AF595062

Celebrating 120 Years of Butantan Institute Contributions for Toxinology

Celebrating 120 Years of Butantan Institute Contributions for Toxinology

Editor

Ana M. Moura-Da-Silva

MDPI • Basel • Beijing • Wuhan • Barcelona • Belgrade • Manchester • Tokyo • Cluj • Tianjin

Editor
Ana M. Moura-Da-Silva
Instituto Butantan
Brazil

Editorial Office
MDPI
St. Alban-Anlage 66
4052 Basel, Switzerland

This is a reprint of articles from the Special Issue published online in the open access journal *Toxins* (ISSN 2072-6651) (available at: http://www.mdpi.com).

For citation purposes, cite each article independently as indicated on the article page online and as indicated below:

LastName, A.A.; LastName, B.B.; LastName, C.C. Article Title. *Journal Name* **Year**, *Volume Number*, Page Range.

ISBN 978-3-0365-3188-5 (Hbk)
ISBN 978-3-0365-3189-2 (PDF)

Cover image courtesy of Butantan Institute.

Contents

MDPI

Editorial

Celebrating 120 Years of Butantan Institute Contributions for Toxinology

Ana M. Moura-da-Silva

Laboratório de Imunopatologia, Instituto Butantan, Av. Vital Brasil, 1500, São Paulo 05503-900, SP, Brazil; ana.moura@butantan.gov.br

A hundred and twenty years ago, the Butantan Institute was founded by the Brazilian physician and scientist Vital Brazil, combining, in the same institution, medical research, and the transfer of results to society in the form of health products. Its foundation was a reaction to the outbreak of bubonic plague in the city of Santos, São Paulo State of Brazil, but the Institution soon also showed its specialization in the study of venomous animals, their venoms, and the production of antivenoms. More than a century after its foundation, the Institute maintains its tradition and initial mission regarding important contributions for collective health. Today, Butantan is an outstanding biomedical research center, which integrates basic research, technological development, the production of immunobiological, and scientific divulgation, seeking the permanent updating and integration of its resources. Butantan is internationally known for its research on venomous animals, and houses one of the largest collections of snakes in the world. Butantan scientists have published countless studies on the characterization of the composition of venoms, the mechanisms of action of their toxins, and the use of toxins as lead molecules for the development of new drugs. The Butantan Institute also operates the "Hospital Vital Brazil", which is specialized in accidents involving venomous animals. The dissemination of scientific knowledge occurs at different levels and media; the institution is a pioneer in the training of graduate students in the field of Toxinology for MSc and PhD degrees.

This Special Issue of *Toxins* celebrates the 120th anniversary of the Butantan Institute, in recognition of their contribution to international Toxinology, highlighting the current production by house scientists and collaborators from other institutions. We selected 19 original articles and 4 reviews approaching several points of Toxinology. The venom of snakes was approached in different aspects as the structural and functional variability in the composition of venoms from individual snakes [1], the mechanisms of action of whole venoms [2,3], the mechanisms of action of individual components as crotoxin [4], phospholipases A2 [5], metalloproteinases [6–8], and the oral immunity induced by whole venoms and their components [9]. Venoms from other animals are also reported; the venoms of centipedes [10], scorpions [11,12], fishes [13,14], and caterpillars [15], including microbial toxins [16]. The great potential for molecules derived from animal venoms as drug leads was also reviewed [17], and the antimicrobial [18], anticoagulant [19,20], and analgesic [21,22] effects have been highlighted in original studies reported here.

In conclusion, this is a small recognition of the institute's contribution to the field of Toxinology that demonstrates its continued and relevant role in this field. I hope you enjoy reading this, and congratulations to the Butantan Institute, scientists, and collaborators, on their 120th anniversary.

Citation: Moura-da-Silva, A.M. Celebrating 120 Years of Butantan Institute Contributions for Toxinology. *Toxins* **2022**, *14*, 76. https://doi.org/10.3390/toxins14020076

Received: 11 January 2022
Accepted: 19 January 2022
Published: 21 January 2022

Funding: The publication fees of this issue were sponsored by Fundação Butantan.

Conflicts of Interest: The authors declare no conflict of interest.

References

1. Sousa, L.F.; Holding, M.L.; Del-Rei, T.H.M.; Rocha, M.M.T.; Mourão, R.H.V.; Chalkidis, H.M.; Prezoto, B.; Gibbs, H.L.; Moura-da-Silva, A.M. Individual Variability in Bothrops atrox Snakes Collected from Different Habitats in the Brazilian Amazon: New Findings on Venom Composition and Functionality. *Toxins* **2021**, *13*, 814. [CrossRef]
2. Megale, Â.; Magnoli, F.C.; Guidolin, F.R.; Godoi, K.S.; Portaro, F.C.V.; Dias-da-Silva, W.; Snake, V. Kn-Ba, a Snake Venom Serine Protease, Induce the Production of Inflammatory Mediators in THP-1 Macrophages. *Toxins* **2021**, *13*, 906. [CrossRef]
3. Kisaki, C.Y.; Arcos, S.S.S.; Montoni, F.; da Silva Santos, W.; Calacina, H.M.; Lima, I.F.; Ca-jado-Carvalho, D.; Ferro, E.S.; Nishiyama, M.Y., Jr.; Iwai, L.K. Bothrops Jararaca Snake Venom Modulates Key Cancer-Related Proteins in Breast Tumor Cell Lines. *Toxins* **2021**, *13*, 519. [CrossRef] [PubMed]
4. Kato, E.E.; Sampaio, S.C. Crotoxin Modulates Events Involved in Epithelial-Mesenchymal Transition in 3D Spheroid Model. *Toxins* **2021**, *13*, 830. [CrossRef] [PubMed]
5. Moreira, V.; Leiguez, E.; Janovits, P.M.; Maia-Marques, R.; Fernandes, C.M.; Teixeira, C. In-flammatory Effects of Bothrops Phospholipases A2: Mechanisms Involved in Biosynthesis of Lipid Mediators and Lipid Accumulation. *Toxins* **2021**, *13*, 868. [CrossRef] [PubMed]
6. Zychar, B.C.; Clissa, P.B.; Carvalho, E.; Alves, A.S.; Baldo, C.; Faquim-Mauro, E.L.; Gon-çalves, L.R.C. Modulation of Adhesion Molecules Expression by Different Metalloproteases Isolated from Bothrops snakes. *Toxins* **2021**, *13*, 803. [CrossRef] [PubMed]
7. Gimenes, S.N.C.; Sachett, J.A.G.; Colombini, M.; Freitas-de-Sousa, L.A.; Ibiapina, H.N.S.; Costa, A.G.; Santana, M.F.; Park, J.J.; Sherman, N.E.; Ferreira, L.C.L.; et al. Observation of Bothrops atrox Snake Envenoming Blister Formation from Five Patients: Pathophysiological Insights. *Toxins* **2021**, *13*, 800. [CrossRef] [PubMed]
8. Bertholim, L.; Chaves, A.F.A.; Oliveira, A.K.; Menezes, M.C.; Asega, A.F.; Tashima, A.K.; Zelanis, A.; Serrano, S.M.T. Systemic Effects of Hemorrhagic Snake Venom Metalloproteinases: Untargeted Peptidomics to Explore the Pathodegradome of Plasma Proteins. *Toxins* **2021**, *13*, 764. [CrossRef]
9. Tsuruta, L.R.; Moro, A.M.; Tambourgi, D.V.; Sant'Anna, O.A. Oral Tolerance Induction by Bothrops jararaca Venom in a Murine Model and Cross-Reactivity with Toxins of Other Snake Venoms. *Toxins* **2021**, *13*, 865. [CrossRef]
10. De Lucca Caetano, L.H.; Nishiyama, M.Y., Jr.; de Carvalho Lins Fernandes Távora, B.; de Oliveira, U.C.; de Loiola Meirelles Junqueira-de-Azevedo, I.; Faquim-Mauro, E.L.; Magalhães, G.S. Recombinant Production and Characterization of a New Toxin from Cryptops iheringi Centipede Venom Revealed by Proteome and Transcriptome Analysis. *Toxins* **2021**, *13*, 858. [CrossRef]
11. Barbosa, M.O.R.; de Paulo, M.E.F.D.; Nencioni, A.L.A. Scorpion Envenomation of Lactating Rats Decreases the Seizure Threshold in Offspring. *Toxins* **2021**, *13*, 853. [CrossRef] [PubMed]
12. Duzzi, B.; Silva, C.C.F.; Kodama, R.T.; Cajado-Carvalho, D.; Squaiella-Baptistão, C.C.; Portaro, F.C.V. New Insights into the Hypotensins from Tityus serrulatus Venom: Pro-Inflammatory and Vasopeptidases Modulation Activities. *Toxins* **2021**, *13*, 846. [CrossRef] [PubMed]
13. Lopes-Ferreira, M.; Sosa-Rosales, I.; Silva-Junior, P.I.; Conceicao, K.; Maleski, A.L.A.; Balan-Lima, L.; Disner, G.R.; Lima, C. Molecular Characterization and Functional Analysis of the Nattectin-like Toxin from the Venomous Fish Thalassophryne maculosa. *Toxins* **2022**, *14*, 2. [CrossRef]
14. Campos, F.V.; Fiorotti, H.B.; Coitinho, J.B.; Figueiredo, S.G. Fish Cytolysins in All Their Complexity. *Toxins* **2021**, *13*, 877. [CrossRef]
15. Alvarez-Flores, M.P.; Gomes, R.N.; Trevisan-Silva, D.; Oliveira, D.S.; Batista, I.F.C.; Buri, M.V.; Alvarez, A.M.; DeOcesano-Pereira, C.; de Souza, M.M.; Chudzinski-Tavassi, A.M. Lonomia obliqua: Envenoming and Innovative Research. *Toxins* **2021**, *13*, 832. [CrossRef]
16. Luz, D.; Gómez, F.D.; Ferreira, R.L.; Melo, B.S.; Guth, B.E.C.; Quintilio, W.; Moro, A.M.; Presta, A.; Sacerdoti, F.; Ibarra, C.; et al. The Deleterious Effects of Shiga Toxin Type 2 Are Neu-tralized In Vitro by FabF8:Stx2 Recombinant Monoclonal Antibody. *Toxins* **2021**, *13*, 825. [CrossRef]
17. Coelho, G.R.; da Silva, D.L.; Beraldo-Neto, E.; Vigerelli, H.; de Oliveira, L.A.; Sciani, J.M.; Pimenta, D.C. Neglected Venomous Animals and Toxins: Underrated Biotechnological Tools in Drug Development. *Toxins* **2021**, *13*, 851. [CrossRef]
18. Oguiura, N.; Corrêa, P.G.; Rosmino, I.L.; Souza, A.L.; Pasqualoto, K.F.M. Antimicrobial Ac-tivity of Snake β-Defensins and Derived Peptides. *Toxins* **2022**, *14*, 1. [CrossRef]
19. Aounallah, H.; Fessel, M.R.; Goldfeder, M.B.; Carvalho, E.; Bensaoud, C.; Chud-zinski-Tavassi, A.M.; Bouattour, A.; M'ghirbi, Y.; Faria, F. rDromaserpin: A Novel Anti-Hemostatic Serpin, from the Salivary Glands of the Hard Tick Hyalomma dromedarii. *Toxins* **2021**, *13*, 913. [CrossRef]
20. Linhares, D.D.C.; Faria, F.; Kodama, R.T.; Amorim, A.M.X.P.; Portaro, F.C.V.; Trevisan-Silva, D.; Ferraz, K.F.; Chudzinski-Tavassi, A.M. Novel Cysteine Protease Inhibitor Derived from the Haementeria vizottoi Leech: Recombinant Expression, Purification, and Characterization. *Toxins* **2021**, *13*, 857. [CrossRef]
21. De Freitas, B.G.; Hösch, N.G.; Pereira, L.M.; Barbosa, T.C.; Picolo, G.; Cury, Y.; Zambelli, V.O. PKCζ-Mitogen-Activated Protein Kinase Signaling Mediates Crotalphine-Induced Antinociception. *Toxins* **2021**, *13*, 912. [CrossRef] [PubMed]
22. Giardini, A.C.; Evangelista, B.G.; Sant'Anna, M.B.; Martins, B.B.; Lancellotti, C.L.P.; Ciena, A.P.; Chacur, M.; Pagano, R.L.; Ribeiro, O.G.; Zambelli, V.O.; et al. Crotalphine Attenuates Pain and Neuroinflammation Induced by Experimental Autoimmune Encephalomyelitis in Mice. *Toxins* **2021**, *13*, 827. [CrossRef] [PubMed]

 toxins

Article

Molecular Characterization and Functional Analysis of the Nattectin-like Toxin from the Venomous Fish *Thalassophryne maculosa*

Monica Lopes-Ferreira [1,*], Ines Sosa-Rosales [2], Pedro Ismael Silva Junior [3], Katia Conceicao [4], Adolfo Luis Almeida Maleski [1,5], Leticia Balan-Lima [1], Geonildo Rodrigo Disner [1] and Carla Lima [1]

1 Immunoregulation Unit of the Laboratory of Applied Toxinology (CeTICs/FAPESP), Butantan Institute, Vital Brasil Avenue, 1500 Butantan, Sao Paulo 05503-009, Brazil; adolfo.maleski@esib.butantan.gov.br (A.L.A.M.); leticia.lima@esib.butantan.gov.br (L.B.-L.); disner.rodrigo@gmail.com (G.R.D.); carla.lima@butantan.gov.br (C.L.)
2 Escuela de Ciências Aplicadas del Mar, Universidad de Oriente, Boca de Rio 6304, Venezuela; sosaines@gmail.com
3 Protein Chemistry Unit of the Laboratory of Applied Toxinology (CeTICs/FAPESP), Butantan Institute, Vital Brasil Avenue, 1500 Butantan, Sao Paulo 05503-009, Brazil; pedro.junior@butantan.gov.br
4 Peptide Biochemistry Laboratory, UNIFESP, Sao Jose dos Campos 12247-014, Brazil; katia.conceicao@unifesp.br
5 Post-Graduation Program of Toxinology, Butantan Institute, Vital Brasil Avenue, 1500 Butantan, Sao Paulo 05503-009, Brazil
* Correspondence: monica.lopesferreira@butantan.gov.br

Abstract: TmC4-47.2 is a toxin with myotoxic activity found in the venom of *Thalassophryne maculosa*, a venomous fish commonly found in Latin America whose envenomation produces an injury characterized by delayed neutrophil migration, production of major pro-inflammatory cytokines, and necrosis at the wound site, as well as a specific systemic immune response. However, there are few studies on the protein structure and functions associated with it. Here, the toxin was identified from the crude venom by chromatography and protein purification systems. TmC4-47.2 shows high homology with the Nattectin from *Thalassophryne nattereri* venom, with 6 cysteines and QPD domain for binding to galactose. We confirm its hemagglutinating and microbicide abilities independent of carbohydrate binding, supporting its classification as a nattectin-like lectin. After performing the characterization of TmC4-47.2, we verified its ability to induce an increase in the rolling and adherence of leukocytes in cremaster post-capillary venules dependent on the $\alpha 5\beta 1$ integrin. Finally, we could observe the inflammatory activity of TmC4-47.2 through the production of IL-6 and eotaxin in the peritoneal cavity with sustained recruitment of eosinophils and neutrophils up to 24 h. Together, our study characterized a nattectin-like protein from *T. maculosa*, pointing to its role as a molecule involved in the carbohydrate-independent agglutination response and modulation of eosinophilic and neutrophilic inflammation.

Keywords: *Thalassophryne*; nattectin; reverse-phase HPLC; MALDI-ToF; hemagglutinating activity; antibacterial activity; inflammation; toxinology; animal toxins

Key Contribution: Nattectin-like lectin regulates eosinophils and neutrophils recruitment.

Citation: Lopes-Ferreira, M.; Sosa-Rosales, I.; Silva Junior, P.I.; Conceicao, K.; Maleski, A.L.A.; Balan-Lima, L.; Disner, G.R.; Lima, C. Molecular Characterization and Functional Analysis of the Nattectin-like Toxin from the Venomous Fish *Thalassophryne maculosa*. *Toxins* **2022**, *14*, 2. https://doi.org/10.3390/toxins14010002

Received: 14 October 2021
Accepted: 15 November 2021
Published: 21 December 2021

Publisher's Note: MDPI stays neutral with regard to jurisdictional claims in published maps and institutional affiliations.

1. Introduction

Lectins are a comprehensive group of proteins with carbohydrate-binding properties. Due to their binding singularity, many natural lectins, purified from several sources such as plants, algae, and fungi, have prospective applications in biotechnology, medical research, and crop protection. Additionally, many groups have identified lectins in organs and tissues (gills, eggs, electric organ, stomach, intestine, liver, skin, mucus, and plasma) of various fish species [1]. Lectins have been reported to have a wide array of functions, including immune-relevant ones such as pathogen recognition, agglutination, opsonization, complement activation, phagocytosis, and other functions such as splicing of RNA, protein folding, trafficking of molecules, control of cell proliferation, and roles in development [2,3].

Interestingly, our group has previously described a lectin, named Nattectin, in the venom gland of the fish *Thalassophryne nattereri* [4]. Nattectin is a basic monomeric protein, non-glycosylated, galactose-specific lectin from the C-type family, presenting remarkable pro-inflammatory activity. *T. nattereri* and *Thalassophryne maculosa* are venomous toadfish that belong to the family Batrachoididae. They are benthic ambush predators that favor from sandy or muddy substrates where their cryptic coloration, or the habit of burying under sand and mud, helps them avoid detection by their prey. Their omnivorous diet is composed of sea worms, crustaceans, mollusks, and other fish. They are found in temperate and tropical waters throughout the coast of America, Europe, Africa, and India. These species present one of the most adapted teleost venom apparatus composed of four canaliculated spines coupled to venom glands at their base capable of delivering a painful wound to predators. Thus, the venom apparatus has more of a defense function rather than predation.

Studies on *T. nattereri* envenomation and its venom composition have been carried out in Brazil by our group since 1998 at the Butantan Institute [5,6]. *T. nattereri* stands out among the venomous animals of medical importance in Brazil for the number of accidents it causes in the North and Northeast regions and the seriousness of the cases [7–9]. One of the main symptoms of *T. nattereri* envenomation is the immediate, intense pain that persists over 24 h. Erythema and edema are also shortly noticed with the efflorescence of bubbles with serous content. These lesions evolve to long-remaining necrosis with a delayed healing process devoid of specific drug treatment [10–13].

However, few studies have been conducted with *T. maculosa*, mainly found in Venezuela, Colombia, and the islands of Aruba, Curaçao, and Trinidad and Tobago. It is known that its venom presents a mixture of bioactive toxins differently expressed in females and males [14]. *T. maculosa* venom induces a significant necrotic lesion characterized by delayed neutrophil influx to the footpad of mice. An acute production of IL-1β, IL-6, and later secretion of TNF-α, MCP-1, KC, and mediators from arachidonic acid metabolism such as LTB4 and PGE2 were detected in the exudate of inflamed footpads [15]. In addition, we demonstrated that bone marrow-derived macrophages and dendritic cells were strongly stimulated by the venom, which demonstrates its ability to stimulate a specific and systemic immune response [16]. Furthermore, Sosa-Rosales et al. [17] performed a partial identification of two proteins with myotoxic activity in *T. maculosa* venom using reverse-phase HPLC and named them according to column retention time as TmC4-47.2. SDS-Page analysis of the crude venom showed a few weighty bands (one located above 97 Mw, one between 68 and 97 Mw, one major band between 29 and 43 Mw, and the last one located below 18.4 Mw). Then, it seems that the isolated nattectin-like protein is one of the main components of the venom, which includes a significant mixture of bioactive molecules involved in the local inflammatory lesion [17].

We understand that the advance in the knowledge of the pattern of ischemic and necrotic lesions induced by *T. maculosa* and the description of the main toxins can lead to the identification of targets for therapeutic intervention. Therefore, the study of the TmC4-47.2 toxin becomes an important tool for understanding the mechanisms of action involved in envenomation. In the present study, we performed the characterization of the TmC4-47.2 protein sequence using sequential C18 and C8 affinity column chromatography and mass spectrometry and determined its functions as a lectin.

2. Results

2.1. *Purification of TmC4-47.2 Toxin from Thalassophryne maculosa Venom*

Initially, the solution of *T. maculosa* venom chromatographed on a C18 column coupled to a high-pressure liquid chromatography system (Figure S1) generated 3 fractions, with the presence of the toxin mainly in fraction 2 that was subsequently re chromatographed 3 times on a reverse-phase C8 column. The first passage of fraction 2 through the C8 column generated 5 new fractions, with the toxin present in fractions 3, 4, and 5 (Figure S2). The second passage through the C8 column of the 3 previous pooled fractions generated 6 more fractions with the protein present only in fractions 3 to 5 (Figure S3).

These 3 pooled fractions were finally rechromatographed on C8 and generated 3 fractions with the toxin with a molecular mass of 15 kDa as shown on the 12% SDS-PAGE gel (Figure 1A). To determine the exact mass and purity of the toxin obtained, we collected the 3 fractions and applied 10 µL of the toxin to an LC/MS system by direct infusion and the deconvolution of the mass spectrum revealed two distinct proteins with very close molecular masses, one with 15,135.9 and the other with 15,634.3 Da, as observed in Figure 1B. Next, the number of cysteines in each protein was determined in the reduced and alkylated samples, a process that adds 57 Da to each cysteine residue present in the native proteins. We can observe in Figure 1C that the treatment generated proteins with masses of 15,484.8 and 15,982.8. By the difference of the masses of the pure samples and the reduced/alkylated samples, we prove the presence of 6 cysteine residues in each of the proteins according to Yang, Liu, and Liu [18].

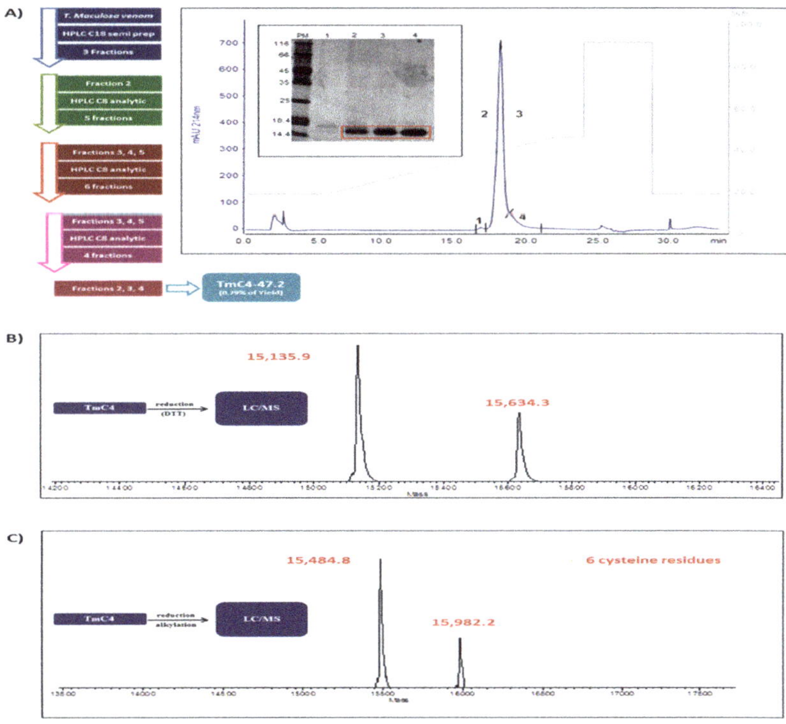

Figure 1. The raw venom of *Thalassophryne maculosa* was submitted to a sequence of fractionation in a high-pressure liquid chromatography (HPLC) system using a C18 followed by C8 reversed phase HPLC columns, according to the scheme in (**A**) (top left). The gradient applied was 20–80% of buffer B in 35 min, in a 1 mL min^{-1} flux. The absorbance was measured at 214 nm and the last pooled fractions containing the isolated toxin are presented in the chromatogram and highlighted in the 12% SDS-PAGE gel with a molecular mass of 15 kDa. 10 µL of the toxin collected from the last 3 fractions were applied to an Liquid Chromatography Mass Spectrometry (LC/MS) system by direct infusion to determine the mass and purity (**B**). The number of cysteines in each protein was determined by LC/MS in the reduced and alkylated samples (**C**).

2.2. *TmC4-47.2 Is a Galactose-Binding Nattectin-Like Lectin*

Subsequently, the pool of the 3 fractions containing the toxin obtained after the last C8 column chromatography was also used for mass determination by MALDI-ToF spectrometry. After digestion, we found 9 peptides with masses ranging from 867.439 to 2065.025 (Figure 2A), which were sequenced and compared to the sequence of Nattectin present in the venom of the fish *T. nattereri* (GenBank LECG_THANI Galactose-specific lectin Q66S03). In Figure 2B, in red, we can observe the sequences of the internal peptides overlapping in the Nattectin sequence with 100% homology and the conserved galactose-binding domain QPD (Gln-Pro-Asp) [19].

A)

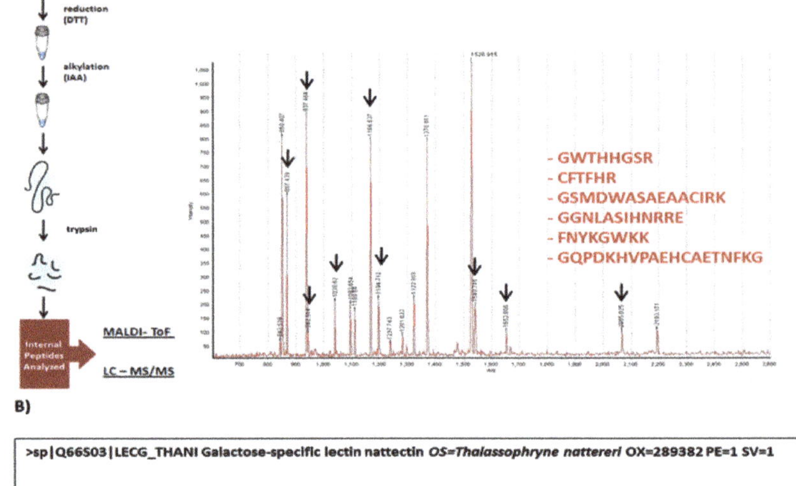

B)

>sp|Q66S03|LECG_THANI Galactose-specific lectin nattectin *OS=Thalassophryne nattereri* OX=289382 PE=1 SV=1

1 MASVPHFTVF LFLACALGIG ANVTRRATSS CPKGWTHHGS RCFTFHRGSM
51 DWASAEAACI RKGGNLASIH NRREQNFITH LIHKLSGENR RTWIGGNDAV
101 KEGMWFWSDG GSKFNYKGWKK GQPDKHVPAE HCAETNFKGA FWNNALCKVK
151 RSFLCAKNL

C)

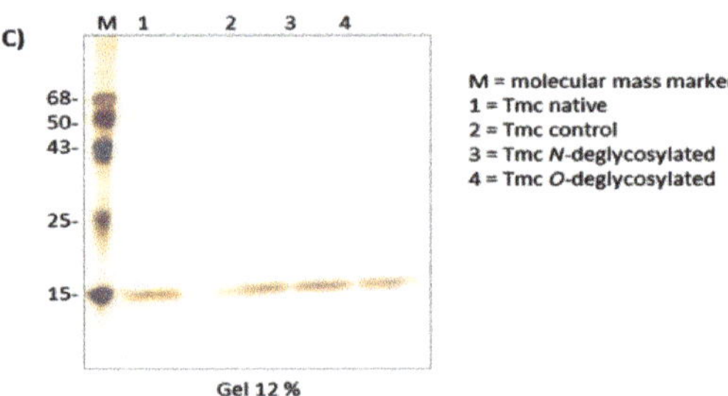

Figure 2. The last pool of the fractions containing the toxin obtained in the C8 column chromatography was used for mass determination and identification of internal peptides by MALDI-ToF spectrometry (**A**). The found peptides were sequenced and overlapped to the *Thalassophryne nattereri* Nattectin (GenBank LECG_THANI Galactose-specific lectin Q66S03) (**B**). Enzymatic deglycosylation was tested to check the effect on the electrophoretic mobility of the native protein in a 12% gel (**C**).

The glycan structure of nattectin-like lectin was further studied by enzymatic deglycosylation [20]. We confirmed that treatment with O-glycosidase or N-glycosidase did not alter the electrophoretic mobility of the proteins compared to the native protein, which shows a single band of 15 kDa (Figure 2C).

After identifying the sequences similarities of the internal peptides of the toxin to those of Nattectin from *T. nattereri*, we confirmed the ability of serum from mice immunized with *T. nattereri* venom or Nattectin to recognize *T. maculosa* toxin (Figure 3A) as the recognition of Nattectin itself from *T. nattereri* venom, which indicates that it is a nattectin-like protein. The high affinity and narrow specificity of the nattectin-like lectin of *T. maculosa* for defined oligosaccharide structures were evaluated using the digoxigenin (DIG) Glycan Differentiation Kit, a competition assay (Figure 3B). The toxin/carbohydrate binding revealed by incubation with different DIG-labeled lectins demonstrated a weak interaction of both Nattectin and nattectin-like protein of *T. maculosa* with PNA, indicative that both Nattectin proteins recognized Gal-β(1–3)-N-acetylgalactosamine, which forms the core 1 structure of many O-glycans [21].

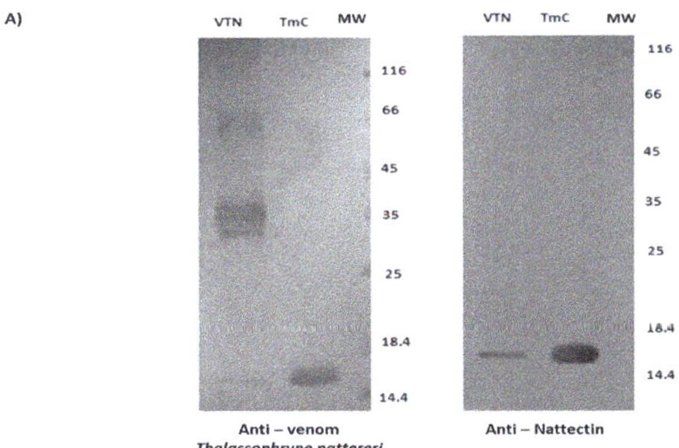

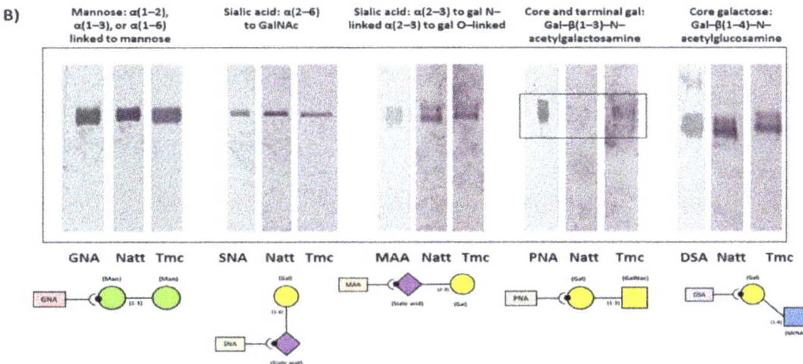

Figure 3. The nattectin-like toxin from *Thalassophryne maculosa* is specifically recognized by Anti-venom and Anti-Natectitn antibodies from *Thalassophryne nattereri*. Serum from mice immunized with *T. nattereri* venom-VTN or Nattectin were tested to check their ability in recognizing *T. maculosa* toxin-TmC (**A**). 10 μg of nattectin-like toxin and *T. nattereri* venom were subjected to 12% SDS-PAGE gel and transferred to a nitrocellulose membrane. The membrane was incubated with *T. nattereri* anti-Natectin anti-venom serum. They were subsequently incubated with peroxidase-labeled mouse anti-IgG and revealed with 4-α-chloro-naphtol. A digoxigenin (DIG) glycan differentiation competition assay kit (Roche Applied Science, Germany) was used to identify the binding specificity of the toxin. Toxin/carbohydrate binding was revealed by incubation with different DIG-labeled lectins and alkaline phosphatase-conjugated anti-DIG antibody (**B**).

2.3. Hemagglutinating and Antimicrobial Activities of *T. maculosa* Nattectin-Like Toxin

To confirm the lectin activity of the nattectin-like protein, hemagglutinating activity was evaluated (Figure 4). Nattectin-like protein agglutinated A type-tested human erythrocytes at a dose of 10 μg. Moreover, when this dose of nattectin-like toxin was previously incubated with D-galactosamine or D-Mannose, the agglutination capacity of erythrocytes by nattectin-like was preserved.

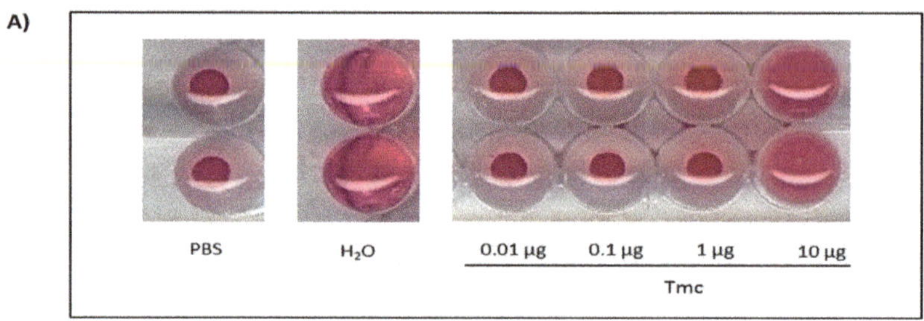

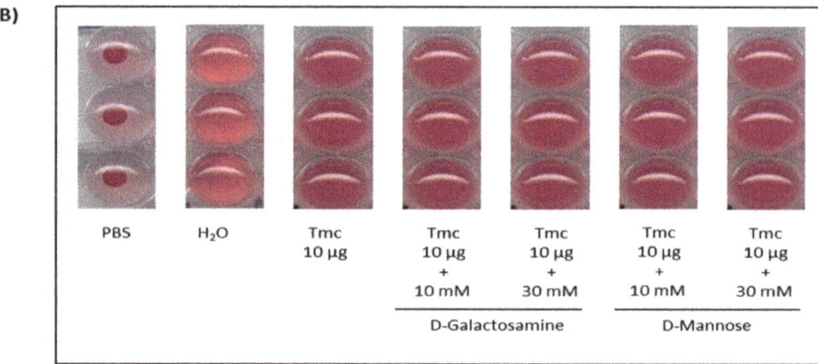

Figure 4. Assessment of the hemagglutinating activity of *Thalassophryne maculosa* toxin-TmC on human erythrocytes (**A**). TmC pre-incubation with D-galactosamine or D-Mannose was tested to confirm the permanence of the hemagglutinating pattern (**B**). Negative and positive controls were made by the respective addition of phosphate-buffered saline—PBS and distilled water—H_2O.

Antimicrobial activity of nattectin-like toxin was performed against *Micrococcus luteus* A270, *Escherichia coli* SBS 363, and *Candida albicans* strains and the Nattectin from *T. nattereri* was tested in parallel as an intern control. We found that 10 μg of nattectin-like toxin did not inhibit the growth of the Gram-negative bacteria (*E. coli*) tested. Furthermore, corroborating the Nattectin effect (that inhibited the growth of all three strains evaluated at 10 μg), this dose of nattectin-like lectin showed an inhibitory effect on *M. luteus* and *C. albicans*.

2.4. TmC4-47.2 Toxin-Induced Alterations in the Microcirculation

The ability of nattectin-like toxin to induce changes in the microcirculation was evaluated using intravital microscopy assay in cremaster muscle of mice using the intra-scrotal application of 10 μg of the toxin and evaluation after 3 h of the injection. We observed in Figure 5 an intense leukocyte recruitment and rolling in the post-capillary venules immediately after the 3 h rest period (0 min.) that increased with time or stayed intense up to 30 min, as it can see from Figure 5B–E. Additionally, we registered a decrease in vessel flow after 10 min, followed by a complete stop of flow in venules and arterioles, possibly due to fibrin thrombus formation after 20 min. Nattectin-like lectin did not induce changes in the caliber of arterioles or damage to muscle fibers.

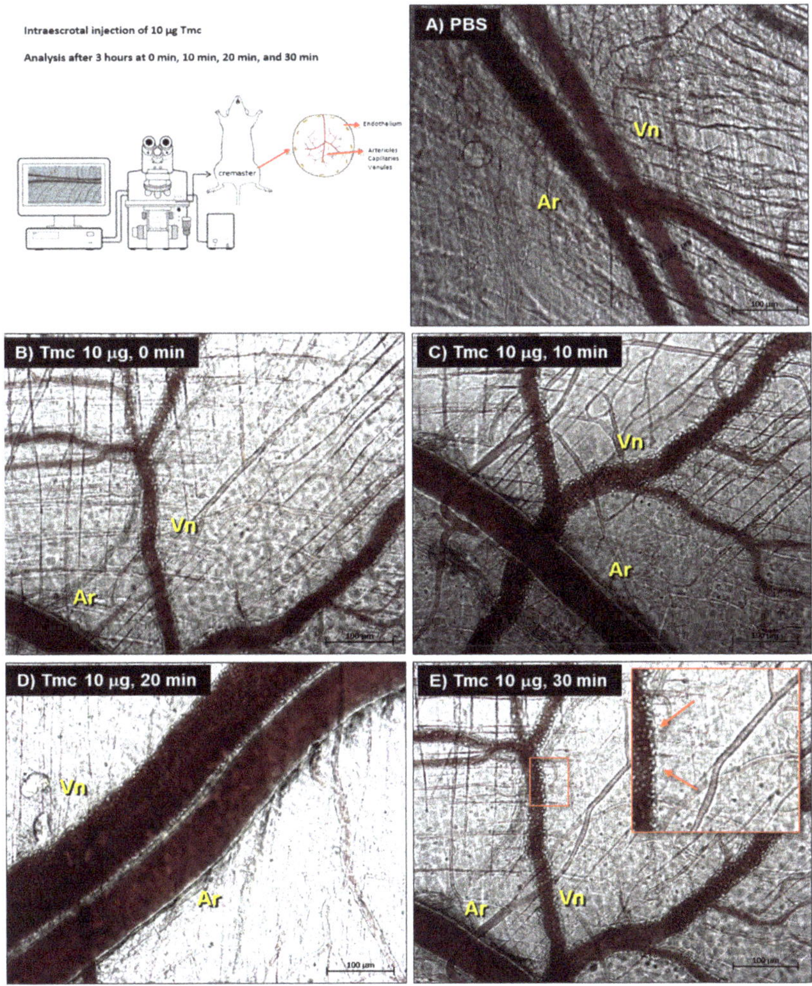

Figure 5. Evaluation of changes in the cremaster muscle microcirculation by intravital microscopy after 3 h of the application of 10 µg of TmC4-47.2 toxin, according to the summarized protocol illustrated in the top-left corner. The tissue microvasculature was evaluated by an optical microscope coupled to a photographic camera in the control group treated with PBS (**A**) and in the set times of 0, 10, 20, and 30 min after the 3 h of exposure (**B–E**). An intense migration and rolling of leukocytes have been observed. In the 6E inset, arrows evidence the leukocytes in venules. Ar: arterioles; Vn: venules.

Our results presented in Figure 6A demonstrate that the intense and prolonged rolling of leukocytes induced by the nattectin-like toxin was followed by adhesion in the post-capillary venules, indicating the toxin's ability to promote extravasation of leukocytes into the surrounded interstitial tissue [22].

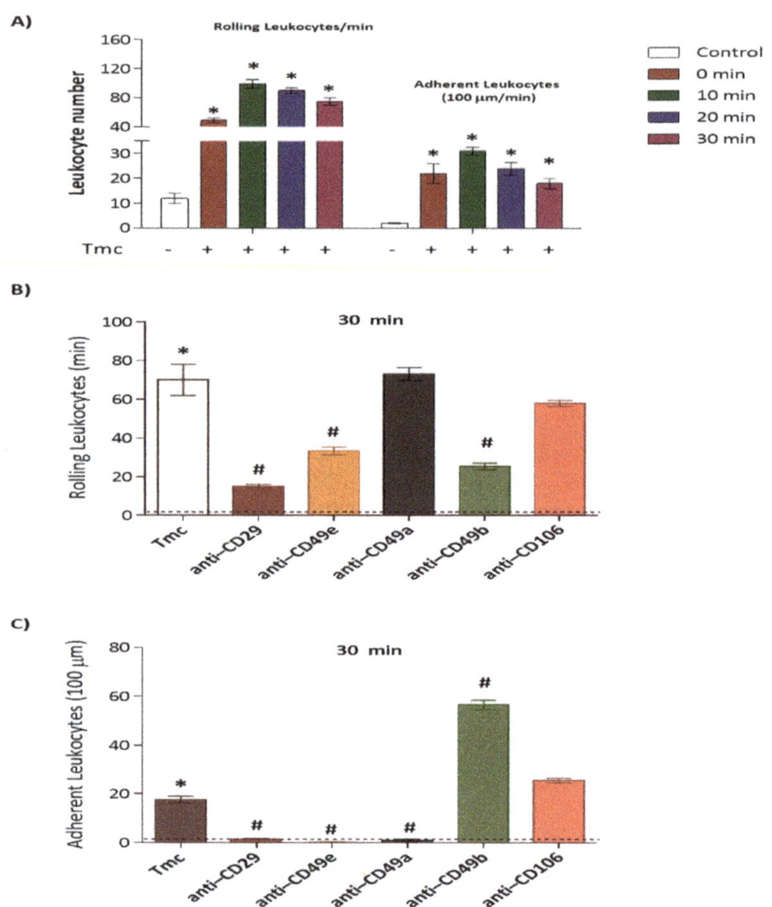

Figure 6. The number of rolling and adherent leukocytes controlled by 10 µg of the nattectin-like toxin TmC4-47.2 (Tmc) were counted in the post-capillary venules of mice at 0 to 30 min after 3 h of exposure using bright field intravital microscopy (**A**). The process of leukocyte recruitment, rolling (**B**), and adherence (**C**) was visualized in the cremaster vasculature of mice with inhibited alpha and beta integrins by neutralizing antibodies before intrascrotal toxin injection. The "*" represents a statistically significant difference with negative control (non-treated) represented by the dashed basal line (**B**,**C**), and the "#" represents statistically significant difference of integrins-treated groups with the *Thalassophryne maculosa* toxins (Tmc), $p < 0.05$.

Integrins are a family of ubiquitous αβ heterodimeric receptors which combine with various ligands in physiological processes and disease, playing a crucial role in cell proliferation, tissue repair, inflammation, infection, and angiogenesis [23]. Next, using bright field intravital microscopy, we visualized the process of leukocyte recruitment in the cremaster vasculature of mice with alpha and beta integrins inhibited by neutralizing antibodies before intra-scrotal toxin injection. We found that treatment of mice with anti-CD29 (beta 1 integrin), anti-CD49e (alpha 5 integrin), and anti-CD49b (alpha 2 integrin) blocked 83%, 57%, and 69%, respectively, of the rolling leukocytes compared to untreated mice (Figure 6B). No inhibition was induced in mice pre-treated with anti-CD49a (alpha 1 integrin) or anti-CD106 (VCAM-1) neutralizing Abs (Figure 6B). Furthermore, the adherent leukocytes induced by nattectin-like lectin were entirely inhibited by anti-CD29, anti-CD49e, and anti-CD49a neutralizing Abs (Figure 6C). In contrast, anti-CD49b or anti-CD106 did not inhibit the adherence of leukocytes to venules.

2.5. Induction of Acute Inflammation by Nattectin-Like Protein

We used a mouse model of peritonitis to evaluate the inflammatory response profile induced by the nattectin-like toxin TmC4-47.2. Balb/c mice received 10 µg of the toxin intraperitoneally diluted in 500 µL of sterile phosphate-buffered saline (PBS), and control mice received only sterile PBS. Six, 16, and 24 h after the exposure, the animals were sacrificed, and the peritoneal cavity was washed to obtain the cell suspension. We analyzed the leukocyte influx recruited to the peritoneum by labeling surface molecules typical for each cell population and the dosage of cytokines (IL-1β, IL-6, and TNF-α) and chemokines (MCP-1, KC, and eotaxin) involved in the inflammatory process.

Our results in Figure 7A show that mice injected with the nattectin-like toxin exhibited intense leukocyte extravasation into the peritoneal cavity after 6 h of injection that was persistent for 24 h. The acute phase (6 h) of inflammation was characterized by the influx of eosinophils (Figure 7B) and mainly neutrophils (Figure 7C). After 16 h, macrophages (Figure 7D) entered the inflamed peritoneal cavity and remained for 24 h in the presence of a large number of both granulocytes.

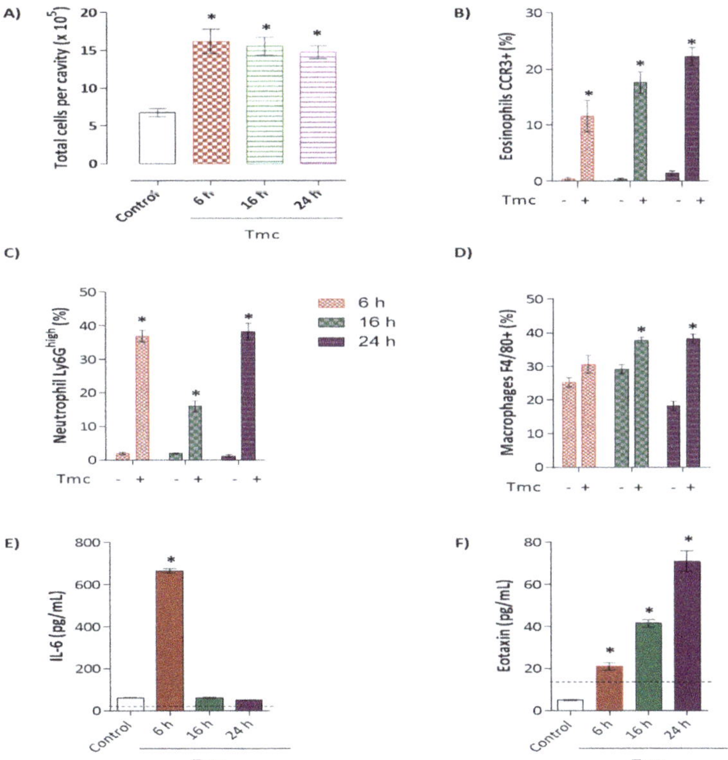

Figure 7. The inflammatory response profile induced by the *Thalassophryne maculosa* toxin TmC4-47.2 (Tmc) was evaluated in a Balb/c mice model of peritonitis. The toxin was applied intraperitoneally at 10 µg in 500 µL of sterile phosphate-buffered saline (PBS). Control mice received sterile PBS. At six, 16, and 24 h after the exposure, the cell suspension from the peritoneal cavity was collected. The total cell number (**A**), the leukocyte influx to the peritoneum (**B–D**), and the dosage of IL-6 (**E**) and eotaxin (**F**) involved in the inflammatory process was analyzed.The "*" represents a statistically significant difference with negative control (non-treated), $p < 0.05$.

Finally, we observed the production of IL-6, an important cytokine involved in the inflammatory process, only within 6 h after the 10 μg toxin injection (Figure 7E). Eotaxin, a chemotactic for eosinophils, was produced 6 h after injection, and increasing levels were observed up to 24 h (Figure 7F). In contrast, the injection did not promote the secretion of the cytokines IL-1β and TNF-α, as well as the neutrophil chemotactic factor, KC (data not shown).

3. Discussion

The *T. nattereri* envenomation in humans and the recapitulation of the injury in mice have been extensively studied by our group [24–26]. We identified in the Natterin group of venom toxins the aerolysin proteins being responsible for the main effects of envenomation [27,28], and their immunopharmacological activities were determined [29–32]. Besides the Natterin proteins, *T. nattereri* venom also contains Nattectin, a lectin with Ca^{2+}-independent hemagglutinating and immunomodulatory activities [4]. Nattectin shows the particular potential to bind types I and V collagen and improve integrin-mediated HeLa cell adhesion and apoptosis protection by its binding to RGD-dependent integrins, mainly the β1 subunit [33]. We further demonstrated its ability to activate antigen-presenting cells [34] and trigger Th1-type immune response with IgG1 production [35].

On the other hand, just a few studies have been conducted with *T. maculosa*, one of the *T. nattereri*'s closest relatives. Sosa-Rosales et al. [17] partly identified a biologically active protein by chromatography. To obtain the toxin TmC4-47.2 from the venom of the *T. maculosa*, we used reverse-phase HPLC-based workflow with the application of the crude venom to the chromatography system coupled to a semi-preparative reverse-phase C18 column, followed by subsequent chromatography of the fractions containing the toxin with a molecular weight of 15 kDa on a reverse-phase C8 analytical column. Finally, the protein was purified into three lyophilized fractions for verification of purity and exact mass by LC/MS. Interestingly we confirmed the presence of two proteins with very close molecular weights, one with a calculated molecular weight of 15,135.9 Da and the other with 15,634.3 Da, unlike the migrated single band ~15 kDa via reducing SDS-PAGE.

To determine the number of cysteine residues and, consequently, of disulfide bridges, we proceeded with the reduction and alkylation of the toxins. After applying the sample to the same LC/MS system, we once more detected the presence of two proteins, now with 15,484.8 Da and 15,982.8 Da, respectively. From the difference in the masses of the pure proteins, we inferred that both proteins have 6 cysteines each and consequently form 3 disulfide bridges, as described in the sequence of Nattectin from *T. nattereri* [28]. Our findings that the anti-venom serum and the anti-Nattectin serum from *T. nattereri* recognize the TmC4-47.2 toxin from *T. maculosa* support the similarity of the sequences.

The sequencing of internal fragments of the toxins generated by enzymatic digestion with trypsin and analysis by MALDI-ToF revealed 100% similarity of the sequenced fragments with the Nattectin sequence, which correspond to 40% of its entire sequence. The high homology rate might reflect a relatively small evolutionary divergence between the two congeneric species and the fact that they harbor the same biological properties. Batrachoididae, the only family in the ray-finned fish order Batrachoidiformes, arose on Earth in the Miocene, which is the first geological epoch of the Neogene Period that extends from about 23 to 5 million years ago (mya). There are about 83 species of toadfishes grouped into 21 genera. The subfamily Thalassophryninae has six phenotypically alike species in the genus *Thalassophryne*, which suggests the time of divergence among species within this genus must be little, although there is no precise description of the time of divergence in the literature. Using peptide mass fingerprinting, we identified a unique carbohydrate recognition domain composed of the amino acids glutamine, proline, and aspartate (QPD), a specific galactose-binding site [36]. We have designated the TmC4-47.2 toxin as a galactose-binding protein.

Glycan structures are common post-translational modifications of proteins, assisting in protein folding and defining the timing for protein disposal as a part of the quality control function in the early secretory pathway [37]. Our data shows that the nattectin-like protein of *T. maculosa* is non-glycosylated since we determined the absence of GlcNAc-β-Ser/Thr or N-linked glycan residues.

Among the distinct post-translational modifications, glycosylation is the most recurrent, and virtually 50% of all known proteins are thought to be glycosylated. In contrast, the galectins are generally small, soluble, non-glycosylated proteins and, unlike the C-Type lectins, do not require structural stabilization via Ca^{2+} complexing for their activity. However, many C-type lectins are Ca^{2+}-independent and may not necessarily bind to sugar ligands [38]. Our findings showing the nattectin-like lectin as a non-glycosylated protein with the ability to interact with the core and terminal galactose-type ligands point to a functional similarity with proto-type galectins that contains only one CRD. Galectins, a family of animal lectins with an affinity for β-galactosides, can form multivalent complexes with cell surface glycoconjugates and trigger several intracellular signals to modulate cell activation, differentiation, and survival [39].

Literature has portrayed the protective role of galectins against infections when binding to the glycoconjugates on the surface of invasive microbes (virus, bacteria, fungi, and parasites) acting as pattern recognition receptors in innate immunity [40]. In fish, they have also been demonstrated to participate in pathogen recognition. Gal-1, a proto-type galectin, from rock bream (*Oplegnathus fasciatus*) bound and agglutinate many bacteria [41] and the Gal-1 of zebrafish bound infectious hematopoietic necrosis virus (IHNV) in a carbohydrate-dependent way [42]. AJL-1, a proto-type galectin of Japanese eel (*Anguilla japonica*), was reported to hinder the biofilm formation of *Aggregatibacter actinomycetemcomitans* [43]. The galectin-8 and -9 of mandarin fish (*Siniperca chuatsi*) were shown to inhibit the growth of some pathogens [44]. Our results corroborate these findings, and through functional analogies we confirmed the ability of nattectin-like protein to induce carbohydrate-independent agglutination of human erythrocytes and exert antimicrobial activity against gram-negative and gram-positive bacterial pathogens as well as against *C. albicans*.

A substantial number of lectins have been identified in other fish. Notably the galectins, which is unusual compared to venom from other animals, like snakes, that commonly present C-type lectins [45]. Probably due to the complex environment, fish have evolved natural immune modulators, such as galectins and C-type lectins. In the fish lectins, evolution likely led to the acquisition of galectin-like properties. Thus, the functional similarities between Nattectins and Galectins studied herein might be the result of convergent evolution and because Nattectin belongs, from an evolutionary and structural perspective, to the C-type lectin family. This is also corroborated by the similarity of *T. maculosa*'s and *T. nattereri*'s Nattectin that present highly conserved residues of carbohydrate-recognition domains found in the C-type lectins.

In addition to the role of galectins in microbial death, several reports show their importance in the transendothelial migration of neutrophils and eosinophils [46,47]. Intravital imaging has yielded relevant understandings into the regulation of inflammatory responses to pathogens and sterile insults in a variety of tissues [48]. Such imaging approaches were useful in characterizing individual steps of the leukocyte recruitment cascade. Moreover, our data show that nattectin-like lectin modulates leukocyte rolling and adherence to vascular endothelium dependent on α5β1 integrin. This crucial adhesion molecule mediates the adherence of many cell types to the extracellular matrix by recognizing its classic ligand fibronectin [49]. It has been reported that the increased expression of α5β1 integrin and fibronectin enrichment in the epidermis are seemingly features of chronic neutrophilic inflammation [50]. Furthermore, polymorphonuclear neutrophils (PMNs) express several surface integrins, including α5β1 [51].

Additionally, we noted that nattectin-like toxin increased eosinophil recruitment to the peritoneal cavity of mice and the production of the pro-inflammatory chemokine, eotaxin. The trafficking of primed mature eosinophils from the bloodstream into inflamed tissues is finely regulated by adhesion molecules, several cytokines, and chemokines with overlapping functions. Studies with mouse and human eosinophils have shown that rolling along the vascular endothelium and the activation-dependent stable adhesion to the vascular endothelium are supported by β1 integrins. Trans-endothelial migration is under the control of eotaxin and its receptor, CCR3 [52].

Together, our findings show that *T. maculosa* possesses two non-glycosylated nattectin-like proteins with similar galactose-binding specificity. Nattectin-like lectin retained their ability to induce carbohydrate-independent agglutination of human erythrocytes. Furthermore, it exerted potent antimicrobial activity against gram-negative and gram-positive bacterial pathogens, as well *C. albicans*. The non-glycosylation status of nattectin-like proteins did not impact protein function since lectins showed a substantial effect on leukocytes rolling and adherence to the endothelial wall of the mice's cremaster muscle. This interaction was dependent on β1 integrin binding. The inflammation of the peritoneal cavity was characterized by eosinophils and neutrophils influx recruited in response to IL-6 and eotaxin mediators produced by nattectin-like protein. Our results point to an immune function of nattectin-like toxin participating in the antimicrobial response through the recognition of pathogens via carbohydrate-binding and modulation of eosinophil and neutrophil infiltration in inflammation.

4. Materials and Methods

4.1. *Thalassophryne maculosa Venom and Isolation of TmC4-47.2 Toxin*

All necessary permits for the capture of *T. maculosa* and to collect the venom were obtained from the Escuela de Ciencias Aplicadas del Mar, Núcleo Nueva Esparta, Universidad de Oriente, Isla Margarita; Venezuela (Permit Number: 30/07). *T. maculosa* individuals were transported to the laboratory and were anesthetized with 2-phenoxyethanol before sacrifice. Venom was shortly obtained from the tip of the spines by applying pressure at their bases. After centrifugation, venom was pooled and stored at −80 °C before use. Protein content was examined using bovine serum albumin (Sigma Chemical Co., St Louis, MO, USA) as standard. Endotoxin content was evaluated (resulting in a total dose < 0.8 pg LPS) with QCL-1000 chromogenic Limulus amoebocyte lysate assay (Bio-Whittaker) according to the manufacturer's instructions. Aliquots of crude venom from *T. maculosa* (5 mg mL^{-1}) was submitted to a reversed-phase (RP) LC10A high-performance liquid chromatography system (HPLC) Shimadzu (Äkta, Amersham Biosciences, Sweden) starting in a C18 semi-preparative column (Shimadzu-Shim Pack-CLC-ODS: 4.6 mm ID × 25 cm, 5 μm, and 100 Å). For elution, a binary gradient of 0.1% TFA aqueous solution (A) and 0.1% TFA 90% acetonitrile solution (B) was used, with a flow rate of 5.0 mL/min for 60 min, and detection at 214 nm absorbance. The chromatographic fractions were manually collected, vacuum dried, and stored at −20 °C. All fractions containing the toxin were identified by 12% SDS-PAGE [53]. The chromatographic fraction containing the toxin was submitted to 3 chromatographic steps using a Jupiter C8 4.60 mm × 250 mm, 10 μm, 300 Å, column (Phenomenex). The elution method was 20% to 80% B solution from 0 to 35 min at 1 mL/min, followed by a gradient of 10% B solution per minute until reaching a maximum concentration of 50% acetonitrile. The presence of the toxin in the eluates was confirmed by absorbance and SDS-PAGE.

4.2. *TmC4-47.2 Toxin Mass Determination*

To determine the toxin mass, we used 10 μL of the sample injected into an LC-MS Surveyor MSQ Plus system (Thermo Finnigan) by direct infusion in a flow of 50 μL/min of aqueous acetonitrile solution (1:1) containing 0.1% formic acid under capillary voltage conditions 3.1 kV, 75 V cone, scan lasting 1 s in the m/z range 200–2000. The equipment was previously calibrated with sodium iodide (m/z range 22.98 to 1971.61). To obtain the molecular weight of the sample, the deconvolution of the mass spectrum obtained was performed using the Mag Tran program version 1.0 beta 8 [54]. For MALDI-ToF mass spectrometry, analyzes were performed in a MALDI-ToF/PRO instrument (Amersham). The samples in solution were mixed 1:1 (v/v) with a supersaturated matrix solution for proteins (sinapic acid), deposited on the sampling plate (0.4–0.8 µL), and allowed to dry at room temperature. The spectrometer was operated in reflection mode, and P14R ($[M+H+]^+$ 1533.85) and angiotensin II ($[M+H+]^+$ 1046.54) (Sigma, St. Louis, MO, USA) were used as external calibrants.

4.3. Reduction, Alkylation, and Digestion of TmC4-47.2 Toxin

The purified toxin was used for the identification of disulfide bridges, according to Yang, Liu, and Liu [18]. The toxin was reduced by 10 µL of 5 mM DTT in 25 mM ammonium bicarbonate (Sigma-Aldrich, St. Louis, MO, USA) and heated at 60 °C for 30 min. For alkylation, 10 µL of 100 mM iodoacetamide in 25 mM ammonium bicarbonate (Sigma-Aldrich, St. Louis, MO, USA) was added and kept for 30 min in the dark. For enzymatic digestion, 2 µL of ultra-grade trypsin (Proteomics grade; Sigma, St. Louis, MO, USA) solution at 40 ng·µL^{-1} was added to 8 µL of the toxin solution and incubated 18 h at 37 °C. The samples were analyzed using a nanoflow LC/MS/MS system customized with a PepFinder Kit. Aliquots of 10 µL were initially charged onto a reversed-phase peptide trap column in a of 10 µL/min flow rate for 3 min. Then peptides were eluted and partitioned on a reversed-phase capillary column (PicoFritTM; 5 µm BioBasic® C18, 300 Å pore size; 75 µm × 10 cm; tip 15 µm, New Objective, Woburn, MA, USA). Solution A comprised 0.1% formic acid, and Solution B acetonitrile and 0.1% formic acid. The flow rate began at 100% A at 10 µL/min for 3 min. Next, it was raised to 70 µL/min for 6.9 min at 100% A and the gradient commenced at 100% A and 0% B. The gradient was increased to 50% B in 60 min, then to 90% B in 5 min and later diminished to 0% B in 5 min and sustained at 100% A for 10 min. The complete program duration was 110 min. By the PepFinder Kit, the flow was separated in a 1:100 ratio. Hence, the factual flow rate injected into the mass spectrometer was 0.5 µL/min. The HPLC was coupled to a Finnigan LCQ Deca XP Plus ion trap mass spectrometer supplied with a nanospray ionization font. Spray voltage was established at 2.5 kV, and the equipment was managed in a data-dependent method, in which one MS scan was captured in the 300−1600 m/z range followed by MS/MS acquisition using collision-induced segregation of the 10 most intense ions from the MS scan. Dynamic peak exclusion was used to avoid the same m/z being chosen for the following 120 s. Tandem mass spectra were extracted by Xcalibur software (version 2.0; Thermo scientific, Waltham, MA). The subsequent MS/MS spectra were searched through a MASCOT search engine (Matrix Science, London, UK) against the NCBI NR database in the taxa Chordata with a 1.20 Da parent tolerance and 0.60 Da fragment tolerance. Iodoacetamide derivatives of cysteine and oxidation of methionine were detailed in MASCOT as fixed and variable modifications, respectively. Scaffold (version Scaffold_2_04_00, Proteome Software Inc., Portland, OR) was used to confirm MS/MS-based peptide and protein description. Peptide distinguishing were credited if they exceeded certain database search engine thresholds. MASCOT descriptions required ion scores higher than the associated identity scores and 20, 30, 40, and 40 for singly, doubly, triply, and quadruply charged peptides. X! Tandem identifications needed at least –Log (Expect Scores) scores of greater than 2.0. Protein identifications were considered if they had at least 2 identified peptides.

4.4. Enzymatic Deglycosylation

In denaturing conditions, for protein deglycosylation, toxin (3.4 µg) were set in 20% SDS for 1 min at 95 °C. After adding 0.2 M sodium phosphate buffer, 0.08% sodium azide, 0.5 M EDTA pH 8, 10% nonidet P-40, incubation was prolonged for 2 min at 95 °C. After cooling, 1 U.µL^{-1} of N-glycosidase F (Roche, Mannheim, Germany) or 25 mU.50 µL^{-1} O-glycosidase was added, and the mixture was incubated for 18 h at 37 °C. The deglycosylation profiles were evaluated by SDS-PAGE.

4.5. Immunoblotting

The purified protein (10 µg) of *T. maculosa* venom was analyzed by 12% SDS-PAGE under non-reducing conditions. Venom from *T. nattereri* (VT*n*) was used as a control.

After SDS-PAGE and transfer to nitrocellulose membrane (pore size = 0.2 μm, Schleicher and Schüll, Dassel, Germany), the toxin was detected using plasma from mice sensitized with *T. nattereri* venom (1:20 dilution) or from Nattectin-sensitized mice followed by Goat anti-mouse IgG HRP (sc-2005 Santa Cruz, at 1:2000) as a secondary antibody (Ab).

4.6. Determination of TmC4-47.2 Toxin Binding Specificity to Carbohydrates

The toxin was cut from 12% SDS-PAGE gel and eluted in PBS to identify toxin binding specificity using a competition assay performed according to Haab [55] using the DIG Glycan Differentiation Kit (#11210238001, Roche Applied Science, Germany). The pure control glycoproteins: carboxypeptidase Y, transferrin, fetuin, and asialofetuin or after pre-incubated with 1 μg of the toxin TmC4-47.2 for 1 h at 37 °C were administered in a 12% SDS–PAGE gel and transferred to a nitrocellulose membrane. Toxin/carbohydrate binding was revealed by incubation with different DIG-labeled lectins and alkaline phosphatase-conjugated anti-DIG antibodies (Roche Molecular Biosciences). Characteristics of specific binding of lectins to carbohydrate moieties used to identify these structures in this study: GNA: *Galanthus nivalis* agglutinin specific to mannose: α(1–2), α(1–3), or α(1–6) linked to mannose; SNA: *Sambucus nigra* agglutinin specific to Sialic acid: α(2–6) to GalNAc; MAA: *Maackia amurensis* agglutinin specific to Sialic acid: α(2–3) to galactose N-linked α(2–3) to galactose O-linked; PNA: *Arachis hypogaea* Peanut agglutinin specific to Core and terminal galactose: Gal-β(1–3)-N-acetylgalactosamine; and DSA: *Datura stramonium* agglutinin specific to Core galactose: Gal-β(1–4)-N-acetylglucosamine.

4.7. Hemagglutinating and Antimicrobial Activities

Human erythrocytes (type A) were collected in 0.15 M citrate buffer, pH 7.4, and washed 3 times by centrifugation with 0.15 M PBS, pH 7.4. To assess the hemolytic activity, aliquots of 10 μL of the selected toxin at 0.01, 0.1, 1, or 10 μg were put in 50 μL in a 3% suspension of erythrocytes in wells of U-shaped bottom plates and incubated for 3 h at room temperature. Solutions of isolated toxin at 10 μg previously incubated with D-Galactosamine (12662, Sigma) or D-Mannose (M2069, Sigma) both at 10 or 30 mM for 1 h were also usedThe hemagglutinating activity was verified by reading the absorbance at 595 nm of each well in a plate reader. Erythrocytes incubated with water were used as positive control (100% hemolysis). Antimicrobial potential was monitored by a liquid growth inhibition assay against *M. luteus* A270, *E. coli* SBS 363, and *C. albicans* MDM8, as described by Rossi et al. [56]. Pre-inoculum of the strains was prepared in Poor Broth (PB broth, 1.0 g peptone in 100 mL of H_2O containing 86 mM NaCl at pH 7.4; 217 mOsM for *M. luteus* and *E. coli* and 1.2 g potato dextrose in 100 mL of H_2O at pH 5.0; 79 mOsM for *C. albicans*), at 37 °C under agitation. The absorbance at 595 nm was set on, and one aliquot of this solution was taken to get cells in logarithmic growth ($A_{595\ nm}$ ~0.6) and diluted 600 times ($A_{595\ nm}$ = 0.0001). The isolated toxin or Nattectin, both at 10 μg, was dissolved in sterile Milli-Q water to 100 μL of PB broth. Tetracycline and Gomesin were used as inhibitor controls. After 18 h of incubation at 30 °C, the inhibition of bacterial growth was determined by measuring absorbance at 595 nm.

4.8. In Vivo Experimental Protocol for Intravital Microscopy

Initially, Balb/c mice were injected by the intrascrotal route with 50 μL of TmC4-47.2 at 10 μg and rested for 3 h. Independent groups of mice were pre-treated with 500 μL of intraperitoneal (i.p.) injection of the purified rat anti-mouse CD29 (beta 1 integrin, 14-0299-82, eBioscience), hamster anti-mouse CD49a (alpha 1 integrin, PA5-95563, Thermo Fisher eBioscience), rat anti-mouse CD49b (alpha 2 integrin, ab238665, Abcam), rat anti-mouse CD49e (alpha 5 integrin, ab221606, Abcam), and rat anti-mouse CD106—VCAM-1, 14-1061-82, Thermo Fisher eBioscience) monoclonal antibodies at 10 μg mL^{-1}, 30 min before TmC4-47.2 injection. Control animals received the same amount of control isotype IgG. Negative control was injected with intrascrotal sterile PBS. The study of the microvascular system was performed with an optical microscope (Axio Imager A.1, Carl-Zeiss,

Oberkochen, DE) coupled to a photographic camera (IcC 1, Carl-Zeiss, Germany) through a 10/025 longitudinal distance objective/numerical aperture and 1.6 optovar. The surgical cremaster preparation was handled as described previously [12]. Mice were anesthetized by an i.p. injection of 2% Xylazine—(Calmiun®, Agener União, São Paulo, Brazil) and with 0.5 g Kg^{-1} of ketamine (Holliday-Scott SA, Buenos Aires, Argentina). The scrotum was exposed, and the cremaster muscle reached. Following the incision with cautery and spreading the muscle over a cover glass, the epididymis and testis were mobilized and pinned aside, allowing the microscopic access to the muscle microcirculation. The exposed tissue was superfused with 37 °C warmed bicarbonate-buffered salines, pH 7.4. The post-capillary venules with a diameter of 25–40 μm were chosen, and the interaction of leukocytes with the luminal surface of the venular endothelium was evaluated, counting the number of rolling leukocytes every 10 min after application of inflammatory agent for 30 min. Rolling leukocytes were defined as those moving at a velocity less than erythrocytes and demonstrated a clear rolling motion. The number of adherent cells was expressed as the number per 100 μm length of venule. The experiments were carried out under the National Council for Animal Experiment Control (CONCEA) and approved by the Butantan Institute's Animal Use Ethics Commission (CEUAIB #275/06).

4.9. Acute Inflammation Induced by TmC4-47.2

Balb/c mice were intraperitoneally injected with TmC4-47.2 at 10 μg in 500 μL, according to Lima et al. [32]. As a negative control, mice were injected i.p. with PBS. After 2, 16, and 24 h, the peritoneum exudates were harvested for total and differential cell count and protein determination. IL-6, TNF-α, IL-1β, KC, eotaxin, and MCP-1 were analyzed using a specific two-site sandwich ELISA with OpEIA Kits (BD-Pharmingen, San Diego, CA, USA). The total leukocyte count was performed in Turk solution, and for differential counts, neutrophils ($Ly6G^+$), eosinophils ($CCR3^+$), and macrophages ($F4/80^+$) were identified by flow cytometry and based on staining and morphologic characteristics using a light microscope Axio Imager A1 (Carl Zeiss, Germany) with an AxioCam ICc1 digital camera (Carl Zeiss).

4.10. Statistical Analysis

All values were expressed as mean ± SEM. Experiments using 3 to 5 mice per group were performed independently two times. Parametric data were evaluated using analysis of variance, followed by the Bonferroni test for multiple comparisons. Non-parametric data were assessed using the Mann-Whitney test. Differences were considered statistically significant at $p < 0.05$ using GraphPad Prism (Graph Pad Software, v6.02, 2013, La Jolla, CA, USA).

Supplementary Materials: The following are available online at https://www.mdpi.com/article/10.3390/toxins14010002/s1. Figure S1: (A) Fractionation process of the *Thalassophryne maculosa* venom. A total of 5 mg of venom was applied to a semi-preparative reversed-phase C18 column coupled to a high-pressure liquid chromatography (HPLC) system. The gradient used was from 20 to 80% buffer B in 60 min under 5 mL/min flow rate. The absorbance was monitored at 214 nm. (B) The protein content was evaluated by 12% SDS-PAGE electrophoresis in polyacrylamide gel (10 μg/well). Samples 1, 2, and 3 correspond to the fractions obtained by chromatography. MW corresponds to molecular mass markers, and the band circled in red refers to the TmC4-47.2.; Figure S2: (A) Second fractioning step for toxin purification. Fraction 2 (1 mg mL^{-1}) obtained in the first chromatography was applied to a reversed-phase analytical C8 column coupled to a high-pressure liquid chromatography (HPLC) system. The gradient used was 20 to 80% of buffer B over 35 min under a flow rate of 1 mL/min. The absorbance was monitored at 214 nm. (B) The protein content of fractions 1, 2s, 2d, 3 and 4 was evaluated by 12% SDS-PAGE electrophoresis on polyacrylamide gel (10 μg/well). MW corresponds to molecular mass markers. The bands highlighted in red refer to TmC4-47.2 toxin; Figure S3: (A) Third fractionation for purification of *Thalassophryne maculosa* TmC4-47.2 toxin. Fractions 2s, 2d and 3 were mixed and 1 mg mL^{-1} of the pool was applied to a reversed-phase analytical C8 column coupled to a high-pressure liquid chromatography (HPLC) system. The gradient used was 20 to 80% buffer B over 35 min under 1 mL min^{-1} flow rate. The absorbance was monitored at 214 nm. (B) The protein content of fractions 1, 2s and 2d was evaluated by 12% SDS-PAGE electrophoresis on polyacrylamide gel (10 μg/well). MW corresponds to molecular mass markers and the bands highlighted in red refer to TmC4-47.2 toxin.

Author Contributions: Methodology, validation, formal analysis, investigation, data curation, writing—original draft preparation, P.I.S.J., K.C., A.L.A.M., L.B.-L., G.R.D.; Conceptualization, resources, data curation, formal analysis, writing—review and editing, visualization, supervision, project administration, funding acquisition, M.L.-F., I.S.-R. and C.L. All authors have read and agreed to the published version of the manuscript.

Funding: This research was funded by the São Paulo Research Foundation—FAPESP (#2013/07467-1) and in part by the Coordination for the Improvement of Higher Education Personnel—CAPES (#001) and National Council for Scientific and Technological Development—CNPq (#305414/2019-4).

Institutional Review Board Statement: The study was conducted according to the animal welfare standards, and all necessary permits for the capture of *T. maculosa* and to collect the venom were obtained from the Escuela de Ciencias Aplicadas del Mar, Núcleo Nueva Esparta, Universidad de Oriente, Isla Margarita; Venezuela (Permit Number #30/07). Additional experiments were carried out under the National Council for Animal Experiment Control (CONCEA) and approved by the Butantan Institute's Animal Use Ethics Commission (CEUAIB #275/06).

Informed Consent Statement: Not applicable.

Acknowledgments: We thank the São Paulo Research Foundation—FAPESP, CAPES, and CNPq for the financial support. Additional thanks to Douglas Boletini-Santos for the assistance in the data collection and to the Butantan Institute Immunoregulation Unit staff. We also thank the anonymous reviewers for their careful reading and insightful comments and suggestions.

Conflicts of Interest: The authors declare no conflict of interest. The funders had no role in the design of the study; in the collection, analyses, or interpretation of data; in the writing of the manuscript, or in the decision to publish the results.

References

1. Ng, T.B.; Cheung, R.C.F.; Wing Ng, C.C.; Fang, E.f.; Wong, J.H. A review of fish lectins. *Curr. Protein Pept. Sci.* **2015**, *16*, 337–351. [CrossRef] [PubMed]
2. Vasta, G.R.; Nita-Lazar, M.; Giomarelli, B.; Ahmed, H.; Du, S.; Cammarata, M.; Parrinello, N.; Bianchet, M.A.; Amzel, L.M. Structural and functional diversity of the lectin repertoire in teleost fish: Relevance to innate and adaptive immunity. *Dev. Comp. Immunol.* **2011**, *35*, 1388–1399. [CrossRef] [PubMed]
3. Brinchmann, M.F.; Patel, D.M.; Pinto, N.; Iversen, M.H. Functional aspects of fish mucosal lectins-interaction with non-self. *Molecules* **2018**, *23*, 1119. [CrossRef] [PubMed]
4. Lopes-Ferreira, M.; Magalhães, G.S.; Fernandez, J.H.; Junqueira-de-Azevedo, I.L.; Le Ho, P.; Lima, C.; Valente, R.H.; Moura-da-Silva, A.M. Structural and biological characterization of Nattectin, a new C-type lectin from the venomous fish. *Thalass. Nattereri Biochim.* **2011**, *93*, 971–980. [CrossRef] [PubMed]
5. Lopes-Ferreira, M.; Barbaro, K.C.; Cardoso, D.F.; Moura-da-Silva, A.M.; Mota, I. *Thalassophryne nattereri* fish venom: Biological and biochemical characterization and serum neutralization of its toxic activities. *Toxicon* **1998**, *36*, 405–410. [CrossRef]
6. Lopes-Ferreira, M.; Moura-da-Silva, A.M.; Mota, I.; Takehara, H.A. Neutralization of *Thalassophryne nattereri* (niquim) fish venom by an experimental antivenom. *Toxicon* **2000**, *38*, 1149–1156. [CrossRef]
7. Almeida, V.G.; Rocha, C.M. Registro de acidentes com peixes peçonhentos e/ou venenosos. *Rev. Soc. Bras. Toxicol* **1989**, *2*, 49–51.
8. Haddad Junior, V.; Pardal, P.P.; Cardoso, J.L.; Martins, I.A. The venomous toadfish *Thalassophryne nattereri* (niquim or miquim): Report of 43 injuries provoked in fishermen of Salinopolis (Para state) and Aracaju (Sergipe state), Brazil. *Rev. Do Inst. De Med. Trop. De São Paulo* **2003**, *45*, 221–223. [CrossRef] [PubMed]
9. Faco, P.E.; Bezerra, G.P.; Barbosa, P.S.; Martins, A.M.; Guimaraes, J.A.; Ferreira, M.L.; Monteiro, H.S. Epidemiology of the injuries caused by *Thalassophryne nattereri* (niquim) in Ceara State (1992–2002). *Rev. Soc. Bras. Med. Trop.* **2005**, *38*, 479–482. [CrossRef] [PubMed]
10. Fonseca, L.A.; Lopes-Ferreira, M. Clinical and experimental studies regarding poisoning caused by a fish *Thalassophryne nattereri* (Niquim). *An. Bras. De Dermatol.* **2000**, *75*, 435–443.
11. Lopes-Ferreira, M.; Nunez, J.; Rucavado, A.; Farsky, S.H.; Lomonte, B.; Ângulo, Y.; Moura Da Silva, A.M.; Gutierrez, J.M. Skeletal muscle necrosis and regeneration after injection of *Thalassophryne nattereri* (niquim) fish venom in mice. *Int. J. Exp. Pathol.* **2001**, *82*, 55–64. [CrossRef] [PubMed]

12. Lopes-Ferreira, M.; Moura-Da-Silva, A.M.; Piran-Soares, A.A.; Ângulo, Y.; Lomonte, B.; Gutierrez, J.M.; Farsky, S.H. Hemostatic effects induced by *Thalassophryne nattereri* fish venom: A model of endothelium-mediated blood flow impairment. *Toxicon* **2002**, *40*, 1141–1147. [CrossRef]
13. Lopes-Ferreira, M.; Emim, J.A.; Oliveira, V.; Puzer, L.; Cezari, M.H.; Araujo, M.S.; Juliano, L.; Lapa, A.J.; Souccar, C.; Moura-Da-Silva, A.M. Kininogenase activity of *Thalassophryne nattereri* fish venom. *Biochem. Pharm.* **2004**, *68*, 2151–2157. [CrossRef] [PubMed]
14. Lopes-Ferreira, M.; Sosa-Rosales, I.; Bruni, F.M.; Ramos, A.D.; Vieira Portaro, F.C.; Conceição, K.; Lima, C. Analysis of the intersexual variation in *Thalassophryne maculosa* fish venoms. *Toxicon* **2016**, *115*, 70–80. [CrossRef] [PubMed]
15. Sosa-Rosales, J.; D'suze, G.; Salazar, V.; Fox, J.; Sevcik, C. Purification of a myotoxin from the toadfish *Thalassophryne maculosa* (günter) venom. *Toxicon* **2005**, *45*, 147–153. [CrossRef] [PubMed]
16. Pareja-Santos, A.; Oliveira Souza, V.M.; Bruni, F.M.; Sosa-Rosales, J.I.; Lopes-Ferreira, M.; Lima, C. Delayed polymorphonuclear leukocyte infiltration is an important component of *Thalassophryne maculosa* venom pathogenesis. *Toxicon* **2008**, *52*, 106–114. [CrossRef] [PubMed]
17. Sosa-Rosales, J.I.; Piran-Soares, A.A.; Farsky, S.H.; Takehara, H.A.; Lima, C.; Lopes-Ferreira, M. Important biological activities induced by *Thalassophryne maculosa* fish venom. *Toxicon* **2005**, *45*, 155–161. [CrossRef] [PubMed]
18. Yang, H.; Liu, N.; Liu, S. Determination of peptide and protein disulfide linkages by maldi mass spectrometry. *Top. Curr. Chem.* **2013**, *331*, 79–116.
19. Zelensky, A.N.; Gready, J.E. The c-type lectin-like domain superfamily. *Febs J.* **2005**, *272*, 6179–6217. [CrossRef] [PubMed]
20. Cummings, R.D. The repertoire of glycan determinants in the human glycome. *Mol. Biosyst.* **2009**, *5*, 1087–1104. [CrossRef]
21. Goldstein, I.J.; Hayes, C.E. The lectins: Carbohydrate-binding proteins of plants and animals. *Adv. Carbo. Chem. Biochem.* **1978**, *35*, 127–340.
22. Chou, R.C.; Kim, N.D.; Sadik, C.D.; Seung, E.; Lan, Y.; Byrne, M.H.; Haribabu, B.; Iwakura, Y.; Luster, A.D. Lipid-cytokine-chemokine cascade drives neutrophil recruitment in a murine model of inflammatory arthritis. *Immunity* **2010**, *33*, 266–278. [CrossRef] [PubMed]
23. Mezu-Ndubuisi, O.J.; Maheshwari, A. The role of integrins in inflammation and angiogenesis. *Pediatr. Res.* **2021**, *89*, 1619–1626. [CrossRef] [PubMed]
24. Lopes-Ferreira, M.; Grund, L.Z.; Lima, C. *Thalassophryne nattereri* fish venom: From the envenoming to the understanding of the immune system. *J. Venom. Anim. Toxins Incl. Trop. Dis.* **2014**, *20*, 35. [CrossRef] [PubMed]
25. Grund, L.Z.; Souza, V.M.; Faquim-Mauro, E.L.; Lima, C.; Lopes-Ferreira, M. Experimental immunization with *Thalassophryne nattereri* fish venom: Striking IL-5 production and impaired of B220+ cells. *Toxicon* **2006**, *48*, 499–508. [CrossRef] [PubMed]
26. Bruni, F.M.; Coutinho, E.; Andrade-Barros, A.I.; Grund, L.Z.; Lopes-Ferreira, M.; Lima, C. Anaphylaxis induced by *Thalassophryne nattereri* venom in mice is an IgE/IgG1-mediated, IL-4-dependent phenomenon. *Sci. Rep.* **2020**, *10*, 584. [CrossRef]
27. Magalhaes, G.S.; Lopes-Ferreira, M.; Junqueira-De-Azevedo, I.L.; Spencer, P.J.; Araujo, M.S.; Portaro, F.C.; Ma, L.; Valente, R.H.; Juliano, L.; Fox, J.W.; et al. Natterins, a new class of proteins with kininogenase activity characterized from *Thalassophryne nattereri* fish venom. *Biochimie* **2005**, *87*, 687–699. [CrossRef]
28. Magalhaes, G.S.; Junqueira-De-Azevedo, I.L.; Lopes-Ferreira, M.; Lorenzini, D.M.; Ho, P.L.; Moura-Da-Silva, A.M. Transcriptome analysis of expressed sequence tags from the venom glands of the fish *Thalassophryne nattereri*. *Biochimie* **2006**, *88*, 693–699. [CrossRef]
29. Komegae, E.N.; Grund, L.Z.; Lopes-Ferreira, M.; Lima, C. The longevity of Th2 humoral response induced by proteases natterins requires the participation of long-lasting innate-like B cells and plasma cells in spleen. *PLoS ONE* **2013**, *8*, e67135. [CrossRef]
30. Komegae, E.N.; Grund, L.Z.; Lopes-Ferreira, M.; Lima, C. TLR2, TLR4 and the MyD88 signaling are crucial for the in vivo generation and the longevity of long-lived antibody-secreting cells. *PLoS ONE* **2013**, *8*, e71185. [CrossRef]
31. Ferreira, M.J.; Lima, C.; Lopes-Ferreira, M. Anti-inflammatory effect of Natterins, the major toxins from the *Thalassophryne nattereri* fish venom is dependent on TLR4/MyD88/PI3K signaling pathway. *Toxicon* **2014**, *87*, 54–67. [CrossRef] [PubMed]
32. Lima, C.; Disner, G.R.; Falcão, M.A.P.; Seni-Silva, A.C.; Maleski, A.L.A.; SouzA, M.M.; Reis Tonello, M.C.; Lopes-Ferreira, M. The natterin proteins diversity: A review on phylogeny, structure, and immune function. *Toxins* **2021**, *13*, 538. [CrossRef]
33. Komegae, E.N.; Ramos, A.D.; Oliveira, A.K.; Serrano, S.M.; Lopes-Ferreira, M.; Lima, C. Insights into the local pathogenesis induced by fish toxins: Role of natterins and nattectin in the disruption of cell-cell and cell-extracellular matrix interactions and modulation of cell migration. *Toxicon* **2011**, *58*, 509–517. [CrossRef] [PubMed]
34. Ishizuka, E.K.; Ferreira, M.J.; Grund, L.Z.; Coutinho, E.M.; Komegae, E.N.; Cassado, A.A.; Bortoluci, K.R.; Lopes-Ferreira, M.; Lima, C. Role of interplay between IL-4 and IFN-γ in the in regulating M1 macrophage polarization induced by Nattectin. *Int. Immunopharmacol.* **2012**, *14*, 513–522. [CrossRef] [PubMed]
35. Saraiva, T.C.; Grund, L.Z.; Komegae, E.N.; Ramos, A.D.; Conceição, K.; Orii, N.M.; Lopes-Ferreira, M.; Lima, C. Nattectin a fish C-type lectin drives Th1 responses in vivo: Licenses macrophages to differentiate into cells exhibiting typical DC function. *Int. immunopharmacol.* **2011**, *11*, 1546–1556. [CrossRef] [PubMed]
36. Drickamer, K.; Fadden, A.J. Genomic analysis of c-type lectins. *Biochem. Soc. Symp.* **2002**, *69*, 59–72. [CrossRef]
37. Moremen, K.W.; Tiemeyer, M.; Nairn, A.V. Vertebrate protein glycosylation: Diversity, synthesis and function. *Nat. Rev. Mol. Cell Biol.* **2012**, *13*, 448–462. [CrossRef] [PubMed]
38. Drickamer, K. C-type lectin-like domains. *Curr. Opin. Struct. Biol.* **1999**, *9*, 585–590. [CrossRef]

39. Yang, R.; Rabinovich, G.; Liu, F. Galectins: Structure, function and therapeutic potential. *Expert Rev. Mol. Med.* **2008**, *10*, E17. [CrossRef]
40. Vasta, G.R. Roles of galectins in infection. *Nat. Rev. Microbiol.* **2009**, *7*, 424–438. [CrossRef] [PubMed]
41. Thulasitha, W.S.; Umasuthan, N.; Whang, I.; Nam, B.H.; Lee, J. Antimicrobial response of galectin-1 from rock bream *Ople-gnathus fasciatus: Molecular*, transcriptional, and biological characterization. *Fish Shellfish. Immunol.* **2016**, *50*, 66–78. [CrossRef] [PubMed]
42. Nita-Lazar, M.; Mancini, J.; Feng, C.; González-Montalbán, N.; Ravindran, C.; Jackson, S.; de Las Heras-Sánchez, A.; Giomarelli, B.; Ahmed, H.; Haslam, S.M.; et al. The zebrafish galectins Drgal1-L2 and Drgal3-L1 bind in vitro to the infectious hematopoietic necrosis virus (IHNV) glycoprotein and reduce viral adhesion to fish epithelial cells. *Dev. Comp. Immunol.* **2016**, *55*, 241–252. [CrossRef] [PubMed]
43. Takayama, S.; Saitoh, E.; Kimizuka, R.; Yamada, S.; Kato, T. Effect of eel galectin ajl-1 on periodontopathic bacterial biofilm formation and their lipopolysaccharide-mediated inflammatory cytokine induction. *Int. J. Antimicrob. Agents* **2009**, *34*, 355–359. [CrossRef] [PubMed]
44. Liang, Z.G.; Li, L.; Chen, S.N.; Mao, M.G.; Nie, P. Expression and antibacterial analysis of galectin-8 and -9 genes in mandarin fish, *Siniperca Chuatsi. Fish Shellfish. Immunol.* **2020**, *107*, 463–468. [CrossRef] [PubMed]
45. Vasta, G.R.; Ahmed, H.; Shao, J.D.; Henrikson, D. Galectins in teleost fish: Zebrafish (*Danio rerio*) as a model species to address their biological roles in development and innate immunity. *Glycoconj. J.* **2004**, *21*, 503–521. [CrossRef]
46. Nishi, N.; Shoji, H.; Seki, M.; Itoh, A.; Miyanaka, H.; Yuube, K.; Hirashima, M.; Nakamura, T. Galectin-8 modulates neutrophil function via interaction with integrin Alpha, M. *Glycobiology* **2003**, *13*, 755–763. [CrossRef]
47. Matsushita, N.; Nishi, N.; Seki, M.; Matsumoto, R.; Kuwabara, I.; Liu, F.T.; Hata, Y.; Nakamura, T.; Hirashima, M. Requirement of divalent galactoside-binding activity of ecalectin/galectin-9 for eosinophil chemoattraction. *J. Biol. Chem.* **2000**, *275*, 8355–8360. [CrossRef] [PubMed]
48. McDonald, B.; Pittman, K.; Menezes, G.B.; Hirota, S.A.; Slaba, I.; Waterhouse, C.C.M.; Beck., P.L.; Muruve, D.A.; Kubes, P. Intravascular danger signals guide neutrophils to sites of sterile inflammation. *Science* **2010**, *330*, 362–366. [CrossRef]
49. Mcdonald, J.A.; Quade, B.J.; Broekelmann, T.J.; Lachance, R.; Forsman, K.; Hasegawa, E.; Akiyama, S. Fibronectin's cell-adhesive domain and an amino-terminal matrix assembly domain participate in its assembly into fibroblast pericellular matrix. *J. Biol. Chem.* **1987**, *262*, 2957–2967. [CrossRef]
50. Carroll, J.M.; Romero, M.R.; Watt, F.M. Suprabasal integrin expression in the epidermis of transgenic mice results in developmental defects and a phenotype resembling psoriasis. *Cell* **1995**, *83*, 957–968. [CrossRef]
51. Lishko, V.K.; Yakubenko, V.P.; Ugarova, T.P. The interplay between integrins alphambeta2 and Alpha5beta1 during cell migration to fibronectin. *Exp. Cell Res.* **2003**, *283*, 116–126. [CrossRef]
52. Rao, S.P.; Ge, X.N.; Sriramarao, P. Regulation of eosinophil recruitment and activation by galectins in allergic asthma. *Front. Med.* **2017**, *4*, 68. [CrossRef] [PubMed]
53. Laemmli, U.K. Cleavage of structural proteins during assembly of the head of bacteriophage T4. *Nature* **1970**, *227*, 680–685. [CrossRef] [PubMed]
54. Zhang, Z.; Marshall, A.G. A universal algorithm for fast and automated charge state deconvolution of electrospray mass-to-charge ratio spectra. *J. Am. Soc. Mass Spectrom.* **1998**, *3*, 225–233. [CrossRef]
55. Haab, B.B. Antibody-lectin sandwich arrays for biomarker and glycobiology studies. *Expert Rev. Proteom.* **2010**, *7*, 9–11. [CrossRef] [PubMed]
56. Rossi, D.C.; Muñoz, J.E.; Carvalho, D.D.; Belmonte, R.; Faintuch, B.; Borelli, P.; Miranda, A.; Taborda, C.P.; Daffre, S. Therapeutic use of a cationic antimicrobial peptide from the spider *Acanthoscurria gomesiana* in the control of experimental candidiasis. *BMC Microbiol.* **2012**, *12*, 28. [CrossRef]

 *toxins*

MDPI

Article

Antimicrobial Activity of Snake β-Defensins and Derived Peptides

Nancy Oguiura [1,*], Poliana Garcia Corrêa [1], Isabella Lemos Rosmino [1], Ana Olívia de Souza [2] and Kerly Fernanda Mesquita Pasqualoto [3]

1 Ecology and Evolution Laboratory, Instituto Butantan, Sao Paulo 05503-900, SP, Brazil; poliana.correa@butantan.gov.br (P.G.C.); isabellarosmino@hotmail.com (I.L.R.)

2 Development and Innovation Laboratory, Instituto Butantan, Sao Paulo 05503-900, SP, Brazil; ana.souza@butantan.gov.br

3 Alchemy–Innovation, Research & Development, University of Sao Paulo, Sao Paulo 05508-000, SP, Brazil; kfmpasqualoto@alchemydrugs.com.br

* Correspondence: nancy.oguiura@butantan.gov.br

Abstract: β-defensins are antimicrobial peptides presenting in vertebrate animals. They participate in innate immunity, but little is known about them in reptiles, including snakes. Although several β-defensin genes were described in Brazilian snakes, their function is still unknown. The peptide sequence from these genes was deduced, and synthetic peptides (with approximately 40 amino acids and derived peptides) were tested against pathogenic bacteria and fungi using microbroth dilution assays. The linear peptides, derived from β-defensins, were designed applying the bioisosterism strategy. The linear β-defensins were more active against *Escherichia coli*, *Micrococcus luteus*, *Citrobacter freundii*, and *Staphylococcus aureus*. The derived peptides (7–14 mer) showed antibacterial activity against those bacteria and on *Klebsiella pneumoniae*. Nonetheless, they did not present activity against *Candida albicans*, *Cryptococcus neoformans*, *Trychophyton rubrum*, and *Aspergillus fumigatus* showing that the cysteine substitution to serine is deleterious to antifungal properties. Tryptophan residue showed to be necessary to improve antibacterial activity. Even though the studied snake β-defensins do not have high antimicrobial activity, they proved to be attractive as template molecules for the development of antibiotics.

Keywords: β-defensins; snakes; antimicrobial activity; bioisosterism; peptides

Key Contribution: We tested many snake β-defensins against bacteria from oral flora, *Micrococcus luteus*, and *Escherichia coli*. This work is the first testing of the antibacterial activity of snake β-defensins besides crotamine. A bioisosterism approach was used to design the derived peptides from β-defensins from *Lachesis muta*, *Bothrops jararaca*, and *Crotalus durissus* snakes. Our results have shown that (i) the defensin's C-terminal portion seems to be crucial against bacteria; (ii) the presence of tryptophan residue into the derived peptide's sequence plays an important role for the antibacterial activity; and (iii) the Cys to Ser substitution has abolished the derived peptides' antifungal activity.

Citation: Oguiura, N.; Corrêa, P.G.; Rosmino, I.L.; de Souza, A.O.; Pasqualoto, K.F.M. Antimicrobial Activity of Snake β-Defensins and Derived Peptides. *Toxins* **2022**, *14*, 1. https://doi.org/10.3390/toxins14010001

Received: 15 October 2021
Accepted: 29 November 2021
Published: 21 December 2021

1. Introduction

With the frightening advent of the global increase of microbial resistance to conventional antibiotics, the search for alternatives has become of utmost importance, and the industry, as well as the regulatory authorities, are realizing the potential of antimicrobial peptides. Since the last decade, antimicrobial peptides have been trialed in clinical phases [1].

In drug development, peptides' properties have been considered more convenient due to their high affinity for the target and selective biological activities [2]. Bacterial resistance is a global health problem due to the indiscriminate use of antibiotics in humans, as well as in animals and in agricultural production that needs multipronged solutions [3]. To contribute to this campaign, many antimicrobial peptides have been discovered and some are in the clinical development phase [4]. For instance, molecules from innate immunity, such as cathelicidins present in snake venoms, have been reported as potentially active against some bacterial strains [5–10], including antibiofilm activity [11,12].

Snakes have developed many active peptides for predation and in their venom composition, many classes of proteins as phospholipases A_2 (PLA_2) [13], L-amino-acid oxidase (LAAO) [14], metalloproteases [15], cathelicidins [8] and crotamine [16–18] are also described as showing antimicrobial activity. Many attempts have been made to shorten these proteins and discover the active site [19–22]. Crotamine is a small basic myotoxin present in the venom of the rattlesnake *C. durissus terrificus* and has a β-defensin structure [23,24] and antimicrobial activity [16–18]. Crotamine and small basic myotoxins from this family are unique and only present in rattlesnake venom [23].

β-defensins from various vertebrates have been studied, and little is known about these molecules and the innate immunity in snakes. β-defensin-like genes with unknown functions have also been described in Brazilian snakes using a PCR approach [25–27]. In the current study, we evaluated the antimicrobial activity of snake β-defensins, both from rattlesnake venom (crotamine) and non-venom β-defensins. Its primary sequences were deduced from genomic sequences and synthesized, except for crotamine purified from rattlesnake venom. Moreover, the amino acid sequences were used to design shorter derived peptides, which can be obtained using simpler and more economical procedures. The snake β-defensins and derived peptides were assayed against microorganisms relevant to the snake's biology and human health because they can cause opportunistic infections.

2. Results

The sequences of mature peptides were deduced from gene codifying sequences and synthesized, except crotamine purified from venom. They were tested in linear form because the linearization could not affect the antibacterial activity, and the linear form facilitates the development of shorter peptides. The alkylation was done to avoid the dimerization and the intramolecular cyclization of peptides with free thiol groups. Besides, the linear form facilitates the development of shorter peptides. Alkylated peptides were purified by high-performance liquid chromatography (HPLC) and analyzed by MALDI-TOF-MS.

The β-defensins were tested using a microbroth dilution assay. It consists of incubating the bacteria (4.105 colony-forming unit—CFU/mL) and the peptide (from 1 to 512 μg/mL) in liquid broth. After the incubation at 37 °C during the night, the bacterial growth was detected by spectrophotometry at 600 nm. If the measure is similar to the broth, it indicates bacterial growth. The molarity was calculated only to defensins that showed antibacterial activity to compare with the derived peptides. The β-defensins were tested against *B. jararaca* oral flora because it would be of biological and medical interest. In addition, we tested on *E. coli* ATCC 25922 and *M. luteus* that are common Gram-negative and Gram-positive bacteria used in antibacterial tests.

Table 1 shows the antibacterial effect of the snake β-defensins from venom (crotamine) and tissues indicating the minimal inhibition concentration that is the lowest concentration of peptide that results in no visible bacterial growth. It was possible to observe that the linearization of crotamine did not modify its antibacterial activity significantly. These β-defensins did not inhibit the bacterial growth of *Providencia rettgeri*, *Serratia marcescens*, *Morganella morganii*, and *Klebsiella pneumonia* up to 512 μg/mL, so the results were not shown in the table. *M. luteus* was the most sensitive bacterium to these peptides. The most active defensins present the highest cationic net charge, DefBm02 (+11), DefbBm03 (+10), DefbLm02 (+8), and crotamine (+7). On the contrary, the β-defensins DefbBju01 (+7), DefbBn02 (+2), and crotasin (−1) did not show antibacterial activity against any strains used in this study. In addition, the substitution in 32nd position, Arg (defbBm02) for Gln (defbBm03), has caused the loss of activity against *Staphylococcus aureus*.

Table 1. Antibacterial effect expressed by the minimal inhibitory concentration (MIC) of snake β-defensins on bacteria.

	MIC (μM)			
Samples	*E. coli*	*M. luteus*	*C. freundii*	*S. aureus*
Crotamine (+7)	13.3	1.7	6.7	>105
Linear crotamine	26.6	1.7	26.7	>105
DefbBm02 (+11)	3.2	0.8	1.6	12.8
DefbBm03 (+10)	1.6	1.6	3.2	>103
DefbBd03 (+6)	>115	0.9	>115	>115
DefbBj01 (+4)	>105	3.2	>105	>105
DefbBju01 (+7)	>111	>111	>111	>111
DefbBn02 (+2)	>113	>113	>113	>113
DefbLm01 (+2)	>113	28.4	>113	>113
DefbLm02 (+8)	3.4	0.9	3.4	6.8
DefbPm (+6)	>109	6.8	>109	>109
DefbTs (+3)	>122	15.3	>122	>122
hBd02 (+6)	15.6	2	>124	>124
Crotasin (−1)	>108	>108	>108	>108

The MIC was determined to be the lowest concentration, resulting in no visible bacterial growth in microbroth dilution assay [28]. The assays were performed in three independent experiments. All the β-defensins tested did not inhibit the growth of *Providencia rettgeri*, *Serratia marcescens*, *Morganella morganii*, and *Klebsiella pneumonia*.

The snake β-defensins crotamine, defbBm02, defbBm03, and defbLm03 were chosen to design the derived peptides for the next step due to their performance in the antibacterial assay. The crotamine 3D model was used as a template for designing the derived peptides because the 3D structure deposited in PDB presented a better resolution (1.70 Å). The software used to build the peptides has a tool to mutate, grow or delete amino acid residues, based on the sequence to be constructed. Furthermore, many algorithms clean the geometry to get a minimum energy structure to start the properties' calculations. These peptides are presented in Table 2. They derived from the C-terminal portion of chosen β-defensins (Table 1). The design considered the bioisosterism strategy [29], for instance, the substitution of SH group (Cys) to OH group (Ser). Linear peptide constructions of different sizes (7 to 14 aa) were considered to verify which minor sequence would retain the antimicrobial activity. The peptides were commercially purchased, and their purities were tested by HPLC and mass spectrometry.

Table 2. Peptides derived from β-defensins applying the bioisosterism strategy.

Peptide	Derived from	Sequence	Hidrofobicity (Kcal/mol)	Net Charge	MW (Da)
PS1 (13 aa)	Crotamine	SRWRWKSSKKGSG	+19.88	+5	1549
PS2 (14 aa)	defbBm02	SQMGRMSSRRRFGK	+19.34	+5	1683
PS3 (12 aa)	defbLm02	SGPGRRSSRRRK	+23.57	+6	1399
PS4 (14 aa)	defbBm03	SQMGQMSSRRRFGK	+18.30	+4	1655
PS5 (13 aa)	PS3	SGPGRRSSRRRWK	+21.48	+6	1585
PS6 (7 aa)	PS1	SRWRWKS	+11.06	+3	1005

The primary structure does not illustrate a structure of interaction. The molecular surfaces translate the molecular form since the calculations consider a water molecule running through the van der Waals radius of each atom of the molecule. The molecular form property, consequently, is dependent on the 3D structure and can be used to visualize those differences.

The coordinates of the polypeptide crotamine from the Brazilian rattlesnake *C. durissus terrificus* [24], retrieved from Protein Data Bank (PBD ID 4GV5; resolution at 1.70 Å) [30], were used as starting geometry to construct the three-dimensional (3D) molecular models of the defensins derived peptides. The crotamine C-terminal portion, containing the fragments Cys30-Gly42 (13 aa), was used to build up the peptides' 3D molecular models (see Material and Methods section). For instance, the Cys residues (SH group) were mutated into Ser (OH group), generating the peptide PS1, and the PS6 peptide was obtained by extracting Ser37-Gly42 from PS1. PS6 (7 aa) corresponds to the minor sequence assayed in this study. The molecular model of each peptide and its respective calculated solvent-accessible molecular surface are shown in Figure 1.

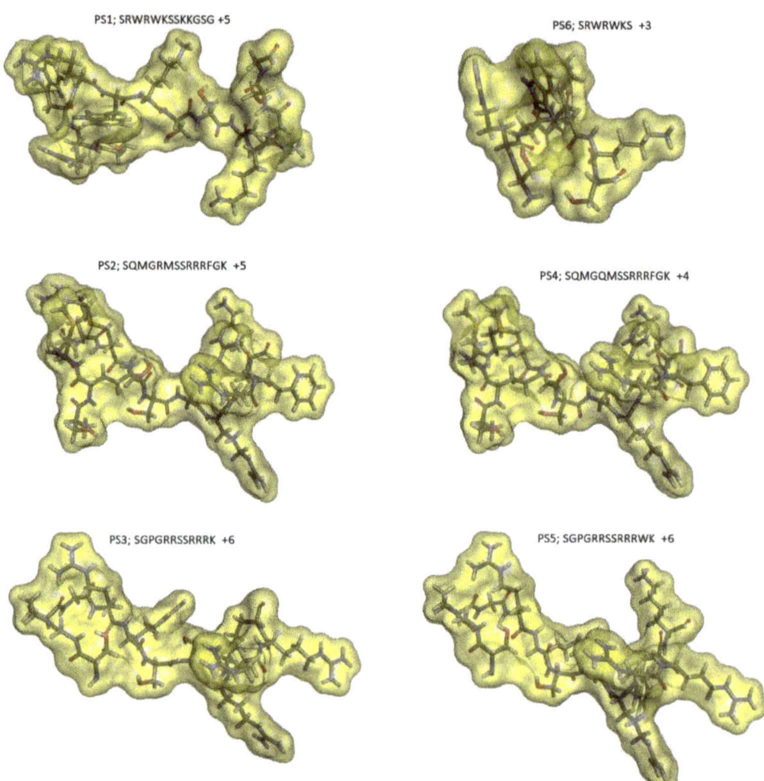

Figure 1. Solvent accessible molecular surfaces using a probe of 1.4 Å (water molecule radius). The peptides are presented in stick models, where hydrogen atoms are in white, oxygens in red, nitrogen atoms in blue, sulfur in orange, and carbon atoms in gray. The molecular surfaces are translucid and presented in yellow color (Discovery Studio Visualizer 4.0 program, Accelrys Software Inc., 2005–2013, San Diego, CA, USA).

The peptide PS1 (SRWRWKSSKKGSG) is a bioisoster of the crotamine terminal portion (substitution of SH(Cys) to OH(Ser) group), and PS6 (SRWRWKS) shares the same sequence of PS1 from the 1st to 7th positions. The calculated solvent accessible surface area and molecular volume values were 1295. 61 Å^2 and 1551.53 Å^3 for PS1; 838.26 Å^2 and 1013.05 Å^3 for PS6, respectively. As mentioned above, PS6 is the minor peptide of this set. The peptides PS2 (SQMGRMSSRRRFGK) and PS4 (SQMGQMSSRRRFGK) have 14 amino acid residues and differ from one another only in the fifth position. PS2 has an arginine (positively charged) in the fifth position, whereas PS4 has a glutamine (polar; non-charged residue). The solvent-accessible surface area and molecular volume values found for PS2 were 1385.25 $Å^2$ and 1672.60 $Å^3$; and for PS4 were 1356.84 $Å^2$ and 1633.52 $Å^3$, respectively. The peptides PS3 (SGPGRRSSRRRK) and PS5 (SGPGRRSSRRRWK) share the same amino acid sequence from the 1st to 11th positions. PS5 has a tryptophan residue before the last residue, lysine (K). The solvent-accessible surface area and molecular volume values found for PS3 were 1179.24 $Å^2$ and 1387.56 $Å^3$; and for PS5 were 1355.29 $Å^2$ and 1562.72 $Å^3$, respectively. Concerning the solvent-accessible molecular surface or surface area values, the peptides can be classified in the following crescent order: PS6 < PS1 < PS3 < PS5 < PS4 < PS2. This property is related to both molecular shape and solvation process, which are important in the ligand–target recognition process.

These β-defensin-derived peptides were tested against bacteria using the microbroth dilution assay from 2.75 to 700 μg/mL. The results showed that these peptides were not active against *M. morganii* and *P. rettgeri* but presented various antibacterial properties against Gram-positive and Gram-negative bacteria (Table 3).

Table 3. Antibacterial effect expressed by the minimal inhibitory concentration (MIC) of short peptides derived from snake β-defensins on bacteria.

	MIC (μM)				
Peptides	***E. coli***	***S. aureus***	***M. luteus***	***K. pneumoniae***	***C. freundii***
PS1	56.5	56.5	28.4	>452	113
PS2	415	415	26.1	>415	>415
PS3	>500	125	31.5	>500	>500
PS4	>423	>423	26.6	>423	>423
PS5	55.2	27.7	13.9	27.7	110
PS6	697	349	43.8	>697	>697

The MIC was determined as the lowest concentration that results in no visible bacterial growth in microbroth dilution assay [28]. The assays were performed in three independent experiments. All the short peptides tested did not inhibit the growth of *Providencia rettgeri, Serratia marcenses*, and *Morganella morganii*.

The derived peptides were also tested against fungi species using the classical resazurin microtiter assay plate method. It is based on the color change of the rezazurin dye that in blue indicates the absence of growth in at least 90%, and pink refers to microorganism growth. MIC is determined as the minimal concentration of the sample that can prevent the color change from blue to pink and refers to the inhibition of 90% of microorganisms. The samples did not inhibit the fungal growth and were considered ineffective until the highest assayed concentrations (250 μM) (Table 4).

Observing the molecular shapes (Figure 1) expressed by the obtained solvent surface area values, peptides PS2 and PS4 (14 aa) are the most related and showed similar activities. The replacement of amino acid residues at the fifth position, R5Q, to a polar non-charged residue showed to be deleterious to maintain the antibacterial activity against *E. coli* and *S. aureus* (Table 3). These results indicate the importance of the basic residue to retain the activity. On the other hand, the PS2 and PS4 inhibitory activity values on *M. luteus* were practically the same (Table 3), suggesting the R5Q substitution is not crucial for that interaction profile.

The PS3 (12 aa) and PS5 (13 aa) peptides differ from one another by the tryptophan (W) insertion at the 12th position, and the PS5 sequence modification can be visualized through the molecular shape (Figure 1). The sequence change has provided a significant improvement in the PS5 activity profile in comparison to PS3. PS5 has presented interesting MIC values against *E. coli* (55.2 μM), *S. aureus* (27.7 μM), *M. luteus* (13.9 μM), and *K. pneumoniae* (27.7 μM). Regarding *C. freundi*, the MIC value was higher (110 μM). The map of electrostatic potential (MEP) was calculated onto the molecular surface of both peptides, and the changes related to the electronic density distribution are presented in Figure 2. The insertion of tryptophan has contributed to exposing the positively charged region of PS5 (intense blue, lower electronic density distribution; arrow in Figure 2).

Table 4. Antifungal effect expressed by the minimal inhibitory concentration (MIC90) of short peptides derived from snake β-defensins on yeasts and filamentous fungi.

	MIC90 (µM)			
Peptides	*Candida albicans*	*Cryptococcus neoformans*	*Trichophyton rubrum*	*Aspergillus fumigatus*
PS1	>250	>250	>250	>250
PS2	>250	>250	>250	>250
PS3	>250	>250	>250	>250
PS4	>250	>250	>250	>250
PS5	>250	>250	>250	>250
PS6	>250	>250	>250	>250

MIC The minimal inhibitory concentration (MIC90) was defined as the lowest concentration that prevents the rezazurin's change in color from blue to pink due to the inhibition of at least 90% of the microorganism's growth [31]. The assays were performed in three independent experiments.

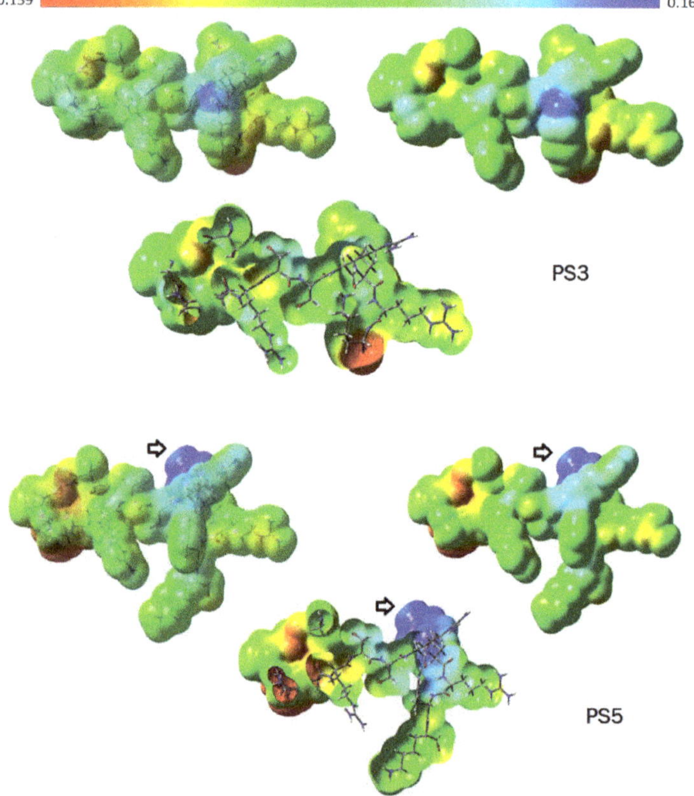

Figure 2. Maps of electrostatic potential (MPEs) were calculated onto the molecular surface of PS3 and PS5 peptides. Regarding the color range, higher electronic density distribution regions are displayed as intense red color (−0.159), and lower electronic density distribution regions are shown in intense blue color (+0.169) (Gaussian 03W, Gaussian, Inc., Pittsburgh, PA, USA, 2003; GaussView 05, Gaussian, Inc., Pittsburgh, PA, USA, 2002–2008).

The PS1 peptide (13 aa) has presented interesting antibacterial activity against *E. coli* (56.5 μM), *S. aureus* (56.5 μM), and *M. luteus* (28.4 μM). As for PS5, regarding *C. freundi*, the MIC value was higher (113 μM). In comparison to PS1, the PS6 peptide (7 aa), having the sequence core SRWRWKS, did not show any activity against *C. freundi*. The minor sequence has retained indeed some antibacterial activity.

Unfortunately, none of the derived peptides have shown antifungal activity against the tested fungi, suggesting that the presence of Cys residues in the sequence is important to that kind of activity profile.

3. Discussion

The β-defensins were tested against *B. jararaca* oral flora because it would be of biological and medical interest. It is known that microorganism infection is one of the snakebite complications due to bacteria or fungi found in the snake's oral cavity [32,33]. In Brazil, over 80% of snake envenomations are caused by *Bothrops* snakes, and around 10% of the snakebites evolve to infections [34]. Among bacteria isolated from abscesses, the most reported were *M. morganii*, *E. coli*, *Providencia* sp., *Klebsiella* sp. [34]. Interestingly, no β-defensins tested in this study showed antibacterial activity against *M. morganii* nor *K. pneumoniae*. Although bacteria were also isolated from the venom of *C. durissus terrificus* [35], envenomations by this species (*C. durissus terrificus*) do not usually cause infection or macroscopic necrosis in the bite site [34]. The toxins present in *C. durissus terrificus* venom: crotamine, PLA_2, LAAO [36], and cathelicidins [22] could help the asepsis snakebite wound.

The linearization did not abolish the antibiotic activity of crotamine and the majority of snake β-defensins studied herein. The MIC against Gram-positive and -negative bacteria is shown in Table 1. The β-defensins presenting lower net charge values (crotasin, −1, defbBn02, +2) did not show antibacterial activity against any bacteria tested or showed a weak antibacterial effect with high MIC (defbLm01, +2). On the other side, the β-defensins defbBm02 (+11) and defbLm02 (+8) showed the best antibacterial activity against the bacteria *E. coli*, *C. freundii*, *M. luteus*, and *S. aureus* with MIC in the range of 0.8 to 12.8 μM. Interestingly, DefbBju, with a net charge of +7, did not inhibit any bacteria tested in this study, indicating that the 3D structure could be essential to its antibacterial activity. The Gram-positive bacteria *M. luteus* was the most sensitive to snake β-defensins while *S. aureus* was inhibited only by defbBm02 and defbLm02. The Gram-negative bacteria *E. coli* and *C. freundii* were inhibited by β-defensins with a net charge higher than +7. The basicity of AMPs is related to antibacterial activity, the higher the positive charge higher the inhibition of bacteria growth, but there are indications that the disulfide bridges modulate the antimicrobial activity [37] and the balance of hydrophilic and hydrophobic surfaces [38]. The net charge of β-defensins was thoroughly discussed by Huang et al. [39]. The activity of linear crotamine corroborates the antibacterial activity of reduced crotamine [17] as well as antifungal activity [18].

Although *S. aureus* is sensitive to several toxins from snake venoms such as PLA_2 [13,40–45], LAAO [46–48], and cathelicidins [8]; crotamine did not inhibit the growth of this bacterium; however, the β-defensins from snakes defbLm02 and defbBm02 showed antibacterial activity with MIC of 32 and 64 μg/mL respectively. Although native crotamine did not show activity against *S. aureus*, fragments of this toxin inhibited the growth of this bacterium [18].

Although *K. pneumoniae* is inhibited by cathelicidins [5,49,50], PLA_2 [40–43,45], and LAAO [51–53], no β-defensin tested in this work showed antibacterial activity against this species. The same was true for *S. marcensens*, sensitive to LAAO [53,54], but not to cathelicidins [5,50]. *M. morganii*, sensitive to the venom of *Montivipera bornmuelleri* [55], was also resistant to snake β-defensins and crotamine.

As most of the bacteria tested were isolated from the oral flora of *B. jararaca*, the lack of antibacterial activity or even weak activity of these molecules may be because these β-defensins only control the coexistence of animals with these microorganisms and the tested snake β-defensin could have other functions in the animal. Many biological functions are described to antimicrobial peptides such as modulation of inflammatory responses, wound healing, angiogenesis promotion [56], regeneration of a lizard tail [57,58], and sperm function in mammals [59], but these biological activities depend on the 3D structure of the β-defensins [60].

The goal of the design of short peptides based on our β-defensins is to get more accessible and cheaper manufacturing as the advantages of small molecules [61]. Peptides P1 to P6 (14 to 7 residues) were tested against bacteria and fungi.

The use of β-defensin fragments caused the decrease of antibacterial activity (see MIC values in Table 3), which was expected since the isolated fragments did not have the same behavior as a conformational organized protein structure. However, it also led to positive results: PS1 showed antibacterial activity against *C. freundii* and *S. aureus* differently from crotamine, as well PS5 that inhibited the growth of *K. pneumoniae*, unlike defbLm02. The decreased antibacterial activity presented by smaller analogs in comparison to the original was also reported to the β-defensin HBD3 [60,62]. A probable cause of the decrease may be the proteolysis that these peptides may be subject to since they were not structurally protected [63]. The peptide with the highest performance was PS5, its only difference to PS3 being an introduction of Trp between basic residues, which may have improved the charge net facilitating the interaction with the bacterial membrane, as happened with the decamer derived from HBD28 [64]. Wessolowski et al. [65] observed that introducing Trp residue in the sequence increases the antibacterial activity, and the cyclization increases the antibacterial activity and the selectivity of peptides. Interestingly, the substitution of Arg for Gln (defbBm02 to defbBm03) did not alter the activity against *E. coli*, *M. luteus*, and *C. freundii*, but it seemed essential to inhibit *S. aureus* by defbBm02. On the other hand, the same substitution was critical for PS4 (derived from defbBm03) to inhibit *S. marcenses*.

It is known that fungi are sensitive to toxins from snake venoms, and LAAO [48], PLA_2 [43], cathelicidins [5,50], and crotamine [18] were already evaluated on *C. albicans*. Other sensitive fungi are *Aspergillus niculans* [5] and *Cryptococcus neoformans* [18]. In contrast, *C. neoformans* was resistant to cathelicidins [51]. Despite the crotamine fragment (residues 27–39) having inhibited the growth of *C. neoformans* [18], the substitution of Cys by Ser in crotamine fragments was deleterious for the antifungal activity [66] in the same way this substitution also abolished the activity against *C. albicans* [67]. Unfortunately, this was true to all the peptides designed in this work showing that this substitution is deleterious to antifungal activity.

We observed that the C-terminal segment of β-defensins is essential to antimicrobial activity, as also observed by Mandal et al. [68]. Additional molecular modifications seem to be necessary to improve the conformational arrangement and to aid the establishment of the structure–property/activity relationships. Based on that, novel promising peptides can be designed to have better antibacterial activity (higher potency). However, to improve the antibacterial activity of derived peptides, it is necessary to protect them from proteolysis by chemical alteration, C-terminal amidation [64], or cyclizing the peptide by disulfide bonds [63], as also the inclusion of Trp in the sequence to ameliorate the balance between hydrophobicity and hydrophilic or to stabilize the peptide structure favoring the membrane interaction. The substitution of Cys residue by Ala should be also considered, since it has been reported to increase the antibacterial activity of short peptides, as well as the substitution of Leu and Ile by Trp [69].

A feature of the β-defensin family is the 3D structure conserved by the disulfide bridges of the cysteine motif and a significant variation of amino acid sequences [70]. These molecules are good templates for development studies since biological targets have already been selected by nature [71].Furthermore, its mechanism of action, which causes the rupture of the bacterial membrane and can bind to different targets as DNA, makes it difficult for the bacteria to develop resistance [56,70].

In summary, the studied snake β-defensins do not have optimum antimicrobial activity, but they proved to be attractive as template molecules for the development of antibiotics.

Our data indicate that the short peptides show specific activity on prokaryote cells but not on eukaryote cells. Based on that, the snake β-defensins C-terminal portion, if optimized, could be indeed used as a bioactive agent.

4. Material and Methods

Twelve peptides (GenBank Accession number is shown in Table 5) were synthesized by Biomatik (Wilmington, NC, USA) and coded as crotasin, DefbBd03, DefbBj01, DefbBju01, DefbBm02, DefbBm03, DefbBn02, DefbLm01, DefLm02, DefbPm, DefbTs and hBd02. The amino acid sequences were deduced from genes and are presented in Table 3. The synthetic peptides were treated with 45 mM DTT by 15 min at 50 °C [72] followed by alkylation using 100 mM IAA (iodoacetamide) 15 min, at room temperature. Alkylated peptides were purified by high performance liquid chromatography (HPLC) and analyzed by MALDI-TOF-MS. This step was performed at the Laboratory of Applied Toxinology—Instituto Butantan with the supervision of Dr. Pedro I. da Silva Jr.

Crotamine was purified from the *C. durissus terrificus* venom (purchased from CEVAP, Botucatu, SP, Brazil) and purified as described [73]. Briefly, crotamine was purified from crude venom by size exclusion on a Sephacryl S200 column (GE Healthcare, Uppsala, Sweden), followed by cation exchange chromatography on a 1-mL Resource S on FPLC system (Akta Purifier System, GE Healthcare). The identity and purity of crotamine were confirmed by MS analysis. The reduction and alkylation were proceed as described above.

Table 5. β-defensins sequences deduced from genes and biochemical characteristics.

Peptide/ Snake	GenBank Accession Number	Amino Acid Sequence	Hidrofobicity Kcal/mol	Net Charge at pH 7	MW Da
Crotamine/ *C. durissus*		YKQCHKKGGHCFPKEKICLPPSSDFGKMDCRWRWKCCKKGSG	+49.14	+8	4886
Crotasin/ *C. durissus*	AF250212	QPQCRWLDGFCHSSPCPSGTTSIGQQDCLWYESCCIPRYEK	+28.91	−1	4708
defbBd03/ *B. diporus*	KC117160	QPECLRQGGMCRPRLCPYVSLGQLDCQNGHVCCRKKPRK	+37.08	+5	4456
defbBj/ *B. jararaca*	KC117163	QEECLQQGGFCRLIRCPFGYDSLEQQDCRKGQRCCIRKPRK	+45.05	+4	4874
defbBju/ *B. jararacussu*	KC117165	QRRCHQKGGMCLPGPCPPGYDSLGQQDCRRGQKCCIKRFGK	+43.85	+7	4591
defbBm02/ *B. mattogrossensis*	KC117167	QRRCRQRRGICRPRPCPPENFSLGRLDCQMGRMCCRRRFGK	+39.97	+11	4978
defbBm03/ *B. mattogrossensis*	KC117168	QRRCRQRRGICRPRPCPPENFSLGRLDCQMGQMCCRRRFGK	+38.93	+10	4950
defbBn02/ *B. neuwiedi*	KC117169	QPECCQEGGICHSKQCPLGYSSLGRLDCQLGQRCCIRIFGK	+33.65	+2	4513
defbLm01/ *L. muta*	KC117171	QEWCRGLGGFCSFYQCRPGHDLGPQDCWPERRCCRWGK	+33.69	+2	4515
defbLm02/ *L. muta*	KC117172	QGQCHQQRGRCFLHQCPLSHYFLGRLDCGPGRRCCRRRK	+36.90	+8	4655
defbPm/ *P. mertensis*	KX664436	QRICLGGRGFCHSTPCPRSTIDYGKKDCWGSLRCCEPKRPGK	+42.13	+6	4695
defbTs/ *T. strigatus*	KX664429	QDLCHNLGGRCFRNRCSWSLRNHGGQDCPWGSVCCKP	+31.16	+3	4188
hBD02/ Human	AF071216	DPVTCLKSGAICHPVFCPRRYKQIGTCGLPGTKCCKKP	+31.07	+6	4104

Brazilian pitvipers *Crotalus* (C.), *Bothrops* (B.), *Lachesis* (L.) and the colubrides *Phalotris* (P.), and *Thamnophis* (T.). The codifying sequences of genes were used to deduce the amino acid sequence and the Signal P software [74] used to determine the mature β-defensins. The hidrofobicity, net charge, and Molecular Weight were theoretical calculated using PepDraw software [http://pepdraw.com/ by Thomas C. Freeman, Jr. Accessed on 27 September 2021].

Crotamine, defbm02, defBm03, and defbLm02 were chosen to design the short peptides due to the best antibacterial activity among the β-defensins tested. Previous studies have indicated the C-terminal of the human β-defensin HBD3 and crotamine as the mandatory region of antimicrobial activity [18,22,60,66,67].

The short peptides, derived from β-defensins (Table 2), were designed applying the bioisosterism strategy [29]; for instance, the substitution of SH group (Cys) by OH group (Ser), regarding the C-terminal portion of the β-defensins (crotamine, defbBm02, defbLm02, defbBm03). Bioisoster groups or substituents share chemical or physical similarities, producing similar biological properties. It was considered linear peptide constructions having different sizes (7 to 14 aa) to verify which would be the minor sequence able to retain the defensins biological activity exploited. The designed peptides were synthesized by GenOne Biotechnologies. The amino acid sequences are presented in Table 1.

4.1. Molecular Modeling and Molecular Properties Calculation

The coordinates of the polypeptide crotamine from the Brazilian rattlesnake *C. durissus terrificus* [24] were retrieved from Protein Data Bank (PBD ID 4GV5; resolution at 1.70 Å) [30] and used as starting geometry to extract the fragment Cys30-Gly42 (13 aa), which were employed to build up the peptides' 3D molecular models. The Cys residues were mutated into Ser, generating the peptide PS1. The PS6 peptide was obtained by extracting Ser37-Gly42 from PS1. PS6 (7 aa) corresponds to the minor sequence designed and assayed in this study. The other fragments were constructed using the tool build-mutate or build-grow available in the Discovery Studio Visualizer 4.0 software (Accelrys Software Inc., 2005–2013). The geometries were optimized, and partial atomic charges were assigned using the CHARMM force field [63], included in Discovery Studio (Accelrys Software Inc., 2005–2013).

Furthermore, the electrostatic potential (EP) property was calculated for the PS3 and PS5 peptides to visualize the changes in electronic density distribution concerning the amino acid substitution patterns (PS5 has one more residue, Trp, in comparison to PS3). The charges from electrostatic potential using a grid-based method (CHELPG [64]) were calculated employing the ab initio method Hartree-Fock/3-21G* basis set (Gaussian 03W software; Gaussian, Inc., Pittsburgh, PA, USA, 2003). The EP maps were calculated onto the peptides' molecular surfaces using GaussView 05 software (Gaussian, Inc., Pittsburg, PA, USA, 2002–2008). The interpretation of EP maps is based on a color scheme, where regions having higher electronic density distribution are presented as an intense red color (negatively charged), whereas regions with lower electronic density distribution are shown as an intense blue color (positively charged). Since the EP property has been calculated onto the PS3 and PS5 molecular surfaces, their molecular shapes were also assessed. Moreover, the molecular volume (intrinsic molecular property) of each peptide considering the van der Waals radii was also calculated employing Discovery Studio Visualizer 4.0 software (Accelrys Software Inc., 2005–2013). Of note, the molecular shape and electronic properties are among the primary molecular properties in the ligand–receptor recognition process.

4.2. Antibacterial Activity

The antibacterial activity of the alkylated peptides was tested against Gram-negative (G−) bacteria (*Klebsiella pneumonia, Serratia marcescens, Morganella morganii, Providencia rettgeri, Citrobacter freundii, Escherichia coli* ATCC 25922 and Gram-positive (G+) bacteria (*Micrococcus luteus* A270 and *Staphylococcus aureus*) using microbroth assay [28]. In a 96-well plate, 90 μL of 10% tryptone soy broth (TSB) with 4×10^5 colony-forming unit (CFU/mL) were mixed with serial twofold dilutions of β-defensins duplicate in concentrations from 512 to 1 μg/mL. The derived peptides coded as PS1, PS2, PS3, PS5, and PS6 (Table 1) were tested against these bacteria in serial twofold dilutions with concentrations bellow 700 μg/mL. After overnight incubation at 37 °C, the turbidity of the bacterial culture was read at 600 nm (Epoch microplate reader, Biotek). The minimal inhibitory concentration (MIC) of each peptide was determined as the lowest concentration that results in no visible bacterial growth. The MIC resulted from three independent experiments. The bacteria *K. pneumonia*, *S. marcescens*, *M. morganii*, *P. rettgeri*, *C. freundii*, *M. luteus* A270 and *S. aureus* were isolated from *Bothrops jararaca* buccal flora and kindly provided by Dr. Márcia R. Franzolin (Laboratory of Bacteriology—Instituto Butantan). *M. luteus* was kindly provided by Dr. Pedro I. da Silva Jr. (Laboratory of Applied Toxinology—Instituto Butantan).

4.3. Antifungal Activity

The antifungal activity was evaluated by the microdilution assay as previously described [31] on clinical strains of *Candida albicans* (IOC 4525), *Cryptococcus neoformans* (IOC 4528), *Trichophyton rubrum* (IOC 4527) and *Aspergillus fumigatus* (IOC 4526). The fungi were cultivated in potato dextrose agar at 28 °C according to the Clinical and Laboratory Standards Institute recommendations [75]. Fungal suspensions were prepared in RPMI 1640 culture media; the CFU/mL was adjusted to 0.5–2.5 $\times$ 10^3 for yeasts (*C. albicans* and *C. neoformans*) and 0.4–5 $\times$ 10^4 CFU/mL for filamentous fungi (*T. rubrum* e *A. fumigatus*), by comparison with a standard curve previously stablished in our laboratory.

The peptides PS1, PS2, PS, PS3, PS4, PS5 and PS6 at 1 mM in acetic acid at 0.01% were diluted in RPMI 1640 culture media at concentrations ranging from 1 to 250 µM. Amphotericin B (AMB) at concentrations below 15 µg/mL (16 µM) and acetic acid from 0.0002 to 0.0025% was used as control.

The fungi suspensions were added to a 96 well plate (100 µL/well) and the samples (100 µL) in different concentrations were added to each well. The plate was incubated at 28 °C for 24–72 h, and 24 h before the end of the assay 25 µL of the rezazurin dye at 0.02% were added to each well. The minimal inhibitory concentration (MIC90) was defined as the lowest concentration that prevents the rezazurin's change in color from blue to pink due to the inhibition of at least 90% of the microorganism's growth. The assays were performed in three independent experiments.

Author Contributions: Conceptualization: K.F.M.P. and N.O.; Investigation: P.G.C., I.L.R., A.O.d.S.; Formal analysis: K.F.M.P., N.O.; Funding acquisition: N.O.; Writing—original draft: N.O.; Writing—review & editing: A.O.d.S., K.F.M.P., N.O. All authors have read and agreed to the published version of the manuscript.

Funding: This research was funded by São Paulo Research Foundation (www.fapesp.br)—FAPESP: 2015/00003-5 (N.O.), 2017/11735-2 (fellowship of I.L.R.); 2008/06524-3 (A.O.S).

Institutional Review Board Statement: Not applicable.

Informed Consent Statement: Not applicable.

Conflicts of Interest: The authors declare no conflict of interest.

References

1. Mahlapuu, M.; Håkansson, J.; Ringstad, L.; Björn, C. Antimicrobial Peptides: An Emerging Category of Therapeutic Agents. *Front. Cell. Infect. Microbiol.* **2016**, *6*, 194. [CrossRef]
2. Uhlig, T.; Kyprianou, T.; Martinelli, F.G.; Oppici, C.A.; Heiligers, D.; Hills, D.; Calvo, X.R.; Verhaert, P. The emergence of peptides in the pharmaceutical business: From exploration to exploitation. *EuPA Open Proteom.* **2014**, *4*, 58–69. [CrossRef]
3. Gil-Gil, T.; Laborda, P.; Sanz-García, F.; Hernando-Amado, S.; Blanco, P.; Martínez, J.L. Antimicrobial resistance: A multifaceted problem with multipronged solutions. *Microbiologyopen* **2019**, *8*, e945. [CrossRef] [PubMed]
4. Andrès, E. Cationic antimicrobial peptides in clinical development, with special focus on thanatin and heliomicin. *Eur. J. Clin. Microbiol. Infect. Dis.* **2012**, *31*, 881–888. [CrossRef]
5. Wang, Y.; Hong, J.; Liu, X.; Yang, H.; Liu, R.; Wu, J.; Wang, A.; Lin, D.; Lai, R. Snake Cathelicidin from *Bungarus fasciatus* Is a Potent Peptide Antibiotics. *PLoS ONE* **2008**, *3*, e3217. [CrossRef]
6. Blower, R.J.; Barksdale, S.M.; van Hoek, M.L. Snake Cathelicidin NA-CATH and Smaller Helical Antimicrobial Peptides Are Effective against *Burkholderia thailandensis*. *PLoS Negl. Trop. Dis.* **2015**, *9*, e0003862. [CrossRef] [PubMed]
7. Zhao, F.; Lan, X.-Q.; Du, Y.; Chen, P.-Y.; Zhao, J.; Zhao, F.; Lee, W.-H.; Zhang, Y. King cobra peptide OH-CATH30 as a potential candidate drug through clinic drug-resistant isolates. *Zool. Res.* **2018**, *39*, 87–96. [PubMed]
8. Barros, E.; Gonçalves, R.M.; Cardoso, M.H.; Santos, N.C.; Franco, O.L.; Cândido, E.S. Snake Venom Cathelicidins as Natural Antimicrobial Peptides. *Front. Pharmacol.* **2019**, *10*, 1415. [CrossRef] [PubMed]

9. Carlile, S.R.; Shiels, J.; Kerrigan, L.; Delaney, R.; Megaw, J.; Gilmore, B.F.; Weldon, S.; Dalton, J.P.; Taggart, C.C. Sea snake cathelicidin (Hc-cath) exerts a protective effect in mouse models of lung inflammation and infection. *Sci. Rep.* **2019**, *9*, 6071. [CrossRef]
10. Pérez-Peinado, C.; Dias, A.S.; Mendonça, D.A.; Castanho, M.A.R.B.; Veiga, A.S.; Andreu, D. Structural determinants conferring unusual long life in human serum to rattlesnake-derived antimicrobial peptide Ctn [15–34]. *J. Pep. Sci.* **2019**, *25*, e3195. [CrossRef] [PubMed]
11. Dean, S.N.; Bishop, B.M.; van Hoek, M.L. Natural and synthetic cathelicidin peptides with anti-microbial and anti-biofilm activity against *Staphylococcus aureus*. *BMC Microbiol.* **2011**, *11*, 114. [CrossRef] [PubMed]
12. Tajbakhsh, M.; Akhavan, M.M.; Fallah, F.; Karimi, A. A Recombinant Snake Cathelicidin Derivative Peptide: Antibiofilm Properties and Expression in *Escherichia coli*. *Biomolecules* **2018**, *8*, 118. [CrossRef]
13. Samy, R.P.; Gopalakrishnakone, P.; Stiles, B.G.; Girish, K.S.; Swamy, S.N.; Hemshekhar, M.; Tan, K.S.; Rowan, E.G.; Sethi, G.; Chow, V.T.K. Snake Venom Phospholipases A2: A Novel Tool against Bacterial Diseases. *Curr. Med. Chem.* **2012**, *19*, 6150–6162. [CrossRef]
14. Izidoro, L.F.M.; Sobrinho, J.C.; Mendes, M.M.; Costa, T.R.; Grabner, A.N.; Rodrigues, V.M.; da Silva, S.L.; Zanchi, F.B.; Zuliani, J.P.; Fernandes, C.F.C.; et al. Snake Venom L-Amino Acid Oxidases: Trends in Pharmacology and Biochemistry. *Biomed. Res. Int.* **2014**, *2014*, 196754. [CrossRef]
15. Samy, R.P.; Gopalakrishnakone, P.; Chow, V.T.K.; Ho, B. Viper Metalloproteinase (*Agkistrodon halys* Pallas) with Antimicrobial Activity against Multi-Drug Resistant Human Pathogens. *J. Cell. Physiol.* **2008**, *216*, 54–68. [CrossRef] [PubMed]
16. Yount, N.Y.; Kupferwasser, D.; Spisni, A.; Dutz, S.M.; Ramjan, Z.H.; Sharma, S.; Waring, A.J.; Yeaman, M.R. Selective reciprocity in antimicrobial activity versus cytotoxicity of hBD-2 and crotamine. *Proc. Natl. Acad. Sci. USA* **2009**, *106*, 14972–14977. [CrossRef]
17. Oguiura, N.; Boni-Mitake, M.; Affonso, R.; Zhang, G. In vitro antibacterial and hemolytic activities of crotamine, a small basic myotoxin from rattlesnake *Crotalus durissus*. *J. Antibiot.* **2011**, *64*, 327–331. [CrossRef] [PubMed]
18. Yamane, E.S.; Bizerra, F.C.; Oliveira, E.B.; Moreira, J.T.; Rajabi, M.; Nunes, G.L.C.; de Souza, A.O.; da Silva, I.D.C.G.; Yamane, T.; Karpel, R.L.; et al. Unraveling the antifungal activity of a South American rattlesnake toxin Crotamine. *Biochimie* **2013**, *95*, 231–240. [CrossRef] [PubMed]
19. Lomonte, B.; Ângulo, Y.; Moreno, E. Synthetic Peptides Derived from the C-Terminal Region of Lys49 Phospholipase A_2 Homologues from Viperidae Snake Venoms: Biomimetic Activities and Potential Applications. *Curr. Pharm. Des.* **2010**, *16*, 3224–3230. [CrossRef]
20. Chen, W.; Yang, B.; Zhou, H.; Sun, L.; Dou, J.; Qian, H.; Huang, W.; Mei, Y.; Han, J. Structure–activity relationships of a snake cathelicidin-related peptide, BF-15. *Peptides* **2011**, *32*, 2497–2503. [CrossRef] [PubMed]
21. Almeida, J.R.; Mendes, B.; Lancellotti, M.; Marangoni, S.; Vale, N.; Passos, O.; Ramos, M.J.; Fernandes, P.A.; Gomes, P.; Da Silva, S.L. A novel synthetic peptide inspired on Lys49 phospholipase A_2 from *Crotalus oreganus abyssus* snake venom active against multidrug resistant clinical isolates. *Eur. J. Med. Chem.* **2018**, *149*, 248–256. [CrossRef] [PubMed]
22. Falcao, C.B.; Radis-Baptista, G. Crotamine and crotalicidin, membrane active peptides from *Crotalus durissus terrificus* rattlesnake venom, and their structurally-minimized fragments for applications in medicine and biotechnology. *Peptides* **2020**, *126*, 170234. [CrossRef] [PubMed]
23. Oguiura, N.; Boni-Mitake, M.; Rádis-Baptista, G. New view on crotamine, a small basic polypeptide myotoxin from South American rattlesnake venom. *Toxicon* **2005**, *46*, 363–370. [CrossRef] [PubMed]
24. Coronado, M.A.; Gabdulkhakov, A.; Georgieva, D.; Sankaran, B.; Murakami, M.T.; Arni, R.K.; Betzel, C. Structure of the polypeptide crotamine from the Brazilian rattlesnake *Crotalus durissus terrificus*. *Acta Cryst.* **2013**, *D69*, 1958–1964. [CrossRef]
25. Rádis-Baptista, G.; Kubo, T.; Oguiura, N.; Silva, A.R.B.P.; Hayashi, M.A.F.; Oliveira, E.B.; Yamane, T. Identification of crotasin, a crotamine-related gene of *Crotalus durissus terrificus*. *Toxicon* **2004**, *43*, 751–759. [CrossRef]
26. Corrêa, P.G.; Oguiura, N. Phylogenetic analysis of β-defensin-like genes of *Bothrops*, *Crotalus* and *Lachesis* snakes. *Toxicon* **2013**, *69*, 65–74. [CrossRef] [PubMed]
27. Oliveira, Y.S.; Corrêa, P.G.; Oguiura, N. Beta-defensin genes of the Colubridae snakes *Phalotris mertensi*, *Thamnodynastes hypoconia*, and *T. strigatus*. *Toxicon* **2018**, *146*, 124–128. [CrossRef]
28. Xiao, Y.; Cai, Y.; Bommineni, Y.R.; Fernando, S.C.; Prakash, O.; Gilliland, S.E.; Zhang, G. Identification and functional characterization of three chicken cathelicidins with potent antimicrobial activity. *J. Biol. Chem.* **2006**, *281*, 2858–2867. [CrossRef]
29. Patani, G.A.; LaVoie, E.J. Bioisosterism: A Rational Approach in Drug Design. *Chem. Rev.* **1996**, *96*, 3147–3176. [CrossRef]
30. Berman, H.M.; Westbrook, J.; Feng, Z.; Gilliland, G.; Bhat, T.N.; Weissig, H.; Shindyalov, I.N.; Bourne, P.E. The Protein Data Bank. *Nucl. Acids Res.* **2000**, *28*, 235–242. [CrossRef] [PubMed]
31. Palomino, J.C.; Martin, A.; Camacho, M.; Guerra, H.; Swings, J.; Portaels, F. Resazurin microtiter assay plate: Simple and inexpensive method for detection of drug resistance in *Mycobacterium tuberculosis*. *Antimicrob. Agents Chemother.* **2002**, *46*, 2720–2722. [CrossRef] [PubMed]
32. Dehghani, R.; Sharif, M.R.; Moniri, R.; Sharif, A.; Kashani, H.H. The identification of bacterial flora in oral cavity of snakes. *Comp. Clin. Pathol.* **2016**, *25*, 279–283. [CrossRef]
33. Dehghani1, R.; Sharif, A.; Assadi, M.A.; Kashani, H.H.; Sharif, M.R. Fungal flora in the mouth of venomous and non-venomous snakes. *Comp. Clin. Pathol.* **2016**, *25*, 1207–1211. [CrossRef]

34. Jorge, M.T.; Ribeiro, L.A. Infections in the bite site after envenoming by snakes of the *Bothrops* genus. *J. Venom. Anim. Toxins* **1997**, *3*, 264–272. [CrossRef]
35. Garcia-Lima, E.; Laure, C.J. A study of bacterial contamination of rattlesnake venom. *Rev. Soc. Bras. Med. Trop.* **1987**, *20*, 19–21. [CrossRef] [PubMed]
36. Bercovici, D.; Chudzinski, A.M.; Dias, W.O.; Esteves, M.I.; Hiraichi, E.; Oishi, N.Y.; Picarelli, Z.P.; Rocha, M.C.; Ueda, C.M.P.M.; Yamanouye, N.; et al. A systematic fraction of *Crotalus durissus terrificus* venom. *Mem. Inst. Butantan* **1987**, *49*, 69–78.
37. Sharma, H.; Nagaraj, R. Human β-Defensin 4 with Non-Native Disulfide Bridges Exhibit Antimicrobial Activity. *PLoS ONE* **2015**, *10*, e0119525. [CrossRef]
38. Liu, S.; Zhou, L.; Li, J.; Suresh, A.; Verma, C.; Foo, Y.H.; Yap, E.P.H.; Tan, D.T.H.; Beuerman, R.W. Linear Analogues of Human b-Defensin 3: Concepts for Design of Antimicrobial Peptides with Reduced Cytotoxicity to Mammalian Cells. *ChemBioChem* **2008**, *9*, 964–973. [CrossRef]
39. Huang, X.-X.; Gao, C.-Y.; Zhao, Q.-J.; Li, C.-L. Antimicrobial Characterization of Site-Directed Mutagenesis of Porcine Beta Defensin 2. *PLoS ONE* **2015**, *10*, e0118170. [CrossRef]
40. Samy, R.P.; Kandasamy, M.; Gopalakrishnakone, P.; Stiles, B.G.; Rowan, E.G.; Becker, D.; Shanmugam, M.K.; Sethi, G.; Chow, V.T.K. Wound Healing Activity and Mechanisms of Action of an Antibacterial Protein from the Venom of the Eastern Diamondback Rattlesnake (*Crotalus adamanteus*). *PLoS ONE* **2014**, *9*, e80199. [CrossRef]
41. Samy, R.P.; Stiles, B.G.; Chinnathambi, A.; Zayed, M.E.; Alharbi, S.A.; Franco, O.L.; Rowan, E.G.; Kumar, A.P.; Lim, L.H.K.; Sethi, G. Viperatoxin-II: A novel viper venom protein as an effective bactericidal Agent. *FEBS Open Bio.* **2015**, *5*, 928–941. [CrossRef]
42. Sudarshan, S.; Dhananjaya, B.L. Antibacterial Potential of a Basic Phospholipase A2 (VRV_PL_V) of *Daboia russellii pulchella* (Russell's Viper) Venom. *Biochemistry* **2014**, *79*, 1237–1244. [CrossRef]
43. Sudarshan, S.; Dhananjaya, B.L. Antibacterial Potential of a Basic Phospholipase A2 (VRV_PL_VIIIa) of *Daboia russellii pulchella* (Russell's Viper) Venom. *J. Venom. Anim. Toxins Incl. Trop. Dis.* **2015**, *21*, 17. [CrossRef]
44. Sudarshan, S.; Dhananjaya, B.L. The Antimicrobial Activity of an Acidic Phospholipase A2 (NN-XIa-PLA2) from the Venom of *Naja naja naja* (Indian Cobra). *Appl. Biochem. Biotechnol.* **2015**, *176*, 2027–2038. [CrossRef]
45. Corrêa, E.A.; Kayano, A.M.; Diniz-Sousa, R.; Setúbal, S.S.; Zanchi, F.B.; Zuliani, J.P.; Matos, N.B.; Almeida, J.R.; Resende, L.M.; Marangoni, S.; et al. Isolation, structural and functional characterization of a new Lys49 phospholipase A_2 homologue from *Bothrops neuwiedi urutu* with bactericidal potential. *Toxicon* **2016**, *115*, 13–21. [CrossRef] [PubMed]
46. Guo, C.; Liu, S.; Yao, Y.; Zhang, Q.; Sun, M.-Z. Past decade study of snake venom L-amino acid oxidase. *Toxicon* **2012**, *60*, 302–311. [CrossRef] [PubMed]
47. Muñoz, L.J.V.; Estrada-Gomez, S.; Núñez, V.; Sanz, L.; Calvete, J.J. Characterization and cDNA sequence of *Bothriechis schlegelii* l-aminoacid oxidase with antibacterial activity. *Int. J. Biol. Macromol.* **2014**, *69*, 200–207. [CrossRef]
48. Costa, T.R.; Menaldo, D.L.; da Silva, C.P.; Sorrechia, R.; de Albuquerque, S.; Pietro, R.C.L.R.; Ghisla, S.; Antunes, L.M.G.; Sampaio, S.V. Evaluating the microbicidal, antiparasitic and antitumor effects of CR-LAAO from *Calloselasma rhodostoma* venom. *Int. J. Biol. Macromol.* **2015**, *80*, 489–497. [CrossRef] [PubMed]
49. Falcao, C.B.; de La Torre, B.G.; Perez-Peinado, C.; Barron, A.E.; Andreu, D.; Radis-Baptista, G. Vipericidins: A novel family of cathelicidin-related peptides from the venom gland of South American pit vipers. *Amino Acids* **2014**, *46*, 2561–2571. [CrossRef]
50. Wei, L.; Gao, J.; Zhang, S.; Wu, S.; Xie, Z.; Ling, G.; Kuang, Y.-Q.; Yang, Y.; Yu, H.; Wang, Y. Identification and Characterization of the First Cathelicidin from Sea Snakes with Potent Antimicrobial and Antiinflammatory Activity and Special Mechanism. *J. Biol. Chem.* **2015**, *290*, 16633–16652. [CrossRef]
51. Lee, M.L.; Tan, N.H.; Fung, S.Y.; Sekaran, S.D. Antibacterial action of a heat-stable form of L-amino acid oxidase isolated from king cobra (*Ophiophagus hannah*) venom. *Comp. Biochem. Physiol. C* **2011**, *153*, 237–242. [CrossRef] [PubMed]
52. Okubo, B.M.; Silva, O.N.; Migliolo, L.; Gomes, D.G.; Porto, W.F.; Batista, C.L.; Ramos, C.S.; Holanda, H.H.S.; Dias, S.C.; Franco, O.L.; et al. Evaluation of an Antimicrobial L-Amino Acid Oxidase and Peptide Derivatives from *Bothropoides mattogrosensis* Pitviper Venom. *PLoS ONE* **2012**, *7*, e33639. [CrossRef] [PubMed]
53. Phua, C.S.; Vejayan, J.; Ambu, S.; Ponnudurai, G.; Gorajana, A. Purification and antibacterial activities of an L-amino acid oxidase from king cobra (*Ophiophagus hannah*) venom. *J. Venom. Anim. Toxins Incl. Trop. Dis.* **2012**, *18*, 198–207. [CrossRef]
54. Skarnes, R.C. L-amino acid oxidase, a bacterial system. *Nature* **1970**, *225*, 1072. [CrossRef]
55. Accary, C.; Hraoui-Bloquet, S.; Hamze, M.; Mallem, Y.; El Omar, F.; Sabatier, J.-M.; Desfontis, J.-C.; Fajloun, Z. Protein Content Analysis and Antimicrobial Activity of the Crude Venom of *Montivipera bornmuelleri*; a Viper from Lebanon. Infect. *Disord. Drug Targets* **2014**, *14*, 49–55. [CrossRef]
56. Lai, Y.; Gallo, R.L. AMPed up immunity: How antimicrobial peptides have multiple roles in immune defense. *Trends Immunol.* **2009**, *30*, 131–141. [CrossRef]
57. Alibardi, L. Granulocytes of Reptilian Sauropsids Contain Beta-Defensin-like Peptides: A Comparative Ultrastructural Survey. *J. Morphol.* **2013**, *274*, 877–886. [CrossRef]
58. Alibardi, L.; Celeghin, A.; Dalla Valle, L. Wounding in lizards results in the release of beta-defensins at the wound site and formation of an antimicrobial barrier. *Dev. Comp. Immunol.* **2012**, *36*, 557–565. [CrossRef] [PubMed]
59. Dorin, J.R.; Barratt, C.L.R. Importance of β-defensins in sperm function. *Mol. Hum. Reprod.* **2014**, *20*, 821–826. [CrossRef]
60. Craik, D.J.; Fairlie, D.P.; Liras, S.; Price, D. The Future of Peptide-based Drugs. *Chem. Biol. Drug Des.* **2013**, *81*, 136–147. [CrossRef] [PubMed]

61. Hoover, D.M.; Wu, Z.; Tucker, K.; Lu, W.; Lubkowski, J. Antimicrobial Characterization of Human β-Defensin 3 Derivatives. *Antimicrob. Agents Chemother.* **2003**, *47*, 2804–2809. [CrossRef] [PubMed]
62. Sudheendra, U.S.; Dhople, V.; Datta, A.; Kar, R.K.; Shelburne, C.E.; Bhunia, A.; Ramamoorthy, A. Membrane disruptive antimicrobial activities of human β-defensin-3 analogs. *Eur. J. Med. Chem.* **2015**, *91*, 91–99. [CrossRef]
63. Scudiero, O.; Nigro, E.; Cantisani, M.; Colavita, I.; Leone, M.; Mercurio, F.A.; Galdiero, M.; Pessi, A.; Daniele, A.; Salvatore, F.; et al. Design and activity of a cyclic mini-β-defensin analog: A novel antimicrobial tool. *Int. J. Nanomed.* **2015**, *10*, 6523–6539.
64. Saravanan, R.; Li, X.; Lim, K.; Mohanram, H.; Peng, L.; Mishra, B.; Basu, A.; Lee, J.-M.; Bhattacharjya, S.; Leong, S.S.J. Design of Short Membrane Selective Antimicrobial Peptides Containing Tryptophan and Arginine Residues for Improved Activity, Salt-Resistance, and Biocompatibility. *Biotechnol. Bioeng.* **2014**, *111*, 37–49. [CrossRef]
65. Wessolowski, A.; Bienert, M.; Dathe, M. Antimicrobial activity of arginine- and tryptophan-rich hexapeptides: The effects of aromatic clusters, d-amino acid substitution and cyclization. *J. Pept. Res.* **2004**, *64*, 159–169. [CrossRef] [PubMed]
66. Dal Mas, C.; Pinheiro, D.A.; Campeiro, J.D.; Mattei, B.; Oliveira, V.; Oliveira, E.B.; Miranda, A.; Perez, K.R.; Hayashi, M.A.F. Biophysical and biological properties of small linear peptides derived from crotamine, a cationic antimicrobial/antitumoral toxin with cell penetrating and cargo delivery abilities. *Biochim. Biophys. Acta Biomembr.* **2017**, *1859*, 2340–2349. [CrossRef]
67. Ponnappan, N.; Budagavi, D.P.; Chugh, A. CyLoP-1: Membrane-active peptide with cell-penetrating and antimicrobial properties. *Biochim. Biophys. Acta* **2017**, *1859*, 167–176. [CrossRef] [PubMed]
68. Mandal, M.; Jagannadham, M.V.; Nagaraj, R. Antibacterial activities and conformations of bovine β-defensin BNBD-12 and analogs:structural and disulfide bridge requirements for activity. *Peptides* **2002**, *23*, 413–418. [CrossRef]
69. Hilpert, K.; Elliott, M.R.; Volkmer-Engert, R.; Henklein, P.; Donini, O.; Zhou, Q.; Winkler, D.F.H.; Hancock, R.E.W. Sequence requirements and an optimization strategy for short antimicrobial peptides. *Chem. Biol.* **2006**, *13*, 1101–1107. [CrossRef] [PubMed]
70. Mookherjee, N.; Anderson, M.A.; Haagsman, H.P.; Davidson, D.J. Antimicrobial host defence peptides: Functions and clinical potential. *Nat. Rev. Drug Discov.* **2020**, *19*, 311–332. [CrossRef]
71. Stone, K.L.; Williams, K.R. Enzymatic digestion of proteins in solutions and in SDS polyacrylamide gel. In *The Protein Protocol Handbook*; Walker, J.M., Ed.; Human Press Inc.: Totowa, NJ, USA, 1996; pp. 415–421.
72. Boni-Mitake, M.; Costa, H.; Spencer, P.J.; Vassillieff, V.S.; Rogero, J.R. Effects of ^{60}Co gamma radiation on crotamine. *Braz. J. Med. Biol. Res.* **2001**, *34*, 1531–1538. [CrossRef] [PubMed]
73. Petersen, T.N.; Brunak, S.; von Heijne, G.; Nielsen, H. SignalP 4.0: Discriminating signal peptides from transmembrane regions. *Nat. Methods* **2011**, *8*, 785–786. [CrossRef] [PubMed]
74. Breneman, C.M.; Wiberg, K.B. Determining atom-centered monopoles from molecular electrostatic potentials.The need for high sampling density in formamide conformational analysis. *J. Comput. Chem.* **1990**, *11*, 361–373. [CrossRef]
75. Clinical and Laboratory standards Institute (CLSI). *Reference Method for Broth Dilution Antifungal Susceptibility Testing of Yeasts*, 4th ed.; CLSI Standard M27; Clinical and Laboratory standards Institute: Wayne, PA, USA, 2017.

 *toxins*

Article

rDromaserpin: A Novel Anti-Hemostatic Serpin, from the Salivary Glands of the Hard Tick *Hyalomma dromedarii*

Hajer Aounallah [1,2], Melissa Regina Fessel [1], Mauricio Barbugiani Goldfeder [1], Eneas Carvalho [3], Chaima Bensaoud [4], Ana Marisa Chudzinski-Tavassi [1,5], Ali Bouattour [2], Youmna M'ghirbi [2,*] and Fernanda Faria [1,*]

1 Innovation and Development Laboratory, Innovation and Development Center, Instituto Butantan, São Paulo 05503-900, Brazil; hajer.aounallah@esib.butantan.gov.br (H.A.); melissa.fessel@butantan.gov.br (M.R.F.); mauricio.goldfeder@butantan.gov.br (M.B.G.); ana.chudzinski@butantan.gov.br (A.M.C.-T.)

2 Laboratory of Viruses, Vectors and Hosts (LR20IPT02), Pasteur Institute of Tunis, University of Tunis El Manar, Tunis 1002, Tunisia; ali.bouattour@pasteur.tn

3 Laboratory of Bacteriology, Butantan Institute, São Paulo 05503-900, Brazil; eneas.carvalho@butantan.gov.br

4 Institute of Parasitology, Biology Centre, Czech Academy of Sciences, 37005 Ceske Budejovice, Czech Republic; chayma.bensaoud@paru.cas.cz

5 Centre of Excellence in New Target Discovery (CENTD), Butantan Institute, Butantã, São Paulo 05503-900, Brazil

* Correspondence: youmna.mghirbi@pasteur.tn (Y.M.); fernanda.faria@butantan.gov.br (F.F.)

Abstract: Hemostatic disorders are caused either by platelet-related dysfunctions, defective blood coagulation, or by a combination of both, leading to an increased susceptibility to cardiovascular diseases (CVD) and other related illnesses. The unique specificity of anticoagulants from hematophagous arthropods, such as ticks, suggests that tick saliva holds great promise for discovering new treatments for these life-threatening diseases. In this study, we combined in silico and in vitro analyses to characterize the first recombinant serpin, herein called Dromaserpin, from the sialotranscriptome of the *Hyalomma dromedarii* tick. Our in silico data described Dromaserpin as a secreted protein of ~43 kDa with high similarities to previously characterized inhibitory serpins. The recombinant protein (rDromaserpin) was obtained as a well-structured monomer, which was tested using global blood coagulation and platelet aggregation assays. With this approach, we confirmed rDromaserpin anticoagulant activity as it significantly delayed plasma clotting in activated partial thromboplastin time and thrombin time assays. The profiling of proteolytic activity shows its capacity to inhibit thrombin in the micromolar range (0.2 to 1 μM) and in the presence of heparin this inhibition was clearly increased. It was also able to inhibit Kallikrein, FXIa and slightly FXIIa, with no significant effect on other factors. In addition, the rDromaserpin inhibited thrombin-induced platelet aggregation. Taken together, our data suggest that rDromaserpin deserves to be further investigated as a potential candidate for developing therapeutic compounds targeting disorders related to blood clotting and/or platelet aggregation.

Keywords: *Hyalomma dromedarii*; salivary glands; serpin; anticoagulants; thrombin inhibitor

Key Contribution: rDromaserpin is a new anti-hemostatic serpin from the salivary glands of the *H. dromedarii* tick. It was able to inhibit blood coagulation mainly through its action on thrombin, in addition to modulating thrombin activity on platelet aggregation.

Citation: Aounallah, H.; Fessel, M.R.; Goldfeder, M.B.; Carvalho, E.; Bensaoud, C.; Chudzinski-Tavassi, A.M.; Bouattour, A.; M'ghirbi, Y.; Faria, F. rDromaserpin: A Novel Anti-Hemostatic Serpin, from the Salivary Glands of the Hard Tick *Hyalomma dromedarii*. *Toxins* **2021**, *13*, 913. https://doi.org/10.3390/toxins13120913

Received: 28 October 2021
Accepted: 8 December 2021
Published: 20 December 2021

1. Introduction

Despite the emergence of new diseases, cardiovascular diseases (CVD) such as stroke, heart attack, and deep vein thrombosis remain the leading causes of death worldwide [1], and are increasing exponentially given today's modern unhealthy lifestyle [2]. They are multifactorial and are caused mainly by hemostasis disorders [3]. In a normal state, hemostasis serves to maintain the fluidity of blood within the veins and arteries, preventing excessive blood loss after injury through clot formation [4]. This system involves a set of well-regulated processes, including blood coagulation, platelet aggregation, and vasoconstriction, led by a complex series of cascade enzymatic reactions [5]. These reactions involve several serine proteases controlled by endogenous molecules including serine protease inhibitors (serpins) [6,7]. Several pathogeneses can ensue when serine protease activity or serpin-mediated regulation becomes unbalanced or dysfunctional [8]. For instance, a failure in the regulation of the blood coagulation may cause abnormal bleeding, causing either hemorrhage or thrombosis [9,10]. Anticoagulants and antiplatelet compounds are commonly used as antithrombotic drugs for the prevention and treatment of many cardiovascular disorders [11]. Apart from the CVD, some cases of severe thrombosis and vascular disorders have been related to the coronavirus disease 2019 (COVID-19) complications [12]. More recently, anticoagulant therapy has been applied to patients with coagulopathy associated with severe COVID-19, and the treatment has shown a decrease in the mortality rate [12]. Today's treatment options for blood clotting dysfunction include naturally derived compounds [13]. With their high target specificity, they are not expected to have many side effects [13]. The search for alternative anticoagulants is still ongoing. Currently, several studies have screened natural resources, aiming to find potential antithrombotic biomolecules, worthy of competition or improvement to those currently available, with higher target specificity and/or lower side effects [14].

Serine protease inhibitors, termed serpins, are the most widely represented protein superfamily of protease inhibitors [15]. Admittedly, serpins have a pleiotropic function, regulating wide spectrums of proteolytic processes from thrombosis, thrombolysis, inflammation, and immune response, to cellular invasion in tissue remodeling, hormone transport, and even tumor growth and development [16]. Despite their multiple functions, serpins share a structurally conserved fold, including three β-sheets and seven to nine α-helices with a reactive center loop (RCL), recognized by target enzymes [17]. The RCL is a solvent-exposed adjustable stretch of 21–22 amino acid residues in length, which acts as a target for individual proteases [17]. Structural dynamic studies of serpins have unveiled the distinct conformational states adopted by the RCL, including the native intact RCL [18], the cleaved inactive form [19], the partially inserted RCL forms [20], and the latent form [21]. These studies have proved that the RCL must change its configuration to bind successfully to target proteases. While most serpins inhibit serine proteases, some of them inhibit cysteine proteases as well [17]. They function in a suicide inhibitory mechanism in which, once the serpin binds to the target protease, both molecules are permanently inactivated [15]. Through this inhibition, serpins toggle between normal physiology and pathology, guaranteeing normal biological function or leading to diseases, respectively [22]. Indeed, serpinopathies, where genetic mutations lead to inactive or aberrant serpins, and other disorders where serpin levels become unbalanced, cause thrombotic or thrombolytic cascades, triggering excess clotting or bleeding, respectively [23]. Restoring overall stability can be managed through the application of serpin protein treatments as therapeutics [24]. Given their central involvement, both in normal physiology and in pathological conditions, studies are increasing to better understand the roles of serpins in some physiological processes and to develop new therapeutic approaches.

Hematophagous animals, such as leeches, mosquitoes, and ticks employ serpins as a weapon to counter the host defense system and guarantee a successful blood meal [25]. These species represent an attractive natural source for the development of novel anti-hemostatic serpins. Particularly, proteins from tick salivary glands have been proven to target several blood coagulation components [26]. Several tick serpins have been described in numerous sequencing projects [27]; however, roughly a dozen of them are functionally described [28]. More specifically, to our knowledge, there is no previous characterization of serpins from the sialotranscriptome of *H. dromedarii*.

In the present study, we combined bioinformatics analyses and experimental assays to characterize a new serpin, named Dromaserpin, from the sialotranscriptome of *H. dromedarii*. Its recombinant form (rDromaserpin) was obtained by expression in *Escherichia coli*, was characterized using biochemical and biophysical tools, and tested in various in vitro assays. Its anticoagulant effect was demonstrated by its ability to significantly prolong the intrinsic coagulation pathway, acting primarily on thrombin and Kallikrein, two trypsin-like serine proteases central in the processes of hemostasis and thrombosis. Its antiplatelet activity was proven by inhibiting platelet aggregation induced by thrombin. Overall, our results suggest that rDromaserpin is a novel protease inhibitor that deserves further in-depth investigation, primarily its molecular and kinetic mechanism. Further in vivo studies could be the object of future studies initiating a path for the development of a new potential antithrombotic drug.

2. Results

2.1. Bioinformatic and Phylogenetic Analysis on Dromaserpin Sequence

The composite full-length Dromaserpin cDNA sequence was 1371 nucleotides long with a single open reading frame (ORF) of 1209 bases. The 5′ non-coding region was 27 bp, and the 3′ non-coding region was 135 bp (data not shown). The ORF coded for a protein with 403 amino acid residues, including a 21 residues signal peptide, which suggests the protein is secreted and acts extracellularly (Figure 1).

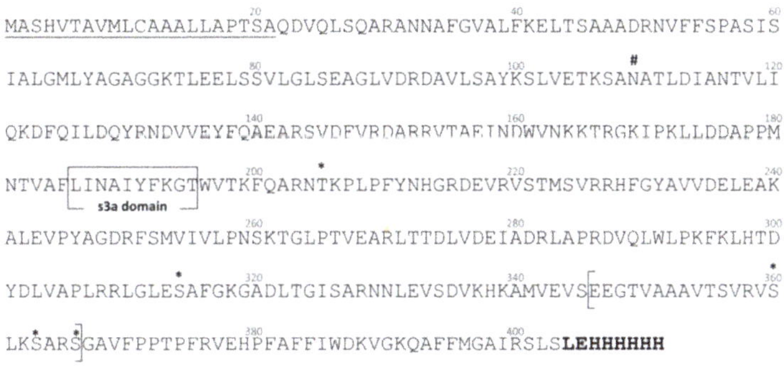

```
MASHVTAVMLCAAALLAPTSAQDVQLSQARANNAFGVALFKELTSAAADRNVFFSPASIS
IALGMLYAGAGGKTLEELSSVLGLSEAGLVDRDAVLSAYKSLVETKSANATLDIANTVLI
QKDFQILDQYRNDVVEYFQAEARSVDFVRDARRVTAEINDWVNKKTRGKIPKLLDDAPPM
NTVAFLINAIYFKGTWVTKFQARNTKPLPFYNHGRDEVRVSTMSVRRHFGYAVVDELEAK
ALEVPYAGDRFSMVIVLPNSKTGLPTVEARLTTDLVDEIADRLAPRDVQLWLPKFKLHTD
YDLVAPLRRLGLESAFGKGADLTGISARNNLEVSDVKHKAMVEVSEEGTVAAAVTSVRVS
LKSARSGAVFPPTPFRVEHPFAFFIWDKVGKQAFFMGAIRSLSLEHHHHHH
```

s3a domain

Figure 1. Analysis of the predicted amino acid sequence of Dromaserpin. The predicted signal peptide is underlined. N-glycosylation and O-glycosylation sites are identified by hashtag and asterisk, respectively. Conserved s3a domain is boxed and conserved reactive center loop (RCL) domain is indicated between square brackets. A poly-histidine tag and two residues, indicated in bold, were added to the C-terminal of the sequence. Numbers indicate amino acid residue positions in the sequence. The rDromaserpin sequence extends from residue 22 to residue 409. In the rDromaserpin, the signal peptide was removed and changed by a methionine.

The predicted rDromaserpin has 391 amino acid residues, including the histidine tag sequence, with a theoretical molecular weight of 43,159.27 Da. The predicted protein has a single potential N-glycosylation site at the position N_{109}, and five O-glycosylation sites at the positions T_{205}, S_{314}, S_{360}, S_{363} and S_{366} (Figure 1). The predicted protein presents the serine protease inhibitor-associated domains s3a and RCL (Figure 1) and, indeed, its membership in the serpin superfamily was confirmed by SMART software (Figure 1). Since the RCL consensus critical residues for serpin inhibitory activity were previously described [29], we aligned the predicted RCL of the newly identified serpin with the corresponding amino acid sequence of inhibitory and non-inhibitory serpins (Figure 2).

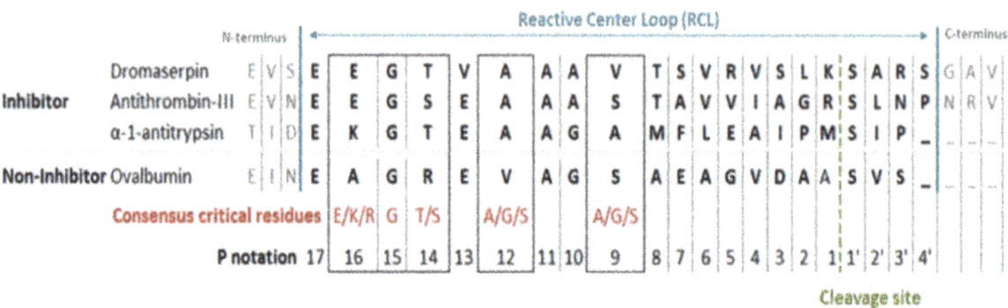

Figure 2. Comparison between RCL sequence in Dromaserpin and inhibitory and non-inhibitory serpins. RCL amino acid sequences alignment between α-Antitrypsin (Uniprot: P01009) and Antithrombin-III (P01008) from human, Ovalbumin from chicken (Uniprot: P01012), and Dromaserpin was obtained using ClustalW. The consensus critical residues for inhibitory activity have been described elsewhere [29]. P notation was applied according to a pervious study [30].

The RCL sequence extends from P17 to P4′. P′ residues are numbered from the cleavage site to the C-terminus. P residues are numbered from the cleavage site to the N-terminus. Dromaserpin's RCL was similar to both inhibitory serpins α-1 Antitrypsin and Antithrombin-III, unlike the non-inhibitory serpin ovalbumin (Figure 2). Particularly, the consensus restudies described as crucial for the inhibitory activity were present in Dromaserpin among these: Glu_{347}, Gly_{348}, Thr_{349}, Ala_{351}, and were perfectly conserved, except Val_{354}. Comparison of the aligned RCLs in Figure 2 hypothesizes that Dromaserpin may be an inhibitory serpin.

As proteins sharing sequence similarities are quite likely to have similar or close activity [31], we aligned the Dromaserpin amino acid sequence with 25 tick serpins (Table S1). These serpins have >30% homology with the Dromaserpin and most have been experimentally proven to act as anti-hemostatic proteins [32]. Sequence comparison showed that amino acid patterns related to the inhibitory propriety (mainly the RCL domain and s3a domain) were present in Dromaserpin and were conserved among all the compared tick serpins (Figure 3b). In the evolutionary analyses, the resulting phylogenetic tree revealed two separate groups of serpins (Figure 3a). Dromaserpin was clustered together with the larger one, composed of 18 serpins (Figure 3a). Dromaserpin was closer to the serpins of genus *Rhipicephalus*. Indeed, the BLASTP search identified two tick serpins, from *Rhipicephalus* genus, whose sequences were very similar to the Dromaserpin sequence.

Dromaserpin shares the highest sequence identity 83.33% (99% coverage) with RHS-1 (accession: AFX65224.1), an anticoagulant serpin from *Rh. haemaphysaloides* presenting anti-chymotrypsin activity [32]. Dromaserpin presents 81.94% identity (97% coverage) with RmS5 (accession: AHC98656.1), from *Rh. microplus*, which has not been functionally characterized [34]. In the tree, Dromaserpin formed a small branch with RHS-1 and RmS5, both presenting a basic amino acid (lysine) in the position P1, similar to Dromaserpin (Figure 3a). The aligned RCLs of the serpins used in the phylogenetic analysis are described in (Figure 3b). The serpins grouped in the same cluster had highly conserved RCLs compared to other tick serpins, which showed fewer similarities.

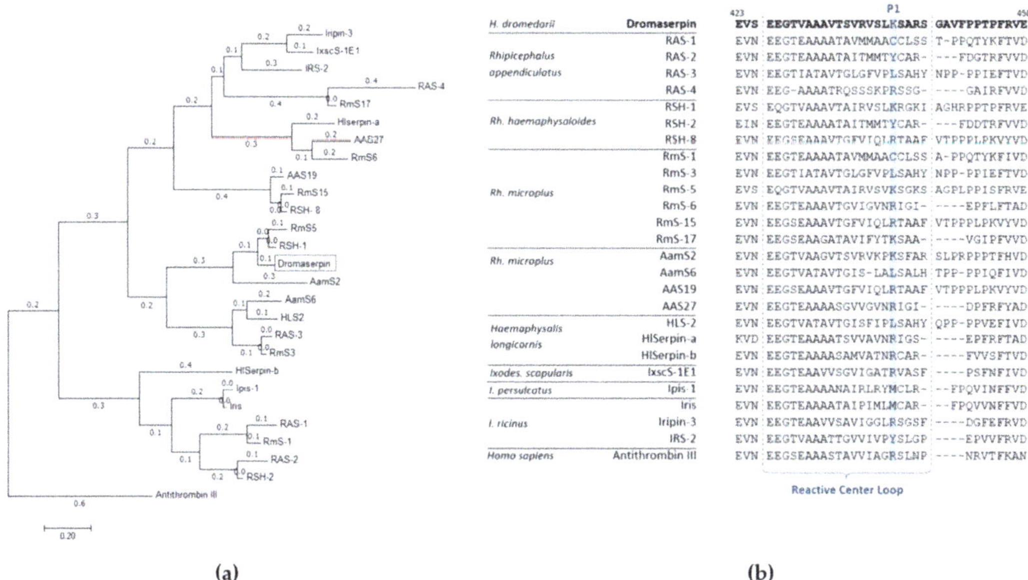

(a) **(b)**

Figure 3. Molecular phylogenetic analysis of 25 tick serpins and Dromaserpin. (**a**) The evolutionary phylogenetic tree of Dromaserpin and the chosen tick serpins. Antithrombin III was utilized as an outgroup. The tree with the highest log likelihood (−12,179.3074) is shown, drawn to scale, with branch lengths measured in the number of substitutions per site (next to the branches). Evolutionary analyses were conducted in MEGA7 [33]. (**b**) Alignment of the RCL regions of the selected serpins was carried out using MEGA7.

2.2. Expression and Purification of the Recombinant Dromaserpin (rDromaserpin)

rDromaserpin has 391 amino acid residues, including the histidine tag sequence, a predicted molecular weight of 43,159.27 Da, and a predicted isoelectric point (pI) of 8.03. These features dictated the purification strategy adopted. rDromaserpin was successfully expressed using the *E. coli* BL21 (DE3) system in the optimal expression condition (1 mM IPTG, 30 °C, 3 h incubation). It was obtained in both soluble and insoluble forms (Figure 4a).

Although rDromaserpin mostly aggregates as inclusion bodies, it was possible to use the soluble fraction for the purification procedure and still obtain an acceptable yield (1.73 mg/L). As expected, purified rDromaserpin migrates at an apparent molecular weight of ~43 kDa on 12.5% SDS-PAGE (Figure 4b). After the purification steps, rDromaserpin was obtained with a high level of purity, observed after Coomassie Blue staining. The purified protein was used for subsequent experiments.

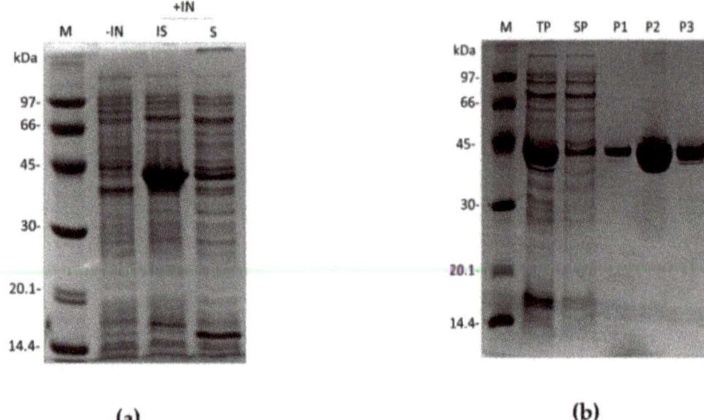

Figure 4. Expression and purification of rDromaserpin. cDNA coding for rDromaserpin was cloned into pET28a expression vector and the recombinant protein was expressed in *E. coli* BL21 (DE3) in 2YT medium. Whole cell lysates of non-induced or induced (IPTG 1 mM, 3 h, 30 °C) cultures were analyzed by SDS-PAGE (12.5%). (**a**) Both soluble and insoluble fractions were analyzed. Lanes M, −IN, +IN, IS, and S represent protein marker, not induced, induced, insoluble and soluble fractions, respectively; (**b**) SDS-PAGE of fractions from three chromatography steps. Lanes M, TP, and SP represent protein marker, total protein and soluble protein respectively. Lanes P1, P2, and P3 represent eluted purified proteins pooled from IMAC chromatography, Q sepharose ion-exchange chromatography, and size exclusion chromatography, respectively.

2.3. Structural Characterization of the rDromaserpin

2.3.1. Analysis of the Secondary Structure of rDromaserpin

The purified rDromaserpin was obtained as a monomer, according to an analytical gel filtration analysis (Figure 5). According to the calibration curve obtained with standards, (Figure 5), the calculated molecular weight of rDromaserpin was 37.025 kDa.

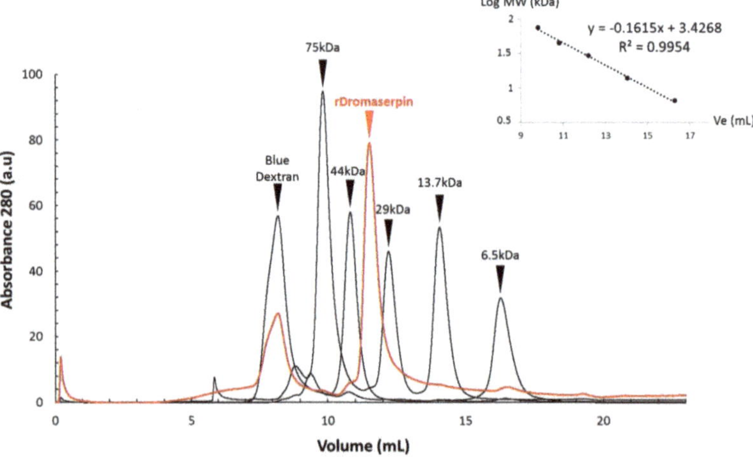

Figure 5. Separation of rDromaserpin and five standard proteins by size exclusion chromatography. Loading of the 100 µL of rDromaserpin (labeled in red) or the following standards (labeled in black): Conalbumin (75 kDa), Ovalbumin (44 kDa), Carbonic anhydrase (29 kDa), Ribonuclease A (13.7 kDa), Aprotinin (6.5 kDa), as well as blue dextran (for the void volume V_0). A calibration curve was obtained with standards to calculate the molecular weight of rDromaserpin using the equation (log (MW) = −0.1615 Ve + 3.4268).

Its secondary structure content was investigated by carrying out circular dichroism (CD) analyses in the Far-UV region, which makes possible estimations of the protein's α-helical, β-sheet and random coil content (Figure 6).

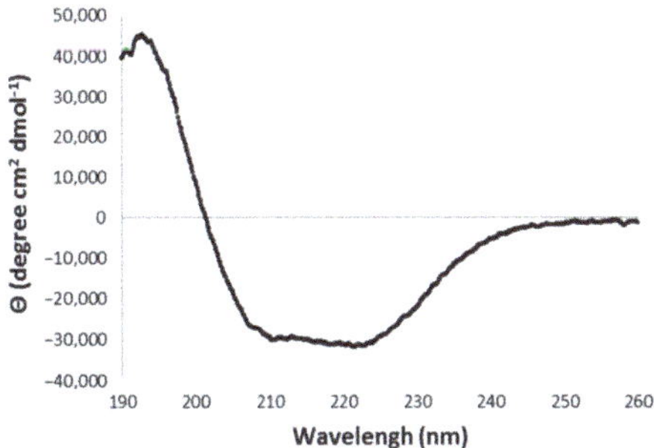

Figure 6. Far-UV CD spectra obtained for rDromaserpin. The ellipticity was expressed as the mean-residue molar ellipticity (θ) in degree cm^2 dmol^{-1}.

The Far-UV CD spectrum presented characteristics of a structured protein, and showed one positive peak at 194 nm and two minimums at 211 nm and 220 nm (Figure 6), a pattern related to an α/β rich protein [35]. The deconvolution of the CD spectrum, using the BestSel program, indicates the secondary structure content of rDromaserpin as approximately 54% α-helices, 21% β-strands and 25% random coils (Figure S1).

2.3.2. Comparative Modeling

The predicted amino acid sequence of Dromaserpin was used to search for homologous proteins, with experimentally-solved 3D structures to serve as templates for structure homology modeling, using the Swiss-Model workspace. Conserpin, which shares 40.62% identity (92% coverage) with Dromaserpin, was the protein selected as a template, and had its 3D structure solved in both RCL open/uncleaved (PDB code: 5cdx) and closed/cleaved (PDB code: 5cdz) conformations. Two three-dimensional model structures of Dromaserpin were thus constructed using each of the conserpin conformations as templates (Figure 7).

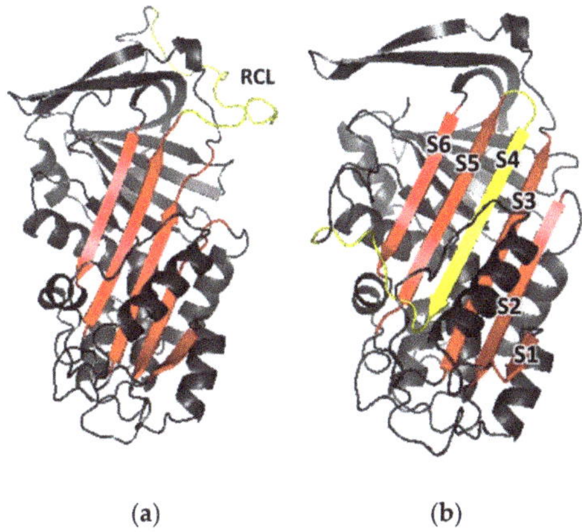

Figure 7. Cartoon representations of the three-dimensional models of Dromaserpin. (**a**) A cartoon representation of Dromaserpin 3D model 1 with an exposed RCL, based on the structure of Conserpin (PDB: 5cdx). (**b**) A cartoon representation of the Dromaserpin 3D model 2 with an inserted RCL, based on the structure of Conserpin (PDB: 5cdz). The RCL (in yellow) is inserted in the 5-sheet β-strand (in red). Loops are colored grey.

According to the obtained models, the overall structure of Dromaserpin adopts a typical serpin fold composed of eight α-helices (Figure S2a) and three large β-sheets (A, B, and C) (Figure S2b). An uncleaved RCL model, based on the structure of Conserpin (5cdx), shows an intact RCL (Figure 7a) (yellow). As observed in other serpins [36], RCL in this configuration is located as a flexible loop outside the Dromaserpin structure core and can act as bait for target proteases. Additionally, a second model was built based on the structure of Conserpin in the latent state (PDB code: 5cdz) (Figure 7b). In this model, the RCL is cleaved and inserted into the central β-sheet A (Figure 7b) (S1, S2, S3, S5, and S6 indicated in red) as a strand, S4 (indicated in yellow on Figure 7b). Thus, according to Ramachandran plot, 97.6% (5cdz) (Figure S3) and 98.5% (5cdx) (Figure S4) of the residues from Dromaserpin 3D models are in favored and/or allowed regions. Neither model had residues located in the disallowed regions of the Φ, Ψ angle pairs of the Ramachandran plot, indicating correct stereochemistry. The percentage of secondary elements in Dromaserpin models was predicted by STRIDE, and was compared to those obtained experimentally by analyzing CD spectra, using the BestSel tool (Table 1).

Table 1. Comparison of the secondary structure content obtained theoretically (exposed RCL and inserted RCL) and experimentally (CD data).

	Elements from the Models		**Elements from CD Data**
	Exposed RCL	**Inserted RCL**	
Total aligned residues (100%)	372	372	391
Helix (%)	29.83	30.11	54.20
Strand (%)	30.65	33.60	21.40
Other (Coil, bridge, Turn) (%)	39.52	36.29	24.40

2.4. *Functional Characterization of the rDromaserpin*

2.4.1. Global Blood Coagulation Assays

The effect of rDromaserpin was investigated on blood clotting using assays that independently target the extrinsic, intrinsic and common coagulation pathways. Prothrombin Time (PT) assay, designed to monitor the extrinsic coagulation [37], detected no significant effect upon the addition of 1 μM of rDromaserpin, and only a very slight increase of clotting time with 5 μM of rDromaserpin (2 s) in the tested conditions (data not shown). Thus, the protein, under these conditions, has minimal or no effect on the extrinsic coagulation pathway. As well, the Thrombin Time (TT) assay conducted on PPP (Platelet-poor plasma) did not show any significant differences under these conditions (data not shown). However, when tested with prothrombin deficient plasma (FII-DP), 5 μM of rDromaserpin significantly increased TT by ~10.97 s, when 2 mU of FIIa was added without pre-incubation (Figure 8a). Similarly, clotting time was increased by ~9.83 s, when rDromaserpin was pre-incubated with FII-DP for 1 h (Figure 8a). The plasma became incoagulable when rDromaserpin was pre-complexed with thrombin, preventing thrombin from converting the fibrinogen to fibrin (Figure 8a).

On the other hand, by using the activated partial thromboplastin time (APTT) assay, a statistically significant delay was shown in blood clot formation when plasma was treated during 20 min with 0.5, 1 or 5 μM rDromaserpin, prolonging the intrinsic coagulation time by 7.4 s, 13.06 s and 42.56 s, respectively (Figure 8b). Our results therefore suggest that rDromaserpin is probably able to inhibit serine protease(s) involved in the intrinsic pathway of blood coagulation.

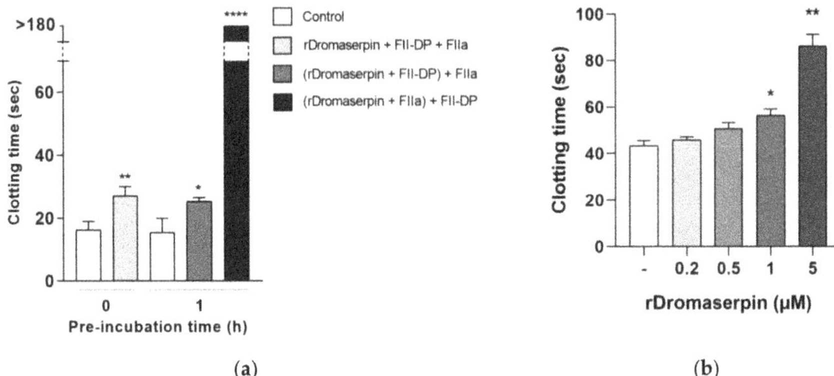

Figure 8. Effect of rDromaserpin on the intrinsic pathway of blood coagulation. (**a**) In vitro TT assay performed on FII-DP in the presence of 5 μM rDromaserpin and 2 mU FIIa. The assay was tested without pre-incubation or by pre-incubating samples in brackets for 1 h at room temperature. (**b**) In vitro APTT evaluation in isolated human plasma incubated for 20 min with 0.2, 0.5, 1, or 5 μM rDromaserpin. (-) refers to plasma used as control in the experiments containing the same volume of buffer. The results correspond to the mean ± SEM values acquired in three independent experiments; the asterisk indicates a significant difference between groups; * ($p < 0.05$) ** ($p < 0.005$) **** ($p < 0.0005$).

2.4.2. Protease Inhibition Profiling of rDromaserpin

As evidenced by its capacity to prolong the clotting time of human plasma, the ability of purified rDromaserpin to inhibit different serine proteases derived from human blood (FIIa, Kallikrein, FIXa, FXIa, FXIIa and FXa) involved in coagulation pathways was tested (Figure 9).

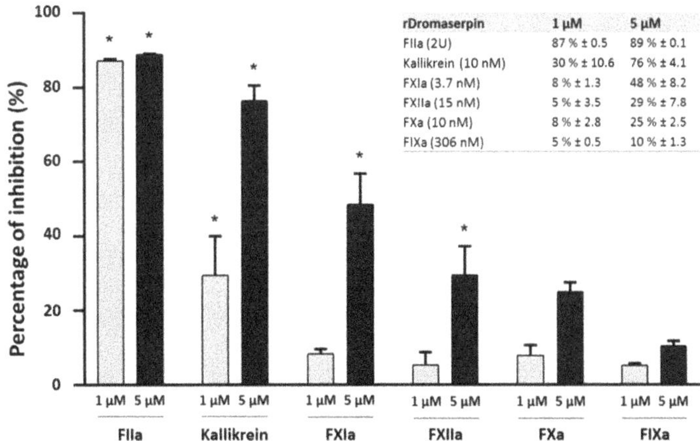

rDromaserpin	1 μM	5 μM
FIIa (2U)	87 % ± 0.5	89 % ± 0.1
Kallikrein (10 nM)	30 % ± 10.6	76 % ± 4.1
FXIa (3.7 nM)	8 % ± 1.3	48 % ± 8.2
FXIIa (15 nM)	5 % ± 3.5	29 % ± 7.8
FXa (10 nM)	8 % ± 2.8	25 % ± 2.5
FIXa (306 nM)	5 % ± 0.5	10 % ± 1.3

Figure 9. Inhibition of serine proteases by rDromaserpin. The ability of rDromaserpin (1 μM and 5 μM) to inhibit the activity of FIIa (2U), Kallikrein (10 nM), FXIa (3.7 nM), FXIIa (15 nM), FXa (10 nM), and FIXa (306 nM), using correspondent synthetic substrates (0.2 μM) was evaluated at 37 °C for 15 min. The results are acquired in three independent experiments. Proteases labelled with an asterisk were inhibited with statistical significance ($p < 0.05$).

Residual proteolytic activity of the tested proteases was measured using the appropriate synthetic chromogenic substrates (see materials and methods section). Statistically significant reductions in enzymatic activity among the tested proteases were observed only for thrombin (FIIa), Kallikrein, and FXIa. At concentrations of 1 µM and 5 µM, rDromaserpin significantly decreased thrombin catalytic activity by 87% ($p < 0.001$) and 89% ($p < 0.001$), respectively. At the concentration of 5 µM, rDromaserpin significantly decreased Kallikrein catalytic activity by ~76% ($p < 0.001$), while 1 µM slightly inhibited the enzyme by ~30% ($p < 0.01$). Five µM of rDromaserpin inhibited FXIa (~48%, $p < 0.01$) and slightly FXIIa (~29%, $p = 0.014$). However, 1 µM of the inhibitor was unable to affect the activity of both factors. As opposed to FXa and FIXa which showed no significant activity in presence of rDromaserpin up to 5 µM. These data suggest that rDromaserpin is a functional thrombin inhibitor, able to target Kallikrein and FXIa, and it might play multiple roles in the tick-host interface. We further recorded thrombin activity using different concentrations of the inhibitor (ranging from 0.02 to 1 µM) (Figure 10a).

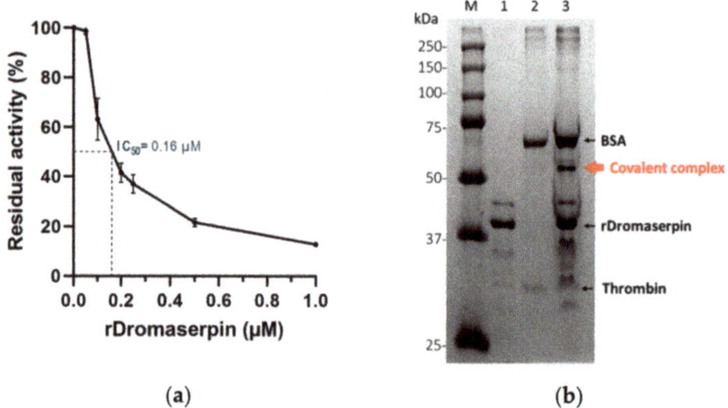

Figure 10. The effect of rDromaserpin on thrombin. (**a**) Residual activity of thrombin in the presence of increasing concentrations of rDromaserpin. Error bars represents the mean ± S.E.M values registered in three independent experiments. An estimate is given of the half maximal inhibitory concentration (IC_{50} = 0.16 µM). (**b**) A covalent complex was formed between rDromaserpin and thrombin. Lanes 1 and 2 represent rDromaserpin and thrombin, respectively. The major band in lane 2 corresponds to BSA (Bovine Serum Albumin) present in thrombin buffer. Lane 3 corresponds to rDromaserpin and thrombin loaded after 1 h incubation. Lane M represents the protein marker (Precision Plus Protein™ Dual Color Standars, BioRad). The covalent complex between rDromaserpin and thrombin is indicated with red arrow. Proteins were resolved on 5–10% SDS-PAGE and visualized by Coomasie Blue staining.

We observed a decrease in thrombin residual activity as the amount of rDromaserpin increased, inferring a dose-dependent inhibition (Figure 10a). At concentrations less than 0.1 µM, rDromaserpin did not affect thrombin (FIIa) activity. Indeed, it was able to significantly inhibit thrombin with an IC_{50} = 0.16 µM.

For a better understanding of its mechanism of action, we investigated the possibility of forming a covalent inhibitor-enzyme complex. With thrombin (36 kDa), rDromaserpin forms an SDS- and heat-resistant complex which migrates above 50 kDa (Figure 10b), proving a mechanism of suicide inhibition.

2.4.3. Effect of Heparin on the Rate of Thrombin Inhibition by rDromaserpin

Glycosaminoglycans are known to improve the inhibitory activity of plasma serpins [38]. We therefore investigated whether heparin, a common serpin activator, is able to enhance rDromaserpin activity toward thrombin. To locate the heparin binding site in the Dromaserpin model in its predicted active state (exposed RCL) (Figure 11a), we examined possible basic residues in revealed binding sites of other heparin-binding serpins. Our in silico analysis showed a single extended basic patch located on the top of the β-sheet A, near β-sheet C, and surrounding the RCL region in the active model of Dromaserpin (Figure 11a). Under the tested conditions, rDromaserpin did not bind to the heparin-sepharose resin (Figure 11c). In contrast, when it was pre-complexed with thrombin, the covalent complex was able to bind to the resin, indicating ternary complex formation (Figure 11c).

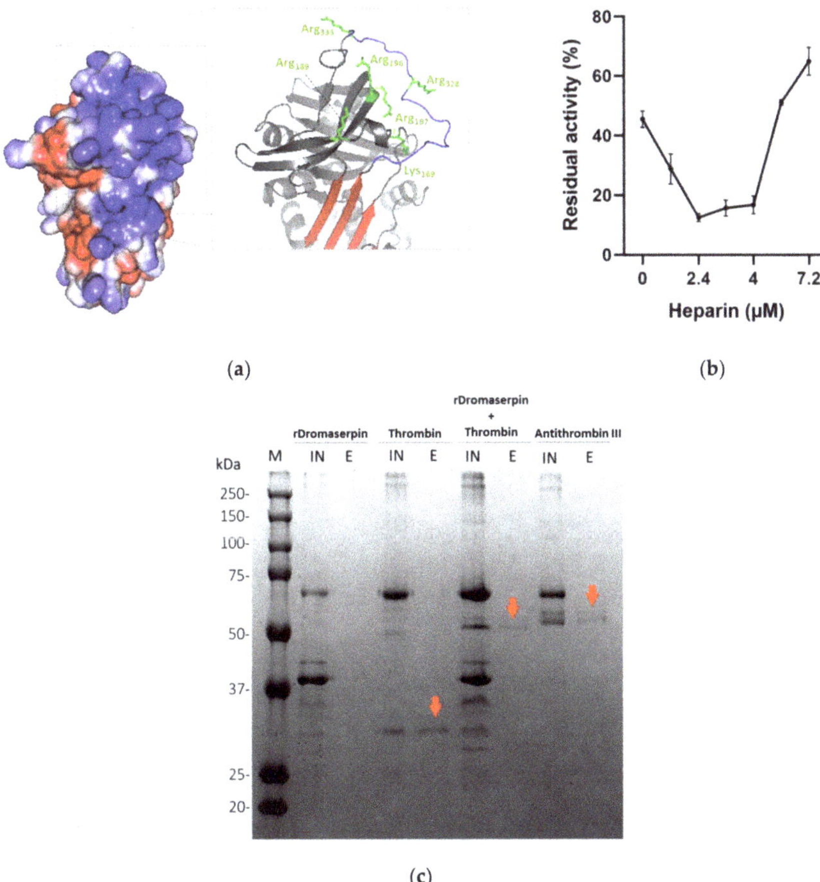

Figure 11. Effect of heparin on thrombin inhibition by rDromaserpin. (**a**) Electrostatic surface of Dromaserpin in the exposed RCL model. Colors relate to the electrostatic surface potential (blue is positive, and red is negative, −2 to 2 k_BT) calculated by APBS (37). The surface charge distribution reveals a prominent basic patch located on top of the β-sheet A and in RCL region. Side chains of basic residues located and surrounding the basic path ($Arg_{196–197}$, Arg_{189}, Arg_{328}, Arg_{335} and Lys_{169}) are labeled in green and shown as sticks. (**b**) Thrombin residual activity was assessed in the presence of 0.2 µM Dromaserpin and an increased concentration of heparin in µM range. Bar error represents the mean ± S.E.M values registered in three independent experiments. (**c**) Binding of rDromaserpin-thrombin complex to heparin. IN, input. E, eluted. Proteins eluted with 1 M NaCl are indicated by the red arrows. Antithrombin III was used as control, known to bind to heparin-sepharose resin.

Experimentally, rDromaserpin induces accelerated thrombin inhibition in the presence of heparin, as evidenced by a U-shaped dose-response curve (Figure 11b). In the absence of heparin, thrombin was inhibited by 0.2 µM of rDromaserpin, showing a residual activity of 48.11% (Figure 11b). Using a range of heparin, the rate of inhibition appears to increase by around 25-fold in an optimal heparin concentration of 2.4 µM, reaching 12.5% of residual activity (Figure 11b). However, heparin-mediated acceleration of thrombin inhibition by rDromaserpin was progressively reduced upon a concentration of 5.6 μM (Figure 11b).

2.4.4. Dromaserpin Inhibits the Function of Thrombin in Activating Platelet Aggregation

Platelet aggregation is one of the host's first lines of anti-tick defense [5]. Knowing that ticks use serpins to hijack host hemostasis [39], and after discovering that rDromaserpin is an efficient inhibitor of thrombin in its action on blood clotting, it was necessary to verify whether it would also be able to inhibit platelet aggregation when thrombin is the agonist. Washed platelets were used in platelet aggregation assays in the presence of 1 μg/mL of thrombin as an agonist, and pre-incubated with 10 μM of rDromaserpin (Figure 12). Our results showed that rDromaserpin was able to inhibit the platelet aggregation function of thrombin by reducing platelet aggregation by 88.16% (±2.84%), compared to the control (Figure 12).

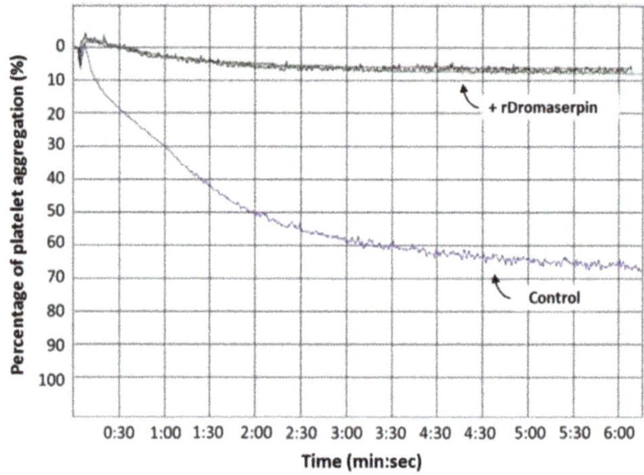

Figure 12. Effect of rDromaserpin on thrombin-induced platelet aggregation. The platelet aggregation assay was performed using washed platelet approach described in Material and Methods section. The trace below (labeled "control") corresponds to platelet activated by thrombin in the presence of rDromaserpin buffer. The traces above (labeled "+ rDromaserpin") correspond to triplicates of 10 μM rDromaserpin pre-incubated with thrombin prior to platelet activation. Data are presented as mean ± S.E.M. of triplicate platelet aggregation assays.

3. Discussion

Tick serpins attract considerable attention in the field of drug discovery and development because of their anti-hemostatic and immunomodulatory properties [27]. The current study was prompted by our previous investigations describing the sialome of *H. dromedarii* tick, highlighting a variety of putative proteins, including serpins, with potentially therapeutic features [40,41]. Our strategy was to combine computational and experimental analysis to characterize a new serpin that could target and modulate the hemostasis system.

Our bioinformatics analysis described Dromaserpin as a serpin of ~43 kDa and 403 amino acids long with a signal peptide, suggesting its secretion and extracellular activity in *H. dromedarii*. The inhibitory activity of serpins is mainly executed by its RCL domain [36]. Noting that the RCL is not the only region to predict inhibitory serpins, other consensus sequences that are well preserved during the evolution have been studied elsewhere [29]. Here, we focus on the RCL region as it is directly involved in the serpin's inhibition mechanism. The RCL sequence in Dromaserpin is located near its C-terminus. It is composed of 21 amino acids, as are most serpins [15]. It is noteworthy that some serpins do not have classical inhibitory activity [15]. These non-inhibitory serpins contain the RCL domain, but there is no evidence that it is involved in their physiological activity. For instance, they may serve in hormone transport [42] and corticosteroid-binding globulin production [43]. Given this information, we found it important to compare RCL sequences in Dromaserpin, inhibitory and non-inhibitory serpins. The RCL of Dromaserpin shares the highest homology with inhibitory serpins α-1 Antitrypsin and Antithrombin-III. These in silico predictions lead us to believe that Dromaserpin might act as an inhibitory serpin, which we subsequently proved with functional assays. While revealing the first high-resolution structure of the serpin α-1 Antitrypsin, Engh et al. unveiled striking evidence about residues in RCL that are proposed to interact with target proteases [19]. Interestingly, Dromaserpin preserves residues essential for inhibitory activity (Figure 2). Moreover, the RCL sequence includes a cleavable reactive center, defined as the P1-P1′ bond, in which the protease cleaves and forms a covalent complex with the serpin [15]. Lys-Ser is the scissile P1-P1′sequence in Dromaserpin. Where the serpin's target protease may not be accurately predicted based on their P1 amino acid residue [8], several previous studies have provided relevant information about the nature of P1 residue in relation to targeted protease [44,45]. Most related investigations report that serpins with inhibitory functions against trypsin or thrombin have polar basic (Arg and Lys) residues at the P1 site, whereas those with aromatic (Phe, Tyr and Trp) residues at the P1 site are more likely to inhibit chymotrypsin [8]. Dromaserpin has a lysine residue at the P1 site, identical to RHS-1 and RmS5, the closest serpins in the evolutionary tree, which inhibit chymotrypsin, thrombin and FXa [32]. Similar observations were reported for AAS19, which has Arg at the P1 site, and targets numerous protease including FXa, FXIIa, thrombin, tryptase, and chymotrypsin [46]. HLserpin-A has also a basic (Arg) in that position; however, it is reported to inhibit the serine proteases cathepsin G, and FXa, and the cystein proteases papain and Cathepsin B, with no effect on thrombin [47]. Experimentally, we proved that rDromaserpin can inhibit thrombin, Kallikrein and, slightly, FXIIa and FXIa, but neither FXa nor FIXa. Overall, we consider that in silico analyses are not sufficient to predict Dromaserpin targets. Apart from the P1-P1′ sequence, dynamic studies on serpins demonstrate that the RCL domain changes its state in order to bind to target proteases [48,49]. Using comparative modeling, we showed Dromaserpin in two conformations: an exposed RCL and an inserted RCL in the β sheet A. Both conformations have been described for other tick serpins in previous structural analyses [50,51]. In these analyses, they assume the use of a suicide inhibition mechanism to inhibit proteases. Experimentally, rDromaserpin forms an SDS and heat-stable complex, with thrombin proving a suicide inhibition mechanism. Apparently, thrombin cleaves the RCL of rDromaserpin in its P1 amino acid residue, forming the obtained covalent complex in the SDS-PAGE gel (Figure 11c). Through this inhibition mechanism Dromaserpin binds to thrombin, and both molecules are permanently inactivated.

To gain functional insight on this new serpin, a recombinant form (rDromaserpin) was overexpressed using an *E. coli* expression system, to prove its potential inhibitory activity in vitro. The feasibility of serpin interaction with targeted proteases depends on their 3D structure and conformation in the solution [16]. A serpin's inhibition mechanism is defined by their ability to switch between distinct structural configurations and interplay between kinetic stability and thermodynamic instability [8]. As described above, we investigated, computationally, the conformational stability of Dromaserpin by predicting its tertiary structure. Here we consider that active rDromaserpin adopts the conformation of the model where the RCL is exposed to the solvent, ready to interact with the target protease. We have deduced from this model 29.83% α-helices, 30.65% β-strands and 39.52% random coils. CD spectroscopy was used for secondary structural content analysis of the predicted models and to determine the rDromaserpin conformation. According to CD data, the active rDromaserpin is well folded; however, the experimental secondary structural content findings differed slightly to that from the theoretical prediction. This discrepancy between the CD results and the model might be explained by the fact that the model was built based on only 372 residues, while CD measures consider the entire protein sequence (391 residues). In fact, 4.9% of the recombinant protein (19 residues: 9 C-terminal and

10 N-terminal) analyzed by CD, were not aligned during the modeling. Moreover, 3.8% of the model sequence (two inserts of 14 residues) were not modeled by Swiss-MODEL and were assigned as random coils by the software. These two percentages account for 8.7% of the residues randomly indicated as coil whereas they may be α helices. This assumption may explain the observed difference in our results.

Serpins generally function as serine proteinase inhibitors, hence their name. In mammals, serpins participate in the regulation of many complex proteolytic pathways [36]. Arthropod serpins are characterized by their hemostatic and anti-inflammatory effects in mammalian blood [52]. Accordingly, we investigated the implication of rDromaserpin in hemostasis. We first tested the effect of rDromaserpin on blood coagulation pathways. Initially, blood clotting was described as two converging enzymatic cascades, the extrinsic and the intrinsic coagulation pathways, stimulated either by exposure of blood to a damaged vessel surface or by blood-borne components of the vascular system, respectively [5]. Over the last century, major advances have been made in the field of hemostasis, promoting cell-based models of coagulation and explaining the hemostatic process as it occurs in vivo [53]. However, the cascade model has helped improve the understanding of coagulation in plasma-based in vitro assays and has allowed for clinically useful interpretations of laboratory tests for plasma coagulation anomalies [53]. We described our in vitro results using the cascade model to explain the involvement of rDromaserpin in blood clotting (Figure 13).

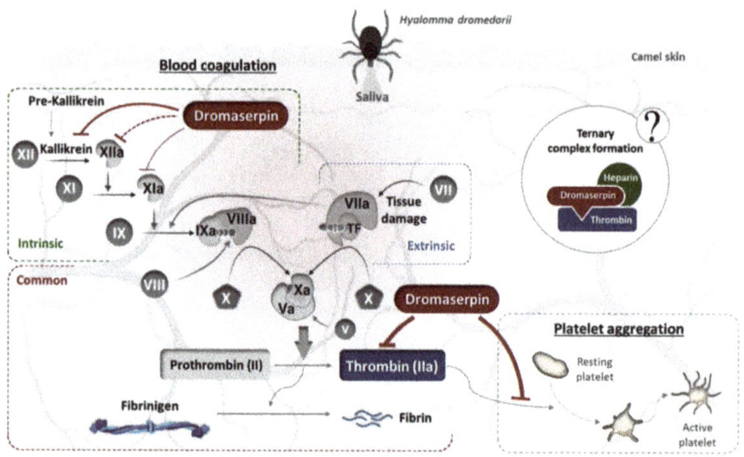

Figure 13. Schematic representation of rDromaserpin effect on hemostasis in the host interface.

Here, we showed that rDromaserpin significantly prolonged the intrinsic pathway by 7.4 s, 13.06 s, and 42.56 s upon adding 0.5, 1, or 5 µM of inhibitor, respectively. In contrast, under the same condition, these amounts do not affect the extrinsic coagulation pathway (data not shown). In TT assay, the rDromaserpin prolonged the common coagulation pathway and significantly increased clotting time by ~10.97 s. It also increased clotting time by 9.83 s when pre-incubated with prothrombin deficient plasma. In contrast, the plasma became incoagulable when rDromaserpin was pre-incubated with thrombin. During the incubation time, the covalent complex formed and rDromaserpin prevented thrombin from converting fibrinogen to fibrin. We suggest, therefore, that rDromaserpin is likely to inhibit serine protease(s) involved in the intrinsic and the common pathway of blood coagulation. The anti-coagulation property of the camel tick salivary glands is already well documented [54]. Indeed, in an earlier investigation, five anticoagulants prolonging APTT were resolved from the salivary gland crude extract of *H. dromedarii* with no effect on PT [54]. However, the molecular identity of the main anti-coagulant molecules in *H. dromedarii* saliva is not fully known. Given its potential anticoagulant activity, we found it imperative to identify the rDromaserpin target(s) among certain factors involved in the intrinsic pathway of blood clotting. Interestingly, in vitro profiling of rDromaserpin's proteolytic activity shows its capacity to inhibit thrombin and Kallikrein, to slightly affect FXIa and FXIIa activity without any relevant effect on FIXa or FXa. The intrinsic pathway is triggered by the activation of FXII when blood comes in contact with a negatively charged surface [6]. FXIIa (activated FXII) promotes its own activation in turn by stimulating Kallikrein formation and activating the upstream factor, FXIa [6]. Under the tested conditions, the effect of rDromaserpin on Kallikrein, FXIa, and FXIIa activity does not appear to be physiologically relevant. To verify its effectiveness in vivo, further experiments should be conducted under various conditions. Once these serine proteases (Kallikrein, FXIIa, and FXIa) are shown to be inhibited in vivo, coagulation time is prolonged, which might assist *H. dromedarii* in the initial stages of the feeding process. By targeting plasma Kallikrein, Dromaserpin could also help the tick avoid the formation of edema by inhibiting local bradykinin releases [55]. Nevertheless, it appears that Dromaserpin has a better efficacy on the common pathway of blood clotting, as it inhibits thrombin significantly. Indeed, among the serine proteases tested in this study, the strongest inhibition was observed with thrombin where the IC_{50} = 0.16 μM. In hemostasis, thrombin plays a key role in blood clotting, involved in the last steps of the common pathway, and in activating platelet aggregation. Indeed, due to its inhibitory activity on thrombin, we investigated whether rDromaserpin would affect platelet aggregation in addition to plasma clotting. Platelet aggregation is a complex process stimulated by several agonists including thrombin, cathepsin G, collagen and ADP [56]. Previous studies considered thrombin to be the most efficient platelet aggregation agonist [57,58]. In the present study, we showed that rDromaserpin significantly reduced platelet aggregation induced by thrombin.

Overall, these results suggest that Dromaserpin may play a crucial role during the tick fixation on the camel. Camels have a particularly active hemostatic mechanism with a short bleeding time and thrombocytosis [59]. They are known for their high level of FVIII in plasma (eight times more than humans), which is a very resistant factor to high temperatures [60]. This promotes hypercoagulability, where the intrinsic pathway is directly involved in high rates of thrombin generation. In this work, we present a tick molecule whose direct host is the camel. The camel's parasites are likely to have powerful compounds in their saliva, such as Dromaserpin, that can keep its blood incoagulable to ensure successful feeding.

The exact mechanism of thrombin inhibition by Dromaserpin remains unclear. Nevertheless, Dromaserpin seems to damage their target protease after forming a stable complex, as proven with the rDromaserpin-thrombin covalent complex (Figure 11c). Usually, the visible consequence of this interaction is a dose-dependent reduction of the enzymatic activity of candidate proteases [61]. As a suicide inhibitor, rDromaserpin reduced residual enzyme activity of thrombin in a dose-dependent manner. On the other hand, serpin activity can be enhanced by cofactors, mainly glycosaminoglycans [38]. For several serpins, heparin interacts and modulates their activity by increasing the level of inhibition toward the targeted proteases [62]. Accordingly, we found it necessary to study whether heparin modulates the activity of Dromaserpin towards thrombin. Structurally, these interactions are mediated by defined and specific amino acid residues present in most heparin-binding proteins (BPH) [63]. In our in silico analysis, the Dromaserpin active model (exposed RCL) reveals an important basic patch located on top of the β-sheet A, near β-sheet C, and surrounding the RCL region. Indeed, electrostatic interactions play a major role in the binding process of heparin to serpins [63]. Molecular docking and computational approaches have revealed the usual consensus sequences, rich of basic residues, suitable to bind heparin [64–66]. As heparin is highly anionic, the basic amino acids, such as $Arg_{196\text{-}197}$, Arg_{189}, Arg_{328}, Arg_{335} and Lys_{169} (respecting the numbering in the model), in the region of the basic patch in Dromaserpin could be included in the heparin-binding site. It important to emphasize that these data are not sufficient to confirm the exact heparin-binding site in Dromaserpin, but the presence of this basic patch might support our hypotheses. Our preliminary investigations have shown that rDromaserpin alone is not able to bind heparin under the conditions tested. However, when complexed with thrombin, heparin-sepharose resin could capture the complex, proving the possibility of forming a ternary complex. The structural aspect of this interaction between rDromaserpin, thrombin, and heparin is under investigation.

The in vitro effect of heparin on thrombin inhibition by rDromaserpin was demonstrated by the increased inhibition rate in a certain range of heparin concentrations. Two mechanisms have been suggested to explain how heparin enhances protease activity and whether or not it binds to serpin only [38]. When only serpin binds to heparin, a saturation effect is usually observed with a range of heparin concentrations [38]. In contrast to our case, at high heparin concentrations, the curve obtained (Figure 11b) is probably consistent with the formation of a ternary complex (Figure 13). Heparin enhances several plasma serpins by using this template mechanism, such as SCCA1 [38], nexin-1 [67], and Antithrombin III [68]. By binding to antithrombin III, heparin acts as an anticoagulant, causing a conformational change in serpin which acts more effectively on coagulation factors and thrombin [68]. Similarly, a tertiary complex, including heparin, is probably formed when thrombin is inhibited by rDromaserpin. However, in-depth structural investigations are necessary to verify this hypothesis.

4. Conclusions

Gathering computational and experimental results, we were able to characterize a novel anti-hemostatic serpin, Dromaserpin, from the salivary glands of *H. dromedarii*. We believe that this is the first recombinant serpin identified from the *Hyalomma* genus which inhibits thrombin in both blood coagulation and platelet aggregation. We also suggested an inhibition mechanism implicating heparin as a cofactor. Future research will explore its molecular mechanisms and kinetics as an anticoagulant candidate. Besides, its effectiveness must be verified in vivo to testify its feasibility as a potential of medicinal applications mainly in treatment of blood clotting disorders.

5. Materials and Methods

5.1. Reagents and Chemicals

All chemicals and reagents were purchased from Sigma-Aldrich, Merck or Fisher, unless otherwise indicated, and were analytical grade or better. Human thrombin and factor Xa were purchased from Promega (Madison, WI, USA), human Kallikrein plasma was purchased from Sigma Aldrich (Saint Louis, MO, USA), and factor XIIa was purchased from HTI (Haematologic Technologies, Essex Junction, VT, USA). The activity of both factor XIa and factor IXa was tested using BIOPHENTM factor XIa (HYPHEN biomed, Neuville-sur-Oise, France) and BIOPHEN factor IXa, respectively. The chromogenic substrates S-2238 (H-D-Phe-Pip-Arg-pNA•2HCl) for thrombin, S-2765 (Z-D-Arg-Gly-Arg-pNA•2HCl) for factor Xa and S-2302 (H-D-Pro-Phe-Arg-pNA•2HCl) for FXIIa and Kallikrein were purchased from Chromogenix (Chromogenix, Milano, Italy). All solutions and buffers were prepared with Milli-Q water (Merck Millipore, Darmstadt, Germany).

5.2. Bioinformatics Analysis on Pedicted Dromaserpin Sequence

The sialotranscriptome of the *H. dromedarii* tick was recently described and annotated [41]. Among the deposited sequences, a cDNA sequence that codes for Dromaserpin was selected for expression. The nucleotide sequence was translated into an amino acid sequence using a TransDecoder, confirmed later with Expasy translate (https://web.expasy.org/translate/ (accessed on 1 April 2020)), and the predicted sequence was used for the following in silico analysis. A homologous search of the full-length Dromaserpin sequence was performed using BLAST programs (https://blast.ncbi.nlm.nih.gov/Blast.cgi (accessed on 5 October 2021)), through which the conserved domain was also identified. Localization of the identified RCL domain was done by SMART software (http://smart.embl-heidelberg.

de/ (accessed on 1 April 2020)). Later, we performed a comparison between the RCL amino acid sequence in Dromaserpin and the RCL sequence of the inhibitory serpins α-1-antitrypsin (P01009) and Antithrombin-III (P01008), and a non-inhibitory serpin Ovalbumin (P01012), available in Uniprot. Full-length amino acid sequence was submitted to the signal 4.1 server (http://www.cbs.dtu.dk/services/SignalP-4.1/ (accessed on 1 April 2020)) and the signal peptide was removed from the protein sequence for subsequent analysis. The predicted molecular weight (MW) and isoelectric point (pI) were determined using ProtParam tools from ExPASy server (https://web.expasy.org/protparam/ (accessed on 1 April 2020)). To determine potential N- or O-linked glycosylation sites, Dromaserpin amino acid was scanned using NetNGlyc 1.0 (http://www.cbs.dtu.dk/services/NetNGlyc/ (accessed on 1 April 2020)) and NetOGlyc 4.0 servers (http://www.cbs.dtu.dk/services/NetOGlyc/ (accessed on 1 April 2020)). For phylogenetic analysis, protein sequences of 25 serpins from tick species and one human serpin were retrieved from GenBank. Accession numbers of these sequences, the percentage of identity and coverage with Dromaserpin are described in Supplementary Table S1. Multiple sequence alignments were performed using ClustalW algorithm (https://www.genome.jp/tools-bin/clustalw (accessed on 5 October 2021)) and sequences were edited manually. The evolutionary history was inferred by using the Maximum Likelihood method based on the JTT matrix-based model [69]. Human Antithrombin III was utilized as outgroup to root the tree. The initial tree was obtained automatically by applying Neighbor-Join and BioNJ algorithms to a matrix of pairwise distances, estimated using a JTT model, and then selecting the topology with superior log likelihood value. The tree is drawn to scale, with branch lengths measured in the number of substitutions per site. Evolutionary analyses were conducted in MEGA7 [33].

5.3. Plasmid Construction-Expression and Purification of the Recombinant Dromaserpin

All manipulation of microorganisms was developed in the BSL-2 area, as authorized by CIBio-IBu (Internal Biosafety Committee of Butantan Institute, Brazil) and CTNBio (National Technical Biosafety Commission, Brazil) (Registration number CQB-039/98 of 13 February 2014). The chosen full coding cDNA sequence was used for plasmid construction after the deletion of the sequence coding for signal peptide. Codon optimization, gene synthesis and molecular cloning of rDromaserpin coding sequence into plasmid pET28a (Novagen/EMD Millipore, Darmstadt, Hesse, Germany) were conducted by GenOne (GenOne Inc. Rio de Janeiro, RJ, Brazil). Recombinant pET28a_Dromaserpin plasmid, coding for the C-terminal His-tagged protein, was transformed into chemically competent *E. coli* BL21 (DE3) strain. The *E. coli* cultures were inoculated in 1 L of 2 YT culture medium (supplemented with 20 μg/mL kanamycin) at 30 °C with 250 rpm agitation. Protein expression was induced with 1 mM isopropyl β-D-Thiogalactoside (IPTG), at OD_{600} 0.6, and cells remained incubated for 3 h at 30 °C following the induction. Expression was confirmed by resolving samples on a 12.5% SDS-PAGE. Bacterial cells were harvested, washed with 150 mM NaCl and re-suspended in lysis buffer (50 mM Tris-HCl pH = 8; 500 mM NaCl, 1% Triton; 10 mM β-mercaptoethanol). Whole cell extracts were obtained through a Panda Plus 2000 (GEA NIRO, Erie, PA, USA) high pressure homogenizer disrupter, three times at 1000 bars, and the suspension was clarified through centrifugation at 16,000× *g* rpm for 1 h at 4 °C. Histidine tagged protein was subsequently purified from the soluble fraction using a HisTrap HP (5 mL; GE Healthcare, GE Healthcare, Uppsala, Uppland, Sweden) column of Immobilized Metal Affinity Chromatography (IMAC) (AKTA AVANT; GE Healthcare GE Healthcare, Uppsala, Uppland, Sweden). Fractions containing the eluted protein were pooled and subjected to Q-sepharose ion-exchange chromatography (AKTA AVANT; GE Healthcare, GE Healthcare, Uppsala, Uppland, Sweden) using a HiTrap Q FF (1 mL; GE Healthcare, GE Healthcare, Uppsala, Uppland, Sweden) column. The purified protein was applied to a Superdex 75 (1 mL; GE Healthcare, GE Healthcare, Uppsala, Uppland, Sweden) column used to exchange the buffer to 1 × Phosphate-buffered saline (PBS), pH 7.4 supplied with 10% glycerol. To evaluate its purity, the purified protein was resolved on a 12.5% SDS-PAGE and stained with Coomassie Brilliant Blue. Finally, rDromaserpin concentration was determined by its absorbance at 280 nm using a Biodrop spectrophotometer (Biochrom™ BioDrop µLite Micro-Volume, Braeside, Australia).

5.4. *Structural Characterization of the rDromaserpin*

5.4.1. Determination of Molecular Size by Gel Filtration

Gel Filtration experiments were performed with purified rDromaserpin using Superdex™ 75 HR 10/300 column (GE Healthcare, Uppsala, Uppland, Sweden) on an ÄKTA purifier liquid-chromatography system. Chromatography was carried out at 4 °C in 1× PBS pH 7.4 and 10% glycerol, at a flow of 0.5 mL/min. Protein elution was monitored by measuring absorbance at 280 nm. Apparent molecular masses of protein eluted from the column were deduced from a calibration curve obtained by loading 100 µL of the following standards: Conalbumin (75,000 Da), Ovalbumin (44,000 Da, 30.5 Å), Carbonic anhydrase (29,000 Da), Ribonuclease A (13,700 Da, 16.4 Å), Aprotinin (6500 Da), as well as blue dextran (for the void volume V_0). According to the calibration curve obtained with standards, we calculated the molecular weight of rDromaserpin as described in the Equation (1).

$$\log(\mathrm{MW}) = -0.1615\ \mathrm{Ve} + 3.4268 \tag{1}$$

5.4.2. Circular Dichroism (CD) Spectroscopy

CD spectra were recorded on a JASCO J-810 spectropolarimeter, equipped with a thermoelectric sample temperature controller (Peltier system) equilibrated to 20 °C. The rDromaserpin samples at 0.73 µM were diluted 10 × in Milli-Q water. The diluted buffer without protein was used to calibrate the equipment. The scans were collected at Far-UV region (from λ = 190 nm to λ = 260 nm) after seven accumulations, using a pathlength quartz cuvette of 1.0 mm. Spectra were corrected by subtracting a buffer blank, normalized to residue molar absorption, measured in mdeg (M^{-1} cm^{-1}), and adjusted to the input buffer. The mean molar residual ellipticity θλ (deg cm^2 $dmol^{-1}$) was calculated as described in the Equation (2), based on a molecular mass of 43,159.27 Da, where θ_λ^{Raw} is ellipticity in millidegrees, C is molar rDromaserpin concentration in M, N is number of amino acid residues, and l is the path length of the cuvette in mm.

$$\theta_\lambda = \frac{\theta_\lambda^{Raw} \times 10^6}{C \times N \times L} \tag{2}$$

The estimation of secondary structure was performed by BeStSel website algorithm (http://bestsel.elte.hu/index.php (accessed on 9 April 2021)) using the CD spectra values ranging from 190 nm to 250 nm.

5.4.3. Comparative Modeling and Structural Analysis of Dromaserpin

The three-dimensional structure (3D) model of Dromaserpin was predicted using a comparative modeling approach. Using its predicted amino acid sequence, we performed a homology search among the 3D structures available in PDB database (https://www.rcsb.org/ (accessed on 22 March 2020)) using BLAST from NCBI (https://blast.ncbi.nlm.nih.gov/Blast.cgi (accessed on 22 March 2020)). For homology modeling, we selected the crystal structure of Conserpin in an uncleaved (PDB code: 5cdx) and a latent (PDB code: 5cdz) state, which presents 40.62% identity with Dromaserpin, covering 94% of the sequence. Aligned regions were selected using 5cdx as template on Swiss-Model online tool (https://swissmodel.expasy.org/ (accessed on 22 March 2020)). A 3D model was generated by PyMOL software (http://www.pymol.org/ (accessed on 22 March 2020)). Subsequently, the overall quality of the Serpin 3D-structure models was evaluated by analyzing each correspondent Ramachandran plot, calculated by MolProbity program (http://molprobity.biochem.duke.edu/ (accessed on 30 September 2021)). From these generated models, we determined the content of the secondary structure using Stride (http://webclu.bio.wzw.tum.de/cgi-bin/stride/stridecgi.py (accessed on 22 March 2020)).

5.5. Functional Characterization of the rDromaserpin

5.5.1. Global blood Coagulation Assays

All experimental procedures were conducted as per the guidelines for the care and use of laboratory subjects of the Butantan Institute. The study on human blood was approved by the National Human Research Ethics Committee, Plataforma Brasil, CAAE: 33013220.4.0000.0086. Informed consent was obtained from the volunteers conferring to the declaration of Helsinki Convention and the Brazilian Department of Health. Blood was drawn from healthy human volunteers and collected in a test tube preloaded with 3.2% sodium citrate. Plasma was harvested immediately by centrifugation of blood at 3000× g rpm at 19 °C for 15 min. APTT, PT, and TT assays were performed by incubating rDromaserpin with the plasma for 20 min as described elsewhere [70]. These assays were performed using commercially available kits (STA-PTT Automate 5 for APTT, STA-Neoplastine® CI Plus for PT and STA-Thrombin 2 for TT–STAGO, Asnières-sur-Seine, France), using Semi-automatic coagulation analyzer START STAGO®. Alternatively, TT assay was also performed using FII-DP (Haematologic Technologies, Essex Junction, VT, USA). TT was measured after pre-incubating rDromaserpin with the plasma or with 2 mU thrombin for 1 h at room temperature. Clotting times were determined in triplicate in the absence or presence of 0.2, 0.5, 1, and 5 µM of rDromaserpin for APTT assay and 1 and 5 µM of rDromaserpin for PT and TT assays.

5.5.2. Protease Inhibition Assays

Inhibitory activity of rDromaserpin was tested against six mammalian serine proteases related to host defense pathways against tick feeding. The mammalian proteases tested (per reaction) included: human factor XIIa (15 nM), human thrombin (2U), human factor Xa (10 nM), human plasma Kallikrein (10 nM), human factor XIa (3.7 nM) and human factor IXa (306 nM) (Enzyme Research Laboratories). Substrates were used at 0.2 mM final concentration, including S-2238 for thrombin, S-2302 for factor XIIa and Kallikrein, and substrate S-2765 for factor Xa. Reagents were mixed at room temperature in technical triplicates. Two concentrations of rDromaserpin (1 µM and 5 µM) were pre-incubated separately with indicated amounts of the proteases listed above for 15 min at 37 °C in 20 mM Tris-HCl, 150 mM NaCl, BSA 0.1%, pH 7.4 buffer. The corresponding substrate for each protease was added in a 100 µL final reaction volume, and substrate hydrolysis was measured at OD_{405} nm every 11 s for 15 min at 37 °C using Epoch™ Microplate Spectrophotometer (BioTeck). The inhibitory activity on factor XIa and factor IXa was tested according to the guidelines of their correspondent kits (HYPHEN biomed, Neuville-sur-Oise, France). Acquired OD_{405} nm data are subjected to one-phase decay analysis in Prism 8 software to determine plateau values as proxies for initial velocity of substrate hydrolysis Vmax or residual enzyme activity. The percent enzyme activity inhibition level is determined using the Equation (3), where $Vmax_{Vi}$ is activity in presence of rDromaserpin and $Vmax_{V0}$ is activity in absence of rDromaserpin. Data are presented as mean ± S.E.M of three independent replicate readings.

$$100 - \left(\frac{Vmax_{Vi}}{Vmax_{V0}}\right) \times 100 \quad (3)$$

As thrombin was the most strongly inhibited protease, its activity was later tested with rDromaserpin at different concentrations (1, 0.5, 0.25, 0.2, 0.1, 0.05, and 0.02 µM) using the protocol described above. The covalent complex was visualized by incubating 1.7 µM rDromaserpin and 2U Thrombin for 1 h at room temperature. After one hour of incubation, 10 µL of 5 × loading buffer (62.5 mM Tris –HCl pH 6.8, 3% SDS, 10% glycerol, 1.25% β-mercaptoethanol, 0.001% blue of bromophenol) were added to the mixture, and samples were boiled for 10 min. Finally, samples were resolved on a 5–10% SDS-PAGE and stained with Coomassie Brilliant Blue.

5.5.3. Thrombin Inhibition by rDromaserpin in the Presence of Heparin

For assays in the absence or presence of heparin, the concentration of thrombin was held at 2U, and the concentration of rDromaserpin was maintained at 0.2 µM. The stock of unfractionated heparin concentration of 10 µg/mL used in this assay represents a concentration of ~400 µM (taking an average molecular mass of 25,000 g/mol). Heparin was added to the reaction in a range of 0.8–7.2 µM. Reactions were incubated for 15 min at 37 °C in thrombin assay buffer. The chromogenic substrate S-2238 was used at a final concentration of 0.2 mM, and residual activity was followed at OD_{405} nm.

5.5.4. Evaluation of Binding to Heparin-Sepharose Resin

The ability of rDromaserpin, thrombin and the covalent complex to bind to heparin was evaluated using Heparin-Sepharose 6 Fast Flow affinity resin (Cytiva, Marlborough, MA, USA). First, 50 µL of the resin was equilibrated with 200 µL of buffer A (20 mM Tris pH 7.4, 150 mM NaCl, 0.1% BSA). In four different tubes, 32 µg rDromaserpin, 133 U thrombin, rDromaserpin-thrombin complex, or 50 µg antithrombin III were added to the resins in a final volume of 250 µL. The samples were incubated for 30 min with gentle shaking. The resins were washed three times with 200 µL of buffer A to remove unbound proteins. For elution, 1 M NaCl was added and incubated for 10 min. The resins were pelleted by centrifugation at 1000× *g* rpm for 5 min, and the supernatants were analyzed by 10% SDS-PAGE.

5.5.5. Effect of rDromaserpin on Platelet Aggregation

The effect of rDromaserpin on the function of thrombin in platelets was also investigated on washed platelets. To prepare washed platelets, approximately 20 mL of freshly human blood collected from healthy donors, in a tube containing 3.8% sodium citrate, was centrifuged at 900× *g* rpm for 20 min at room temperature. Platelet rich plasma (PRP) was recuperated and 2% EDTA was added at a ratio of 1:100, followed by 2000 rpm centrifugation for 15 min at room temperature. The pellet containing platelets was resuspended in 10 mL of wash buffer (140 mM NaCl, 10 mM $NaHCO_3$, 2.5 mM KCl, 0.49 mM Na_2HPO_4, 1 mM $MgCl_2$, 22 mM Sodium Citrate, 52.23 mM BSA-dissolved in 100 mL of water, pH adjusted to 6.5 with HCl), and centrifuged at 2000× *g* rpm for 15 min at room temperature. The wash step was repeated twice under the same conditions. The supernatant was discarded and the pellet resuspended in 2 mL of Tyrode buffer (1 mM $CaCl_2$, 134 mM NaCl, 12 mM $NaHCO_3$, 2.9 mM KCl, 0.34 mM Na_2HPO_4, 1 mM $MgCl_2$, 10 mM HEPES-pH adjusted to 7.4 with HCl. Add glucose 1% at the time of use). The aggregometer (CHRONO-LOG® Model 700, Havertown, PA, USA), was calibrated with Tyrode buffer containing glucose. To determine the effect of rDromaserpin on platelet aggregation, 10 µM rDromaserpin or equal volume of 1 × PBS buffer pH 7.4, 1, 10% glycerol were pre-incubated for 10 min at 37 °C, with agonist 1 µg /mL NIH-U thrombin in a 50 µL reaction. Platelet aggregation was induced following the addition of pre-warmed washed platelets in a final volume of 500 µL. Platelet aggregation was monitored for 6 min and the percentage of platelet aggregation inhibition was deduced using The AGGRO/LINK®8 program. Data are presented as mean ± S.E.M. of triplicate platelet aggregation assays.

5.6. Statistical Analysis

Data are presented as the mean ± S.E.M unless indicated otherwise and statistical significance was calculated for at least three independent experiments employing the Student's *t*-test, using GraphPad Prism 8.0 software (GraphPad Software, Inc., San Diego, CA, USA). *p*-values < 0.05 were considered to be statistically significant.

Supplementary Materials: The following are available online at https://www.mdpi.com/article/10.3390/toxins13120913/s1. Table S1: GenBank accession numbers of serpins used in the phylogenetic analysis; Figure S1: CD spectrum analysis by the BesSel program; Figure S2: Cartoon representations of the secondary elements in the active model of Dromaserpin; Figure S3: Ramachandran plot of Dromaserpin Model with an exposed RCL, generated by MolProbity program; Figure S4: Ramachandran plot of Dromaserpin Model with an inserted RCL, generated by MolProbity program. References [32,34,47,50,71–78] are cited in the supplementary materials.

Author Contributions: Conceptualization, H.A., C.B., Y.M. and F.F.; Data curation, H.A. and M.R.F.; Formal analysis, H.A., M.R.F., E.C. and F.F.; Funding acquisition, A.M.C.-T. and F.F.; Investigation, H.A.; Methodology, M.B.G.; Project administration, A.M.C.-T., A.B. and F.F.; Resources, C.B.; Software, M.R.F.; Supervision, Y.M. and F.F.; Validation, M.B.G. and F.F.; Writing–original draft, H.A.; Writing–review & editing, H.A., M.R.F., M.B.G., Y.M. and F.F. All authors have read and agreed to the published version of the manuscript.

Funding: This research was funded by the Coordenação de Aperfeiçoamento de Pessoal de Nível Superior—Brazil (CAPES)—Finance Code 001; by FAPESP (CETICS: 2013/07467-1) and Fundação Butantan. C.B. received the European Union within ESIF in frame of Operational Programme Research, Development and Education (project no. CZ.02.2.69/0.0/0.0/20_079/0017809).

Institutional Review Board Statement: Not applicable.

Informed Consent Statement: Not applicable.

Acknowledgments: The Factor IXa Kit was generously provided by Elizabeth Angelica Martins from the Laboratory of Recombinant Biologics, Butantan Institute. Kallikrein was kindly provided by Benedito Prezotto from the Laboratory of Pharmacology, Butantan Institute. Heparin was kindly provided by Durvanei Augusto Maria from the Laboratory of Development and Innovation, Butantan Institute. We would like to thank Deborah Glassman for her English corrections and comments on the manuscript. We are grateful to Karem Jridi, a graphic designer at the NGC Company (New Generation Concept) in Tunisia, for the figure editing.

Conflicts of Interest: The authors declare no conflict of interest.

References

1. The World Health Organization Web Site. Available online: https://www.who.int/health-topics/cardiovascular-diseases/#tab=tab_1 (accessed on 17 September 2020).
2. Stoney, C.M.; Kaufmann, P.G.; Czajkowski, S.M. Cardiovascular disease: Psychological, social, and behavioral influences: Introduction to the special issue. *Am. Psychol.* **2018**, *73*, 949–954. [CrossRef]
3. Chang, J.C. Hemostasis based on a bovel "two-path unifying theory" and classification of hemostatic disorders. *Blood Coagul. Fibrinolysis* **2018**, *29*, 573–584. [CrossRef] [PubMed]
4. Ogedegbe, H.O. An overview of hemostasis. *Lab. Med.* **2002**, *33*, 948–953. [CrossRef]
5. Versteeg, H.H.; Heemskerk, J.W.M.; Levi, M.; Reitsma, P.H. New fundamentals in hemostasis. *Physiol. Rev.* **2013**, *93*, 327–358. [CrossRef]
6. Mussbacher, M.; Kral-Pointner, J.B.; Salzmann, M.; Schrottmaier, W.C.; Assinger, A. Mechanisms of hemostasis: Contributions of platelets, coagulation factors, and the vessel wall. In *Fundamentals of Vascular Biology*; Geiger, M., Ed.; Learning Materials in Biosciences; Springer International Publishing: Cham, Switzerland, 2019; pp. 145–169. ISBN 978-3-030-12270-6.
7. Page, M.J.; Macgillivray, R.T.A.; Di Cera, E. Determinants of specificity in coagulation proteases. *J. Thromb. Haemost.* **2005**, *3*, 2401–2408. [CrossRef]
8. Gettins, P.G.W. Serpin structure, mechanism, and function. *Chem. Rev.* **2002**, *102*, 4751–4804. [CrossRef]
9. Crawley, J.T.B.; Zanardelli, S.; Chion, C.K.N.K.; Lane, D.A. The central role of thrombin in hemostasis. *J. Thromb. Haemost.* **2007**, *5* (Suppl. S1), 95–101. [CrossRef] [PubMed]
10. Siller-Matula, J.M.; Schwameis, M.; Blann, A.; Mannhalter, C.; Jilma, B. Thrombin as a multi-functional enzyme. Focus on in vitro and in vivo effects. *Thromb. Haemost.* **2011**, *106*, 1020–1033. [CrossRef] [PubMed]
11. Mega, J.L.; Simon, T. Pharmacology of antithrombotic drugs: An assessment of oral antiplatelet and anticoagulant treatments. *Lancet* **2015**, *386*, 281–291. [CrossRef]
12. Tang, N.; Bai, H.; Chen, X.; Gong, J.; Li, D.; Sun, Z. Anticoagulant treatment is associated with decreased mortality in severe coronavirus disease 2019 patients with coagulopathy. *J. Thromb. Haemost.* **2020**, *18*, 1094–1099. [CrossRef] [PubMed]
13. Koehn, F.E.; Carter, G.T. The evolving role of natural products in drug discovery. *Nat. Rev. Drug Discov.* **2005**, *4*, 206–220. [CrossRef] [PubMed]
14. Carvalhal, F.; Cristelo, R.R.; Resende, D.I.S.P.; Pinto, M.M.M.; Sousa, E.; Correia-da-Silva, M. Antithrombotics from the sea: Polysaccharides and beyond. *Mar. Drugs* **2019**, *17*, 170. [CrossRef] [PubMed]
15. Lucas, A.; Yaron, J.R.; Zhang, L.; Ambadapadi, S. Overview of serpins and their roles in biological systems. *Methods Mol. Biol.* **2018**, *1826*, 1–7. [CrossRef]
16. Lucas, A.; Yaron, J.R.; Zhang, L.; Macaulay, C.; McFadden, G. Serpins: Development for therapeutic applications. *Methods Mol. Biol.* **2018**, *1826*, 255–265. [CrossRef]

17. Van Gent, D.; Sharp, P.; Morgan, K.; Kalsheker, N. Serpins: Structure, Function and Molecular Evolution. *Int. J. Biochem. Cell Biol.* **2003**, *35*, 1536–1547. [CrossRef]
18. Stein, P.E.; Leslie, A.G.; Finch, J.T.; Carrell, R.W. Crystal structure of uncleaved ovalbumin at 1.95 A Resolution. *J. Mol. Biol.* **1991**, *221*, 941–959. [CrossRef]
19. Engh, R.; Löbermann, H.; Schneider, M.; Wiegand, G.; Huber, R.; Laurell, C.B. The S variant of human alpha 1-antitrypsin, structure and implications for function and metabolism. *Protein Eng.* **1989**, *2*, 407–415. [CrossRef] [PubMed]
20. Skinner, R.; Abrahams, J.P.; Whisstock, J.C.; Lesk, A.M.; Carrell, R.W.; Wardell, M.R. The 2.6 A structure of antithrombin indicates a conformational change at the heparin binding site. *J. Mol. Biol.* **1997**, *266*, 601–609. [CrossRef] [PubMed]
21. Tucker, H.M.; Mottonen, J.; Goldsmith, E.J.; Gerard, R.D. Engineering of plasminogen activator inhibitor-1 to reduce the rate of latency tansition. *Nat. Struct. Biol.* **1995**, *2*, 442–445. [CrossRef]
22. Chen, H.; Davids, J.A.; Zheng, D.; Bryant, M.; Bot, I.; Van Berckel, T.J.C.; Biessen, E.; Pepine, C.; Ryman, K.; Progulski-Fox, A.; et al. The serpin solution; targeting thrombotic and thrombolytic serine proteases in inflammation. *Cardiovasc. Hematol. Disord. Drug Targets* **2013**, *13*, 99–110. [CrossRef]
23. Irving, J.A.; Ekeowa, U.I.; Belorgey, D.; Haq, I.; Gooptu, B.; Miranda, E.; Pérez, J.; Roussel, B.D.; Ordóñez, A.; Dalton, L.E.; et al. The serpinopathies studying serpin polymerization in vivo. *Meth. Enzymol.* **2011**, *501*, 421–466. [CrossRef]
24. Rau, J.C.; Beaulieu, L.M.; Huntington, J.A.; Church, F.C. Serpins in thrombosis, hemostasis and fibrinolysis. *J. Thromb. Haemost.* **2007**, *5* (Suppl. S1), 102–115. [CrossRef] [PubMed]
25. Bang, N.U. Leeches, snakes, ticks, and vampire bats in today's cardiovascular drug development. *Circulation* **1991**, *84*, 436–438. [CrossRef]
26. Jmel, M.A.; Aounallah, H.; Bensaoud, C.; Mekki, I.; Chmelař, J.; Faria, F.; M'ghirbi, Y.; Kotsyfakis, M. Insights into the role of tick salivary protease inhibitors during ectoparasite-host crosstalk. *Int. J. Mol. Sci.* **2021**, *22*, 892. [CrossRef] [PubMed]
27. Chmelař, J.; Kotál, J.; Kovaříková, A.; Kotsyfakis, M. The use of tick salivary proteins as novel therapeutics. *Front. Physiol.* **2019**, *10*, 812. [CrossRef] [PubMed]
28. Chmelař, J.; Kotál, J.; Langhansová, H.; Kotsyfakis, M. Protease inhibitors in tick saliva: The role of serpins and cystatins in tick-host-pathogen interaction. *Front. Cell Infect. Microbiol.* **2017**, *7*, 216. [CrossRef] [PubMed]
29. Irving, J.A.; Pike, R.N.; Lesk, A.M.; Whisstock, J.C. Phylogeny of the serpin superfamily: Implications of patterns of amino acid conservation for structure and function. *Genome Res.* **2000**, *10*, 1845–1864. [CrossRef]
30. Schechter, I.; Berger, A. On the size of the active site in proteases. I. Papain. 1967. *Biochem. Biophys. Res. Commun.* **2012**, *425*, 497–502. [CrossRef]
31. Nguyen, K.D.; Pan, Y. A knowledge-based multiple-sequence alignment algorithm. *IEEE/ACM Trans. Comput. Biol. Bioinform.* **2013**, *10*, 884–896. [CrossRef] [PubMed]
32. Yu, Y.; Cao, J.; Zhou, Y.; Zhang, H.; Zhou, J. Isolation and characterization of two novel serpins from the tick *Rhipicephalus haemaphysaloides*. *Ticks Tick Borne Dis.* **2013**, *4*, 297–303. [CrossRef] [PubMed]
33. Kumar, S.; Stecher, G.; Tamura, K. MEGA7: Molecular evolutionary genetics analysis version 7.0 for bigger datasets. *Mol. Biol. Evol.* **2016**, *33*, 1870–1874. [CrossRef] [PubMed]
34. Tirloni, L.; Seixas, A.; Mulenga, A.; Vaz, I.d.S.; Termignoni, C. A Family of serine protease inhibitors (serpins) in the cattle tick *Rhipicephalus (boophilus) microplus*. *Exp. Parasitol.* **2014**, *137*, 25–34. [CrossRef] [PubMed]
35. Martin, S.R.; Schilstra, M.J. Circular dichroism and its application to the study of biomolecules. *Methods Cell Biol.* **2008**, *84*, 263–293. [CrossRef] [PubMed]
36. Sanrattana, W.; Maas, C.; De Maat, S. SERPINs—From trap to treatment. *Front. Med.* **2019**, *6*, 25. [CrossRef] [PubMed]
37. Ng, V.L. Prothrombin time and partial thromboplastin time assay considerations. *Clin. Lab. Med.* **2009**, *29*, 253–263. [CrossRef]
38. Higgins, W.J.; Fox, D.M.; Kowalski, P.S.; Nielsen, J.E.; Worrall, D.M. Heparin enhances serpin inhibition of the cysteine protease cathepsin L. *J. Biol. Chem.* **2010**, *285*, 3722–3729. [CrossRef] [PubMed]
39. Chmelar, J.; Calvo, E.; Pedra, J.H.F.; Francischetti, I.M.B.; Kotsyfakis, M. Tick salivary secretion as a source of antihemostatics. *J. Proteom.* **2012**, *75*, 3842–3854. [CrossRef]
40. Bensaoud, C.; Aounallah, H.; Sciani, J.M.; Faria, F.; Chudzinski-Tavassi, A.M.; Bouattour, A.; M'ghirbi, Y. Proteomic informed by transcriptomic for salivary glands components of the camel tick *Hyalomma dromedarii*. *BMC Genom.* **2019**, *20*, 675. [CrossRef] [PubMed]
41. Bensaoud, C.; Nishiyama, M.Y.; Ben Hamda, C.; Lichtenstein, F.; Castro de Oliveira, U.; Faria, F.; Loiola Meirelles Junqueira-de-Azevedo, I.; Ghedira, K.; Bouattour, A.; M'Ghirbi, Y.; et al. *De novo* assembly and annotation of *Hyalomma dromedarii* Tick (Acari: Ixodidae) sialotranscriptome with regard to gender differences in gene expression. *Parasit. Vectors* **2018**, *11*, 314. [CrossRef]
42. Schussler, G.C. The thyroxine-binding proteins. *Thyroid* **2000**, *10*, 141–149. [CrossRef]
43. Simard, M.; Underhill, C.; Hammond, G.L. Functional implications of corticosteroid-binding globulin N-glycosylation. *J. Mol. Endocrinol.* **2018**, *60*, 71–84. [CrossRef]
44. Polderdijk, S.G.I.; Huntington, J.A. Identification of serpins specific for activated protein C using a lysate-based screening assay. *Sci. Rep.* **2018**, *8*, 8793. [CrossRef] [PubMed]
45. Roberts, T.H.; Hejgaard, J.; Saunders, N.F.W.; Cavicchioli, R.; Curmi, P.M.G. Serpins in unicellular eukarya, archaea, and bacteria: Sequence analysis and evolution. *J. Mol. Evol.* **2004**, *59*, 437–447. [CrossRef] [PubMed]

46. Kim, T.K.; Radulovic, Z.; Mulenga, A. Target validation of highly conserved *Amblyomma americanum* tick saliva serine protease inhibitor 19. *Ticks Tick Borne Dis.* **2016**, *7*, 405–414. [CrossRef] [PubMed]
47. Wang, F.; Song, Z.; Chen, J.; Wu, Q.; Zhou, X.; Ni, X.; Dai, J. The immunosuppressive functions of two novel tick serpins, HlSerpin-a and HlSerpin-b, from *Haemaphysalis longicornis*. *Immunology* **2019**, *159*, 109–120. [CrossRef] [PubMed]
48. Mahon, B.P.; Ambadapadi, S.; Yaron, J.R.; Lomelino, C.L.; Pinard, M.A.; Keinan, S.; Kurnikov, I.; Macaulay, C.; Zhang, L.; Reeves, W.; et al. Crystal structure of cleaved Serp-1, a Myxomavirus-derived immune modulating serpin: Structural design of serpin reactive center loop peptides with improved therapeutic function. *Biochemistry* **2018**, *57*, 1096–1107. [CrossRef]
49. Pongprayoon, P.; Niramitranon, J.; Kaewhom, P.; Kaewmongkol, S.; Suwan, E.; Stich, R.W.; Jittapalapong, S. Dynamic and structural insights into tick serpin from *Ixodes ricinus*. *J. Biomol. Struct. Dyn.* **2019**, *38*, 2296–2303. [CrossRef]
50. Chmelar, J.; Oliveira, C.J.; Rezacova, P.; Francischetti, I.M.B.; Kovarova, Z.; Pejler, G.; Kopacek, P.; Ribeiro, J.M.C.; Mares, M.; Kopecky, J.; et al. A Tick salivary protein targets cathepsin G and chymase and inhibits host inflammation and platelet aggregation. *Blood* **2011**, *117*, 736–744. [CrossRef] [PubMed]
51. Kovářová, Z.; Chmelař, J.; Sanda, M.; Brynda, J.; Mareš, M.; Rezáčová, P. Crystallization and diffraction analysis of the serpin IRS-2 from the hard tick *Ixodes ricinus*. *Acta Crystallogr. Sect. F Struct. Biol. Cryst. Commun.* **2010**, *66*, 1453–1457. [CrossRef]
52. Meekins, D.A.; Kanost, M.R.; Michel, K. Serpins in arthropod biology. *Semin. Cell Dev. Biol.* **2017**, *62*, 105–119. [CrossRef] [PubMed]
53. Smith, S.A. The Cell-based model of coagulation. *J. Vet. Emerg. Crit. Care* **2009**, *19*, 3–10. [CrossRef] [PubMed]
54. Ibrahim, M.A.; Masoud, H.M.M. Thrombin inhibitor from the salivary gland of the camel Tick *Hyalomma dromedarii*. *Exp. Appl. Acarol.* **2018**, *74*, 85–97. [CrossRef]
55. Francischetti, I.M.B.; Sa-Nunes, A.; Mans, B.J.; Santos, I.M.; Ribeiro, J.M.C. The role of saliva in tick feeding. *Front. Biosci.* **2009**, *14*, 2051–2088. [CrossRef] [PubMed]
56. Rumbaut, R.E.; Thiagarajan, P. *Platelet Aggregation*; Morgan & Claypool Life Sciences, Texas Medical Center: Houston, TX, USA, 2010.
57. Davì, G.; Patrono, C. Platelet activation and atherothrombosis. *N. Engl. J. Med.* **2007**, *357*, 2482–2494. [CrossRef] [PubMed]
58. Furman, M.I.; Liu, L.; Benoit, S.E.; Becker, R.C.; Barnard, M.R.; Michelson, A.D. The cleaved peptide of the thrombin receptor is a strong platelet agonist. *Proc. Natl. Acad. Sci. USA* **1998**, *95*, 3082–3087. [CrossRef]
59. Al Ghumlas, A.K.; Gader, A.G.M.A. Characterization of the aggregation responses of camel platelets. *Vet. Clin. Pathol.* **2013**, *42*, 307–313. [CrossRef] [PubMed]
60. Abdel Gader, A.G.M.; Al Momen, A.K.M.; Alhaider, A.; Brooks, M.B.; Catalfamo, J.L.; Al Haidary, A.A.; Hussain, M.F. Clotting factor VIII (FVIII) and thrombin generation in camel plasma: A comparative study with humans. *Can. J. Vet. Res.* **2013**, *77*, 150–157.
61. Schechter, N.M.; Plotnick, M.I. Measurement of the kinetic parameters mediating protease-serpin inhibition. *Methods* **2004**, *32*, 159–168. [CrossRef]
62. Gray, E.; Hogwood, J.; Mulloy, B. The anticoagulant and antithrombotic mechanisms of heparin. In *Heparin—A Century of Progress*; Springer: Berlin/Heidelberg, Germany, 2012; pp. 43–61. [CrossRef]
63. Varki, A.; Cummings, R.D.; Esko, J.D.; Freeze, H.H.; Stanley, P.; Bertozzi, C.R.; Hart, G.W.; Etzler, M.E. *Essentials of Glycobiology*, 2nd ed.; Cold Spring Harbor Laboratory Press: Woodbury, NY, USA, 2009; 784p.
64. Forster, M.; Mulloy, B. Computational approaches to the identification of heparin-binding sites on the surfaces of proteins. *Biochem. Soc. Trans.* **2006**, *34*, 431–434. [CrossRef] [PubMed]
65. Handel, T.M.; Johnson, Z.; Crown, S.E.; Lau, E.K.; Sweeney, M.; Proudfoot, A.E. Regulation of protein function by glycosaminoglycans—As exemplified by chemokines. *Annu. Rev. Biochem.* **2005**, *74*, 385–410. [CrossRef]
66. Fromm, J.R.; Hileman, R.E.; Caldwell, E.E.O.; Weiler, J.M.; Linhardt, R.J. Pattern and spacing of basic amino acids in heparin binding sites. *Arch. Biochem. Biophys.* **1997**, *343*, 92–100. [CrossRef] [PubMed]
67. Li, W.; Huntington, J.A. Crystal structures of protease Nexin-1 in complex with heparin and thrombin suggest a 2-Step recognition mechanism. *Blood* **2012**, *120*, 459–467. [CrossRef] [PubMed]
68. Li, W.; Johnson, D.J.D.; Esmon, C.T.; Huntington, J.A. Structure of the antithrombin-thrombin-heparin ternary complex reveals the antithrombotic mechanism of heparin. *Nat. Struct. Mol. Biol.* **2004**, *11*, 857–862. [CrossRef]
69. Jones, D.T.; Taylor, W.R.; Thornton, J.M. The Rapid generation of mutation data matrices from protein sequences. *Comput. Appl. Biosci.* **1992**, *8*, 275–282. [CrossRef]
70. Batista, I.F.C.; Ramos, O.H.P.; Ventura, J.S.; Junqueira-de-Azevedo, I.L.M.; Ho, P.L.; Chudzinski-Tavassi, A.M. A new Factor Xa inhibitor from *Amblyomma cajennense* with a Unique Domain Composition. *Arch. Biochem. Biophys.* **2010**, *493*, 151–156. [CrossRef] [PubMed]
71. Bock, S.C.; Wion, K.L.; Vehar, G.A.; Lawn, R.M. Cloning and Expression of the cDNA for Human Antithrombin III. *Nucleic Acids Res.* **1982**, *10*, 8113–8125. [CrossRef] [PubMed]
72. Mulenga, A.; Khumthong, R.; Blandon, M.A. Molecular and Expression Analysis of a Family of the *Amblyomma americanum* Tick Lospins. *J. Exp. Biol.* **2007**, *210*, 3188–3198. [CrossRef] [PubMed]
73. Porter, L.; Radulović, Ž.; Kim, T.; Braz, G.R.C.; Da Silva Vaz, I.; Mulenga, A. Bioinformatic Analyses of Male and Female *Amblyomma americanum* Tick Expressed Serine Protease Inhibitors (Serpins). *Ticks Tick Borne Dis.* **2015**, *6*, 16–30. [CrossRef]

74. Imamura, S.; Da Silva Vaz Junior, I.; Sugino, M.; Ohashi, K.; Onuma, M. A Serine Protease Inhibitor (Serpin) from *Haemaphysalis longicornis* as an Anti-Tick Vaccine. *Vaccine* **2005**, *23*, 1301–1311. [CrossRef]
75. Chlastáková, A.; Kotál, J.; Beránková, Z.; Kaščáková, B.; Martins, L.A.; Langhansová, H.; Prudnikova, T.; Ederová, M.; Kutá Smatanová, I.; Kotsyfakis, M.; et al. Iripin-3, a New Salivary Protein Isolated From *Ixodes ricinus* Ticks, Displays Immunomodulatory and Anti-Hemostatic Properties In Vitro. *Front. Immunol.* **2021**, *12*, 626200. [CrossRef]
76. Schwarz, A.; Von Reumont, B.M.; Erhart, J.; Chagas, A.C.; Ribeiro, J.M.C.; Kotsyfakis, M. De Novo *Ixodes ricinus* Salivary Gland Transcriptome Analysis Using Two Next-Generation Sequencing Methodologies. *FASEB J.* **2013**, *27*, 4745–4756. [CrossRef] [PubMed]
77. Ibelli, A.M.G.; Kim, T.K.; Hill, C.C.; Lewis, L.A.; Bakshi, M.; Miller, S.; Porter, L.; Mulenga, A. A Blood Meal-Induced *Ixodes scapularis* Tick Saliva Serpin Inhibits Trypsin and Thrombin, and Interferes with Platelet Aggregation and Blood Clotting. *Int. J. Parasitol.* **2014**, *44*, 369–379. [CrossRef] [PubMed]
78. Mulenga, A.; Tsuda, A.; Onuma, M.; Sugimoto, C. Four Serine Proteinase Inhibitors (Serpin) from the Brown Ear Tick, *Rhiphicephalus appendiculatus*; cDNA Cloning and Preliminary Characterization. *Insect Biochem. Mol. Biol.* **2003**, *33*, 267–276. [CrossRef]

 toxins

MDPI

Article

PKCζ-Mitogen-Activated Protein Kinase Signaling Mediates Crotalphine-Induced Antinociception

Bárbara G. de Freitas †, Natália G. Hösch †, Leandro M. Pereira, Tereza C. Barbosa, Gisele Picolo, Yara Cury and Vanessa O. Zambelli *

Laboratory of Pain and Signaling, Butantan Institute, São Paulo 05503-900, Brazil; bguimaf@gmail.com (B.G.d.F.); natalia.hosh@esib.butantan.gov.br (N.G.H.); le.marciopereira@yahoo.com.br (L.M.P.); tereza.barbosa@butantan.gov.br (T.C.B.); gisele.picolo@butantan.gov.br (G.P.); yara.cury@esib.butantan.gov.br (Y.C.)

* Correspondence: vanessa.zambelli@butantan.gov.br; Tel.: +55-11-2627-9765

† These authors have contributed equally to this work.

Abstract: Crotalphine (CRP) is a structural analogue to a peptide that was first identified in the crude venom from the South American rattlesnake *Crotalus durissus terrificus*. This peptide induces a potent and long-lasting antinociceptive effect that is mediated by the activation of peripheral opioid receptors. The opioid receptor activation regulates a variety of intracellular signaling, including the mitogen-activated protein kinase (MAPK) pathway. Using primary cultures of sensory neurons, it was demonstrated that crotalphine increases the level of activated ERK1/2 and JNK-MAPKs and this increase is dependent on the activation of protein kinase Cζ (PKCζ). However, whether PKCζ-MAPK signaling is critical for crotalphine-induced antinociception is unknown. Here, we biochemically demonstrated that the systemic crotalphine activates ERK1/2 and JNK and decreases the phosphorylation of p38 in the lumbar spinal cord. The in vivo pharmacological inhibition of spinal ERK1/2 and JNK, but not of p38, blocks the antinociceptive effect of crotalphine. Of interest, the administration of a PKCζ pseudosubstrate (PKCζ inhibitor) prevents crotalphine-induced ERK activation in the spinal cord, followed by the abolishment of crotalphine-induced analgesia. Together, our results demonstrate that the PKCζ-ERK signaling pathway is involved in crotalphine-induced analgesia. Our study opens a perspective for the PKCζ-MAPK axis as a target for pain control.

Keywords: analgesic peptide; protein kinase C; hyperalgesia; cell-signaling

Key Contribution: Crotalphine-induced analgesia is mediated by down-regulation of p38-MAPK and up-regulation of PKCζ-MAPK signaling pathway.

Citation: de Freitas, B.G.; Hösch, N.G.; Pereira, L.M.; Barbosa, T.C.; Picolo, G.; Cury, Y.; Zambelli, V.O. PKCζ-Mitogen-Activated Protein Kinase Signaling Mediates Crotalphine-Induced Antinociception. *Toxins* **2021**, *13*, 912. https://doi.org/10.3390/toxins13120912

Received: 30 October 2021
Accepted: 6 December 2021
Published: 20 December 2021

Publisher's Note: MDPI stays neutral with regard to jurisdictional claims in published maps and institutional affiliations.

1. Introduction

Crotalphine, a structural analogue to an antinociceptive peptide that was first identified in the crude venom from the South American rattlesnake *Crotalus durissus terrificus*, induces a potent and long-lasting antinociceptive effect, mediated by the activation of peripheral opioid receptors [1]. Unlike opioids, treatment with crotalphine for several days does not induce tolerance and withdrawal symptoms. Crotalphine is not an opioid receptor agonist, however, this peptide induces the release of dynorphin A that activates peripheral kappa opioid receptors (KOR) [2]. More recently, attempting to identify the direct molecular targets of crotalphine in pain pathways, it was demonstrated that this peptide partially activates and desensitizes the TRPA1 ion channel at subnanomolar concentrations and this effect is critical for the peptide's analgesic effect [3].

Mitogen-activated protein kinases (MAPKs) transduce a multiple extracellular stimulus into intracellular effects by modifying the transcription, as well as inducing post-translational changes in target proteins. There are three major members in the MAPK family: extracellular signal-regulated kinases (ERK) 1/2, C-Jun N-terminal kinase (JNK) and p38, which represent three different signaling pathways [4]. The emerging evidence suggests that MAPKs are involved in nociception and in the development of side effects of drugs, such as morphine. In this regard, peripheral inflammation or nerve injury are followed by spinal activation of ERK1/2, JNK and p38 that seems to increase neuronal excitability and up-regulate transcriptional factors involved in nociception [5]. However, the ERK1/2-signaling activation is essential for morphine and KOR agonists-mediated analgesia in rodents, suggesting that ERK activation may also positively affect the pain outcome [6,7].

It is known that crotalphine increases the level of activated ERK and JNK in cultured sensory neurons from the dorsal root ganglia (DRG) and the ERK levels increase is dependent on the activation of KOR and the protein kinase Cζ (PKCζ) [8]. However, whether the PKCζ-MAPK signaling pathway is involved in crotalphine-induced antinociception has not yet been evaluated in vivo. Considering that crotalphine has a potent and long-lasting analgesic effect in rodents, identifying the intracellular signaling mechanisms responsible for the peptide effects may provide insights to guide the development of better analgesics. Thus, using biochemical and pharmacological strategies, we sought to investigate the role of MAPK in crotalphine-induced antinociception.

2. Results

2.1. PGE_2 Is Responsible for the Long-Lasting Antinociceptive Effect Induced by Crotalphine

The data show that the magnitude of the crotalphine antinociceptive effect depends on tissue inflammation [8]. Therefore, we first characterized the effect of one single systemic administration of crotalphine on naïve rats and rats sensitized with an intraplantar injection of Prostaglandin E_2 (PGE_2) at day one. The mechanical threshold was assessed 1 h after crotalphine (3 h after PGE_2) and every day for 8 days (192 h) (Figure 1A). The intraplantar injection of PGE_2 (100 ng/paw) induced a significant decrease in the mechanical threshold with a peak response occurring 3 h after the administration, compared with the basal values obtained before any treatment, which characterize hypersensitivity. The systemic administration of crotalphine (20 ng/kg or 1 µg/kg, p.o.) blocked the PGE_2-induced hypersensitivity and increased the nociceptive threshold of the animals when compared with the basal values (analgesia). This effect is detected at 3 h and lasts until 120 h (5 days) after one single administration of crotalphine (Figure 1B). Next, we tested the crotalphine effect in non-sensitized (naïve) rats. A single administration of crotalphine (20 ng/kg and 1 µg/kg, p.o.) induces a significant increase in the mechanical nociceptive threshold (analgesia), starting at 3 h and lasting until 5 h after treatment, when compared with the baseline. A lower dose of crotalphine (8 pg/kg, p.o.) did not change the mechanical threshold (Figure 1C). As expected, no difference in the nociceptive threshold was detected in rats treated with saline (control). Together, these results demonstrate that a single administration of PGE_2 enhances and prolongs crotalphine-induced analgesia.

2.2. Crotalphine Increases the Spinal ERK and JNK and Decreases p38 Activation

In order to investigate the potential mechanisms involved in the antinociception induced by systemic crotalphine, we sought to determine whether crotalphine interferes with the expression and phosphorylation of ERK, JNK and p38 (Figure 2A). Importantly, since the inhibition of MEK, an up-stream MAPK kinase, blocks PGE_2 induced hyperalgesia, the following experiments were conducted in naïve rats, i.e., without sensitization to focus on the mechanism involved in crotalphine-induced analgesia [7]. Moreover, we selected the 1 µg/kg dose of crotalphine for the next experiments.

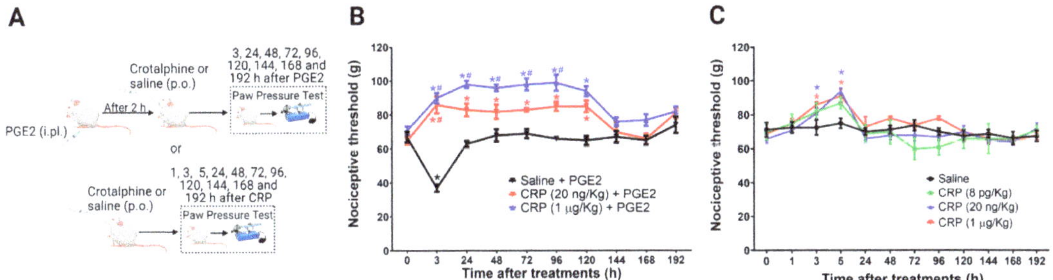

Figure 1. Effect of crotalphine in the mechanical nociceptive threshold of sensitized and non-sensitized rats. (**A**) Schematic representation of treatment and rat paw pressure test. (**B**) PGE_2 (100 ng/paw) was injected 2 h before the oral crotalphine administration and nociceptive threshold was assessed before (0) and 3, 24, 48, 72, 96, 120, 144, 168 and 198 h after the PGE_2 injection. Data are presented as mean ± SEM. * significantly different from baseline. # significantly different from the saline + PGE_2 group, $n = 6$ per group ($p < 0.05$). (**C**) Nociceptive threshold was obtained before (0) and 1, 3, 5, 24, 48, 72, 96, 120, 144, 168 and 198 h after the oral administration of crotalphine. Data are presented as mean ± SEM. * significantly different from the saline group, $n = 5$ per group ($p < 0.05$). The observer was blinded to the experimental conditions.

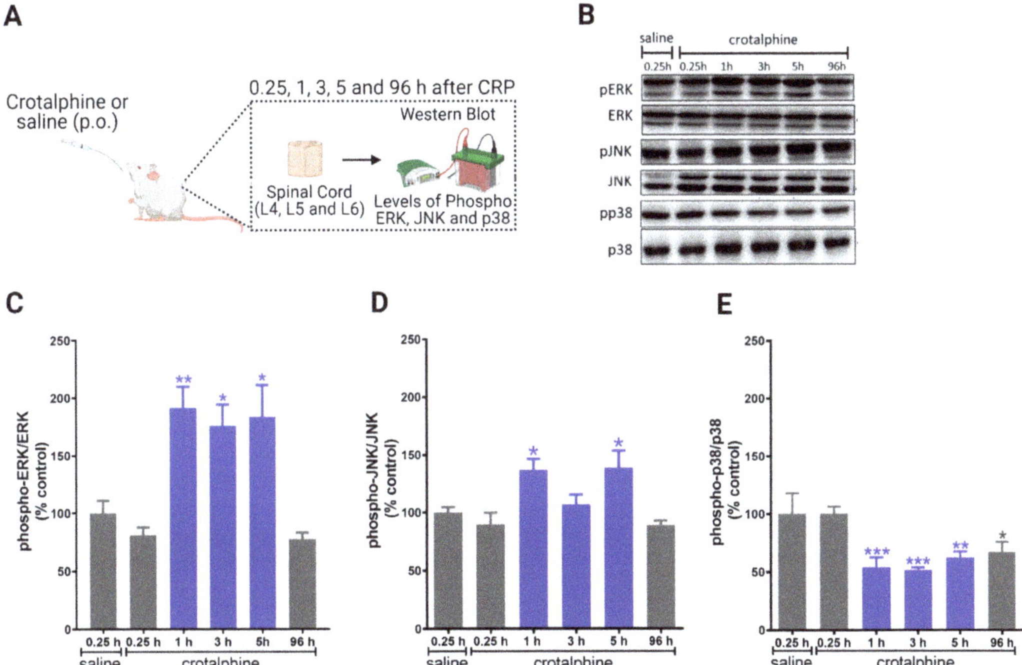

Figure 2. Systemic crotalphine administration increases ERK1/2 and JNK and decreases p38 phosphorylation levels in the spinal cord. (**A**) Schematic representation of treatment and immunoblot. (**B**) Representative blots showing the phosphorylated and total ERK1/2, JNK and p38 levels in the total lysate of spinal cord. Changes in protein expression of ERK 1/2 (**C**), JNK (**D**) and p38 (**E**) MAPKs at different time points were determined by Western blot analysis in lumbar spinal cord extracts from crotalphine or saline-treated rats. Graphs represent the ratio between the phosphorylated protein and the total amount of the targeted protein. Data are presented as mean ± SEM and expressed as % of control (saline) animals. * significantly different from mean values of saline treated animals, $n = 6$ per group ($p < 0.05$). ** significantly different from mean values of saline treated animals, $n = 6$ per group ($p < 0.01$). *** significantly different from mean values of saline treated animals, $n = 6$ per group ($p < 0.001$). The observer was blinded to the experimental conditions.

Crotalphine increases the activation of ERK1/2 in the lumbar spinal cord at 1, 3 and 5 h after administration, when compared with saline-treated rats (Figure 2B,C). Crotalphine also increases the levels of JNK phosphorylation at 1 and 5 h (Figure 2B,D). Conversely, crotalphine decreased the phosphorylation of p38 for up to 96 h after administration (Figure 2B,E). Together, these data indicate that ERK1/2 and JNK are activated in the period in which crotalphine is inducing analgesia, whereas p38 is repressed for a longer period.

2.3. Spinal ERK and JNK Activation Participates in Crotalphine-Induced Analgesia

To evaluate the functional significance of the increased MAPKs activation in the antinociceptive effect of crotalphine, we used a pharmacological MEK inhibitor (MAPK-I, PD98059) (Figure 3A). First, to investigate whether the peripheral MAPKs participate in crotalphine-induced analgesia, the inhibitor was injected by an intraplantar route in two different doses (15 and 30 μg/paw). MAPK-I did not interfere with crotalphine-induced antinociception (Figure 3B). However, the intrathecal MAPK-I injection (30 μg) completely reversed crotalphine-induced mechanical analgesia, showing that spinal MAPKs, but not intraplantar, are involved in crotalphine's beneficial effects (Figure 3C). We next examined which MAPK would be responsible for the crotalphine effects.

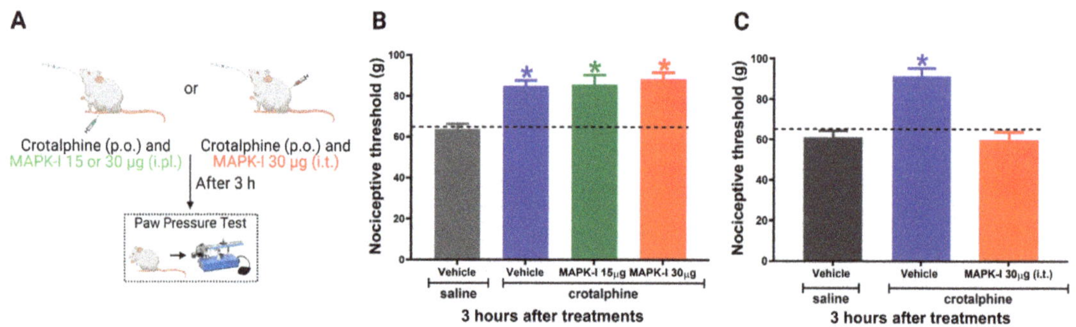

Figure 3. Spinal MAPKs are involved in crotalphine-induced analgesia. (**A**) Schematic representation of treatment and behavior assay. Nociceptive threshold was obtained in the rat paw pressure test before (0) and 3 h after systemic administration of crotalphine (CRP, 1 μg/kg). MEK inhibitor (MAPK-I) was administered immediately after crotalphine (**B**) intraplantarly or (**C**) intrathecally. Data represent mean values ± SEM. * significantly different from baseline (dotted line), $n = 6$ per group ($p < 0.05$). The observer was blinded to the experimental conditions.

The ERK-I and JNK-I intrathecal administration prevented the antinociceptive effect of crotalphine. Nevertheless, the p38 inhibitor did not interfere with crotalphine-induced analgesia (Figure 4B). The control animals injected with the inhibitors alone did not display changes in the nociceptive threshold [7]. Together, these results indicate that crotalphine-induced ERK and JNK activation is responsible, at least in part, for the analgesic effect of this peptide.

2.4. Spinal PKCζ Is Involved in the Antinociceptive Effect of Crotalphine

The activation of the κ-opioid receptor increases the phosphorylation of MAPKs (ERK1/2 and JNK) in neuronal and non-neuronal cells [6,9]. In addition, pretreatment of sensory neurons (DRG cells) with a PKCζ pseudosubstrate abolishes crotalphine-mediated ERK1/2 and JNK phosphorylation [8]. Based on this data and on the results showing that spinal ERK1/2 and JNK are involved in crotalphine-induced analgesia, we further investigated whether PKCζ plays a role in crotalphine effects. To examine the PKCζ participation on the mechanical nociceptive threshold, we used the PKCζ pseudosubstrate, which selectively inhibits the atypical PKCζ isozyme [8,10]. The PKCζ pseudosubstrate was injected by an intrathecal route (3 μg) [11] before the crotalphine administration (Figure 5A). As shown in Figure 5B, the analgesia induced by crotalphine was completely abolished by the PKCζ pseudosubstrate.

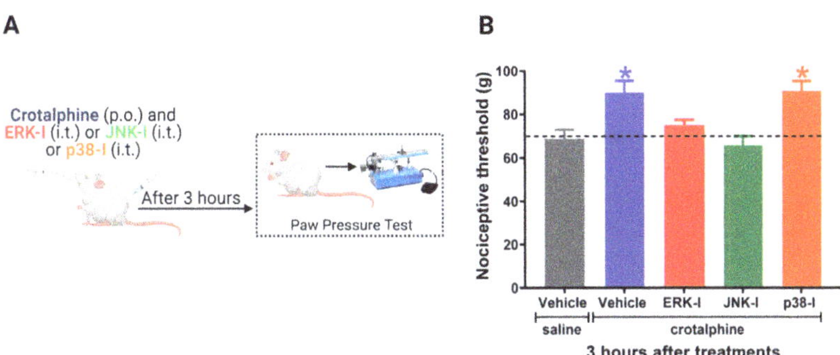

Figure 4. ERK1/2 and JNK inhibitors prevent crotalphine-induced analgesia. (**A**) Schematic representation of the experimental procedure. (**B**) Nociceptive threshold was assessed in the rat paw pressure test before (0) and 3 h after systemic administration of crotalphine (1 μg/kg) with concomitant intrathecal injection of the ERK inhibitor (ERK-I, 30 μg/30 μL), JNK inhibitor (JNK-I, 30 μg/30 μL) or p38 inhibitor (p38-I, 30 μg/30 μL). Data represents mean values ± SEM. * significantly different from baseline (dotted line), $n = 6$ per group ($p < 0.05$). The observer was blinded to the experimental conditions.

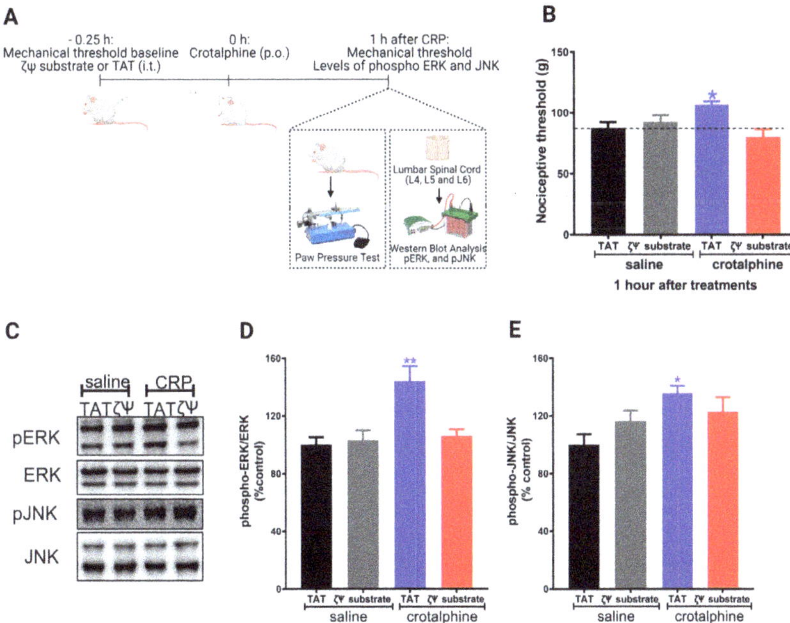

Figure 5. PKCζ mediates crotalphine-induced ERK1/2 activation and antinociception. (**A**) Schematic representation of paw pressure test and experimental procedure. (**B**) PKCζ pseudosubstrate (ζΨ substrate; 3 μg/30 μL) was intrathecally injected 15 min before systemic crotalphine administration (1 μg/kg) and nociceptive threshold was assessed before PKCζ pseudosubstrate injection and 1 h after crotalphine treatment. Data represents mean values ± SEM. *significantly different from baseline (dotted line), $n = 6$ per group ($p < 0.05$). (**C**) Representative blots showing the phosphorylated and total ERK1/2 and JNK levels in the total lysate of the spinal cord. ERK1/2 (**D**) and JNK (**E**) MAPKs were determined by Western blot analysis in lumbar spinal cord extracts from rats treated with crotalphine and the PKCζ pseudosubstrate. Graphs represent the ratio between the phosphorylated protein and the total amount of the targeted protein. Data are presented as mean ± SEM and expressed as % of control (TAT + saline) animals. * significantly different from mean values of TAT + saline treated animals, $n = 6$ per group ($p < 0.05$). ** significantly different from mean values of TAT + saline treated animals, $n = 6$ per group ($p < 0.01$). The observer was blinded to the experimental conditions.

Finally, to examine the direct role that PKCζ plays in the activation of the MAPKs by crotalphine, we used the PKCζ pseudosubstrate and performed the biochemical studies. Our results show that PKCζ inhibition prevents crotalphine-induced ERK1/2 activation (Figure 5D), without interfering with the JNK (Figure 5E). These data suggest that spinal PKCζ activation mediates crotalphine-induced ERK1/2 activation that culminates in analgesia.

3. Discussion

Crotalphine, a 14-amino acid peptide isolated from *C. d. terrificus* venom, has a potent, long-lasting and KOR-mediated antinociceptive effect [1,8,12,13]. In the present study, we showed that a single systemic administration of crotalphine induces analgesia for 5 h in non-sensitized rats. Moreover, when the peptide is administered in PGE_2 sensitized rats, a potent antinociceptive effect is detected for 5 days. Importantly, the peptide did not induce delayed hypersensitivity, which is a decrease in the nociceptive threshold that follows the analgesic effect, a side effect frequently produced by systemic morphine [7,14]. Together, these results confirm the previous finding showing that a peripheral sensitization enhances the antinociceptive effects of opioid-like drugs [8].

Several preclinical studies have shown that opioids activate the MAPK pathway [9,15–17] and this activation is usually associated with cellular stress, inflammation and activation of the sensory neuron that ultimately contribute to nociception [18,19]. However, the use of MAPK inhibitors as analgesics in humans is controversial [20] and the clinical trials involve mainly p38 inhibitors [21,22]. Here, we revealed that spinal ERK1/2 and JNK are activated at the same time period that crotalphine induces analgesia. Importantly, the pharmacological disruption of these kinases is sufficient to blunt the antinociceptive effect. Although several studies have shown that the activation of opioid receptors leads to ERK1/2 and JNK activation [23,24], there are few studies correlating the MAPK pathway with analgesia. Recently, Abraham and co-workers (2018) showed that the ERK1/2 signaling is required by KOR agonists to induce analgesia, since the agonist efficacy was reduced in females with estrogen-induced ERK1/2 impairment [6]. Of interest, one systemic dose of morphine in rats promotes analgesia through spinal ERK activation [7], however, whether this effect is mediated by KOR is unknown. Together, these studies suggest that regardless of its up-stream signaling, the same MAPK can activate different signaling pathways simultaneously (essential and detrimental) and therefore affects the effectiveness of MAPK inhibitor as analgesics.

As mentioned before, crotalphine activates ERK1/2 and JNK in cultured sensory neurons mediated by KOR since the selective opioid receptor antagonist, Nor-BNI, prevents this effect [8]. In the present study, we did not check whether KOR is involved in crotalphine-induced ERK1/2 and JNK activation; however, crotalphine induces the release of dynorphin A that activates KOR in vivo [2].

Unlike what was observed for ERK and JNK, the basal levels of spinal p38 phosphorylation were decreased for at least 96 h after crotalphine administration. As expected, the p38 inhibitor did not interfere with the peptide effect. The phosphorylation of p38 is crucial for the activation of the transcription factors that are involved with the synthesis of pro-inflammatory cytokines and neuromediators, such as calcitonin gene-related peptide (CGRP) and substance P [25–27]. Aiming to investigate the mechanisms involved in the crotalphine long-lasting effect, we tracked the MAPK levels in the spinal cord for 96 h. Our biochemical data confirm the behavioral results showing that crotalphine may have a prolonged inhibitory effect that lasts up to 96 h. However, there is a limitation in our study since the phosphorylation levels in PGE_2-sensitized animals was not determined. However, we can speculate from the results obtained with naïve rats that p38 inhibition may contribute to the long-lasting crotalphine-induced analgesia in sensitized rats. Further studies are needed to address this interesting effect.

PKC activation represents an early signaling element in the opioid pathways to ERK and the PKCζ isoform is responsible for mediating KOR-induced ERK phosphorylation [9,15,28]. Additionally, cell culture assays have shown that crotalphine-induced phosphorylation of ERK1/2 and JNK requires PKCζ activation [8]. Thus, to confirm the contribution of PKCζ in crotalphine-induced antinociception, we injected the PKCζ pseudosubstrate intrathecally, which selectively inhibits the atypical PKCζ isozyme, in non-sensitized rats. Our results demonstrate that PKCζ activation is involved in crotalphine-induced analgesia since the PKCζ pseudosubstrate prevented the antinociception. Moreover, Western blot analyses showed that the PKCζ pseudosubstrate decreased the levels of ERK1/2 phosphorylation, but not of JNK, which was increased in the lumbar spinal cord of rats treated with crotalphine. These results are consistent with the previous data showing that crotalphine activates ERK via PKCζ in cultured sensory neurons [8]. The absence of effect on JNK phosphorylation does not exclude that this pathway is modulated by PKCζ, since in sensitized sensory neuron cultures this kinase is inhibited by crotalphine. Moreover, it is important to mention that, given the short half-life of the PKCζ pseudosubstrate, the crotalphine effects were assessed at 1h after treatment.

Additional studies are required to understand which molecules and/or signaling pathways are being modulated by these kinases. However, it is possible that crotalphine and maybe KOR-agonists use PKCζ/ERK1/2 signaling pathway to regulate transcription factors that are essential for the expression of receptors/channels and/or activation of others signaling pathways, which are involved with antinociception [19,23,29,30]. For example, MAPK phosphorylation activates transcription factors, such as cAMP responsive element binding protein (CREB), which regulates dynorphin gene expression [30,31].

A number of limitations to our study should be noted; for example, we did not investigate whether crotalphine crosses the blood–brain barrier and directly acts on the spinal cord. We also did not demonstrate whether the effect detected in spinal cord tissue is a consequence of a peripheral action, for example, the peptide acting on sensory fiber synapses with the spinal cord neurons. Further studies are necessary to address these questions.

Taken together, our findings show that crotalphine induces analgesia by down-regulating p38 signaling pathways and activating ERK and JNK in the spinal cord via PKCζ. These results contribute to the understanding of the molecular mechanisms that are involved in the analgesic effect of drugs with opioid activity and opens a perspective for the PKCζ-MAPK axis as a target for pain control.

4. Conclusions

In conclusion, our findings demonstrate that a single analgesic dose of crotalphine results in activation of ERK and JNK, and this phenomenon is mediated by PKCζ activation. Targeting the PKCζ-MAPK axis may become an interesting therapeutic alternative to induce analgesia.

5. Materials and Methods

5.1. Animals

In this study, we used male Wistar rats weighting between 170 and 190 g. The rats were housed in an environment with temperatures of 21 ± 2 °C, and light-controlled with a 12 h/12 h light/dark cycle. Standard food and water were available ad libitum. The procedures were performed according to the guidelines for the ethical use of conscious animals in pain research, following the International Association for the Study of Pain [32]. This study was approved by the Institutional Animal Care Committee of the Butantan Institute (CEUAIB, protocol number 9766020419 (date of approval 19 March 2014) and 1245/14 (date of approval 17 April 2019)). The ARRIVE guidelines was followed while conducting the study.

5.2. Chemicals and DrugAadministration

Crotalphine (<E-F-S-P-E-N-C-Q-G-E-S-Q-P-C, where <E is pyroglutamic acid and a disulfide bond between 7C-14C) was synthesized as described by Konno et al. (2008) [1] by the American Peptide Co. (Sunnyvale, CA, USA). The peptide was stored lyophilized

at −20 °C until use. For the stock solutions, the peptide was dissolved in sterile distilled water and then diluted in sterile saline immediately before the experiments. Crotalphine at doses of 8 pg/kg, 20 ng/kg or 1μg/kg (oral route) was administered 60 min before the first nociceptive evaluation. PGE_2 was purchased from Sigma Chemical Co. (St. Louis, MO, USA) and was prepared as previously described [7]. The PGE_2 was injected (100 ng/paw) in a volume of 100 μL with 0.2% of ethanol [8]. ERK inhibitor (ERK-I), p38 inhibitor (SB20358; p38-I) and JNK inhibitor (SP660125; JNK-I), were dissolved in 5% dimethyl sulfoxide (DMSO) and administered intrathecally (i.t.) at the dose of 30 μg. PD98059, a MEK inhibitor, was dissolved in 5% DMSO and administered intraplantarly (i.pl.) (10 or 30 μg/paw) or intrathecally (30 μg) [33]. The peptide ζ-pseudosubstrate ([C]SIYRRGARRWRKLYRAN; amino acids 105-121 in ζ-PKC) was synthesized by Zhejiang Ontores Biotechnologies Co. (Hangzhou, Zhejiang, China) and fused to the cell permeable TAT peptide (YGRKKRRQRRR) transduction domain peptide. The control group received the TAT peptide (American Peptide Co., Sunnyvale, CA, USA). Stock solutions were made in sterile distilled water and diluted in sterile saline. This peptide was administered intrathecally at the dose of 3 μg. In summary, for these intrathecal injections, the animals were anesthetized with 2% isoflurane and the needle was placed in the subarachnoid space on the midline between L4 and L5 vertebrae [34], with a maximum volume of 50 μL. The animals recovered consciousness approximately 1 min after discontinuing the anesthetic.

5.3. *Behavioral Assessment*

Evaluation of the Antinociceptive Effect

The behavioral tests were performed between 9:00 am and 4:00 pm. To assess the mechanical nociceptive threshold, we used the rat paw pressure test [35] (Ugo Basile, VA, Italy). The pressure was recorded before and for up to 192 h after the crotalphine treatment or PGE_2 injection. Testing was conducted blind to the group designation. To reduce stress, the rats were habituated to the testing procedure a day before experimentation. In this test, an increasing amount of force (measured in g) was applied to the hind paw of the rat and interrupted when the animal withdrew the paw. The force necessary to induce this reaction was recorded as the mechanical nociceptive threshold. A maximum pressure of 250 g (i.e., cut-off) was established to avoid damage to the paw.

5.4. *Biochemical Studies*

Western Blot Analysis

Spinal cord tissues (L4–L6) were collected and homogenized in a lysis buffer containing a protease and phosphatase inhibitor cocktail (Sigma-Aldrich). Protein concentrations of the samples were determined using a Bradford assay [36]. Total protein of 30 μg was separated on SDS-PAGE gel (10% gradient gel) and transferred to nitrocellulose membranes (BioRad, Santo Amaro, SP, Brazil). The membranes were blocked for 120 min with 5% bovine serum albumin (BSA) or non-fatty milk and incubated overnight at 4 °C with a primary antibody against phospho-ERK, phospho-JNK, phospho-p38 or non-phosphorylated forms of these proteins (1:1000; Cell Signaling Technology, Danvers, MA, USA). The membranes were then incubated in the correspondent peroxidase-conjugated secondary antibody (1:5000; antirabbit or antimouse, Sigma-Aldrich, St. Louis, MO, USA, cat. numbers A0545 and A8919, respectively) for 120 min at room temperature. The enhanced chemiluminescence method was used to develop the filters (Amersham GE Healthcare Bio-Sciences Corp.; Piscataway, NJ, USA). The densitometric data were analyzed using the UVITEC Cambridge software (UVITEC Cambridge, Cambridge, UK) and normalized to the total protein (not phosphorylated).

5.5. *Statistical Analyses*

Statistical analysis was performed using GraphPad Prism 8 program (GraphPad Software Inc., San Diego, CA, USA). The results are presented as the mean ± SEM. The statistical evaluation of the data was performed using two-way analysis of variance (ANOVA) with post hoc testing by Tukey for Figures 1, 3, 4 and 5B. Figures 2 and 5D,E were analyzed using One-way ANOVA followed by Tukey's post hoc test. A value of $p < 0.05$ was considered significant.

Author Contributions: Conceptualization, V.O.Z. and Y.C.; behavior experiments, B.G.d.F., N.G.H. and L.M.P., western blot experiments T.C.B., N.G.H. and B.G.d.F.; data analysis, B.G.d.F., N.G.H. and V.O.Z.; writing—original draft preparation, N.G.H. and V.O.Z.; writing—review and editing, V.O.Z. and G.P.; supervision, V.O.Z.; project administration, V.O.Z.; funding acquisition, V.O.Z. and Y.C. All authors have read and agreed to the published version of the manuscript.

Funding: This work was supported by funds from the Conselho Nacional de Desenvolvimento Científico e Tecnológico (CNPq) grants numbers 305345/2011-7 and 444299/2014-9; Fundação de Amparo à Pesquisa do Estado de São Paulo, Brazil (FAPESP) grants numbers 2011/04459-2 and 2013/07467-1 (CETICs program); and the Instituto Nacional de Ciência e Tecnologia em Toxinologia (INCTTOX PROGRAM) (CNPq) grant number 573790/2008-6/FAPESP (grant number 2008/57898-0). BGF and LMP received a fellowship from the CNPq/PIBIC (144228/2014-9 and 107340/2005-4, respectively). Coordenação de Aperfeiçoamento de Pessoal de Nível Superior—Brasil (CAPES)—Finance Code 001. Fundação Butantan.

Institutional Review Board Statement: The study was conducted according to the guidelines Ethics Committee on the Use of Experimental Animals at the Butantan Institute (protocol number 1245/14 (date of approval 17 April 2019) and 9766020419 (date of approval 19 March 2014)).

Informed Consent Statement: Not applicable.

Acknowledgments: Schematic images created with BioRender.com.

Conflicts of Interest: Authors declare no conflict of interest.

References

1. Konno, K.; Picolo, G.; Gutierrez, V.P.; Brigatte, P.; Zambelli, V.O.; Camargo, A.C.M.; Cury, Y. Crotalphine, a novel potent analgesic peptide from the venom of the South American rattlesnake Crotalus durissus terrificus. *Peptides* **2008**, *29*, 1293–1304. [CrossRef] [PubMed]
2. Machado, F.C.; Zambelli, V.O.; Fernandes, A.C.O.; Heimann, A.S.; Cury, Y.; Picolo, G. Peripheral interactions between cannabinoid and opioid systems contribute to the antinociceptive effect of crotalphine. *Br. J. Pharmacol.* **2014**, *171*, 961–972. [CrossRef] [PubMed]
3. Bressan, E.; Touska, F.; Vetter, I.; Kistner, K.; Kichko, T.I.; Teixeira, N.B.; Picolo, G.; Cury, Y.; Lewis, R.J.; Fischer, M.J.M.; et al. Crotalphine desensitizes TRPA1 ion channels to alleviate inflammatory hyperalgesia. *Pain* **2016**, *157*, 2504–2516. [CrossRef]
4. Wei, Z.; Liu, H.T. MAPK signal pathways in the regulation of cell proliferation in mammalian cells. *Cell Res.* **2002**, *12*, 9–18. [CrossRef]
5. Obata, K.; Noguchi, K. MAPK activation in nociceptive neurons and pain hypersensitivity. *Life Sci.* **2004**, *74*, 2643–2653. [CrossRef]
6. Abraham, A.D.; Schattauer, S.S.; Reichard, K.L.; Cohen, J.H.; Fontaine, H.M.; Song, A.J.; Johnson, S.D.; Land, B.B.; Chavkin, C. Estrogen regulation of GRK2 inactivates Kappa opioid receptor signaling mediating analgesia, but not aversion. *J. Neurosci.* **2018**, *38*, 8031–8043. [CrossRef] [PubMed]
7. De Freitas, B.G.; Pereira, L.M.; Santa-Cecília, F.V.; Hösch, N.G.; Picolo, G.; Cury, Y.; Zambelli, V.O. Mitogen-Activated Protein Kinase Signaling Mediates Morphine Induced-Delayed Hyperalgesia. *Front. Neurosci.* **2019**, *13*, 1018. [CrossRef] [PubMed]
8. Zambelli, V.O.; Fernandes, A.C.D.O.; Gutierrez, V.P.; Ferreira, J.C.B.; Parada, C.A.; Mochly-Rosen, D.; Cury, Y. Peripheral sensitization increases opioid receptor expression and activation by crotalphine in rats. *PLoS ONE* **2014**, *9*, e90576. [CrossRef] [PubMed]
9. Belcheva, M.M.; Clark, A.L.; Haas, P.D.; Serna, J.S.; Hahn, J.W.; Kiss, A.; Coscia, C.J. μ and κ opioid receptors activate ERK/MAPK via different protein kinase C isoforms and secondary messengers in astrocytes. *J. Biol. Chem.* **2005**, *280*, 27662–27669. [CrossRef] [PubMed]
10. Berra, E.; Diaz-Meco, M.T.; Dominguez, I.; Municio, M.M.; Sanz, L.; Lozano, J.; Chapkin, R.S.; Moscat, J. Protein kinase C ζ isoform is critical for mitogenic signal transduction. *Cell* **1993**, *74*, 555–563. [CrossRef]
11. He, Y.; Wang, Z.J. Nociceptor beta II, delta, and epsilon isoforms of PKC differentially mediate paclitaxel-induced spontaneous and evoked pain. *J. Neurosci.* **2015**, *35*, 4614–4625. [CrossRef]
12. Gutierrez, V.P.; Konno, K.; Chacur, M.; Sampaio, S.C.; Picolo, G.; Brigatte, P.; Zambelli, V.O.; Cury, Y. Crotalphine induces potent antinociception in neuropathic pain by acting at peripheral opioid receptors. *Eur. J. Pharmacol.* **2008**, *594*, 84–92. [CrossRef]
13. Gutierrez, V.P.; Zambelli, V.O.; Picolo, G.; Chacur, M.; Sampaio, S.C.; Brigatte, P.; Konno, K.; Cury, Y. The peripheral L-arginine-nitric oxide-cyclic GMP pathway and ATP-sensitive K^+ channels are involved in the antinociceptive effect of crotalphine on neuropathic pain in rats. *Behav. Pharmacol.* **2012**, *23*, 14–24. [CrossRef]

14. Van Elstraete, A.C.; Sitbon, P.; Trabold, F.; Mazoit, J.X.; Benhamou, D. A single dose of intrathecal morphine in rats induces long-lasting hyperalgesia: The protective effect of prior administration of ketamine. *Anesth. Analg.* **2005**, *101*, 1750–1756. [CrossRef]
15. Bohn, L.M.; Belcheva, M.M.; Coscia, C.J. Mitogenic signaling via endogenous κ-opioid receptors in C6 glioma cells: Evidence for the involvement of protein kinase C and the mitogen- activated protein kinase signaling cascade. *J. Neurochem.* **2000**, *74*, 564–573. [CrossRef] [PubMed]
16. Bruchas, M.R.; Macey, T.A.; Lowe, J.D.; Chavkin, C. Kappa opioid receptor activation of p38 MAPK is GRK3- and arrestin-dependent in neurons and astrocytes. *J. Biol. Chem.* **2006**, *281*, 18081–18089. [CrossRef]
17. Jensen, K.B.; Lonsdorf, T.B.; Schalling, M.; Kosek, E.; Ingvar, M. Increased sensitivity to thermal pain following a single opiate dose is influenced by the COMT val158 met polymorphism. *PLoS ONE* **2009**, *4*, e6016. [CrossRef]
18. Zhuang, Z.Y.; Wen, Y.R.; Zhang, D.R.; Borsello, T.; Bonny, C.; Strichartz, G.R.; Decosterd, I.; Ji, R.R. A peptide c-Jun N-terminal kinase (JNK) inhibitor blocks mechanical allodynia after spinal nerve ligation: Respective roles of JNK activation in primary sensory neurons and spinal astrocytes for neuropathic pain development and maintenance. *J. Neurosci.* **2006**, *26*, 3551–3560. [CrossRef]
19. Minden, A.; Lin, A.; Smeal, T.; Dérijard, B.; Cobb, M.; Davis, R.; Karin, M. c-Jun N-terminal phosphorylation correlates with activation of the JNK subgroup but not the ERK subgroup of mitogen-activated protein kinases. *Mol. Cell. Biol.* **1994**, *14*, 6683–6688. [CrossRef]
20. Ostenfeld, T.; Krishen, A.; Lai, R.Y.; Bullman, J.; Green, J.; Anand, P.; Scholz, J.; Kelly, M. A randomized, placebo-controlled trial of the analgesic efficacy and safety of the p38 MAP kinase inhibitor, losmapimod, in patients with neuropathic pain from lumbosacral radiculopathy. *Clin. J. Pain* **2015**, *31*, 283–293. [CrossRef] [PubMed]
21. Espírito Santo, V.; Passos, J.; Nzwalo, H.; Carvalho, I.; Santos, F.; Martins, C.; Salgado, L.; Silva, C.E.; Vinhais, S.; Vilares, M.; et al. Selumetinib for plexiform neurofibromas in neurofibromatosis type 1: A single-institution experience. *J. Neurooncol.* **2020**, *147*, 459–463. [CrossRef] [PubMed]
22. Tong, S.E.; Daniels, S.E.; Black, P.; Chang, S.; Protter, A.; Desjardins, P.J. Novel p38α mitogen-activated protein kinase inhibitor shows analgesic efficacy in acute postsurgical dental pain. *J. Clin. Pharmacol.* **2012**, *52*, 717–728. [CrossRef]
23. Bruchas, M.R.; Yang, T.; Schreiber, S.; DeFino, M.; Kwan, S.C.; Li, S.; Chavkin, C. Long-acting κ opioid antagonists disrupt receptor signaling and produce noncompetitive effects by activating c-Jun N-terminal kinase. *J. Biol. Chem.* **2007**, *282*, 29803–29811. [CrossRef]
24. Komatsu, T.; Sakurada, S.; Kohno, K.; Shiohira, H.; Katsuyama, S.; Sakurada, C.; Tsuzuki, M.; Sakurada, T. Spinal ERK activation via NO-cGMP pathway contributes to nociceptive behavior induced by morphine-3-glucuronide. *Biochem. Pharmacol.* **2009**, *78*, 1026–1034. [CrossRef]
25. Fiebich, B.L.; Schleicher, S.; Butcher, R.D.; Craig, A.; Lieb, K. The Neuropeptide Substance P Activates p38 Mitogen-Activated Protein Kinase Resulting in IL-6 Expression Independently from NF-κB. *J. Immunol.* **2000**, *165*, 5606–5611. [CrossRef]
26. Ji, R.R.; Gereau IV, R.W.; Malcangio, M.; Strichartz, G.R. MAP kinase and pain. *Brain Res. Rev.* **2009**, *60*, 135–148. [CrossRef]
27. Zhang, Y.; Song, N.; Liu, F.; Lin, J.; Liu, M.; Huang, C.; Liao, D.; Zhou, C.; Wang, H.; Shen, J. Activation of mitogen-activated protein kinases in satellite glial cells of the trigeminal ganglion contributes to substance P-mediated inflammatory pain. *Int. J. Oral Sci.* **2019**, *11*, 24. [CrossRef]
28. Belcheva, M.M.; Szùcs, M.; Wang, D.; Sadee, W.; Coscia, C.J. μ-Opioid Receptor-mediated ERK Activation Involves Calmodulin-dependent Epidermal Growth Factor Receptor Transactivation. *J. Biol. Chem.* **2001**, *276*, 33847–33853. [CrossRef]
29. Kam, A.Y.F.; Chan, A.S.L.; Wong, Y.H. Phosphatidylinositol-3 kinase is distinctively required for μ-, but not κ-opioid receptor-induced activation of c-Jun N-terminal kinase. *J. Neurochem.* **2004**, *89*, 391–402. [CrossRef]
30. Kreibich, A.S.; Blendy, J.A. cAMP response element-binding protein is required for stress but not cocaine-induced reinstatement. *J. Neurosci.* **2004**, *24*, 6686–6692. [CrossRef]
31. Carlezon, W.A.; Duman, R.S.; Nestler, E.J. The many faces of CREB. *Trends Neurosci.* **2005**, *28*, 436–445. [CrossRef]
32. Zimmermann, M. Ethical guidelines for investigations of experimental pain in conscious animals. *Pain* **1983**, *16*, 109–110. [CrossRef]
33. Ma, W.; Zheng, W.H.; Powell, K.; Jhamandas, K.; Quirion, R. Chronic morphine exposure increases the phosphorylation of MAP kinases and the transcription factor CREB in dorsal root ganglion neurons: An in vitro and in vivo study. *Eur. J. Neurosci.* **2001**, *14*, 1091–1104. [CrossRef]
34. Milligan, E.D.; Twining, C.; Chacur, M.; Biedenkapp, J.; O'Connor, K.; Poole, S.; Tracey, K.; Martin, D.; Maier, S.F.; Watkins, L.R. Spinal glia and proinflammatory cytokines mediate mirror-image neuropathic pain in rats. *J. Neurosci.* **2003**, *23*, 1026–1040. [CrossRef]
35. Randall, L.O.; Selitto, J.J. A method for measurement of analgesic activity on inflamed tissue. *Arch. Int. Pharmacodyn. Ther.* **1957**, *111*, 409–419.
36. Bradford, M.M. A rapid and sensitive method for the quantification of microgram quantities of protein utilizing the principle of protein-dye binding. *Anal. Biochem.* **1976**, *72*, 248–254. [CrossRef]

 *toxins*

Article

Bitis arietans Snake Venom and Kn-Ba, a Snake Venom Serine Protease, Induce the Production of Inflammatory Mediators in THP-1 Macrophages

Ângela Alice Amadeu Megale [1,*], Fabio Carlos Magnoli [1], Felipe Raimondi Guidolin [1], Kemily Stephanie Godoi [1], Fernanda Calheta Vieira Portaro [2,*] and Wilmar Dias-da-Silva [1,*]

[1] Immunochemistry Laboratory, Butantan Institute, São Paulo 05503-900, Brazil; fabio.magnoli@butantan.gov.br (F.C.M.); guidolin.felipe@gmail.com (F.R.G.); kemily.godoi@esib.butantan.gov.br (K.S.G.)
[2] Laboratory of Structure and Function of Biomolecules, Butantan Institute, São Paulo 05503-900, Brazil
* Correspondence: angela.amadeu@butantan.gov.br (Â.A.A.M.); fernanda.portaro@butantan.gov.br (F.C.V.P.); wilmar.silva@butantan.gov.br (W.D.-d.-S.)

Abstract: *Bitis arietans* is a snake of medical importance found throughout sub-Saharan Africa and in savannas and pastures of Morocco and western Arabia. The effects of its venom are characterized by local and systemic alterations, such as inflammation and cardiovascular and hemostatic disturbances, which can lead to victims' death or permanent disability. To better characterize the inflammatory process induced by this snake's venom, the participation of eicosanoids and PAF (platelet- activating factor) in this response were demonstrated in a previous study. In addition, edema and early increased vascular permeability followed by an accumulation of polymorphonuclear (PMN) cells in the peritoneal cavity were accompanied by the production of the eicosanoids LTB_4, LTC_4, TXB_2, and PGE_2, and local and systemic production of IL-6 and MCP-1. In this context, the present study focused on the identification of inflammatory mediators produced by human macrophages derived from THP-1 cells in response to *Bitis arietans* venom (BaV), and Kn-Ba, a serine protease purified from this venom. Here, we show that Kn-Ba, and even the less intensive BaV, induced the production of the cytokine TNF and the chemokines RANTES and IL-8. Only Kn-Ba was able to induce the production of IL-6, MCP-1, and IP-10, whereas PGE_2 was produced only in response to BaV. Finally, the release of IL-1β in culture supernatants suggests the activation of the inflammasomes by the venom of *Bitis arietans* and by Kn-Ba, which will be investigated in more detail in future studies.

Keywords: *Bitis arietans* venom (BaV); Kn-Ba; inflammation; cytokines and chemokines; PGE_2; THP-1 macrophages

Key Contribution: Inflammation is closely related to the development of local and systemic deleterious effects due to snake envenomation. Therefore, comprehension of the molecular mechanisms involved in this process is a valuable tool to establish efficient complementary therapies. In this context, this study showed, for the first time, that BaV and Kn-Ba induced the production of different inflammatory mediators, such as cytokines, chemokines, and lipid mediators by THP-1 macrophages. Among these mediators, the production and release of the cytokine IL-1β must be highlighted, as it indicates the possible involvement of inflammasomes in *Bitis arietans* envenomation.

Citation: Megale, Â.A.A.; Magnoli, F.C.; Guidolin, F.R.; Godoi, K.S.; Portaro, F.C.V.; Dias-da-Silva, W. *Bitis arietans* Snake Venom and Kn-Ba, a Snake Venom Serine Protease, Induce the Production of Inflammatory Mediators in THP-1 Macrophages. *Toxins* **2021**, *13*, 906. https://doi.org/10.3390/toxins13120906

Received: 29 October 2021
Accepted: 10 December 2021
Published: 16 December 2021

Publisher's Note: MDPI stays neutral with regard to jurisdictional claims in published maps and institutional affiliations.

1. Introduction

The global burden of snake envenoming is estimated at around 400,000 cases/year, with 20,000 deaths/year. Due to poor notifications, 1,800,000 envenomations/year and 94,000 deaths/year may occur [1]. It is estimated that, on the African continent, more than 60% of these accidents happen in sub-Saharan Africa, of which 95% occur in rural regions, leading to about 12,000 to 32,000 deaths and more than 9000 amputations related to post-envenoming complications [1,2].

Except in some islands, at high altitudes, and in regions with long-lasting snow seasons, venomous snake species are distributed worldwide, living naturally in intimate contact with human populations [3]. In the different global regions, the number of venomous snake species is large and varied. In the Middle East and North Africa, 17 snake species are found, and in sub-Saharan Africa, encompassing the Central, East, South, and West regions of the African continent, 26 species are found. Among these snakes, the *Bitis* genus includes six species, which, due to the number and severity of envenomation incidents, are important: *B. arietans*, *B. somalica*, *B. parviocula*, *B. gabonica*, *B. rhinoceros*, and *B. nasicornis* [4–6]. *B. arietans* is found in North Africa and in all sub-Saharan regions and is considered one of the most relevant snakes of medical importance in the African continent [4,7–9].

Local and systemic symptoms are observed in human victims of *Bitis arietans* bite. Intense pain, blistering, edema, ecchymosis, hemorrhage, draining lymph node hypertrophy, and necrosis are the common local symptoms. The systemic symptoms are fever, neutrophilic leukocytosis, thrombocytopenia, hemolysis, and bleeding. These systemic symptoms result in anemia, lower resistance to infections, diffuse hemorrhages, myocardial injury, coagulopathy, blood hypotension, and sometimes permanent injury of the affected body regions. All of these can cause death [9–13].

The *Bitis arietans* venom (BaV) is a complex mixture of proteins (±90–95%), peptides, carbohydrates, nucleic-acid-derived segments, metallic cations, biogenic amines, lipids, and free amino acids. Among the proteins, there are substantial quantities of enzymes, such as snake venom serine proteases (SVSPs), snake venom metalloproteases (SVMPs), and phospholipase-A_2. Aside from some non-enzymatic proteins such as disintegrins, type-C lectins, cystatins, and type-Kunitz protease inhibitors are also found [14–17]. More recently, our group showed the presence of some new and already known proline-rich peptides in BaV, also known as bradykinin-potentiating peptides (BPPs) [18].

Among the most common components of BaV are proteases, specially SVSPs and SVMPs, which represent 58% of the venom composition [15–17]. These two protease classes, already studied in other snake venoms, such as *Bothrops atrox* [19–21] and *Bothrops jararaca* [22–24], have been identified as important toxins for the local and systemic effects observed in envenomation.

Due to the importance of these molecules, our group has been working on the purification and characterization of SVSPs and SVMPs present in the BaV. Thus, initially, a serine protease with fibrinogenolytic and kinin-releasing activities, named Kn-Ba, was purified and characterized in vitro. Kn-Ba is also recognized by antibodies (Abs) present in the horse-serum anti-*Bitis* spp. venoms [25].

Based on the importance of the inflammatory process in snake envenomation [11,23,26–29], the in vivo inflammation induced by BaV was studied in mice peritoneal cavities. BaV induced local tissue vessel dilatation followed by plasma infiltration; erythrocytes and neutrophils accumulation with simultaneous production of the eicosanoids LTB_4, LTC_4, TXB_2, and PGE_2; as well as the local and systemic production of IL-6 and MCP-1. In general, this study demonstrated that the BaV contains toxins that trigger the inflammatory process, which is partially dependent on lipid mediators [30].

In this context, aiming to better understand the inflammatory mechanisms involved in *Bitis arietans* envenomation, the present study focused on the identification of inflammatory mediators induced by BaV and Kn-Ba in human macrophages differentiated from THP-1 cells (THP-1 macrophages).

2. Results

2.1. THP-1 Macrophage Differentiation

Human non-adherent pre-monocytes of the THP-1 lineage were cultured in supplemented RPMI medium at a density of 2–8 × 10^5 cells/mL (Supplementary Figure S1A). In these conditions, the cells grew in clumps and showed a high rate of proliferation, doubling the number of cells within 2–3 days. When the cells reached high concentrations, not exceeding 1×10^6 cells/mL, live cells were differentiated in macrophages using PMA (Supplementary Figure S1B). In contrast to monocytes, differentiated macrophages become adherent and relatively larger than undifferentiated cells, stopping proliferation and spreading.

To confirm the differentiation, the expression of CD11b was evaluated by flow cytometry. CD11b, which together with CD18 forms the inactive C3b (iC3b) receptor, known as CR3 receptor, has been reported to be induced during differentiation of monocytes into macrophages. CR3 has important functions both as an adhesion molecule and a membrane receptor mediating recognition of different ligands [31–33]. As showed in Supplementary Figure S1C, only differentiated macrophages expressed CD11b, confirming the differentiation of these cells for further treatments.

2.2. Total Protein and Endotoxin Contents Determination in BaV and Kn-Ba

For THP-1 macrophage treatments, stimuli stock solutions were prepared at a concentration of 5 mg/mL (BaV) and 26 µg/mL (Kn-Ba) in sterile PBS, pH 7.2. The determined endotoxin concentrations were <0.1 UE/µg in both samples, which can be considered as an acceptable endotoxin level [34].

2.3. Effects of BaV and Kn-Ba on THP-1 Macrophage Viabilities

To evaluate the cytotoxic effects caused by treatments with BaV and Kn-Ba, the release of lactate dehydrogenase (LDH) in culture supernatants was evaluated. After the incubation with 0.5–15 µg of BaV and 0.5–1 µg of Kn-Ba for 24 h, 48 h, and 72 h, cell viability was greater than 90% (Supplementary Figure S2).

2.4. BaV Induces the Production of TNF and IL-1β in THP-1 Macrophages

The concentration of cytokines IL-1β, IL-6, IL-10, IL-12, and TNF were evaluated on cell-free supernatants by CBA (Cytometric Bead Array) after 24 h, 48 h, and 72 h of treatments with BaV. As shown in Figure 1, BaV induced the production of TNF and IL-1β in a concentration-dependent manner. The production of TNF was detected only after 24 h in response to the two highest doses of BaV (10 µg and 15 µg/well). On the other hand, the maximum production of IL-1β, also in response to the two highest doses of BaV, was reached in 24 h, then decreased after 48 h and 72 h of treatment. BaV did not induce the production of IL-6, IL-10, or IL-12.

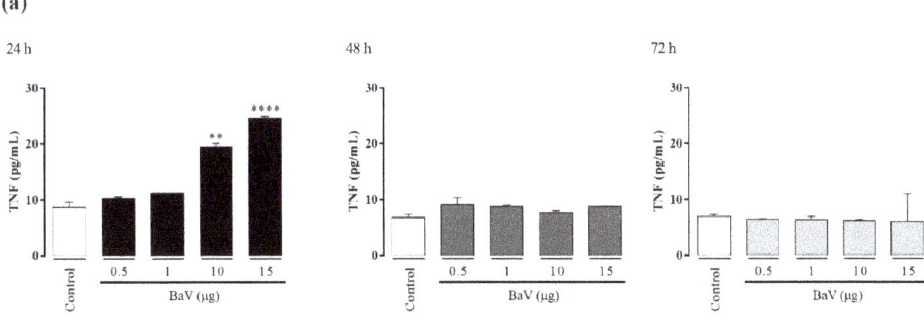

Figure 1. *Cont.*

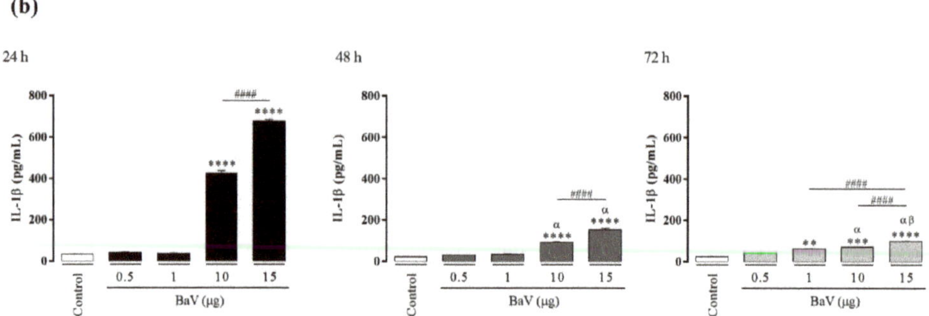

Figure 1. Cytokine production in THP-1 macrophages induced by BaV. THP-1 macrophages (2×10^5 cells/well) were treated with different concentrations of BaV (0.5 to 15 μg/well) to induce cytokine production. Cells incubated only with culture medium were used as control. After 24 h, 48 h, and 72 h, the cytokine production was evaluated in culture supernatant by CBA. (**a**) TNF and (**b**) IL-1β. Results expressed as mean of duplicates ± SEM and analyzed by one-way ANOVA followed by Tukey's post-test (each period compared with respective control) or two-way ANOVA followed by Tukey's post-test (comparison between different periods of treatment). (*) Significant difference in relation to the respective control. (#) Significant difference between concentrations of BaV on each period. Symbols indicate differences between periods of treatment: (α) difference of 48 h or 72 h compared to 24 h; (β) difference from 72 h to 48 h. $p < 0.01$ (**); $p < 0.001$ (***); and $p < 0.0001$ (**** and ####).

2.5. BaV Induces the Production of RANTES and IL-8 in THP-1 Macrophages

As the two highest doses of BaV (10 and 15 μg/well) were able to induce the cytokines production without affecting the cell viability, these doses were selected to evaluate the production of chemokines CXCL8/IL-8, CCL5/RANTES, CXCL9/MIG, CCL2/MCP-1, and CXCL10/IP-10 in cell-free supernatants by CBA after 24 h, 48 h, and 72 h of treatments with BaV. BaV induced the production of RANTES and IL-8. RANTES was produced in a dose-dependent manner after 24 h, decaying thereafter. IL-8, in contrast, started to be produced—also in a dose-dependent manner—mainly after 48 h and continuing after 72 h (Figure 2). BaV did not induce the production of CXCL9/MIG, CCL2/MCP-1, and CXCL10/IP-10.

2.6. Kn-Ba Induces the Production of TNF, IL-6, and IL-1β in THP-1 Macrophages

The concentrations of cytokines IL-1β, IL-6, IL-10, IL-12, and TNF were evaluated on cell-free supernatants by CBA after 24 h, 48 h, and 72 h of treatments with Kn-Ba. Kn-Ba induced the production of TNF and IL-1β, and, unlike BaV, it was also able to induce IL-6 production. At the two doses used (0.5 and 1 μg), Kn-Ba induced the production of these chemokines in concentration-dependent manners. The TNF and IL-6 production were detected after 24 h, decreasing thereafter, whereas the production of IL-1 β was accentuated after 48 h and maintained after 72 h (Figure 3). Kn-Ba did not induce the production of IL-10 or IL-12.

2.7. Kn-Ba Induces the Production of IP-10, MCP-1, RANTES, and IL-8 in THP-1 Macrophages

The concentrations of chemokines CXCL8/IL-8, CCL5/RANTES, CXCL9/MIG, CCL2/MCP-1, and CXCL10/IP-10 were evaluated in cell-free supernatants by CBA after 24 h, 48 h, and 72 h of treatments with Kn-Ba (0.5 and 1 μg). Just like BaV, Kn-Ba induced the production of RANTES and IL-8, but was also able to induce the production of IP-10 and MCP-1, in time- and concentration-dependent manners (Figure 4).

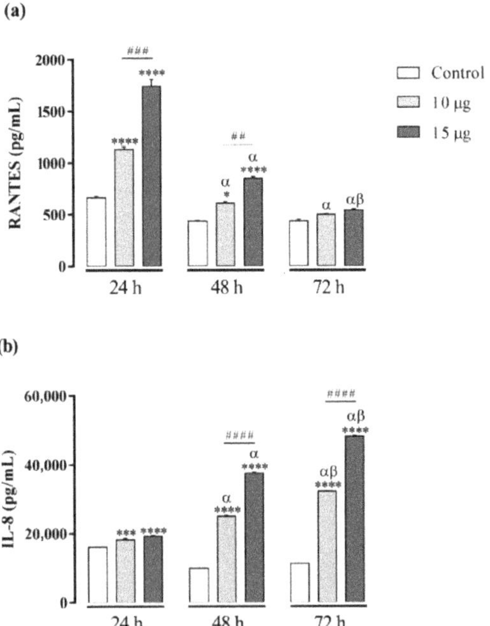

Figure 2. Chemokine productions in THP-1 macrophages induced by BaV. THP-1 macrophages (2×10^5 cells/well) were treated with two concentrations of BaV (10 and 15 µg/well). Cells incubated only with culture medium were used as control. After 24 h, 48 h, and 72 h, the production of chemokines were evaluated in culture supernatant by CBA. (**a**) RANTES and (**b**) IL-8. Results expressed as mean of duplicates ± SEM and analyzed by one-way ANOVA followed by Tukey's post-test (each period compared with respective control) or two-way ANOVA followed by Tukey's post-test (comparison between different periods of treatment). (*) Significant difference in relation to the respective control. (#) Significant difference between concentrations of BaV on each period. Symbols indicate differences between periods of treatment: (α) difference of 48 h or 72 h compared to 24 h; (β) difference from 72 h to 48 h. $p < 0.05$ (*); $p < 0.01$ (##); $p < 0.001$ (*** and ###); and $p < 0.0001$ (**** and ####).

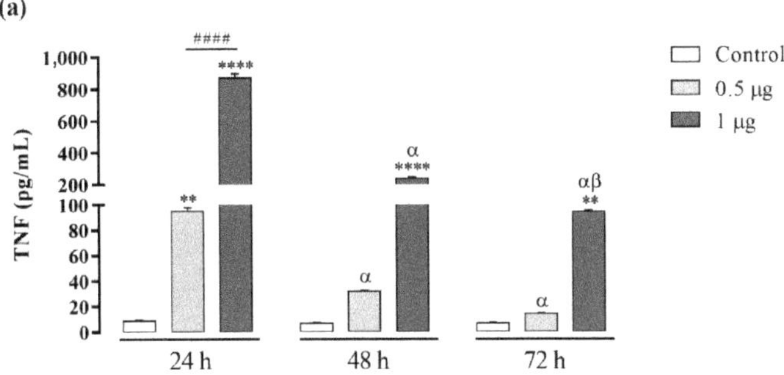

Figure 3. *Cont.*

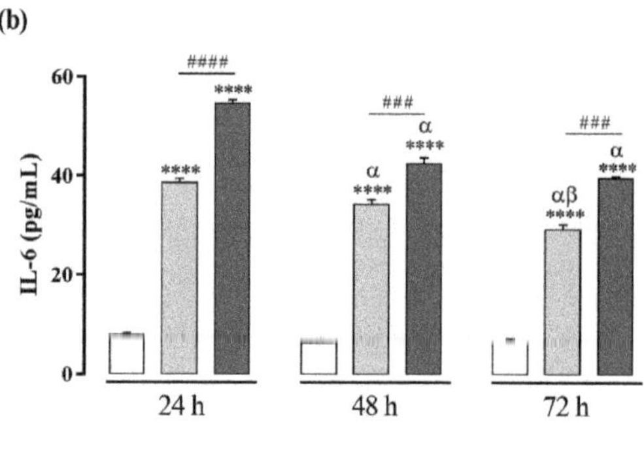

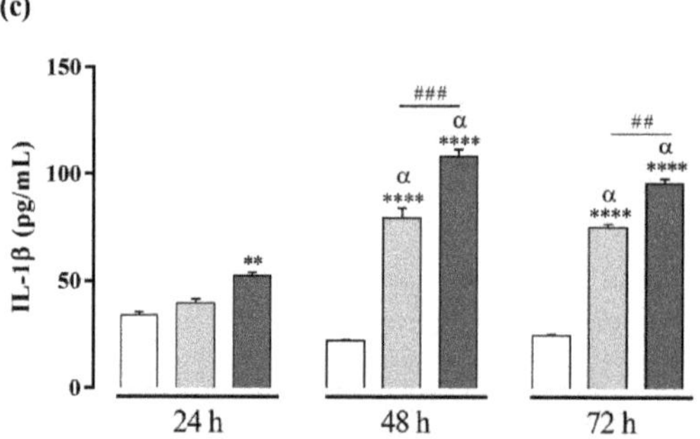

Figure 3. Cytokine production in THP-1 macrophages induced by Kn-Ba. THP-1 macrophages (2×10^5 cells/well) were treated with two concentrations of Kn-Ba (0.5 and 1 μg/well). Cells incubated only with culture medium were used as control. After 24 h, 48 h, and 72 h the production of cytokines were evaluated in culture supernatant by CBA. (**a**) TNF, (**b**) IL-6 and (**c**) IL-1β. Results expressed as mean of duplicates ± SEM and analyzed by one-way ANOVA followed by Tukey's post-test (each period compared with respective control) or two-way ANOVA followed by Tukey's post-test (comparison between different periods of treatment). (*) Significant difference in relation to the respective control. (#) Significant difference between concentrations of Kn-Ba on each period. Symbols indicate differences between periods of treatment: (α) difference of 48 h or 72 h compared to 24 h; (β) difference from 72 h to 48 h. $p < 0.01$ (** and ##); $p < 0.001$ (###); and $p < 0.0001$ (**** and ####).

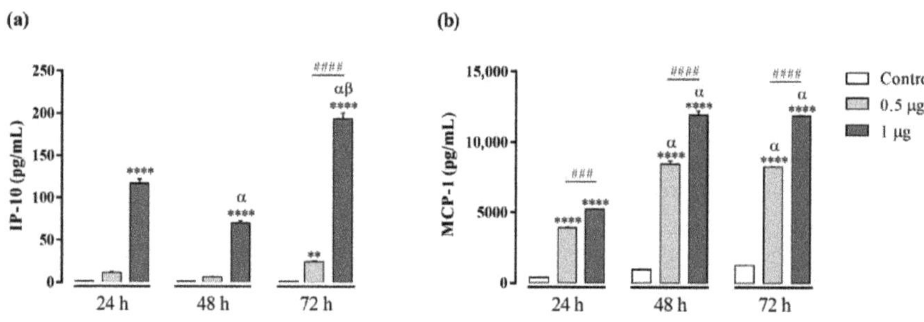

Figure 4. *Cont.*

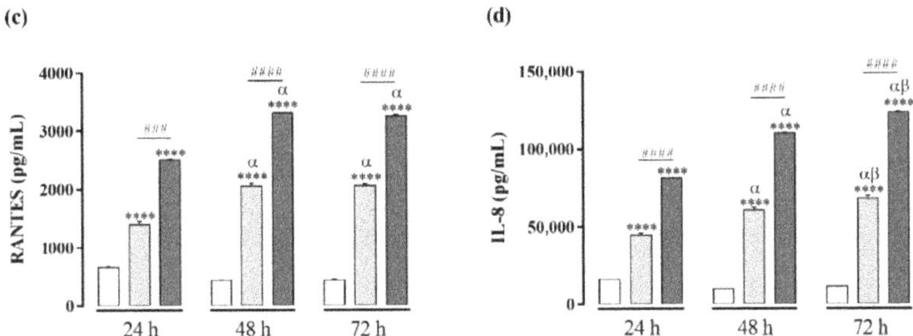

Figure 4. Chemokine productions in THP-1 macrophages induced by Kn-Ba. THP-1 macrophages (2 × 10^5 cells/well) were treated with two concentrations of Kn-Ba (0.5 and 1 μg/well). Cells incubated only with culture medium were used as control. After 24 h, 48 h, and 72 h, the production of chemokines were evaluated in culture supernatant by CBA. (**a**) IP-10, (**b**) MCP-1, (**c**) RANTES and (**d**) IL-8. Results expressed as mean of duplicates ± SEM and analyzed by one-way ANOVA followed by Tukey's post-test (each period compared with respective control) or two-way ANOVA followed by Tukey's post-test (comparison between different periods of treatment). (*) Significant difference in relation to the respective control. (#) Significant difference between concentrations of Kn-Ba on each period. Symbols indicate differences between periods of treatment: (α) difference of 48 h or 72 h compared to 24 h; (β) difference from 72 h to 48 h. $p < 0.01$ (**); $p < 0.001$ ($^{###}$); and $p < 0.0001$ (**** and $^{####}$).

2.8. Kn-Ba Induced High Levels of TNF and IL-6, Whereas BaV Is Involved in the IL-1β Production in THP-1 Macrophages

The cytokine concentrations produced in response to highest doses of BaV (15 μg/well) and Kn-Ba (1 μg/well) were compared. As shown in Figure 5, the increased productions of TNF and IL-6 in response to Kn-Ba were evident. However, the highest concentration of IL-1β was detected mainly in response to BaV, suggesting that this important inflammatory cytokine is produced in response to other components present in BaV.

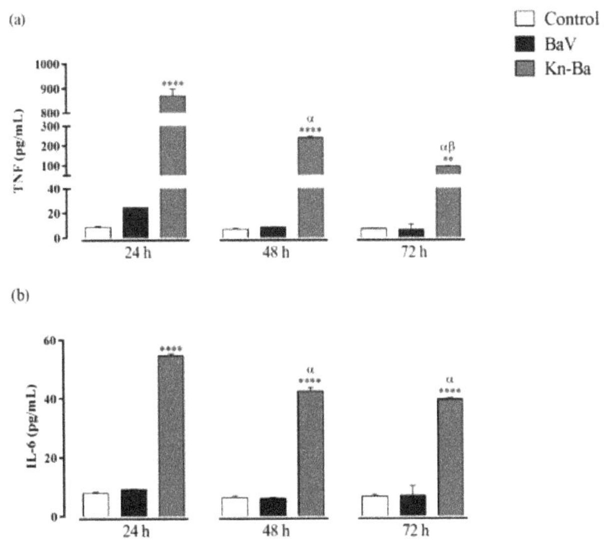

Figure 5. *Cont.*

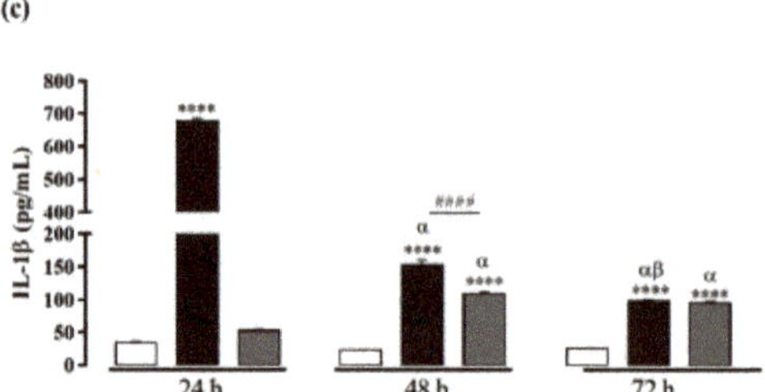

Figure 5. Comparison between cytokine productions induced by higher doses of BaV and Kn-Ba. THP-1 macrophages (2×10^5 cells/well) were treated with higher concentrations of BaV (15 µg/well) and Kn-Ba (1 µg/well). Cells incubated only with culture medium were used as control. After 24 h, 48 h, and 72 h, the production of cytokines was evaluated in culture supernatant by CBA. (**a**) TNF, (**b**) IL-6 and (**c**) IL-1β. Results expressed as mean of duplicates ± SEM and analyzed by one-way ANOVA followed by Tukey's post-test (each period compared with respective control) or two-way ANOVA followed by Tukey's post-test (comparison between different periods of treatment). (*) Significant difference in relation to the respective control. (#) Significant difference between stimuli on each period. Symbols indicate differences between periods of treatment: (α) difference of 48 h or 72 h compared to 24 h; (β) difference from 72 h to 48 h. $p < 0.01$ (**); and $p < 0.0001$ (**** and ####).

2.9. *Kn-Ba Induced the Production of Highest Levels of all Evaluated Chemokines in THP-1 Macrophages*

The chemokine concentrations produced in response to the highest doses of BaV (15 µg/well) and Kn-Ba (1 µg/well) were compared. As depicted in Figure 6, the higher productions of the chemokines IP-10, MCP-1, RANTES, and IL-8 were detected in response to Kn-Ba.

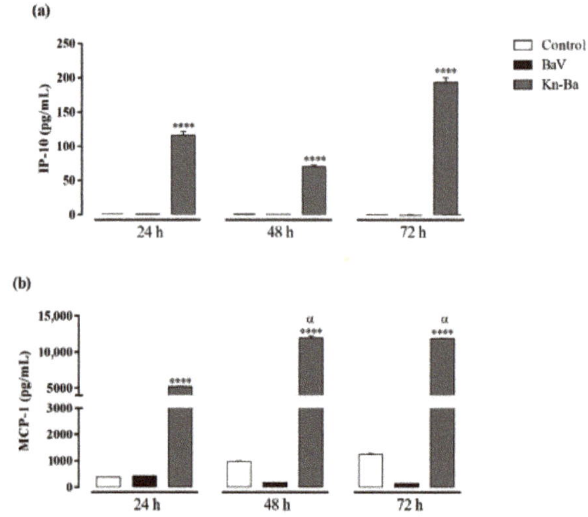

Figure 6. *Cont.*

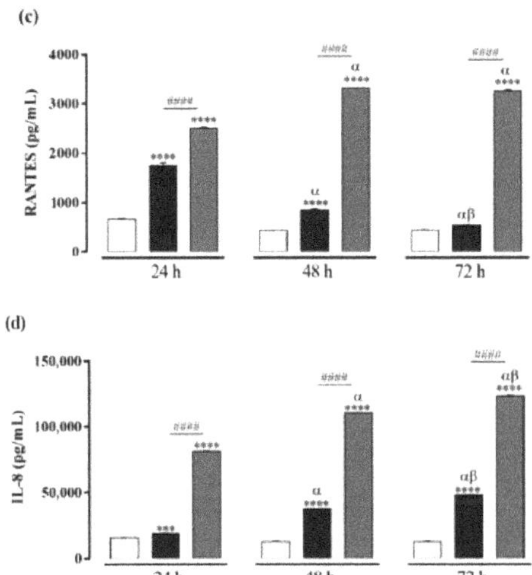

Figure 6. Comparison between chemokine productions induced by higher doses of BaV and Kn-Ba. THP-1 macrophages (2 × 10^5 cells/well) were treated with higher concentrations of BaV (15 µg/well) and Kn-Ba (1 µg/well). Cells incubated only with culture medium were used as control. After 24 h, 48 h, and 72 h, the production of chemokines was evaluated in culture supernatant by CBA. (**a**) IP-10, (**b**) MCP-1, (**c**) RANTES and (**d**) IL-8. Results expressed as mean of duplicates ± SEM and analyzed by one-way ANOVA followed by Tukey's post-test (each period compared with respective control) or two-way ANOVA followed by Tukey's post-test (comparison between different periods of treatment). (*) Significant difference in relation to the respective control. (#) Significant difference between stimuli on each period. Symbols indicate differences between periods of treatment: (α) difference of 48 h or 72 h compared to 24 h; (β) difference from 72 h to 48 h. $p < 0.001$ (***); and $p < 0.0001$ (**** and ####).

2.10. BaV, But Not Kn-Ba, Induced the Production of PGE_2

The concentrations of the eicosanoids PGE_2 and LTB_4 were evaluated in cell-free supernatants by competitive ELISA after 30 min of treatments with higher doses of BaV (15 µg/well) and Kn-Ba (1 µg/well). BaV, but not Kn-Ba, induced the production of PGE_2 (Figure 7). No stimulus was able to induce the production of LTB_4 (data not shown).

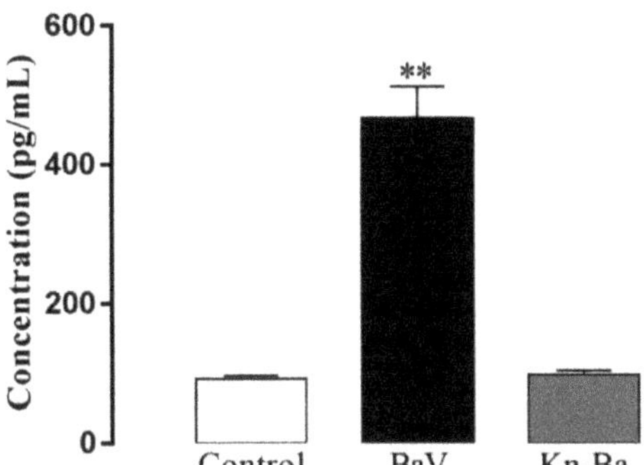

Figure 7. Production of PGE_2. THP-1 macrophages (2 × 10^5 cells/well) were treated with higher concentrations of BaV (15 µg/well) and Kn-Ba (1 µg/well). Cells incubated only with culture medium were used as control. After 30 min, the production of PGE_2 was evaluated by ELISA. Results expressed as mean of duplicates ± SEM and analyzed by one-way ANOVA followed by Tukey's post-test. (Asterisk) Significant difference from control and Kn-Ba. $p < 0.01$ (**).

3. Discussion

Inflammation is the hallmark of envenomation by snakes in the Viperidae family. The activation of the inflammatory process and its cascade of events play an important role in the pathogenesis of envenomation, in the clinical picture, and in the outcome of the accident. In this sense, the envenomation by *Bothrops asper* [35,36], *B. atrox* [37,38], and *B. jararaca* [39,40], which, among other characteristic changes such as hemostatic and cardiovascular disorders, induce a prominent inflammatory response, which has been associated not only with local damage but also with systemic disturbances caused by Viperidae venoms [11,29]. Therefore, given the importance in understanding the inflammatory response induced by the BaV as a tool for the development of complementary therapies, or the improvement of *B. arietans* antivenoms, we show that this venom was capable of inducing, in vitro, the production of several inflammatory mediators, including cytokines, chemokines, and lipid mediators.

Our previous in vivo results demonstrated that BaV induces inflammation with the participation of diverse endogenous inflammatory mediators produced by residents or derivatives of recently migrated blood cells [30]. Among these cells, macrophages became candidates for target cells in the present study of cell inflammation induced by BaV and Kn-Ba, since macrophages play a key role in inflammation [41] and also participate in the inflammatory response related to other snake envenomations [22,23,42]. In this sense, studies have shown that in the absence of macrophages, the recruitment of immune cells to the site of inflammation, especially neutrophils, is impaired [22,23].

Macrophages express, on their cell surface, receptors named pattern-recognizing receptors (PRRs), which recognize the molecular domains such as pathogen-associated molecular patterns (PAMPs) expressed by pathogens, virus, fungus, bacteria; and damage-associated molecular patterns (DAMPs), which are expressed by host-damaged molecules [43–45]. In addition, the PRRs can also recognize venom-associated molecular patterns, also known as VAMPs. The term VAMP was coined by Brazilian researchers, who showed that mice deficient in TLR (Toll-like receptor)-2 and TLR-4, two important PRRs, as well as the adapter molecule CD14, produce fewer inflammatory mediators in response to the *Tityus serrulatus* scorpion venom, including decreased levels of IL-6 and TNF-α [46].

In view of the important role of macrophages in the inflammatory process, we used THP-1 macrophages to evaluate the in vitro pro-inflammatory properties of BaV and Kn-Ba. To support the homogeneity of the experiments, the human monocytes of the THP-1 lineage were differentiated in vitro into macrophages with PMA, as described in the literature [33,47]. There are different protocols using PMA for the differentiation of THP-1 monocytes to macrophages, but in some cases the differentiation profile may be not comparable to primary monocyte-derived macrophages [48]. To overcome this inconvenience, macrophages treated with PMA were kept at rest for 4 days in culture media without PMA, which was shown to be a good alternative to induce a differentiation pattern similar to primary monocyte-derived macrophages [33]. This differentiation was evaluated by morphological analyzes such as adhesion to the support surface, spreading and emission of pseudopods, and loss of proliferation. In addition to the morphological alterations, the differentiated macrophages, but no undifferentiated THP-1 monocytes, expressed CD11b, an important surface marker of monocyte-to-macrophage maturation [31–33]. CD11b forms the molecular complex CD11b-CD18, known as CR3 receptor, which is expressed in macrophages and in other leucocytes and binds to several ligands, such as iC3b adhered to pathogens. These promote phagocytosis [49] and expanding the macrophages' biological activities [33]. After differentiation, the THP-1 macrophages were used as target cells in the studies with BaV and its purified serine protease, Kn-Ba.

In sequence, the effects of BaV and Kn-Ba as inflammation-inducers were demonstrated by the production of the cytokines TNF, IL-6, IL-1β; and the chemokines IP-10, MCP-1, RANTES, and IL-8; besides the lipid mediator PGE_2.

The role of purified SVSP on the inflammation induced by snake venoms has been extensively studied [50–52]. However, although SVSPs presents a high degree of similarity between their amino acids sequences, their functions may differ [53], so that not all are involved with the inflammatory response. In this sense, two SVSPs purified from *Bothrops pirajai* venom, named BpirSP27 and BpirSP41, seem to be not involved in the inflammatory events related to *B. pirajai* envenomation, such as edema, pain, and leukocyte recruitment [54]. In contrast, it was showed that SVSPs from the venoms of *Bothrops alternatus* and *B. moojeni* are able to promote edema and pain, two classic signs of inflammation [50]. SVSPs from *Bothrops asper* can also activate endogenous matrix metalloproteases [51], while SVSPs from *Crotalus durissus terrificus* venom induced edema and the increased expression of COX-2 and PGE_2 production [52]. In agreement with these finds, here we show that Kn-Ba induces the production of inflammatory cytokines and chemokines by THP-1-derived human macrophages, indicating its participation in BaV-induced inflammation.

Our results show differences between the type and concentration of some inflammatory mediators produced in response to treatments with BaV or purified Kn-Ba. In this sense, it can be suggested that the mediators produced only after stimuli with BaV, but not with Kn-Ba, as in the case of PGE_2, were generated by other components present in the venom. However, here we also show that some mediators, more precisely IL-6, IP-10, and MCP-1, were produced only after stimulation with Kn-Ba. This is a complex issue involving both the composition of the venom itself and the immune response triggered against the components of the whole venom. It is well known that BaV is a complex mixture of proteins (±90–95%), which includes, besides enzymes such as SVSP and SVMPs, non-enzymatic proteins as Kunitz-type proteases inhibitors [14–17]. Therefore, it can be suggested that the activity of some proteases could be hindered by endogenous inhibitors present in the venom itself. Thus, it is possible that purified Kn-Ba inducing the production of different inflammatory mediators, in a greater degree than BaV, may be due to the absence of these inhibitors. In addition, the different components of BaV can act in a different way in immune cells, making the inflammatory response diverse from the one developed only against purified Kn-Ba.

Among the inflammatory mediators produced in response to BaV and Kn-Ba, the cytokine IL-1β is noteworthy. According to literature data, after the activation of macrophages via TLRs, a cascade of intracellular signaling begins. This cascade may culminate in the activation of an important transcription factor named NF-κB, which is responsible for the expression of diverse inflammatory genes, including TNF-α, IL-6, iNOS, and also pro IL-1β [45,55,56]. Pro IL-1β is a zymogen whose activation and later secretion depends on the inflammasomes' activation, a multiprotein complex formed due to cell activation by the recognition of patterns via cytosolic PRRs of the NLR family (NOD-like receptors) [57,58]. An important subfamily of NLRs, called NALP, is involved in the induction of the inflammatory response via cytokines of the IL-1 family, including IL-1β, IL-8, and IL-33. Recognition via NALP promotes the formation of inflammasomes, which are responsible for the activation of inflammatory caspases 1, 4, and 5 in humans, and caspases-1, 11, and 12 in mice, which can convert pro-IL-1β into mature IL-1β [59,60]. IL-1β is a highly inflammatory cytokine produced during various inflammatory conditions, which mediates innate and adaptative immune responses by promoting acute phase response and recruiting inflammatory cells [61,62]. The overproduction of IL-1β is harmful and can trigger autoimmune diseases [63,64].

Here, we show that both BaV and Kn-Ba induced the production and secretion of IL-1β in the supernatant of macrophage cultures, suggesting the participation of inflammasomes in the inflammatory process related to the *Bitis arietans* envenomation. In addition, as higher concentrations of this cytokine were detected after treatment with BaV, our results indicate there are other components present in BaV, besides Kn-Ba, which may be involved in the possible inflammasome activation.

IL-1β can act synergistically with other cytokines produced by human macrophages to amplify *Bitis-arietans*-related inflammations, such as with TNF-α, activating the endothelium and inducing vasodilation and increased vascular permeability, and, with IL-6, activating hepatocytes and inducing the production of acute phase proteins, which can activate the complement system and act as opsonins, facilitating phagocytosis by macrophages and neutrophils [45].

Summing up, our study showed that BaV and Kn-Ba are able to induce the activation and the production of inflammatory mediators in THP-1-derived human macrophages, and this is the first suggestion that the inflammasomes may play a role in *Bitis arietans* envenomation.

4. Conclusions

Our study showed, for the first time, the pro-inflammatory effects of BaV and Kn-Ba upon THP-1 derived human macrophages. Both stimuli are responsible for the production of the cytokines TNF and the chemokines RANTES and IL-8. However, the significant production of these mediators occurred in response to Kn-Ba. Only Kn-Ba was able to induce the production of IL-6, MCP-1, and IP-10, whereas PGE_2 was produced only in response to BaV. Finally, although both stimuli induced the production of IL-1β, suggesting that the inflammasomes may play a role in BaV envenomation, the highest production of this cytokine in response to BaV suggests the participation of other venom components in this process (Figure 8). These results together with previous data published by our group describing the biological actions of Kn-Ba and BaV-induced inflammation in vivo help us to better understand the inflammatory mechanisms involved in this envenomation, and to list important toxins for the future development of antivenoms.

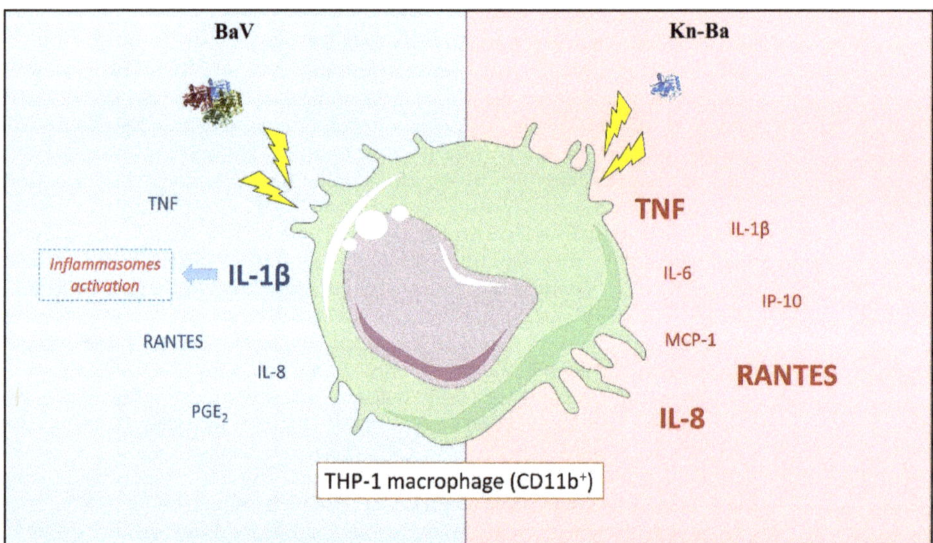

Figure 8. Comparative overview of the inflammatory profiles induced by BaV and Kn-Ba in THP-1 macrophages. THP-1 macrophages were treated with BaV (left; blue) and Kn-Ba (right; red) resulting in the synthesis and secretion of the indicated cytokines, chemokines, and PGE_2. The mediators highlighted in larger size were produced by both stimuli, but at higher levels at the evidenced side. Both stimuli induced the production and release of IL-1β, indicating the inflammasomes pathway activation during BaV envenomation. However, the highest production of this cytokine in response to BaV indicated the participation of other venom components in this process. The schematic art pieces used in this figure were provided by Servier Medical art. Servier Medical Art by Servier is licensed under a Creative Commons Attribution 4.0 Unported License.

5. Materials and Methods

5.1. *BaV*

Lyophilized *B. arietans* venom (BaV) was purchased from Venom Supplies, Tanunda, Australia. The venoms were obtained from male and female snakes of different ages, captured in South Africa and maintained in captivity. Stock solutions were prepared in sterile phosphate-buffered saline (PBS, 8.1 mM Na_2HPO_4; 1.5 mM KH_2PO_4; 137 mM NaCl; 2.7 mM KCl, pH 7.2) and stored at −80 °C.

5.2. *Kn-Ba*

Kn-Ba was purified from BaV as previously described [25]. Purified Kn-Ba was diluted in sterile PBS pH 7.2 and stored at −80 °C.

5.3. *Total Protein Quantification*

Total protein concentration of BaV and Kn-Ba were measured by the bicinchoninic acid method [65] using the Pierce BCA Protein Assay kit (Pierce Biotechnology, Waltham, IL, USA), according to manufacture instructions. A standard curve was prepared using increased concentrations (0–2000 μg/mL) of pure bovine serum albumin (BSA, Sigma Aldrich, St. Louis, MO, USA) diluted in sterile PBS pH 7.4. Absorbances were obtained in the plate spectrophotometer (ELX 800, Biotek Instruments, Winooski, VT, USA) at λ540 nm.

5.4. *Endotoxin Contents Determination*

Samples of BaV (20 μg/mL) and Kn-Ba (1 μg/mL) were diluted in sterile PBS, pH 7.2, and the endotoxin presence was analyzed by the Microbiological Quality Control of Butantan Foundation using the PYROGENT TM Gel clot LAL Assays kit (Lonza, Walkersville, MD, USA), according to manufacture instructions. Endotoxin concentration was estimated by comparing with an *Escherichia coli* LPS standard (0.125 UE/mL)

5.5. *Human THP-1 Pre-Monocyte Culture*

The non-adherent human pre-monocyte of the THP-1 cell lineage was acquired from the Rio de Janeiro Cell Bank (Rio de Janeiro, RJ, Brazil, cat.: BCRJ: 0234) and cultured according to the manufacture instructions. The suspension cells were cultured in 75 cm^3 culture flasks (Corning Inc., New York, NY, USA) with RPMI-1640 medium (Gibco, Invitrogen Corp., Waltham, MA, USA) supplemented with 23 mM $NaHCO_3$, 13 mM $C_6H_{12}O_6$, 10 mM Hepes, 2 mM L-glutamine, 1 mM sodium pyruvate, 10% FBS (Fetal Bovine Serum, Cultilab, São Paulo, SP, Brazil), 100 U/mL of penicillin, and 100 μg/mL of streptomycin (Gibco, Invitrogen Corp., Waltham, MA, USA). The cell was maintained at 37 °C in an atmosphere containing 5% CO_2 at a density of 2–8 × 10^5 cells/mL, not exceeding the concentration of 1 × 10^6 cells/mL. Total cell number and cell viability were periodically monitored in a Neubauer chamber by Trypan blue exclusion (Trypan blue 0.4% in PBS pH 7.2; 1:1; *v*/*v*).

5.6. *Human THP-1 Pre-Monocyte Differentiation into THP-1 Macrophages*

Human pre-monocytes were differentiated into macrophages using Phorbol 12-Myristate 13-Acetate (PMA, Sigma Aldrich, St. Louis, MO, USA), according to protocol described by Daigneault and colleagues (2010) [33]. Briefly, live cells were transferred to 24 culture wells plates containing 1 mL/well of supplemented RPMI medium containing PMA (100 ng/mL) at a density of 2 × 10^5 cells/well. The cell was incubated at 37 °C in an atmosphere containing 5% CO_2 during 72 h. After the fourth day, PMA-containing medium was replaced by medium without PMA (2 mL/well), and the cells were kept at rest for 4 more days. At the end of the rest period, macrophage adhesion was visualized under a phase contrast microscope (Leica DM2500, Wetzlar, Germany) and photographed.

5.7. CD11b Expression in Differentiated THP-1 Macrophages

The CD11b expression by differentiated THP-1 macrophages was evaluated by flow cytometry, according to the literature [32], with some modifications, as described below. THP-1 monocytes were cultured in 75 cm^3 culture flasks until reaching a density of 8×10^5 cells/mL, according to what is described in 5.5. At this time, the culture medium was replaced with medium containing 100 ng/mL of PMA, and differentiation was conducted as described in 5.6. The adherent macrophages were carefully detached with cell scraper and centrifuged (260× *g*) for 10 min at 10 °C. Contents of viable cells were determined by Trypan blue exclusion, and the cell pellet was resuspended in FACS buffer (PBS with 1% BSA and 0.01% sodium azide) at the density of 20×10^6 cells/mL. Cells were transferred to 96 round-well-bottom microplates (50 µL/well) and were incubated protected from light for 30 min at 4 °C with anti-CD11b PE-labelled antibody (IgG1κ PE, Clone VIM 12, BD, Becton Dickinson, San Jose, CA, USA) or with isotypic control (IgG1κ PE, BD) diluted at 1:5 (*v*/*v*) in FACS buffer. After incubation, the cell suspensions were centrifuged (305× *g*) for 5 min at 4 °C, washed three times with FACS buffer and resuspended in 300 µL/well of FACS buffer with 1% paraformaldehyde. The data acquisition was performed in flow cytometer (FACS Aria III, Becton Dickinson, San Jose, CA, USA). In parallel, the CD11b expression was also evaluated in THP-1 monocytes as control.

5.8. Incubation of THP-1 Macrophages with BaV and Kn-Ba

THP-1 differentiated in 24-well plates, according to what is described in 5.4, were treated with BaV (0.5 µg, 1 µg, 10 µg, and 15 µg/well) or with Kn-Ba (0.5 µg and 1 µg/well) diluted in 500 µL/well of supplemented RPMI medium without FBS and incubated at 37 °C in an atmosphere containing 5% CO_2 during 24 h, 48 h, and 72 h [66]. The supernatants were collected, centrifuged (405× *g*) for 5 min at 4 °C for removal of cell debris, aliquoted, and stored at −80 °C.

5.9. Release of Lactate Dehydrogenase (LDH)

The cytotoxic effects after BaV and Kn-Ba treatments were evaluated by the quantification of lactate dehydrogenase (LDH) enzyme in cell-free supernatants using the kit CytoTox 96® Non-Radioactive Cytotoxicity Assay (Promega, Madison, WI, USA) [67], according to the manufacturer's instructions. Cell-free supernatants of THP-1 macrophages treated during 45 min with lysis buffer were used as a positive control of LDH release, and cells treated only with culture medium were used as LDH release background.

5.10. Quantification of Cytokines and Chemokines Produced by THP-1 Macrophages

The cytokines and chemokines produced by THP-1 macrophages in response to the treatments with BaV and Kn-Ba were evaluated in cell-free supernatants. For quantification of inflammatory cytokines IL-1β, IL-6, IL-10, IL-12, and TNF-α, the CBA Human Inflammatory Cytokines kit was used (BD, Bioscience, Franklin Lakes, NJ, USA), and for quantification of the chemokines CXCL8/IL-8, CCL5/RANTES, CXCL9/MIG, CCL2/MCP-1, and CXCL10/IP-10, the CBA Human Chemokines kit was used (BD, Bioscience, Franklin Lakes, NJ, USA), according to the manufacturer's instructions. The data acquisition was performed in a flow cytometer (FACS Aria III, Becton Dickinson, San Jose, CA, USA) and analyzed by the software FACS Diva, version 6.1.3.

5.11. Quantification of Lipid Inflammatory Mediators Produced by THP-1 Macrophages

The concentration of the lipid mediators LTB_4 and PGE_2 in cell-free supernatants was quantified by competitive ELISA using specific kits (EIA, Cayman Chemical, Ann Arbor, MI, USA), according to the manufacturer's instructions. The absorbances were determined in a spectrophotometer at λ 405/420 nm (VersaMax, Molecular Devices, San Jose, CA, USA). The eicosanoids concentrations were calculated by the interpolation on the kits' standard curve.

5.12. Statistical Analysis

Data are presented as means ± standard error (SEM) and analyzed by Graph Pad Prism, version 6.0 for Windows (Graph Pad Software, San Diego, CA, USA). One-way ANOVA test and multiple comparisons by Tukey HSD were used for comparisons of one variable in more than two groups. For comparisons of two or more variables, two-way ANOVA followed by Tukey HSD were used. For all tests, the values $p < 0.05$ were considered significant. The assays were conducted in duplicate and repeated at least twice in independent days.

Supplementary Materials: The following are available online at https://www.mdpi.com/article/10.3390/toxins13120906/s1, Figure S1. Differentiation of THP-1 macrophages. (A) Human pre-monocytes of the THP-1 lineage were cultured in 75 cm^3 bottles containing supplemented RPMI medium. The cells were kept in an incubator at 37 °C and 5% CO_2, at a concentration of 2–8 × 10^5 cells/mL. (B) Macrophages were differentiated from THP-1 pre-monocytes with 100 ng/mL PMA for 3 days, followed by a 4-day rest period in the absence of PMA. 200× magnification. (C) CD11b expression in differentiated THP-1 macrophages. Adherent THP-1 macrophages differentiated with PMA and pre-monocytes in suspension were analyzed for CD11b expression by flow cytometry. Result expressed as mean of duplicates ± SEM and analyzed by two-way ANOVA followed by Tukey's post-test. (*) Significant difference in relation to the respective controls ($p < 0.0001$). Figure S2. Cell cytotoxicity. Differentiated THP-1 macrophages was treated with 0.5–15 µg/well of BaV or with 0.5 and 1 µg/well Kn-Ba for 24 h, 48 h, and 72 h, and cell viability was evaluated by the quantification of lactate dehydrogenase (LDH) enzyme in cell-free supernatants. Results expressed as mean of duplicates ± SEM and analyzed by one-way ANOVA followed by Tukey's post-test. (*) Significant difference in relation to the respective control. $p < 0.01$ (**), and $p < 0.0001$ (****).

Author Contributions: Conceptualization, Â.A.A.M. and W.D.-d.-S.; methodology, Â.A.A.M., F.C.V.P. and W.D.-d.-S.); validation, Â.A.A.M., F.C.V.P. and W.D.-d.-S.; formal analysis, Â.A.A.M. and W.D.-d.-S.; investigation, Â.A.A.M. and F.C.M.; resources, W.D.-d.-S. and F.C.V.P.; data curation, Â.A.A.M., F.C.M., F.C.V.P. and W.D.-d.-S.; writing—original draft preparation, Â.A.A.M., F.C.V.P. and W.D.-d.-S.; writing—review and editing, Â.A.A.M., F.R.G., K.S.G., F.C.V.P. and W.D.-d.-S.; supervision, W.D.-d.-S. and F.C.V.P.; project administration, W.D.-d.-S.; funding acquisition, W.D.-d.-S. and F.C.V.P. All authors have read and agreed to the published version of the manuscript.

Funding: This research was funded by Coordination for the Improvement of Higher-Level Education Personnel—CAPES (Â.A.A.M.: fellowship Ph.D. PROEX; W.D.S.: 23038.000814/2011–83); São Paulo Research Foundation (FAPESP 2013/07467-1 and Fapesp 2019/20832-7) and the National Council for Scientific and Technological Development (*PROÁFRICA*, CNPq: 490048/2005-6).

Institutional Review Board Statement: Not applicable.

Informed Consent Statement: Not applicable.

Acknowledgments: The authors thank Denise Tambourgi for the Immunochemistry Laboratory structure and to Martin Wesley for English reviewing in the article.

Conflicts of Interest: The authors declare no conflict of interest.

References

1. Kasturiratne, A.; Wickremasinghe, A.R.; de Silva, N.; Gunawardena, N.K.; Pathmeswaran, A.; Premaratna, R.; Savioli, L.; Lalloo, D.G.; de Silva, H.J. The global burden of snakebite: A literature analysis and modelling based on regional estimates of envenoming and deaths. *PLoS Med.* **2008**, *5*, 1591–1604. [CrossRef]
2. Chippaux, J.P. Estimate of the burden of snakebites in sub-Saharan Africa: A meta-analytic approach. *Toxicon* **2011**, *57*, 586–599. [CrossRef] [PubMed]
3. Swaroop, S.; Grab, B. Snakebite mortality in the world. *Bull. World Health Organ.* **1954**, *10*, 35–76. [PubMed]
4. WHO. *Guidelines for the Production, Control and Regulation of Snake Antivenom Immunoglobulins*; World Health Organization: Geneva, Switzerland, 2016; pp. 1–146.
5. Shupe, S. *Venomous Snakes of the World: A Manual for Use by US Amphibious Forces*; Skyhorse Publishing Inc.: New York, NY, USA, 2013.
6. Shupe, S. *Generic and Species Descriptions*; Skyhorse Publishing: New York, NY, USA, 2013.

7. WHO. *Guidelines for the Prevention and Clinical Management of Snakebite in Africa*; World Health Organization: Brazzaville, Congo, 2010; pp. 1–145.
8. Currier, R.B.; Harrison, R.A.; Rowley, P.D.; Laing, G.D.; Wagstaff, S.C. Intra-specific variation in venom of the African Puff Adder (*Bitis arietans*): Differential expression and activity of snake venom metalloproteinases (SVMPs). *Toxicon* **2010**, *55*, 864–873. [CrossRef] [PubMed]
9. Warrell, D.A.; Ormerod, L.D.; Davidson, N.M. Bites by puff-adder (*Bitis arietans*) in Nigeria, and value of antivenom. *Br. Med. J.* **1975**, *4*, 697–700. [CrossRef]
10. Bey, T.A.; Boyer, L.V.; Walter, F.G.; McNally, J.; Desai, H. Exotic snakebite: Envenomation by an African puff adder (*Bitis arietans*). *J. Emerg. Med.* **1997**, *15*, 827–831. [CrossRef]
11. Langhorn, R.; Persson, F.; Åblad, B.; Goddard, A.; Schoeman, J.P.; Willesen, J.L.; Tarnow, I.; Kjelgaard-Hansen, M. Myocardial injury in dogs with snake envenomation and its relation to systemic inflammation. *J. Vet. Emerg. Crit. Care* **2014**, *24*, 174–181. [CrossRef]
12. Lavonas, E.J.; Tomaszewski, C.A.; Ford, M.D.; Rouse, A.M.; Kerns, W.P., 2nd. Severe puff adder (*Bitis arietans*) envenomation with coagulopathy. *J. Toxicology. Clin. Toxicol.* **2002**, *40*, 911–918. [CrossRef]
13. Theakston, R.; Wyatt, G. Venom antibody levels in a patient bitten by a young puff adder (*Bitis arietans)* during a world record attempt. *Ann. Trop. Med. Parasitol.* **1985**, *79*, 305–307. [CrossRef]
14. Markland, F.S. Snake venoms and the hemostatic system. *Toxicon* **1998**, *36*, 1749–1800. [CrossRef]
15. Juarez, P.; Wagstaff, S.C.; Oliver, J.; Sanz, L.; Harrison, R.A.; Calvete, J.J. Molecular cloning of disintegrin-like transcript BA-5A from a *Bitis arietans* venom gland cDNA library: A putative intermediate in the evolution of the long-chain disintegrin bitistatin. *J. Mol. Evol.* **2006**, *63*, 142–152. [CrossRef]
16. Fasoli, E.; Sanz, L.; Wagstaff, S.; Harrison, R.A.; Righetti, P.G.; Calvete, J.J. Exploring the venom proteome of the African puff adder, *Bitis arietans*, using a combinatorial peptide ligand library approach at different pHs. *J. Proteom.* **2010**, *73*, 932–942. [CrossRef]
17. Archer, J.; Whiteley, G.; Casewell, N.R.; Harrison, R.A.; Wagstaff, S.C. VTBuilder: A tool for the assembly of multi isoform transcriptomes. *BMC Bioinform.* **2014**, *15*, 1–11. [CrossRef]
18. Kodama, R.T.; Cajado-Carvalho, D.; Kuniyoshi, A.K.; Kitano, E.S.; Tashima, A.K.; Barna, B.F.; Takakura, A.C.; Serrano, S.M.; Dias-Da-Silva, W.; Tambourgi, D.V.; et al. New proline-rich oligopeptides from the venom of African adders: Insights into the hypotensive effect of the venoms. *Biochim. Biophys. Acta* **2015**, *1850*, 1180–1187. [CrossRef] [PubMed]
19. Petretski, J.; Kanashiro, M.; Silva, C.; Alves, E.; Kipnis, T. Two related thrombin-like enzymes present in *Bothrops atrox* venom. *Braz. J. Med Biol. Res. Rev. Bras. Pesqui. Med. E Biol.* **2000**, *33*, 1293–1300. [CrossRef]
20. Kanashiro, M.M.; Rita de Cássia, M.E.; Petretski, J.H.; Prates, M.V.; Alves, E.W.; Machado, O.L.; da Silva, W.D.; Kipnis, T.L. Biochemical and biological properties of phospholipases A2 from *Bothrops atrox* snake venom. *Biochem. Pharmacol.* **2002**, *64*, 1179–1186. [CrossRef]
21. Stocker, K.; Barlow, G. The coagulant enzyme from *Bothrops atrox* venom (batroxobin). *Methods Enzymol.* **1976**, *45*, 214–223.
22. Clissa, P.B.; Laing, G.D.; Theakston, R.D.G.; Mota, I.; Taylor, M.J.; Moura-da-Silva, A.M. The effect of jararhagin, a metalloproteinase from *Bothrops jararaca* venom, on pro-inflammatory cytokines released by murine peritoneal adherent cells. *Toxicon* **2001**, *39*, 1567–1573. [CrossRef]
23. Costa, E.; Clissa, P.; Teixeira, C.; Moura-da-Silva, A. Importance of metalloproteinases and macrophages in viper snake envenomation–induced local inflammation. *Inflammation* **2002**, *26*, 13–17. [CrossRef]
24. Serrano, S.M.; Sampaio, C.A.; Mentele, R.; Camargo, A.C.; Fink, E. A novel fibrinogen-clotting enzyme, TL-BJ, from the venom of the snake *Bothrops jararaca*: Purification and characterization. *Thromb. Haemost.* **2000**, *83*, 438–444.
25. Megale, Â.A.A.; Magnoli, F.C.; Kuniyoshi, A.K.; Iwai, L.K.; Tambourgi, D.V.; Portaro, F.C.; da Silva, W.D. Kn-Ba: A novel serine protease isolated from *Bitis arietans* snake venom with fibrinogenolytic and kinin-releasing activities. *J. Venom. Anim. Toxins* **2018**, *24*, 1–11. [CrossRef]
26. Teixeira, C.F.; Cury, Y.; Moreira, V.; Picolo, G.; Chaves, F. Inflammation induced by *Bothrops asper* venom. *Toxicon* **2009**, *54*, 67–76. [CrossRef]
27. Teixeira, C.F.; Fernandes, C.M.; Zuliani, J.P.; Zamuner, S.F. Inflammatory effects of snake venom metalloproteinases. *Mem. Do Inst. Oswaldo Cruz* **2005**, *100*, 181–184. [CrossRef]
28. Voronov, E.; Apte, R.; Sofer, S. The systemic inflammatory response syndrome related to the release of cytokines following severe envenomation. *J. Venom. Anim. Toxins* **1999**, *5*, 5–33. [CrossRef]
29. Szold, O.; Ben-Abraham, R.; Frolkis, I.; Sorkine, M.; Sorkine, P. Tumor necrosis factor as a mediator of cardiac toxicity following snake envenomation. *Crit. Care Med.* **2003**, *31*, 1449–1453. [CrossRef]
30. Megale, Â.A.A.; Portaro, F.C.; Da Silva, W.D. *Bitis arietans* snake venom induces an inflammatory response which is partially dependent on lipid mediators. *Toxins* **2020**, *12*, 594. [CrossRef]
31. Schwende, H.; Fitzke, E.; Ambs, P.; Dieter, P. Differences in the state of differentiation of THP-1 cells induced by phorbol ester and 1, 25-dihydroxyvitamin D3. *J. Leukoc. Biol.* **1996**, *59*, 555–561. [CrossRef]
32. Mittar, D.; Paramban, R.; McIntyre, C. Flow cytometry and high-content imaging to identify markers of monocyte-macrophage differentiation. *J. BD Biosci.* **2011**, *1*, 1–20.
33. Daigneault, M.; Preston, J.A.; Marriott, H.M.; Whyte, M.K.; Dockrell, D.H. The identification of markers of macrophage differentiation in PMA-stimulated THP-1 cells and monocyte-derived macrophages. *PLoS ONE* **2010**, *5*, 1–10. [CrossRef]

34. Zoccal, K.F. A Peçonha do Escorpião *Tityus serrulatus* é Reconhecida por Receptores de Reconhecimento Padrão e Induz Ativação Celular e Inflamação. Ph.D. Thesis, Universidade de São Paulo, São Paulo, Brazil, 2014.
35. Gutiérrez, J.M.; Chaves, F.; Cerdas, L. Inflammatory infiltrate in skeletal muscle injected with *Bothrops asper* venom. *Rev. Biol. Trop.* **1986**, *34*, 209–214.
36. Zamuner, S.R.; Zuliani, J.P.; Fernandes, C.M.; Gutiérrez, J.M.; Teixeira, C.d.F.P. Inflammation induced by *Bothrops asper* venom: Release of proinflammatory cytokines and eicosanoids, and role of adhesion molecules in leukocyte infiltration. *Toxicon* **2005**, *46*, 806–813. [CrossRef] [PubMed]
37. Barros, S.; Friedlanskaia, I.; Petricevich, V.; Kipnis, T. Local inflammation, lethality and cytokine release in mice injected with *Bothrops atrox* venom. *Mediat. Inflamm.* **1998**, *7*, 339–346. [CrossRef] [PubMed]
38. Moreira, V.; Dos-Santos, M.C.; Nascimento, N.G.; da Silva, H.B.; Fernandes, C.M.; Lima, M.R.D.I.; Teixeira, C. Local inflammatory events induced by *Bothrops atrox* snake venom and the release of distinct classes of inflammatory mediators. *Toxicon* **2012**, *60*, 12–20. [CrossRef]
39. Farsky, S.H.; Walber, J.; Costa-Cruz, M.; Curry, Y.; Teixeira, C.F. Leukocyte response induced by *Bothrops jararaca* crude venom: In vivo and in vitro studies. *Toxicon* **1997**, *35*, 185–193. [CrossRef]
40. Zamuner, S.R.; Gutiérrez, J.M.; Muscará, M.N.; Teixeira, S.A.; Teixeira, C.F. *Bothrops asper* and *Bothrops jararaca* snake venoms trigger microbicidal functions of peritoneal leukocytes in vivo. *Toxicon* **2001**, *39*, 1505–1513. [CrossRef]
41. Murray, P.J.; Wynn, T.A.J. Protective and pathogenic functions of macrophage subsets. *Nat. Rev. Immunol.* **2011**, *11*, 723–737. [CrossRef]
42. Laing, G.D.; Clissa, P.B.; Theakston, R.D.G.; Moura-da-Silva, A.M.; Taylor, M.J. Inflammatory pathogenesis of snake venom metalloproteinase-induced skin necrosis. *Eur. J. Immunol.* **2003**, *33*, 3458–3463. [CrossRef]
43. Lentschat, A.; Karahashi, H.; Michelsen, K.S.; Thomas, L.S.; Zhang, W.; Vogel, S.N.; Arditi, M. Mastoparan, a G protein agonist peptide, differentially modulates TLR4- and TLR2-mediated signaling in human endothelial cells and murine macrophages. *J. Immunol.* **2005**, *174*, 4252–4261. [CrossRef]
44. Liu, X.; Zhan, Z.; Li, D.; Xu, L.; Ma, F.; Zhang, P.; Yao, H.; Cao, X. Intracellular MHC class II molecules promote TLR-triggered innate immune responses by maintaining activation of the kinase Btk. *Nat. Immunol.* **2011**, *12*, 416–424. [CrossRef] [PubMed]
45. Medzhitov, R. Recognition of microorganisms and activation of the immune response. *Nature* **2007**, *449*, 819–826. [CrossRef]
46. Zoccal, K.F.; Bitencourt Cda, S.; Paula-Silva, F.W.; Sorgi, C.A.; de Castro Figueiredo Bordon, K.; Arantes, E.C.; Faccioli, L.H. TLR2, TLR4 and CD14 recognize venom-associated molecular patterns from *Tityus serrulatus* to induce macrophage-derived inflammatory mediators. *PLoS ONE* **2014**, *9*, 1–12. [CrossRef]
47. Takashiba, S.; Van Dyke, T.E.; Amar, S.; Murayama, Y.; Soskolne, A.W.; Shapira, L. Differentiation of monocytes to macrophages primes cells for lipopolysaccharide stimulation via accumulation of cytoplasmic nuclear factor κB. *J. Infect. Immun.* **1999**, *67*, 5573–5578. [CrossRef]
48. Kohro, T.; Tanaka, T.; Murakami, T.; Wada, Y.; Aburatani, H.; Hamakubo, T.; Kodama, T. A comparison of differences in the gene expression profiles of phorbol 12-myristate 13-acetate differentiated THP-1 cells and human monocyte-derived macrophage. *J. Atheroscler. Trhromb.* **2004**, *11*, 88–97. [CrossRef]
49. Aderem, A.; Underhill, D.M. Mechanisms of phagocytosis in macrophages. *Annu. Rev. Immunol.* **1999**, *17*, 593–623. [CrossRef]
50. Mamede, C.C.N.; de Sousa, B.B.; da Cunha Pereira, D.F.; Matias, M.S.; de Queiroz, M.R.; de Morais, N.C.G.; Vieira, S.A.P.B.; Stanziola, L.; de Oliveira, F. Comparative analysis of local effects caused by *Bothrops alternatus* and *Bothrops moojeni* snake venoms: Enzymatic contributions and inflammatory modulations. *Toxicon* **2016**, *117*, 37–45. [CrossRef]
51. Saravia-Otten, P.; Frisan, T.; Thelestam, M.; Gutierrez, J.M. Membrane independent activation of fibroblast proMMP-2 by snake venom: Novel roles for venom proteinases. *Toxicon* **2004**, *44*, 749–764. [CrossRef]
52. Costa, C.; Belchor, M.; Rodrigues, C.; Toyama, D.; de Oliveira, M.; Novaes, D.; Toyama, M. Edema induced by a *Crotalus durissus terrificus* venom serine protease (Cdtsp 2) involves the PAR pathway and PKC and PLC activation. *Int. J. Mol. Sci.* **2018**, *19*, 2405. [CrossRef]
53. Matsui, T.; Fujimura, Y.; Titani, K. Snake venom proteases affecting hemostasis and thrombosis. *Biochim. Biophys. Acta* **2000**, *1477*, 146–156. [CrossRef]
54. Menaldo, D.L.; Bernardes, C.P.; Pereira, J.C.; Silveira, D.S.; Mamede, C.C.; Stanziola, L.; de Oliveira, F.; Pereira-Crott, L.S.; Faccioli, L.H.; Sampaio, S.V. Effects of two serine proteases from *Bothrops pirajai* snake venom on the complement system and the inflammatory response. *Int. Immunopharmacol.* **2013**, *15*, 764–771. [CrossRef]
55. Lawrence, T. The nuclear factor NF-kappaB pathway in inflammation. *Cold Spring Harb. Perspect. Biol.* **2009**, *1*, 1–10. [CrossRef]
56. Arias-Salvatierra, D.; Silbergeld, E.K.; Acosta-Saavedra, L.C.; Calderon-Aranda, E.S. Role of nitric oxide produced by iNOS through NF-κB pathway in migration of cerebellar granule neurons induced by lipopolysaccharide. *Cell. Signal.* **2011**, *23*, 425–435. [CrossRef]
57. Eder, C. Mechanisms of interleukin-1β release. *J. Immunobiol.* **2009**, *214*, 543–553. [CrossRef]
58. Martín-Sánchez, F.; Diamond, C.; Zeitler, M.; Gomez, A.; Baroja-Mazo, A.; Bagnall, J.; Spiller, D.; White, M.; Daniels, M.; Mortellaro, A. Inflammasome-dependent IL-1 β release depends upon membrane permeabilisation. *J. Cell Death Differ.* **2016**, *23*, 1219–1231. [CrossRef]
59. Martinon, F.; Tschopp, J. Inflammatory caspases: Linking an intracellular innate immune system to autoinflammatory diseases. *Cell* **2004**, *117*, 561–574. [CrossRef]

60. Meylan, E.; Tschopp, J.; Karin, M. Intracellular pattern recognition receptors in the host response. *Nature* **2006**, *442*, 39–44. [CrossRef]
61. Zheng, H.; Fletcher, D.; Kozak, W.; Jiang, M.; Hofmann, K.J.; Corn, C.A.; Soszynski, D.; Grabiec, C.; Trumbauer, M.E.; Shaw, A. Resistance to fever induction and impaired acute-phase response in interleukin-1β-deficient mice. *Immunity* **1995**, *3*, 9–19. [CrossRef]
62. Rider, P.; Carmi, Y.; Guttman, O.; Braiman, A.; Cohen, I.; Voronov, E.; White, M.R.; Dinarello, C.A.; Apte, R.N. IL-1α and IL-1β recruit different myeloid cells and promote different stages of sterile inflammation. *J. Immunol.* **2011**, *187*, 4835–4843. [CrossRef]
63. Dinarello, C.A. Interleukin-1 in the pathogenesis and treatment of inflammatory diseases. *Blood* **2011**, *117*, 3720–3732. [CrossRef]
64. Broderick, L.; De Nardo, D.; Franklin, B.S.; Hoffman, H.M.; Latz, E. The inflammasomes and autoinflammatory syndromes. *Annu. Rev. Pathol.* **2015**, *10*, 395–424. [CrossRef] [PubMed]
65. Smith, P.K.; Krohn, R.I.; Hermanson, G.; Mallia, A.; Gartner, F.; Provenzano, M.; Fujimoto, E.; Goeke, N.; Olson, B.; Klenk, D. Measurement of protein using bicinchoninic acid. *Anal. Biochem.* **1985**, *150*, 76–85. [CrossRef]
66. Delafontaine, M.; Villas-Boas, I.; Mathieu, L.; Josset, P.; Blomet, J.; Tambourgi, D. Enzymatic and pro-inflammatory activities of *Bothrops lanceolatus* venom: Relevance for envenomation. *Toxins* **2017**, *9*, 244. [CrossRef] [PubMed]
67. Rucavado, A.; Flores-Sánchez, E.; Franceschi, A.; Magalhaes, A.; Gutiérrez, J.M. Characterization of the local tissue damage induced by LHF-II, a metalloproteinase with weak hemorrhagic activity isolated from *Lachesis muta muta* snake venom. *Toxicon* **1999**, *37*, 1297–1312. [CrossRef]

Article

Oral Tolerance Induction by *Bothrops jararaca* Venom in a Murine Model and Cross-Reactivity with Toxins of Other Snake Venoms

Lilian Rumi Tsuruta [1,*], Ana Maria Moro [1], Denise V. Tambourgi [2] and Osvaldo Augusto Sant'Anna [2]

1 Biopharmaceuticals Laboratory, Butantan Institute, São Paulo 05503-900, Brazil; ana.moro@butantan.gov.br
2 Immunochemistry Laboratory, Butantan Institute, São Paulo 05503-900, Brazil; denise.tambourgi@butantan.gov.br (D.V.T.); osvaldo.santanna@butantan.gov.br (O.A.S.)
* Correspondence: lilian.tsuruta@butantan.gov.br

Abstract: Oral tolerance is defined as a specific suppression of cellular and humoral immune responses to a particular antigen through prior oral administration of an antigen. It has unique immunological importance since it is a natural and continuous event driven by external antigens. It is characterized by low levels of IgG in the serum of animals after immunization with the antigen. There is no report of induction of oral tolerance to *Bothrops jararaca* venom. Here, we induced oral tolerance to *B. jararaca* venom in BALB/c mice and evaluated the specific tolerance and cross-reactivity with the toxins of other *Bothrops* species after immunization with the snake venoms adsorbed to/encapsulated in nanostructured SBA-15 silica. Animals that received a high dose of *B. jararaca* venom (1.8 mg) orally responded by showing antibody titers similar to those of immunized animals. On the other hand, mice tolerized orally with three doses of 1 μg of *B. jararaca* venom showed low antibody titers. In animals that received a low dose of *B. jararaca* venom and were immunized with *B. atrox* or *B. jararacussu* venom, tolerance was null or only partial. Immunoblot analysis against the venom of different *Bothrops* species provided details about the main tolerogenic epitopes and clearly showed a difference compared to antiserum of immunized animals.

Keywords: oral tolerance; *Bothrops jararaca*; snake venom; ELISA

Key Contribution: We describe the induction of oral tolerance by administration of *Bothrops jararaca* venom through an original approach, showing the main tolerogenic epitopes in *B. jararaca* venom and other species of *Bothrops*.

Citation: Tsuruta, L.R.; Moro, A.M.; Tambourgi, D.V.; Sant'Anna, O.A. Oral Tolerance Induction by *Bothrops jararaca* Venom in a Murine Model and Cross-Reactivity with Toxins of Other Snake Venoms. *Toxins* **2021**, *13*, 865. https://doi.org/10.3390/toxins13120865

Received: 15 October 2021
Accepted: 30 November 2021
Published: 3 December 2021

Publisher's Note: MDPI stays neutral with regard to jurisdictional claims in published maps and institutional affiliations.

1. Introduction

Snake venoms are composed of a high diversity of proteins and peptides with biological activities, allowing these animals to defend themselves and immobilize their prey [1]. The composition of snake venoms among species displays high variability, both in qualitative and quantitative aspects and complexity [2]. Accidents with snakebite envenoming cause local and systemic effects and represent a public health problem in developing countries, where they reach lower socio-economic segments and kill >100,000 people each year [3]. The primary treatment for the systemic effects of snake envenoming is the intravenous administration of antivenom against specific venoms. Antivenoms specifically neutralize the venoms used in their production and those of related species, which means that antivenoms are produced regionally depending on demand [3]. Indeed, there is a crisis related to the supply of antivenoms, especially in sub-Saharan Africa and parts of Asia; the development of new treatments for patients with snakebite envenoming should be promoted on the basis of recent scientific knowledge related to snake venoms [3]. Recently, several studies have reported that antivenom serum antibodies, generated against specific snake venoms, are cross-reactive with venoms from other species, considering homologous and heterologous snake venoms [4–7].

Most snakebites in Brazil occur because of the genus *Bothrops* and are considered a serious public health problem. *Bothrops* venom components mainly cause local damage and systemic effects targeting blood hemostasis, endothelial microcirculation, extracellular matrix, and the cardiovascular system [1,8].

Oral tolerance is the induction of peripheral immune tolerance by the oral administration of the antigen and is characterized by the inhibition of the specific immune response to this antigen due to prior oral exposure [9–14]. It is a natural and continuous process driven by external antigens. It has a unique immunological importance, as it develops unresponsiveness to ingested food and potential insults from the environment to maintain host homeostasis by protecting against food allergies and colitis caused by autoimmunity [12,13,15]. The gut is regularly exposed to multiple types of antigens, and the associated immune system has specialized immune cells and lymph nodes to balance responses to commensal bacteria (microbiome), innocuous antigens, and harmful microorganisms [11]. Depending on the properties of the antigen, such as size and solubility, the orally administrated antigen that reaches the intestinal epithelium is transported by different routes and can lead to the induction of tolerance or immunity [14]. The oral tolerance induction mechanism has been extensively studied using animal models, mainly for food allergens [11]. It involves multiple factors, and it is known that the dose of the administered antigen and the consumption time are decisive. Administration of a single high dose of antigen leads to the mechanisms of anergy or depletion, whereas exposure to multiple low doses favors the development of regulatory T cells [11,16]. Anergy induction means obtaining antigen-unresponsive T cells, while depletion induction refers to apoptosis of antigen-specific T cells [14]. Previous studies have shown that genetic and environmental factors are involved in the induction of oral tolerance, demonstrating that this characteristic is a process under the influence of multiple factors [17–19].

Oral tolerance induction by the administration of one kind of antigen/allergen has been extensively investigated, as has been the mechanism of this process involving immune cells and pathways [11,13,20]. This process has not been explored by the administration of a complex mixture of proteins. Snake envenomation by the oral route does not occur in nature; instead, snakes inject their venoms when there is a dangerous situation and/or they need to defend themselves. Oral antigen application of this kind to induce oral tolerance represents a novel experimental approach. To our knowledge, there is no report of oral tolerance induction using *B. jararaca* venom as an antigen. We propose a method for inducing oral tolerance to *B. jararaca* venom in mice, followed by evaluation of serum cross-reactivity with the toxins from other *Bothrops* species compared to the serum generated by the immunization process.

2. Results

2.1. Induction of Oral Tolerance by B. jararaca Venom Is Dependent on Dose Administed

Oral tolerance is characterized by low IgG levels in the antiserum of animals after immunization with the antigen previously administered orally. Initially, the induction of oral tolerance by administering *B. jararaca* venom was tested using two conditions and the results are summarized in Figure 1. Group I did not receive venom orally. Briefly, one group of mice orally received a single dose of 1.8 mg of the venom (Group II) and other groups received 1 μg of snake venom on the first, third and fifth day (Groups III and IV). At 12 days after the last low dose of 1 μg, all groups were immunized with snake venom combined with silica adjuvant. Groups I, II and III received venom from *B. jararaca*, while Group IV was challenged with venom from *B. atrox*. Antisera corresponding to 7 days after immunization were analyzed, and no difference in antibody titers between all groups of mice was observed (data not shown). Antisera were then collected on the 35th day after immunization and titers determined (Figure 1), followed by analysis based on ANOVA. According to our results, animals that received high doses of *B. jararaca* venom orally were not tolerized (Group II), with antibody titers being similar to those of mice immunized only (Group I). No statistically significant differences were found between Groups I and II and Groups I and IV. Interestingly, mice given a low dose of venom orally and immunized with the same antigen (Group III) had the lowest antibody titer showing the induction of oral tolerance with a statistically significant difference compared to other groups. On the other hand, animals that received a low oral dose of *B. jararaca* and were immunized with *B. atrox* venom (Group IV) showed no tolerance induction. Our results suggest that oral tolerance to snake venom appears to be dose-dependent and is specific for the venom of each snake species.

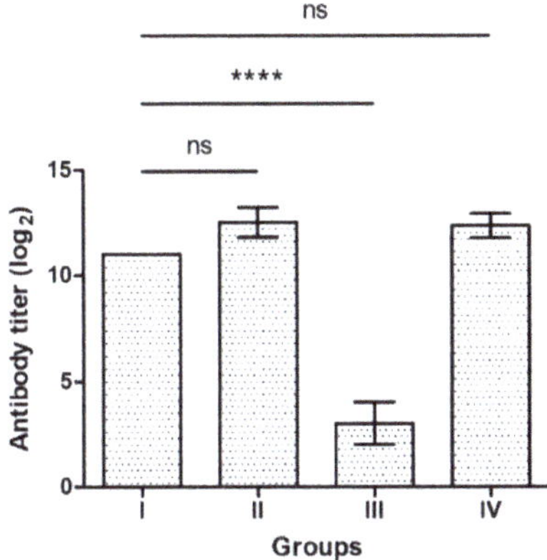

Group I: mice immunized with 1 μg of *B. jararaca* venom
Group II: mice orally tolerized with 1.8 mg of *B. jararaca* venom and immunized with 1 μg of *B. jararaca* venom
Group III: mice orally tolerized with 3 doses of 1 μg of *B. jararaca venom* and immunized with 1 μg of *B. jararaca* venom
Group IV: mice orally tolerized with 3 doses of 1 μg of *B. jararaca* and immunized with 1 μg of *B. atrox venom*

Figure 1. Induction of oral tolerance by administration of *B. jararaca* venom evaluated by antibody titer of mouse antiserum after immunization. Group I was not subjected to oral tolerance. The mice were orally tolerized with a single high dose of venom (Group II) or with low doses of venom on the first, third and fifth day (Group III and IV). Twelve days after receiving the last oral dose, Groups I, II and III were immunized with *B. jararaca* venom, and group IV received *B. atrox* venom. Serum antibody titers corresponding to 35 days after immunization were determined by ELISA. Error bars represent the standard deviation of an experiment ($n = 3$). ANOVA with 95% confidence intervals was used to determine significant differences between each group of mice. ns: not significant; ****: p-value was <0.0001.

2.2. Characterization of Antisera from Orally Tolerized Animals Reveals Specificity for Other Snake Venoms

To characterize the specificity and cross-reactivity of the antiserum after induction of oral tolerance with *B. jararaca* venom, another experiment was carried out with induction of oral tolerance by administering three doses of 1 μg of the *B. jararaca* venom (Groups I and III). Eight days after the last dose, all groups of mice were immunized intraperitoneally with snake venom combined to adjuvant. Groups I and II were immunized with *B. jararaca*, while Groups III and IV received *B. jararacussu* venom. Antiserum was collected 12, 25 and 45 days after immunization to analyze the specificity of response to *B. jararaca* and *B. jararacussu* venoms. The specificity of antiserum for the venom of *B. jararaca*, using this venom as the antigen is shown in Figure 2, while Figure 3 shows the specificity for the venom of *B. jararacussu*. Our results showed that Group I, orally tolerized and immunized with *B. jararaca* venom, was tolerized for snake venom in view of the low IgG level in all analyzed antisera (Figures 2A and 3A). Group III was tolerized orally and immunized with *B. jararacussu* venom and the antibody titer increased on the 45th day after venom injection when specificity for *B. jararaca* venom was evaluated (Figure 2C), while the same phenomenon was observed on day 25 for the specificity of *B. jararacussu* venom (Figure 3C). Antiserum from groups of mice immunized with snake venom only (Groups II and IV) had high antibody titers for both snake venoms 25 days after immunization

(Figures 2B,D and 3B,D). Furthermore, antisera from mice immunized with *B. jararacussu* venom (Group IV) showed reactivity with *B. jararaca* venom (Figure 2B) and antisera from mice immunized with *B. jararaca* venom (Group II) showed reactivity with *B. jararacussu* venom (Figure 3B). In the case of specificity for *B. jararacussu* venom, mice immunized with *B. jararaca* venom (Group II) showed a reduction in antibody titer 45 days after immunization (Figure 3B).

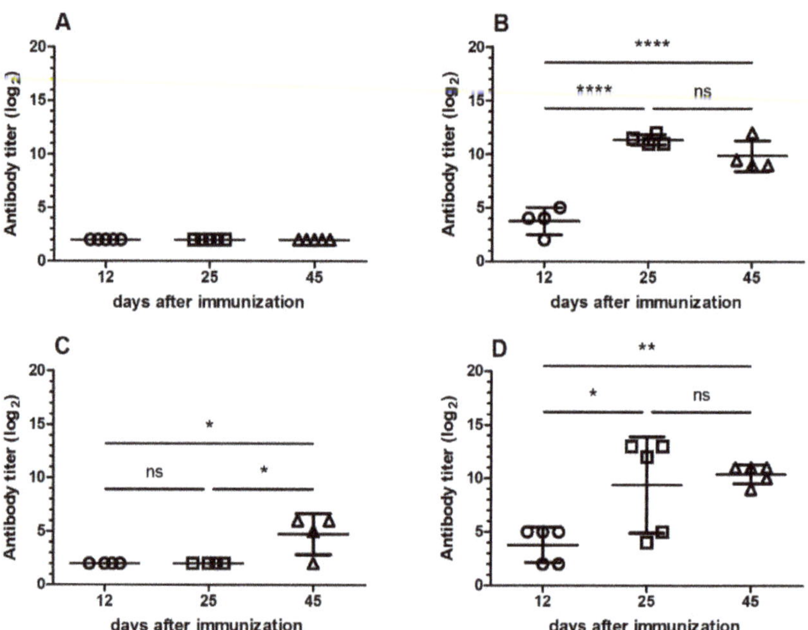

Figure 2. Antiserum specificity for *B. jararaca* venom from mice subjected or not to oral tolerance followed by immunization with snake venom. Microplate wells were coated with 1 µg of *B. jararaca* venom, and serial dilutions of antiserum of Groups I–IV corresponding to 12, 25 and 45 days after snake venom immunization were applied; antibody titer was determined by ELISA. (**A**) Group I: mice tolerized with *B. jararaca* venom and immunized with the same snake venom; (**B**) Group II: mice immunized with *B. jararaca* venom; (**C**) Group III: mice tolerized with *B. jararaca* venom and immunized with *B. jararacussu* venom; (**D**) Group IV: mice immunized with *B. jararacussu* venom. The mean and standard deviation ($n = 5$) of each antibody titer obtained were plotted. ANOVA with 95% confidence intervals was used to determine significant differences between each group of mice group. ns: not significant; *: p-value was 0.01–0.05; **: p-value was 0.001–0.01; ****: p-value was <0.0001.

IgG1 and IgG2a titers were determined for antisera obtained 45 days after immunization for Groups I, II, III and IV (Figure 4). Orally tolerized mice had lower antibody titers than immunized mice and no difference between IgG1 and IgG2a titers was observed. IgG1 antibody titers from mice immunized with snake venoms (*B. jararaca* or *B. jararacussu* venom) were higher than IgG2a titers. IgE was not found in any of the four groups (data not shown).

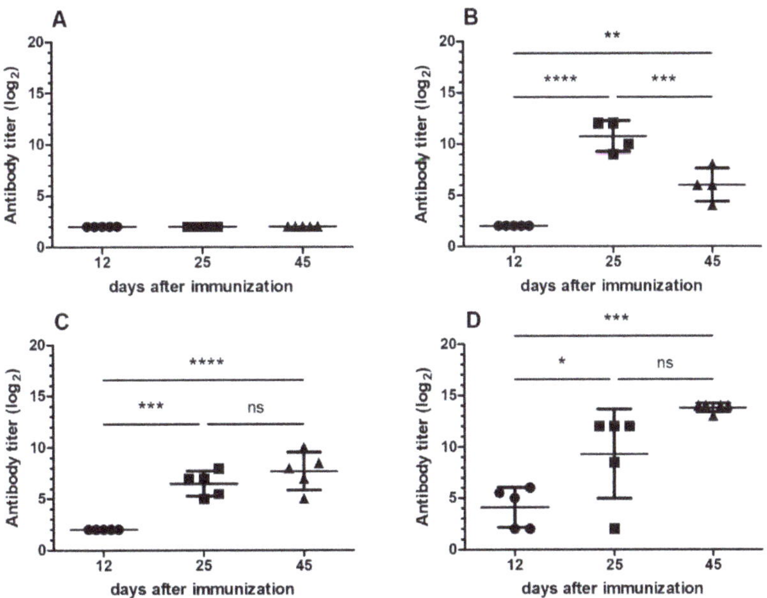

Figure 3. Antiserum specificity for *B. jararacussu* venom from mice subjected or not to oral tolerance followed by immunization with snake venom. Microplate wells were coated with 1 μg of *B. jararacussu* venom, and serial dilutions of antiserum of Groups I–IV corresponding to 12, 25 and 45 days after snake venom immunization were applied; antibody titer was determined by ELISA. (**A**) Group I: mice tolerized with *B. jararaca* venom and immunized with the same snake venom; (**B**) Group II: mice immunized with *B. jararaca* venom; (**C**) Group III: mice tolerized with *B. jararaca* venom and immunized with *B. jararacussu* venom; (**D**) Group IV: mice immunized with *B. jararacussu* venom. The mean and standard deviation (n = 5) of each antibody titer obtained were plotted. ANOVA with 95% confidence intervals was used to determine significant differences between each mice group. ns: not significant; *: p-value was 0.01–0.05; **: p-value was 0.001–0.01; ***: p-value was 0.0001–0.001; ****: p-value was <0.0001.

Next, we evaluated the reactivity of antisera from mice subjected to oral tolerance with venoms of some *Bothrops* species and *Bitis anetans* by Western blotting. Protein profiles of the snake venoms under non-reducing conditions are shown in Figure 5; venoms differed in composition and band intensity.

Snake venoms subjected to electrophoresis were transferred to PVDF membranes, the membrane was incubated with a pool of antisera corresponding to 45 days after the immunization of Groups I, II, III and IV and revealed by chemiluminescent reagent (Figure 6). Antisera from orally tolerized animals showed lower reactivity with snake venom proteins compared to antisera obtained from mice immunized with snake venom. Antisera of mice tolerized with *B. jararaca* venom and immunized with the same snake venom (Group I) recognized some proteins from the venom of *B. jararaca*, and a protein band from the venom of *B. alternatus*, *B. jararacussu* and *B. atrox amazonia* (Figure 6A). On the other hand, the antisera from mice immunized with the venom of *B. jararaca* recognized venom proteins of the *Bothrops* species analyzed (*Bothrops jararaca*, *B. alternatus*, *B. jararacussu* and *B. atrox amazonia*), and a wide range of reactivity was found for the venoms of *B. jararaca* and *B. alternatus* (Figure 6B). Interestingly, some protein bands recognized by both tolerized and non-tolerized mouse antisera (Groups I and II) were very similar (Figure 6A,B). When antisera from mice orally tolerized or not and immunized with *B. jararacussu* venom (Groups III and IV) were evaluated, at least one protein of the snake venom was recognized by antisera from Group III (Figure 6C), and antisera from immunized mice (Group IV) showed higher reactivity with *B. jararacussu* and *B. jararaca* venoms and low reactivity with other snake venoms (Figure 6D). The reactivity of Groups III and IV antisera was tested for *Bitis anetans* venom, and antisera from tolerized or not mice showed reactivity with this venom. Each antiserum recognized a distinct protein (Figure 6C,D). Our results showed that antisera from mice tolerized orally with *B. jararaca* venom and immunized with snake venoms had reactivity with venoms from other species at different levels. Interestingly, antisera from orally tolerized mice demonstrated cross-reactivity with different venom epitopes in comparison to the antisera of immunized animals.

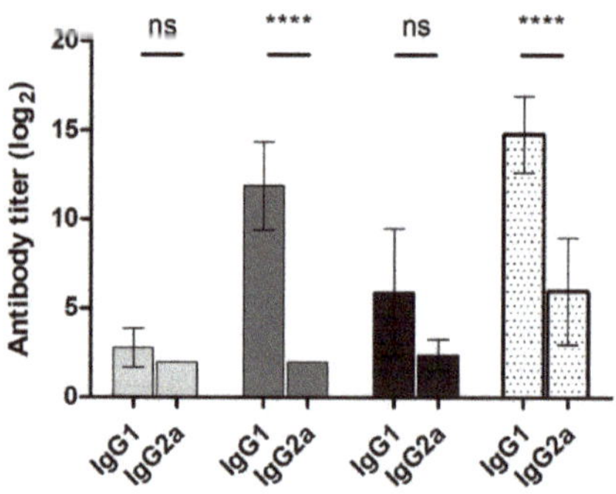

Figure 4. IgG1 and IgG2a antibody titers of groups of mice subjected to oral tolerance induction with *B. jararaca* venom followed by immunization. Microplate wells were coated with 1 μg of *B. jararaca* venom, and serial dilutions of Groups I-IV antiserum, as previously described in "Section 4", corresponding to 45 days after immunization with snake venom, were applied; antibody titer was determined by ELISA using mouse anti-IgG1 or anti-IgG2a antibodies conjugated to HRP. Error bars represent the standard deviation of one experiment (n = 5). 2-way ANOVA with 95% confidence intervals was used to determine significant difference between IgG1 and IgG2a antibody titers for each group. ns: not signigficant; ****: p-value was < 0.0001.

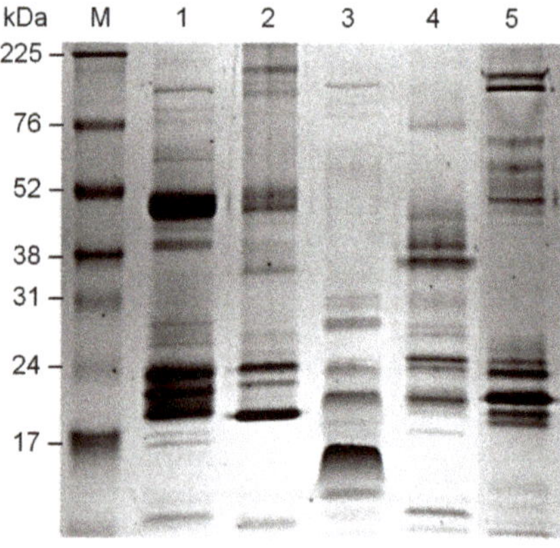

Figure 5. Electrophoretic profiles of snake venoms under non-reducing conditions. 12% SDS-PAGE of 5 μg of each snake venom. Coomassie blue staining. M: Rainbow marker high range (Cytiva); snake venom from 1: *B. jararaca*; 2: *B. alternatus*; 3: *B. jararacussu*; 4: *B. atrox amazonia*; 5: *Bitis anetans*.

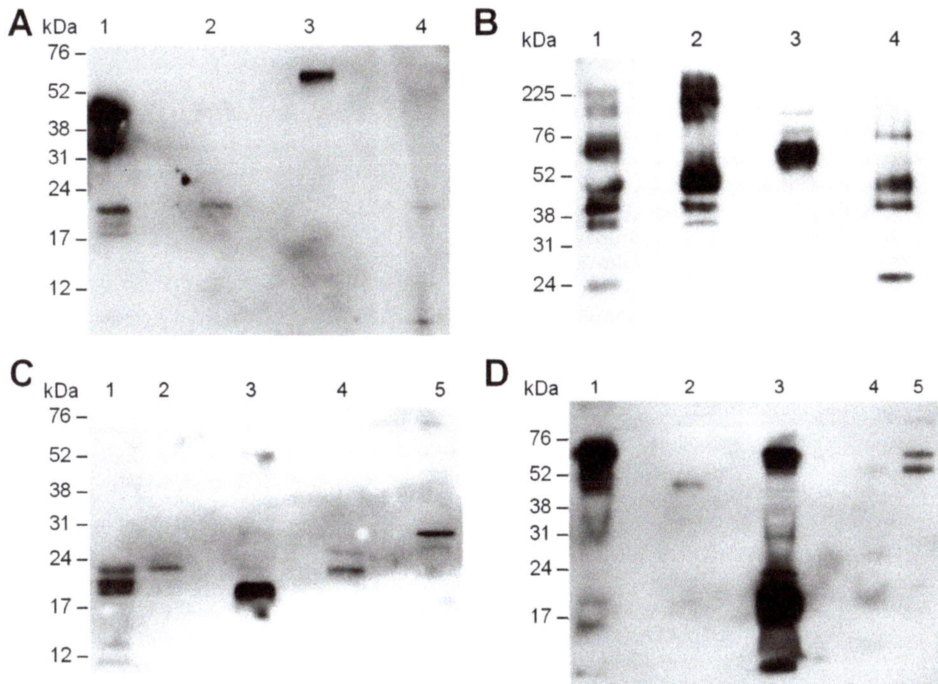

Figure 6. Analysis of the reactivity of the pool of antisera from animals subjected or not to oral tolerance induction with *B. jararaca* venom followed by immunization with snake venom. Snake venoms were subjected to electrophoresis and transferred to PVDF membrane according to Figure 5. The membrane was incubated with a pool of antisera corresponding to 45 days after immunization. (**A**): 1/400 dilution of Group I antisera; (**B**): 1/1000 dilution of Group II antisera; (**C**): 1/400 dilution of Group III antisera; (**D**) 1/1500 dilution of Group IV antisera. Peroxidase-conjugated anti-mouse IgG was then added, and the membrane was revealed with ECL reagent.

3. Discussion

Oral tolerance is an immunological process in which the specific immune response is inhibited by prior oral administration of antigen. The induction of this process can be assessed after antigen immunization and measured by determination of antibody titer or by the decrease in allergic symptoms after allergen challenge [14].

The ability of the immune system to adapt to dietary antigens and commensal bacteria is important and prevents the development of food allergies, celiac disease, and autoimmune diseases [20]. Most studies related to oral tolerance have been carried out in animal models to establish the safe dose and duration of the process and to understand the immune cells and pathways involved, because of the risk of testing in humans [20]. These studies are of crucial importance for understanding the mechanisms involved and can promote strategies for the development of natural oral tolerance and prevent intoxication, allergies, and autoimmune diseases.

There is a report of daily oral administration for 14 days of *Crotalus durissus* terrificus snake venom (200 µg/kg) in male Swiss mice (17–21 g) that induced tolerance to the antinociceptive effect, and no antibody titers were measurable after prolonged treatment [21]. The expected effect was obtained by administering a dose of venom higher than that in our study, and mice were not immunized after receiving the oral dose. The cited study had a different purpose in relation to the present work and a different approach was taken.

Most incidences of snakebites in Brazil, considering all regions, are related to *Bothrops* species [22] and, therefore, constituted the objective of this study. They are also extensively investigated because they are responsible for more fatality cases in Central and South America than other groups of snakes [1]. So far, oral tolerance by administering *B. jararaca* venom in animal models has not been published and we have successfully established a protocol in BALB/c mice. The antibody titers were measured in the antisera after immunization with snake venoms to understand the mechanism involved.

We observed that mice receiving a single dose of 1.8 mg of *B. jararaca* venom orally did not develop tolerance and that the antibody titer was similar to that of the group of mice that was only immunized (Figure 1). These results are in agreement with those found in the literature for other antigens [11,16]. We have shown that repeated exposure to a dose of 1 μg of *B. jararaca* venom induced tolerance (Figure 1). We also found that tolerance was more effectively induced when animals received the same snake venom during oral administration and immunization (Figures 1–3), showing that it is specific for one type of antigen, even when the antiserum showed cross-reactivity with venom from other snakes.

There was no reference concerning an immunization protocol with snake venom after oral administration, and we established that immunization would be performed about 7 days after the last dose. In the first protocol, mice were immunized 12 days after the last dose, while in the second protocol it was 8 days, and under both conditions, the oral tolerance induction was verified by low antibody titer. We think that the immunization protocol could be applied in this period of time to induce oral tolerance with *B. jararaca* venom.

In the first protocol of oral tolerance induction, mice received a low dose of *B. jararaca* venom and were immunized with *B. atrox* venom, and no tolerance induction was observed (Figure 1). On the basis of this result, we excluded this condition in the next protocol of oral tolerance induction and challenged the immunization with the venom of another *Bothrops* species (*B. jararacussu*). Partial oral tolerance induction was observed (Figures 2 and 3) with *B. jararacussu* venom immunization, showing that this tolerance can be induced with another *Bothrops* venom. Variations in venom composition and biological activities of Brazilian snakes from *Bothrops* genus were observed [23] and these differences could be influenced on oral tolerance induction when immunization with heterologous venom was applied. Further studies involving different conditions for establishing oral tolerance induction with *B. jararaca* venom and other *Bothrops* species could elucidate these observations.

We also performed immunoblot analysis of the venom of different species of *Bothrops* incubating with antisera from animals tolerized with administration of *B. jararaca* venom or only immunized with snake venom. The epitopes recognized by each antiserum were clearly different (Figure 6), showing that the reactivity profile of antisera in relation to venom components changed according to the protocol used to induce tolerance. To see if proteins in snake venom from species other than *Bothrops* are recognized by antisera from mice orally tolerized or not, reactivity with *Bitis anetans* venom was evaluated. Antisera from mice partially tolerized (Group III) and from animals immunized with *B. jararacussu* venom (Group IV) were tested and each antiserum recognized different proteins in *Bitis anetans* venom, showing that reactivity with other snake venoms could be explored in future studies.

The application of oral tolerance has advantages because it is non invasive and uses a simple route. New knowledge and in-depth understanding of this process could contribute to the application of oral tolerance in prophylaxis and treatment of diseases [20].

Most of the studies related to oral tolerance induction involve the administration of one antigen [11,13,20]. The present work is an experimental study of oral tolerance induction developed with *B. jararaca* venom that is composed of a complex of proteins with biological activity. This process does not happen in the nature, and thus, the observation of oral tolerance induction represented a novel experimental approach. To demonstrate oral tolerance induction in our approach, the immune response was evaluated by the determination of the antibody titer, and then, the epitopes recognized by antiserum from orally tolerized or not tolerized mice were compared. Our results showed that it is possible to induce oral tolerance by administration of the complex mixture of the proteins, such as *B. jararaca* venom. Investigations related to the modulation of immune response and the mechanisms involved in this process were not part of the scope of the present work. The administration of a single high dose of *B. jararaca* venom was not lethal to the animals but did not induce oral tolerance. Further experimental studies should be conducted with other snake venoms to confirm this phenomenon as this approach would be used to develop vaccines by oral administration. The protocol established for oral tolerance induction should also be applied to other snake venoms in future studies together with the investigation of the mechanisms involved in this process. Therefore, the present work opened new perspectives to explore the process of oral tolerance induction.

4. Materials and Methods

4.1. Animals

BALB/c mice, female, aged 2–4 months, were used and maintained at the animal facilities of the Immunochemistry Laboratory, Butantan Institute, and they were caged and handled according to the International Animal Welfare Recommendations and in line with the guidelines for the use of animals in biomedical research [24]. Ethics Committee on Animal Use of the Butantan Institute approved the experiment protocol (Protocol IBU 454/08; 9 April 2008).

4.2. Snake Venoms

Lyophilized venoms (*Bothrops jararaca, B. alternatus, B. jararacussu, B. atrox amazonia* and *Bitis anetans*) were obtained from the Laboratory of Herpetology, Butantan Institute, São Paulo, Brazil, and stored at −20 °C. Venoms were resuspended in phosphate-buffered saline (PBS).

4.3. Protein Quantification

Protein concentration of the snake venom samples was determined by a microtiter-based Bradford Protein Assay (BioRad) in microplates using bovine serum albumin (BSA—Sigma–Aldrich, St. Louis, MO, USA) for the standard curve [25].

4.4. Induction of Oral Tolerance and Immunization

To establish the oral tolerance induction protocol, BALB/c mice were subdivided into 4 groups of animals (n = 3). Group I did not receive venom orally. Oral tolerance was induced by gavage administration of *B. jararaca* venom, and the mice received a single high dose of 1.8 mg (Group II) or 3 doses of 1 μg on the first, third and fifth day (Groups III and IV). At 12 days after the last low dose (Groups III and IV), all groups were intraperitoneally immunized with 1 μg of snake venom adsorbed to/encapsulated in nanostructured SBA-15 silica [26]. Groups I, II and III received *B. jararaca* venom, while Group IV was immunized with *B. atrox* venom. Antiserum corresponding to the 7th and 35th day after immunization was collected by bleeding from the retroorbital plexus and the antibody titer was determined by ELISA. Animals that did not undergo any procedure (n = 3) were considered the control group, and sera were added to the antibody titer assay.

Another oral tolerance induction protocol was then carried out. BALB/c mice were subdivided into 4 groups of animals (n = 5). Oral tolerance was induced by gavage administration of 1 μg of the *B. jararaca* venom on the first, third and fifth day (Groups I and III). Groups II and IV did not receive venom orally. Eight days after the last dose, mice were intraperitoneally immunized with 1 μg of snake venom adsorbed to/encapsulated in nanostructured SBA-15 silica. Groups I and II were immunized with *B. jararaca*, while Groups III and IV received *B. jararacussu* venom. Antiserum was collected by bleeding from the retroorbital plexus at 12, 25 and 45 days after immunization, and the antibody titers were determined by ELISA. Sera from mice of control group were also added. The cross-reactivity of serum after induction of oral tolerance with *B. jararaca* venom was analyzed by Western blotting.

4.5. ELISA Procedure

IgG titer of mouse sera was determined by ELISA. *B. jararaca* or *B. jararacussu* venom was diluted to 10 µg/mL in carbonate buffer (50 mM $Na_2CO_3/NaHCO_3$, pH 9.6), and 100 µL/well were added to a 96-well EIA/RIA Clear Flat Bottom Polystyrene High Bind Microplate (3590, Corning, NY, USA). The plate was incubated at 4 °C overnight, washed 3 times with PBS/0.05% Tween and then incubated with the PBS/5% BSA blocking solution at 37 °C for 2 h. The blocking solution was replaced with a serial twofold dilution of antiserum in PBS/1% BSA, starting at an appropriate dilution for each one. Sera from animals that did not receive any venom was used as negative control, and samples were incubated at 37 °C for 1 h. Wells were washed 5 times with PBS/0.05% Tween to remove unbound antibodies. HRP-conjugated anti-mouse IgG (H + L) (Promega, Madison, WI, USA) was used as secondary antibody at 1/2500 dilution, and incubation was at 37 °C for 1 h. The plate was washed again, and the substrate used was OPD (o-phenylenediamine) (Sigma–Aldrich) in phosphate-citrate buffer with H_2O_2. After 5 min incubation, the reaction was stopped by adding 0.2 M citric acid. The absorbance at 450 nm was measured with a microplate reader. Antibody titers were expressed as log2 maximum serum dilution giving a positive reaction at which the absorbance was equal to two times the control value of the control serum.

Immunoglobulin isotypes, IgG1 and IgG2a, were evaluated in mouse antisera. For this purpose, HRP rat anti-mouse IgG1 (BD Biosciences, San Jose, CA, USA) at 1/400 dilution and HRP rat anti-mouse IgG2a (BD Biosciences) at 1/1000 dilution were used as secondary antibody. Antibody titer was determined as described above.

4.6. Electrophoresis and Western Blotting

Venom samples of 5 µg were prepared by adding non-reducing sample buffer and were separated by electrophoresis on 12% SDS-PAGE gel. Amersham ECL High-Range Rainbow Molecular Weight Markers (Cytiva, Marlborough, MA, USA) was also included. The gel was stained with Coomassie Blue or transferred to Amersham Hybond-P PVDF Membrane (Cytiva) for 1 h at 10 V using a BioRad Trans-blot SD semi-dry transfer cell. The membranes were stained with 0.5% Ponceau S, washed with ultrapure water and then blocked with PBS/5% BSA at 37 °C for 2 h. Next, the membranes were washed twice with PBS/0.1% Tween 20 for 20 s and were incubated overnight with an appropriate dilution of pooled antisera (Groups I, II, III and IV corresponding to 45 days after immunization) at 4 °C. The membranes were washed 4 times for 5 min and incubated with anti-mouse IgG (whole molecule)-HRP (Sigma) at 1/30,000 dilution in PBS/1% BSA for 1 h at room temperature. After washing, the blots were developed using Amersham ECL Prime Western Blotting Detection Reagent (Cytiva).

4.7. Statistical Analysis

Statistical analysis was performed by GraphPad Prism software. ANOVA (Tukey's multiple comparison test) was used to compare average antibody titer obtained between antisera of animal groups, orally tolerized or not with *B. jararaca* venom. The sera obtained at different times after immunization were submitted to analyses. Two-way ANOVA with 95% confidence intervals was used to determine significant differences between IgG1 and IgG2a antibody titers for each mouse group.

Author Contributions: Conceptualization, O.A.S.; methodology, L.R.T. and O.A.S.; validation, L.R.T. and O.A.S.; formal analysis, L.R.T., D.V.T. and O.A.S.; investigation, L.R.T. and O.A.S.; resources, A.M.M., D.V.T. and O.A.S.; data curation, L.R.T.; writing—original draft preparation, L.R.T.; writing—review and editing, L.R.T., A.M.M. and D.V.T.; visualization, L.R.T., A.M.M., D.V.T. and O.A.S.; supervision, O.A.S.; project administration, O.A.S.; funding acquisition, A.M.M., D.V.T. and O.A.S. All authors have read and agreed to the published version of the manuscript.

Funding: This research was funded by Brazilian National Council for Scientific and Technological Development (CNPq), grant number 474435/2007-5, by INCTTOX (Instituto Nacional de Ciência e Tecnologia em Toxinas) from the São Paulo Research Foundation (FAPESP) and CNPq, and by FAPESP (grant number 2017/17844-8). A.M.M. is grateful for CNPq (process 307405/2020-0).

Institutional Review Board Statement: The methodology of this work was approved by Ethic Committee on Animal Use of the Butantan Institute (Protocol IBU 454/08, 9 April 2008).

Informed Consent Statement: Not applicable.

Data Availability Statement: The data presented in this study are available on request from corresponding author.

Acknowledgments: We thank Stephanie Murari do Nascimento for technical support.

Conflicts of Interest: The authors declare no conflict of interest.

References

1. Tashima, A.K.; Zelanis, A.; Kitano, E.S.; Ianzer, D.; Melo, R.L.; Rioli, V.; Sant'anna, S.S.; Schenberg, A.C.; Camargo, A.C.; Serrano, S.M. Peptidomics of three Bothrops snake venoms: Insights into the molecular diversification of proteomes and peptidomes. *Mol. Cell Proteom.* **2012**, *11*, 1245–1262. [CrossRef]
2. Tasoulis, T.; Isbister, G.K. A Review and Database of Snake Venom Proteomes. *Toxins* **2017**, *9*, 290. [CrossRef]
3. Gutiérrez, J.M.; Calvete, J.J.; Habib, A.G.; Harrison, R.A.; Williams, D.; Warrell, D.A. Snakebite envenoming. *Nat. Rev. Dis. Primers* **2017**, *3*, 17063. [CrossRef]
4. Lipps, B.V.; Khan, A.A. Antigenic cross reactivity among the venoms and toxins from unrelated diverse sources. *Toxicon* **2000**, *38*, 973–980. [CrossRef]
5. Prieto da Silva, A.R.; Yamagushi, I.K.; Morais, J.F.; Higashi, H.G.; Raw, I.; Ho, P.L.; Oliveira, J.S. Cross reactivity of different specific Micrurus antivenom sera with homologous and heterologous snake venoms. *Toxicon* **2001**, *39*, 949–953. [CrossRef]
6. Stábeli, R.G.; Magalhães, L.M.; Selistre-de-Araujo, H.S.; Oliveira, E.B. Antibodies to a fragment of the *Bothrops moojenil*-amino acid oxidase cross-react with snake venom components unrelated to the parent protein. *Toxicon* **2005**, *46*, 308–317. [CrossRef] [PubMed]
7. Rocha, M.M.; Paixão-Cavalcante, D.; Tambourgi, D.V.; Furtado, M.F. Duvernoy's gland secretion of *Philodryas olfersii* and *Philodryas patagoniensis* (Colubridae): Neutralization of local and systemic effects by commercial bothropic antivenom (Bothrops genus). *Toxicon* **2006**, *47*, 95–103. [CrossRef] [PubMed]
8. Mamede, C.C.N.; Simamoto, B.B.S.; Pereira, D.F.C.; Costa, J.O.; Ribeiro, M.S.M.; de Oliveira, F. Edema, hyperalgesia and myonecrosis induced by Brazilian bothropic venoms: Overview of the last decade. *Toxicon* **2020**, *187*, 10–18. [CrossRef] [PubMed]
9. Faria, A.M.; Ficker, S.M.; Speziali, E.; Menezes, J.S.; Stransky, B.; Verdolin, B.A.; Lahmann, W.M.; Rodrigues, V.S.; Vaz, N.M. Aging and immunoglobulin isotype patterns in oral tolerance. *Braz. J. Med. Biol. Res.* **1998**, *31*, 35–48. [CrossRef]
10. Sun, J.B.; Czerkinsky, C.; Holmgren, J. Mucosally induced immunological tolerance, regulatory T cells and the adjuvant effect by cholera toxin B subunit. *Scand. J. Immunol.* **2010**, *71*, 1–11. [CrossRef] [PubMed]
11. Commins, S.P. Mechanisms of Oral Tolerance. *Pediatr. Clin. N. Am.* **2015**, *62*, 1523–1529. [CrossRef]
12. Anagnostou, K.; Clark, A. What do we mean by oral tolerance? *Clin. Exp. Allergy* **2016**, *46*, 782–784. [CrossRef]
13. Sricharunrat, T.; Pumirat, P.; Leaungwutiwong, P. Oral tolerance: Recent advances on mechanisms and potential applications. *Asian Pac. J. Allergy Immunol.* **2018**, *36*, 207–216. [CrossRef]
14. Tordesillas, L.; Berin, M.C. Mechanisms of Oral Tolerance. *Clin. Rev. Allergy Immunol.* **2018**, *55*, 107–117. [CrossRef]
15. Smith, K.M.; Eaton, A.D.; Finlayson, L.M.; Garside, P. Oral tolerance. *Am. J. Respir. Crit. Care Med.* **2000**, *162*, S175–S178. [CrossRef] [PubMed]
16. Weiner, H.L.; da Cunha, A.P.; Quintana, F.; Wu, H. Oral tolerance. *Immunol. Rev.* **2011**, *241*, 241–259. [CrossRef]
17. Da Silva, A.C.; Massa, S.; Sant'Anna, O.A. Preliminary results corroborating the polygenic control of immunological tolerance. *Braz. J. Med. Biol. Res.* **1990**, *23*, 581–584. [PubMed]
18. Da Silva, A.C.; de Souza, K.W.; Machado, R.C.; da Silva, M.F.; Sant'Anna, O.A. Genetics of immunological tolerance: I. Bidirectional selective breeding of mice for oral tolerance. *Res. Immunol.* **1998**, *149*, 151–161. [CrossRef]
19. Da Silva, M.F.; Nóbrega, A.; Ribeiro, R.C.; Levy, M.S.; Ribeiro, O.G.; Tambourgi, D.V.; Sant'Anna, D.A.; da Silva, A.C. Genetic selection for resistance or susceptibility to oral tolerance imparts correlation to both Immunoglobulin E level and mast cell number phenotypes with a profound impact on the atopic potential of the individual. *Clin. Exp. Allergy* **2006**, *36*, 1399–1407. [CrossRef]
20. Wambre, E.; Jeong, D. Oral Tolerance Development and Maintenance. *Immunol. Allergy Clin. N. Am.* **2018**, *38*, 27–37. [CrossRef] [PubMed]
21. Brigatte, P.; Hoffmann, F.A.; Bernardi, M.M.; Giorgi, R.; Fernandes, I.; Takehara, H.A.; Barros, S.B.; Almeida, M.G.; Cury, Y. Tolerance to the antinociceptive effect of Crotalus durissus terrificus snake venom in mice is mediated by pharmacodynamic mechanisms. *Toxicon* **2001**, *39*, 1399–1410. [CrossRef]

22. Chippaux, J.P. Incidence and mortality due to snakebite in the Americas. *PLoS Negl. Trop. Dis.* **2017**, *11*, e0005662. [CrossRef]
23. Queiroz, G.P.; Pessoa, L.A.; Portaro, F.C.; Furtado Mde, F.; Tambourgi, D.V. Interspecific variation in venom composition and toxicity of Brazilian snakes from Bothrops genus. *Toxicon* **2008**, *52*, 842–851. [CrossRef]
24. Giles, A.R. Guidelines for the use of animals in biomedical research. *Thromb. Haemost.* **1987**, *58*, 1078–1084. [CrossRef] [PubMed]
25. Bradford, M.M. A rapid and sensitive method for the quantitation of microgram quantities of protein utilizing the principle of protein-dye binding. *Anal. Biochem.* **1976**, *72*, 248–254. [CrossRef]
26. Mercuri, L.P.; Carvalho, L.V.; Lima, F.A.; Quayle, C.; Fantini, M.C.; Tanaka, G.S.; Cabrera, W.H.; Furtado, M.F.; Tambourgi, D.V.; Matos, J.d.R.; et al. Ordered mesoporous silica SBA-15: A new effective adjuvant to induce antibody response. *Small* **2006**, *2*, 254–256. [CrossRef] [PubMed]

Article

Recombinant Production and Characterization of a New Toxin from *Cryptops iheringi* Centipede Venom Revealed by Proteome and Transcriptome Analysis

Lhiri Hanna De Lucca Caetano [1], Milton Yutaka Nishiyama-Jr [2], Bianca de Carvalho Lins Fernandes Távora [1], Ursula Castro de Oliveira [2], Inácio de Loiola Meirelles Junqueira-de-Azevedo [2], Eliana L. Faquim-Mauro [1] and Geraldo Santana Magalhães [1,*]

[1] Laboratório de Imunopatologia, Instituto Butantan, São Paulo 05503-900, Brazil; lhiri.hanna@gmail.com (L.H.D.L.C.); bianca.tavora@butantan.gov.br (B.d.C.L.F.T.); eliana.faquim@butantan.gov.br (E.L.F.-M.)

[2] Laboratório de Toxinologia Aplicada, Instituto Butantan, São Paulo 05503-900, Brazil; milton.nishiyama@butantan.gov.br (M.Y.N.-J.); ursula.oliveira@butantan.gov.br (U.C.d.O.); inacio.azevedo@butantan.gov.br (I.d.L.M.J.-d.-A.)

* Correspondence: geraldo.magalhaes@butantan.gov.br

Abstract: Among the Chilopoda class of centipede, the *Cryptops* genus is one of the most associated with envenomation in humans in the metropolitan region of the state of São Paulo. To date, there is no study in the literature about the toxins present in its venom. Thus, in this work, a transcriptomic characterization of the *Cryptops iheringi* venom gland, as well as a proteomic analysis of its venom, were performed to obtain a toxin profile of this species. These methods indicated that 57.9% of the sequences showed to be putative toxins unknown in public databases; among them, we pointed out a novel putative toxin named Cryptoxin-1. The recombinant form of this new toxin was able to promote edema in mice footpads with massive neutrophils infiltration, linking this toxin to envenomation symptoms observed in accidents with humans. Our findings may elucidate the role of this toxin in the venom, as well as the possibility to explore other proteins found in this work.

Keywords: *Cryptops iheringi*; centipede; venom; toxin; transcriptome; proteome; recombinant protein; venomics; chilopoda

Key Contribution: The first transcriptome profile of the venom gland of *Cryptops iheringi* species and the characterization of a new recombinant toxin named Cryptoxin-1, which may play an important role in envenomation.

Citation: De Lucca Caetano, L.H.; Nishiyama-Jr, M.Y.; de Carvalho Lins Fernandes Távora, B.; de Oliveira, U.C.; de Loiola Meirelles Junqueira-de-Azevedo, I.; Faquim-Mauro, E.L.; Magalhães, G.S. Recombinant Production and Characterization of a New Toxin from *Cryptops iheringi* Centipede Venom Revealed by Proteome and Transcriptome Analysis. *Toxins* **2021**, *13*, 858. https://doi.org/10.3390/toxins13120858

Received: 8 October 2021
Accepted: 18 November 2021
Published: 2 December 2021

Publisher's Note: MDPI stays neutral with regard to jurisdictional claims in published maps and institutional affiliations.

1. Introduction

The Centipedes of the Chilopoda class are venomous arthropods that are widely distributed throughout the world [1–3]. The pair of glands, located in each jaw, produce venom that is used to kill or immobilize its prey by inoculation [4–6]. These animals are well adapted to urban areas and are commonly found in backyards and other home areas, and because of this, they often pose a danger to humans by injecting their venom as a defense [2]. The symptoms and complications induced by the envenomation caused by centipedes indicate that its venom comprises a natural set of proteins, peptides, and enzymes with a rich diversity of biological activities [7]. Most of the recent studies of the genus *Scolopendra* have indicated that the venom of a single centipede contains more than 500 proteins [8–10].

Centipedes' venoms have been used for hundreds of years in traditional Chinese medicine, as well as in Korea and other countries in East Asia to treat many disorders such as stroke, hemiplegia, epilepsy, cough, tetanus, burns, cardiovascular diseases, and myocutaneous disease, among others [11,12]. These historical and ethnopharmacological practices indicate that these animals' toxins could be explored for therapeutic uses and drug development. Despite this, the pharmacological properties of the toxins and the accidental envenomation of humans have not been studied extensively.

In Brazil, epidemiological data on accidents with centipedes are also very scarce. However, two retrospective studies that include occurrences recorded at the Vital Brazil Hospital of the Butantan Institute, São Paulo, Brazil, showed that the majority of accidents with centipedes were caused by the *Cryptops* and *Otostigmus* genus, with the first being responsible for more than 60% of the cases reported [2,13]. The envenomation symptoms are characterized by burning pain, paresthesia, edema, and local hemorrhage, and can develop into superficial necrosis [2,13,14]. A systemic reaction, although rare, may occur [15–20].

The toxicology of centipede venom has been understudied in Brazil, and the scarce literature that does exist generally refers to species of the Scolopendridae family, especially the genus *Scolopendra* [21–23]; this is mainly due to the difficulties of obtaining sufficient amounts of venom to conduct biological activities. In this context, the extraction of centipede venom can be time-consuming, and the yields are typically very low, even when it is extracted through electrostimulation [24].

To date, only Malta, et al. (2008) [25] have explored this class of venom in the literature, demonstrating nociception induction, edema, and myotoxicity in mice. However, this study was unable to further characterize the venom due to the difficulty of isolating the venom's toxins. Therefore, the identification of proteins and peptides responsible for the symptoms in human envenomation is highly important for the development of better treatments. In addition, these molecules may have applications in toxinology, immunology, ecology, agriculture, and pharmacy. Thus, the present study, based on the transcriptome and proteome approaches, reports the gene expression profile of the venom gland, identifies novel toxins and characterizes a new toxin that has been named Cryptoxin-1.

2. Results

2.1. Identification of Toxins from Transcriptomic and Proteomic Analysis

In this study, we used a proteotranscriptomic approach to characterize the venom from *C. iheringi*, since no protein or gene sequence was available in public databases. Therefore, the venom was submitted for proteome analysis while the venom gland mRNA was extracted and submitted for transcriptome investigation.

The *C. iheringi*'s venom gland mRNA was extracted and sequenced by Illumina HiSeq 1500 technology (Figures S1 and S2 of the supplementary material). A total of 15,904,398 paired-end reads were generated. The relevant pre-processing quality control, filtering, and trimming steps were applied, resulting in 14,964,551 (94.1%) high-quality reads. The transcriptomic profile of the *C. iheringi* venom gland generated 88,774 assembled transcripts with an average length of 766 bp, a Transcript N50 of 1104 and contained 16,266 (18.3%) transcripts with a length of greater than 1 Kb (Table 1). We evaluated the completeness of the *C. iheringi* transcriptome assembly using BUSCO (Benchmarking Universal Single-Copy Orthologs), searching against the 954 metazoa ortholog groups, and identified 934 (97.8%) of the conserved groups in metazoa; of these, 885 (92.7% of total) were complete, and 49 (5.1%) genes were fragmented.

For the alignment of *C. iheringi* transcriptome assembly against the 106,197 transcripts from 10 species from the Scolopendromorpha orders (Table 2) (*C. anomalans, H. marginata, S. alternans, S. cingulata, S. dehaani, S. morsitans, S. subspinipes, S. virirdis, S. rubiginosus, S. sexspinosus*) we obtained a total of 5328 (6%) *C. iheringi* hits, with the *Cryptops anomalans* having the highest rate of identification, of 4272 (4.83%). The sequence similarity surveys, by BLASTx alignment, resulted in 71.4% of unknown transcripts. Therefore only 28.6% of all transcripts presented at least one protein homolog against the Uniprot and TSA databases.

To further characterize the toxins sequences, the crude venom was analyzed by LC-MS/MS, and then, we performed automatic peptide matching against the predicted proteins from the *C. ihering*'s transcriptome. The sequences identified by this approach were labeled as putative known toxins if they were present in a public database, and if not, they were referred to as putative unknown toxins.

Table 1. Description of Transcriptome sequencing and Assembly of *Cryptops iheringi* and the transcriptome completeness analysis by BUSCO.

Description	Number (%)
Total Raw Paird-end Reads	15,904,398
Total High-quality Paird-end Reads	14,964,551
Total transcripts	88,774
Percent GC Content	42.76%
Transcript N50	1104
Median transcript length	416
Average transcript length	766.27
Longest transcript length	23,855
Number of transcripts >1 kb	16,266 (18.3%)
Shortest transcript length	209
Total assembled bases	68,024,656
Complete BUSCOs	885 (92.7%)
Complete and single-copy BUSCOs	713 (74.7%)
Complete and duplicated BUSCOs	172 (18%)
Fragmented BUSCOs	49 (5.1%)
Missing BUSCOs	20 (2.2%)
Total BUSCO groups searched	954 (100%)

Table 2. The number of transcripts from TSA/NCBI for each species from Scolopendromorpha orders and the number of hits from *C. iheringi* transcriptome assembly against the orders.

Description	Total Transcripts	Number (%)
Total *Cryptops iheringi* Hits	-	5328 (6%)
Cryptops anomalans	33,662	4272
Scolopocryptops rubiginosus	28,965	575
Scolopendra cingulata	23,301	283
Scolopocryptops sexspinosus	1540	117
Scolopendra subspinipes	648	32
Scolopendra viridis	520	29
Hemiscolopendra marginata	764	17
Scolopendra morsitans	662	3
Scolopendra alternans	51	0
Scolopendra dehaani	16,084	0

Furthermore, the predicted proteins from the *C. ihering*'s transcriptome that were not identified in the approach above were labeled as non-toxins, aligned with the Gene Ontology (GO) database, and classified according to their main biological category, in accordance with the GO nomenclature. The remaining sequences that were not identified as a match within the searched databases, and which were not identified through the association between the transcriptome and total venom proteome, were called unknown transcripts and they were no longer explored.

Among the non-toxins transcripts, around 6877.9 transcripts (27%) belong to the Biological Process, 10,953.7 transcripts (43%) to the Cellular Component, and 7642.12 transcripts (30%) to the Molecular Function, the five most representative categories for each GO term were represented as the percentage of transcripts in Figure 1.

The proteomic analysis of the crude venom revealed that 454 predicted proteins of the transcriptome could be classified as unknown venom components or as putative venom toxins, which were further classified into 24 different protein families (Figure 2). In terms of relative expression in TPM (transcript per million), putative unknown toxins and putative known toxins represented 24.97% of the transcriptome.

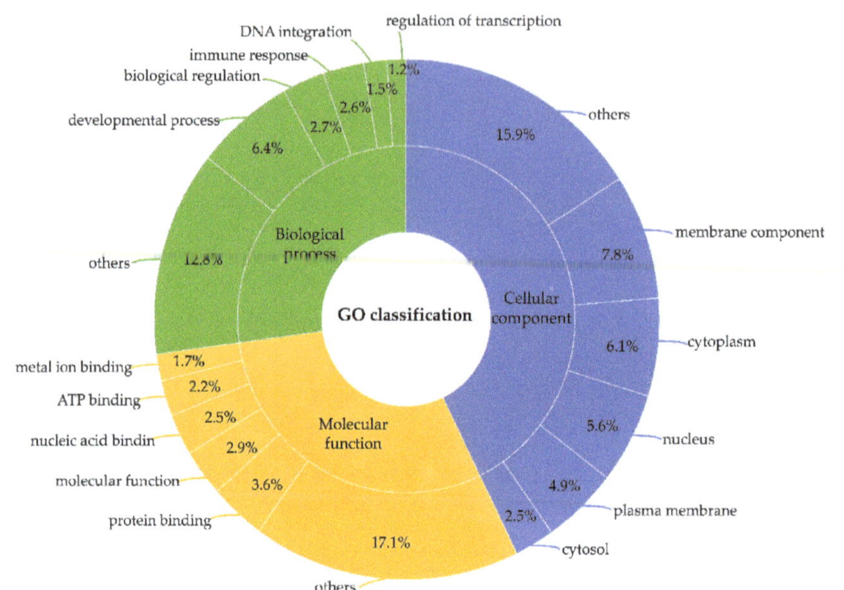

Figure 1. Non-toxins distribution of the five most representative categories of ontologies in the total number of transcripts from the transcriptome analysis of *C. ihering*'s venom gland. Annotation was performed according to the Gene Ontology terms for cellular component, biological process, and molecular function categories.

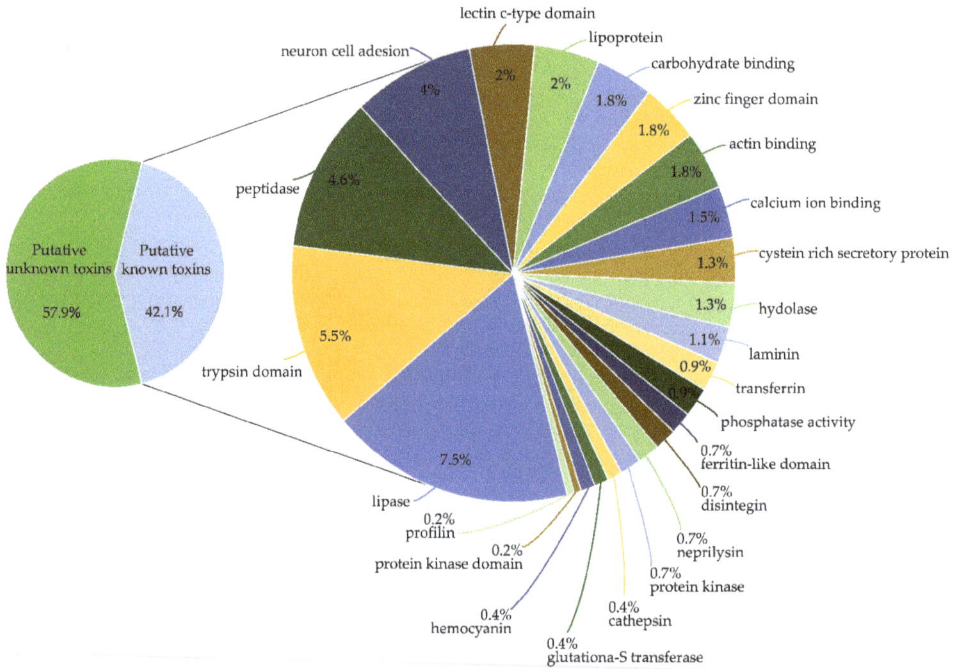

Figure 2. Distribution of the diversity of transcripts encoding putative known toxins found in the proteotranscriptomic approach of the venom gland of *C. iheringi*. Percentages correspond to the relative expression in TPM of each category.

2.2. Novel Toxins Identified by the Mass Spectrometry

In order to identify the major proteins of C. *ihering*'s venom, their approximate molecular masses, and relative abundances, the crude venom was submitted for an SDS-PAGE, followed by a mass spectrometry analysis of the selected gel bands. The results revealed bands with apparent molecular masses ranging from 15 to 200 kDa, with proteins of the high molecular mass showing a higher intensity (Figure 3).

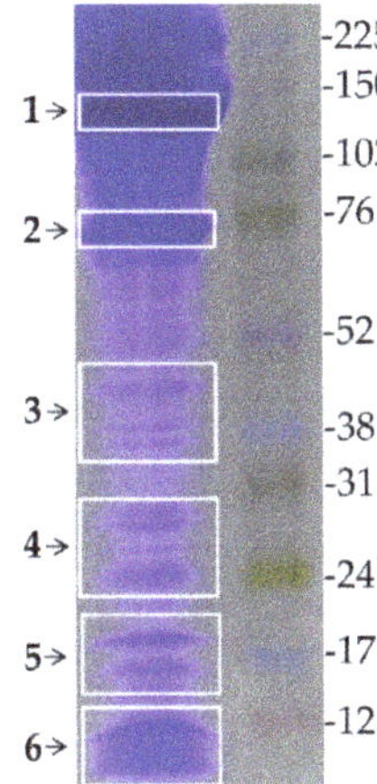

Figure 3. SDS-PAGE of the crude venom extracted from *C. iheringi*. Venom proteins were separated on a 12.5% SDS-PAGE gel and stained with Coomassie Brilliant Blue. The molecular mass (kDa) marker is shown on the right. The Selected groups of bands (from 1 to 6, on the left) were excised from the gel and processed for LC/MS-MS analysis.

Six groups of protein bands were selected and removed from the gel to be analyzed. The proteins were identified by matching the resulting mass spectra peptides to the deduced molecular masses of tryptic peptides, derived from full-length proteins as predicted from the transcriptome assembly. Matches of at least three peptides were considered valid. To identify a robust set of abundant venom toxins we focused on those sequences that exceeded the threshold value of 100 TPM. This allowed us to generate a list of 11 putative venom toxins that are highly abundant and likely play an important role in the venom (Table 3 and supplementary material Excel Table S1).

Table 3. Toxin identification of the major bands of *C. iheringi* venom based on the transcriptome and proteomic data. The numerical identifications correspond to the group of bands where the protein was found.

Band Group	Accession Number	Unique Peptides	Proteome Coverage	Molecular Weigth	Best Hit	Species	Identity and Acession Number
1	Ciheringi01366	32	53%	152 kDa	uncharacterized protein	*Centruroides sculpturatus*	25.07% XP_023226371.1
2	Ciheringi05450	78	77%	76 kDa	Hemocyanin	*Scolopendra dehaani*	55.99% SMH67860.1
3	Ciheringi11581	10	36%	37 kDa	No hit	-	-
3	Ciheringi14246	8	29%	27 kDa	Lipase	*Centruroides sculpturatus*	29.74% XP_023229615.1
3	Ciheringi16405	4	14%	28 kDa	Venom allergen	*Scolopendra subspinipes*	42.92% QEE04219.1
3	Ciheringi21566	5	34%	22 kDa	No hit	-	-

Table 3. *Cont.*

Band Group	Accession Number	Unique Peptides	Proteome Coverage	Molecular Weigth	Best Hit	Species	Identity and Acession Number
4	Ciheringi38643	12	56%	11 kDa	No hit	-	-
4	Ciheringi10323	4	53%	10 kDa	Lipase	*Branchiostoma floridae*	37.39% XP_035699465.1
5	Ciheringi24930	3	22%	15 kDa	No hit	-	-
5	Cryptoxin-1	12	68%	12 kDa	No hit	-	-
6	Ciheringi05125	7	74%	14 kDa	Profilin	*Orussus abietinus*	80.95% XP_012283556.1

2.3. Cloning and Expression of Cryptoxin-1

After the venom proteotranscriptomic analysis, three putative unknown toxins that showed best proteome coverage (Ciheringi14246, Ciheringi38643, and Cryptoxin-1) were selected to be cloned and expressed in the recombinant form to study their biological activities. However, after initial evaluations, only the protein Cryptoxin-1 was obtained in a soluble form and with a satisfactory yield (8.5 mg/L of culture) after its expression in *E. coli*. This putative toxin, Cryptoxin-1, was characterized as described below.

Cryptoxin-1 is composed of 119 amino acids (Figure 4), with a predicted molecular mass of 12,769.33 Da, and a theoretical pI of 5.76. It also showed a predicted signal peptide, indicating that this toxin is secreted. In addition, it showed a GRAVY (Grand average of hydropathicity) index of −0.392, indicating its hydrophilic characteristics and the absence of a predicted glycosylation site.

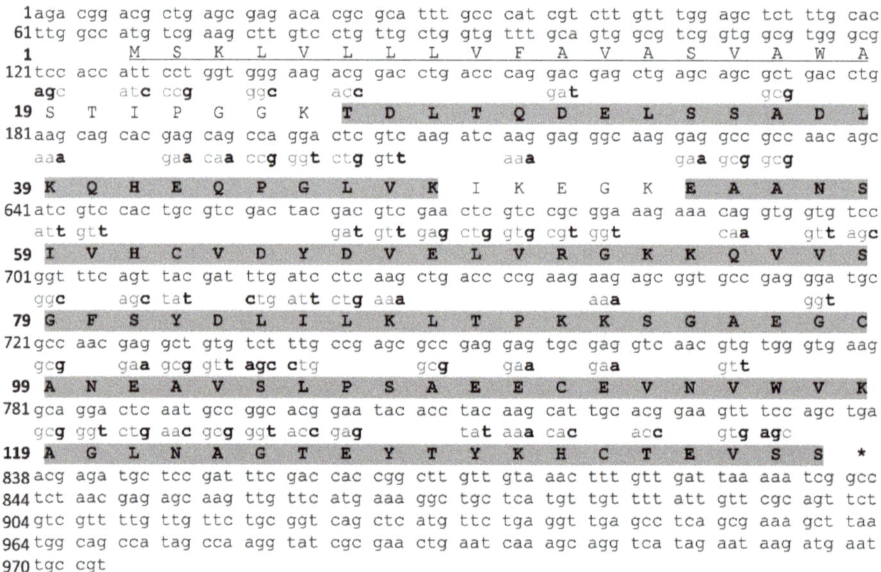

Figure 4. Nucleotide sequence of Cryptoxin-1 found in the transcriptome and its amino acid translation. The nucleotides in bold were changed to optimize expression in *E. coli*. The coverage of peptides found in the proteome is highlighted in grey. Underlined amino acids indicate the predicted signal peptide (SignalP-5.0), (*) indicates the stop codon.

The cDNA of Cryptoxin-1 was codon-optimized and cloned into the pET-24b (+) expression vector and then transformed into *E. coli* BL21 (DE3). The SDS-PAGE protein expression analysis revealed a single major band at around 16 kDa (Figure 5a, line 3). The mass spectrometry analysis (MALDI-TOF-MS) of purified Cryptoxin-1 showed a molecular mass of 14,138.5 Da (Figure 5c), which corresponds to the combination of Cryptoxin-1 (12,769.33 Da), a C-terminal tail of six histidines for IMAC purification, and additional residues encoded by the cloning vector (1360.67 Da). Its expression was also confirmed by polyclonal anti-histidine antibody immunoblotting, as shown in Figure 5b.

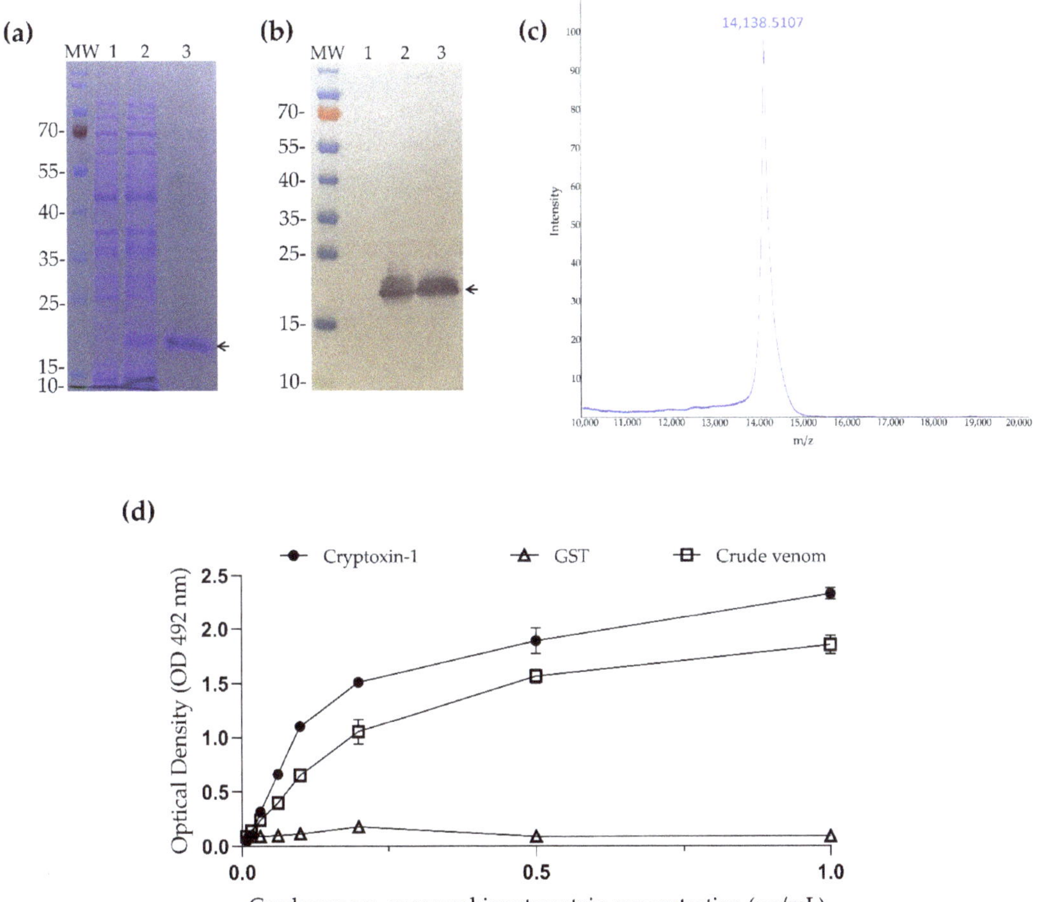

Figure 5. (**a**) 12% SDS-PAGE gel stained with Coomassie Brilliant blue. MW—Molecular mass marker in kDa; 1—*E. coli* sediment before IPTG induction; 2—*E. coli* sediment after IPTG induction; 3—Cryptoxin-1 purified by nickel-sepharose's affinity. (**b**) Recognition of the recombinant protein by immunoblotting using the polyclonal antibody anti-histidine (Sigma–Aldrich, St. Louis, MO, USA). Arrow indicates protein height; (**c**) MALDI-TOF—Mass spectrometry analysis of purified Cryptoxin-1 showing its molecular mass of 14,138.51 Da. (**d**) ELISA, IgG anti—*C. iheringi* venom against crude venom, Cryptoxin-1, and GST (unrelated recombinant protein). Fixed antibody dilution used was 1:200 (7.5 ug/mL) versus serial protein dilution starting at 1 µg/mL.

To further confirm the presence of Cryptoxin-1 in the venom, its recombinant form, as well as the whole venom and a recombinant non-related protein (negative control) glutathione protein S-Transferase (GST) from *Schistosoma mansoni* were tested by ELISA, using a purified polyclonal IgG anti- *C. ihering*'s venom. As can be seen in Figure 5d, the anti-venom IgG recognized both the venom and Cryptoxin-1, while the negative control (GST) was not recognized.

2.4. *Crypotoxin-1 Induces Edema in Mice Footpad*

As previously demonstrated, local tissue inflammation is one of the deleterious effects in *C. iheringi* envenomation [25]. Thus, we evaluated the local injury induced by Cryptoxin-1 injection in the footpad of BALB/c mice.

Mice were injected, through the right footpad, with either PBS (negative control), 45 μM of recombinant proteins Cyptoxin-1, or GST (negative protein control). The edema was measured through the thickness of the footpad at different time intervals, including: 1, 24, 48, and 72 h. The group injected with Cryptoxin-1 experienced a marked presence of edema during all the measurement times with a statistical difference when compared to the control groups (Figure 6a).

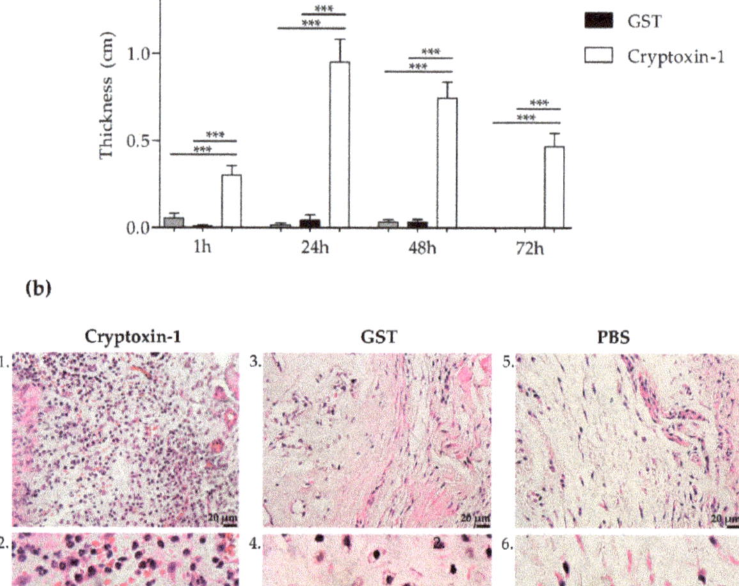

Figure 6. (**a**) Cryptoxin-1 induced footpad edema in mice. Groups of BALB/c mice were injected with 30 μL (45 μM) of Cryptoxin-1, GST (negative protein control), or 30 uL PBS (negative control). Edema was determined by thickness difference, at times 1, 24, 48, and 72 h. The results represent the ± S.E.M compared with the negative control group (Cryptoxin-1 vs PBS and Cryptoxin-1 vs GST), (n = 5). Statistical analysis was performed by ANOVA, followed by the Bonferroni test, *** $p < 0.0001$. (**b**) Histological analysis of the footpad of mice at 24 h after protein injection or PBS. All samples were analyzed with hematoxylin and eosin staining. 1. and 2.: Cryptoxin-1, bar 20 μm (40×) and 10 μm (100×) respectively; 3. and 4.: GST, bar 20 and 10 μm respectively; 5. and 6.: PBS, bar 20 and 10 μm, respectively. Neutrophilic inflammatory infiltrates (arrow). The images are representative of five mice/groups.

The cellular infiltration was then analyzed using the histological sections. Twenty-four hours after the injection, the GST and PBS groups presented normal tissue without an excess of inflammatory infiltration (Figure 6b, images 3, 4, 5, and 6). In contrast, 24 h after the Cryptoxin-1 injection, we observed the predominance of neutrophilic inflammatory infiltration (Figure 6b, images 1. and 2.).

2.5. Cryptoxin-1 Induces Potent Neutrophil Migration in Mice Footpad

Since we verified the peak of the edema induced by Cryptoxin-1 injection, as well as neutrophil infiltration in the histological analysis at 24 h, we confirmed this cellular profile by flow cytometry. Thus, at the peak of the edema (24 h), cellular suspensions were prepared from the footpad of the different mice groups and stained with anti-CD45, anti-CD11b, and anti-Ly6G mAbs conjugated to fluorochromes followed by flow cytometry. As shown in Figure 7, Cryptoxin-1 induced a significant level of neutrophils infiltration compared to that achieved in the other groups.

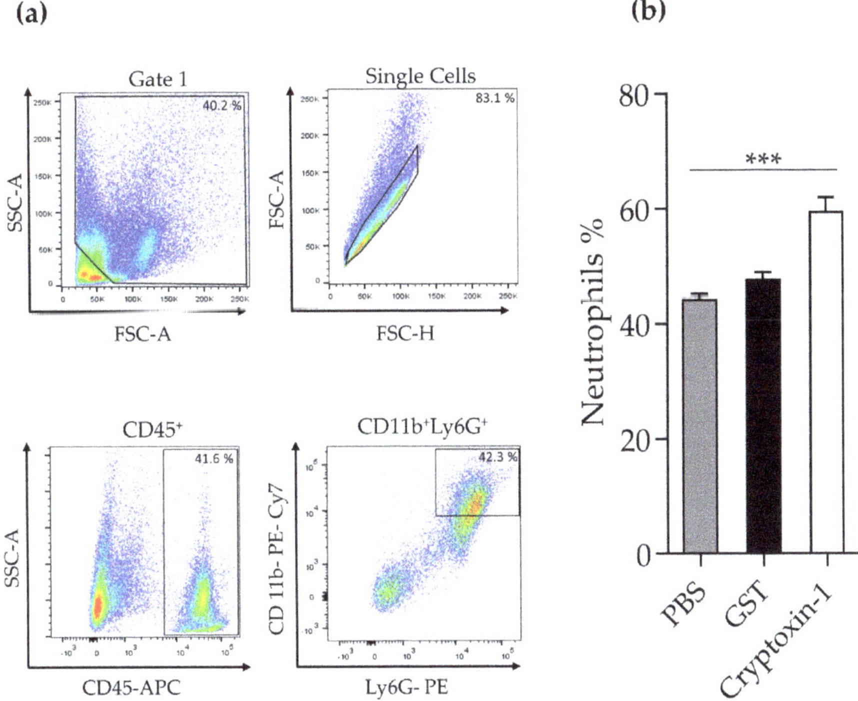

Figure 7. Neutrophil migration in the footpad of BALB/c mice injected with Cryptoxin-1, GST (45 μM), or PBS. (**a**) Flow cytometry gate strategy. Cells suspensions were prepared from footpad macerates after 24 h of the injection. Samples of cells (1×10^6 cells) were incubated with anti-CD 45- APC, anti-CD 11b-PE-Cy7, and anti-Ly6G (PE) antibodies followed by flow cytometry analysis. (**b**) The mean of the percentage of $CD45^+CD11b^+Ly6G^+$ cells of individual mice/group (n = 5) ± S.E.M. Statistical analyses was performed by ANOVA, followed by Bonferroni test, *** $p < 0.05$ Cryptoxin-1 group compared with PBS or GST groups.

3. Discussion

Centipedes are well adapted to urban areas and are very commonly found in gardens and other residential areas. As a consequence, there is a great risk of accidents occurring for humans [2,13,26]. Although their venom may cause undesirable effects, centipedes have been used in traditional eastern medicine for centuries [11]. However, individual substances have rarely been refined [27,28].

The vast majority of studies of centipede venom are restricted to the *Scolopendra* genus [8–10,27,29–33]. In addition, some studies use the whole centipede instead of the whole venom for their proteomics analyses, making a more specified comparison inviable [34,35]. Five comparative studies have demonstrated that centipede venoms are complex cocktails, encompassing more than 60 phylogenetically distinct protein families [10,32,36–38]. Among them, there exist, enzymes, protease inhibitors, a great diversity of cysteine-rich proteins, and unknown proteins that are yet to be functionally characterized. Therefore, in this study, we aimed to contribute to the understanding of the toxin genes present in centipedes by generating a gene expression profile of the venom gland of *Cryptops iheringi* species.

Since literature for this species is scarce, we followed the transcriptome and proteomic approaches that were effective to identify toxins for other related species. In this regard, Ward, et al. (2018) [33], using these techniques, were able to identify 39 new toxins in the venom gland of the *Scolopendra viridis*, while Liu, et al. (2020) [8] found more than 400 toxin-like unknown sequences in the venom gland of *Scolopendra mojiangica.* Similarly, we found as high as 57.9% of the proteins to be uncharacterized from the *C. iheringi* centipede and 454 protein sequences that could only be characterized as putative unknown toxins or known toxins due to the proteomic approach. Among them, 263 proteins showed no similarity with the available sequences in public databases, indicating a great diversity of components with an unknown structure and function.

The putative venom toxins of *C. iheringi* revealed diversely distributed proteins with novel structures and biological activities that need to be further investigated. The majority of the venom proteins are putatively functional enzymes. Most notably, lipases and other hydrolases (8.8%), which include a large group of different proteins, such as phospholipases, are frequently reported as venom components of several other arthropods, such as centipedes, spiders, and scorpions [25,34,39–43], contributing to prey digestion and venom toxicity [42].

Trypsin domain proteins were also found in this venom (5.8%). Food protein degradation is crucial for digestion and is catalyzed by trypsin enzymes. Trypsin appeared early in evolution, and it became the most abundant proteinase in the digestive systems of invertebrates [44]. Trypsin performs two main functions, namely, the hydrolysis of protein and the activation of other digestive proteases, although it also plays a role in the innate immunity of these animals [45]. Some trypsin domains proteins have also been found in the centipede *S. subspinipes dehaani* venom gland transcriptome [29].

Peptidases (4.6%) were found to be another relevant group, comprised of endopeptidases, carboxypeptidases, and esterases that are among the reported protein components of some centipedes [30]. These kinds of proteins have an effect on amino acid production for digestive purposes and may be responsible for the tissue deleterious effects of the envenomation [46]. Several classes of peptidases, for which activities were not yet clarified, have also been found in the venom proteome and transcriptome of the scorpion *Hadrurus spadix* [47].

Putative neuron cell adhesion toxins were also present in this venom (4%). Findings on black widow spider venom indicate that such toxins can modulate a neuronal adhesion receptor, which stimulates strong neuronal exocytosis in vertebrates, and, interestingly, may perform functions in synapse development [48,49]. In the context of venom activity, interesting studies exist showing that a toxin from the snakes *Bothrops atrox* and *Bothrops moojeni* is capable of improving spatial memory disorder in temporal ischemic rats through its effects on the neural cell adhesion molecule [50]. Regarding the other putative venom toxin found here, more studies are necessary to propose and define its function in the venom of arthropods, especially from *C. iheringi.*

In addition to the whole venom proteome, *C. ihering*'s crude venom was subjected to a protein separation by SDS-PAGE to better visualize the main bands and their relative expression. In this gel, the venom showed an electrophoretic profile with a wide range of proteins between 15 and 200 kDa, with a large amount located above 70 kDa. The main bands of the venom were excised from the gel and subjected to LC-MS/MS mass spectrometry, in a strategy successfully utilized for the proteome decomplexation of other venoms [51]. The proteomic analysis returned several peptides with good spectra quality, which allowed us to classify 11 toxins, five of which ranged in size between 17 and 37 kDa, whose sequences did not show any similarity to the public databases, and therefore, represent new *C. iheringi* specific toxins. In order to unravel the function of unknown putative toxins, one, named Cryptoxin-1, was cloned and expressed in *E. coli*. The presence of this toxin in the venom was further confirmed by ELISA, where it was strongly recognized by IgG anti- *C. iheringi*'s venom, indicating its presence in the venom.

As it is known from the literature, envenomation by centipedes usually causes pain, erythema, and edema formation in humans and mice [13]. However, the characterization of the inflammatory activities induced by the venom is poorly described in the literature. For the centipede *C. iheringi*, there is only one published article showing that the venom induces strong pro-inflammatory activity able to induce edema and nociception, in addition to being myotoxic for mice [25]. Similarly, previous studies showed that the crude venom of the centipede species *S. viridicornis*, and *O. pradoi* induced edema in mice's footpads, which progressively diminished by 72 h [25,52]. In this respect, the injection of Cryptoxin-1 into mice's footpads was able to cause an edema of rapid evolution and progressive decay after 72 h. In addition, the injected animals were prostrate, bristly, with low temperature, and showed erythema at the injection site (data not shown).

After considering the results obtained with the edematogenic activity, we performed a histological analysis of the mice footpad injected with Cryptoxin-1 to characterize the cellular influx in the peak of the edema. The histological sections demonstrated the predominance of neutrophil infiltration, and flow cytometry analysis confirmed this result. Following these findings, Fung et al. (2011) [26] reported that 40% of patients who have been admitted to Hong Kong Emergency Hospital with centipede bites (species not specified), showed an increased neutrophil-predominant leukocytosis in their blood tests with an edema and erythema at the bite site, and strong pain. We also observed that the neutrophil infiltration lasted up to 72 h, which was also reported for the crude venom of *S. viridicornis* [53]. Taking these observations together, the results indicate that Cryptoxin-1 may contribute to the symptoms observed in envenomation. It is important to point out that in all the experiments, the recombinant GST, which was subjected to the same expression and purification procedures as Cryptoxin-1, was used as a non-related protein control to exclude any effect related to the protein purification steps.

Kinetics cellular infiltrate studies show that neutrophils are the first inflammatory cells to reach the lesion site and that edematogenic activity may occur due to the neutrophils release of cytokines, prostaglandin, myeloperoxidase, bradykinin, and histamine, causing increased vasodilation and the permeability of small vessels, resulting in the migration of other cells to the local tissue [54,55]. Although it is already known that innate immune cells participate in the local inflammatory response [27], the correlation between the local edema induced by *C. iheringi* venom and cellular infiltration is not completely understood. Therefore, further investigation is necessary to elucidate the complex interplay of the toxins present in its venom.

In this work, we described the profile of toxins present in the *C. iheringi* venom gland using transcriptome and proteome approaches that may contribute to understanding the venom composition and its effects in envenomation. In addition, new toxin genes were identified that may allow for the characterization of their role in this venom, and possibly for other toxins in related species. Furthermore, a new recombinant toxin named Cryptoxin-1 was also characterized as showing a proinflammatory activity, suggesting that it is likely to be one of the components responsible for the envenomation symptoms observed in accidents with humans. Additional studies are being conducted with this toxin as well as the other unknown toxins to understand their role in envenomation. Keeping this in mind, we understand the potential of novel developments for further studies concerning this centipede species and its venom.

4. Materials and Methods

4.1. Specimen Collection and Venom Extraction

Seven *C. iheringi* adult specimens were collected in the metropolitan area of the city of São Paulo, Brazil with the permission of SISBIO (15222-2) and kept in the Arthropod Laboratory of the Butantan Institute. To obtain the venom, the animals were anesthetized by anoxia, and the venom was extracted through electrical discharges (12 V) in the ventral region of the head (coxo sternum) with an electroshock device. The venom obtained through the forcipules was aspirated with an automatic micropipette and deposited in a microcentrifuge tube in an ice bath. The venom obtained was stored at −80 °C for a subsequent proteome analysis. The extraction was performed every 30 days.

4.2. RNA Isolation, Library Preparation, and Illumina Sequencing

The heads of seven specimens of *Cryptops iheringi* were submitted for the dissection the of venom glands for transcriptomics. The total RNA was extracted with TRIZOL Reagent (Invitrogen, Life Technologies Corp., Carlsbad, CA, USA), a method based on the procedure described by Chomczynski et al. (1987) [56]. The total RNA was quantified by its absorbance at a wavelength of 260 nm in a NanoDrop 2000 device (Thermo Fisher Scientific, Waltham, MA, USA). Beginning with an amount of total RNA ranging from 75 to 77 µg for each sample, the purification of mRNA was performed through an affinity to magnetic microspheres containing oligo (dT), using the protocol of the Dynabeads® mRNA DIRECT kit (Invitrogen, Life Technologies Corp.), with reagents to reduce the number of ribosomal RNA (rRNA). The quantification of mRNA was performed using the Quant-iT RiboGreen® reagent (Invitrogen, Life Technologies Corp.), according to the manufacturer's specifications. All RNA procedures were performed using RNAse-free tubes and tips with a filter and water, treated with diethylpyrocarbonate (DEPC, Sigma–Aldrich, St. Louis, MO, USA). After the extraction of mRNA, its integrity was assessed using the 2100 Bioanalyzer, pico chip series (Agilent Technologies Inc. Santa Clara, CA, USA). The mRNA was then subjected to a purification and concentration step using the MinElute® PCR Purification Kit (Qiagen) protocol. To confirm that mRNA was not lost of during this purification and concentration step, a further quantification of the mRNA was performed through its absorbance at a wavelength of 260 nm in a NanoDrop 2000 device (Thermo Fisher Scientific, Waltham, MA, USA).

A cDNA library was generated by TruSeq RNA Sample Prep Kit protocol (Illumina, San Diego, CA, USA). The cDNA was synthesized from fragmented mRNA using random hexamer primers, followed by ligation with appropriate sequencing adaptors. The size distribution of the cDNA libraries was measured with a 2100 Bioanalyzer using DNA1000 assay (Agilent Technologies Inc. Santa Clara, CA, USA). An ABI StepOnePlus Real-Time PCR System with *KAPA* Library *Quantification* was used for library sample quantification before sequencing. The cDNA library was then sequenced on Illumina HiSeq 1500 System, in a Rapid Run mode in a 2-lane paired-end flowcell, run for 300 cycles, generating 2 × 151 bp paired-end reads for each fragment, according to the manufacturer's protocol (Illumina).

4.3. RNA-Seq Raw Data Pre-Processing, De Novo Assembly, and Functional Annotation

After large-scale sequencing of the cDNA, using Illumina HiSeq1500 equipment, bioinformatics analyses were performed. Thus, the sequencing platform generated sequencing images, which were converted to BCL format, after the CASAVA software was used to demultiplex the samples through the identification of the indexes (barcodes). The demultiplexing step generates the FASTQ file format, with a quality control of Q30.

For the pre-processing of the reads, an in-house pipeline was used to analyze the raw reads with a read filter by quality, eliminating reads with homopolymer and low complexity regions, poly-A/T/N tails, removal of adapters, indexes, and low-quality edges using the software FASTQ-mcf, [57] and bowtie2 [58]. The criteria used for filtering were as follows: the removal of homopolymer regions and a low complexity above 90% of the sequence, trimming tip regions with an average quality lower than 25. Only reads at a minimum size of 40 bp were kept. The raw reads were filtered by PhiX contaminants using the software Bowtie2 [58] standard parameters.

The transcriptome was assembled using the rnaSPAdes [59] with a K-mer size of 55.

The TransDecoder software version 3.0.1 (http://transdecoder.sourceforge.net/; accessed on 15 January 2018) was used to identify Open Reading Frames (ORFs) from the assembled transcripts with protein lengths higher than 60 amino acids. The program SignalP version 5.0 [60] was used for signal peptide predictions.

The completeness of the transcriptome was also estimated by the presence of sequences belonging to the set of ultraconserved eukaryotic proteins, tested using the BUSCO approach based on metazoa database [61].

Using TSA/NCBI, we downloaded the transcriptome assemblies from 10 species from the Scolopendromorpha orders (Table 2) (*Cryptops anomalans* (GERT01.1), *Hemiscolopendra marginata* (GHBY01.1), *Scolopendra alternans* (GASK01.1), *Scolopendra cingulate* (GCAP01.1), *Scolopendra dehaani* (GBIM01.1), *Scolopendra morsitans* (GHKQ01.1), *Scolopendra subspinipes* (GGDW01.1), *Scolopendra virirdis* (GGNE01.1, *Scolopocryptops rubiginosus* (GCIY01.1), *Scolopocryptops sexspinosus* (GHBZ01.1)) summarizing 106197 transcripts used to create the database for Blast alignment. The C. iheringi were aligned against the Scolopendromorpha database using the BlastN alignment tool with a cutoff of 1×10^{-15}.

The predicted amino acid sequences were aligned using the BLASTx and BLASTp programs [62] against NCBI's Uniprot/Swissprot protein databases, and Transcriptome Shotgun Assembly (TSA), to access sequence similarity with proteins in other species with a cutoff e-value of 1×10^{-5}. The hmm search tool [63] allowed us to identify the conserved PFAM domains [64], with a cut-off e-value $< 1 \times 10^{-3}$. The priority order of the UniProt/Swissprot, PFAM, and TSA-NCBI protein hits was used to select the best candidate for each transcript.

The sequencing reads were aligned against the *C. iheringi* transcriptome with the bowtie2 program [58]. The method was used to estimate the transcript abundance. Further computing of the abundance for each transcript was performed by RSEM [65], along with a Maximum Likelihood abundance estimate, using the Expectation-Maximization algorithm for its statistical model. Final abundance estimates were calculated as Expected counts, Fragments Per Kilobase of exons per Million fragments mapped (FPKM) and Transcripts Per Million (TPM) values. Functional annotation was performed using the Blast2GO program [66], which is a tool used for analyzing a set of sequencing tags that makes it possible to understand the physiological meaning of a large number of genes. Transcript sequences were used as input sequences for the Blast2GO program. BLASTx was used to find counterparts in the NCBI database NR with a cut-off value of 1×10^{-5}. Furthermore, the analysis was performed using the first 20 hits, a minimum alignment length of 33 amino acids, and a low complexity filter activation. The program then extracted the Gene Ontology (GO) terms for each hit obtained by mapping the existing annotation associations, after an annotation rule assigns the GO term to the sequence in question. After the BLAST, mapping, and annotation steps, the graphs, tables, and organization charts provided by the program were analyzed. For the distribution data of the GO terms provided by the program, tables with raw data were used instead of the graphs provided, since this allowed for greater formatting flexibility for the presentation of the data.

Bioinformatics analyses were performed using the computational infrastructure of the Center of Toxin, Immune response and cell signaling (CeTICS), and the Bioinformatics and Computational Biology Core in the Butantan Institute. The raw data generated in this project was deposited in the NCBI BioProject section under the accession code PRJNA763193, BioSample SAMN21432369 and SRA SRR1608688.This Transcriptome Shotgun Assembly was deposited in NCBI TSA under the accession GJOG00000000.

4.4. SDS-PAGE and LC-MS/MS Analysis of *C. iheringi*

C. iheringi venom was analyzed by SDS-PAGE (12% acrylamide resolution and Pierce, USA) under reducing conditions [67]. After protein separation by electrophoresis, the gels were stained with Coomassie Brilliant Blue R-250 (GE Healthcare) (0.1% coomassie R-250, 40% ethanol; 10% acetic acid) and bleached with a bleach solution (20% methanol, 5% acetic acid) for which a molecular weight standard for proteins (Middle Range protein Ladder, ready-to-use—Thermo Fisher Scientific, Waltham, MA, USA) was used. Electrophoresis occurred at room temperature, using a voltage of 150 V and a current of 40 mA per gel. The regions of interest were cut out from the gel manually with the aid of a sterile scalpel on a clean surface and placed in Eppendorf tubes washed with methanol and Milli-Q water. To remove the Coomassie dye, the gel bands containing the venom proteins were incubated under agitation for 10 min in a bleaching solution (50% acetonitrile; 25 mM NH_4HCO_3; pH 8), followed by a 10 min rest; the procedure was repeated until complete dye removal. After removing the dye, the fragments were washed with 100% acetonitrile and dried in a vacuum centrifugation system for 25 min.

During enzymatic digestion, the gels were rehydrated with trypsin solution (10–15 µg/mL trypsin; 25 mM NH_4HCO_3). The enzymatic solution was incubated (4 °C), the samples remained in an ice bath for 30 min, then 25 mM NH_4HCO_3 was added and kept at 37 °C for 20 h. After this period, the digestion tube supernatant was transferred to a methanol treated Eppendorf tube. To extract the peptides, the gel fragments were covered with a 50% acetonitrile solution; 5% TFA, and gently homogenized for 30 min, the supernatants were reduced to a volume of 5 µL in a vacuum centrifugation system and then purified in C18 ZIP TIPs micro columns following the manufacturer's instructions.

The total venom was filtered through a 0.22-micron filter solubilized in ammonia bicarbonate (Sigma–Aldrich, St. Louis, MO, USA # A-6141) pH 8.0, containing a phosphatase inhibitor. The protein content of the samples was quantified using the BCA kit (Thermo Fisher Scientific, Waltham, MA, USA). The protein disulfide bridges were reduced by adding 10 mM DTT (Sigma–Aldrich, St. Louis, MO, USA # D-5545), followed by incubation for 30 min at 56 °C. The proteins were alkylated with 55 mM IAA (Sigma–Aldrich, St. Louis, MO, USA # I-1149), at room temperature, in the dark, for 30 min. Protein digestion was performed using trypsin (Promega # V5111), in a 1:20 ratio, at 37 °C, for 18 h. Columns C 18 (Harvard Apparatus, Holliston, MA, USA) were used for cleaning and the desalination of the sample. Digestion products were loaded onto a tandem system containing a pre-column that separates the products and passes the effluent automatically into an LTQ-Orbitrap Velos mass spectrometer (Thermo Fisher Scientific, Waltham, MA, USA). The enzymatic digestion with trypsin and desalination of the gel bands, as well as the crude venom were carried out at CEFAP: Center for research support facilities at the University of São Paulo, USP.

4.5. LC-MS/MS Data Analysis

After analyzing the samples by LC-MS/MS, the raw data were collected and passed through the Mascot platform (Matrix Science, Boston, MA, USA), using carbamidomethylation and methionine oxidation as variable modifications. The resulting files were exported in the .dat extension and processed in the Scaffold Q + software version 4.0 (Proteome Software, Portland, OR, USA), using as selection criteria of the presence of at least three peptide fragments, a probability rate of 95% for protein identification and a false discovery rate (FDR) of 5%. In addition, for the analysis carried out on the Scaffold Q + and Mascot platforms, a FASTA database was built containing the amino-acid sequences obtained through the predicted proteins from assembled transcripts of the transcriptome venom gland, together with sequences of possible sample contaminants such as trypsin and human keratin, to avoid alignment and coverage errors. Based on the data on the probability of protein identification and percentage of coverage, a contig for each band of venom was identified.

4.6. Protein Expression and Purification

The nucleotide sequence of Cryptoxin-1 was optimized for expression in *E. coli* and the construction pET24b-Cryptoxin-1 was performed using Invitrogen™ Gene Synthesis (GeneArt™ Thermo Fisher Scientific, Waltham, MA, USA).

For their expression, chemically competent *E. coli* BL21 Star™ (DE3) (Invitrogen®) were transformed with pET24b Cryptoxin-1 construction or pET-42a (Novagen) containing the sequence of glutathione S-transferase (GST). For each experiment, a cell colony grown overnight from LB-agar plates was transferred into a liquid LB medium and grown overnight at 30 °C in the presence of 50 µg/mL kanamycin. This culture was diluted 1:50 into 200 mL of fresh LB broth/kanamycin. When the cell suspension reached an optical density of 0.6 at 30 °C (OD 600 nm) it was induced with a final concentration of 1 mM of isopropyl-b-D-thiogalactoside (IPTG). Cells were then grown for four hours after which the cells were collected by centrifugation at 10,000× *g* for 10 min at 10 °C (ultracentrifuge Beckman). The whole-cell pellets were then resuspended in a binding buffer (20 mM sodium phosphate pH 7.4 and 500 mM NaCl) and lysed on ice by an ultrasonication device (amplitude of 20% with 3 s pulse and 4 s interval between each pulse) for 60 s, and the process was repeated 5 times. Cell debris were removed from the protein solution by centrifugation at 10,000× *g* for 10 min in a Beckman ultracentrifuge. The entire amount of the supernatant containing the soluble protein was purified with high-performance immobilized metal affinity chromatography (IMAC) using HisTrapHP 5 mL column pre-packed (Cytiva™, Marlborough, MA, USA) coupled to the ÄKTA™ start protein purification system and then desalted into phosphate-buffered saline (PBS) buffer pH 7.4 with HiTrap® Desalting Columns (Cytiva™, Marlborough, MA, USA). The endotoxin was removed with Pierce™ High-Capacity Endotoxin Removal Spin Columns (Thermo Fisher Scientific, Waltham, MA, USA) following the manufacturer's protocol. Quantification of recombinant proteins was performed using the BCA Pierce™ Protein Assay Kit (Thermo Fisher Scientific, Waltham, MA, USA) following the manufacturer's protocol.

4.7. Mass Spectrometry Analysis

Mass spectrometry of Cryptoxin-1 was performed on the MALDI-TOF Autoflex Speed (Bruker Corporation, Billerica, MA, USA) equipment following pre-established protocols for protein analysis. In summary, 0.5 µL of a saturated solution of sinapinic acid in ethanol was mixed with 100 ng of Cryptoxin-1. After drying, 1 µL of a TA30 (0.1% trifluoroacetic acid/acetonitrile in a proportion of 70/30) was added to the mixture and applied to the Ground Steel plate for analysis. Data acquisition was performed in a linear mode with positive polarity, with the following parameters: Ion Source 1—19.50 kV, Ion Source 2—17.60 kV, Lens—9.0 kV, Pulsed Ion Extraction 170 ns, Mass Range 5—70 kDa, Laser Frequency 500 Hz, Gain Detector 10.0×. The results were analyzed using the online software mMass version 5.5.0, 2013 (Martin Strohalm© Open Source Mass Spectrometry Tool).

4.8. Rabbit Specific Antivenom Production

The anti-venom serum from the *C. iheringi* centipede was obtained by immunization of rabbits. Two hundred micrograms of the venom and 2.5 mg of aluminum hydroxide (Brenntag Specialties, Inc., South Plainfield, NJ, USA) were added to a final volume of 1 mL of PBS. After this, 250 µL of this mixture was injected intramuscularly. After 1 month, the rabbits received five consecutive boosters of antigen with 15-day intervals. Blood was collected and sera were separated and stored at 20 °C until use. Antibodies present in the hyperimmune serum were purified on HiTrapProtein G HP 5 mL column pre-packed with high-performance protein G-Sepharose (GE Healthcare, Little Chalfont, UK) and quantified by BCA (QuantiProBCA Assay Kit, Sigma-Aldrich, St. Louis, MO, USA). To characterize the polyclonal anti-venom, ELISA and immunoblotting assays were performed. The experimental protocols were approved by the Butantan Institute Ethical Committee for Animal Research (certified by CEUAIB n° 8172250816).

4.9. ELISA Immunoassay

Microplates of 96-well (Sarstedt, Germany) were sensitized with 100 µL/well of the heterologous proteins in the serial 1:2 dilution from 1 to 0.008 µg/mL and incubated in a humid chamber at 4 °C for 18 h. Crude venom and GST protein were used as controls under the same conditions. Subsequently, blocking was performed with PBS containing 1% bovine serum albumin (BSA) for 30 min. After blocking, the addition of 1:200 (7.5 µg) of polyclonal IgG anti-*C. iheringi* venom antibody diluted in PBS + 1% BSA at 37 °C for 1 h. Subsequently, the microplates were incubated with a peroxidase-conjugated anti-rabbit IgG antibody (1:5000) at 37 °C for 45 min. After this, a revelation solution of OPD (ortho-phenylenediamine) was added (1 mg of OPD, 2 mL of citrate/phosphate buffer, and 1 µL of hydrogen peroxide). Then, the microplates were statically incubated, in the dark, at 24 °C for 15 min, and sulfuric acid (H_2SO_4) 2 N was used to stop the reaction and the plate was read in an ELISA reader (Labsystems Multiskan, Thermo Fisher Scientific, Waltham, MA, USA) at 492 nm.

4.10. Mice

For the experiments, BALB/c male mice (between 18 and 20 g) were bred from the animal house facilities of the Butantan Institute, São Paulo, Brazil. The animals were kept in a controlled temperature, 12/12 light/dark cycle, and were provided with standard food and water *ad libitum*. The experimental protocols were approved by the Butantan Institute Ethical Committee for Animal Research (certified by CEUAIB n° 4300061120).

4.11. Evaluation of Paw Edema

Mice (n = 6) were injected (30 µL) with Cryptoxin-1 (45 µM), GST (45 µM), or PBS (negative control) in the right hind paw. The edema-forming activity was studied after 1, 24, 48, and 72 h, by pachymeter. The results were expressed as the difference in paw thickness before (control) and after (experimental) injection (mean ± S.E.M).

4.12. Histological Analysis

Mice (n = 5) were injected in the right paw with Cryptoxin-1 45 µM/30 µL, GST 45 µM/30 µL or PBS (negative control) and 24 h after the injection, the animals were euthanized, and the right paws were collected for footpad skin removal. The samples were then fixed in 4% paraformaldehyde in PBS, pH 7.2, for 24 h. After dehydration in a crescent ethanol series up to 95%, the samples were embedded in glycol methacrylate (Leica Microsystems Nussloch GmbH, Heidelberg, Germany). Sections of 4 µm were obtained in a Microm HM340 microtome and stained with the hematoxylin-eosin solution for morphological studies of tissues.

4.13. Analysis of the Neutrophil Infiltrate in Footpad Tissue by Flow Cytometry

The Neutrophil migration in the footpad after 24 h of Cryptoxin-1 injection was analyzed by flow cytometry. Groups of mice (n = 5) were euthanized, and the right paws were removed at the tibiotarsal joint and macerated. The debris was resuspended into 1 mL of PBS and then centrifuged (5 min/1200 rpm/4 °C). The cell pellets were recovered and counted by Trypan blue exclusion (Sigma–Aldrich, St. Louis, MO, USA) using a hemocytometer. The cells that were resuspended in an RPMI-1640 cell culture media (Gibco Thermo Fisher Scientific, Waltham, MA, USA), were then incubated with anti-FcγRII/III mAb for 30 min at 4 °C. Afterward, the cell suspensions were centrifuged (5 min/1200 rpm/4 °C) and resuspended in the culture medium (10^6 cells/well) and incubated with anti-leukocyte (CD45-APC) and anti-neutrophil (Ly6G-PE/CD11b-PeCy7) monoclonal antibodies (BD Biosciences, Franklin Lakes, NJ, USA) for 30 min at 4 °C. The cells were then washed and resuspended in PBS containing 0.1% paraformaldehyde (Merck, Darmstadt, Germany). All samples were acquired in the flow cytometer (FACS Canto II, BD Biosciences, Franklin Lakes, NJ, USA). Around 20,000 events were collected for each sample The data were analyzed using FlowJo software 7.5 (BD Biosciences, Franklin

Lakes, NJ, USA). The forward and side scatter density plots (FSC × SSC) were used to exclude the debris and select the cell population, followed by the selection of the single cells. After this, the $CD45^+$ cells were selected and then, using the fluorescent minus one (FMO) methodology, the $CD45^+CD11b^+Ly6G^+$ cells were determined. The results were expressed as the mean of the percentage of $CD45^+CD11b^+Ly6G^+$ cell population of individual mice/group ± standard error of the mean (S.E.M).

4.14. Statistical Analysis

All statistical analyses and graphical representations were analyzed using the GraphPad Prism 9.1.2 program. Statistical tests performed using ANOVA followed by the Bonferroni test and t Student's test. The p values followed the pattern recommended by the software: * $p < 0.05$; ** $p < 0.001$; *** $p < 0.0001$; **** $p < 0.00001$.

Supplementary Materials: The following are available online at https://www.mdpi.com/article/10.3390/toxins13120858/s1, Figure S1: Capillary electrophoresis of the RNA sample extracted from *C. iheringi*; Figure S2: Size distribution of *C. iheringi* cDNA library evaluated in Agilent 2100 Bioanalyzer; Table S1: *Cryptops iheringi* transcriptome data.

Author Contributions: Conceptualization: G.S.M., L.H.D.L.C., I.d.L.M.J.-d.-A. and M.Y.N.-J. Methodology: G.S.M., M.Y.N.-J., E.L.F.-M., I.d.L.M.J.-d.-A. and L.H.D.L.C. Validation: E.L.F.-M., G.S.M., M.Y.N.-J., L.H.D.L.C. and B.d.C.L.F.T. Formal analysis: E.L.F.-M., G.S.M., M.Y.N.-J., L.H.D.L.C. and B.d.C.L.F.T. Investigation: E.L.F.-M., G.S.M., M.Y.N.-J., L.H.D.L.C., B.d.C.L.F.T. and U.C.d.O. Resources: G.S.M., L.H.D.L.C., I.d.L.M.J.-d.-A. Data curation: M.Y.N.-J., L.H.D.L.C. and G.S.M. Writing—original draft preparation: G.S.M., L.H.D.L.C. Writing—review and editing: G.S.M., L.H.D.L.C., M.Y.N.-J. Visualization: G.S.M., L.H.D.L.C., E.L.F.-M., I.d.L.M.J.-d.-A., M.Y.N.-J., B.d.C.L.F.T. and U.C.d.O. Supervision: G.S.M. and M.Y.N.-J. Project administration: G.S.M. and L.H.D.L.C. Funding acquisition: G.S.M. and L.H.D.L.C. All authors have read and agreed to the published version of the manuscript.

Funding: This research was supported by São Paulo Research Foundation (FAPESP) (2017/16999-8) granted to Geraldo Santana Magalhães. A fellowship, from the National Coordination of High Education Personnel Formation Programs (CAPES- Demanda Social) and FAPESP (2017/13812-4), was granted to Lhiri Hanna De Lucca Caetano. A fellowship from National Council for Scientific and Technological Development (CNPq) (312096/2018-6) was granted to Eliana L. Faquim-Mauro. Computational Infrastructure from the Bioinformatics and Computational Biology Core from Laboratório de Toxinologia Aplicada, Butantan Institute was also funded by FAPESP (2013/07467-1).

Institutional Review Board Statement: Ethics Committee Name: Committee on Animal Use of the Butantan Institute (CEUAIB); Approval code: CEUAIB n° 4300061120; Approval Date: 22 December 2020; Approval code: CEUAIB n° 886/12; Approval Date: 19 November 2014.

Data Availability Statement: Not applicable.

Acknowledgments: We would like to thank the staff of the Butantan Institute and Butantan Foundation for their technical support, as well as the availability of the institute's facilities, especially the Immunopathology Laboratory, the Laboratory of Applied Toxinology Laboratory (LETA), the computational infrastructure from Bioinformatics and Computational Biology Core (NBBC) and the graduate course of Butantan Institute (ESIB).

Conflicts of Interest: The authors declare no conflict of interest.

References

1. Bücherl, W.B.; Buckley, E.E. *Venomous Chilopods or Centipeds*; Academic Press: New York, NY, USA, 1971; Volume 3.
2. Knysak, I.; Martins, R.; Bertim, C.R. Epidemiological aspects of centipede (Scolopendromorphae: Chilopoda) bites registered in greater S. Paulo, SP, Brazil. *Revista Saúde Pública* **1998**, *32*, 514–518. [CrossRef]
3. Attems, C.G. Myriapoda II *Scolopendromorpha*. *Berl. Leipz.* **1930**, *54*, 1–308.
4. Bucherl, W. Ação do veneno dos escolopendromorfos do Brasil sobre alguns animais de laboratório. *Mem. Inst. Butantan* **1947**, *19*, 181–198.

5. Antoniazzi, M.M.; Pedroso, C.M.; Knysak, I.; Martins, R.; Guizze, S.P.G.; Jared, C.; Barbaro, K.C. Comparative mor-phological study of the venom glands of the centipede Cryptops iheringi, Otostigmus pradoi and Scolopendra viridicomis. *Toxicon* **2009**, *53*, 367–374. [CrossRef] [PubMed]
6. Stankiewicz, M.; Hamon, A.; Benkhalifa, R.; Kadziela, W.; Hue, B.; Lucas, S.; Mebs, D.; Pelhate, M. Effects of a centipede venom fraction on insect nervous system, a native Xenopus oocyte receptor and on an expressed Drosophila muscarinic receptor. *Toxicon* **1999**, *37*, 1431–1445. [CrossRef]
7. Undheim, E.A.B.; King, G.F. On the venom system of centipedes (Chilopoda), a neglected group of venomous animals. *Toxicon* **2011**, *57*, 512–524. [CrossRef]
8. Liu, Z.-C.; Liang, J.-Y.; Lan, X.-Q.; Li, T.; Zhang, J.-R.; Zhao, F.; Li, G.; Chen, P.-Y.; Zhang, Y.; Lee, W.-H.; et al. Comparative analysis of diverse toxins from a new pharmaceutical centipede. *Zool. Res.* **2020**, *41*, 138.
9. Yang, S.; Liu, Z.H.; Xiao, Y.; Li, Y.; Rong, M.Q.; Liang, S.P.; Zhang, Z.Y.; Yu, H.N.; King, G.F.; Lai, R. Chemical Punch Packed in Venoms Makes Centipedes Excellent Predators. *Mol. Cell. Proteom.* **2012**, *11*, 640–650. [CrossRef]
10. Undheim, E.A.B.; Jones, A.; Clauser, K.R.; Holland, J.W.; Pineda, S.S.; King, G.F.; Fry, B.G. Clawing through Evolution: Toxin Diversification and Convergence in the Ancient Lineage Chilopoda (Centipedes). *Mol. Biol. Evol.* **2014**, *31*, 2124–2148. [CrossRef]
11. Pemberton, R.W. Insects and other arthropods used as drugs in Korean traditional medicine. *J. Ethnopharmacol.* **1999**, *65*, 207–216. [CrossRef]
12. Kong, Y.; Shao, Y.; Chen, H.; Ming, X.; Wang, J.B.; Li, Z.Y.; Wei, J.F. A Novel Factor Xa-Inhibiting Peptide from Cen-tipedes Venom. *Int. J. Pept. Res. Ther.* **2013**, *19*, 303–311. [CrossRef]
13. Medeiros, C.R.; Susaki, T.T.; Knysak, I.; Cardoso, J.L.; Málaque, C.M.S.; Fan, H.W.; Santoro, M.L.; França, F.O.S.; Barbaro, K.C. Epidemiologic and clinical survey of victims of centipede stings admitted to Hospital Vital Brazil (São Paulo, Brazil). *Toxicon* **2008**, *52*, 606–610. [CrossRef]
14. Barroso, E.; Hidaka, A.; Santos, A.; França, J.; Sousa, A.; Valente, J.; Magalhães, A.; Pardal, P. Centipede accidents notified by "Centro de Informações Toxicológicas de Belém", over a two-year period. *Rev. Soc. Bras. Med. Trop.* **2001**, *34*, 527–530. [CrossRef] [PubMed]
15. Mumcuoglu, K.Y.; Leibovici, V. Centipede (Scolopendra) bite: A case report. *Isr. J. Med. Sci.* **1989**, *25*, 47–49.
16. Harada, K.; Asa, K.; Imachi, T.; Yamaguchi, Y.; Yoshida, K.-I. Centipede inflicted postmortem injury. *J. Forensic Sci.* **1999**, *44*, 849–850. [CrossRef] [PubMed]
17. Wang, I.-K.; Hsu, S.-P.; Chi, C.-C.; Lee, K.-F.; Lin, P.Y.; Chang, H.-W.; Chuang, F.-R. Rhabdomyolysis, acute renal failure, and multiple focal neuropathies after drinking alcohol soaked with centipede. *Ren. Fail.* **2004**, *26*, 93–97. [CrossRef]
18. Ozsarac, M.; Karcioglu, O.U.; Ayrik, C.; Somuncu, F.; Gumrukcu, S. Acute coronary ischemia following centipede envenomation: Case report and review of the literature. *Wilderness Environ. Med.* **2004**, *15*, 109–112. [CrossRef]
19. Yildiz, A.; Biceroglu, S.; Yakut, N.; Bilir, C.; Akdemir, R.; Akilli, A. Acute myocardial infarction in a young man caused by centipede sting. *Emerg. Med. J.* **2006**, *23*, 3. [CrossRef]
20. Hasan, S.; Hassan, K. Proteinuria associated with centipede bite. *Pediatr. Nephrol.* **2004**, *20*, 550–551. [CrossRef]
21. Mohamed, A.H.; Zaid, E.; El-Beih, N.M.; Abd El-Aal, A.A. Effects of an extract from the centipede Scolopendra moristans on intestine, uterus and heart contractions and on blood glucose and liver and muscle glycogen levels. *Toxicon* **1980**, *18*, 581–589. [CrossRef]
22. Mohamed, A.H.; Abu-Sinna, G.; El-Shabaka, H.A.; Abd El-Aal, A. Proteins, lipids, lipoproteins and some enzyme char-acterizations of the venom extract from the centipede Scolopendra morsitans. *Toxicon* **1983**, *21*, 371–377. [CrossRef]
23. Gomes, A.; Datta, A.; Sarangi, B.; Kar, P.K.; Lahiri, S.C. Isolation, purification & pharmacodynamics of a toxin from the venom of the centipede Scolopendra subspinipes dehaani Brandt. *Indian J. Exp. Boil.* **1983**, *21*, 203–207.
24. Cooper, A.M.; Fox, G.A.; Nelsen, D.R.; Hayes, W.K. Variation in venom yield and protein concentration of the centipedes Scolopendra polymorpha and Scolopendra subspinipes. *Toxicon* **2014**, *82*, 30–51. [CrossRef] [PubMed]
25. Malta, M.B.; Lira, M.S.; Soares, S.L.; Rocha, G.C.; Knysak, I.; Martins, R.; Guizze, S.P.G.; Santoro, M.L.; Barbaro, K.C. Toxic activities of Brazilian centipede venoms. *Toxicon* **2008**, *52*, 255–263. [CrossRef] [PubMed]
26. Fung, H.T.; Lam, S.K.; Wong, O.F. Centipede bite victims: A review of patients presenting to two emergency departments in Hong Kong. *Hong Kong Med. J.* **2011**, *17*, 381–385.
27. Undheim, E.A.B.; Fry, B.G.; King, G.F. Centipede Venom: Recent Discoveries and Current State of Knowledge. *Toxins* **2015**, *7*, 679–704. [CrossRef]
28. Undheim, E.A.; Jenner, R.A.; King, G.F. Centipede venoms as a source of drug leads. *Expert Opin. Drug Discov.* **2016**, *11*, 1139–1149. [CrossRef]
29. Liu, Z.-C.; Zhang, R.; Zhao, F.; Chen, Z.-M.; Liu, H.-W.; Wang, Y.-J.; Jiang, P.; Zhang, Y.; Wu, Y.; Ding, J.-P.; et al. Venomic and Transcriptomic Analysis of Centipede Scolopendra subspinipes dehaani. *J. Proteome Res.* **2012**, *11*, 6197–6212. [CrossRef]
30. González-Morales, L.; Pedraza-Escalona, M.; Diego-Garcia, E.; Restano-Cassulini, R.; Batista, C.V.F.; Gutiérrez, M.D.C.; Possani, L.D. Proteomic characterization of the venom and transcriptomic analysis of the venomous gland from the Mexican centipede Scolopendra viridis. *J. Proteom.* **2014**, *111*, 224–237.
31. Rong, M.; Yang, S.; Wen, B.; Mo, G.; Kang, D.; Liu, J.; Lin, Z.; Jiang, W.; Li, B.; Du, C.; et al. Peptidomics combined with cDNA library unravel the diversity of centipede venom. *J. Proteom.* **2015**, *114*, 28–37. [CrossRef]

32. Smith, J.J.; Undheim, E.A.B. True Lies: Using Proteomics to Assess the Accuracy of Transcriptome-Based Venomics in Centi-pedes Uncovers False Positives and Reveals Startling Intraspecific Variation in Scolopendra Subspinipes. *Toxins* **2018**, *10*, 96. [CrossRef] [PubMed]
33. Ward, M.J.; Rokyta, D.R. Venom-gland transcriptomics and venom proteomics of the giant Florida blue centipede, Scolopendra viridis. *Toxicon* **2018**, *152*, 121–136. [CrossRef] [PubMed]
34. Yoo, W.G.; Lee, J.H.; Shin, Y.; Shim, J.-Y.; Jung, M.; Kang, B.-C.; Oh, J.; Seong, J.; Lee, H.K.; Kong, H.S.; et al. Antimicrobial peptides in the centipede Scolopendra subspinipes mutilans. *Funct. Integr. Genom.* **2014**, *14*, 275–283. [CrossRef]
35. Zhao, F.; Lan, X.; Li, T.; Xiang, Y.; Zhao, F.; Zhang, Y.; Lee, W.-H. Proteotranscriptomic Analysis and Discovery of the Profile and Diversity of Toxin-like Proteins in Centipede. *Mol. Cell. Proteom.* **2018**, *17*, 709–720. [CrossRef]
36. Undheim, E.A.; Sunagar, K.; Hamilton, B.R.; Jones, A.; Venter, D.J.; Fry, B.G.; King, G.F. Multifunctional warheads: Diversification of the toxin arsenal of centipedes via novel multidomain transcripts. *J. Proteom.* **2014**, *102*, 1–10. [CrossRef]
37. Undheim, E.A.; Grimm, L.L.; Low, C.-F.; Morgenstern, D.; Herzig, V.; Zobel-Thropp, P.; Pineda, S.S.; Habib, R.; Dziemborowicz, S.; Fry, B.G.; et al. Weaponization of a Hormone: Convergent Recruitment of Hyperglycemic Hormone into the Venom of Arthropod Predators. *Structure* **2015**, *23*, 1283–1292. [CrossRef]
38. Jenner, R.A.; von Reumont, B.M.; Campbell, L.I.; Undheim, E.A.B. Parallel Evolution of Complex Centipede Venoms Revealed by Comparative Proteotranscriptomic Analyses. *Mol. Biol. Evol.* **2019**, *36*, 2748–2763, Erratum in **2021**, *38*, 4057. [CrossRef]
39. Undheim, E.A.; Sunagar, K.; Herzig, V.; Kely, L.; Low, D.H.W.; Jackson, T.N.W.; Jones, A.; Kurniawan, N.; King, G.F.; Ali, S.A.; et al. A Proteomics and Transcriptomics Investigation of the Venom from the Barychelid Spider Trittame loki (Brush-Foot Trapdoor). *Toxins* **2013**, *5*, 2488–2503. [CrossRef] [PubMed]
40. Meng, X.; Li, C.; Xiu, C.; Zhang, J.; Li, J.; Huang, L.; Zhang, Y.; Liu, Z. Identification and Biochemical Properties of Two New Acetylcholinesterases in the Pond Wolf Spider (Pardosa pseudoannulata). *PLoS ONE* **2016**, *11*, e0158011. [CrossRef]
41. Alex, A.B.; Deshpande, S.B. Indian red scorpion venom modulates spontaneous activity of rat right atria through the in-volvement of cholinergic and adrenergic systems. *Indian J. Exp. Biol.* **1999**, *37*, 455–460. [PubMed]
42. Krayem, N.; Gargouri, Y. Scorpion venom phospholipases A2: A minireview. *Toxicon* **2020**, *184*, 48–54. [CrossRef]
43. Gonzalez-Morales, L.; Diego-Garcia, E.; Segovia, L.; Gutierrez, M.D.; Possani, L.D. Venom from the centipede Scolo-pendra viridis Say: Purification, gene cloning and phylogenetic analysis of a phospholipase A2. *Toxicon* **2009**, *54*, 8–15. [CrossRef]
44. Li, C.; Wang, F.; Aweya, J.J.; Yao, D.; Zheng, Z.; Huang, H.; Li, S.; Zhang, Y. Trypsin of Litopenaeus vannamei is required for the generation of hemocyanin-derived peptides. *Dev. Comp. Immunol.* **2018**, *79*, 95–104. [CrossRef] [PubMed]
45. Muhlia-Almazán, A.; Sánchez-Paz, A.; García-Carreño, F.L. Invertebrate trypsins: A review. *J. Comp. Physiol. B* **2008**, *178*, 655–672. [CrossRef]
46. Bottrall, J.L.; Madaras, F.; Biven, C.D.; Venning, M.G.; Mirtschin, P.J. Proteolytic activity of Elapid and Viperid Snake venoms and its implication to digestion. *J. Venom Res.* **2010**, *1*, 18–28. [PubMed]
47. Rokyta, D.R.; Ward, M.J. Venom-gland transcriptomics and venom proteomics of the black-back scorpion (Hadrurus spadix) reveal detectability challenges and an unexplored realm of animal toxin diversity. *Toxicon* **2017**, *128*, 23–37. [CrossRef]
48. Silva, J.P.; Ushkaryov, Y.A. The latrophilins, "split-personality" receptors. *Adv. Exp. Med. Biol.* **2010**, *706*, 59–75. [PubMed]
49. Jackson, V.A.; del Toro, D.; Carrasquero, M.; Roversi, P.; Harlos, K.; Klein, R.; Seiradake, E. Structural basis of latro-philin-FLRT interaction. *Structure* **2015**, *23*, 774–781. [CrossRef]
50. Wu, W.; Guan, X.; Jiang, S.; Yang, J.; Sui, N.; Chen, A.; Kuang, P.; Zhang, X. Effect of batroxobin on expression of neural cell adhesion molecule in temporal infarction rats and spatial learning and memory disorder. *J. Tradit. Chin. Med. = Chung i Tsa Chih Ying Wen Pan* **2001**, *21*, 294–298.
51. Eichberg, S.; Sanz, L.; Calvete, J.J.; Pla, D. Constructing comprehensive venom proteome reference maps for integrative venomics. *Expert Rev. Proteom.* **2015**, *12*, 557–573. [CrossRef] [PubMed]
52. Tavora, B.; Kimura, L.F.; Antoniazzi, M.M.; Chiariello, T.M.; Faquim-Mauro, E.L.; Barbaro, K.C. Involvement of mast cells and histamine in edema induced in mice by Scolopendra viridicornis centipede venom. *Toxicon* **2016**, *121*, 51–60. [CrossRef]
53. Kimura, L.F.; Prezotto-Neto, J.P.; Távora, B.D.C.L.F.; Antoniazzi, M.M.; Knysak, I.; Guizze, S.P.G.; Santoro, M.L.; Barbaro, K.C. Local inflammatory reaction induced by Scolopendra viridicornis centipede venom in mice. *Toxicon* **2013**, *76*, 239–246. [CrossRef] [PubMed]
54. Akdis, C.A.; Blaser, K. Histamine in the immune regulation of allergic inflammation. *J. Allergy Clin. Immunol.* **2003**, *112*, 15–22. [CrossRef]
55. Jutel, M.; Akdis, M.; Akdis, C.A. Histamine, histamine receptors and their role in immune pathology. *Clin. Exp. Allergy* **2009**, *39*, 1786–1800. [CrossRef]
56. Chomczynski, P.; Sacchi, N. Single-step method of RNA isolation by acid guanidinium thiocyanate-phenol-chloroform ex-traction. *Anal. Biochem.* **1987**, *162*, 156–159. [CrossRef]
57. Aronesty, E. *Command-Line Tools for Processing Biological Sequencing Data*; Ea-Utils, Expression Analysis: Durham, NC, USA, 2011.
58. Langmead, B.; Salzberg, S.L. Fast gapped-read alignment with Bowtie 2. *Nat. Methods* **2012**, *9*, 357–359. [CrossRef]
59. Bankevich, A.; Nurk, S.; Antipov, D.; Gurevich, A.A.; Dvorkin, M.; Kulikov, A.S.; Lesin, V.M.; Nikolenko, S.I.; Pham, S.; Prjibelski, A.D.; et al. SPAdes: A New Genome Assembly Algorithm and Its Applications to Single-Cell Sequencing. *J. Comput. Biol.* **2012**, *19*, 455–477. [CrossRef] [PubMed]

60. Thomas Nordahl, P.; Søren, B.; Gunnar Von, H.; Henrik, N. SignalP 4.0: Discriminating signal peptides from transmem-brane regions. *Nat. Methods* **2011**, *8*, 785.
61. Simão, F.A.; Waterhouse, R.M.; Ioannidis, P.; Kriventseva, E.V.; Zdobnov, E.M. BUSCO: Assessing genome assembly and annotation completeness with single-copy orthologs. *Bioinformatics* **2015**, *31*, 3210–3212. [CrossRef]
62. Altschul, S.F.; Madden, T.L.; Schäffer, A.A.; Zhang, J.; Zhang, Z.; Miller, W.; Lipman, D.J. Gapped BLAST and PSI-BLAST: A new generation of protein database search programs. *Nucleic Acids Res.* **1997**, *25*, 3389–3402. [CrossRef] [PubMed]
63. Mistry, J.; Finn, R.D.; Eddy, S.R.; Bateman, A.; Punta, M. Challenges in homology search: HMMER3 and convergent evolution of coiled-coil regions. *Nucleic Acids Res.* **2013**, *41*, e121. [CrossRef]
64. Finn, R.D.; Bateman, A.; Clements, J.; Coggill, P.; Eberhardt, R.Y.; Eddy, S.R.; Heger, A.; Hetherington, K.; Holm, L.; Mistry, J.; et al. Pfam: The protein families database. *Nucleic Acids Res.* **2014**, *42*, D222–D230. [CrossRef] [PubMed]
65. Li, B.; Dewey, C.N. RSEM: Accurate transcript quantification from RNA-Seq data with or without a reference genome. *BMC Bioinform.* **2011**, *12*, 323. [CrossRef] [PubMed]
66. Conesa, A.; Götz, S.; García-Gómez, J.M.; Terol, J.; Talón, M.; Robles, M. Blast2GO: A universal tool for annotation, visualization and analysis in functional genomics research. *Bioinformatics* **2005**, *21*, 3674–3676. [CrossRef] [PubMed]
67. Laemmli, U.K. Cleavage of Structural Proteins during the Assembly of the Head of Bacteriophage T4. *Nature* **1970**, *227*, 680–685. [CrossRef] [PubMed]

Article

Novel Cysteine Protease Inhibitor Derived from the *Haementeria vizottoi* Leech: Recombinant Expression, Purification, and Characterization

Débora do Carmo Linhares [1], Fernanda Faria [2,*], Roberto Tadashi Kodama [3], Adriane Michele Xavier Prado Amorim [2], Fernanda Calheta Vieira Portaro [3], Dilza Trevisan-Silva [4], Karla Fernanda Ferraz [2] and Ana Marisa Chudzinski-Tavassi [2,4,*]

1 Laboratory of Industrial Biotechnology, Institute for Technological Research, Av. Prof. Almeida Prado, 532, São Paulo 05508-901, SP, Brazil; deboradclinhares@gmail.com
2 Laboratory of Development and Innovation, Butantan Institute, Av. Vital Brasil, 1500-Butantã, São Paulo 05503-900, SP, Brazil; adriane_mi@yahoo.com.br (A.M.X.P.A.); karla.ferraz@esib.butantan.gov.br (K.F.F.)
3 Laboratory of Biomolecule Structure and Function, Butantan Institute, Av. Vital Brasil, 1500-Butantã, São Paulo 05503-900, SP, Brazil; pararoberval@gmail.com (R.T.K.); fernanda.portaro@butantan.gov.br (F.C.V.P.)
4 Centre of Excellence in New Target Discovery—CENTD, Butantan Institute, Av. Vital Brasil, 1500-Butantã, São Paulo 05503-900, SP, Brazil; dilza.silva@butantan.gov.br
* Correspondence: fernanda.faria@butantan.gov.br (F.F.); ana.chudzinski@butantan.gov.br (A.M.C.-T.)

Abstract: Cathepsin L (CatL) is a lysosomal cysteine protease primarily involved in the terminal degradation of intracellular and endocytosed proteins. More specifically, in humans, CatL has been implicated in cancer progression and metastasis, as well as coronary artery diseases and others. Given this, the search for potent CatL inhibitors is of great importance. In the search for new molecules to perform proteolytic activity regulation, salivary secretions from hematophagous animals have been an important source, as they present protease inhibitors that evolved to disable host proteases. Based on the transcriptome of the *Haementeria vizzotoi* leech, the cDNA of Cystatin-Hv was selected for this study. Cystatin-Hv was expressed in *Pichia pastoris* and purified by two chromatographic steps. The kinetic results using human CatL indicated that Cystatin-Hv, in its recombinant form, is a potent inhibitor of this protease, with a K_i value of 7.9 nM. Consequently, the present study describes, for the first time, the attainment and the biochemical characterization of a recombinant cystatin from leeches as a potent CatL inhibitor. While searching out for new molecules of therapeutic interest, this leech cystatin opens up possibilities for the future use of this molecule in studies involving cellular and in vivo models.

Keywords: leech; *Haementeria vizottoi*; cysteine proteases inhibitor; recombinant cystatin; cathepsin L

Key Contribution: Cystatin-Hv is the first cysteine protease inhibitor described from leeches and obtained in its recombinant form, allowing its biochemical characterization. It was discovered in the transcriptome of the salivary complexes from *Haementeria vizottoi* leeches, obtained in recombinant form and characterized biochemically. It was able to strongly inhibit the human cathepsin L with a K_i of 7.9 nM.

Citation: Linhares, D.d.C.; Faria, F.; Kodama, R.T.; Amorim, A.M.X.P.; Portaro, F.C.V.; Trevisan-Silva, D.; Ferraz, K.F.; Chudzinski-Tavassi, A.M. Novel Cysteine Protease Inhibitor Derived from the *Haementeria vizottoi* Leech: Recombinant Expression, Purification, and Characterization. *Toxins* **2021**, *13*, 857. https://doi.org/10.3390/toxins13120857

Received: 1 October 2021
Accepted: 11 November 2021
Published: 2 Decemeber 2021

1. Introduction

Human cysteine proteases participate in several physiological processes, such as the degradation of peptides and proteins [1], and constitute the major components of lysosomes [2]. In this group of enzymes, cathepsin L (CatL) is an endopeptidase that degrades intracellular and endocytosed proteins in the lysosome. Recent studies have suggested that this protease plays many critical roles in diverse cellular settings. Thus, overexpression of CatL has been reported in several human diseases, such as liver fibrosis, Type I and II diabetes, cardiac, bone, immune, and kidney disorders [3–5]. Additionally, membrane-bound or released CatL mediates the cleavage of the S1 subunit of the coronavirus surface spike glycoprotein, participating in the invasion into human host cells via the so-called late pathway [6,7]. Although the preferred pathway for infection is via the serine protease TMPRSS2 (early pathway), CatL, together with TMPRSS2, present tempting targets for pharmacological inhibition [7].

The endogenous inhibitors named cystatins are the most effective mechanism for controlling the activity of cathepsin L [1,2]. It is interesting to note that CatL inhibitors can also be found in several organisms, such as hematophagous animals, where they play an important role in their survival. Cystatins from hematophagous animals participate in the immunological modulations [8], reducing the processing capacity and presentation of antigens by the host's antigen-presenting cells. The production of cytokines by the host's macrophages is also affected, resulting in an anti-inflammatory response [9]. Tick cystatins have been characterized as capable of inhibiting several cathepsins involved in blood digestion, embryonic development of the tick, and the immune response of the host [10,11]. Cystatins that specifically inhibit CatL have also been described, such as sialostatin L, present in the *Ixodes scapularis* tick. Sialostatin L inhibits the protective proteolytic activity of host cells at infestation sites, thus promoting tick survival [12]. In addition, it also holds anti-inflammatory and immunosuppressive activities through the inhibition of killer T cells [13].

Tick cystatins have been described as potent and specific protease inhibitors, expanding their potential to be used further as new vaccine antigens and anti-tick drugs of medical importance [10,14]. Furthermore, studies of cystatins in cell models, such as cancer [15], inflammation [16], and immunomodulation [17], demonstrate the significant potential of cystatins as new drugs or prototypes for developing new drugs for veterinary and human use.

Similar to ticks, leeches are hematophagous parasites that possess interesting compounds in their saliva capable of assisting the maintenance of blood flow for successful feeding. Thus, many anticoagulant and antiplatelet molecules have been characterized for this purpose [18–21]. However, cysteine protease inhibitors have rarely been studied in the literature, and, until now, no cystatin—native or recombinant—was characterized for leeches apart from the cystatin B gene first characterized in *Theromyzon tessulatum* leeches. It was demonstrated that the innate immune response in the leech involves a cysteine protease inhibitor not previously detected in other invertebrate models, highlighting the need for further study of the innate immunity mechanism in these animals [22].

The present work is the first of its kind to characterize the recombinant cystatin of leeches attained through the sequence of the Hviz00340 transcript from the transcriptome of *Haementeria vizottoi* [23]. Cystatin-Hv was successfully expressed in *Pichia pastoris*, and after purification, was characterized for its ability to inhibit cathepsin L. The present study of the first recombinant cystatin derived from leeches will allow a more detailed investigation of its role in feeding the parasite. In addition, the molecule itself can be further investigated in cellular and in vivo models to better understand its potential role in the search for new molecules of therapeutic interest.

2. Results

2.1. Selection and Purification of Cystatin-Hv

The cystatin-Hv cDNA sequence, coding for the predicted cysteine protease inhibitor derived from the leech *Haementeria vizottoi*, is composed of 396 nucleotides, resulting in 131 amino acids. The signal peptidase cleavage site of Cystatin-Hv predicted by SignalP and Expasy was located between the 19th and 20th amino acid residue. To express and secrete the recombinant protein in *Pichia pastoris*, the DNA fragment corresponding to the mature protein was cloned into pD912-AK. The calculated molecular mass of this predicted protein is 12,487.92 Da, and the pI is 5.23.

This protein presents domain and typical cystatin active sites, such as the highly conserved first hairpin loop QVVAG and the second hairpin loop PW [24], associated with cystatins type C (Figure 1). The highest identity is found with a hypothetical protein of the leech *Helobdella robusta* (47% identity, access number: XP_009012188.1), presenting low identity with the preliminary characterized cystatin B of the leech *Theromyzon tessulatum* (21% identity, access number: AAN28679) [25], the Sialostasin of the tick *Ixodes scapularis* (19% identity, access number: Q8MVB6) [26], the Iristasin of the tick *Ixodes ricinus* (14% identity, access number: 5O46_A) [16], the OmC2 of the tick *Ornithodoros moubata* (21% identity, access number: 3L0R_B) [27], and with the cystatin 2a of the tick *Rhipicephalus (Boophilus) microplus* (21% identity, access number: AGW80657.1) [28].

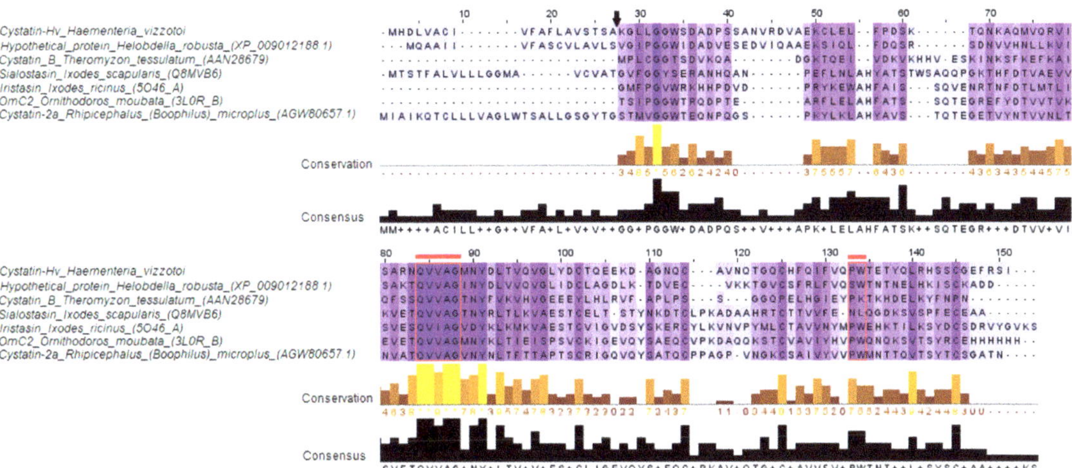

Figure 1. Multiple alignments of cystatin-Hv with other cystatins from leeches and ticks. Regions highlighted in red show the first hairpin loop and the second hairpin loop. The black arrow indicates the signal peptide cleavage site. The conservation of the sequences is shown in yellow bars, and the score values 9 and 10 (or asterisk) indicate total conservation of the aligned sequences; the consensus sequence is shown in black bars (Clustal Omega).

For the expression of the recombinant Cystatin-Hv, the selected expression vector carries a secretion signal derived from *S. cerevisiae* (SS alpha-factor), located upstream of the insert, which fused to the recombinant protein, promotes its secretion out of the cell. The production of Cystatin-Hv using *P. pastoris* (X33) was performed as described, with four 100 mL replicates, beginning the expression step with OD_{600nm} around 5, going up to OD_{600nm} 69 after 44 h of assay (average values), as shown in Figure 2.

Culture supernatant after 44 h of expression was recovered, concentrated, dialyzed, submitted to ion-exchange chromatography, and pooled fractions were analyzed on SDS-PAGE and used for inhibition assays against papain (Figure 3). Inhibitory activity was detected relative to pool 2 (eluted within 10% to 15% of NaCl 1 M buffer), evidenced by the decrease in fluorescence emission (the result of proteolysis) when compared to other pooled fractions and positive control (reaction without pooled fractions). Inhibition of papain occurred in a dose–response manner, as shown in Figure 4.

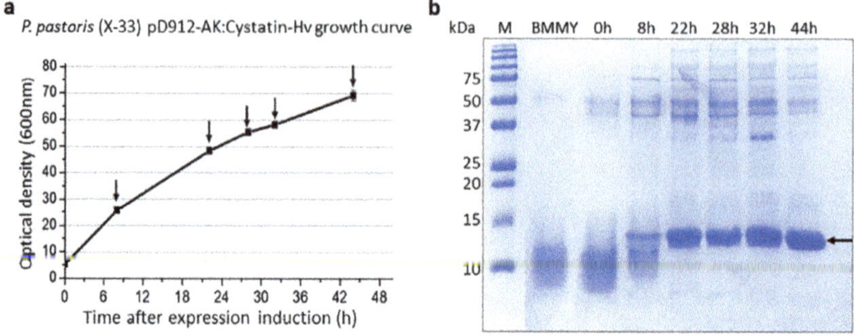

Figure 2. Growth curve of *P. pastoris* (X-33) pD912-AK: Cystatin-Hv during the assay of recombinant protein expression (**a**) and Coomassie-stained SDS-PAGE (15%) of proteins recovered from culture supernatant collected during the experiment (**b**). Vertical arrows: points of methanol feeding and supernatant collection; lane M: low molecular weight protein markers "Precision Plus Protein™ Dual Color Standards" (Bio-Rad); lane BMMY: fresh culture medium; lanes 0 h to 44 h: proteins recovered at different incubation times; horizontal arrow: expected sized protein band.

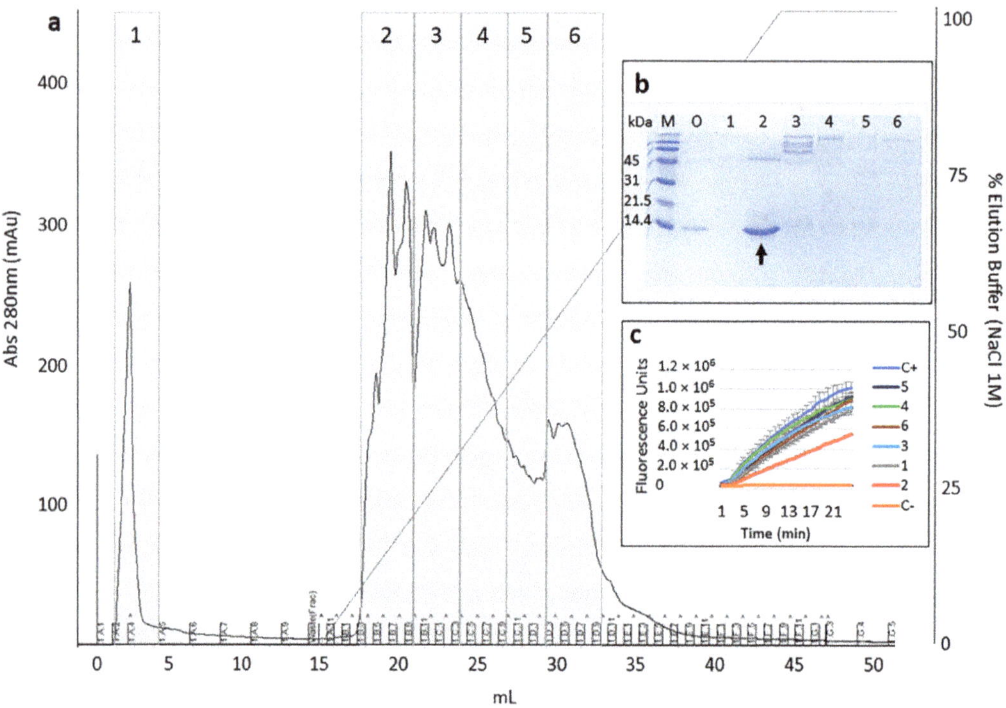

Figure 3. Chromatogram showing the elution profile of culture supernatant proteins from MonoQ resin, using 1 M NaCl as elution buffer, whose fractions were pooled 1–6 for analysis (**a**) SDS-PAGE of pooled fractions (**b**) and inhibition assay of papain activity (papain 10.7 nM, zFR-MCA substrate 5 µM) in the presence of protein pools (250 ng) (**c**). (**b**) Lane M: low molecular weight protein markers "SDS standards low range" (Bio-Rad); O: culture supernatant; lanes 1 to 6: proteins recovered from pools; vertical arrow indicates protein band compatible with Cystatin-Hv. (**c**) Here C+ indicates the positive control (papain and substrate), and C− indicates negative control (substrate).

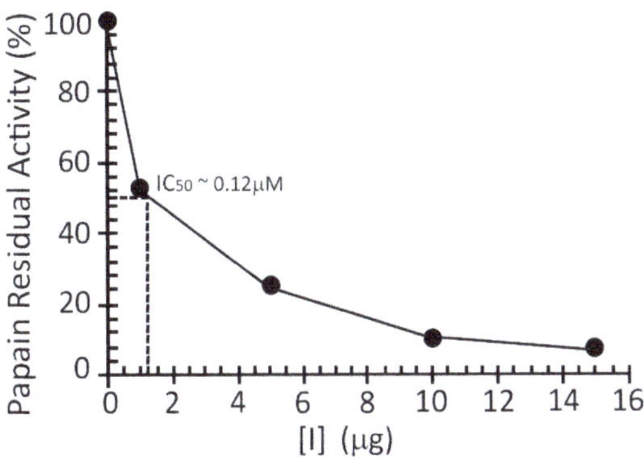

Figure 4. Inhibition assay of Cystatin-Hv (1 μg, 5 μg, 10 μg e 15 μg of pool 2 from ion-exchange chromatography) against papain (21.4 nmol/L or 50 ng) using 5 μM of Z-FR-MCA substrate. An estimate of the half maximal inhibitory concentration (IC_{50}) is given, considering the dominance of Cystatin-Hv in pool 2.

Once the inhibitory activity was detected, pooled fractions (referred to as pool 2) were further purified by size-exclusion chromatography, leading to a single band named Cystatin-Hv (Figure 5). Mass spectrometry analysis (LC-MS/MS) was performed to confirm the accuracy of the molecular mass and allowed the identification of Cystatin-Hv with eight unique peptides, covering 89% of the mature protein sequence (Supplementary Figure S1).

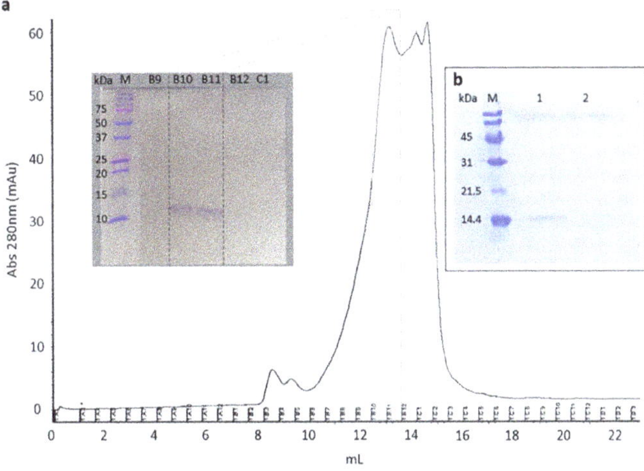

Figure 5. Size exclusion chromatography of pool 2 (from previous ion-exchange chromatography), highlighting protein bands into fractions B10 and B11 (purified Cystatin-Hv), on SDS-PAGE, related to inhibitory activity against papain and cathepsin L (**a**), SDS-PAGE analysis of purified Cystatin-Hv under reducing (1) and nonreducing (2) conditions (**b**). Molecular markers used were "Precision Plus Protein™ Dual Color Standards" (Bio-Rad) (**a**) and "SDS standards low range" (Bio-Rad) (**b**).

2.2. Inhibition Studies

Enzymatic kinetics assays were performed with three different concentrations of Cystatin-Hv (8 nM, 16 nM, and 24 nM respectively), cathepsin L (0.4 nM) and two concentrations of Z-FR-MCA substrate (1 K_m and 2 K_m). Experimental data of reaction rates were linearized and treated further as proposed by Dixon [29] (Figure 6), thus allowing us to determine the inhibition constant, K_i, of 7.9 nM.

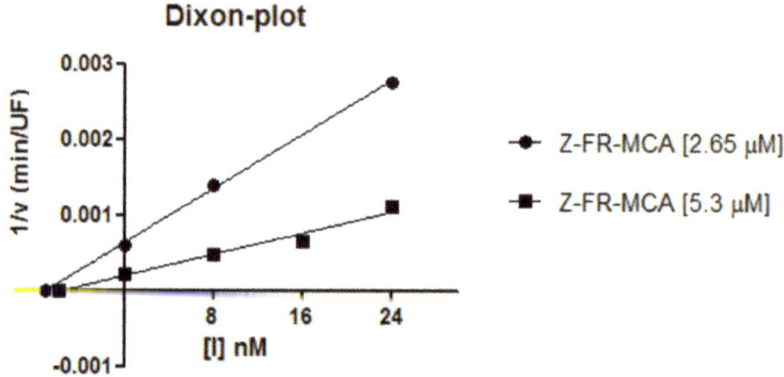

Figure 6. Dixon diagram presenting kinetic enzymatic assays of cathepsin L inhibition by Cystatin-Hv. Two substrate concentrations (1 K_m or 2.65 µM and 2 K_m or 5.3 µM) and three cystatin-Hv concentrations (8 nM, 16 nM, and 24 nM) were used. The inhibition constant was set as 7.9 nM. The graphic was made using the software GraphPad Prism 5. Slope value: 7.964×10^{-5} to 9.909×10^{-5} (2.65 µM) and 2.496×10^{-5} to 4.547×10^{-5} (5.3 µM) for a 95% confidence interval. UF: Units of Fluorescence.

3. Discussion

The present work was carried out to assess biodiversity and contribute to developing new molecules that can generate and inspire new therapeutic possibilities. In this context, the search for molecules from animal secretions related to feeding is rather compelling since, from the evolutionary perspective, the proteins present in such secretions have been subjected to selective pressure for better efficiency to ultimately facilitate the animal's survival and perpetuation [30]. Hence, it is expected that proteins present in the saliva of hematophagous animals should have a specific action on the host or prey, and it is up to the researchers to isolate these components, identify their actions, and study how these molecules can be used for our benefit.

Cystatins present in hematophagous have the key function in inhibiting endogenous cysteine proteases of the animal and in helping the feeding process as well. The saliva of these animals contains not only inhibitors that reduce host blood clotting and premature blood clotting inside the gut but also molecules that interfere and inhibit the performance of the host's immune system [17,24] and allow the hematophagous to keep feeding for an extended period.

An important dimension in the discussion on cystatins present in the salivary complexes of leeches relates to the issue of the innate immune response. It is reported that symbiotic bacteria, antimicrobial peptides, and phagocytic immune cells play a protective role in defending from harmful agents and preventing premature degradation of the ingested blood meal, which is concentrated and maintained over a period of many weeks inside the digestive tract [31]. Although the leech defense system has been poorly investigated, studies with cystatin B have demonstrated the involvement of this cysteine protease inhibitor in the innate immunity of *Theromyzon tessulatum* leeches since an increase in cystatin B gene expression has been shown in large circulating coelomic cells after bacterial challenge [22,25]. While more studies are needed to further elucidate the function of cysteine protease inhibitors for leeches, it is likely that these molecules also work as immunoregulators, given the major implication of cathepsins in immunity [22], similarly to what has been described for ticks, a better characterized group of hematophagous.

In ticks, this group of inhibitors has been extensively explored. It was noted that in tick saliva, the majority (84%) of cystatin transcripts belong to a group that is secreted extracellularly, suggesting a predominantly immunoregulation function [32]. Cystatin OmC2, from the *Ornithodoros moubata* tick, for example, targets two lysosomal cathepsins, S and C, which perform the function of processing antigens in antigen-presenting cells, apart from affecting the maturation of dendritic cells [17]. Cystatin Iristatin, identified from the tick *Ixodes ricinus*, inhibited the proteolytic activity of cathepsins L and C and decreased the production of several inflammation inducers (IL-1, IL-4, IL-9, IFN-γ) by different populations of T cells, among other anti-inflammatory activities [16].

Cystatins are also present in humans, where, as in other animal species, they act as inhibitors of endogenous cysteine proteases, such as cathepsins. Overexpression of these enzymes has been observed in a number of tumorous cells, such as breast, lung, brain, head, neck, and melanoma cancers, where they act on the degradation of the extracellular matrix enabling tumor growth, invasion of other tissues, and migration into the bloodstream [2,8]. In particular, cathepsin L is a lysosomal endopeptidase widely expressed and involved in the degradation of intracellular or phagocyted proteins that can also be found in a variety of extracellular media as well as in the cell nucleus [33,34]. In this way, positive regulation of the lysosomal endopeptidase cathepsin L has often been observed in a number of human cancers, and its levels of expression in tumor tissues or their presence in the environment adjacent to the tumors is considered to be largely correlated with their aggressiveness [2,34–36].

There is little information available about cystatins regarding leeches, most of which are the results of transcriptomic analyses suggesting the participation of these molecules in the immune response [25]. Functional studies with cystatins present in the leeches have not yet been reported in the literature.

The present study started with the library of transcripts of the salivary complexes of the leech *Haementeria vizottoi*, where 1204 Isotigs were obtained, and among them, 123 were identified as related to feeding [23]. After further screening, one Isotig was selected for this study, starting with the gene sequence, through the cloning and recombinant production of the protein, Cystatin-Hv, to its functional characterization.

In general, the benefits of protein production by *P. pastoris* system include appropriate folding, especially for cysteine-rich proteins (in the endoplasmic reticulum) and secretion (by Kex2 as signal peptidase) of recombinant proteins to the supernatant environment of the expression [37]. In the case of Cystatin-Hv, a protein with five cysteines, the expression occurred satisfactorily, as expected, with compatible quality acceptable to the scalability of the process. Furthermore, the use of the *P. pastoris* expression system, due to its limited production of endogenous secretory proteins, is known to favor an easy purification protein process [37]. In this sense, the isolation in two chromatography steps was sufficient to achieve a pure form of recombinant Cystatin-Hv, similar to the purification process performed by Cardoso [38], characterizing a tick cystatin that presented an inhibitory effect against the activity of a hemoglobin lytic enzyme.

The inhibition assays allowed us to confirm the activity of Cystatin-Hv, in its recombinant form, as a strong inhibitor of cathepsin L. Further, results plotted in the Dixon diagram (Figure 6), with curves intercept on the X-axis, suggest a noncompetitive mechanism of action for this inhibitor. Although cystatins are usually described as competitive inhibitors, the noncompetitive mechanism was observed for soybean [39], corn [40], and chestnut seed [41] plant cystatins, as well as for human Cystatin SA [42]. In order to improve Cystatin-Hv characterization and understanding of its mechanism, complementary assays are to be performed, also against other known cathepsins. The inhibition constant (K_i) in the order of nM (7.9 nM) is compatible with the one found in the literature for the dissociation constant of cathepsin L with human cystatins [1]. Similar K_i values were obtained related to cystatins of hematophagous animals such as the bovine ectoparasite *Rhipicephalus microplus*, whose protein identified as Rmcystatin-4 was cloned, expressed, and purified, and has demonstrated inhibitory activity against cathepsin L with a K_i of 11.1 nM [38]. The cystatin OmC2 from the tick genus *Ornithodoros* also presented similar K_i values in the range of nM against lysosomal cathepsins S and C [17].

Although the K_i value in relation to papain has not been obtained, the IC_{50} value of approximately 0.12 μM, considering Cystatin-Hv dominant in pool 2, indicates a greater potency of cystatin-Hv for the inhibition of cathepsin L. However, future studies should be carried out with papain and other cathepsins to assess the specificity of cystatin-Hv in relation to a particular protease.

The character of recombinant Cystatin-Hv as an inhibitor of cysteine proteases, especially human cathepsin L, opens interesting possibilities for its potential biological function as an immunoregulator and an anti-inflammatory molecule, justifying our efforts to study this protein in its recombinant form. Inhibition of CatL has also been recognized as having a significant role in the prevention of cell invasion by viruses of the coronavirus family in vitro. Given the recent emergence of the novel SARS-CoV-2, calls for more attention to inhibitors of this cysteine protease are well justified [7]. Thus, the first recombinant cystatin from leeches will allow a more detailed investigation of its role in feeding the parasite. In addition, the molecule itself can be investigated in cellular and in vivo models to understand its significance in the possible search for new molecules of therapeutic interest.

4. Conclusions

The present work is the first of its kind to characterize the recombinant cystatin of leeches attained through the sequence of the transcript Hviz00340 from the transcriptome of *Haementeria vizottoi* [23]. Cystatin-Hv was successfully expressed in *Pichia pastoris*, and after purification, it was characterized for its ability to inhibit cathepsin L. Kinetic studies have indicated that recombinant Cystatin-Hv is a potent inhibitor of cathepsin L, with a K_i of 7.9 nM. Thus, a rigorous study of this molecule could be promising, and future work will be carried out in this direction to better characterize the therapeutic potential of Cystatin-Hv.

5. Materials and Methods

5.1. Strains, Plasmids, Enzymes

The expression vector pD912-AK with the synthesized insert of interest (codon-optimized) was purchased from ATUM 2.0 (Newark, NJ, USA), and *P. pastoris* strain X-33 (Invitrogen, Waltham, MA, USA) was used as the expression host. The *Escherichia coli* strain DH5α and restriction enzyme SacI were purchased from Thermo Fisher (Waltham, MA, USA). *E. coli* cells with plasmids were cultured at 37 °C in Luria–Bertani medium (yeast extract, 5 g/L; tryptone, 10 g/L; NaCl, 10 g/L; agar, 15 g/L) containing 25 μg/mL Zeocin (Invitrogen, Waltham, MA, USA). Papain was purchased from Sigma-Aldrich (St. Louis, MO, USA) and human cathepsin L from R&D Systems (Minneapolis, MN, USA).

5.2. Sequence Source and In Silico Characterization

The Cystatin-Hv cDNA sequence (transcript Hviz00340) was obtained from the sialotranscriptome of *Haementeria vizzotoi* leech [23]. The signal peptide sequence, determined by SignalP 4.0 [43], was excluded from further analysis and from the insert synthesis. Theoretical pI and Mw were determined using the Expasy platform. Identity and similarity percentages of the full-length amino acid sequence were obtained by BLAST search (NCBI database), and multiple sequence alignments were performed on a sequence of Cystatin-Hv versus known cystatins, using Clustal Omega [44].

5.3. Expression and Purification of Recombinant Protein

The vector pD912-AK: Cystatin-Hv was linearized with SacI and electroporated into competent *P. pastoris* X-33 cells. Transformants were screened on YPD medium plates containing 25 μg/mL Zeocin, and the presence of Cystatin-Hv insert was confirmed by PCR. Expression was carried out in replicates inoculating 50 mL of BMGY medium [1.0% yeast extract, 2.0% peptone, 100 mM potassium phosphate pH 6.0, 1.34% YNB, 4×10^{-5}% D-biotin (w/v), and 1% glycerol (v/v)] and cultivated under the influence of 350 rpm orbital shaking at 28 °C for 24 h. Cells were harvested by centrifugation at 450× g for 5 min at 4 °C and resuspended in 50 mL of BMMY medium [1.0% yeast extract, 2.0% peptone, 100 mM potassium phosphate pH 6.0, 1.34% YNB, 4×10^{-5}% D-biotin (w/v), and 0.5% methanol (v/v)] to absorbance at 600 nm of 5.0. Incubation was carried out at 30 °C and 350 rpm orbital shaking for 44 h, with further additions of methanol to a final concentration of 0.5% every 12 h, approximately. Samples (1 mL) were taken during the assay, submitted to protein precipitation with methanol/chloroform [45], and analyzed by SDS-PAGE.

Cells were removed from the supernatant by centrifugation (3500× *g* for 15 min at 4 °C) and filtration (0.45 µm). The supernatant was dialyzed (5 kDa molecular exclusion) and concentrated with 20 mM Tris-HCl pH 8.0 using Cogent µScale TFF System (Merck, Darmstadt, Germany) and submitted to ion-exchange chromatography in a Mono Q 5/50 GL (GE Healthcare) 1 mL column connected to an AKTA Avant system (GE Healthcare), equilibrated with 20 mM Tris-HCl pH 8.0. The sample was added, and the column was washed with 15 CV (column volume) of 20 mM Tris-HCl pH 8.0 (0.5 mL/min), followed by the elution step supported by a crescent linear gradient of 20 mM Tris-HCl pH 8.0, 1.0 M NaCl along 30 CV (0.5 mL/min). Fractions (~500 µL) were pooled, analyzed by SDS-PAGE, and the one presenting inhibitory activity towards papain was applied on Superdex 75 10/300 column (GE Healthcare), being eluted with 20 mM Tris-HCl pH 8.0 along 2 CV (1 mL/min). Fractions with expected molecular weight, single band, were pooled, quantified using the bicinchoninic acid (BCA) Protein Assay Kit (Pierce, WA, USA) and further analyzed for inhibitory activity against papain and cathepsin L.

5.4. Mass Spectrometry for Sequence Confirmation

A purified sample of the recombinant Cystatin-Hv was submitted to in-solution trypsin digestion prior to mass spectrometry analysis by LC-MS/MS. The generated tryptic peptides were desalted, dried, and dissolved in 20 µL of 0.1% (*v*/*v*) formic acid, and 2 µL were automatically injected into a 2 cm C-18 trap column (3 µm particle size, 100 Å pore size, 75 µm I.D., Thermo Fisher Scientific, Waltham, MA, USA) by an Easy nanoLC 1200 coupled to a QExactive plus (Thermo Fisher Scientific, Waltham, MA, USA) mass spectrometer. Chromatographic separation of tryptic peptides was performed on a 15 cm long analytical column (Acclaim PepMap, 2 µm particle size, 100 Å pore size, 50 µm I.D.—Thermo Fisher Scientific, Waltham, MA, USA). Peptides were eluted with a linear gradient of 5–100% Buffer B (80% acetonitrile in 0.1% formic acid) at 200 nL/min for 30 min. The spray voltage was set to 2.4 kV, and the mass spectrometer was operated in positive, data-dependent mode, in which one full MS scan was acquired in the m/z range of 300–1500 followed by MS/MS acquisition using high-energy collisional dissociation (HCD) of the seven most intense ions from the MS scan using an isolation window of 2.0 *m*/*z*.

The obtained MS and MS/MS spectra were analyzed using PEAKS Studio X, and the searches were performed against a customized database. Briefly, the database used included all *Pichia pastoris* protein sequences downloaded from UniProt (a total of 16,348 sequences, downloaded on 14 October 2021) and the translated amino acid sequence of Cystatin-Hv (without the signal peptide sequence). This reference database was concatenated with common contaminants for mass spectrometry experiments (116 sequences), and the decoy sequences were used for false discovery (FDR) rate control. The search engine was set to detect specific tryptic peptides at an FDR of 1%, allowing two missed cleavages. Methionine oxidation, acetylation of the protein N-termini, and deamidation of asparagine and guanidine were set as variable modifications, and carbamidomethylation of cysteine was set as a fixed modification.

5.5. Inhibitory Assays

Papain was utilized to select the chromatographic fractions that contained Cystatin-Hv, and, after obtaining the inhibitor in its homogeneous form, cathepsin L was used to determine the value of the inhibition constant (K_i). The assays were implemented according to Portaro et al. (2000) [46] with some minor modifications. Enzymes were preactivated for 15 min at room temperature with 6 mM DTT in 50 mM sodium phosphate, 200 mM NaCl, 5 mM EDTA, and pH 5.5 (final volume 100 µL). For the pool selection steps, 10 ng of papain, 5 µM of fluorogenic substrate Z-FR-AMC (Sigma-Aldrich, St. Louis, MO, USA) and 250 ng of protein from pooled purification fractions were used. Cathepsin L (0.4 nM) was employed against three concentrations of purified Cystatin-Hv (8 nM, 16 nM and 24 nM) and two concentrations of the fluorogenic substrate Z-FR-AMC (1 K_m and 2 K_m, where the K_m = 2.6 μM, ref. [47]) to determine the K_i value [48]. Control reactions were carried out in the same conditions but without Cystatin-Hv. The activity was measured (fluorescence at λ_{EM} 480 nm and λ_{EX} 360 nm) in a Victor 3 (Perkin Elmer, Boston, MA, USA) plate reader. The temperature remained constant at 37 °C, and one reading per minute was performed for 15 min, the plates being shaken before each measurement. The residual activity of human cathepsin L in the presence of Cystatin-Hv in different amounts was determined, and the inhibition constant (K_i) of Cystatin-Hv towards human cathepsin L was determined by the Dixon Plot equation (1/V vs. [I]) [49], using the software GraphPad Prism 5.

Supplementary Materials: The following are available online at https://www.mdpi.com/article/10.3390/toxins13120857/s1, Figure S1: Amino acid sequence of the mature Hviz340 recombinant protein and the tryptic peptides identified by LC-MS/MS analysis using a QExactive plus mass spectrometer and database search in PEAKS Studio X.

Author Contributions: Conceptualization, D.d.C.L., F.F. and A.M.C.-T.; methodology, D.d.C.L., F.F., R.T.K., F.C.V.P. and K.F.F.; investigation, D.d.C.L. and R.T.K.; formal analysis, D.d.C.L., R.T.K., A.M.X.P.A. and D.T.-S.; writing—original draft preparation, D.d.C.L., F.F. and F.C.V.P.; writing—review and editing, D.d.C.L., F.F., R.T.K., F.C.V.P., A.M.X.P.A., D.T.-S. and A.M.C.-T.; supervision, F.F. and A.M.C.-T.; funding acquisition, D.d.C.L., F.F. and A.M.C.-T. All authors have read and agreed to the published version of the manuscript.

Funding: This research was funded by Support Foundation for the Technological Research Institute: N°500115A/PTC1043/15; São Paulo Research Foundation (FAPESP): #2013/07467-1, #2015/13124-5 and #2019/20832-7; FAPESP and GlaxoSmithKline: #2016/06026-0 and #2015/50040-4.

Acknowledgments: Authors acknowledges Tatiana Joly Siquini for the assistance with culture procedures and Viviane Barbosa Portas and Melissa Regina Fessel for the assistance with protein purification techniques and equipment handling.

Conflicts of Interest: The authors declare no conflict of interest.

References

1. Grzonka, Z.; Jankowska, E.; Kasprzykowski, F.; Kasprzykowska, R.; Lankiewicz, L.; Wiczk, W.; Wieczerzak, E.; Ciarkowski, J.; Drabik, P.; Janowski, R.; et al. Structural studies of cysteine proteases and their inhibitors. *Acta Biochim. Pol.* **2001**, *48*, 1–20. [CrossRef]
2. Sudhan, D.; Siemann, D.W. Cathepsin L targeting in cancer treatment. *Pharmacol. Ther.* **2015**, *155*, 105–116. [CrossRef] [PubMed]
3. Dana, D.; Pathak, S.K. A Review of Small Molecule Inhibitors and Functional Probes of Human Cathepsin L. *Molecules* **2020**, *25*, 698. [CrossRef] [PubMed]
4. Manchanda, M.; Das, P.; Gahlot, G.P.S.; Singh, R.; Roeb, E.; Roderfeld, M.; Gupta, S.D.; Saraya, A.; Pandey, R.M.; Chauhan, S.S. Cathepsin L and B as Potential Markers for Liver Fibrosis: Insights From Patients and Experimental Models. *Clin. Transl. Gastroenterol.* **2017**, *8*, e99. [CrossRef]
5. Liu, C.-L.; Guo, J.; Zhang, X.; Sukhova, G.; Libby, P.; Shi, G.-P. Cysteine protease cathepsins in cardiovascular disease: From basic research to clinical trials. *Nat. Rev. Cardiol.* **2018**, *15*, 351–370. [CrossRef]
6. Takeda, M. Proteolytic activation of SARS-CoV-2 spike protein. *Microbiol. Immunol.* **2021**, 1–9. [CrossRef]
7. Liu, T.; Luo, S.; Libby, P.; Shi, G.-P. Cathepsin L-selective inhibitors: A potentially promising treatment for COVID-19 patients. *Pharmacol. Ther.* **2020**, *213*, 107587. [CrossRef] [PubMed]
8. Ochieng, J.; Chaudhuri, G. Cystatin Superfamily. *J. Health Care Poor Underserved* **2010**, *21*, 51–70. [CrossRef] [PubMed]
9. Schönemeyer, A.; Lucius, R.; Sonnenburg, B.; Brattig, N.; Sabat, R.; Schilling, K.; Bradley, J.; Hartmann, S. Modulation of Human T Cell Responses and Macrophage Functions by Onchocystatin, a Secreted Protein of the Filarial Nematode *Onchocerca volvulus*. *J. Immunol.* **2001**, *167*, 3207–3215. [CrossRef]
10. Schwarz, A.; Valdés, J.J.; Kotsyfakis, M. The role of cystatins in tick physiology and blood feeding. *Ticks Tick-Borne Dis.* **2012**, *3*, 117–127. [CrossRef]
11. Lu, S.; da Rocha, L.A.; Torquato, R.J.; Junior, I.D.S.V.; Florin-Christensen, M.; Tanaka, A.S. A novel type 1 cystatin involved in the regulation of *Rhipicephalus microplus* midgut cysteine proteases. *Ticks Tick-Borne Dis.* **2020**, *11*, 101374. [CrossRef] [PubMed]
12. Kotsyfakis, M.; Sa-Nunes, A.; Francischetti, I.M.B.; Mather, T.N.; Andersen, J.F.; Ribeiro, J. Antiinflammatory and Immunosuppressive Activity of Sialostatin L, a Salivary Cystatin from the Tick *Ixodes scapularis*. *J. Biol. Chem.* **2006**, *281*, 26298–26307. [CrossRef]

13. Horka, H.; Staudt, V.; Klein, M.; Taube, C.; Reuter, S.; Dehzad, N.; Andersen, J.F.; Kopecky, J.; Schild, H.; Kotsyfakis, M.; et al. The Tick Salivary Protein Sialostatin L Inhibits the Th9-Derived Production of the Asthma-Promoting Cytokine IL-9 and Is Effective in the Prevention of Experimental Asthma. *J. Immunol.* **2012**, *188*, 2669–2676. [CrossRef]
14. Parizi, L.F.; Rangel, C.K.; Sabadin, G.A.; Saggin, B.F.; Kiio, I.; Xavier, M.A.; Matos, R.D.S.; Camargo-Mathias, M.I.; Seixas, A.; Konnai, S.; et al. *Rhipicephalus microplus* cystatin as a potential cross-protective tick vaccine against *Rhipicephalus appendiculatus*. *Ticks Tick-Borne Dis.* **2020**, *11*, 101378. [CrossRef]
15. Wei, N.; Lin, Z.; Xu, Z.; Cao, J.; Zhou, Y.; Zhang, H.; Gong, H.; Zhou, J.; Li, G. A Tick Cysteine Protease Inhibitor RHcyst-1 Exhibits Antitumor Potential. *Cell. Physiol. Biochem.* **2018**, *46*, 2385–2400. [CrossRef]
16. Kotál, J.; Stergiou, N.; Buša, M.; Chlastáková, A.; Beránková, Z.; Řezáčová, P.; Langhansová, H.; Schwarz, A.; Calvo, E.; Kopecký, J.; et al. The structure and function of Iristatin, a novel immunosuppressive tick salivary cystatin. *Cell. Mol. Life Sci.* **2019**, *76*, 2003–2013. [CrossRef] [PubMed]
17. Zavašnik-Bergant, T.; Vidmar, R.; Sekirnik, A.; Fonović, M.; Salat, J.; Grunclová, L.; Kopáček, P.; Turk, B. Salivary Tick Cystatin OmC2 Targets Lysosomal Cathepsins S and C in Human Dendritic Cells. *Front. Cell. Infect. Microbiol.* **2017**, *7*, 288. [CrossRef] [PubMed]
18. Faria, F.; Kelen, E.M.A.; Sampaio, C.A.M.; Bon, C.; Duval, N.; Chudzinski-Tavassi, A.M. A New Factor Xa Inhibitor (Lefaxin) from the *Haementeria depressa* Leech. *Thromb. Haemost.* **1999**, *82*, 1469–1473. [CrossRef]
19. Chudzinski-Tavassi, A.M.; Bermej, E.; Rosenstein, R.E.; Faria, F.; Sarmiento, M.I.K.; Alberto, F.; Sampaio, M.U.; Lazzari, M.A. Nitridergic Platelet Pathway Activation by Hementerin, a Metalloprotease from the Leech *Haementeria depressa*. *Biol. Chem.* **2003**, *384*, 1333–1339. [CrossRef]
20. Salzet, M.; Chopin, V.; Baert, J.-L.; Matias, I.; Malecha, J. Theromin, a Novel Leech Thrombin Inhibitor. *J. Biol. Chem.* **2000**, *275*, 30774–30780. [CrossRef]
21. Oliveira, D.G.L.; Alvarez-Flores, M.P.; Lopes, A.R.; Chudzinski-Tavassi, A.M. Functional characterisation of Vizottin, the first factor Xa inhibitor purified from the leech *Haementeria vizottoi*. *Thromb. Haemost.* **2012**, *108*, 570–578. [CrossRef] [PubMed]
22. Lefebvre, C.; Cocquerelle, C.; Vandenbulcke, F.; Hot, D.; Huot, L.; Lemoine, Y.; Salzet, M. Transcriptomic analysis in the leech *Theromyzon tessulatum*: Involvement of cystatin B in innate immunity. *Biochem. J.* **2004**, *380*, 617–625. [CrossRef]
23. Amorim, A.M.X.P.; De Oliveira, U.C.; Faria, F.; Pasqualoto, K.F.M.; Junqueira-De-Azevedo, I.D.L.; Chudzinski-Tavassi, A.M. Transcripts involved in hemostasis: Exploring salivary complexes from *Haementeria vizottoi* leeches through transcriptomics, phylogenetic studies and structural features. *Toxicon* **2015**, *106*, 20–29. [CrossRef]
24. Zi, M.; Xu, Y. Involvement of cystatin C in immunity and apoptosis. *Immunol. Lett.* **2018**, *196*, 80–90. [CrossRef]
25. Lefebvre, C.; Vandenbulcke, F.; Bocquet, B.; Tasiemski, A.; Desmons, A.; Verstraete, M.; Salzet, M.; Cocquerelle, C. Cathepsin L and cystatin B gene expression discriminates immune cœlomic cells in the leech *Theromyzon tessulatum*. *Dev. Comp. Immunol.* **2008**, *32*, 795–807. [CrossRef]
26. Valenzuela, J.G.; Francischetti, I.M.B.; Pham, V.M.; Garfield, M.K.; Mather, T.N.; Ribeiro, J.M.C. Exploring the sialome of the tick *Ixodes scapularis*. *J. Exp. Biol.* **2002**, *205*, 2843–2864. [CrossRef]
27. Salát, J.; Paesen, G.C.; Řezáčová, P.; Kotsyfakis, M.; Kovářová, Z.; Šanda, M.; Majtán, J.; Grunclová, L.; Horká, H.; Andersen, J.F.; et al. Crystal structure and functional characterization of an immunomodulatory salivary cystatin from the soft tick *Ornithodoros moubata*. *Biochem. J.* **2010**, *429*, 103–112. [CrossRef] [PubMed]
28. Parizi, L.F.; Githaka, N.W.; Acevedo, C.; Benavides, U.; Seixas, A.; Logullo, C.; Konnai, S.; Ohashi, K.; Masuda, A.; da Silva Vaz, I., Jr. Sequence characterization and immunogenicity of cystatins from the cattle tick *Rhipicephalus* (*Boophilus*) *microplus*. *Ticks Tick-Borne Dis.* **2013**, *4*, 492–499. [CrossRef]
29. Copeland, R.A. *Enzymes: A Practical Introduction to Structure, Mechanism, and Data Analysis*, 2nd ed.; Viley-VCH: New York, NY, USA, 2000.
30. King, G.F. Venoms as a platform for human drugs: Translating toxins into therapeutics. *Expert Opin. Biol. Ther.* **2011**, *11*, 1469–1484. [CrossRef] [PubMed]
31. Silver, A.; Graf, J. Innate and procured immunity inside the digestive tract of the medicinal leech. *Invertebr. Surviv. J. ISJ* **2011**, *8*, 173–178.
32. Chmelař, J.; Kotál, J.; Langhansová, H.; Kotsyfakis, M. Protease Inhibitors in Tick Saliva: The Role of Serpins and Cystatins in Tick-host-Pathogen Interaction. *Front. Cell. Infect. Microbiol.* **2017**, *7*, 216. [CrossRef]
33. Reiser, J.; Adair, B.; Reinheckel, T. Specialized roles for cysteine cathepsins in health and disease. *J. Clin. Investig.* **2010**, *120*, 3421–3431. [CrossRef] [PubMed]
34. Goulet, B.; Sansregret, L.; Leduy, L.; Bogyo, M.; Weber, E.; Chauhan, S.S.; Nepveu, A. Increased Expression and Activity of Nuclear Cathepsin L in Cancer Cells Suggests a Novel Mechanism of Cell Transformation. *Mol. Cancer Res.* **2007**, *5*, 899–907. [CrossRef]
35. Berdowska, I. Cysteine proteases as disease markers. *Clin. Chim. Acta* **2004**, *342*, 41–69. [CrossRef]
36. Tamhane, T.; Lllukkumbura, R.; Lu, S.; Maelandsmo, G.M.; Haugen, M.H.; Brix, K. Nuclear cathepsin L activity is required for cell cycle progression of colorectal carcinoma cells. *Biochimie* **2016**, *122*, 208–218. [CrossRef] [PubMed]
37. Karbalaei, M.; Rezaee, S.A.; Farsiani, H. *Pichia pastoris*: A highly successful expression system for optimal synthesis of heterologous proteins. *J. Cell. Physiol.* **2020**, *235*, 5867–5881. [CrossRef]
38. Cardoso, T.H.; Lu, S.; Gonzalez, B.R.; Torquato, R.J.; Tanaka, A.S. Characterization of a novel cystatin type 2 from *Rhipicephalus microplus* midgut. *Biochimie* **2017**, *140*, 117–121. [CrossRef] [PubMed]

39. Zhao, Y.; Botella, M. Ángel; Subramanian, L.; Niu, X.; Nielsen, S.S.; Bressan, R.A.; Hasegawa, P.M. Two Wound-Inducible Soybean Cysteine Proteinase Inhibitors Have Greater Insect Digestive Proteinase Inhibitory Activities than a Constitutive Homolog. *Plant Physiol.* **1996**, *111*, 1299–1306. [CrossRef]
40. Abe, M.; Abe, K.; Iwabuchi, K.; Domoto, C.; Arai, S. Corn Cystatin I Expressed in *Escherichia coli*: Investigation of Its Inhibitory Profile and Occurrence in Corn Kernels. *J. Biochem.* **1994**, *116*, 488–492. [CrossRef]
41. Pernas, M.; Sánchez-Monge, R.; Gómez, L.; Salcedo, G. A chestnut seed cystatin differentially effective against cysteine proteinases from closely related pests. *Plant Mol. Biol.* **1998**, *38*, 1235–1242. [CrossRef]
42. Eliyahu, E.; Shtraizent, N.; He, X.; Chen, D.; Shalgi, R.; Schuchman, E.H. Identification of Cystatin SA as a Novel Inhibitor of Acid Ceramidase. *J. Biol. Chem.* **2011**, *286*, 35624–35633. [CrossRef] [PubMed]
43. Petersen, T.N.; Brunak, S.; von Heijne, G.; Nielsen, H. SignalP 4.0: Discriminating signal peptides from transmembrane regions. *Nat. Methods* **2011**, *8*, 785–786. [CrossRef] [PubMed]
44. Sievers, F.; Wilm, A.; Dineen, D.; Gibson, T.J.; Karplus, K.; Li, W.; López, R.; McWilliam, H.; Remmert, M.; Söding, J.; et al. Fast, scalable generation of high-quality protein multiple sequence alignments using Clustal Omega. *Mol. Syst. Biol.* **2011**, *7*, 539. [CrossRef]
45. Wessel, D.; Flügge, U. A method for the quantitative recovery of protein in dilute solution in the presence of detergents and lipids. *Anal. Biochem.* **1984**, *138*, 141–143. [CrossRef]
46. Portaro, F.C.V.; Santos, A.B.F.; Cezari, M.H.S.; Juliano, M.A.; Juliano, L.; Carmona, E. Probing the specificity of cysteine proteinases at subsites remote from the active site: Analysis of P4, P3, P2′ and P3′ variations in extended substrates. *Biochem. J.* **2000**, *347*, 123–129. [CrossRef]
47. Kodama, R.T.; Kuniyoshi, A.K.; Da Silva, C.C.F.; Cajado-Carvalho, D.; Duzzi, B.; Mariano, D.C.; Pimenta, D.C.; Borges, R.; Da Silva, W.D.; Portaro, F.C.V. A Kunitz-type peptide from *Dendroaspis polylepis* venom as a simultaneous inhibitor of serine and cysteine proteases. *J. Venom. Anim. Toxins Incl. Trop. Dis.* **2020**, *26*, 1–13. [CrossRef] [PubMed]
48. Portaro, F.C.V.; Cezari, M.H.S.; Juliano, M.A.; Juliano, L.; Walmsley, A.R.; Prado, E.S. Design of kallidin-releasing tissue kallikrein inhibitors based on the specificities of the enzyme's binding subsites. *Biochem. J.* **1997**, *323*, 167–171. [CrossRef] [PubMed]
49. Segel, I. *Enzyme Kinetics Behavior and Analysis of Rapid Equilibrium and Steady State Enzyme Systems*; John Wiley: New York, NY, USA, 1976.

MDPI

Article

Scorpion Envenomation of Lactating Rats Decreases the Seizure Threshold in Offspring

Marina de Oliveira Rodrigues Barbosa, Maria Eliza F. do Val de Paulo and Ana Leonor Abrahão Nencioni *

Laboratory of Pharmacology, Butantan Institute, Av. Dr. Vital Brazil 1500, São Paulo 05503-900, Brazil; orb.marina@gmail.com (M.d.O.R.B.); maria.paulo@butantan.gov.br (M.E.F.d.V.d.P.)
* Correspondence: ana.nencioni@butantan.gov.br; Tel.: +55-11-2627-9761

Abstract: Few data are available in the literature describing the long-term effects of envenoming in the perinatal period. In this study, the relationship between envenoming of lactating rats and possible behavioral changes in the mother and in her offspring were investigated. Lactating Wistar rats received a single dose of *T. serrulatus* crude venom on postnatal days 2 (V2), 10 (V10) or 16 (V16), and had their maternal behavior evaluated. The seizure threshold was evaluated in adulthood offspring. A decrease in maternal care during envenoming was observed in V2 and V10 groups. The retrieval behavior was absent in the V2 group, and a lower seizure threshold in the adult offspring of all groups was observed. During envenoming, mothers stayed away from their offspring for a relatively long time. Maternal deprivation during the early postnatal period is one of the most potent stressors for pups and could be responsible, at least in part, for the decrease in the convulsive threshold of the offspring since stress is pointed to as a risk factor for epileptogenesis. Furthermore, the scorpionic accident generates an intense immune response, and inflammation in neonates increases the susceptibility to seizures in adulthood. Therefore, maternal envenoming during lactation can have adverse effects on offspring in adulthood.

Keywords: scorpion accidents; lactation; maternal care; seizure threshold

Key Contribution: Maternal envenoming during lactation can have adverse effects on offspring. Maternal deprivation and inflammation can decrease the convulsive threshold of the offspring in adulthood.

Citation: Barbosa, M.d.O.R.; de Paulo, M.E.F.d.V.; Nencioni, A.L.A. Scorpion Envenomation of Lactating Rats Decreases the Seizure Threshold in Offspring. *Toxins* **2021**, *13*, 853. https://doi.org/10.3390/toxins13120853

Received: 15 October 2021
Accepted: 19 November 2021
Published: 30 November 2021

1. Introduction

Scorpions are terrestrial arthropods, which inhabit different biomes and are distributed worldwide, except in Antarctica [1].

Scorpionism is the main cause of accidents with venomous animals in Brazil having overcome snakebite since 2004 [2]. *Tityus serrulatus*, popularly known as yellow scorpion, has been described in 1922 by Lutz and Mello and is responsible for most of the serious accidents with scorpions in Brazil [3–5].

Scorpion accidents are classified according to symptoms into: mild, characterized by local signs such as edema, erythema, sweating, numbness and twitching; moderate, in which, in addition to the previous symptoms, vomiting, abdominal pain, tachypnea, tachycardia or bradycardia, hypertension, agitation, hypersalivation and priapism can also occur; and severe, in which the main symptoms are cardiovascular and pulmonary complications such as heart failure and pulmonary edema, and neurological symptoms such as encephalopathy, coma and convulsion [6].

The number of scorpion accidents has increased in the last few decades and most of the victims consist of people of reproductive age, so pregnant and lactating women are becoming possible targets [4,5]. In the state of São Paulo, between 2007 and 2019, pregnant women comprised 3% of female victims of scorpionism, with the majority during the second trimester of gestation [7]. However, there are no data on the number of lactating victims, and it is difficult to estimate the real proportion of victims in the perinatal period.

Experimentally, some effects of pre- or post-natal injection of Brazilian scorpion venoms in rats have been demonstrated. Prenatal injection of *T. serrulatus* venom in mothers increased the number of post-implantation losses, altered some reflex and physical parameters of the pups, and caused an increase in the weight of the placentas, liver and lungs of the pups [8–10]. Prenatal injection of *T. bahiensis* venom also increased the weight of some organs in the pups and altered physical and behavioral parameters both in childhood and in adulthood [11–13] and, when injected during lactation, it caused a delay in physical and reflex development in childhood, and reduced anxiety in adulthood [14].

Pregnancy and breastfeeding are very important for the adequate development of the pups, both from a physical and a behavioral point of view. Particularly, the quality of care offered to the newborns is important for the maturation of cerebral architecture, especially the hippocampal areas responsible for cognitive functions and stress responsiveness [15–17]. Clinical studies have demonstrated that adverse conditions such as stress early in life predispose the individual to developing several psychiatric disorders such as anxiety, depression, and epilepsy [18–20]. Experimentally, a relationship was observed between stress in perinatal period and a decrease in neurogenesis [21,22]. In addition, stress is relevant to the process of epileptogenesis, both in childhood and in adult life [23]. Hormones and neurotransmitters mediating the influence of early-life stress on excitability may create a permanent vulnerability for the development of epilepsy [20].

It is of utmost importance that the short and long term effects of scorpionism during pregnancy and lactation are well elucidated, in order to minimize the damage to the health of affected mothers and children.

Previous empirical observations conducted in our laboratory revealed a change in the pattern of care provided by mothers envenomed by scorpions during breastfeeding.

Maternal care is responsible for the proper development of the central nervous system [15,17]. Therefore, it is necessary to investigate whether the correct development of the offspring is affected by the envenomation of the mothers. We believe that the stress caused by a single dose of scorpion venom can cause changes in the behavior of the mothers, resulting directly or indirectly in altered development of the offspring.

Therefore, the present study aimed to evaluate the effect of moderate maternal envenoming on the development of the offspring's central nervous system, with particular attention paid to the susceptibility to seizures.

Pentylenetetrazole (PTZ), widely used as seizure-inducing drug, is a $GABA_A$ antagonist and suppresses the function of inhibitory synapses, leading to increased neuronal activity and consequently causes generalized seizures in animals [24]. In small doses, PTZ has been used as a model for absence seizures and in higher doses it produces convulsive seizures [25]. Here, it was used to test the convulsive threshold of the offspring of envenomed rats.

2. Results

2.1. Maternal Behavior

The mothers in the Ct group spent most of the time in an arched-back nursing posture and the care offered to the pups decreased over the days (Figure 1). In the V2 and V10 groups, the time spent in the arched-back nursing posture is significantly reduced on the day of envenoming (Figure 1). In the V16 group, there was no change, because at this stage this behavior is almost extinguished (Figure 1).

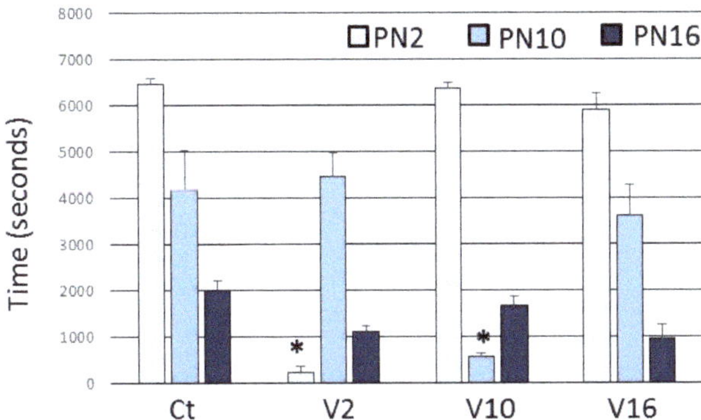

Figure 1. Time spent in arched-back nursing posture after the treatment. Ct injected with 0.9% NaCl in PN2, PN10 and PN16. V2 injected with *T. serrulatus* venom in PN2 and 0.9% NaCl in PN10 and PN16. V10 injected with *T. serrulatus* venom in PN10 and 0.9% NaCl in PN2 and PN16. V16 injected with *T. serrulatus* venom in PN16 and 0.9% NaCl in PN2 and PN10. (n = 6 females per group). Data are expressed as means ± SEM. (*) $p < 0.001$ compared to the control group (Two-way ANOVA).

2.2. Retrieval Test

The test was performed only on PN2 because, over the first week postpartum, the pups are able to move around and the mother ceases to exhibit the behavior of picking them up, and the frequency of licking/grooming and arched-back nursing decreases [15,26].

In the V2 group, dams that failed to group the litter during the test period demonstrated higher latencies to retrieve pups. In V10 and V16 groups, the mothers behaved similarly to the control group (Figure 2).

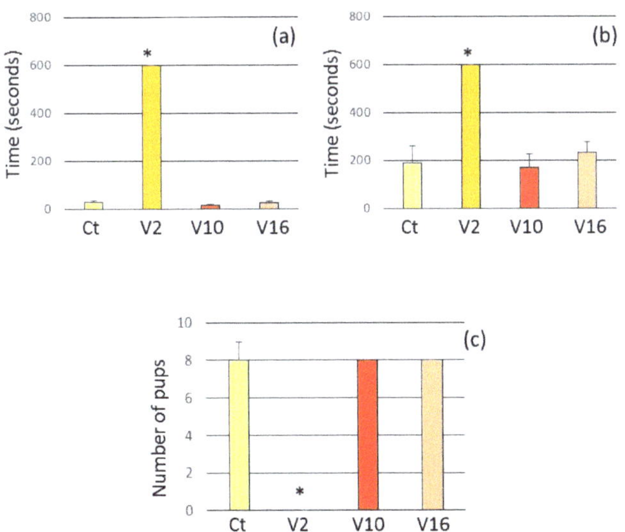

Figure 2. Assessment of mothers in retrieval test after the treatment. Ct injected with 0.9% NaCl in PN2, PN10 and PN16. V2 injected with *T. serrulatus* venom in PN2 and 0.9% NaCl in PN10 and PN16. V10 injected with *T. serrulatus* venom in PN10 and 0.9% NaCl in PN2 and PN16. V16 injected with *T. serrulatus* venom in PN16 and 0.9% NaCl in PN2 and PN10. (n = 6 females per group). (**a**) Latency to retrieve the first pup. (**b**) Latency to retrieve the last pup. (**c**) Number of retrieved pups. Data are expressed as means ± SEM. (*) $p < 0.001$ compared to the control group (Two-way ANOVA).

2.3. Seizure Threshold

In all groups, females and males obtained similar results, with no gender distinction. The V2 group developed seizure behavior earlier in relation to the control group, requiring fewer booster doses (Figure 3).

In addition, in all experimental groups the animals showed more intense convulsions than in the control group (Figure 3).

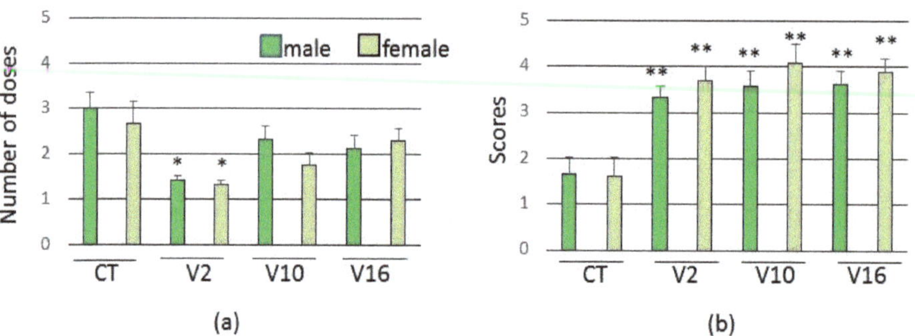

Figure 3. Assessment of seizure threshold (n = 6 females and 6 males per group). (**a**) Number of doses needed to reach the seizure threshold. (**b**) Average intensity of seizure behavior. Data are expressed as means ± SEM. (*) $p < 0.01$ and (**) $p < 0.05$ compared to the control group (Kruskal Wallis test).

3. Discussion

Despite the increase in the number of scorpion accidents in the last few years, there is insufficient information on the effects of scorpion envenomation on pregnancy, lactation, and neonatal outcomes, and unfortunately the available data are controversial.

Experimental studies provide clues as to what might happen in an accident in humans. Thus, in the present study, we used lactating rats to evaluate the possible consequences for the offspring after a scorpion sting in the mother during breastfeeding.

A possible scorpion accident was simulated through a single subcutaneous injection of *T. serrulatus* venom, which is the most common method of inoculation in accidental bites. The dose of 4.0 mg/kg was determined in previous experiments in our laboratory (unpublished data) as causing symptoms observed in moderate to severe envenoming cases such as severe local pain, piloerection, respiratory perturbation and increased lacrimal and salivary secretions. The days of injection were chosen based on the different periods of brain maturation [27]. The development of the mammalian brain begins in embryogenesis, and its maturation continues in the postnatal period [27,28]. In rats, the neurogenesis in cortical regions starts on the ninth day of gestation and extends until the fifteenth postnatal day. The growth spurt of the brain corresponds to the period in which this organ increases in weight most rapidly. In rodents this event peaks around postnatal day 7, reviewed by [29].

Our results demonstrated a decrease in maternal care on the days of the venom injection, and offspring that were more susceptible to convulsion in adulthood. It is relevant to note that the earlier the envenoming, the lower the seizure threshold in the offspring.

In mammals, maternal care is the care that the mother provides to the offspring, and it is essential for the adequate development of the nervous system [15,17,30]. Maternal care modulates the expression of various genes and neurochemical content in one or more regions of the brain and changes in this process can cause variations in behavioral responses and the development of mood disorders [17,31].

Studies in animals and humans have shown that, during early childhood, the brain is particularly sensitive to stress [32], probably because in this period the system is still developing [19].

Maternal deprivation during the early postnatal period is one of the most potent stressors for pups [32] and it was experimentally demonstrated that it can affect behavior,

ACTH and neurotrophin levels in rats, which persist into adulthood [33]. A stressful status can lead to permanent neurobehavioral alterations and an increased susceptibility to psychiatric disorders [34].

Epidemiological studies point to stress as a risk factor for epileptogenesis in adults and young people [20,35]. Experimentally, it was demonstrated that early-life stress in rats has long-lasting effects on brain excitability and may promote age-specific seizures and epilepsy [36]. On the other hand, increased maternal care makes mice genetically predisposed to epilepsy less susceptible to seizures [37].

In our experiments, we observed that mothers from groups V2, V10 and V16 remained away from the pups for a long period after venom injection. This is probably due to the fact that they were experiencing the symptoms of envenoming, mainly pain, as we could see from the vocalization that occurs with minimal contact by either the pups or the observers, with the mother, as well as frequent licking at the injection site. The symptoms of envenoming last for a few hours, creating a long period of separation between mothers and pups, which remained without maternal care.

In addition, mothers in group V2 did not collect the offspring during the observation period in the retrieval test. Retrieval is a common behavior displayed by rodents [38] and, usually, the assessment of maternal behavior is performed in the first week after delivery, as the offspring during this period are exclusively dependent on maternal care [15]. The time spent licking/grooming the pups decreases over the days, as does the time spent in contact with the pups. In the first 10 days after the delivery, the mothers stay longer in the nest and the pups grow, and maternal care tends to gradually decrease and the mother becomes less responsive towards her offspring [15,30]. Thus, the test was performed only in the V2 group, and was not performed in V10 and V16 groups.

Based on the above considerations, we believe that the stress of maternal deprivation could be responsible, at least in part, for the decrease in the convulsive threshold of the offspring.

However, we cannot disregard the possibility that, due to the physiological effects of the venom on the mother, some alteration occurs in the composition of the milk (milk components that could be in higher or lower concentrations), directly affecting the nervous system of the offspring, and studies are being developed in this regard in our laboratory.

It is also possible that some component of the venom or some cytokines produced by the mother are directly passed on, since scorpion accidents are capable of inducing an intense immune response in the injured individual. Several cytokines are increased in the plasma of envenomed patients, such as IL1-α, IL1-β, IL-6, IL-8, IL-10 TNF-α and IFN-γ [39–41]. In addition, it has been shown that scorpion venom can also alter some cytokines in pups of envenomed mothers [14].

These cytokines could affect the nervous system of the pups and be responsible for the decreased seizure threshold, as it is believed that inflammatory processes may be related to epileptogenesis [42]. Inflammation in neonates has already been shown to increase their susceptibility to seizures in adulthood and the induction of an inflammatory response was able to decrease the seizure threshold when performed between days 7 and 14 postnatally [43].

Evidence supports the hypothesis that cytokines not only act as inflammatory mediators, but also have neuromodulatory action. IL1-β is described as having excitatory effects in several brain regions [44]. In the hippocampus, TNF-α can increase the expression of AMPA receptors [45]. Concomitantly, this cytokine may be associated with decreased expression of $GABA_A$ receptors [46]. Epidemiological analyses have demonstrated that central nervous system infections are a major cause of acquired epilepsy revised by [47]. It is due to changes in the physiological properties of neurons within the hippocampus [48].

Thus, the inflammatory process resulting from the envenoming could also be responsible for the decrease in the convulsive threshold of the offspring.

4. Conclusions

This study provides evidence of the interference caused by scorpion envenoming on maternal behavior and on the development of the central nervous system of the pups. A decrease in the care provided to immature pups is clear. Furthermore, it was possible to observe a lower seizure threshold in the offspring of injured mothers. Our results highlight the importance of the perinatal context in the individual's health, even in adulthood.

5. Materials and Methods

5.1. Venom and Drugs

Dried venom of *T. serrulatus* obtained from the Strategic Nucleus of Venoms and Antivenoms of Butantan Institute (São Paulo, Brazil) was dissolved in 0.9% NaCl and injected (4.0 mg/kg, s.c.) into the back of lactating rats. The control group was injected with 0.9% NaCl (1.0 mL/kg s.c.).

Pentylenotetrazole (PTZ; Sigma-Aldrich™, St. Louis, MO, USA) was dissolved in 0.9% NaCl and injected in the adult offspring (initial dose of 20 mg/kg i.p., and booster doses of 10 mg/kg every 10 min).

5.2. Animals

Twelve male and twenty-four female Wistar rats (250–300 g), maintained under controlled conditions (food and water was permitted ad libitum, and the animals were maintained on a 12:12 light/dark schedule with lights on at 7 a.m.), were used for mating. All the experimental procedures were approved by our Institutional Ethics Committee for Experiments on Animals (No. 1927030818).

5.3. Animal Mating and Pregnancy Diagnosis

For mating, two females and one male were housed overnight. The impregnation was confirmed the next morning by the presence of spermatozoa in the vaginal smear. Pregnant females were housed individually until delivery.

5.4. Weaning

At weaning, the littermates were separated and housed by sex until 2 months of age, when the animals were submitted to the convulsive threshold test (two males and two females from each litter).

5.5. Experimental Groups

Lactating rats were divided into four groups (n = 6) that received the following treatments on post-natal (PN) day 2 (PN2), 10 (PN10) and PN16:

- CT group: 0.9% NaCl on PN 2, PN10, and PN16 (1 mL/kg, s.c.);
- V2 group: venom (4 mg/kg, s.c.) on PN2 and 0.9% NaCl on PN10, and PN16 (1 mL/kg, s.c.);
- V10 group: venom (4 mg/kg, s.c.) on PN10 and 0.9% NaCl on PN2, and PN16 (1 mL/kg, s.c.);
- V16 group: venom (4 mg/kg, s.c.) on PN16 and 0.9% NaCl on PN2, and PN10 (1 mL/kg, s.c.).

All the mothers were submitted to injections in PN2, PN10 and PN16, with 0.9% NaCl or venom according to the treatment group, so that all went through the same handling stress.

5.6. Evaluation of Maternal Behavior

Each female was provided with shredded paper one day before delivery for nest making. Maternal care was recorded with a digital camera (Canon Vixia HF R800, Tokyo, Japan) for 2 h after venom or 0.9% NaCl injection on PN2, PN10, and PN16. The time of arched-back nursing posture was recorded. Statistical analyses were performed by two-way ANOVA, and the level of significance was set at $p < 0.001$.

5.7. Retrieval Test

The same animals used in previous experiment were submitted to the retrieval test as described by [49]. Immediately after the maternal behavior observation session, the pups were removed from their mothers for 5 min. After this period, the whole litter was placed back in the housing cage in a dispersed manner. The mother was then observed for 10 min and the latency to pick up the first pup, the latency to pick up all the pups, and the number of pups picked up, were observed.

Statistical analyses were performed by two-way ANOVA, and the level of significance was set at $p < 0.001$.

5.8. Seizure Threshold

At 60 days of age, one male and one female from each litter received an initial dose of 20 mg/kg PTZ, and a booster of 10 mg/kg every 10 min until the occurrence of score 5 or more according to the score scale adapted from that proposed by [50] (Table 1).

Table 1. Score scale adapted from that proposed by Fischer and Kittner (1998).

Score	Behavior
0	no evidence of convulsive activity
1	mouth and facial movements
2	myoclonic body jerking
3	jaw clonus
4	head and forelimb clonus
5	head and forelimb clonus with full rearing
6	head and forelimb clonus and falling
7	run with generalized tonic clonic convulsion
8	death

The total injected volume of PTZ varied approximately between 0.6 mL (initial dose plus a booster) and 1.2 mL (initial dose plus three boosters).

After the observation, the average score in each experimental group was computed.

Statistical analyses were performed by the Kruskal–Wallis test, and the level of significance was set at $p < 0.001$ and $p < 0.05$.

Author Contributions: M.d.O.R.B. and A.L.A.N. conceived and designed the experiments; M.d.O.R.B. and M.E.F.d.V.d.P. performed the experiments and analyzed the data; M.d.O.R.B. and A.L.A.N. wrote the paper. All authors have read and agreed to the published version of the manuscript.

Funding: M.d.O.R.B. was supported by the Coordination for the Improvement of Higher Education Personnel (CAPES, Brazil).

Institutional Review Board Statement: The study was conducted according to the guidelines of the Declaration of Helsinki and approved by the Institutional Review Board (or Ethics Committee) of Butantan Institute (Animal Care and Use Protocol 1927030818, approved 19 December 2018).

Informed Consent Statement: Not applicable.

Data Availability Statement: Not applicable.

Conflicts of Interest: The authors declare no conflict of interest. The funders had no role in the design of the study; in the collection, analyses, or interpretation of data; in the writing of the manuscript, or in the decision to publish the results.

References

1. Lourenço, W.R. Partenoghenesis in Scorpions: Some history-new data. *J. Venom. Anim. Toxins Incl. Trop. Dis.* **2008**, *14*, 19–44. [CrossRef]
2. Reckziegel, G.C.; Pinto, V.L., Jr. Scorpionism in Brazil in the years 2000 to 2012. *J. Venom. Anim. Toxins incl. Trop. Dis.* **2014**, *20*, 2–8. [CrossRef] [PubMed]

3. Bucaretchi, F.; Bacarat, E.C.E.; Nogueira, R.J.N.; Chavez, A.; Zambrone, F.A.D.; Fonseca, M.R.C.C.; Tourinho, F.S. A comparative study of severe scorpion envenomation in children caused by *Tityus bahiensis* and *Tityus serrulatus*. *Rev. Inst. Med. Trop.* **1995**, *37*, 331–336. [CrossRef] [PubMed]
4. Bucaretchi, F.; Fernandes, L.C.R.; Fernandes, C.B.; Branco, M.M.; Vieira, R.J.; Capitani, E.M.; Hyslop, S. Clinical consequences of *Tityus bahiensis* and *Tityus serrulatus* scorpion stings in the region of Campinas, southeastern Brazil. *Toxicon* **2014**, *89*, 17–25. [CrossRef] [PubMed]
5. Brazilian Ministry of Health. *Manual of Diagnosis and Treatment of Accidents Caused by Venomous Animals*; Brazilian Ministry of Health: Brasilia, Brazil, 2009.
6. Santos, M.S.V.; Silva, C.G.L.; Silva Neto, B.; Grangeiro Junior, C.R.P.; Lopes, V.H.G.; Teixeira Junior, A.G.; Bezerra, D.A.; Luna, J.V.C.P.; Cordeiro, J.B.J.; Gonçalves Júnior, J.; et al. Clinical and epidemiological aspects of scorpionism in the world: A systematic review. *Wilderness Environ. Med.* **2016**, *27*, 504–518. [CrossRef] [PubMed]
7. SINAN. Reports for the SINAN. 2020. Available online: http://tabnet.datasus.gov.br/cgi/tabcgi.exe?sinannet/cnv/animaisbr.def (accessed on 22 November 2020).
8. Barão, A.A.S.; Nencioni, A.L.A.; Coronado, V.A. Embryotoxic effects of maternal exposure to *Tityus serrulatus* scorpion venom. *J. Venom. Anim. Toxins Incl. Trop. Dis.* **2008**, *14*, 322–337. [CrossRef]
9. Barão, A.A.S.; Bellot, R.G.; Dorce, V.A.C. Developmental effects of *Tityus serrulatus* scorpion venom on the rat offspring. *Brain Res. Bull.* **2008**, *76*, 499–504. [CrossRef] [PubMed]
10. Cruttenden, K.; Nencioni, A.L.A.; Bernardi, M.M.; Dorce, V.A.C. Reproductive toxic effects of *Tityus serrulatus* scorpion venom in rats. *Reprod. Toxicol.* **2008**, *25*, 497–503. [CrossRef]
11. Dorce, A.L.; Bellot, R.G.; Dorce, V.A.; Nencioni, A.L. Effects of prenatal exposure to *Tityus bahiensis* scorpion venom on rat offspring development. *Reprod. Toxicol.* **2009**, *28*, 365–370. [CrossRef]
12. Dorce, A.L.C.; Dorce, V.A.C.; Nencioni, A.L.A. Effects of in utero exposure to *Tityus bahiensis* scorpion venom in adult rats. *Neurotoxicol. Teratol.* **2010**, *32*, 187–192. [CrossRef]
13. Dorce, A.L.C.; Dorce, V.A.C.; Nencioni, A.L.A. Mild reproductive effects of the *Tityus bahiensis* scorpion venom in rats. *J. Venom. Anim. Toxins Incl. Trop. Dis.* **2014**, *20*, 4. [CrossRef] [PubMed]
14. Martins, A.N.; Nencioni, A.L.A.; Dorce, A.L.C.; Paulo, M.E.F.V.; Frare, E.O.; Dorce, V.A.C. Effect of maternal exposure to *Tityus bahiensis* scorpion venom during lactation on the offspring of rats. *Reprod. Toxicol.* **2016**, *59*, 147–158. [CrossRef] [PubMed]
15. Champagne, F.; Francis, D.; Mar, A.; Meaney, M. Variations in maternal care in the rat as a mediating influence for the effects of environment on development. *Phisiol. Behav.* **2003**, *79*, 359–371. [CrossRef]
16. Bagot, R.C.; Van Hasselt, F.N.; Champagne, D.L.; Meaney, M.J.; Krugers, H.J.; Joëls, M. Maternal care determines rapid effects of stress mediators on synaptic plasticity in adult rat hippocampal dentate gyrus. Neurobiol. *Learn. Mem.* **2009**, *92*, 292–300. [CrossRef]
17. Sequeira-Cordeiro, A.; Massis-Calvo, M.; Mora-Callegos, A.; Fornaguera-Trás, J. Maternal behavior as an early modulator of neurobehavioral offspring responses by Sprague-Dawley rats. *Behav. Brain Res.* **2013**, *237*, 63–70. [CrossRef]
18. Van Campen, J.S.; Jansen, F.E.; Steinbusch, L.C.; Braun, K.P.J. Stress sensitivity of childhood epilepsy is related to experienced negative life events. *Epilepsia* **2012**, *53*, 1554–1562. [CrossRef]
19. Van Campen, J.S.; Jansen, F.E.; Graan, P.N.E.; Braun, K.P.J.; Joels, M. Early life stress in epilepsy: A seizure precipitant and risk factor for epileptogenesis. *Epilepsy Behav.* **2014**, *38*, 160–171. [CrossRef]
20. Huang, L.T. Early-life stress impairs the developing hippocampus and primes seizure occurrence: Cellular, molecular and epigenetic mechanisms. *Front. Mol. Neurosci.* **2014**, *7*, 8. [CrossRef]
21. Brummelte, S.; Pawluski, J.L.; Galea, L.A.M. High post-partum levels of corticosterone given to dams influence postnatal hippocampal cell proliferation and behavior of offspring: A model of post-partum stress and possible depression. *Hormones Behav.* **2006**, *50*, 370–382. [CrossRef]
22. Kawamura, T.; Chen, J.; Takahashi, T.; Ichitani, Y.; Nakahara, D. Prenatal stress suppresses cell proliferation in the early developing brain. *NeuroReport* **2006**, *17*, 1515–1518. [CrossRef]
23. Godoy, L.D.; Garcia-Cairasco, N. Maternal behavior and the neonatal HPA axis in the Wistar Audiogenic Rat (WAR) strain: Early-life implications for a genetic animal model in epilepsy. *Epilepsy Behav.* **2021**, *117*, 107877. [CrossRef] [PubMed]
24. Shimada, T.; Yamagata, K. Pentylenetetrazole-Induced Kindling Mouse Model. *J. Vis. Exp.* **2018**, *136*, e56573. [CrossRef] [PubMed]
25. Lüttjohann, A.; Fabene, P.F.; Van Luijtelaar, G. A revised Racine's scale for PTZ- induced seizures in rats. *Physiol. Behav.* **2009**, *98*, 579–586. [CrossRef] [PubMed]
26. Leon, M.; Moltz, H. Maternal pheromone: Discrimination by pre-weanling albino rats. *Physiol. Behav.* **1971**, *7*, 265–267. [CrossRef]
27. Rice, D.; Barone, S.J. Critical periods of vulnerability for the developing nervous system: Evidence from humans and animal models. *Environ. Health Perspect.* **2000**, *108*, 511–533.
28. Vasung, L.; Turk, E.A.; Ferradal, S.L.; Sutin, J.; Stout, J.N.; Ahtam, B.; Lin, P.Y.; Grant, E. Exploring early human brain development with structural and physicological neuroimaging. *Neuroimage* **2019**, *187*, 226–354. [CrossRef]
29. Semple, B.D.; Blomgren, K.; Gimlin, K.; Ferriero, D.M.; Noble-Haeusslein, L.J. Brain development in rodents and humans: Identifying benchmarks of maturation and vulnerability to injury across species. *Prog. Neurobiol.* **2013**, *106*, 1–16. [CrossRef]
30. Numan, M. Maternal behavior. In *The Physiology of Reproduction*, 2nd ed.; Knobil, E., Neill, J.D., Eds.; Rasen Press: New York, NY, USA, 1994; pp. 221–302.

31. Weinstock, M. Prenatal stressors in rodent: Effects on behavior. *Neurol. Stress* **2017**, *6*, 3–13. [CrossRef]
32. Lupien, S.J.; McEwen, B.S.; Gunnar, M.R.; Heim, C. Effects of stress throughout the lifespan on the brain, behaviour and cognition. *Nat. Ver. Neurosci.* **2009**, *10*, 434–445. [CrossRef]
33. Réus, G.Z.; Stringari, R.B.; Ribeiro, K.F.; Cipriano, A.L.; Panizzutti, B.S.; Stertz, L.; Lersch, C.; Kapeninski, F.; Quevedo, J. Maternal deprivation induces depressive-like behavior and alters neurotrophin levels in the rat brain. *Neurochem. Res.* **2011**, *36*, 460–466. [CrossRef]
34. Agid, O.; Shapira, B.; Zislin, J.; Ritsner, M.; Hanin, B.; Murad, H.; Troudart, T.; Bloch, M.; Heresco-Levy, U.; Lerer, B. Environment and vulnerability to major psychiatric illness: A case control study of early parental loss in major depression bipolar disorder and schizophrenia. *Mol. Psychiatry* **1999**, *4*, 163–172. [CrossRef]
35. Koe, A.S.; Jones, N.C.; Salzberg, M.R. Early life stress as an influence on limbic epilepsy: An hypothesis whose time has come? *Front. Behav. Neurosci.* **2009**, *3*, 24. [CrossRef] [PubMed]
36. Dubé, C.M.; Molet, J.; Singh-Taylor, A.; Ivy, A.; Maras, P.M.; Baram, T.Z. Hyper-excitability and epilepsy generated by chronic early-life stress. *Neurobiol. Stress* **2015**, *2*, 10–19. [CrossRef] [PubMed]
37. Leussis, M.P.; Heirinch, S.C. Quality of rearing guides expression of behavioral and neural seizure phenotypes in el mice. *Brain Res.* **2009**, *1260*, 84–93. [CrossRef] [PubMed]
38. Bridges, R.S. Neuroendocrine regulation of maternal behavior. *Front. Neuroendocrinol.* **2015**, *36*, 178–196. [CrossRef]
39. Meki, A.R.; Monhey el-Dean, Z.M. Serum interleukin-1 beta, interleukin-6, nitric oxide and alpha-antitrypsin in scorpion envenomed children. *Toxicon* **1998**, *36*, 1851–1859. [CrossRef]
40. Fukuhara, Y.D.M.; Dellalibera-Joviliano, R.; Cunha, F.Q.C.; Donadi, E.A. Increased plasma levels of IL-6, IL-8, IL-10 and TNF-alpha in patients moderately or severely envenomed by *Tityus serrulatus* scorpion sting. *Toxicon* **2003**, *41*, 49–55. [CrossRef]
41. Reis, M.B.; Zoccal, K.F.; Gardinassi, L.G.; Faccioli, L.H. Scorpion envenomation and inflammation: Beyond neurotoxic effects. *Toxicon* **2019**, *167*, 174–179. [CrossRef]
42. Caviness, J.N.; Brown, P. Myoclonus: Current concepts and recent advances. *Lancet Neurol.* **2004**, *3*, 598–607. [CrossRef]
43. Riazi, K.; Galic, M.A.; Pittman, Q.J. Contributions of peripheral inflammation to seizure susceptibility: Cytokines and brain excitability. *Epilepsy Res.* **2010**, *89*, 34–42. [CrossRef]
44. Vezzani, A.; Friedman, A.; Digledine, R.J. The role of inflammation in epileptogenesis. *Neuropharmacology* **2012**, *69*, 16–24. [CrossRef]
45. Beattie, E.C.; Stellwagen, D.; Morishita, D.; Bresnahan, J.C.; Ha, B.K.; Von Zastrow, M.; Beattie, M.S.; Malenka, R.C. Control of synaptic strength by glial TNF-alpha. *Science* **2002**, *295*, 2282–2285. [CrossRef]
46. Sawada, M.; Hara, N.; Maeno, T. Tumor necrosis factor reduces the Ach-induced outward current in identified aplysia neurons. *Neurosc. Lett.* **1991**, *131*, 217–220. [CrossRef]
47. Galic, M.A.; Riazi, K.; Henderson, A.K.; Tsutsui, S.; Pittman, Q.J. Viral-like brain inflammation during development causes increased seizure susceptibility in adult rats. *Neurobiol. Dis.* **2009**, *36*, 343–351. [CrossRef] [PubMed]
48. Wu, M.H.; Huang, C.C.; Chen, S.H.; Liang, Y.C.; Tsai, J.J.; Hsieh, C.L. Herpes simplex virus type 1 inoculation enhances hippocampal excitability and seizure susceptibility in mice. *Eur. J. Neurosci.* **2003**, *18*, 3294–3304. [CrossRef] [PubMed]
49. Slamberová, R.; Charousová, P.; Pometlová, M. Maternal behavior is impaired by methamphetamine administered during pre-mating, gestation and lactation. *Reprod. Toxicol.* **2005**, *20*, 103–110. [CrossRef] [PubMed]
50. Fisher, W.; Kittner, H. Influence of ethanol on the pentylenetetrazol-induced kindling in rats. *J. Transm.* **1998**, *105*, 1129–1142. [CrossRef] [PubMed]

Article

New Insights into the Hypotensins from *Tityus serrulatus* Venom: Pro-Inflammatory and Vasopeptidases Modulation Activities

Bruno Duzzi [1,*,†], Cristiane Castilho Fernandes Silva [1,†], Roberto Tadashi Kodama [1], Daniela Cajado-Carvalho [1], Carla Cristina Squaiella-Baptistão [2] and Fernanda Calheta Vieira Portaro [1,*]

1 Laboratory of Structure and Function of Biomolecules, Butantan Institute, São Paulo 05503-900, SP, Brazil; cristiane.silva@esib.butantan.gov.br (C.C.F.S.); pararoberval@gmail.com (R.T.K.); daniela.carvalho@butantan.gov.br (D.C.-C.)

2 Immunochemistry Laboratory, Butantan Institute, São Paulo 05503-900, SP, Brazil; carla.baptistao@butantan.gov.br

* Correspondence: brunoduzzi9@gmail.com (B.D.); fernanda.portaro@butantan.gov.br (F.C.V.P.)

† These authors contributed equally to this study.

Citation: Duzzi, B.; Silva, C.C.F.; Kodama, R.T.; Cajado-Carvalho, D.; Squaiella-Baptistão, C.C.; Portaro, F.C.V. New Insights into the Hypotensins from *Tityus serrulatus* Venom: Pro-Inflammatory and Vasopeptidases Modulation Activities. *Toxins* **2021**, *13*, 846. https://doi.org/10.3390/toxins13120846

Received: 14 September 2021
Accepted: 25 October 2021
Published: 26 November 2021

Publisher's Note: MDPI stays neutral with regard to jurisdictional claims in published maps and institutional affiliations.

Abstract: The *Tityus serrulatus* scorpion is considered the most dangerous of the Brazilian fauna due to the severe clinical manifestations in injured victims. Despite being abundant components of the venom, few linear peptides have been characterized so far, such as hypotensins. In vivo studies have demonstrated that hypotensin I (TsHpt-I) exerts hypotensive activity, with an angiotensin-converting enzyme (ACE)-independent mechanism of action. Since experiments have not yet been carried out to analyze the direct interaction of hypotensins with ACE, and to deepen the knowledge about these peptides, hypotensins I and II (TsHpt-II) were studied regarding their modulatory action over the activities of ACE and neprilysin (NEP), which are the peptidases involved in blood pressure control. Aiming to search for indications of possible pro-inflammatory action, hypotensins were also analyzed for their role in murine macrophage viability, the release of interleukins and phagocytic activity. TsHpt-I and -II were used in kinetic studies with the metallopeptidases ACE and NEP, and both hypotensins were able to increase the activity of ACE. TsHpt-I presented itself as an inhibitor of NEP, whereas TsHpt-II showed weak inhibition of the enzyme. The mechanism of inhibition of TsHpt-I in relation to NEP was defined as non-competitive, with an inhibition constant (Ki) of 4.35 μM. Concerning the analysis of cell viability and modulation of interleukin levels and phagocytic activity, BALB/c mice's naïve macrophages were used, and an increase in TNF production in the presence of TsHpt-I and -II was observed, as well as an increase in IL-6 production in the presence of TsHpt-II only. Both hypotensins were able to increase the phagocytic activity of murine macrophages in vitro. The difference between TsHpt-I and -II is the residue at position 15, with a glutamine in TsHpt-I and a glutamic acid in TsHpt-II. Despite this, kinetic analyzes and cell assays indicated different actions of TsHpt-I and -II. Taken together, these results suggest a new mechanism for the hypotensive effects of TsHpt-I and -II. Furthermore, the release of some interleukins also suggests a role for these peptides in the venom inflammatory response. Even though these molecules have been well studied, the present results suggest a new mechanism for the hypotensive effects of TsHpt-I

Keywords: *Tityus serrulatus*; venom components; hypotensins; NEP inhibition; cytokines

Key Contribution: Despite the great similarity in primary structures between hypotensins, different activities were demonstrated. Both hypotensins increase ACE activity at different levels, while only TsHpt-I has shown a non-competitive inhibition over NEP activity, suggesting other alternative hypotensive mechanisms for this peptide. Furthermore, the release of some cytokines may suggest a role for these peptides in the inflammatory response induced by the venom.

1. Introduction

The image of the scorpion has long been connected to human history, being represented in cults, legends, philosophy and arts, as it is one of the oldest animals on the planet. Dating from the Silurian period, more than 400 million years ago, scorpions are organisms that have long intrigued human beings [1,2].

The order Scorpiones is represented by 2200 species and, through taxonomic studies, have been grouped into 20 families and 165 genera of scorpions. The most dangerous, and capable of causing fatal accidents in humans, belong to the Buthidae family, represented by the following genera: *Androctonus* and *Leiurus* (North Africa and Middle East), *Centruroides* (Mexico and the United States) and *Tityus* (South America and Trinidad) [3].

In Brazil, the scorpions *Tityus serrulatus*, *T. bahiensis* and *T. stigmurus* are the animals responsible for serious accidents. Among these, the *T. serrulatus* scorpion, popularly known as the "Brazilian yellow scorpion", is the one with the highest Medical and Scientific relevance in Brazil. It is mainly distributed among the states of Bahia, Goiás (including the Federal District), Paraná, Espírito Santo, Rio de Janeiro, Minas Gerais and São Paulo [4].

T. serrulatus reproduces by parthenogenesis, and each female is able to generate about 70 offspring during its life. They are commonly found in sewers, cemeteries and wastelands, where they find safe shelter and plenty of food. Therefore, in addition to the potency of its venom, their adaptation to urban centers may explain the significant increase in the number of accidents caused by this scorpion in Brazil [4–6].

In this scenario, accidents caused by *T. serrulatus* stings are considered a public health problem in Brazil due to its potential to cause severe clinical manifestations, which might bring a prognosis of death, especially on children aged from 0 to 14 years. Although most of the cases have been classified as mild, the biggest concern is related to the high number of cases that are reported annually in Brazil, since scorpion stings represent 41% of all venomous animal accidents, including snakes, spiders, bees and others, as reported in 2016 [7].

Generally, the *T. serrulatus* venom (TsV) is composed of mucus, inorganic salts, lipids, amines, nucleotides, enzymes, kallikrein inhibitors, natriuretic peptides, high molecular weight proteins, peptides, amino acids and neurotoxins [8]. Current studies carried out by Oliveira and colleagues [9] involving the transcriptome of the venom glands have shown that more than 30% of the venom is made up of enzymes and, approximately, 40% of peptides. The peptides present in the TsV can be classified as structured—which are stabilized by disulfide bonds—or linear [8]. The so-called structured peptides have been the most studied components, classified as neurotoxins that interact with ion channels (Na^+ and K^+) and are related to the most serious effects caused by the venom. On the other hand, linear peptides, although found with some abundance in the venom, are still poorly characterized. The peptidomic analysis performed by Rates and colleagues [10] demonstrated the existence of a great diversity of peptides in the TsV, all of them not yet characterized. Many could not be found in the database, and as this information is scarce, this requires de novo sequencing. Other studies using "omic" techniques have demonstrated the complexity of the venom in relation to linear peptides from post-translational modifications of larger proteins, generating lists with hundreds of components [11–13].

Among the linear peptides we have the hypotensins (TsHpt), identified from TsV proteomic analyzes. Both peptides are made up of 25 amino acid residues that contain two consecutive prolines in their C-terminal portion and a punctual difference between TsHpt-I and -II, which is the residue at position 15, being a glutamine in TsHpt-I and a glutamic acid in TsHpt-II. Studies with TsHpt-I, also known as Ts14, showed that this peptide was able to exert hypotensive activity in normotensive Wistar rats by potentiating bradykinin. The hypothesis is that the vasodilation effect is related to the release of NO by an independent mechanism of ACE inhibition [14].

In order to increase knowledge about hypotensins and their biological activities, the present work demonstrates, for the first time, the interaction of these peptides with human vasopeptidases, ACE (EC 3.4.15.1) and NEP (EC 3.4.24.11), alongside with cellular assays, which were carried out in order to verify the possible action of hypotensins as inflammatory or anti-inflammatory peptides.

2. Results

2.1. Modulation of ACE and NEP Activities by Hypotensins

For the initial kinetic tests, the synthetic hypotensins were incubated with the metallopeptidases ACE and NEP and their fluorescent substrates, Abz-FRK(Dnp)P-OH and Abz-RGFK (Dnp)-OH, respectively. These enzymes were chosen for the studies because they are considered of medical importance, where their modulations caused by venom peptides may be related to some symptoms present in serious accidents, such as hypotension/hypertension. As shown in Table 1, the hypotensins showed different activities in relation to the modulations of the peptidases studied.

Table 1. Modulations of ACE and NEP activities by interaction with synthetic hypotensins.

	Metallopeptidases			
	ACE		NEP	
	Activation (%)	Inhibition (%)	Activation (%)	Inhibition (%)
TsHpt-I	64	-	-	75
TsHpt-II	44	-	-	11

The results were obtained by incubating the enzymes with their FRET substrates (10 μM) in a final volume of 100 μL. For the ACE assays, the used buffer was Tris HCl 100 mM, NaCl 50 mM and $ZnCl_2$ 10 μM, pH 7.0. For NEP, the assays were made in Tris HCl 50 mM, pH 7.5 buffer. All reactions occurred at 37 °C, in a Victor 3 fluorimeter (Perkin–Elmer) adjusted for excitation and emission readings at 320 and 420 nm, respectively, for 15 min (one reader per minute). Results were obtained in triplicate and analyzed on GraFit 5 software.

Both peptides increased the catalytic activity of ACE, and TsHpt-I was more effective than TsHpt-II in activating the enzyme. Contrary to that observed with ACE, the results involving the peptides tested with NEP revealed the presence of inhibitors of this enzyme. As a highlight, TsHpt-I was able to reduce by 75% the hydrolysis of the FRET substrate used.

As already mentioned, the suggested mechanism of action on hypotension caused by TsHpt-I in vivo is the release of NO, together with the agonistic effect on bradykinin B2 receptors [15]. Taken together, our results with TsHpt-I seem to indicate a second hypotensive mechanism for this peptide, which involves the inhibition of NEP. Interestingly, TsHpt-I showed a strong inhibition of NEP's catalytic activity, in contrast with TsHpt-II. This fact is probably the result of a single difference between the primary structures between the hypotensins; that is, the presence of glutamine at position 15 in TsHpt-I, instead of a glutamic acid present in TsHpt-II.

2.2. Stability of Hypotensins against Metallopeptidases ACE and NEP

As hypotensins showed interactions, even if opposite, in relation to the studied vasopeptidases, tests to determine the susceptibility to hydrolysis of both synthetic peptides were carried out. For this, the hypotensins were incubated with the enzymes, in their respective buffers, and later analyzed in an HPLC-C18 system.

Both hypotensins were resistant to hydrolysis and, therefore, did not behave as substrates for ACE and NEP, even after 4 h of incubation at 37 °C (Figure 1).

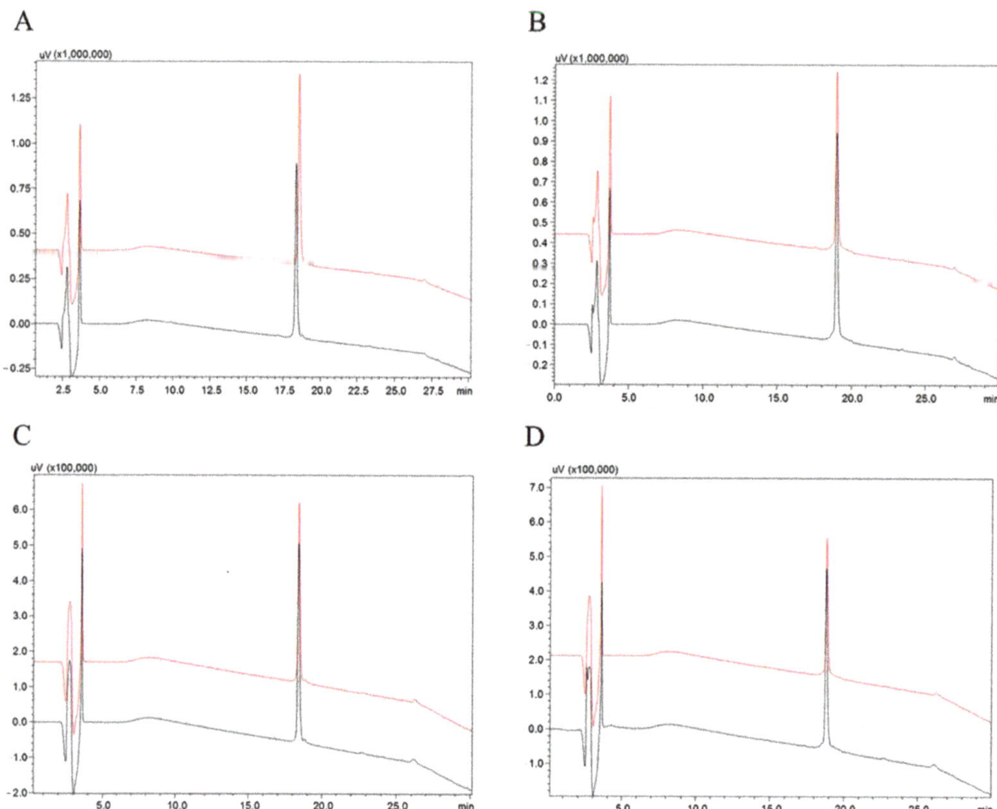

Figure 1. Analysis of the susceptibility to hydrolysis of the hypotensins over the vasopeptidases NEP and ACE. Incubations were carried out in Tris HCl 100 mM, NaCl 50 mM and $ZnCl_2$ 10 µM, pH 7.0, for the ACE assays, and, for the NEP assays, in Tris HCl 50 mM, pH 7.5 buffer, for four hours at 37 °C. (**A**) TsHpt-I control (black line) and after incubation with NEP (red line). (**B**) TsHpt-II control (black line) and TsHpt-II incubated with NEP (red line). (**C**) TsHpt-I control (black line) and incubated with ACE (red line). (**D**) TsHpt-II control (black line) and after incubation with ACE (red line). The samples of negative and experimental controls (**A**) TsHpt-I and (**B**) TsHpt-II are shown on the left and right, respectively. The hydrolysis products were considered ≤5%. The gradient used was 20% to 60% solvent B in 20 min, and 60% to 100% solvent B in 5 min (flow 1 mL/min, absorbance at 214 nm). Restek Ultra C-18 5 µm column, 250 × 4.6 mm.

Since TsHpt-I demonstrated inhibition of NEP catalytic activity, and was resistant to hydrolysis, kinetic analyses to determine the inhibition constant (Ki) and the mechanism of inhibition of NEP by TsHpt-I were carried out.

TsHpt-I behaved as a non-competitive inhibitor of NEP, presenting an inhibition constant (Ki) of 4.35 µM (Figure 2). The determination of the mechanism and Ki of TsHpt-II on neprilysin activity were not performed as this peptide showed weak interaction with the peptidase (Table 1).

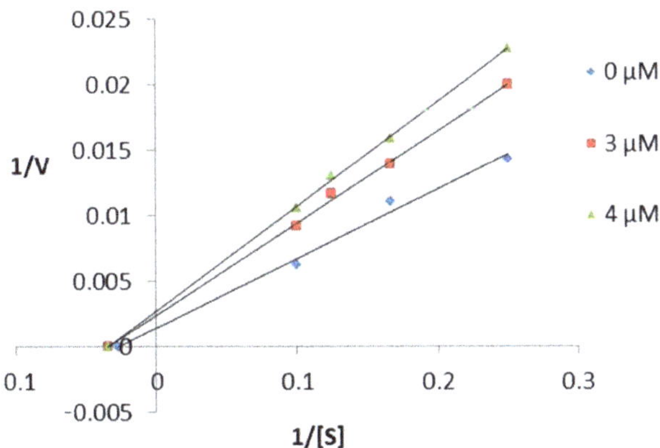

Figure 2. Lineweaver–Burk plot of NEP inhibition by TsHpt-I. The inhibition constant (Ki) was determined using four Abz-RGFK-EDDnp substrate concentrations (4 µM, 6 µM, 8 µM and 10 µM), with variation in the concentration of TsHpt-I (3 µM and 4 µM), keeping the amount of enzyme fixed (1.5 ng). The experiment was performed in triplicate.

TsHpt-I is the second neprilysin inhibitor from the *Tityus serrulatus* venom described in the literature. Previous observations by our group revealed the presence of a peptide capable of inhibiting human NEP and an NEP-like present in cockroaches, the [des-Arg1]-Proctolin [16]. [des-Arg1]-Proctolin was characterized as a competitive inhibitor of NEP, with an inhibition constant of 0.94 µM [16]. Thus, there may be a synergistic action between these two peptides present in TsV, TsHpt-I and [des-Arg1]-Proctolin, for the inhibition of NEP, and this fact may be related to hypotension associated with scorpionism.

2.3. *In Vitro Pro-Inflammatory Effects of TsHpt I and -II on Mouse Macrophages*

After 24 h of incubation, TsHpt-I and -II exerted intermediate cytotoxicity on naïve murine peritoneal macrophages in vitro, as demonstrated by the statistically significant decrease in cell viability, to about 75% and 60%–70% at concentrations of 10 µg/mL and 50 µg/mL, respectively, when compared to 100% in the control group (medium), as determined by the MTT method (Figure 3). Cytotoxicity was not shown to be dose-dependent (Figure 3).

Despite the moderate cytotoxic effect in the concentrations of 10 µg/mL and 50 µg/mL, TsHpt-I and -II were not able to decrease cellular viability in the higher concentration of 100 µg/mL. Interestingly, in this same concentration, both peptides were able to promote the release of significant levels of TNF, and TsHpt-II also induced the release of IL-6 (Figure 4). None of the peptides were able to induce the release of IL-10, IL-12p70, IFN-γ and MCP-1 (data not shown).

In order to confirm the pro-inflammatory effects of both peptides on the biological function of macrophages, the phagocytic activity of macrophages, cultured for 24 h in the presence of TsHpt-I and -II, were analyzed, and compared to the negative and positive controls. Both treatments with TsHpt-I and -II promoted a significant increase in the phagocytic index (Figure 5), demonstrating that the pro-inflammatory action of these peptides also affects the biological function of macrophages.

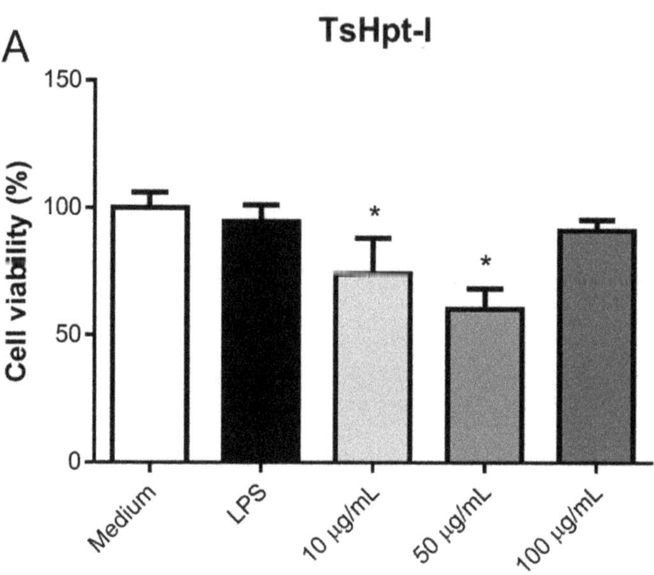

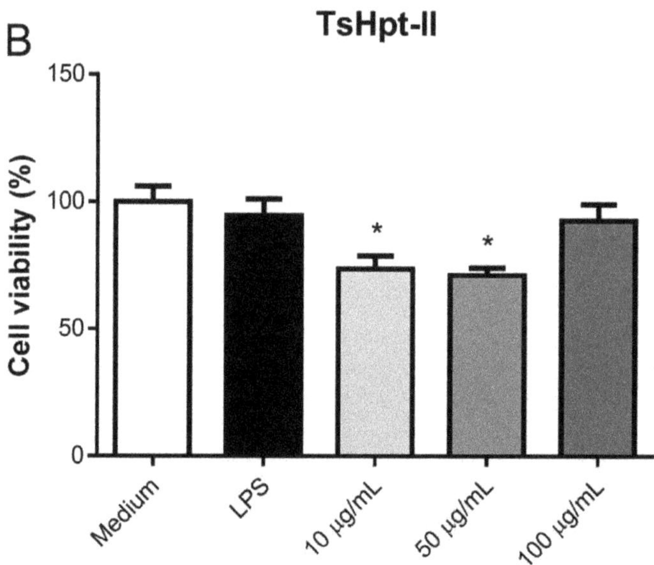

Figure 3. Cytotoxic effect of hypotensins I and II on murine peritoneal macrophages in vitro. Peritoneal exudate macrophages from naïve BALB/c mice were cultured for 24 h in the presence of TsHpt-I (**A**) and -II (**B**) at concentrations of 10 µg/mL, 50 µg/mL and 100 µg/mL. Cells cultured with medium alone or in the presence of 5 µg/mL LPS were used as the controls. Cell viability was determined by the MTT method. The results represent the mean ± SD of two independent experiments performed in triplicate. * $p < 0.05$.

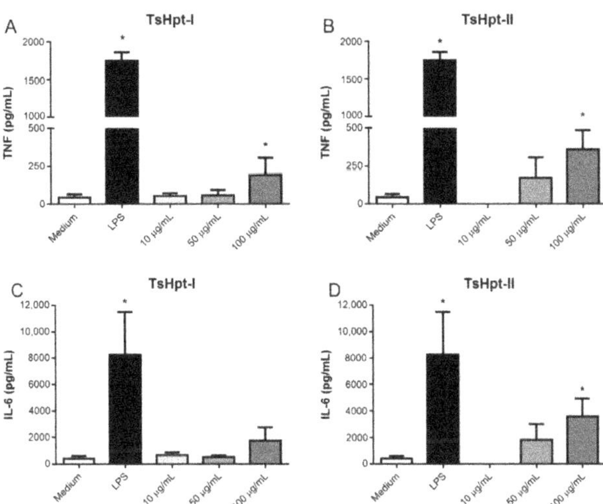

Figure 4. Cytokine production by murine peritoneal macrophages stimulated in vitro with TsHpt-I and -II. Peritoneal exudate macrophages from naïve BALB/c mice were cultured for 24 h in the presence of TsHpt-I (**A**,**C**) and -II (**B**,**D**), at concentrations of 10 µg/mL, 50 µg/mL and 100 µg/mL. Cells cultured with medium alone or in the presence of 5 µg/mL LPS were used as negative and positive controls, respectively. The concentrations of TNF (**A**,**B**) and IL-6 (**C**,**D**) were determined by the CBA method, according to the manufacturer's instructions. The results represent the mean ± SD of three independent experiments performed in triplicate. * $p < 0.05$.

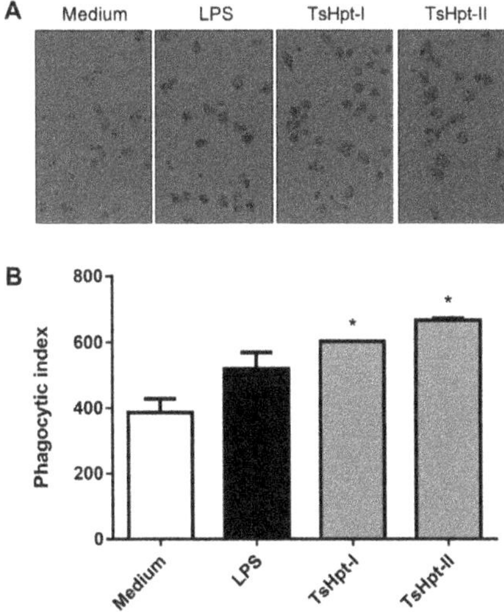

Figure 5. Phagocytic activity of the murine peritoneal macrophages stimulated in vitro with TsHpt-I and -II. Peritoneal exudate macrophages from naïve BALB/c mice were cultured on glass coverslips for 24 h in culture medium or in the presence of LPS (5 µg/mL) or TsHpt-I and -II (100 µg/mL). After this period, the cells were incubated for 1 h with a suspension of *Saccharomyces cerevisiae*. Coverslips were washed, fixed and stained by Giemsa. (**A**) Phagocytosis images in 400× magnification. (**B**) The phagocytic index was calculated based on the percentage of phagocytic cells and on the average number of yeasts per cell, by counting approximately 200 cells per coverslip in 1000× magnification. The results represent one of two reproducible experiments performed in duplicate (mean ± SD). * $p < 0.05$.

3. Discussion

Identified from the proteomic analyses on the venom of *Tityus serrulatus* (TsV), the hypotensins are peptides made up of 25 amino acid residues that contain two consecutive prolines in their C-terminal portion. Hypotensins are a family of peptides with small structural differences between them, with TsHpt-I being the best-studied member—both the natural molecule and its synthetic counterpart. Tests involving the natural or synthetic TsHpt-I demonstrated that this peptide was able to exert hypotensive activity in normotensive Wistar rats through bradykinin potentiation. The hypothesis about the vasodilation effect is related to NO release, which is a mechanism independent of ACE inhibition [14].

Focusing on clinical conditions related to blood pressure alterations observed in accidents caused by *T. serrulatus*, the present study investigated the possible interaction of TsHpt-I and -II with the metallopeptidases angiotensin converting enzyme (ACE) and neprilysin (NEP). ACE is considered an important molecule in the regulation of blood pressure, as it generates angiotensin II (Ang II) from the cleavage of angiotensin I (Ang I), in addition to degrading bradykinin (Bk) [17]. NEP also acts to control blood pressure through the excretion of Na^+ and water, as it is capable of degrading natriuretic peptides (ANP, BNP and CNP) [18,19]. Thus, both ACE and NEP are known as vasopeptidases. The third vasopeptidase is the endothelin-converting-enzyme I (ECE-1, EC 3.4.24.71), a metallopeptidase capable of releasing endothelin-I (ET-I), a vasoconstrictor peptide, from the big-endothelin [20].

Regarding TsV, data from the literature describe the presence of an ACE-like one [9,21], capable of converting Ang I to Ang II, and degrading Bk. Hence, the presence of this vasopeptidase in TsV may collaborate with the hypertension observed in accidents involving humans. The possible presence of an NEP-like in TsV has also been described, as well as an inhibitor of this metallopeptidase, called [des-Arg1]-Proctolin [16]. This peptide was characterized as a competitive inhibitor of human NEP, presenting an inhibition constant of 0.94 μM [16]. Proteomic studies also reported the presence of an ECE-like enzyme in TsV [9]. Moreover, high levels of ET-I were observed in the sera of patients after envenomation with the scorpion *Androctonus australis hector*, indicating that molecules of scorpion venoms also have an effect on the endothelin axis. [22]. Therefore, the presence of vasopeptidases and their inhibitors in scorpion venoms may contribute to acute changes caused in the cardiovascular system observed in cases of envenomation [23].

Although the preferred prey of the *Tityus serrulatus* scorpion are insects, such as crickets and cockroaches, its venom is dangerous to humans. According to data from proteomics and transcriptomics studies [9], the toxic effect of TsV in humans may be the result of evolutionarily preserved molecules present in both insects and mammals. This suggestion may explain the presence of ACE-like, ECE-like and NEP-like enzymes in TsV, together with neprilysin inhibitors and hypotensins. [23].

Interestingly, both hypotensins were able to increase the catalytic activity of ACE, but in different ways. While TsHpt-I activated ACE by 64%, TsHpt-II increased by 46% the hydrolysis of the substrate Abz-RGFK-EDDnp. In fact, in studies on the determination of the hypotensive mechanism of TsHpt-I, ACE activation can also be observed; however, the results were not discussed [14]. Studies with NEP also indicated different interactions with hypotensins, and results with TsHpt-I showed that this peptide is a non-competitive inhibitor of NEP, with a Ki value of 4.35 μM. As TsHpt-I is the second NEP inhibitor described in the *T. serrulatus* venom, it is possible that there is a combined action between hypotensin I and [des-Arg^1]-Proctolin [16], which may be related to the hypotension caused by the envenomation. In contrast, TsHpt-II displayed a low interaction with NEP, and the different results are, probably, the effect of a single difference between the primary structures of the hypotensins. However, future studies of circular dichroism will be needed to clarify this matter.

As hypotensins demonstrated new activities in vitro, cytotoxicity and possible pro- or anti-inflammatory actions were investigated in order to increase our knowledge of these molecules. Both hypotensins have immunomodulatory potential, with pro-inflammatory effects on murine peritoneal macrophages, when used at a concentration of 100 μg/mL. Interestingly, at this concentration, both hypotensins did not exert cytotoxic activity on the tested cells, which makes the two molecules even more interesting, due to their pharmacological potential for the long-term development of new immunostimulants and/or adjuvants. Pucca and collaborators [24] demonstrated the pro-inflammatory effect of three peptides derived from the TsV venom on a strain of murine macrophages, with increased production of IL-6. Similar to the results presented in this work, the effects were subtle compared to LPS, although significant in relation to the negative control. In our study, we observed increased TNF production in the presence of both peptides and increased IL-6 production in the presence of TsHpt-II. As for the mechanism of molecules as hypotensives, it is important to emphasize that both TNF and IL-6 induce vasodilation, and its massive release can even lead to shock.

Cassini-Vieira and colleagues suggested an anti-inflammatory role for TsHpt-I from *T. serrulatus* venom, based on its ability to reduce neutrophil infiltration and TNF production in a murine model of sponge implant-induced inflammation. On the other hand, increased macrophage infiltration was observed in this model, indicating a pro-inflammatory role, which demonstrates the need for further studies on the mechanisms of action of TsHpt-I [25].

Interestingly, both treatments with both peptides promoted a significant increase in the phagocytic index, demonstrating that the pro-inflammatory action of these peptides also affects the macrophages' biological function. This phenomenon is interesting, considering the possible development of immunomodulators. It is known that adjuvant and/or immunostimulant molecules generally induce a pro-inflammatory environment that favors the activation of antigen-presenting cells and, consequently, the development of adaptive immunity against specific antigens. Increased macrophages' phagocytic capacity by hypotensins may reflect increased microbicidal activity and/or antigen presentation. The anti candida and anti-biofilm activities of TistH, a hypotensin present in the venom of the *Tityus stigmurus* scorpion, were recently described and confirmed [26], but functional antigen presentation assays are needed to deepen our knowledge of the immunomodulatory action of these peptides.

Despite the biotechnological potential of hypotensins, the activities already described for these peptides, and the new results showed in the present work, indicate that both molecules do not have a specific target or mechanisms of action. Considering that they are multifunctional toxins present in the *Tityus serrulatus* venom, new studies aiming at drug development should be very carefully carried out in order to minimize unexpected effects.

4. Conclusions

Despite the great similarity between the primary structures of hypotensins, different levels of interactions with the vasopeptidases ACE and NEP were observed. Hypotensins increase ACE activity at different levels while TsHpt-I is a non-competitive inhibitor of NEP, suggesting other hypotensive mechanisms for this peptide in addition to those already described. Furthermore, the release of some interleukins may suggest a role for these peptides in the inflammatory response induced by the venom. Hypotensins are multifunctional toxins, and further studies are needed to clarify the potential of these molecules for biotechnological use.

5. Material and Methods

5.1. Reagents

The synthetic peptides TsHpt-I and TsHpt-II were obtained by the solid-phase peptide synthesis method, and purchased from GenOne Biotechnologies (Rio de Janeiro, Brazil). Angiotensin Converting Enzyme (ACE) from rabbit lung, RPMI 1640 medium, LPS from *E. coli* 0127:B8, Trypan Blue, Giemsa stain and glutaraldehyde solution were purchased from Sigma-Aldrich (St. Louis, MO, USA). Neprilysin and the Fluorescence Resonance Energy Transfer (FRET) substrates, Abz-FRK (Dnp) P-OH and Abz-RGFK (Dnp)-OH, were provided by Prof. Adriana Carmona, from the Department of Biophysics of UNIFESP-EPM, São Paulo, SP, Brazil. Acetonitrile and TFA used in RP-HPLC were acquired from J. T. Baker (Avantor, Radnor, PA, USA). Fetal cow serum (FCS) and penicillin and streptomycin antibiotics were purchased from Cultilab (Campinas, SP, Brazil). Tetrazolium salt 3-(4,5-dimethylthiazol-2-yl)-2,5- diphenyltetrazolium bromide (MTT) was purchased from Invitrogen (Waltham, MA, USA). DMSO was purchased from Merck (Darmstadt, Germany). BD Cytometric Bead Array Mouse Inflammation Kit was purchased from BD Biosciences (San Jose, CA, USA). The *Saccharomyces cerevisiae* suspension was obtained from washing and adjusting the concentration of bread yeast (Fleischmann – Petrópolis, RJ, Brazil) in RPMI.

5.2. Interactions of Hypotensins with Vasopeptidases ACE and NEP

For ACE assays, 10 μM of each peptide was incubated with 3.0 ng of peptidase and 10 μM of Abz-FRK (Dnp) P-OH substrate, in Tris HCl 100 mM, NaCl 50 mM and $ZnCl_2$ 10 μM, pH 7.0 buffer. For NEP, 10 μM of TsHpt-I or -II was incubated with 1.5 ng of peptidase, 3.5 μM Abz-RGFK (Dnp)-OH in Tris HCl 50 mM, pH 7.5 buffer. All reactions occurred at 37 °C, in a final volume of 100 μL, in a Victor 3 fluorimeter (Perkin–Elmer, Waltham, MA, USA) adjusted for excitation and emission readings at 320 and 420 nm, respectively, for 15 min (one reader per minute). Results were obtained in triplicate and analyzed using GraFit 5 (Erithacus software, East Grinstead, West Sussex, UK).

5.3. Stability Test of the Synthetic Peptides

The synthetic peptides TsHpt-I and TsHpt-II (30 μM) were incubated with ACE (3.0 ng), in Tris HCl 100 mM, NaCl 50 mM and $ZnCl_2$ 10 μM, pH 7.0 buffer, and NEP (1.5 ng), in Tris HCl 50 mM, pH 7.5 buffer, at 37 °C for 4 h. Samples containing only the synthetic peptides were used as the negative control. After incubation, samples were analyzed by reverse phase chromatography on RP-HPLC (Shimadzu, Kyoto, Japan), using a Restek Ultra C-18 column (5 μm, 250 × 4.6 mm). Solvents used were 0.1% TFA in water (solvent A), and acetonitrile plus solvent A (9:1) as solvent B. Separations were performed at a flow rate of 1 mL/min and a 20–60% gradient of solvent B over 20 min. In all cases, elution was followed by the measurement of ultraviolet absorption (214 nm).

5.4. Characterization of TsHpt-I as a NEP Inhibitor

To determine the inhibition constant (Ki) of TsHpt-I over NEP, four concentrations of Abz-RGFK (Dnp)-OH (4 μM, 6 μM, 8 μM and 10 μM) and 3 μM and 4 μM of TsHpt-I were tested using 1.5 ng of peptidase in 100 μL of final volume of Tris HCl pH 7.5. The Km value of the substrate used was determined to be 14 μM [27]. Controls without the TsHpt-I were also performed in all assays. The reactions were monitored as described above (item 5.2). The Lineweaver–Burk plot was constructed (1/V × 1/(S)) according to the presented mechanism. The Ki was calculated as described by Segel [28]. All assays were performed in triplicate.

5.5. Cell Assays

5.5.1. Murine Peritoneal Macrophages Obtainment

Male young, between 8 and 12 weeks of age and weighing between 20 and 22 g, BALB/c mice adults were used. Mice were obtained from the Central Animal Facility of the Butantan Institute and housed in the Laboratory of Immunochemistry bioterium, Butantan Institute. The mice were kept in boxes lined with shavings, containing 3 animals per box, under natural light, full-time ventilation and exhaustion, filtered water and commercial feed ad libitum. After a period of 2 to 3 days of acclimatization of the animals, they were euthanized in a CO_2 chamber. All experimental procedures involving animals were in accordance with the ethical principles in animal research adopted by the Brazilian Society of Animal Science and the National Brazilian Legislation no.11.794/08. Animal care and experimental procedures were approved by the Institutional Committee for the Care and

Use of Laboratory Animals from Butantan Institute (CEUAIB protocol number 5396310517, approved on 21 June 2017).

Peritoneal exudate cells (BALB/c naïve mice, $n = 6$) were collected by two washes with 5 mL of RPMI medium. The cells obtained were washed twice with RMPI medium, at 400 g, for 10 min, at 18 °C, and resuspended in R10 medium (RPMI + 10% fetal cow serum). After counting in a Neubauer chamber, in the presence of Trypan blue, the cell concentration was adjusted to 2×10^6 cells/mL. Cells were distributed into 96-well culture plates (2×10^5 cells/100 µL/well) and incubated for 2 h at 37 °C and 5% CO_2. After incubation, non-adherent cells were discarded, and adherent cells (macrophages) were resuspended in R10 medium containing the respective tested stimuli, at different concentrations, in triplicates, in a final volume of 200 µL/well. Cells resuspended in R10 medium were used as a negative control, and cells stimulated with LPS (5 µg/mL) as a positive control. Cells were then incubated for 24 h at 37 °C and 5% CO_2. After this period, the culture supernatants were collected and stored at −80 °C, for subsequent dosage of cytokines, and cell viability was determined by MTT assay.

5.5.2. Effect of Hypotensins on the Cell Viability and Production of Inflammatory Mediators by Murine Peritoneal Macrophages Stimulated In Vitro with Hypotensins

After a 24-h period of incubation of cells with hypotensins (10 µg/mL, 50 µg/mL and 100 µg/mL), MTT assays were performed, consisting of the addition of 100 µL of R10 containing 0.5 mg/mL of tetrazolium salt 3-(4.5-dimethylthiazol-2-yl)-2,5-diphenyltetrazolium bromide (MTT) in all wells, followed by a new incubation, under the same conditions, for 4 h. Then, the supernatants were discarded, and 100 µL/well of DMSO were added to dissolve the formed crystals. Absorbance was determined in a spectrophotometer at 540 nm. Cell viability was calculated based on the absorbances of the samples and the negative control (cells cultivated only with R10).

The concentration of pro- and anti-inflammatory cytokines (IL-6, IL-10, IL-12p70, IFN-α, MCP-1 and TNF) in the samples incubated with both hypotensins (10 µg/mL, 50 µg/mL and 100 µg/mL) was determined by the CBA method (BD Cytometric Bead Array Mouse Inflammation Kit), according to the manufacturer's instructions. The samples were acquired in a BD FACSCanto II flow cytometer, and the data were analyzed using BD FCAP Array version 3.0 software.

5.5.3. Effect of TsHpt-I and -II on Phagocytic Function of Murine Peritoneal Macrophages

Peritoneal macrophages from naïve BALB/c mice ($n = 6$) were obtained as described above (item 5.5.1). Cells were distributed in 24-well culture plates (5×10^5 cells/500 µL/well), with glass coverslips inside, and incubated for 2 h at 37 °C and 5% CO_2. After incubation, non-adherent cells were discarded, and adherent cells (macrophages) were stimulated in vitro with hypotensins, in duplicate, at a concentration of 100 µg/mL, for 24 h at 37 °C and 5% CO_2. After this period, the culture supernatants were discarded, and the cells were incubated for 1 h with a suspension of *Saccharomyces cerevisiae* at a concentration of 1.5×10^6 yeasts/mL/well. Coverslips were washed 10 times with PBS, fixed with 0.5% glutaraldehyde and stained by Giemsa. Phagocytosis was evaluated by immersion optical microscopy (1000×) and quantified by counting approximately 200 cells per cover slip. The percentage of phagocytic cells and the average number of internalized yeasts per phagocytic cell were calculated. The phagocytic index was obtained by multiplying the two values (percentage × mean) and represents the total phagocytic capacity of each cell population.

5.6. Statistical Analysis

The results were statistically analyzed using the GraphPad Prism 5 program, using the one-way ANOVA test followed by Tukey's post-test. Results with a p-value < 0.05 were considered significant.

Author Contributions: Conceptualization, B.D. and F.C.V.P.; methodology, B.D., D.C.-C., C.C.S.-B., C.C.F.S. and R.T.K.; validation, B.D., C.C.S.-B., C.C.F.S. and F.C.V.P.; formal analysis, B.D., D.C.-C.,

C.C.S.-B., C.C.F.S., R.T.K. and F.C.V.P.; investigation, B.D., C.C.S.-B. and C.C.F.S.; resources, F.C.V.P.; data curation, B.D., D.C-C., C.C.S.-B., C.C.F.S. and R.T.K.; writing—original draft preparation, B.D., D.C.-C., C.C.S.-B., C.C.F.S., R.T.K. and F.C.V.P.; writing—review and editing, B.D., D.C.-C., C.C.S.-B., C.C.F.S., R.T.K. and F.C.V.P.; supervision, F.C.V.P.; project administration, F.C.V.P.; funding acquisition, B.D., D.C.-C., R.T.K. and F.C.V.P. All authors have read and agreed to the published version of the manuscript.

Funding: This work was supported by grants from FAPESP (BD: 2014/12976-5, DC-C: 2013/15343-0, RTK: 2015/13124-5, FCVP: 2019/20832-7) and CAPES (CCFS: 88882.377116/2019-01).

Institutional Review Board Statement: The study was conducted according to the guidelines of the Declaration of Helsinki and approved by the Institutional Review Board (or Ethics Committee) of Butantan Institute (CEUAIB protocol number 5396310517, approved on 21 June 2017).

Informed Consent Statement: Not applicable.

Acknowledgments: The authors thank Martin Wesley for English reviewing in the article.

Conflicts of Interest: The authors declare that they have no competing financial interest.

References

1. Bortoluzzi, L.R.; Querol, M.V.M.; Querol, E. Notas sobre a ocorrência de *Tityus serrulatus* Lutz & Mello, 1922 (Scorpiones, Buthidae) no oeste do Rio Grande do Sul, Brasil. *Biota Neotrop.* **2007**, *7*, 357–359. [CrossRef]
2. Cardoso, J.; França, F.S.; Wen, F.H.; Málaque, C.M.S.; Haddad, V., Jr. *Animais Peçonhentos no Brasil. Biologia, Clínica e Terapêutica dos Acidentes*, 2nd ed.; Sarvier: São Paulo, Brazil, 2009.
3. Lourenço, W.R.; Von Eickstedt, V.R.D.; Cardoso, J.L.C.; França, F.O.S.; Wen, F.H.; Málaque, C.M.S. Escorpiões de importância médica. In *Animais Peçonhentos no Brasil. Biologia, Clínica e Terapêutica dos Acidentes*; Sarvier: São Paulo, Brazil, 2003; ISBN 8573781335.
4. Marcussi, S.; Arantes, E.C.; Soares, A.M.; Giglio, J.R.; Mazzi, M.V. *Escorpiões: Biologia, Envenenamento e Mecanismos de Ação de suas Toxinas*, 1st ed.; FUNPEC: Ribeirão Preto, Brazil, 2011.
5. FUNASA. *Manual de Diagnóstico e Tratamento de Acidentes por Animais Peçonhentos*, 2nd ed.; Ministério da Saúde. Fundação Nacional de Saúde: Brasília, Brazil, 2001; ISBN 85-7346-014-8.
6. Matthiesen, F.A. The breeding of *Tityus serrulatus* Lutz & Mello, 1927, in captivity (Scorpions, Buthidae). *Rev. Bras. Pesqui. Med. Biol.* **1971**, *4*, 299–300.
7. Chippaux, J.-P. Epidemiology of envenomations by terrestrial venomous animals in Brazil based on case reporting: From obvious facts to contingencies. *J. Venom. Anim. Toxins Incl. Trop. Dis.* **2015**, *21*. [CrossRef] [PubMed]
8. Pucca, M.B.; Cerni, F.A.; Pinheiro Junior, E.L.; Bordon, K.D.C.F.; Amorim, F.G.; Cordeiro, F.A.; Longhim, H.T.; Cremonez, C.M.; Oliveira, G.H.; Arantes, E.C. *Tityus serrulatus* venom—A lethal cocktail. *Toxicon* **2015**, *108*, 272–284. [CrossRef] [PubMed]
9. de Oliveira, U.C.; Nishiyama, M.Y.; Santos, M.B.V.; Chalkidis, D.M.; Santos-da-Silva, A.D.P.; Souza-Imberg, A.; Candido, D.M.; Yamanouye, N.; Junqueira-de-Azevedo, I.L.M. Proteomic endorsed transcriptomic profiles of venom glands from *Tityus obscurus* and *T. serrulatus* scorpions. *PLoS ONE* **2018**, *13*, e0193739. [CrossRef] [PubMed]
10. Rates, B.; Ferraz, K.K.F.; Borges, M.H.; Richardson, M.; De Lima, M.E.; Pimenta, A.M.C. *Tityus serrulatus* venom peptidomics: Assessing venom peptide diversity. *Toxicon* **2008**, *52*, 611–618. [CrossRef] [PubMed]
11. Verano-Braga, T.; Dutra, A.A.A.; León, I.R.; Melo-Braga, M.N.; Roepstorff, P.; Pimenta, A.M.C.; Kjeldsen, F. Moving pieces in a venomic puzzle: Unveiling post-translationally modified toxins from *Tityus serrulatus*. *J. Proteome Res.* **2013**, *12*, 3460–3470. [CrossRef]
12. Nascimento, D.G.; Rates, B.; Santos, D.M.; Verano-Braga, T.; Barbosa-Silva, A.; Dutra, A.A.A.; Biondi, I.; Martin-Eauclaire, M.F.; De Lima, M.E.; Pimenta, A.M.C. Moving pieces in a taxonomic puzzle: Venom 2D-LC/MS and data clustering analyses to infer phylogenetic relationships in some scorpions from the Buthidae family (Scorpiones). *Toxicon* **2006**, *47*, 628–639. [CrossRef] [PubMed]
13. Rocha-Resende, C.; Leão, N.M.; de Lima, M.E.; Santos, R.A.; de Castro Pimenta, A.M.; Verano-Braga, T. Moving pieces in a cryptomic puzzle: Cryptide from *Tityus serrulatus* Ts3 Nav toxin as potential agonist of muscarinic receptors. *Peptides* **2017**, *98*, 70–77. [CrossRef] [PubMed]
14. Verano-Braga, T.; Rocha-Resende, C.; Silva, D.M.; Ianzer, D.; Martin-Eauclaire, M.F.; Bougis, P.E.; de Lima, M.E.; Santos, R.A.S.; Pimenta, A.M.C. Biochemical and Biophysical Research Communications *Tityus serrulatus* Hypotensins: A new family of peptides from scorpion venom. *Biochem. Biophys. Res. Commun.* **2008**, *371*, 515–520. [CrossRef]
15. Verano-Braga, T.; Figueiredo-Rezende, F.; Melo, M.N.; Lautner, R.Q.; Gomes, E.R.M.; Mata-Machado, L.T.; Murari, A.; Rocha-Resende, C.; Elena de Lima, M.; Guatimosim, S.; et al. Structure–function studies of *Tityus serrulatus* Hypotensin-I (TsHpt-I): A new agonist of B2 kinin receptor. *Toxicon* **2010**, *56*, 1162–1171. [CrossRef] [PubMed]

16. Duzzi, B.; Cajado-Carvalho, D.; Kuniyoshi, A.K.; Kodama, R.T.; Gozzo, F.C.; Fioramonte, M.; Tambourgi, D.V.; Portaro, F.V.; Rioli, V. [des-Arg 1]-Proctolin: A novel NEP-like enzyme inhibitor identified in *Tityus serrulatus* venom. *Peptides* **2016**, *80*, 18–24. [CrossRef] [PubMed]
17. Skidgel, R.A.; Erdos, E.G. Novel activity of human angiotensin I converting enzyme: Release of the NH2- and COOH-terminal tripeptides from the luteinizing hormone-releasing hormone. *Proc. Natl. Acad. Sci. USA* **1985**, *82*, 1025–1029. [CrossRef]
18. Roques, B.P.; Noble, F.; Dauge, V.; Fournie-Zaluski, M.C.; Beaumont, A. Neutral endopeptidase 24.11: Structure, inhibition, and experimental and clinical pharmacology. *Pharmacol. Rev.* **1993**, *45*, 87–146. [PubMed]
19. Rawlings, N.D.; Salvesen, G. *Handbook of Proteolytic Enzymes*, 3rd ed.; Elsevier: Amsterdam, Holland, 2013; ISBN 9780123822192.
20. Daull, P.; Jeng, A.Y.; Battistini, B. Towards Triple Vasopeptidase Inhibitors for the Treatment of Cardiovascular Diseases. *J. Cardiovasc. Pharmacol.* **2007**, *50*, 247–256. [CrossRef] [PubMed]
21. Cajado-Carvalho, D.; Kuniyoshi, A.K.; Duzzi, B.; Iwai, L.K. Insights into the Hypertensive Effects of *Tityus serrulatus* Scorpion Venom: Purification of an angiotensin-converting enzyme-like peptidase. *Toxins* **2016**, *8*, 348. [CrossRef] [PubMed]
22. Sifi, A.; Adi-Bessalem, S.; Laraba-Djebari, F. Involvement of the Endothelin Receptor Type A in the Cardiovascular Inflammatory Response Following Scorpion Envenomation. *Toxins* **2020**, *12*, 389. [CrossRef] [PubMed]
23. Cologna, C.; Marcussi, S.; Giglio, J.; Soares, A.; Arantes, E. *Tityus serrulatus* Scorpion Venom and Toxins: An Overview. *Protein Pept. Lett.* **2009**, *16*, 920–932. [CrossRef] [PubMed]
24. Pucca, M.B.; Cerni, F.A.; Pinheiro-Junior, E.L.; Zoccal, K.F.; Bordon, K.D.C.F.; Amorim, F.G.; Peigneur, S.; Vriens, K.; Thevissen, K.; Cammue, B.P.A.; et al. Non-disulfide-bridged peptides from *Tityus serrulatus* venom: Evidence for proline-free ACE-inhibitors. *Peptides* **2016**, *82*, 44–51. [CrossRef] [PubMed]
25. Cassini-Vieira, P.; Felipetto, M.; Prado, L.B.; Verano-Braga, T.; Andrade, S.P.; Santos, R.A.S.; Teixeira, M.M.; de Lima, M.E.; Pimenta, A.M.C.; Barcelos, L.S. Ts14 from *Tityus serrulatus* boosts angiogenesis and attenuates inflammation and collagen deposition in sponge-induced granulation tissue in mice. *Peptides* **2017**, *98*, 63–69. [CrossRef]
26. Torres-Rêgo, M.; Gláucia-Silva, F.; Rocha Soares, K.S.; de Souza, L.B.F.C.; Damasceno, I.Z.; dos Santos-Silva, E.; Lacerda, A.F.; Chaves, G.M.; da Silva-Júnior, A.A.; de Fernandes-Pedrosa, M.F. Biodegradable cross-linked chitosan nanoparticles improve anti-Candida and anti-biofilm activity of TistH, a peptide identified in the venom gland of the *Tityus stigmurus* scorpion. *Mater. Sci. Eng. C* **2019**, *103*, 109830. [CrossRef] [PubMed]
27. Barros, N.M.T.; Campos, M.; Bersanetti, P.A.; Oliveira, V.; Juliano, M.A.; Boileau, G.; Juliano, L.; Carmona, A.K. Neprilysin carboxydipeptidase specificity studies and improvement in its detection with fluorescence energy transfer peptides. *Biol. Chem.* **2007**, *388*. [CrossRef] [PubMed]
28. Segel, I.H. *Enzyme Kinetics: Behavior and Analysis of Rapid Equilibrium and Steady State Enzyme Systems*; Willey: New York, NY, USA, 1975.

Article

Crotoxin Modulates Events Involved in Epithelial–Mesenchymal Transition in 3D Spheroid Model

Ellen Emi Kato [1] and Sandra Coccuzzo Sampaio [1,2,*]

[1] Laboratory of Pathophysiology, Butantan Institute, São Paulo 05503-900, Brazil; ellen.kato@esib.butantan.gov.br
[2] Department of Pharmacology, Institute of Biomedical Sciences, University of São Paulo, São Paulo 05508-060, Brazil
* Correspondence: sandra.coccuzzo@butantan.gov.br

Abstract: Epithelial–mesenchymal transition (EMT) occurs in the early stages of embryonic development and plays a significant role in the migration and the differentiation of cells into various types of tissues of an organism. However, tumor cells, with altered form and function, use the EMT process to migrate and invade other tissues in the body. Several experimental (in vivo and in vitro) and clinical trial studies have shown the antitumor activity of crotoxin (CTX), a heterodimeric phospholipase A2 present in the *Crotalus durissus terrificus* venom. In this study, we show that CTX modulates the microenvironment of tumor cells. We have also evaluated the effect of CTX on the EMT process in the spheroid model. The invasion of type I collagen gels by heterospheroids (mix of MRC-5 and A549 cells constitutively prepared with 12.5 nM CTX), expression of EMT markers, and secretion of MMPs were analyzed. Western blotting analysis shows that CTX inhibits the expression of the mesenchymal markers, N-cadherin, α-SMA, and αv. This study provides evidence of CTX as a key modulator of the EMT process, and its antitumor action can be explored further for novel drug designing against metastatic cancer.

Keywords: crotoxin; epithelial–mesenchymal transition; spheroid model; tumor stroma

Key Contribution: We demonstrated for the first time the modulatory effect of CTX on tumor–stroma interaction in a 3D model by showing its inhibitory capacity on crucial events of EMT and provide evidences that CTX is a potential modulator of the signaling cascade involved in the tumor progression.

Citation: Kato, E.E.; Sampaio, S.C. Crotoxin Modulates Events Involved in Epithelial–Mesenchymal Transition in 3D Spheroid Model. *Toxins* **2021**, *13*, 830. https://doi.org/10.3390/toxins13110830

Received: 15 October 2021
Accepted: 19 November 2021
Published: 22 November 2021

1. Introduction

Events of metastasis begin with epithelial–mesenchymal transition (EMT), whereby epithelial cells lose their inherent characteristics, including apicobasal polarity, and acquire invasive and infiltrating mesenchymal properties [1–4]. The EMT program involves the loss of expression of E-cadherin. In contrast, the expression of N-cadherin is switched on along with upregulation of other mesenchymal markers, including vimentin, fibronectin, and metalloproteases [5–9]. Cancer-associated fibroblasts (CAFs), activated fibroblasts commonly found in the tumor microenvironment, are characterized by higher expression of myofibroblastic markers, including α-smooth muscle actin (α-SMA). CAFs are among the predominant cells within solid tumors and secrete several soluble growth factors, including transforming growth factor-β1 (TGF-β1), interleukin-6 (IL-6), fibroblast growth factor (FGF), platelet-derived growth factor (PDGF), stromal cell-derived factor-1 (SDF-1), and hepatocyte growth factor (HGF). Some of these growth mediators can induce EMT in carcinoma cells [10–12].

The majority of data that highlight the importance of EMT in carcinogenesis have been obtained using in vitro cell line models. Over the last decade, three-dimensional (3D) cancer cell culture systems in vitro have been developed to understand the interactions and crosstalk between cell–cell and cell–matrix that drives tumor progression [13–15]. The activation of EMT is mostly studied in various cancer spheroid models [16]. Therefore, the development of novel drugs that can inhibit the onset of EMT is one of the goals of cancer research. In this direction, several compounds derived from animal venoms are considered critical scientific tools. Firstly, animal venoms have practical therapeutic applications as they provide structural templates for the development of new drugs. Secondly, studies on animal venoms have contributed significantly to understanding the regulatory mechanisms guiding cell functions under normal and pathophysiological conditions, such as cancer [17].

Snake venoms are complex mixtures of bioactive molecules [17–19]. Phospholipase A2 enzymes (PLA2; EC 3.1.1.4) are among the most well-characterized components of known snake venoms. Among these, crotoxin (CTX), the major toxin from *Crotalus durissus terrificus* venom, is a heterodimer comprised of a basic subunit (CB), responsible for the phospholipase activity, and neurotoxic and myotoxic properties of the molecule. The CB subunit is associated non-covalently with crotapotin, an acidic, non-enzymatic peptide (CA). The CTX complex is a potent neurotoxin, while isolated subunits present low lethality [20–25]. Sixteen CTX isoforms were identified, resulting from a random combination of four CA isoforms (CA1, CA2, CA3, and CA4) and four CB isoforms (CBa2, CBb, CBc, and CBd) [22]. These combinations between the isoforms determine the formation of different complexes, responsible for the different pharmacological and biological properties reported for CTX [21]. Several studies have shown that CTX has in vivo and in vitro anti-inflammatory, immunomodulatory, and antitumoral properties [26–31]. Cura and colleagues (2002) suggested that CTX may have greater selectivity on solid tumors since CTX could inhibit in vivo growth of Lewis lung carcinoma and MX-I human mammary carcinoma. However, it has low antitumor activity against HL-60 leukemia cells [32]. Recent studies have demonstrated that CTX not only inhibits tumor growth but also modulates stromal cells in the tumor microenvironment, such as the reprogramming of endothelial cells and macrophages, thus exhibiting an antiangiogenic phenotype [26,28,29,33].

Based on the properties mentioned above of CTX on the tumor microenvironment we also present, herein, the ability of this toxin to modulate EMT. In this study, we show for the first time, the modulatory effect of CTX on EMT markers in the 3D-spheroid model composed of tumor cells and fibroblasts. CTX is a promising target for the future development of anti-metastasis therapeutics.

2. Results

2.1. Effect of CTX on MRC-5 Cell Differentiation with Different Stimulatory Factors under 2D Condition

We confirmed the expression of α-SMA in differentiating myofibroblasts through immunostaining. MRC-5 cells, previously incubated with 12.5 nM CTX, were cultured in the presence of TGF-β1 (2 ng/mL) or conditioned medium (CM) from A549 cells for 3 days. There was an undetectable level of α-SMA in both control and CTX-treated unstimulated MRC-5 cells (cultured in DMEM with 10% FBS only). We found that TGF-β1 and tumor-CM induced higher α-SMA expression in untreated than in CTX-treated MRC-5 cells (Figure 1A,B). Under the same experimental condition, Calu-3-CM induced α-SMA expression in MRC-5 cells, while in the presence of CTX this marker was undetectable (Figure S1).

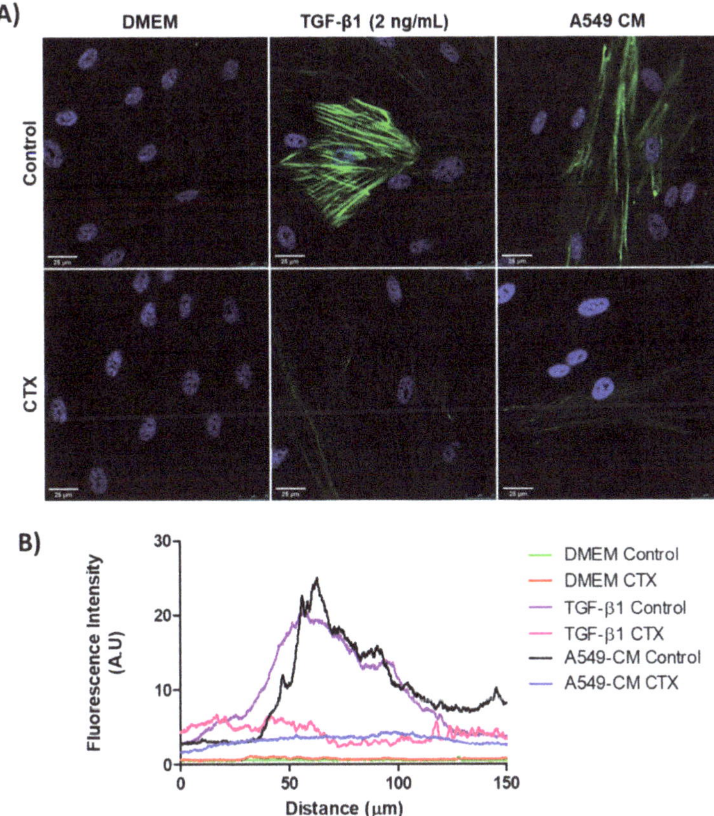

Figure 1. Myofibroblast differentiation with various stimulatory factors. (**A**) Representative immunofluorescent images of MRC-5 cells pretreated with CTX (12.5 nM) for 2 h and then incubated in DMEM (with 10% FBS), TGF-β1 (2 ng/mL), or tumor-conditioned media from A549 cells for 3 days. The control group of cells were untreated and grown in DMEM with 10% FBS only. (**B**) Quantification of α-SMA fluorescence intensity. Green fluorescence indicates α-SMA-containing stress fibers and blue fluorescence indicates the nuclei. Scale bar = 25 µm. The data are presented from three independent experiments.

2.2. CTX Does Not Interfere in the Spheroid Formation and Prevents Non-Spheroid from Forming Cells

The human adenocarcinoma cell line A549 was mixed with MRC-5 cells in a ratio of 1:4 for spheroid formation by the hanging drop method. We observed regular round-shaped spheroids composed only of MRC-5 cells (Figure 2A); the presence of CTX did not affect the spheroid formation (Figure 2B). On the other hand, A549 cells were unable to aggregate to form a spherical structure due to weak intercellular interactions (data not shown). When A549 cells were mixed with MRC-5 cells, compact spheres were observed with a small group of tumor cells around them. These tumor cells were unable to integrate into the spheres (Figure 2C,D). Interestingly, there was a drastic reduction in tumor cells around the spheroids formed by the CTX treated MRC-5 cells (Figure 2E,F). On the other hand, MRC-5/Calu-3 spheroids were less compact compared to MRC-5/A549 spheroids (Figure S2A,B), while in the presence of CTX, fewer tumor cells were observed around the spheroid (Figure S2C,D). In addition, live/dead staining of MRC-5/A549 cells in spheroids confirmed that CTX did not affect cell viability (Figure 2E,F).

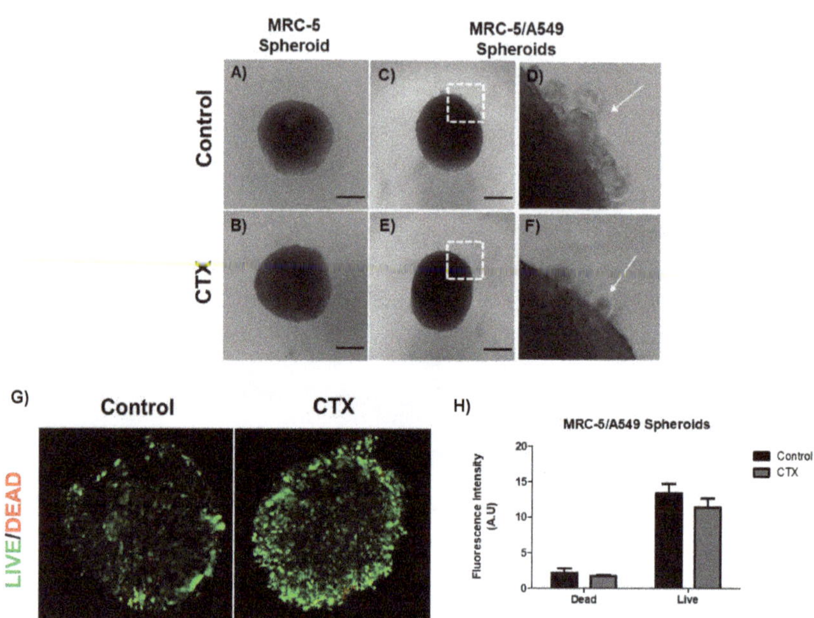

Figure 2. Spheroid formation by MRC-5 cells alone or in combination with A549 tumor cells. (**A**) MRC-5 single spheroid formation by the hanging drop method. (**B**) MRC-5 single spheroid constitutively formed in the presence of 12.5 nM of CTX. (**C**) MRC-5/A549 spheroid formation by the hanging drop method. (**D**) A small number of cancer cells did not incorporate into the cell aggregates (white arrow) (**E**) MRC-5/A549 spheroid constituted in the presence of 12.5 nM CTX—after 24 h, cell aggregates formed compact structures. (**F**) A small number of cancer cells did not incorporate into the cell aggregates (white arrow). Images obtained from inverted microscope 4X. (**E**) Live (green)/Dead (red) image of MRC-5/A549 spheroids. Scale bar = 100 μm. (**F**) Quantitative analysis from live/dead assay measured by relative fluorescence intensity. All the data presented here are from three independent experiments.

2.3. Spheroid Cell Invasion of Collagen Gel

To analyze the effect of CTX on the invasive phenotype, MRC-5/A549 spheroids were embedded into polymerized collagen type I gels for up to 48 h. Invading cells from MRC-5/A549 spheroids were elongated and spindle-shaped (Figure 3A). Spheroids constituted in the presence of CTX showed a 50% reduction in the invaded gel region (Figure 3B,C). On the other hand, the invasion distance of MRC-5/Calu-3 spheroids was shorter and preserved cell–cell interaction compared to MRC-5/A549 spheroids, while in the presence of CTX, the invasion area was reduced by invading cells (Figure S3).

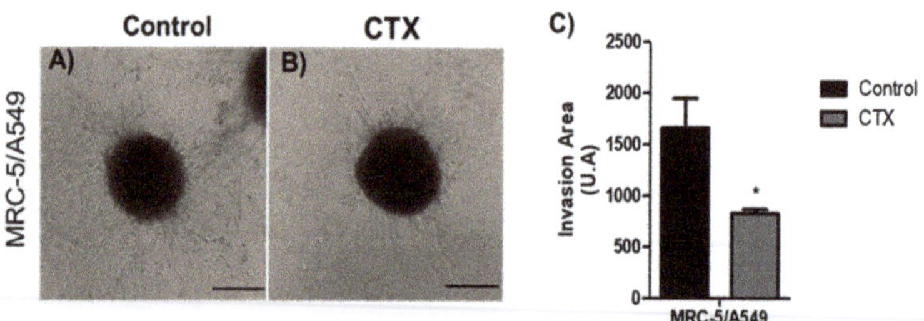

Figure 3. Invasion area of spheroids in 3D collagen gels. Representative images are of invasion into

type I collagen gel (1.2 mg/mL) (**A**) by cells of MRC-5/A549 spheroids or (**B**) MRC-5/A549 constituted with 12.5 nM of CTX. Cell invasion was photographed under phase-contrast microscopy at 48 h. (**C**) Cell invasion area was measured and analyzed on ImageJ software. * $p < 0.05$ compared to the control group. The data presented here are from three independent experiments ($n = 5$).

2.4. Effect of CTX on the Expression of EMT Markers in Composite Spheroids

To investigate the mechanism underlying the migration ability of MRC-5/A549 spheroids, we performed Western blot analysis for epithelial marker E-cadherin and mesenchymal markers—N-cadherin, vimentin, and α-SMA. Our results showed that after 3 days, MRC-5/A549 spheroids lost E-cadherin, but the expression of all EMT markers was elevated. MRC-5/A549 constitutively formed in the presence of CTX showed a significant reduction in mesenchymal markers, α-SMA (43%), N-cadherin (46%), and integrin αv (41%) (Figure 4). However, there was no marked difference in vimentin levels. On the other hand, at the same incubation period, MRC-5/Calu-3 spheroids presented high expression of E-cadherin and low expression of mesenchymal markers, which means that MRC-5/Calu-3 spheroids need a longer period of incubation to induce EMT. However, the presence of CTX on this spheroid inhibited only α-SMA expression compared to the control group, while there was no difference in integrin αv expression (Figure S4).

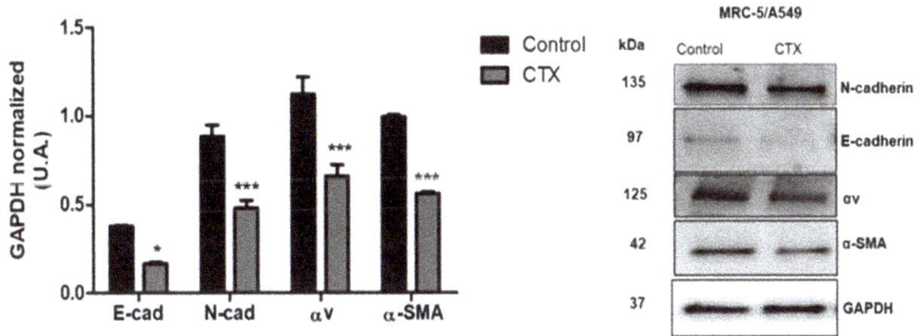

Figure 4. Expression of EMT-associated proteins. Quantitative analysis of expression of EMT markers. After three days in culture, MRC-5/A549 and MRC-5/Calu-3 spheroids were lysed, and Western blot analyses were conducted for E-cadherin, N-cadherin, α-SMA, integrin subunit αv. GAPDH served as the loading control. *** $p < 0.001$ compared to the control group. * $p < 0.01$ compared to the control group. The data presented are from three independent experiments ($n = 4$).

2.5. Effect of Crotoxin on MMP-9, MMP-13, and Cytokine Secretions in 3D Collagen Gel Matrix

To determine the effect of CTX on spheroids in the 3D collagen gel, the release of matrix metalloproteinases, MMP-9 and MMP-13 (ECM-digesting enzymes), in the culture media were measured. Our results showed that a monolayer of MRC-5, grown in serum-free media, produced endogenous MMP-9; the enzyme was undetectable in media harboring A549 cells monolayer (Figure 5A). MMP-13 were produced endogenously by all cell lines. CTX did not interfere with MMP-9 release by MRC-5 cells, even though CTX repressed MMP-13 (69%) release by A549 cells (Figure 5B). Interestingly, MRC-5/A549 spheroids embedded in collagen gel promoted secretions of MMP-9 and MMP-13. However, composite spheroids treated with CTX inhibited MMP-9 and MMP-13 secretions (37% and 39%, respectively).

In addition, cytokines, chemokines, and growth factors released during the spheroid invasion of 3D collagen gel were assessed by a membrane-based cytokine array. Our results showed that the concentrations of 32 out of 80 cytokines had reduced by 1.2- to 5-fold in the culture media containing CTX-treated spheroid cells (Figure 5C). We found that CTX inhibits tumor-related cytokines, particularly IL-6, IL-8, HGF, TGF-β1, and IGFBP-1. CTX also inhibits chemokines that bind to CXCR1 and CXCR2 receptors such as CXCL5,

CXCL1/2/3, CXCL1a, CXCL6, and CXCL8 (IL-8) that are involved in cancer angiogenesis and metastasis (Figure 5C).

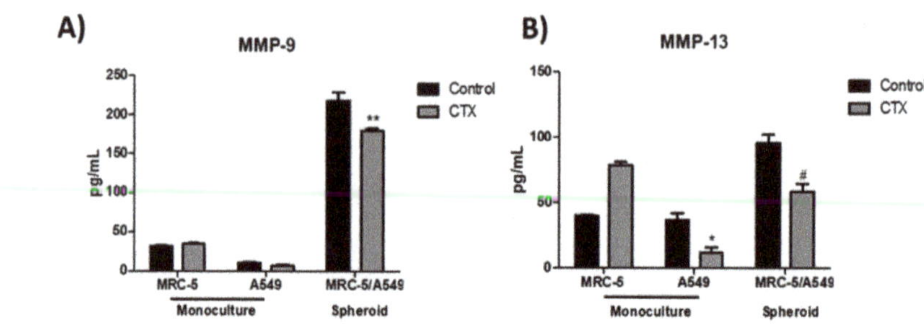

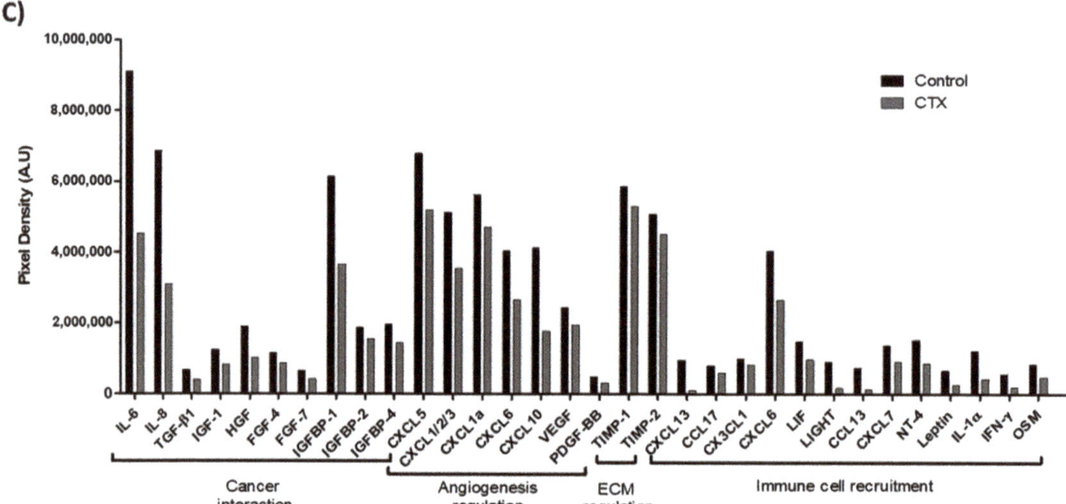

Figure 5. Effect of crotoxin on the production of MMP-9, MMP-13, cytokines, and chemokines in the 3D collagen gel. Monoculture of human lung fibroblasts MRC-5 and human lung adenocarcinoma cell line A549 incubated with 12.5 nM of CTX for three days; invasion of collagen gel by composite spheroids of MRC-5/A549 at 48 h. After this period, spent media were collected to quantify the volume of MMP-9 (**A**) and MMP-13 (**B**) by ELISA. * $p < 0.05$ compared to the control group. # $p < 0.05$ compared to the control group. ** $p < 0.01$ compared to the control group. ($n = 6$). In (**C**), cytokine array analysis using composite spheroid invaded gel of at 48 h. The graph indicates the decrease in the growth factors by more than 1.2-folds. A pool of $n = 4$ was used.

3. Discussion

The present study aimed to investigate the possible modulatory effect of crotoxin, a toxin with PLA2 activity from the venom of the snake, *Crotalus durissus terrificus*, on EMT. Here, we used an in vitro spheroid model composed of human lung adenocarcinoma and human lung fibroblast cell lines. This model resembles early tumor–stroma interactions, mimicking an avascular tumor initiation step [13,34,35]. During tumor spheroid formation, fibroblasts become activated and acquire a myofibroblast-like phenotype referred to cancer-associated fibroblasts. These CAFs express contractile proteins (particularly α-smooth muscle actin), synthesize a large amount of ECM components, and secrete various matrix metalloproteinases [12,36].

In this study, we used a co-culture system to understand the reciprocal crosstalk between NSCLC tumor cell line, A549, and normal human lung fibroblast cell line, MRC-5.

We found that tumor-conditioned media promoted the expression of α-SMA in MRC-5 cells, whereas CTX inhibited it. It has already been reported that the TGF-β pathway is the dominant mediator of crosstalk to initiate the process of CAF activation [37]. Previous studies have demonstrated the inhibitory effect of CTX on the functions of stromal cells such as macrophages [29] and endothelial cells [33] when co-cultivated with tumor-conditioned media. Thus, CTX impairs tumor progression. A recent study demonstrated that when human skin fibroblast cells were incubated with crude venom (CdtV) from *Crotalus durissus terrificus*, cells showed altered protrusions, formed highly polymerized actin filaments, and produced a high amount of fibronectin [38]. Based on these observations, we suggest that CTX plays varied roles in different microenvironments and may regulate the process of tissue repair.

Stadler and colleagues (2018) used different colon cancer cell lines for the spheroid formation and showed that some of the cells did not integrate into the spheroids. They hypothesized that the non-spheroid forming (NSF) cells are a subpopulation of tumor cells that had lost cell–cell adhesion properties and rendered them the ability to migrate [39]. Moreover, Sodek and colleagues (2009) correlated the ability of ovarian cancer cell lines to form compact spheroids with their migratory and invading capacity in 3D matrices. These cells exhibited myofibroblast-like features [40]. Our results showed that the presence of CTX in the composite spheroid prevented the loss of cell–cell adhesion properties of the cells and reduced the invasion area in a 3D collagen matrix.

To confirm our findings, the expression of a well-defined set of EMT-associated markers was analyzed by western blotting. Three-day-old MRC-5/A549 spheroids presented an upregulation of mesenchymal markers (such as N-cadherin, α-SMA, and integrin αv) and downregulation of E-cadherin in consensus with its rapid progression toward EMT. In contrast, as shown in the Supplementary Material, MRC-5/Calu-3 spheroids presented no alterations on EMT markers at the same experimental condition. These findings concur with a previous study that showed EMT progression in A549 and Calu-3 cells in vitro when exposed to TGF-β1 and pro-inflammatory cytokines. The authors suggest that differential cell plasticity and susceptibility to EMT may depend on tissue origin [41]. As fibroblasts become CAFs during spheroid formation, the interaction between tumor cells and CAFs leads to invasion strategies; CAFs turn into primary drivers to help tumor cells migrate by remodeling ECM and creating tracks [42]. Taken together, our data suggest that CTX significantly inhibits expression of N-cadherin, α-SMA, and integrin αv in MRC-5/A549 spheroids, which correlates with the reduced invasion area in the collagen gel. It also suggests an involvement of CTX with actin polymerization via integrin-dependent signaling pathway with subsequent impairment of migratory ability, a finding that was also observed in endothelial cells in the tumor microenvironment [33]. These findings are in agreement with a similar study conducted with PLA2 (BthTX-II) extracted from the venom of *Bothrops jararacussu*. BthTX-II displayed a weak catalytic activity and presented an inhibitory effect on the adhesion, proliferation, invasion, and migration of human breast cancer cells. It also inhibited EMT by modulating epithelial and mesenchymal markers [43]. Many studies have shown the inhibitory effect of other PLA2s on tumor cells, however the findings did not correlate the effect with the enzymatic activity of the molecule. CTX acts on focal adhesion kinases (FAK), a crucial component of integrin-mediated cell signaling in endothelial cells [33], and inhibits tyrosine phosphorylation, consequently inhibiting the activity of proteins involved in the intracellular signaling pathway of macrophages [44].

Our data suggest that CTX affected MMP-9 secretions, which significantly increased during tumor–CAF crosstalk in the spheroid model. Unsurprisingly, monocultures of A549 and MRC-5 cells released endogenous levels of MMP-9. It has been shown that high expression of MMP-9 is associated with the aggressiveness of malignant cells in solid tumors [45]. MMP-9 has also been reported to activate the bioactive form of TGF-β [46] and downregulate the expression of E-cadherin [47], thus initiating the process of EMT. Moreover, Eberlein and colleagues (2015) reported that the activation of normal fibroblast during tumor cell–fibroblast crosstalk occurs through the αvβ6/TGF-β signaling pathway [37].

We hypothesize that CTX regulates TGF-β activation since our findings showed reduced secretions of MMP-9 and αv integrin in the spheroid model. Moreover, human MMP-13 is expressed in skin fibroblasts and has a role in acute wound healing by remodeling fibrillar collagens [48]. In this study, we found an increased secretion of MMP-13 in MRC-5 monoculture incubated with CTX, suggesting its involvement in wound healing. The modulatory activity of CTX during the healing process in an inflammatory environment has already been shown in an earlier study [49]. Conversely, MMP-13 secretions were reduced in A549 monoculture in the presence of CTX as well as in the heterospheroid model. A study using NSCLC from patients showed that both MMP-9 and MMP-13 were associated with metastasis, invasion, and prognosis; MMP-13 mainly activates MMP-9 to participate in the invasion and metastasis of NSCLC [50]. This corroborates our finding of the microenvironment-dependent modulatory activity of CTX on the release of MMP-9 and MMP-13.

Cytokines released during MRC-5/A549 spheroids' invasion of collagen gel have shown that CTX drastically inhibits chemokines that bind to the receptor CXCR1 and CXCR2, such as CXCL5, CXCL-8 (IL-8), CXCL1/2/3, CXCL1a, and CXCL6. It is well established that the CXCL5/CXCR2 and IL-8/CXCR1/CXCR2 axes contribute to carcinogenesis by promoting tumor cell proliferation, migration, and invasion, and the EMT process of many tumor cells, including NSCLC [50–53], hepatocarcinoma cells [54], and papillary thyroid carcinoma cells [55]. Additionally, CTX inhibits primary growth factors involved with EMT such as HGF [56], TGF-β1 [37,57], and VEGF [53].

4. Conclusions

This is the first scientific report about the modulatory effect of CTX on paracrine signaling during tumor–stroma crosstalk as it inhibits CAFs' differentiation. In the 3D model, CTX repressed the expression of mesenchymal markers, chemokines, and growth factors associated with the EMT process. CTX also suppressed the invasion of collagen gel due to a decrease in MMP-9 and MMP-13 secretions and activities. We have provided evidence that CTX is a potential modulator of the signaling cascade involved in the progression of EMT, a critical phase that results in metastasis.

5. Materials and Methods

5.1. Crotoxin

CTX was obtained from the venom collected from several *Crotalus durissus terrificus* and provided by the Laboratory of Herpetology, Butantan Institute. The purification of this toxin was performed as described previously [58]. The purity of CTX was verified by non-reducing sodium dodecyl sulfate-polyacrylamide gel electrophoresis (12.5%) [59] and PLA2 activity was assessed by colorimetric assay using a synthetic chromogenic substrate as described elsewhere [60].

5.2. Cell Culture

The human non-small cell lung adenocarcinoma A549 cell line was kindly provided by Dr Durvanei Augusto Maria, Laboratory of Biochemistry and Biophysics, Butantan Institute. MRC-5 human lung fibroblast cell lines were purchased from Rio de Janeiro Bank Cells (Banco de Células do Rio de Janeiro, BCRJ code-0180) (Rio de Janeiro, Brazil). Monolayer cell was cultured in 75 cm^2 flasks, containing DMEM (Dulbecco's Modified Eagles Medium) supplemented with 10% fetal bovine serum (FBS), 100 units/mL of penicillin, and 0.1 mg/mL of streptomycin (CultiLab) with 5% CO_2 at 37 °C for 48 h. Subconfluent cells were trypsinized with trypsin/EDTA (Gibco®) for subsequent use. A549 cells from passage 32–37, MRC-5 cells from passage 29–32 were used in our experiments.

5.3. Spheroid Preparation and Culture

Composite spheroids containing a mixture of tumor cells A549 and MRC-5 fibroblasts were prepared based on the hanging drop method as described earlier [13,35]. Subconfluent cells were trypsinized and resuspended in DMEM with 10% FBS to a concentration of 1×10^6/mL. Cell suspension of fibroblasts and tumor cells were prepared with or without 12.5 nM of CTX and then mixed at the ratio of 4:1 (4 fibroblasts per 1 tumor cell). Approximately 40 drops (25 mL/drop, 2.5×10^4 cells) were dispensed onto a lid of a cell culture dish. The lid was then inverted and placed over a cell culture dish containing a culture medium for humidity. For spheroid optimization, it was added to suspension cells 0.24% of methylcellulose so that it can contribute to more compact and circular spheroid morphology. The lid was incubated for 24 h, at 37 °C, and 5% CO_2.

5.4. Invasion in Three-Dimensional Spheroid Collagen Gel

Type I collagen gel (1.2 mg/mL) was prepared as described by [13]. Type I collagen gel solution (5 mL of 2X DMEM, 1 mL 10X HEPES (0.2 M, pH 8.0), and 4 mL type I collagen (3 mg/mL, PureCol, Advanced Biomatrix®)) were prepared and kept on ice before and during experiments. For this assay, 100 µL/well of collagen gel was added in a 48-well plate; it was allowed to polymerize for 30 min at 37 °C to form the first layer of the gel. After gel polymerization, the spheroid was embedded into the collagen gel by pipetting it into the second collagen gel layer (100 µL/well). The collagen–spheroid mixtures were then left to polymerize in the cell culture incubator, the 400 µL of serum-free DMEM media were added on top of the wells. After 48 h, the spent media were collected for ELISA assay. Cell migration from spheroids embedded in collagen gels was monitored under an inverted light microscope (Leica DMIL®, Wetzlar, Germany) and photographed at different time points.

5.5. Confocal Immunofluorescence Analysis

A monolayer of MRC-5 (2×10^5/well) was grown in DMEM supplemented with 10% SFB for 24 h. Fibroblasts were then treated with CTX (12.5 nM) for 2 h and, washed with PBS and cultured in either DMEM with 1% SFB in the presence of 2 ng/mL of TGF-β1 (Peprotech, cat number 100–21C, Cranbury, NJ, USA) or conditioned media from A549 cells for three days. After this, cells were fixed in 4% paraformaldehyde for 20 min, rehydrated with PBS (3 × 10 min), permeabilized in 0.2% Triton X-100 in PBS for 10 min, and kept overnight in a blocking solution containing PBS with 1% BSA and 0.1% Triton X-100, at 4 °C. The slides were incubated for 1 h with anti-α-SMA (1:250) in PBS with 1% BSA, and 0.1% Triton X-100 at 37 °C. After three washes with PBS containing 0.05% Tween 20, the slides were incubated with Goat anti-rabbit conjugated with DyLight 549 (1:800) for 1 h at room temperature. The slides were washed twice and then incubated with DAPI (25 mg/mL; Sigma Aldrich, Saint Louis, MO, USA) for 2 min. Mounting medium was applied, and the cells were photographed using a Leica DMi8 confocal microscope (Leica Microsystems, Mannheim, Germany), equipped with a DFC 310 FX digital camera, 63× magnification with oil immersion. Images were captured with the LAS AF and processed with the ImageJ software (National Institutes of Health, Bethesda, MD, USA).

5.6. Western Blotting

Spheroids of MRC-5/A459 were lysed, and total cellular protein was extracted using RIPA lysis (Sigma Aldrich, R0278) with a protease and phosphatase cocktail (Sigma Aldrich, P0044; P5726; P8340). Cell lysates were then centrifuged at 12,000 rpm for 30 min at 4 °C, the supernatant containing the soluble proteins was collected and measured by the BCA protein assay (Novagen®, 71285). The samples containing 30 µg protein were subjected to SDS/PAGE under reducing conditions on a 4–20% gradient polyacrylamide gel (Bio-Rad, cat no. 456–1094, Hercules, CA, USA). Following electrophoresis, proteins were transferred to a nitrocellulose membrane, which was then blocked with TBS-T (20 mM Tris-HCl, 150 mM NaCl, and 0.1% Tween 20) containing 5% nonfat dry milk for 2 h. The membranes were incubated with primary antibodies: anti-E-cadherin (1:1000), anti-N-cadherin (1:1000), anti-vimentin (1:5000), anti-α-SMA (1:2000), anti-integrin αv (1:2000), and anti-GAPDH (1:1000) overnight at 4 °C (Abcam). The membrane was subsequently incubated with peroxidase-conjugated secondary antibody (anti-mouse IgG and anti-goat IgG) for 2 h. Detection was performed with an enhanced chemiluminescence kit (Thermo Scientific, Rockford, IL, USA). The signals were detected with an image acquisition system (Alliance 9.7, Uvitec, Cambridge, UK). Band intensities were measured with Image J software (NIH).

5.7. ELISA Assay

As mentioned earlier, spheroids' spent media were collected for collagen invasion assay. Human MMP-9 (ab100610) and MMP-13 (ab100605) were quantified by using specific ELISA kits following the manufacturer's instructions. ELISA Kits were purchased from Abcam Company (Cambridge, UK).

5.8. Human Cytokine Array Assay

The expression levels of chemokines and cytokines were analyzed using a Human Cytokine Antibody Array (C5) (RayBiotech, Inc., Norcross, GA, USA). As per the manufacturer's instructions, antibody-embedded membranes were incubated with 1 mL of conditioned media (CM) at 4 °C overnight, followed by incubation with a biotin-conjugated detection antibody cocktail and diluted HRP-streptavidin at room temperature. Proteins were then visualized using a chemiluminescent substrate reagent. The intensity of each spot represents cytokines, which were then quantified densitometrically using ImageJ software.

5.9. Statistical Analysis

The GraphPad Prism software was used (version 5.0). Student's *t*-test and one-way ANOVA (followed by Tukey's post test) were used for comparisons, and differences were considered significant at $p < 0.05$.

Supplementary Materials: The following are available online at https://www.mdpi.com/article/10.3390/toxins13110830/s1, Figure S1: Myofibroblast differentiation in Calu-3 conditioned medium. Figure S2: MRC-5/Calu-3 spheroid. Figure S3: Invasion area of MRC-5/Calu-3 spheroids in 3D collagen gels. Figure S4: Expression of EMT-related proteins.

Author Contributions: All authors contributed to data analysis, model development, and manuscript writing. Thus, each author participated sufficiently in the work to take public responsibility for appropriate portions of the content and, therefore, agreed to be accountable for all aspects in ensuring that questions related to the accuracy or integrity of any part of the work are appropriately investigated and resolved. All authors have read and agreed to the published version of the manuscript.

Funding: This work was supported by São Paulo Research Foundation (FAPESP)-Grants 2017/11161-6. This study was financed in part by the Coordenação de Aperfeiçoamento de Pessoal de Nível Superior—Brasil (CAPES)—Finance Code 001. This study was financed in part by the National Council for Scientific and Technological Development—CNPq-Productivity grant (301685/2017-7).

Institutional Review Board Statement: Not applicable.

Informed Consent Statement: Not applicable.

Data Availability Statement: Not applicable.

Acknowledgments: The authors gratefully acknowledge the financial support by São Paulo Research Foundation, Conselho Nacional de Desenvolvimento Científico e Tecnológico (CNPq) and Coordenação de Aperfeiçoamento de Pessoal de Nível Superior (CAPES). The authors also thank, for technical support, Alexsander S. de Souza from the Multiuser Platform of the Scientific Development Center of the Butantan Institute, São Paulo, Brazil, for confocal microscopy.

Conflicts of Interest: The authors declare no conflict of interest. The funders had no role in the design of the study; in the collection, analyses, or interpretation of data; in the writing of the manuscript, or in the decision to publish the results.

References

1. Kalluri, R.; Weinberg, R.A. The Basics of Epithelial-Mesenchymal Transition. *J. Clin. Investig.* **2009**, *119*, 1420–1428. [CrossRef] [PubMed]
2. Kim, K.K.; Wei, Y.; Szekeres, C.; Kugler, M.C.; Wolters, P.J.; Hill, M.L.; Frank, J.A.; Brumwell, A.N.; Wheeler, S.E.; Kreidberg, J.A.; et al. Epithelial Cell Alpha3beta1 Integrin Links Beta-Catenin and Smad Signaling to Promote Myofibroblast Formation and Pulmonary Fibrosis. *J. Clin. Investig.* **2009**, *119*, 213–224. [CrossRef] [PubMed]
3. Thiery, J.P. Epithelial–Mesenchymal Transitions in Tumour Progression. *Nat. Rev. Cancer* **2002**, *2*, 442–454. [CrossRef]
4. Zeisberg, M.; Neilson, E.G. Biomarkers for Epithelial-Mesenchymal Transitions. *J. Clin. Investig.* **2009**, *119*, 1429–1437. [CrossRef]
5. Ding, S.; Chen, G.; Zhang, W.; Xing, C.; Xu, X.; Xie, H.; Lu, A.; Chen, K.; Guo, H.; Ren, Z.; et al. MRC-5 Fibroblast-Conditioned Medium Influences Multiple Pathways Regulating Invasion, Migration, Proliferation, and Apoptosis in Hepatocellular Carcinoma. *J. Transl. Med.* **2015**, *13*, 237. [CrossRef] [PubMed]
6. Ding, S.-M.; Lu, A.-L.; Zhang, W.; Zhou, L.; Xie, H.-Y.; Zheng, S.-S.; Li, Q.-Y. The Role of Cancer-Associated Fibroblast MRC-5 in Pancreatic Cancer. *J. Cancer* **2018**, *9*, 614–628. [CrossRef]
7. Heylen, N.; Baurain, R.; Remacle, C.; Trouet, A. Effect of MRC-5 Fibroblast Conditioned Medium on Breast Cancer Cell Motility and Invasion in Vitro. *Clin. Exp. Metastasis* **1997**, *16*, 193–203. [CrossRef]
8. Lamouille, S.; Xu, J.; Derynck, R. Molecular Mechanisms of Epithelial–Mesenchymal Transition. *Nat. Rev. Mol. Cell Biol.* **2014**, *15*, 178–196. [CrossRef]
9. Shintani, Y.; Abulaiti, A.; Kimura, T.; Funaki, S.; Nakagiri, T.; Inoue, M.; Sawabata, N.; Minami, M.; Morii, E.; Okumura, M. Pulmonary Fibroblasts Induce Epithelial Mesenchymal Transition and Some Characteristics of Stem Cells in Non-Small Cell Lung Cancer. *Ann. Thorac. Surg.* **2013**, *96*, 425–433. [CrossRef] [PubMed]
10. Kalluri, R. EMT: When Epithelial Cells Decide to Become Mesenchymal-like Cells. *J. Clin. Investig.* **2009**, *119*, 1417–1419. [CrossRef]
11. Kalluri, R. The Biology and Function of Fibroblasts in Cancer. *Nat. Rev. Cancer* **2016**, *16*, 582–598. [CrossRef]
12. Kalluri, R.; Zeisberg, M. Fibroblasts in Cancer. Nature reviews. *Cancer* **2006**, *6*, 392–401. [CrossRef] [PubMed]
13. Lu, N.; Karlsen, T.V.; Reed, R.K.; Kusche-Gullberg, M.; Gullberg, D. Fibroblast α11β1 Integrin Regulates Tensional Homeostasis in Fibroblast/A549 Carcinoma Heterospheroids. *PLoS ONE* **2014**, *9*, e103173. [CrossRef]
14. Sung, K.E.; Su, X.; Berthier, E.; Pehlke, C.; Friedl, A.; Beebe, D.J. Understanding the Impact of 2D and 3D Fibroblast Cultures on In Vitro Breast Cancer Models. *PLoS ONE* **2013**, *8*, e76373. [CrossRef]
15. Veelken, C.; Bakker, G.-J.; Drell, D.; Friedl, P. Single Cell-Based Automated Quantification of Therapy Responses of Invasive Cancer Spheroids in Organotypic 3D Culture. *Methods* **2017**, *128*, 139–149. [CrossRef] [PubMed]
16. Forte, E.; Chimenti, I.; Rosa, P.; Angelini, F.; Pagano, F.; Calogero, A.; Giacomello, A.; Messina, E. Cancers EMT/MET at the Crossroad of Stemness, Regeneration and Oncogenesis: The Ying-Yang Equilibrium Recapitulated in Cell Spheroids. *Cancers* **2017**, *9*, 98. [CrossRef] [PubMed]
17. Calderon, L.A.; Sobrinho, J.C.; Zaqueo, K.D.; de Moura, A.A.; Grabner, A.N.; Mazzi, M.V.; Marcussi, S.; Nomizo, A.; Fernandes, C.F.C.; Zuliani, J.P.; et al. Antitumoral Activity of Snake Venom Proteins: New Trends in Cancer Therapy. *BioMed Res. Int.* **2014**, *2014*, 203639. [CrossRef]
18. Koh, D.C.I.; Armugam, A.; Jeyaseelan, K. Snake Venom Components and Their Applications in Biomedicine. *Cell. Mol. Life Sci.* **2006**, *63*, 3030–3041. [CrossRef]
19. Lewis, R.J.; Garcia, M.L. Therapeutic Potential of Venom Peptides. *Nat. Rev. Drug Discov.* **2003**, *2*, 790–802. [CrossRef]
20. Aird, S.D.; Yates, J.R.; Martino, P.A.; Shabanowitz, J.; Hunt, D.F.; Kaiser, I.I. The Amino Acid Sequence of the Acidic Subunit B-Chain of Crotoxin. *Biochim. Biophys. Acta* **1990**, *1040*, 217–224. [CrossRef]
21. Faure, G.; Xu, H.; Saul, F.A. Crystal Structure of Crotoxin Reveals Key Residues Involved in the Stability and Toxicity of This Potent Heterodimeric β-Neurotoxin. *J. Mol. Biol.* **2011**, *412*, 176–191. [CrossRef] [PubMed]
22. Faure, G.; Bon, C. Crotoxin, a Phospholipase A2 Neurotoxin from the South American Rattlesnake *Crotalus Durissus Terrificus*: Purification of Several Isoforms and Comparison of Their Molecular Structure and of Their Biological Activities. *Biochemistry* **1988**, *27*, 730–738. [CrossRef]
23. Fernandes, C.A.H.; Pazin, W.M.; Dreyer, T.R.; Bicev, R.N.; Cavalcante, W.L.G.; Fortes-Dias, C.L.; Ito, A.S.; Oliveira, C.L.P.; Fernandez, R.M.; Fontes, M.R.M. Biophysical Studies Suggest a New Structural Arrangement of Crotoxin and Provide Insights into Its Toxic Mechanism. *Sci. Rep.* **2017**, *7*, 43885. [CrossRef] [PubMed]
24. Fraenkel-Conrat, H.; Singer, B. Fractionation and Composition of Crotoxin. *Arch. Biochem. Biophys.* **1956**, *60*, 64–73. [CrossRef]
25. Sampaio, S.C.; Hyslop, S.; Fontes, M.R.; Prado-Franceschi, J.; Zambelli, V.O.; Magro, A.J.; Brigatte, P.; Gutierrez, V.P.; Cury, Y. Crotoxin: Novel Activities for a Classic Beta-Neurotoxin. *Toxicon* **2010**, *55*, 1045–1060. [CrossRef] [PubMed]
26. de Araújo Pimenta, L.; de Almeida, M.E.S.; Bretones, M.L.; Cirillo, M.C.; Curi, R.; Sampaio, S.C. Crotoxin Promotes Macrophage Reprogramming towards an Antiangiogenic Phenotype. *Sci. Rep.* **2019**, *9*, 4281. [CrossRef] [PubMed]
27. De-Oliveira, R.B.B.; Kato, E.E.; de Lima, T.S.; Alba-Loureiro, T.C.; Curi, R.; Cirillo, M.C.; Sampaio, S.C. Crotoxin-Treated Macrophages Stimulate ROS Production and Killing Activity in Co-Cultured Neutrophils Article Information. ROS Production and Killing Activity in co-cultured Neutrophils. *J. Appl. Microb. Res.* **2018**, *1*, 1–55.

28. Brigatte, P.; Faiad, O.J.; Ferreira Nocelli, R.C.; Landgraf, R.G.; Palma, M.S.; Cury, Y.; Curi, R.; Sampaio, S.C. Walker 256 Tumor Growth Suppression by Crotoxin Involves Formyl Peptide Receptors and Lipoxin A4. *Mediat. Inflamm.* **2016**, *2016*, 2457532. [CrossRef]
29. Costa, E.S.; Faiad, O.J.; Landgraf, R.G.; Ferreira, A.K.; Brigatte, P.; Curi, R.; Cury, Y.; Sampaio, S.C. Involvement of Formyl Peptide Receptors in the Stimulatory Effect of Crotoxin on Macrophages Co-Cultivated with Tumour Cells. *Toxicon* **2013**, *74*, 167–178. [CrossRef]
30. Lima, T.S.; Cataneo, S.C.; Iritus, A.C.C.; Sampaio, S.C.; Della-Casa, M.S.; Cirillo, M.C. Crotoxin, a Rattlesnake Toxin, Induces a Long-Lasting Inhibitory Effect on Phagocytosis by Neutrophils. *Exp. Biol. Med.* **2012**, *237*, 1219–1230. [CrossRef] [PubMed]
31. Sampaio, S.C.; Brigatte, P.; Sousa-e-Silva, M.C.; dos-Santos, E.C.; Rangel-Santos, A.C.; Curi, R.; Cury, Y. Contribution of Crotoxin for the Inhibitory Effect of *Crotalus Durissus Terrificus* Snake Venom on Macrophage Function. *Toxicon* **2003**, *41*, 899–907. [CrossRef]
32. Cura, J.E.; Blanzaco, D.P.; Brisson, C.; Cura, M.A.; Cabrol, R.; Larrateguy, L.; Mendoza, C.; Sechi, J.C.; Silveira, J.S.; Theiller, E.; et al. Phase I and Pharmacokinetics Study of Crotoxin (Cytotoxic PLA(2), NSC-624244) in Patients with Advanced Cancer. *Clin. Cancer Res. Off. J. Am. Assoc. Cancer Res.* **2002**, *8*, 1033–1041.
33. Kato, E.E.; Pimenta, L.A.; Almeida, M.E.S.D.; Zambelli, V.O.; Santos, M.F.D.; Sampaio, S.C. Crotoxin Inhibits Endothelial Cell Functions in Two- and Three-Dimensional Tumor Microenvironment. *Front. Pharmacol.* **2021**, *12*, 713332. [CrossRef] [PubMed]
34. Ivascu, A.; Kubbies, M. Diversity of Cell-Mediated Adhesions in Breast Cancer Spheroids. *Int. J. Oncol.* **2007**, *31*, 1403–1413. [CrossRef] [PubMed]
35. Österholm, C.; Lu, N.; Lidén, Å.; Karlsen, T.V.; Gullberg, D.; Reed, R.K.; Kusche-Gullberg, M. Fibroblast EXT1-Levels Influence Tumor Cell Proliferation and Migration in Composite Spheroids. *PLoS ONE* **2012**, *7*, e41334. [CrossRef]
36. Orimo, A.; Weinberg, R.A. Stromal Fibroblasts in Cancer: A Novel Tumor-Promoting Cell Type. *Cell Cycle* **2006**, *5*, 1597–1601. [CrossRef]
37. Eberlein, C.; Rooney, C.; Ross, S.J.; Farren, M.; Weir, H.M.; Barry, S.T. E-Cadherin and EpCAM Expression by NSCLC Tumour Cells Associate with Normal Fibroblast Activation through a Pathway Initiated by Integrin Avβ6 and Maintained through TGFβ Signalling. *Oncogene* **2015**, *34*, 704–716. [CrossRef]
38. Crispin, L.A.D.C. Caracterização Morfológica de Fibroblastos Cultivados Na Presença de Venenos de Serpentes Com Propriedades Pró Ou Anti-Inflamatória. Bachelor's Thesis, University of Santo Amaro, São Paulo, Brazil, 2017.
39. Stadler, M.; Scherzer, M.; Walter, S.; Holzner, S.; Pudelko, K.; Riedl, A.; Unger, C.; Kramer, N.; Weil, B.; Neesen, J.; et al. Exclusion from Spheroid Formation Identifies Loss of Essential Cell-Cell Adhesion Molecules in Colon Cancer Cells OPEN. *Sci. REPoRtS* **2018**, *8*, 1151. [CrossRef]
40. Sodek, K.L.; Ringuette, M.J.; Brown, T.J. Compact Spheroid Formation by Ovarian Cancer Cells Is Associated with Contractile Behavior and an Invasive Phenotype. *Int. J. Cancer* **2009**, *124*, 2060–2070. [CrossRef]
41. Buckley, S.T.; Medina, C.; Ehrhardt, C. Differential Susceptibility to Epithelial-Mesenchymal Transition (EMT) of Alveolar, Bronchial and Intestinal Epithelial Cells in Vitro and the Effect of Angiotensin II Receptor Inhibition. *Cell Tissue Res.* **2010**, *342*, 39–51. [CrossRef]
42. Labernadie, A.; Kato, T.; Brugués, A.; Serra-Picamal, X.; Derzsi, S.; Arwert, E.; Weston, A.; González-Tarragó, V.; Elosegui-Artola, A.; Albertazzi, L.; et al. A Mechanically Active Heterotypic E-Cadherin/N-Cadherin Adhesion Enables Fibroblasts to Drive Cancer Cell Invasion. *Nat. Cell Biol.* **2017**, *19*, 224–237. [CrossRef] [PubMed]
43. de Vasconcelos Azevedo, F.V.P.; Zóia, M.A.P.; Lopes, D.S.; Gimenes, S.N.; Vecchi, L.; Alves, P.T.; Rodrigues, R.S.; Silva, A.C.A.; Yoneyama, K.A.G.; Goulart, L.R.; et al. Antitumor and Antimetastatic Effects of PLA2-BthTX-II from Bothrops Jararacussu Venom on Human Breast Cancer Cells. *Int. J. Biol. Macromol.* **2019**, *135*, 261–273. [CrossRef]
44. Sampaio, S.C.; Santos, M.F.; Costa, E.P.; Rangel-Santos, A.C.; Carneiro, S.M.; Curi, R.; Cury, Y. Crotoxin induces actin reorganization and inhibits tyrosine phosphorylation and activity of small GTPases in rat macrophages. *Toxicon* **2006**, *47*, 909–919. [CrossRef] [PubMed]
45. Gong, L.; Wu, D.; Zou, J.; Chen, J.; Chen, L.; Chen, Y.; Ni, C.; Yuan, H.; Gong, L.; Wu, D.; et al. Prognostic Impact of Serum and Tissue MMP-9 in Non-Small Cell Lung Cancer: A Systematic Review and Meta-Analysis. *Oncotarget* **2016**, *7*, 18458–18468. [CrossRef] [PubMed]
46. Kobayashi, T.; Kim, H.; Liu, X.; Sugiura, H.; Kohyama, T.; Fang, Q.; Wen, F.-Q.; Abe, S.; Wang, X.; Atkinson, J.J.; et al. Matrix Metalloproteinase-9 Activates TGF-and Stimulates Fibroblast Contraction of Collagen Gels. *Am. J. Physiol. Lung Cell. Mol. Physiol.* **2014**, *306*, 1006–1015. [CrossRef] [PubMed]
47. Nawrocki-Raby, B.; Gilles, C.; Polette, M.; Martinella-Catusse, C.; Bonnet, N.; Puchelle, E.; Foidart, J.-M.; van Roy, F.; Birembaut, P. E-Cadherin Mediates MMP Down-Regulation in Highly Invasive Bronchial Tumor Cells. *Am. J. Pathol.* **2003**, *163*, 653. [CrossRef]
48. Toriseva, M.J.; Ala-Aho, R.; Karvinen, J.; Baker, A.H.; Marjomäki, V.S.; Heino, J.; Kähäri, V.-M. Collagenase-3 (MMP-13) Enhances Remodeling of Three-Dimensional Collagen and Promotes Survival of Human Skin Fibroblasts. *J. Investig. Dermatol.* **2007**, *127*, 49–59. [CrossRef]
49. Crispin, L.A.D.C. Importância Da Atividade Secretora de Macrófagos Induzida Pela Crotoxina Sobre Funções de Fibroblastos Envolvidos Com o Processo de Cicatrização. Estudos in Vitro. Master's Thesis, University of São Paulo, São Paulo, Brazil, 2021.
50. Xiao-yan, X.; Lu-tian, P.; Xiu-xia, L. Expression of MMP-9 and MMP-13 in Non-Small Cell Lung Cancer and Their Significance. *Chin. J. Clin. Exp. Pathol.* **2014**, *12*, 1358–1364.

51. Zhang, W.; Wang, H.; Sun, M.; Deng, X.; Wu, X.; Ma, Y.; Li, M.; Shuoa, S.M.; You, Q.; Miao, L. CXCL5/CXCR2 Axis in Tumor Microenvironment as Potential Diagnostic Biomarker and Therapeutic Target. *Cancer Commun.* **2020**, *40*, 69–80. [CrossRef]
52. Cheng, Y.; Mo, F.; Li, Q.; Han, X.; Shi, H.; Chen, S.; Wei, Y.; Wei, X. Targeting CXCR2 Inhibits the Progression of Lung Cancer and Promotes Therapeutic Effect of Cisplatin. *Mol. Cancer* **2021**, *20*, 62. [CrossRef]
53. Desai, S.; Laskar, S.; Pandey, B.N. Autocrine IL-8 and VEGF Mediate Epithelial–Mesenchymal Transition and Invasiveness via P38/JNK-ATF-2 Signalling in A549 Lung Cancer Cells. *Cell. Signal.* **2013**, *25*, 1780–1791. [CrossRef] [PubMed]
54. Zhou, S.L.; Zhou, Z.J.; Hu, Z.Q.; Li, X.; Huang, X.W.; Wang, Z.; Fan, J.; Dai, Z.; Zhou, J. CXCR2/CXCL5 Axis Contributes to Epithelial–Mesenchymal Transition of HCC Cells through Activating PI3K/Akt/GSK-3β/Snail Signaling. *Cancer Lett.* **2015**, *358*, 124–135. [CrossRef]
55. Cui, D.; Zhao, Y.; Xu, J. Activated CXCL5-CXCR2 Axis Promotes the Migration, Invasion and EMT of Papillary Thyroid Carcinoma Cells via Modulation of β-Catenin Pathway. *Biochimie* **2018**, *148*, 1–11. [CrossRef] [PubMed]
56. Liu, F.; Song, S.; Yi, Z.; Zhang, M.; Li, J.; Yang, F.; Yin, H.; Yu, X.; Guan, C.; Liu, Y.; et al. HGF Induces EMT in Non-Small-Cell Lung Cancer through the HBVR Pathway. *Eur. J. Pharmacol.* **2017**, *811*, 180–190. [CrossRef]
57. Hwang, H.J.; Oh, M.S.; Lee, D.W.; Kuh, H.J. Multiplex Quantitative Analysis of Stroma-Mediated Cancer Cell Invasion, Matrix Remodeling, and Drug Response in a 3D Co-Culture Model of Pancreatic Tumor Spheroids and Stellate Cells. *J. Exp. Clin. Cancer Res.* **2019**, *38*, 258. [CrossRef] [PubMed]
58. Sampaio, S.C.; Rangel-Santos, A.C.; Peres, C.M.; Curi, R.; Cury, Y. Inhibitory Effect of Phospholipase A(2) Isolated from *Crotalus Durissus Terrificus* Venom on Macrophage Function. *Toxicon* **2005**, *45*, 671–676. [CrossRef]
59. Laemmli, U.K. Cleavage of Structural Proteins during the Assembly of the Head of Bacteriophage T4. *Nature* **1970**, *227*, 680–685. [CrossRef]
60. de Araújo, A.L.; Radvanyi, F. Determination of Phospholipase A2 Activity by a Colorimetric Assay Using a PH Indicator. *Toxicon* **1987**, *25*, 1181–1188. [CrossRef]

Article

Crotalphine Attenuates Pain and Neuroinflammation Induced by Experimental Autoimmune Encephalomyelitis in Mice

Aline C. Giardini [1], Bianca G. Evangelista [1], Morena B. Sant'Anna [1], Barbara B. Martins [1], Carmen L. P. Lancellotti [2], Adriano P. Ciena [3], Marucia Chacur [4], Rosana L. Pagano [5], Orlando G. Ribeiro [6], Vanessa O. Zambelli [1] and Gisele Picolo [1,*]

1 Laboratory of Pain and Signaling, Butantan Institute, Sao Paulo 05503-900, SP, Brazil; aline.giardini@butantan.gov.br (A.C.G.); biah.evangelista31@gmail.com (B.G.E.); morena.santanna@butantan.gov.br (M.B.S.); barbara.martins@esib.butantan.gov.br (B.B.M.); vanessa.zambelli@butantan.gov.br (V.O.Z.)
2 Department of Pathological Sciences, Medical Science School Santa Casa of Sao Paulo, Sao Paulo 01221-020, SP, Brazil; luciapl@uol.com.br
3 Laboratory of Morphology, Institute of Biosciences, São Paulo State University, Rio Claro 13506-52, SP, Brazil; adriano.ciena@unesp.br
4 Laboratory of Functional Neuroanatomy of Pain, Instituto de Ciências Biomédicas, Universidade de Sao Paulo, Sao Paulo 05508-900, SP, Brazil; chacurm@icb.usp.br
5 Laboratory of Neuroscience, Hospital Sírio-Libanês, Sao Paulo 01308-060, SP, Brazil; rosana.lpagano@hsl.org.br
6 Laboratory of Immunogenetics, Butantan Institute, Sao Paulo 05503-900, SP, Brazil; orlando.ribeiro@butantan.gov.br
* Correspondence: gisele.picolo@butantan.gov.br; Tel.: +55-11-2627-9897

Citation: Giardini, A.C.; Evangelista, B.G.; Sant'Anna, M.B.; Martins, B.B.; Lancellotti, C.L.P.; Ciena, A.P.; Chacur, M.; Pagano, R.L.; Ribeiro, O.G.; Zambelli, V.O.; et al. Crotalphine Attenuates Pain and Neuroinflammation Induced by Experimental Autoimmune Encephalomyelitis in Mice. *Toxins* **2021**, *13*, 827. https://doi.org/10.3390/toxins13110827

Received: 28 October 2021
Accepted: 17 November 2021
Published: 22 November 2021

Publisher's Note: MDPI stays neutral with regard to jurisdictional claims in published maps and institutional affiliations.

Abstract: Multiple sclerosis (MS) is a demyelinating disease of inflammatory and autoimmune origin, which induces sensory and progressive motor impairments, including pain. Cells of the immune system actively participate in the pathogenesis and progression of MS by inducing neuroinflammation, tissue damage, and demyelination. Crotalphine (CRO), a structural analogue to a peptide firstly identified in *Crotalus durissus terrificus* snake venom, induces analgesia by endogenous opioid release and type 2 cannabinoid receptor (CB2) activation. Since CB2 activation downregulates neuroinflammation and ameliorates symptoms in mice models of MS, it was presently investigated whether CRO has a beneficial effect in the experimental autoimmune encephalomyelitis (EAE). CRO was administered on the 5th day after immunization, in a single dose, or five doses starting at the peak of disease. CRO partially reverted EAE-induced mechanical hyperalgesia and decreased the severity of the clinical signs. In addition, CRO decreases the inflammatory infiltrate and glial cells activation followed by TNF-α and IL-17 downregulation in the spinal cord. Peripherally, CRO recovers the EAE-induced impairment in myelin thickness in the sciatic nerve. Therefore, CRO interferes with central and peripheral neuroinflammation, opening perspectives to MS control.

Keywords: neurodegenerative disease; neurodegeneration; inflammation; IL-17; glial cells

Key Contribution: CRO ameliorates multiple sclerosis symptoms in mice. CRO controls clinical signs in an animal model of multiple sclerosis.

1. Introduction

Multiple sclerosis (MS) is a chronic inflammatory and demyelinating disease which affects more than 2.8 million people worldwide [1]. The autoimmune inflammation that affects the central nervous system (CNS) progressively results in oligodendrocyte injury and demyelination. In the early stages, the axons are preserved, however, with the advance of the disease, the damages are irreversible [2]. The origin of MS is still unclear; however, the combination of environmental, such as lifestyle and viral exposure, and genetic factors may contribute to the development of the disease [3–6].

The literature shows multifocal inflamed regions in gray and white matter, with oligodendrocyte death and myelin sheath disruption [7,8]. The clinical presentation of MS is highly variable, making the diagnosis of the disease difficult [9]. The inflammatory response is mainly mediated by T helper (Th) cells, which play an essential role in the course of MS [10]. Pro-inflammatory cytokines such as interleukin (IL)-6, IL-17, IL-21, IL-22, IL-23, and tumor necrosis factor (TNF)-α are produced by Th17 cells [10–13]. IL-17, which is one of the main cytokines produced by these cells, acts as an inducer of neutrophil infiltration [13–15]. Th1 cells mediate CNS inflammation by stimulating macrophage infiltration and producing important cytokines including interferon (IFN)-γ, which plays a relevant role in spinal neuroinflammation [10]. In addition, B-lymphocytes play a critical role in tissue damage, propagating inflammation, and spinal demyelination in MS [16,17].

In addition to peripheral immune cells, activation of CNS resident cells also adds to the pathogenesis of MS, particularly microglia and astrocytes. Microglia cells are activated when tissue integrity is disturbed, secreting pro-inflammatory or anti-inflammatory cytokines, controlling the expression of anti-inflammatory molecules, regulating phagocytosis of debris and tissue repair, responses that depend on the changes that occurred in their specific microenvironment. Astrocytes strengthen and contribute to the maintenance of the integrity of the blood-brain barrier, restricting the entry of peripheral immune cells into the CNS [18] and modulating synaptic activity and plasticity [19,20]. These activated glial cells produce chemokines and cytokines, leading to the recruitment of additional immune cells to the CNS parenchyma besides producing exacerbated sensitization stages, including neural hypersensitivity [21]. Pain is one of the sensory alterations presented by MS patients, and can be of inflammatory, neuropathic or skeletal muscle origin [22]. It also affects the quality of life, aggravating symptoms such as depression, sleep disturbance, and mood [23–27]. Importantly, like many other progressive diseases, there is no specific treatment for MS, and therapies focus on delaying disease progression and promoting symptom relief, thus improving the patient's quality of life.

Animal models have provided valuable answers regarding the mechanisms involved in the development and progression of MS and the efficacy of new drugs with therapeutic potential [11,28]. These experimental models reproduce CNS inflammation, demyelination of neurons, and some motor changes observed in MS [28–31], in addition to sensory alterations. Among the sensory alterations, the literature has demonstrated alterations in the pain threshold in both humans and animals that, in the MOG_{35-55}-induced experimental autoimmune encephalomyelitis (EAE) model, arise before the onset of motor alterations [32–34], allowing for the study of pain previous to the first evidence of motor alterations, which would compromise pain sensitivity assessment.

Considering that MS is a chronic neurodegenerative disease whose treatment only delays progression and alleviates symptoms without interfering with the pathologic process, new therapeutic strategies are needed. Crotalphine (CRO) is a 14-amino acid peptide (EFSPENCQGESQPC) containing a disulfide bridge and a pyroglutamic acid [35], capable of inducing a potent and durable antinociceptive effect in acute and chronic pain models (including cancer pain and neuropathic pain) [35–37]. CRO induces antinociception by activating CB2 receptors, which, in turn, induces the release of endogenous peptides, particularly dynorphin A, the endogenous agonist of kappa opioid receptors [38]. It was also observed that CRO desensitizes transient receptor potential ankyrin ion channels (TRPA1) [39], a receptor that has a relevant role in the maintenance of inflammatory hyperalgesia [40]. Based on clinical studies showing the effectiveness of cannabinoids on clinical symptoms in patients with MS, particularly muscle stiffness and spasms, sleep disorders, and neuropathic pain [41,42], the CRO effect on clinical symptoms, neuroinflammation, and axonal demyelination of mice submitted to the MOG_{35-55}-induced experimental autoimmune encephalomyelitis, an animal model of MS, is investigated here.

2. Results

2.1. CRO Caused Analgesia in EAE-Induced Pain and Attenuated Clinical Signs

CRO (10, 50, or 100 µg/Kg) or saline was first administered on the fifth day after immunization, in a single dose. The immunized animals (EAE) showed decreased nociceptive threshold (termed hypernociception) on the fourth day after immunization when compared to the complete Freund's adjuvant (CFA) group (control). The hypernociception remained until the last day of evaluation. i.e., 10th day, when the assessment of the pain threshold was stopped due to the onset of motor dysfunctions. The treatment with CRO with doses of 10, 50, and 100 µg/kg, resulted in a partial reversal of the EAE-induced hyperalgesia (Figure 1A). The experiments were performed in the morning and the pain threshold was assessed 1 h after the treatment with CRO.

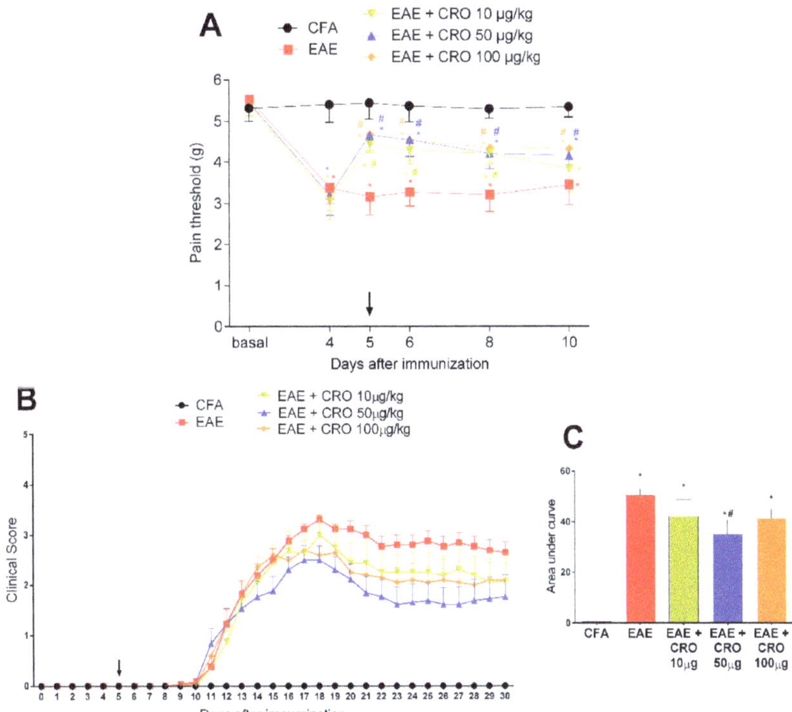

Figure 1. Evaluation of CRO effect in pain sensitivity and clinical signs of animals immunized with MOG_{35-55}. Animals with EAE were treated with a single dose of CRO p.o. on the fifth day, in different doses. Pain sensitivity was measured by the electronic von Frey model (**A**) and motor impairment was evaluated according to a visual scale of clinical signs from 0 to 5 (**B**). For statistical analysis of clinical signs, the area under the curve was evaluated (**C**). Data represent mean ± SEM 8–10 animals per group. # $p < 0.05$ compared to untreated group (EAE). * $p < 0.05$ compared to the control group (CFA). The two-way ANOVA test, followed by Bonferroni post-hoc test was used in the electronic von Frey test, and the One-way ANOVA test was used, followed by Tukey's post-hoc test, was used in the area under the curve.

The first motor impairment symptom appeared on the 11th day after immunization and increased progressively, peaking around the 17–18th day. CRO, at the dose of 50 µg/kg, reduced the disease severity, represented by lower clinical scores, when compared to the untreated group (Figure 1B,C). This dose was chosen for subsequent experiments.

2.2. CRO Reverts EAE-Induced Motor Impairment

To evaluate the efficacy of CRO after the onset of motor impairment, CRO was administered for five consecutive days (one dose by day), starting from the 12th day of immunization. The effectiveness of this protocol was compared with the single-dose protocol. The EAE groups showed a progressive increment in clinical score from the 11th day after immunization, peaking at the 16th day. Both groups treated with CRO showed lower clinical signs scores when compared to the untreated group, showing the potential of CRO to interfere with motor impairment even after its onset as well as to prevent it (Figure 2A,B). At this point, the analgesic effect of CRO was not evaluated, since it is not possible to apply the nociceptive test due to the motor impairment. The body weight of the animals was measured during the course of the disease and, as expected, EAE interfered with the weight gain of the animals (Figure 2C). Crotalphine did not alter the EAE-induced weight loss (Figure 2C).

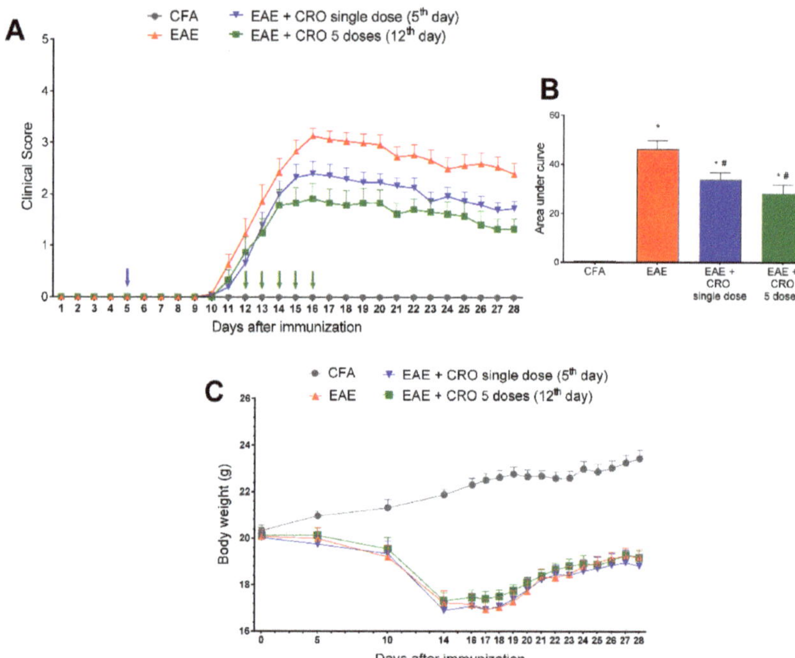

Figure 2. Evaluation of CRO effect on clinical signs and body weight of animals immunized with $MOG_{35–55}$ and treated before or after motor impairment. Animals were immunized with $MOG_{35–55}$ and treated with CRO in a single dose on the fifth day after immunization (blue arrow), or in five doses (green arrows, one daily dose for five consecutive days starting on the 12th day after immunization). Clinical signs were evaluated daily (**A**), and, for statistical analysis, the area under the curve was evaluated (**B**). Body weight (g) was measured during the course of the disease (**C**). Data represent mean ± SEM 8–10 animals per group. * $p < 0.05$ compared to the CFA group. # $p < 0.05$ compared to the untreated group (EAE). A one-way ANOVA test was used, followed by Tukey's post-hoc test in the area under the curve.

2.3. CRO Reduced EAE-Induced Inflammatory Infiltrates in the Spinal Cord 28 Days after Immunization

Central nervous system inflammation is a hallmark of MS pathogenesis [43]. Therefore, the effect of CRO in EAE-induced leukocyte migration to the spinal cord was investigated. The level of cellular infiltrate was scored from 0 (no cellular infiltrate) to 4 (high cellular infiltrate) As shown in panel (Figure 3A), HE staining revealed that there was no cell infiltration in the control group treated with CFA, and the spinal cord of EAE mice displayed mixed cellular infiltrate with numerous perivascular clusters. Of interest, treatment with CRO significantly reduces the EAE-induced cell infiltration at the peak of the disease (Figure 3B), with no difference observed on the 28th day after immunization (Figure 3C).

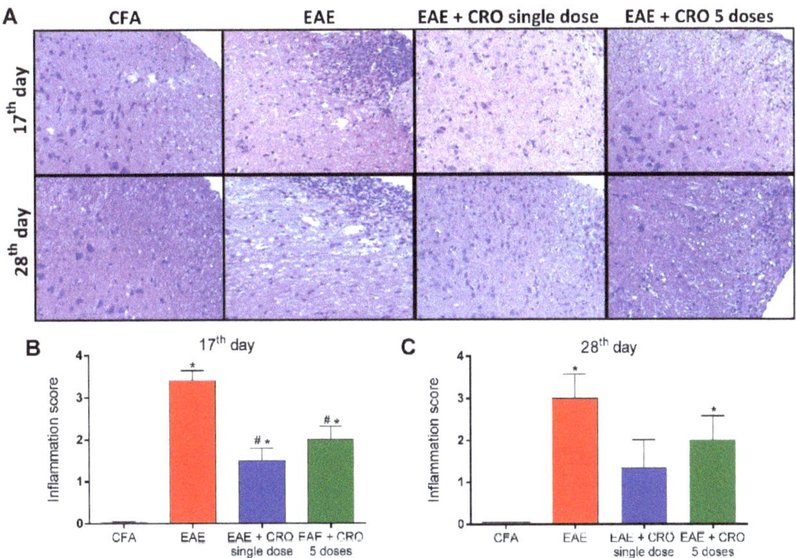

Figure 3. A single dose of CRO reduces cellular infiltration in the spinal cord. The animals were sedated and euthanized for spinal cord collection. Representative 20× microscopic fields of transverse sections of the spinal cord stained with hematoxylin and eosin ((**A**), panel). Quantification of the inflammatory score (0–4) in 17th (**B**) and 28th (**C**) days. Data represent the mean ± SEM of 3 animals per group. * $p < 0.05$ compared to the control group (CFA). # $p < 0.05$ compared to the untreated group (EAE). One-way ANOVA was used, followed by Tukey's post-hoc test.

2.4. CRO Decreased EAE-Induced Glial Cells Immunoreactivity in the Spinal Cord

Besides leucocyte infiltration, activation of resident CNS cells likewise contributes to the pathogenesis of EAE [44,45]. Next, the effect of CRO on EAE-induced astrocytes and microglia activation in the dorsal horn of the spinal cord was evaluated. Immunoreactivity for GFAP (glial fibrillary acidic protein), an astrocyte marker, and for Iba-1, a microglia marker was evaluated on the17th (clinical signs peak) and 28th (period of stabilization of clinical signs) days. The results show that EAE increases the reactivity for GFAP (Figure 4A,B) and Iba-1 (ionized calcium-binding adaptor-1) (Figure 5A,B) when compared with the control group (CFA), in both periods. On the other hand, the single-dose and five doses CRO protocol revert astrocyte and microglia activation induced by EAE (Figures 4 and 5).

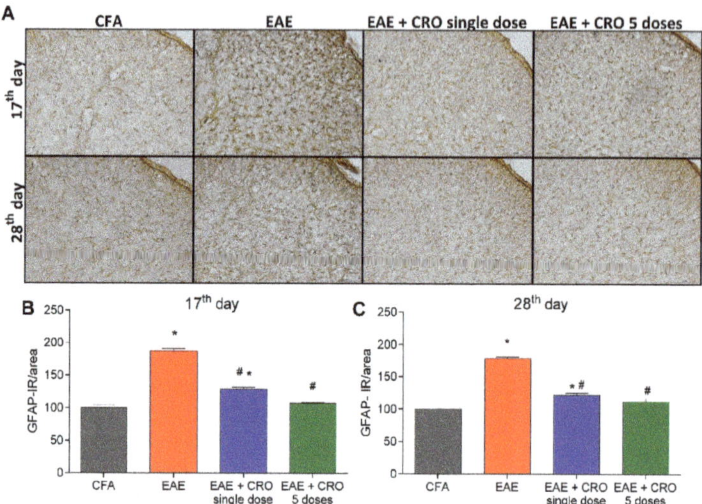

Figure 4. CRO effect on immunoreactivity for astrocytes in the spinal cord of EAE animals. Animals immunized with $MOG_{35–55}$ and treated or untreated with CRO were perfused for the spinal cord collection. Representative 20× microscopic field of transverse sections of the spinal cord (L3-L6) stained with GFAP ((**A**), panel). Immunohistochemistry analysis on 17th (**B**) and 28th days (**C**) after immunization. Immunoreactivity (IR) was quantified using Image J software. Data represent mean ± SEM of 3 animals per group. # $p < 0.05$ compared to EAE group. * $p < 0.05$ compared to control group (CFA). A one-way ANOVA test was used, followed by Tukey's post-hoc test.

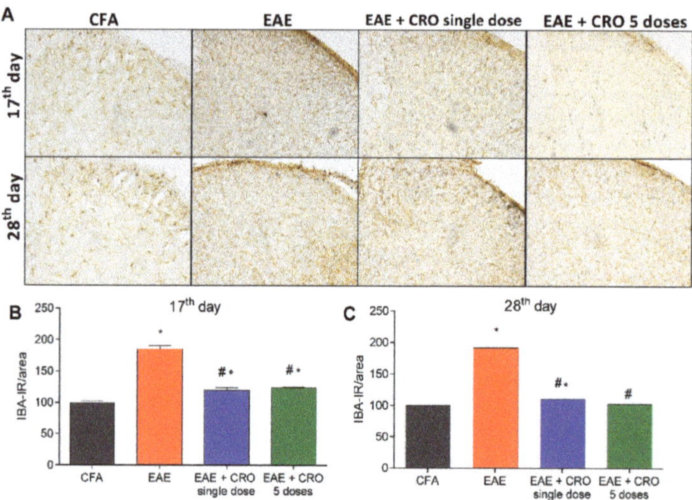

Figure 5. Evaluation of immunoreactivity for microglia in the spinal cord of animals with EAE and treated with CRO. Animals immunized with $MOG_{35–55}$ and treated or untreated with CRO were perfused and had the spinal cords (L3-L6) collected. Representative 20× microscopic field of transverse sections of the spinal cord (L3-L6) stained with Iba-1 ((**A**), panel). Immunohistochemistry analysis on the 17th (**B**) and 28th (**C**) days after immunization. Immunoreactivity was quantified using Image J software. Data represent mean ± SEM of 3 animals per group. # $p < 0.05$ compared to EAE + sal group. * $p < 0.05$ compared to control group (CFA). A one-way ANOVA test was used, followed by Tukey's post-hoc test.

2.5. CRO Mitigate EAE-Induced IL-17 and TNF-α Release in the Spinal Cord at the Peak of the Disease

IL-17 and TNF-α are proinflammatory cytokines that activate T cells and other immune cells to produce and release a variety of cytokines and chemokines, and to express cell adhesion molecules [46]. Figure 6 shows that the levels of IL-17 and TNF-α are increased in the spinal cord of EAE animals at the peak of disease. Of interest, the treatment with CRO on the 5th day prevents EAE-induced IL-17 and TNF-α release. Moreover, the protocol of five doses of CRO does not prevent IL-17 and TNF-α release (Figure 6).

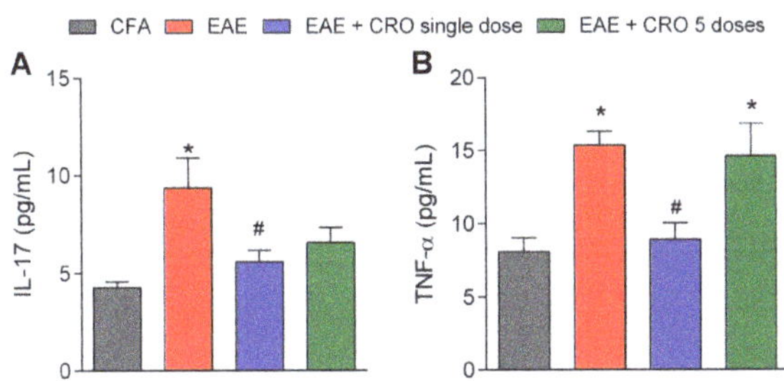

Figure 6. CRO effect on cytokine release in the spinal cord of EAE animals. Evaluation of the expression of IL-17 (**A**) and TNF-α (**B**) cytokines was analyzed by multiplex assay. The animals were immunized, euthanized, and the lumbar portion of the spinal cord was collected on the 17th day after immunization. Data represent the mean ± SEM of 5–7 animals per group. * $p < 0.05$ compared to the CFA group. # $p < 0.05$ compared to the untreated group (EAE). One-way ANOVA was used, followed by Tukey's post-hoc test.

2.6. CRO Not Interfered with the Nerve Growth Factor Expression in the Spinal Cord of Animals with EAE

Nerve growth factor (NGF) is a neurotrophin associated with the differentiation and survival of numerous neurons localized in the peripheral and CNS [47]. It was next investigated if CRO could interfere with the expression of NGF in the lumbar section of the spinal cord (L3–L6) on the 28th day after immunization. Our data confirm the results of the literature showing that NGF expression is reduced in the spinal cord of EAE mice; however, CRO does not change those levels (Figure 7).

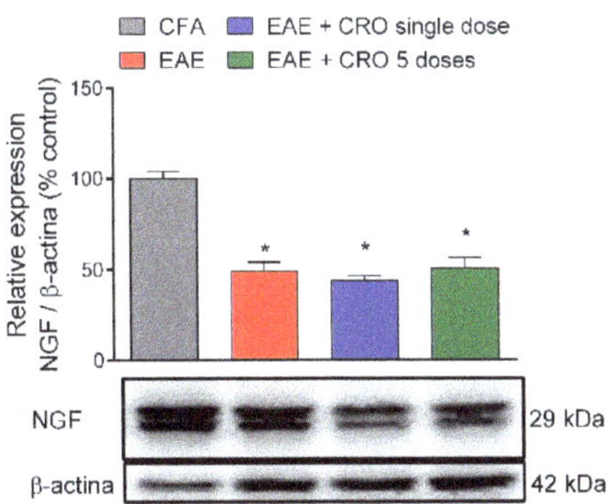

Figure 7. Effect of CRO in neurotrophic factor in the spinal cord of animals immunized with $MOG_{35–55}$. Animals, on the 28th day after immunization, were sedated and euthanized for the spinal cord collection. The expression of neural growth factor (NGF) was evaluated by western blotting. Beta-actin was used as the loading control. Results were normalized by CFA group expression. Results are expressed as the mean ± SEM of 5 animals per gropu. * $p < 0.05$ in relation to the CFA group. One-way ANOVA followed by Tukey's post-hoc test.

2.7. CRO Prevented the Peripheral Demyelination of the Sciatic Nerve Induced by EAE

We have previously demonstrated that EAE decreases sciatic nerve myelin thickness when quantified through the g-ratio estimative (i.e., dividing axon diameter by the fiber diameter) [48]. Here, we checked whether CRO would have a beneficial effect on EAE-induced demyelination.

In order to address this question, the distal portion of the sciatic nerve was collected on the 17th and the 28th days after immunization and its morphology was analyzed (Figure 8A) by transmission electron microscopy. The results demonstrated the presence of intact fibers, with a similar distribution of small and large diameter myelinic fibers, non-myelinic fibers, and Schwann cell nuclei in the control group (CFA). At the peak of the disease (17th day after immunization), EAE animals showed no reduction in myelin sheath thickness (Figure 8C). However, on the 28th day after immunization (Figure 8D), a decrease in the sciatic nerve myelin thickness, defined by an increase in the g-ratio when compared with control animals, was detected. Importantly, animals treated with a single dose of CRO on the fifth day after immunization have a thickness of myelin sheaths similar to the control group. This data shows that CRO prevents EAE-induced demyelination, which may preserve the nerve conduction and contribute to the analgesic effect.

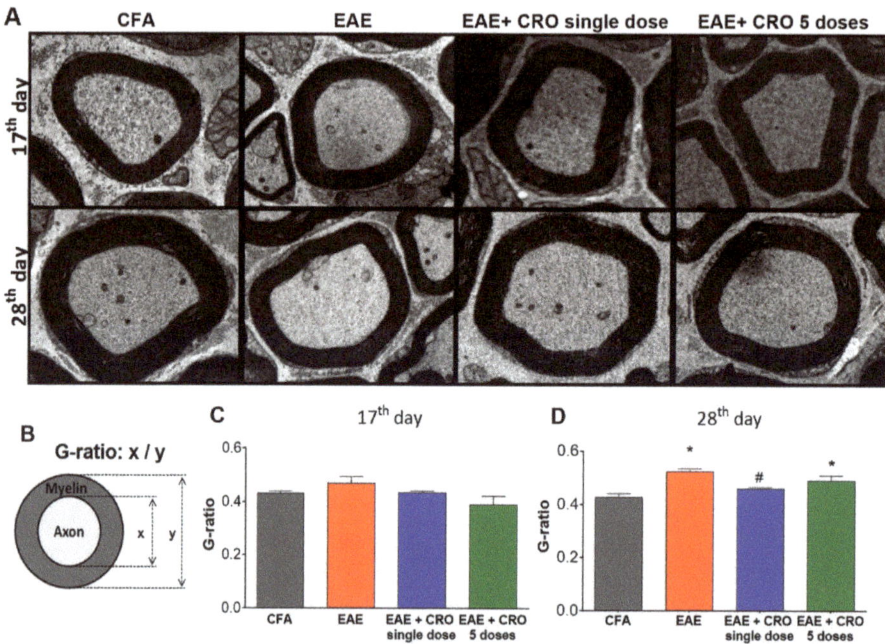

Figure 8. Effect of CRO on the peripheral nerve myelin sheath of animals with EAE. Animals were sedated, euthanized, perfused and the sciatic nerve was collected for evaluation by transmission electron microscopy, 7500× magnification. Representative microscopic fields of transverse sections of the peripheral nerve stained with toluidine blue solution ((**A**), panel). The G-ratio (**B**) was measured using ImageJ software for evaluation of axonal myelin content on the 17th (**C**) and 28th days (**D**) after immunization. Data represent the mean ± SEM of 3 animals per group. * $p < 0.05$ compared to the control group (CFA). # $p < 0.05$ compared to the untreated group (EAE). One-way ANOVA was used, followed by Tukey's post-hoc test.

3. Discussion

MS is a chronic inflammatory disease of autoimmune origin which, demyelinates neurons from the CNS. Components of the CNS are recognized as antigens, more specifically the myelin sheath of the axons. T lymphocyte-mediated CNS aggression is observed in MS [49], as well as in the activation of glial cells (microglia and astrocytes) in both the spinal cord and brain regions [50,51], innate immune cell activation, and cytokines and chemokines release [52]. This inflammatory response greatly alters the properties of neurons, leading to demyelination and axonal loss [53], as well as to several motor, cognitive and sensory alterations [54,55].

Presently, the effect of CRO on disease development was evaluated. CRO is an analgesic peptide, firstly identified in the venom of *Crotalus durissus terrificus* snake [35], which induces long-lasting antinociceptive effect in animals, observed in acute and chronic pain models, which do not induce some of the side effects observed for analgesic drugs, such as alteration in spontaneous motor and general activity or the development of tolerance to its antinociceptive effect after prolonged treatment [35–37]. It was previously demonstrated by our group that CRO induces analgesia mediated by the activation of CB2 cannabinoid receptors [38]. First, we evaluated the disease development regarding motor impairment through the observation of clinical signs. The results showing that CRO ameliorates the EAE-induced clinical signs corroborate data from the literature, where the administration of a phytocannabinoid, an agonist of CB2 receptors, attenuated the clinical severity of the disease in mice, both preventively and after the onset of clinical signs [56].

In addition to the motor impairment, central neuropathic pain is observed in animal models of EAE, occurring due to prolonged inflammation in the spinal cord, resulting in activation of glial cells and aggression to the myelin sheath, causing painful hypersensitivity [31,57]. Thus, activation of spinal astrocytes and microglia seems to be a key element in the generation and maintenance of peripheral neuropathic pain [58,59], reducing the nociceptive threshold of animals before neurological dysfunction occurs [60]. It was presently shown that CRO partially reverses EAE-induced mechanical hypersensitivity. Importantly, the analgesic effect was already observed 1 h after its administration. Considering the rapid analgesic effect observed, as well as the fact that hypernociception is detected before the onset of clinical signs, it is assumed that the observed pain does not depend on the onset of central inflammation and that the observed analgesic effect is due to mechanisms independent of the anti-inflammatory observed effect.

As previously pointed out, central inflammation is a key factor in the progressing of EAE as well as MS [53–55]. Our data clearly demonstrated that crotalphine is capable of preventing the cell influx to the CNS. However, in addition to migrated cells, resident cells also play an important role in the onset of disease development. Microglial reactivity is manifested by morphological changes, modifications in the expression of surface molecules, and secretion of several substances such as cytokines, trophic factors, and chemokines [61–63]. Specifically, in the acute period of EAE, microglia reactivity was detected along with increased clinical symptomatology. Treatment with a microglial inhibitor or microglia depletion has been shown to induce beneficial effects on EAE symptoms, demonstrating that microglial cells play a role in the pathogenesis of EAE [64,65]. In addition, it has been reported that activated microglia may release a large variety of molecules, which may contribute to immune cell recruitment and the spread of inflammatory response [64–67]. Astrocytes also play important roles in the development of chronic pain, producing neurotoxic mediators, cytokines, and chemokines, with proinflammatory activity [21,68,69]. Here, the literature findings were reproduced, showing that EAE increases microglial and astrocytes labeling, both at the peak of the disease and in the late period [31,61]. Importantly, CRO decreased astrocytes and microglia reactivity in both periods, observed in the dorsal horn of the spinal cord, a region related to the nociceptive pathway [70]. In addition to that, histological analysis of the spinal cord showed that CRO decreases cell infiltrates in EAE mice, suggesting a correlation with reduced glial activation, TNF-α cytokine production, and lower clinical signs. These results point to the reduction of the neuroinflammation induced by CRO as a key factor for the antinociceptive effect and the improvement of clinical signs.

The contribution of Th17 cells and its effector cytokine signature, IL-17, is well described as promoters of the induction and progression of MS/EAE [71–73]. The role of

Th1 cells in the process is known, but Th17 cells have a greater proliferative capacity than Th1 cells, in addition to being able to cross the blood-brain barrier (BBB) more easily [71]. Here, it was confirmed that IL-17 levels are up-regulated in the EAE group. Of interest, CRO prevented the IL-17 increase in the spinal cord (17th day). It is evident that IL-17 is a key cytokine in the EAE model [74]. In fact, anti-IL17A treatment showed satisfactory results in attenuating the development of EAE [75], for example, in reducing the clinical signs and improving the histological findings. In addition, IL-17 knockout mice exhibited delayed onset, reduced clinical scores, and early recovery after immunization [76]. Data from our group also demonstrated the importance of the IL-17 to the development of EAE using crotoxin, the main neurotoxin from the venom of the *Crotalus durissus terrificus* snake. In these studies, crotoxin attenuated clinical signs by inhibiting the release of IL-17 and reducing $CD4^{+}IL\text{-}17^{+}$ cell proliferation in lymph nodes [77]. These data point out that the reduction of this pro-inflammatory cytokine is one of the main factors that contribute to CRO attenuation of disease progression. Importantly, although the central release of IL-17 was decreased, the proliferation of Th17 cells in the lymph nodes, at the onset of the immune process, was not altered by CRO (data not shown). In addition to IL-17, the profile of TNF-α was also investigated, since IL-17 and TNF-α are the two major cytokines involved in MS [78–80]. Our results demonstrated that a single dose of CRO reversed EAE-induced TNF-α release in the spinal cord at the peak of the disease. The contribution of TNF-α to the development and progression of MS is well established. However, despite the pro-inflammatory actions of TNF-α, the use of anti-TNF-α monoclonal antibodies in MS patients, in addition to being ineffective, promotes an unexpected worsening of the disease [81], indicating that this cytokine also plays a protective role. Several studies have shown that the pathogenic and homeostatic activities of TNF-α are mediated by distinct cellular and molecular pathways and depend on the type of TNF-α receptor that is activated. Astrocytes, oligodendrocytes, and Treg cells express the TNFR2 receptor; the activation of this receptor mediates neuronal survival, re-myelination, and acts on immunoregulation. On the other hand, activation of TNFR1 receptors induces neuroinflammation and demyelination. Substances that activate CB2 type cannabinoid receptors act on glial cells and neurons, inhibiting TNF-α release and having antioxidant action [82]. Thus, the decrease in TNF-α levels presently observed may be a consequence of CB2 receptor activation, and since this decrease is partial, it is plausible that the remaining TNF-α would continue to exert its neuroprotective effect.

The increased chemokines levels detected in EAE makes the BBB more permeable to inflammatory cells; thus peripheral macrophages infiltrate the CNS [83]. Microglia and macrophages are considered important in the development of EAE and actively contribute to the pathogenesis and progression of the disease [84], causing the release of proinflammatory cytokines and leading to gliosis, inflammation tissue damage, and demyelination, culminating in neuronal death (neurodegeneration) in the CNS [61,85,86]. During the acute stage of EAE, NGF and its tyrosine kinase receptor (TrkA) expression are decreased in the CNS in rats [87]. Furthermore, studies suggest that NGF is responsible for the induction of axonal regeneration, survival, maintenance, and the differentiation of oligodendrocytes, as well as for facilitating the migration and proliferation of oligodendrocyte precursors to the myelin injury sites, a key role in the recovery of demyelinating diseases and stimulating recovery from neuroinflammatory diseases, including EAE [88–90]. On the 28th day after immunization, the expression of neurotrophin NGF was analyzed. In our study, corroborating data from the literature, a reduction of NGF in the spinal cord of the EAE animals was observed; however, it was not prevented by CRO.

It is noteworthy that MS is a disease that affects the CNS and that the literature is scarce regarding the effects of EAE on the demyelination of peripheral nerves. To verify possible peripheral demyelination in this animal model of MS and whether the treatment with CRO can interfere in this process, the morphology of the sciatic nerve was analyzed by transmission electron microscopy. This methodology has been applied for the evaluation of peripheral neurodegeneration, such as demyelination and the effect of neuroprotective therapies on these processes [91]. It was presently observed that this autoimmune disease alters the structure of the sciatic nerve myelin sheath of animals. Our results, together with results previously published [92], indicate that sciatic nerve demyelination may contribute to the neuropathic pain detected in this EAE model. In contrast to previous knowledge about the expression of myelin basic protein (MBP) and MOG exclusively at the CNS, it is currently known that these proteins are expressed in the peripheral nervous system and can therefore be attacked by autoantibodies, contributing to peripheral demyelination [93–96]. Interestingly, treatment with a single dose of CRO induces improvement in myelin fiber thickness. These results are promising and indicate that CRO may prevent demyelination or promote remyelination. Further studies to confirm these hypotheses are necessary.

4. Conclusions

Our results demonstrated that CRO interferes with the normal course of EAE, acting on both central and peripheral sites. This process occurs through the inhibition of CRO with cytokines release and reduction in the activation of glial cells and inflammatory infiltrate, indicating lower neuroinflammation and central sensitization, thus attenuating hyperalgesia and clinical signs. These results show the effectiveness of CRO in this animal model of neurodegenerative disease, indicating important central sites to be controlled in order to interfere with disease development.

5. Materials and Methods

5.1. Animals

Female C57BL/6J mice were used, supplied by Butantan Institute Central Animal Facility. The animals were kept with water and feed ad libitum in an appropriate sound-proof room, controlled temperature (22 °C ± 1), and light-dark cycle (12/12 h), with a maximum of 6 animals per cage. All procedures were performed according to the "Ethical Guide for the Use of Pain-Sensitive Animals in Pain Testing", published by IASP [97], reported following the Animal Research Reporting of In Vivo Experiments (ARRIVE) guidelines [http://www.nc3rs.org.uk/arrive-guidelines (accessed on 8 September 2021)] and approved by the Ethics Committee on Animal Use of the Butantan Institute (CEUAIB protocol number 7334170718).

5.2. Experimental Design

Mice were handled according to Figure 9. Briefly, EAE was induced in female C57BL/6J mice. Pain sensitivity was evaluated before (day 0) and on the fourth, fifth, sixth, eighth and 10th days (as Section 5.4) after immunization (as Section 5.3). Clinical signs were daily evaluated. CRO was administered in a single dose (on the 5th day after immunization–1st protocol) or in five repeated doses from the 12th day (one daily dose for five consecutive days–2nd protocol). At the 17th (peak of disease) and 28th (remission phase) days after immunization, the spinal cord was collected for inflammatory infiltrate determination (as Section 5.6), glial cells immunoreactivity evaluation (as Section 5.7), cytokines release measurement (as Section 5.8), and nerve growth factor expression (as Section 5.9). At the same periods, the sciatic nerve was removed for myelin thickness determination (as Section 5.10).

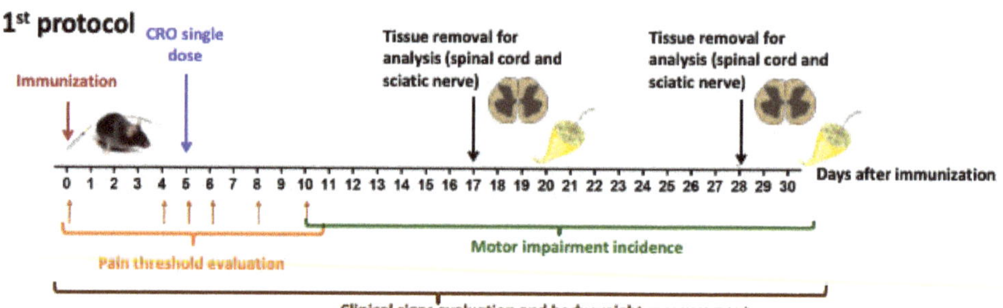

Figure 9. Experimental design for the evaluation of CRO effect on EAE development.

5.3. Induction of the EAE

The EAE model was induced as previously described [98]. Animals were anesthetized (isoflurane 1.5–2% in oxygen) and immunized subcutaneously at the base of the tail (s.c.) with 200 µg of myelin oligodendrocyte glycoprotein ($MOG_{35–55}$) peptide (Proteimax Technology, Sao Paulo, SP, Brazil) emulsified in incomplete Freund's adjuvant (IFA) supplemented with *Mycobacterium tuberculosis* (4 mg/mL, H37Ra, Difco™, Detroit, MI, USA). Intraperitoneal (i.p.) injection of 300 ng of pertussis toxin (Sigma-Aldrich™, St. Louis, MO, USA) was performed immediately after immunization and 48 h later. Control animals were treated with the vehicle (complete Freund's adjuvant—CFA, s.c. and pertussis toxin, i.p.). Animals were daily monitored according to the following clinical scores: 0.0, no symptoms; 0.5, some loss of tail tone; 1.0, loss of tail tone; 1.5, hip weakness upon ambulation; 2.0, hip weakness and partial hind limb paresis; 2.5, partial hind limb paralysis; 3.0, total hind limb paralysis but still moving on the cage; 3.5, complete hind limb paralysis and forelimbs weakness; 4.0, total hind limb paralysis and partial forelimbs paralysis; 4.5, complete hind limb paralysis, partial paralysis of the forelimbs, decreased responsiveness (consider euthanasia); 5, Immobile and unresponsive, moribund (immediate euthanasia). In the occurrence of score 3, food and water were turned available inside the cage. In the occurrence of score 4, animals were evaluated twice a day and if this score persists for three consecutive assessments, euthanasia was applied and the score of 5 was considered to the end of the period of evaluation. Body weight was measured during the course of the disease.

5.4. Evaluation of Pain Sensitivity: Determination of Mechanical Hyperalgesia by Electronic von Frey Test

The animals were habituated in the experimental room 30 min before the measurements. The mice were individually placed in bottomless acrylic boxes. On the bottom of the boxes, a wire mesh allows access to the plantar surface of the hind paw of the animal. A polypropylene disposable tip of 0.5 mm^2 with increasing perpendicular force was applied on the plantar surface of the hind limb until the animal shows the paw withdrawal reaction, thus indicating its nociceptive threshold [99]. The tip is coupled to a transducer, which indicates the value in grams of the force applied to the animal's paw; the accuracy of the device is 0.1 g and the device records the maximum force applied (Insight Equipamentos Ltda., Ribeirao Preto, SP, Brazil). Pain sensitivity was evaluated before (baseline) and at different periods after immunization with $MOG_{35–55}$, before they presented motor alterations.

5.5. Pharmacological Treatment with CRO

Crotalphine (<E-F-S-P-E-N-C-Q-G-E-S-Q-P-C, where <E is pyroglutamic acid and 7C-14C forms a disulfide bond; MW 1534.6 Da) was synthesized by Proteimax Biotecnologia, Brazil (São Paulo, SP, Brazil) as previously described [35], and was administered by oral route, diluted in sterile saline (NaCl 0.9% in distilled water). For dose setting of CRO, animals with EAE were allocated into different groups which received CRO (10, 50 or 100 µg/kg, p.o.) or saline (p.o.), on the fifth day after immunization (1 day after the onset of nociceptive response, the first symptom of the disease). Administration of CRO was based on previous assays showing long-lasting analgesic effects in experimental models of pain [36,37]. CRO (50 µg/kg) was also administered in five repeated doses (one daily dose for five consecutive days) from the 12th day (one day after motor impairment manifestation). The protocol of five consecutive doses was based on previous results demonstrating the greatest difficulty in controlling the symptoms of the disease after the onset of clinical signs [77]; however, it was performed to determine whether crotalphine would be able to interfere with the development of the disease even after the motor impairment trigger.

5.6. Histology of Spinal Cord to Inflammatory Infiltrate Determination (Hematoxylin and Eosin Stain)

To characterize the cellular infiltrate in the spinal cord, histological analysis was performed in EAE animals after CRO or saline treatment. The region comprised between lumbar segments from the spinal cord was removed and fixed in 10% of paraformaldehyde (PFA) solution for 24 h, at room temperature. After the fixation period, the tissues were transferred for a 70% ethyl alcohol solution, dehydrated, embedded in paraffin, and cut into sections of 5 µm thickness. The sections were stained with hematoxylin-eosin (HE). This classical technique in histological diagnosis has two dyes: hematoxylin, which stains all the nuclei of all cells, and eosin, which stains the cytoplasm of all cells [100].

5.7. Glial Cells Immunoreactivity Evaluated by Immunohistochemistry Assay

At the 17th (peak of disease) and 28th (remission phase) days after immunization, animals were anesthetized using ketamine (75 mg/kg i.p.) and xylazine (10 mg/kg, i.p.) and were transcardially perfused with 0.9% saline followed by 4% of PFA solution. The spinal cord was collected, and tissues were included in PFA solution for 4 h to be post-fixed. After that, tissues were kept, for at least 48 h, in a 30% sucrose solution in a refrigerator (2–8 °C). After this period, the regions of the spinal cord corresponding to the L3-L6 were sectioned (30 µm) in a microtome, collected in an anti-freezing solution (500 mL of phosphate buffer 0.05 M, 300 mL of ethylene glycol, and 150 g of sucrose) and kept at −20 °C. After washing three times with phosphate buffer (PB 0.1M, pH 7.4), the material was incubated free-floating for 12–16 h with specific Iba-1 (ionized calcium-binding adaptor-1, 1:2000, Wako Chemicals, Richmond, VA, USA) and GFAP (glial fibrillary acidic protein, 1:1000, Sigma-Aldrich™, St. Louis, MO, USA) antibodies in PB containing 0.3% Triton X-100 and 5% normal donkey serum. After that, sections were washed in PB (3 times), incubated with the biotinylated secondary antibody (anti-rabbit, 1:200 in PB containing 0.3% Triton X-100) for 2 h, washed 3 times in PB, and incubated with the avidin complex -biotin peroxidase (ABC Elite, Vector) for 2 h. After the washing process, tissues were reacted in diaminobenzidine (DAB, Sigma Aldrich™, USA) and 0.01% hydrogen peroxide in PB. The slices were mounted on gelatinized slides and kept at 37 °C for 48 h, dehydrated at room temperature, cleaned, and covered with a coverslip. Immunoreactivity quantification was performed in the dorsal horn of the spinal cord using ImageJ software (NIH/EUA).

5.8. Evaluation of IL-17 and TNF-α Release by Multiplex Assay

To investigate the IL-17 and TNF-α cytokines release, EAE animals treated with CRO or saline were euthanized, and the lumbar portion of the spinal cord (L3-L6) was collected on the 17th day after immunization. Collected tissue was homogenized in RIPA buffer (Sigma-Aldrich™, USA) containing protease inhibitors cocktail (1:100, Sigma-Aldrich™, USA) and phosphatase (1:300, Sigma-Aldrich™, USA). The homogenate was centrifuged for 5 min at 10,000× g and 4 °C. An aliquot of the supernatant was used for protein determination by the Bradford method [39,101]. The samples were normalized (3 µg/µL) and the protein concentration was determined using the commercial kit (Millipore, Burlington, MA, USA) by xMap method (MULTIPLEX), All the samples were done in duplicate and read using the Luminex 200 equipment—xPonent software version 4.2 (LEAC lab, Sao Paulo, SP, Brazil).

5.9. Nerve Growth Factor Expression by Western Blotting Assay

On the 17th day after immunization, animals were euthanized under anesthesia and the L3-L6 regions of the spinal cord were removed. The samples were homogenized in buffer containing Hepes-NaOH (1M, pH 7.9), EGTA (200 mM), NaCl (0.9%), Triton-X 100 (1%), and phosphatase and protease inhibitor (1:300, Sigma-Aldrich™, USA). The samples were centrifuged at 12,000× g rpm for 20 min at 4 °C. After centrifugation, an aliquot of the supernatant was used for protein concentration determination using the method of Bradford (Thermo Fisher Scientific, Waltham, MA, USA) [101]. Aliquots containing 30 µg of protein were boiled in Laemmli buffer 5× and after that subjected to polyacrylamide gel electrophoresis (SDS-PAGE), in an Mini-Protean apparatus for mini gel (BioRad, Hercules, CA, USA). After separation by electrophoresis, the proteins were transferred to a nitrocellulose membrane (BioRad, Hercules, CA, USA), and subsequently, the membrane was blocked in TBST (20 mM Tris-HCl, 150 mM NaCl, and 0.1% Tween 20) containing 5% of BSA for 1 h, then washed in TBST, followed by incubation with antibody NGF (1:1000, Abcam, Boston, MA, USA) overnight at 4 °C. Then, membranes were incubated with anti-IgG from the animal producing the respective peroxidase-conjugated primary antibody (1:5000, Abcam, USA), for 2 h at room temperature. ECL (Pierce) solution with a digital image capture system (UVITEC Cambridge) was used for bands visualization and the UVITEC Cambridge program was used for optical density determination. Results were normalized by membrane incubation with beta-actin (1:5000, Sigma Aldrich™, USA).

5.10. Transmission Electronic Microscopy

After intraperitoneal anesthesia with urethane (3 g/kg), animals were perfused with modified Karnovsky fixative solution containing 2.5% glutaraldehyde and 2% PFA in 0.1M sodium phosphate buffer solution (pH 7.4) [102]. After dissection of the muscles of the posterior thigh, the sciatic nerve was collected. The samples were post-fixed in 1% osmium tetroxide solution at 4 °C and then immersed in 5% aqueous uranyl acetate solution at room temperature. After that, the samples were dehydrated using a series of alcohols, then immersed in propylene oxide, and finally included in Spurr resin. Semi-thin sections (15 µM) were cut using Reichert Ultra Cut® ultra-microtome (Leica, Wetzlar, Germany) and stained with 1% toluidine blue solution. Subsequently, the 60 nm ultrathin sections were collected on 200 mesh copper grids (Sigma-Aldrich™, USA) and contrasted with 4% uranyl acetate solution and 0.4% aqueous lead citrate solution [103]. The grids were examined using the Jeol 1010 transmission electron microscope (NEP/MEPA/ESALQ) at the University of São Paulo.

5.11. Statistical Analysis

Statistical analysis was performed using GraphPad Prism version 6 software (GraphPad, San Diego, CA, USA). Test one-way analysis of variance (ANOVA) followed by Tukey's post-test was used for comparisons among three or more groups. Two-way analysis of variance (ANOVA) followed by Bonferroni post-test was used to compare repeated measures of the electronic von Frey test. The data are expressed as the mean ± SEM and $p < 0.05$ was indicative of a significant statistical difference.

Author Contributions: Conceptualization, G.P., V.O.Z., M.C. and R.L.P.; methodology, A.C.G., B.G.E., M.B.S., B.B.M., histological and immune assays, A.C.G., B.G.E., C.L.P.L., A.P.C., O.G.R., formal analysis, A.C.G., M.B.S., G.P., V.O.Z., M.C., R.L.P.; data curation, A.C.G., B.G.E., B.B.M., writing—original draft preparation, A.C.G.; supervision, G.P.; funding acquisition, G.P., V.O.Z., M.C., R.L.P. All authors have read and agreed to the published version of the manuscript.

Funding: This work was supported by funds from Conselho Nacional de Desenvolvimento Científico e Tecnológico—CNPq (Process number 467211/2014-0), from Coordenação de Aperfeiçoamento de Pessoal de Nível Superior—Brazil [CAPES- Finance Code 001] and from São Paulo Research Foundation (FAPESP) grants 2015/01254-1 and 2013/07467-1 (CETICs program).

Institutional Review Board Statement: All procedures were performed according to the "Ethical Guide for the Use of Pain-Sensitive Animals in Pain Testing", published by IASP [93], reported in accordance with the Animal Research Reporting of In Vivo Experiments (ARRIVE) guidelines (http://www.nc3rs.org.uk/arrive-guidelines, accessed on 27 October 2021) and approved by the Ethics Committee on Animal Use of the Butantan Institute (CEUAIB protocol number 7334170718, approval date 7 August 2018).

Informed Consent Statement: Not applicable.

Data Availability Statement: Not applicable.

Conflicts of Interest: The authors declare no conflict of interest.

References

1. The Multiple Sclerosis International Federation. *Atlas of MS: Mapping Multiple Sclerosis Around of World*, 3rd ed.; The Multiple Sclerosis International Federation: London, UK. Available online: https://www.atlasofms.org (accessed on 30 September 2021).
2. Ohno, N.; Ikenaka, K. Axonal and neuronal degeneration in myelin diseases. *Neurosci. Res.* **2019**, *139*, 48–57. [CrossRef]
3. Perricone, C.; Versini, M.; Ben-Ami, D.; Gertel, S.; Watad, A.; Segel, M.J.; Ceccarelli, F.; Conti, F.; Cantarini, L.; Bogdanos, D.P.; et al. Smoke and autoimmunity: The fire behind the disease. *Autoimmun. Rev.* **2016**, *15*, 354–374. [CrossRef] [PubMed]
4. Fierz, W. Multiple sclerosis: An example of pathogenic viral interaction? *Virol. J.* **2017**, *14*, 42. [CrossRef] [PubMed]
5. Olsson, T.; Barcellos, L.F.; Alfredsson, L. Interactions between genetic, lifestyle and environmental risk factors for multiple sclerosis. *Nat. Rev. Neurol.* **2017**, *13*, 25–36. [CrossRef] [PubMed]
6. Alrouji, M.; Manouchehrinia, A.; Gran, B.; Constantinescu, C.S. Effects of cigarette smoke on immunity, neuroinflammation and multiple sclerosis. *J. Neuroimmunol.* **2019**, *329*, 24–34. [CrossRef]
7. Henderson, A.P.; Barnett, M.H.; Parratt, J.D.; Prineas, J.W. Multiple sclerosis: Distribution of inflammatory cells in newly forming lesions. *Ann. Neurol.* **2009**, *66*, 739–753. [CrossRef]
8. Geurts, J.J.; Calabrese, M.; Fisher, E.; Rudick, R.A. Measurement and clinical effect of grey matter pathology in multiple sclerosis. *Lancet Neurol.* **2012**, *11*, 1082–1092. [CrossRef]
9. Thompson, A.J.; Banwell, B.L.; Barkhof, F.; Carroll, W.M.; Coetzee, T.; Comi, G.; Correale, J.; Fazekas, F.; Filippi, M.; Freedman, M.S.; et al. Diagnosis of multiple sclerosis: 2017 revisions of the McDonald criteria. *Lancet Neurol.* **2018**, *17*, 162–173. [CrossRef]
10. El-behi, M.; Rostami, A.; Ciric, B. Current Views on the Roles of Th1 and Th17 Cells in Experimental Autoimmune Encephalomyelitis. *J. Neuroimmune Pharmacol.* **2010**, *5*, 189–197. [CrossRef]
11. Sloane, E.; Ledeboer, A.; Seibert, W.; Coats, B.; van Strien, M.; Maier, S.F.; Johnson, K.W.; Chavez, R.; Watkins, L.R.; Leinwand, L.; et al. Anti-inflammatory cytokine gene therapy decreases sensory and motor dysfunction in experimental Multiple Sclerosis: MOG-EAE behavioral and anatomical symptom treatment with cytokine gene therapy. *Brain Behav. Immun.* **2009**, *23*, 92–100. [CrossRef]
12. Slavin, A.; Kelly-Modis, L.; Labadia, M.; Ryan, K.; Brown, M.L. Pathogenic mechanisms and experimental models of multiple sclerosis. *Autoimmunity* **2010**, *43*, 504–513. [CrossRef] [PubMed]
13. Jadidi-Niaragh, F.; Mirshafiey, A. Th17 Cell, the New Player of Neuroinflammatory Process in Multiple Sclerosis. *Scand. J. Immunol.* **2011**, *74*, 1–13. [CrossRef] [PubMed]

14. Ivanov, I.I.; McKenzie, B.S.; Zhou, L.; Tadokoro, C.E.; Lepelley, A.; Lafaille, J.J.; Cua, D.J.; Littman, D.R. The orphan nuclear receptor RORgammat directs the differentiation program of proinflammatory IL-17+ T helper cells. *Cell* **2006**, *126*, 1121–1133. [CrossRef]
15. Wojkowska, D.W.; Szpakowski, P.; Ksiazek-Winiarek, D.; Leszczynski, M.; Glabinski, A. Interactions between neutrophils, Th17 cells, and chemokines during the initiation of experimental model of multiple sclerosis. *Mediat. Inflamm.* **2014**, *2014*, 590409. [CrossRef]
16. Arneth, B.M. Impact of B cells to the pathophysiology of multiple sclerosis. *J. Neuroinflamm.* **2019**, *16*, 128. [CrossRef] [PubMed]
17. Ahn, J.J.; Abu-Rub, M.; Miller, R.H. B Cells in Neuroinflammation: New Perspectives and Mechanistic Insights. *Cells* **2021**, *10*, 1605. [CrossRef] [PubMed]
18. Horng, S.; Therattil, A.; Moyon, S.; Gordon, A.; Kim, K.; Argaw, A.T.; Hara, Y.; Mariani, J.N.; Sawai, S.; Flodby, P.; et al. Astrocytic tight junctions control inflammatory CNS lesion pathogenesis. *J. Clin. Investig.* **2017**, *127*, 3136–3151. [CrossRef] [PubMed]
19. Pirttimaki, T.M.; Parri, H.R. Astrocyte plasticity: Implications for synaptic and neuronal activity. *Neuroscientist* **2013**, *19*, 604–615. [CrossRef]
20. Lines, J.; Martin, E.D.; Kofuji, P.; Aguilar, J.; Araque, A. Astrocytes modulate sensory-evoked neuronal network activity. *Nat. Commun.* **2020**, *11*, 3689. [CrossRef] [PubMed]
21. Chiang, C.Y.; Sessle, B.J.; Dostrovsky, J.O. Role of astrocytes in pain. *Neurochem. Res.* **2012**, *37*, 2419–2431. [CrossRef]
22. Urits, I.; Adamian, L.; Fiocchi, J.; Hoyt, D.; Ernst, C.; Kaye, A.D.; Viswanath, O. Advances in the Understanding and Management of Chronic Pain in Multiple Sclerosis: A Comprehensive Review. *Curr. Pain Headache Rep.* **2019**, *23*, 59. [CrossRef]
23. Kalia, L.V.; O'Connor, P.W. Severity of chronic pain and its relationship to quality of life in multiple sclerosis. *Mult. Scler.* **2005**, *11*, 322–327. [CrossRef]
24. Ehde, D.M.; Osborne, T.L.; Hanley, M.A.; Jensen, M.P.; Kraft, G.H. The scope and nature of pain in persons with multiple sclerosis. *Mult. Scler.* **2006**, *12*, 629–638. [CrossRef] [PubMed]
25. Osborne, T.L.; Jensen, M.P.; Ehde, D.M.; Hanley, M.A.; Kraft, G. Psychosocial factors associated with pain intensity, pain-related interference, and psychological functioning in persons with multiple sclerosis and pain. *Pain* **2007**, *127*, 52–62. [CrossRef] [PubMed]
26. Hirsh, A.T.; Turner, A.P.; Ehde, D.M.; Haselkorn, J.K. Prevalence and Impact of Pain in Multiple Sclerosis: Physical and Psychologic Contributors. *Arch. Phys. Med. Rehabil.* **2009**, *90*, 646–651. [CrossRef] [PubMed]
27. Amtmann, D.; Askew, R.L.; Kim, J.; Chung, H.; Ehde, D.M.; Bombardier, C.H.; Kraft, G.H.; Jones, S.M.; Johnson, K.L. Pain affects depression through anxiety, fatigue, and sleep in multiple sclerosis. *Rehabil. Psychol.* **2015**, *60*, 81–90. [CrossRef] [PubMed]
28. Procaccini, C.; De Rosa, V.; Pucino, V.; Formisano, L.; Matarese, G. Animal models of Multiple Sclerosis. *Eur. J. Pharmacol.* **2015**, *759*, 182–191. [CrossRef]
29. Baxter, A.G. The origin and application of experimental autoimmune encephalomyelitis. *Nat. Rev. Immunol.* **2007**, *7*, 904–912. [CrossRef]
30. Basso, A.S.; Frenkel, D.; Quintana, F.J.; Costa-Pinto, F.A.; Petrovic-Stojkovic, S.; Puckett, L.; Monsonego, A.; Bar-Shir, A.; Engel, Y.; Gozin, M.; et al. Reversal of axonal loss and disability in a mouse model of progressive multiple sclerosis. *J. Clin. Investig.* **2008**, *118*, 1532–1543. [CrossRef] [PubMed]
31. Olechowski, C.J.; Parmar, A.; Miller, B.; Stephan, J.; Tenorio, G.; Tran, K.; Leighton, J.; Kerr, B.J. A diminished response to formalin stimulation reveals a role for the glutamate transporters in the altered pain sensitivity of mice with experimental autoimmune encephalomyelitis (EAE). *Pain* **2009**, *149*, 565–572. [CrossRef]
32. Lisi, L.; Navarra, P.; Cirocchi, R.; Sharp, A.; Stigliano, E.; Feinstein, D.L.; Dello Russo, C. Rapamycin reduces clinical signs and neuropathic pain in a chronic model of experimental autoimmune encephalomyelitis. *J. Neuroimmunol.* **2012**, *243*, 43–51. [CrossRef] [PubMed]
33. Rodrigues, D.H.; Sachs, D.; Teixeira, A.L. Mechanical hypernociception in experimental autoimmune encephalomyelitis. *Arq. Neuro-Psiquiatr.* **2009**, *67*, 78–81. [CrossRef] [PubMed]
34. Yuan, S.; Shi, Y.; Tang, S.J. Wnt Signaling in the Pathogenesis of Multiple Sclerosis-Associated Chronic Pain. *J. Neuroimmune Pharmacol.* **2012**, *7*, 904–913. [CrossRef] [PubMed]
35. Konno, K.; Picolo, G.; Gutierrez, V.P.; Brigatte, P.; Zambelli, V.O.; Camargo, A.C.; Cury, Y. Crotalphine, a novel potent analgesic peptide from the venom of the South American rattlesnake Crotalus durissus terrificus. *Peptides* **2008**, *29*, 1293–1304. [CrossRef] [PubMed]
36. Gutierrez, V.P.; Konno, K.; Chacur, M.; Sampaio, S.C.; Picolo, G.; Brigatte, P.; Zambelli, V.O.; Cury, Y. Crotalphine induces potent antinociception in neuropathic pain by acting at peripheral opioid receptors. *Eur. J. Pharmacol.* **2008**, *594*, 84–92. [CrossRef]
37. Brigatte, P.; Konno, K.; Gutierrez, V.P.; Sampaio, S.C.; Zambelli, V.O.; Picolo, G.; Curi, R.; Cury, Y. Peripheral kappa and delta opioid receptors are involved in the antinociceptive effect of crotalphine in a rat model of cancer pain. *Pharmacol. Biochem. Behav.* **2013**, *109*, 1–7. [CrossRef]
38. Machado, F.C.; Zambelli, V.O.; Fernandes, A.C.; Heimann, A.S.; Cury, Y.; Picolo, G. Peripheral interactions between cannabinoid and opioid systems contribute to the antinociceptive effect of crotalphine. *Br. J. Pharmacol.* **2014**, *171*, 961–972. [CrossRef]
39. Bressan, E.; Touska, F.; Vetter, I.; Kistner, K.; Kichko, T.I.; Teixeira, N.B.; Picolo, G.; Cury, Y.; Lewis, R.J.; Fischer, M.J.; et al. Crotalphine desensitizes TRPA1 ion channels to alleviate inflammatory hyperalgesia. *Pain* **2016**, *157*, 2504–2516. [CrossRef]

40. Da Costa, D.S.M.; Meotti, F.C.; Andrade, E.L.; Leal, P.C.; Motta, E.M.; Calixto, J.B. The involvement of the transient receptor potential A1 (TRPA1) in the maintenance of mechanical and cold hyperalgesia in persistent inflammation. *Pain* **2010**, *148*, 431–437. [CrossRef]
41. Leussink, V.I.; Husseini, L.; Warnke, C.; Broussalis, E.; Hartung, H.P.; Kieseier, B.C. Symptomatic therapy in multiple sclerosis: The role of cannabinoids in treating spasticity. *Ther. Adv. Neurol. Disord.* **2012**, *5*, 255–266. [CrossRef]
42. Zajicek, J.P.; Apostu, V.I. Role of cannabinoids in multiple sclerosis. *CNS Drugs* **2011**, *25*, 187–201. [CrossRef] [PubMed]
43. Lassmann, H.; Brück, W.; Lucchinetti, C.F. The immunopathology of multiple sclerosis: An overview. *Brain Pathol.* **2007**, *17*, 210–218. [CrossRef] [PubMed]
44. Brambilla, R. The contribution of astrocytes to the neuroinflammatory response in multiple sclerosis and experimental autoimmune encephalomyelitis. *Acta Neuropathol.* **2019**, *137*, 757–783. [CrossRef] [PubMed]
45. Chu, F.; Shi, M.; Zheng, C.; Shen, D.; Zhu, J.; Zheng, X.; Cui, L. The roles of macrophages and microglia in multiple sclerosis and experimental autoimmune encephalomyelitis. *J. Neuroimmunol.* **2018**, *318*, 1–7. [CrossRef] [PubMed]
46. Griffin, G.K.; Newton, G.; Tarrio, M.L.; Bu, D.X.; Maganto-Garcia, E.; Azcutia, V.; Alcaide, P.; Grabie, N.; Luscinskas, F.W.; Croce, K.J.; et al. IL-17 and TNF-α sustain neutrophil recruitment during inflammation through synergistic effects on endothelial activation. *J. Immunol.* **2012**, *188*, 6287–6299. [CrossRef]
47. Bothwell, M. NGF, BDNF, NT3, and NT4. *Neurotrophic Factors* **2014**, *220*, 3–15. [CrossRef]
48. Teixeira, N.B.; Picolo, G.; Giardini, A.C.; Boumezbeur, F.; Pottier, G.; Kuhnast, B.; Servent, D.; Benoit, E. Alterations of peripheral nerve excitability in an experimental autoimmune encephalomyelitis mouse model for multiple sclerosis. *J. Neuroinflamm.* **2020**, *17*, 266. [CrossRef]
49. Goverman, J. Autoimmune T cell responses in the central nervous system. *Nat. Rev. Immunol.* **2009**, *9*, 393–407. [CrossRef] [PubMed]
50. Petzold, A.; Eikelenboom, M.J.; Gveric, D.; Keir, G.; Chapman, M.; Lazeron, R.H.; Cuzner, M.L.; Polman, C.H.; Uitdehaag, B.M.; Thompson, E.J.; et al. Markers for different glial cell responses in multiple sclerosis: Clinical and pathological correlations. *Brain* **2002**, *125*, 1462–1473. [CrossRef]
51. Huseby, E.S.; Kamimura, D.; Arima, Y.; Parello, C.S.; Sasaki, K.; Murakami, M. Role of T cell-glial cell interactions in creating and amplifying central nervous system inflammation and multiple sclerosis disease symptoms. *Front. Cell Neurosci.* **2015**, *9*, 295. [CrossRef]
52. Szczuciński, A.; Losy, J. Chemokines and chemokine receptors in multiple sclerosis. Potential targets for new therapies. *Acta Neurol. Scand.* **2007**, *115*, 137–146. [CrossRef]
53. Brück, W. The pathology of multiple sclerosis is the result of focal inflammatory demyelination with axonal damage. *J. Neurol.* **2005**, *252*, v3–v9. [CrossRef] [PubMed]
54. Compston, A.; Coles, A. Multiple sclerosis. *Lancet* **2008**, *372*, 1502–1517. [CrossRef]
55. Duffy, S.S.; Lees, J.G.; Perera, C.J.; Moalem-Taylor, G. Managing Neuropathic Pain in Multiple Sclerosis: Pharmacological Interventions. *Med. Chem.* **2018**, *14*, 106–119. [CrossRef] [PubMed]
56. Alberti, T.B.; Barbosa, W.L.; Vieira, J.L.; Raposo, N.R.; Dutra, R.C. (-)-β-Caryophyllene, a CB2 Receptor-Selective Phytocannabinoid, Suppresses Motor Paralysis and Neuroinflammation in a Murine Model of Multiple Sclerosis. *Int. J. Mol. Sci.* **2017**, *18*, 691. [CrossRef]
57. Duffy, S.S.; Perera, C.J.; Makker, P.G.; Lees, J.G.; Carrive, P.; Moalem-Taylor, G. Peripheral and Central Neuroinflammatory Changes and Pain Behaviors in an Animal Model of Multiple Sclerosis. *Front. Immunol.* **2016**, *7*, 369. [CrossRef] [PubMed]
58. Milligan, E.D.; Watkins, L.R. Pathological and protective roles of glia in chronic pain. *Nat. Rev. Neurosci.* **2009**, *10*, 23–36. [CrossRef]
59. Inoue, K.; Tsuda, M. Microglia and neuropathic pain. *Glia* **2009**, *57*, 1469–1479. [CrossRef]
60. Thibault, K.; Calvino, B.; Pezet, S. Characterisation of sensory abnormalities observed in an animal model of multiple sclerosis: A behavioural and pharmacological study. *Eur. J. Pain* **2011**, *15*, 231.e1–16. [CrossRef]
61. Almolda, B.; Gonzalez, B.; Castellano, B. Antigen presentation in EAE: Role of microglia, macrophages and dendritic cells. *Front. Biosci.* **2011**, *16*, 1157–1171. [CrossRef]
62. Moreno, M.; Guo, F.; Ko, E.M.; Bannerman, P.; Soulika, A.; Pleasure, D. Origins and significance of astrogliosis in the multiple sclerosis model, MOG peptide EAE. *J. Neurol. Sci.* **2013**, *333*, 55–59. [CrossRef] [PubMed]
63. Norden, D.M.; Trojanowski, P.J.; Villanueva, E.; Navarro, E.; Godbout, J.P. Sequential activation of microglia and astrocyte cytokine expression precedes increased Iba-1 or GFAP immunoreactivity following systemic immune challenge. *Glia* **2016**, *64*, 300–316. [CrossRef]
64. Heppner, F.L.; Greter, M.; Marino, D.; Falsig, J.; Raivich, G.; Hovelmeyer, N.; Waisman, A.; Rulicke, T.; Prinz, M.; Priller, J.; et al. Experimental autoimmune encephalomyelitis repressed by microglial paralysis. *Nat. Med.* **2005**, *11*, 146–152. [CrossRef] [PubMed]
65. Bhasin, M.; Wu, M.; Tsirka, S.E. Modulation of microglial/macrophage activation by macrophage inhibitory factor (TKP) or tuftsin (TKPR) attenuates the disease course of experimental autoimmune encephalomyelitis. *BMC Immunol.* **2007**, *8*, 10. [CrossRef]
66. Aloisi, F. Immune function of microglia. *Glia* **2001**, *36*, 165–179. [CrossRef]
67. Raivich, G.; Banati, R. Brain microglia and blood-derived macrophages: Molecular profiles and functional roles in multiple sclerosis and animal models of autoimmune demyelinating disease. *Brain Res. Rev.* **2004**, *46*, 261–281. [CrossRef]

68. Ji, R.R.; Donnelly, C.R.; Nedergaard, M. Astrocytes in chronic pain and itch. *Nat. Rev. Neurosci.* **2019**, *20*, 667–685. [CrossRef] [PubMed]
69. Li, T.; Chen, X.; Zhang, C.; Zhang, Y.; Yao, W. An update on reactive astrocytes in chronic pain. *J. Neuroinflamm.* **2019**, *16*, 140. [CrossRef]
70. Basbaum, A.I.; Fields, H.L. Endogenous pain control systems: Brainstem spinal pathways and endorphin circuitry. *Annu. Rev. Neurosci.* **1984**, *7*, 309–338. [CrossRef]
71. Balasa, R.; Barcutean, L.; Balasa, A.; Motataianu, A.; Roman-Filip, C.; Manu, D. The action of TH17 cells on blood brain barrier in multiple sclerosis and experimental autoimmune encephalomyelitis. *Hum. Immunol.* **2020**, *81*, 237–243. [CrossRef]
72. Salehipour, Z.; Haghmorad, D.; Sankian, M.; Rastin, M.; Nosratabadi, R.; Soltan Dallal, M.M.; Tabasi, N.; Khazaee, M.; Nasiraii, L.R.; Mahmoudi, M. Bifidobacterium animalis in combination with human origin of Lactobacillus plantarum ameliorate neuroinflammation in experimental model of multiple sclerosis by altering CD4+ T cell subset balance. *Biomed. Pharmacother.* **2017**, *95*, 1535–1548. [CrossRef]
73. Milovanovic, J.; Arsenijevic, A.; Stojanovic, B.; Kanjevac, T.; Arsenijevic, D.; Radosavljevic, G.; Milovanovic, M.; Arsenijevic, N. Interleukin-17 in Chronic Inflammatory Neurological Diseases. *Front. Immunol.* **2020**, *11*, 947. [CrossRef]
74. Kurte, M.; Luz-Crawford, P.; Vega-Letter, A.M.; Contreras, R.A.; Tejedor, G.; Elizondo-Vega, R.; Martinez-Viola, L.; Fernández-O'Ryan, C.; Figueroa, F.E.; Jorgensen, C.; et al. IL17/IL17RA as a Novel Signaling Axis Driving Mesenchymal Stem Cell Therapeutic Function in Experimental Autoimmune Encephalomyelitis. *Front. Immunol.* **2018**, *9*, 802. [CrossRef] [PubMed]
75. Mardiguian, S.; Serres, S.; Ladds, E.; Campbell, S.J.; Wilainam, P.; McFadyen, C.; McAteer, M.; Choudhury, R.P.; Smith, P.; Saunders, F.; et al. Anti-IL-17A treatment reduces clinical score and VCAM-1 expression detected by in vivo magnetic resonance imaging in chronic relapsing EAE ABH mice. *Am. J. Pathol.* **2013**, *182*, 2071–2081. [CrossRef]
76. Komiyama, Y.; Nakae, S.; Matsuki, T.; Nambu, A.; Ishigame, H.; Kakuta, S.; Sudo, K.; Iwakura, Y. IL-17 plays an important role in the development of experimental autoimmune encephalomyelitis. *J. Immunol.* **2006**, *177*, 566–573. [CrossRef] [PubMed]
77. Teixeira, N.B.; Sant'Anna, M.B.; Giardini, A.C.; Araujo, L.P.; Fonseca, L.A.; Basso, A.S.; Cury, Y.; Picolo, G. Crotoxin down-modulates pro-inflammatory cells and alleviates pain on the MOG. *Brain Behav. Immun.* **2020**, *84*, 253–268. [CrossRef] [PubMed]
78. Neurath, M.F.; Finotto, S. IL-6 signaling in autoimmunity, chronic inflammation and inflammation-associated cancer. *Cytokine Growth Factor Rev.* **2011**, *22*, 83–89. [CrossRef]
79. Gijbels, K.; Van Damme, J.; Proost, P.; Put, W.; Carton, H.; Billiau, A. Interleukin 6 production in the central nervous system during experimental autoimmune encephalomyelitis. *Eur. J. Immunol.* **1990**, *20*, 233–235. [CrossRef]
80. Williams, S.K.; Maier, O.; Fischer, R.; Fairless, R.; Hochmeister, S.; Stojic, A.; Pick, L.; Haar, D.; Musiol, S.; Storch, M.K.; et al. Antibody-mediated inhibition of TNFR1 attenuates disease in a mouse model of multiple sclerosis. *PLoS ONE* **2014**, *9*, e90117. [CrossRef]
81. Lenercept Multiple Sclerosis Study Group and The University of British Columbia MS/MRI Analysis Group. TNF neutralization in MS: Results of a randomized, placebo-controlled multicenter study. *Neurology* **1999**, *53*, 457–465. [CrossRef]
82. Facchinetti, F.; Del Giudice, E.; Furegato, S.; Passarotto, M.; Leon, A. Cannabinoids ablate release of TNFalpha in rat microglial cells stimulated with lypopolysaccharide. *Glia* **2003**, *41*, 161–168. [CrossRef]
83. Engelhardt, B. Molecular mechanisms involved in T cell migration across the blood-brain barrier. *J. Neural Transm.* **2006**, *113*, 477–485. [CrossRef] [PubMed]
84. Rawji, K.S.; Yong, V.W. The benefits and detriments of macrophages/microglia in models of multiple sclerosis. *Clin. Dev. Immunol.* **2013**, *2013*, 948976. [CrossRef] [PubMed]
85. Liu, C.; Li, Y.; Yu, J.; Feng, L.; Hou, S.; Liu, Y.; Guo, M.; Xie, Y.; Meng, J.; Zhang, H.; et al. Targeting the shift from M1 to M2 macrophages in experimental autoimmune encephalomyelitis mice treated with fasudil. *PLoS ONE* **2013**, *8*, e54841. [CrossRef] [PubMed]
86. Jiang, H.R.; Milovanović, M.; Allan, D.; Niedbala, W.; Besnard, A.G.; Fukada, S.Y.; Alves-Filho, J.C.; Togbe, D.; Goodyear, C.S.; Linington, C.; et al. IL-33 attenuates EAE by suppressing IL-17 and IFN-γ production and inducing alternatively activated macrophages. *Eur. J. Immunol.* **2012**, *42*, 1804–1814. [CrossRef] [PubMed]
87. Zhu, L.; Pan, Q.X.; Zhang, X.J.; Xu, Y.M.; Chu, Y.J.; Liu, N.; Lv, P.; Zhang, G.X.; Kan, Q.C. Protective effects of matrine on experimental autoimmune encephalomyelitis via regulation of ProNGF and NGF signaling. *Exp. Mol. Pathol.* **2016**, *100*, 337–343. [CrossRef]
88. Micera, A.; Properzi, F.; Triaca, V.; Aloe, L. Nerve growth factor antibody exacerbates neuropathological signs of experimental allergic encephalomyelitis in adult lewis rats. *J. Neuroimmunol.* **2000**, *104*, 116–123. [CrossRef]
89. Triaca, V.; Tirassa, P.; Aloe, L. Presence of nerve growth factor and TrkA expression in the SVZ of EAE rats: Evidence for a possible functional significance. *Exp. Neurol.* **2005**, *191*, 53–64. [CrossRef]
90. Acosta, C.M.; Cortes, C.; MacPhee, H.; Namaka, M.P. Exploring the role of nerve growth factor in multiple sclerosis: Implications in myelin repair. *CNS Neurol. Disord. Drug Targets* **2013**, *12*, 1242–1256. [CrossRef] [PubMed]
91. Da Silva, J.T.; Santos, F.M.; Giardini, A.C.; Martins, D.e.O.; de Oliveira, M.E.; Ciena, A.P.; Gutierrez, V.P.; Watanabe, I.S.; Britto, L.R.; Chacur, M. Neural mobilization promotes nerve regeneration by nerve growth factor and myelin protein zero increased after sciatic nerve injury. *Growth Factors* **2015**, *33*, 8–13. [CrossRef] [PubMed]
92. Wang, I.C.; Chung, C.Y.; Liao, F.; Chen, C.C.; Lee, C.H. Peripheral sensory neuron injury contributes to neuropathic pain in experimental autoimmune encephalomyelitis. *Sci. Rep.* **2017**, *7*, 42304. [CrossRef]

93. Itoyama, Y.; Webster, H.D.; Richardson, E.P.; Trapp, B.D. Schwann cell remyelination of demyelinated axons in spinal cord multiple sclerosis lesions. *Ann. Neurol.* **1983**, *14*, 339–346. [CrossRef]
94. Puckett, C.; Hudson, L.; Ono, K.; Friedrich, V.; Benecke, J.; Dubois-Dalcq, M.; Lazzarini, R.A. Myelin-specific proteolipid protein is expressed in myelinating Schwann cells but is not incorporated into myelin sheaths. *J. Neurosci. Res.* **1987**, *18*, 511–518. [CrossRef]
95. Griffiths, I.R.; Mitchell, L.S.; McPhilemy, K.; Morrison, S.; Kyriakides, E.; Barrie, J.A. Expression of myelin protein genes in Schwann cells. *J. Neurocytol.* **1989**, *18*, 345–352. [CrossRef] [PubMed]
96. Trotter, J.; DeJong, L.J.; Smith, M.E. Opsonization with antimyelin antibody increases the uptake and intracellular metabolism of myelin in inflammatory macrophages. *J. Neurochem.* **1986**, *47*, 779–789. [CrossRef] [PubMed]
97. Zimmermann, M. Ethical guidelines for investigations of experimental pain in conscious animals. *Pain* **1983**, *16*, 109–110. [CrossRef]
98. Sant'Anna, M.B.; Giardini, A.C.; Ribeiro, M.A.C.; Lopes, F.S.R.; Teixeira, N.B.; Kimura, L.F.; Bufalo, M.C.; Ribeiro, O.G.; Borrego, A.; Cabrera, W.H.K.; et al. The Crotoxin:SBA-15 Complex Down-Regulates the Incidence and Intensity of Experimental Autoimmune Encephalomyelitis Through Peripheral and Central Actions. *Front. Immunol.* **2020**, *11*, 591563. [CrossRef]
99. Cunha, T.M.; Verri, W.A.; Vivancos, G.G.; Moreira, I.F.; Reis, S.; Parada, C.A.; Cunha, F.Q.; Ferreira, S.H. An electronic pressure-meter nociception paw test for mice. *Braz. J. Med. Biol. Res.* **2004**, *37*, 401–407. [CrossRef] [PubMed]
100. Brancroft, J.D. *Manual of Histological Techniques and Their Diagnostic Application*, 2nd ed.; Livingstone: London, UK, 1994; p. 147.
101. Bradford, M.M. A rapid and sensitive method for the quantitation of microgram quantities of protein utilizing the principle of protein-dye binding. *Anal. Biochem.* **1976**, *72*, 248–254. [CrossRef]
102. Ciena, A.P.; de Almeida, S.R.; Alves, P.H.; Bolina-Matos, R.e.S.; Dias, F.J.; Issa, J.P.; Iyomasa, M.M.; Watanabe, I.S. Histochemical and ultrastructural changes of sternomastoid muscle in aged Wistar rats. *Micron* **2011**, *42*, 871–876. [CrossRef]
103. Watanabe, I.; Yamada, E. The fine structure of lamellated nerve endings found in the rat gingiva. *Arch. Histol. Jpn.* **1983**, *46*, 173–182. [CrossRef] [PubMed]

Article

The Deleterious Effects of Shiga Toxin Type 2 Are Neutralized In Vitro by FabF8:Stx2 Recombinant Monoclonal Antibody

Daniela Luz [1,†], Fernando D. Gómez [2,†], Raíssa L. Ferreira [1], Bruna S. Melo [1], Beatriz E. C. Guth [3], Wagner Quintilio [4], Ana Maria Moro [4], Agostina Presta [2], Flavia Sacerdoti [2], Cristina Ibarra [2], Gang Chen [5], Sachdev S. Sidhu [5], María Marta Amaral [2,*] and Roxane M. F. Piazza [1,*]

1 Laboratório de Bacteriologia, Instituto Butantan, Sao Paulo 05503-900, Brazil; daniedaluz@gmail.com (D.L.); raissalozzardo@gmail.com (R.L.F.); brn.smelo@gmail.com (B.S.M.)
2 Laboratorio de Fisiopatogenia, Instituto de Fisiología y Biofísica Bernardo Houssay (IFIBIO Houssay-CONICET), Departamento de Fisiología, Facultad de Medicina, Universidad de Buenos Aires, Buenos Aires 1121, Argentina; gomezfernandoD@gmail.com (F.D.G.); agospresta@hotmail.com (A.P.); flasacerdoti@gmail.com (F.S.); cristinaadrianaibarra@gmail.com (C.I.)
3 Departamento de Microbiologia, Imunologia e Parasitologia, Universidade Federal de São Paulo, Sao Paulo 04023-062, Brazil; bec.guth@unifesp.br
4 Laboratório de Biofármacos, Instituto Butantan, Sao Paulo 05503-900, Brazil; wagner.quintilio@butantan.gov.br (W.Q.); ana.moro@butantan.gov.br (A.M.M.)
5 Banting and Best Department of Medical Research, Terrence Donnelly Centre for Cellular and Biomolecular Research, University of Toronto, Toronto, OT M5S 3E1, Canada; gchen2012@gmail.com (G.C.); sachdev.sidhu@utoronto.ca (S.S.S.)
* Correspondence: mmamaral74@gmail.com (M.M.A.); roxane.piazza@butantan.gov.br (R.M.F.P.)
† The authors equally contributed to the present study.

Citation: Luz, D.; Gómez, F.D.; Ferreira, R.L.; Melo, B.S.; Guth, B.E.C.; Quintilio, W.; Moro, A.M.; Presta, A.; Sacerdoti, F.; Ibarra, C.; et al. The Deleterious Effects of Shiga Toxin Type 2 Are Neutralized In Vitro by FabF8:Stx2 Recombinant Monoclonal Antibody. *Toxins* **2021**, *13*, 825. https://doi.org/10.3390/toxins13110825

Received: 14 October 2021
Accepted: 15 November 2021
Published: 22 November 2021

Abstract: Hemolytic Uremic Syndrome (HUS) associated with Shiga-toxigenic *Escherichia coli* (STEC) infections is the principal cause of acute renal injury in pediatric age groups. Shiga toxin type 2 (Stx2) has in vitro cytotoxic effects on kidney cells, including human glomerular endothelial (HGEC) and Vero cells. Neither a licensed vaccine nor effective therapy for HUS is available for humans. Recombinant antibodies against Stx2, produced in bacteria, appeared as the utmost tool to prevent HUS. Therefore, in this work, a recombinant FabF8:Stx2 was selected from a human Fab antibody library by phage display, characterized, and analyzed for its ability to neutralize the Stx activity from different STEC-Stx2 and Stx1/Stx2 producing strains in a gold standard Vero cell assay, and the Stx2 cytotoxic effects on primary cultures of HGEC. This recombinant Fab showed a dissociation constant of 13.8 nM and a half maximum effective concentration (EC_{50}) of 160 ng/mL to Stx2. Additionally, FabF8:Stx2 neutralized, in different percentages, the cytotoxic effects of Stx2 and Stx1/2 from different STEC strains on Vero cells. Moreover, it significantly prevented the deleterious effects of Stx2 in a dose-dependent manner (up to 83%) in HGEC and protected this cell up to 90% from apoptosis and necrosis. Therefore, this novel and simple anti-Stx2 biomolecule will allow further investigation as a new therapeutic option that could improve STEC and HUS patient outcomes.

Keywords: STEC; Stx2; antibody fragment; monoclonal antibody

Key Contribution: Here we describe the generation, characterization, and efficacy of one Fab antibody fragment anti-Stx2 in protecting cells against Stx2 cytotoxic effects and its ability to neutralize Stx produced by STEC strains, which is demonstrated herein for the first time.

1. Introduction

The hemolytic uremic syndrome (HUS) in children is mostly caused by Shiga toxin-producing *Escherichia coli* (STEC) infection, which is also responsible for outbreaks in the United States, Europe, South America, and Japan [1–3]. In Argentina, where post-diarrheal HUS is endemic, around 300 new cases are reported each year [4]. Since the early 2000s, epidemiologically, the emergence of the non-O157 STEC infection, replacing the traditionally predominant O157 serogroup occurrence [5]. The contamination by STEC strains is usually by contaminated food or water ingestion, person-to-person transmission, or contact with ruminants or its contaminated environment [6]. The primary infection symptom is diarrhea, which is an average incubation phase of three days that could turn bloody in about 60% of patients. However, Shiga toxins (Stx) released by STEC triggers thrombogenic and inflammatory microvascular endothelial cell alterations, leading to HUS in 5–15% of STEC infection cases. HUS is defined by hemolytic anemia, thrombocytopenia, and acute renal injury [7,8]. Besides death, this syndrome can lead to long-term consequences such as hypertension and renal disease because of the high sensitivity to the Stx of the microvascular endothelial cells in the kidney [9].

The Stx toxins produced by STEC are Stx1 and Stx2, they appear to differ significantly in their effectiveness to induce protein synthesis inhibition and cytotoxicity, with some subtypes of Stx2 more potent than Stx1, on the other hand, other subtypes have similar potency [10]. Stxs is AB_5 type toxin, consisting of a homo-pentameric B subunit (7.7 kDa per monomer) which binds to the host receptor globotriaosylceramide (Gb3) and mediate the enzymatically active A subunit (~32 kDa) endocytosis. Once inside the cell, the A subunit depurinates the conserved adenine residue of 28S eukaryotic rRNA, stopping peptide elongation and leading to cell death [11–13]. No specific drug has proved effective as specific therapy for STEC-HUS, which remains as symptomatic care. The antibiotics administration in STEC infection and STEC-HUS remains controversial, with some bacteriostatic antibiotics having a beneficial effect while others can increase the Stx liberation by the bacteria [14]. Proofs of evidence of an advantage from complement blockade therapy in STEC-HUS are also lacking [15]. One alternative treatment for STEC infection and possibly for HUS is neutralizing anti-Stx antibody therapy.

Monoclonal antibodies (mAb) against Stx have been evaluated in animal models (reviewed in [16,17]). Moreover, few mAbs candidates have also been tested in healthy volunteers during phase I studies [18,19]. In addition, a chimeric anti-Stx1 and Stx2 mAb was challenged in a phase II study in South America, but definite evidence of its therapeutic efficacy remains vague [20,21].

In addition to conventional antibodies, recombinant antibodies can be an attractive replacement to avoid animal immunization and other limitations of hybridoma technology, a successful, but cumbersome and costly approach to generate monoclonal antibodies [22,23]. In this context, we may include a family of Stx2B-binding VHHs that neutralize Stx2 in vitro at a nanomolar to the subnanomolar range [24] and the FabC11:Stx2 generated by phage display technology and produced very efficiently using bacterial protein synthesis systems which were able to prevent Stx2 toxicity to human kidney cells and in mice [25,26]. Therefore, the generation of such molecules and studies concerning their applicability will provide new therapeutic options for treating STEC infections to prevent or ameliorate HUS outcomes.

Herein, also employing phage display antibody library F [27], a monovalent FabF8:Stx2 was generated, and efficiently produced in the bacterial system with neutralizing qualities against Stx. We introduce a novel and simple antitoxin agent as a new therapeutic option for STEC infections therapy.

2. Results

2.1. Selection of FabF8:Stx2 from a Human Antibody Fragment Phage Display Library

The FabF8:Stx2 was generated from the selection using purified Stx2a toxin and a human synthetic antibody phage display library (library F) developed by Persson et al. [27]. The cloning was confirmed by sequencing (Figure 1A). The 48 kDa fragment corresponds to the purified Fab fragment, however, a 25 kDa protein also appears, which corresponds to non-assembled variable chains (Figure 1B). As determined by surface plasmon resonance, the purified FabF8:Stx2 showed an affinity constant (K_D) of 13.8 nM (Figure S1). The half-maximum effective concentration (EC_{50}) was determined as being 160 ng/mL (calculated as described in the material and methods) as well as, specificity just for the selected toxin, with no significant cross-reactivity to Stx1 toxin (Figure 1C).

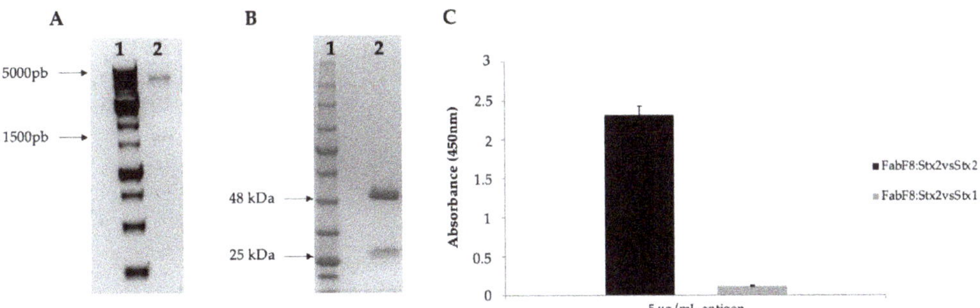

Figure 1. The FabF8:Stx2 generation. (**A**) FabF8:Stx2 gene cloning. Electrophoretic profile on 1.5% agarose gel stained with SYBR (1:1000) of restriction analyzes of FabF8:Stx2 clone. (1) 1Kb molecular weight marker (Invitrogen); (2) Clone F8 anti-Stx2 (FabF8:Stx2); (**B**) FabF8:Stx2 purification. Electrophoretic profile on 15% non-denaturing polyacrylamide gel stained with Coomassie blue of sample eluted from the purifications of Fab fragment. (1) Blueyed molecular weight marker (GE); (2) Clone F8 anti-Stx2. (**C**) ELISA assay to assess cross-reaction of ligands against Stx toxins (5 µg/mL) using EC_{50} concentration of FabF8:Stx2.

2.2. *FabF8:Stx2 Neutralizes the Cytotoxic Effect of Supernatants from Different Stx2-Producing Strains*

The FabF8:Stx2 was employed in a gold standard Vero cell assay (VCA) to test its neutralization ability to the toxicity of the supernatants from different STEC strains producing Stx2 or Stx1/2. The ability of this antibody in neutralizing the purified Stx2 was 84% (Table 1). Bacterial supernatant cytotoxicity was tested in the absence and presence of FabF8:Stx2 (Figure 2). It was observed that 85% of the tested supernatants were neutralized (from 7 to 100%), in 20 strains this rate was above 20% and this ability was superior to 40% in 16 strains. Therefore, the FabF8:Stx2 neutralizing ability ranged from 0 to 100%. No significant differences were observed in its neutralizing ability against Stx2 or Stx1/Stx2-producing strains. This recombinant antibody failed to neutralize only four strains, none of them producing just Stx2a (Figure 2, Table 1).

Table 1. Stx-producing strains features.

Strain	Serotype	Source	Stx	Subtype	Neutralization Rate (%)
EPM 50	O87:H16	Animal	Stx2	2b	90
EPM 96	O93:H19	Food	Stx2	2a, 2d	86
EPM 82	O112:H21	Animal	Stx2	2c	90
EPM 1	O157:H7	Human	Stx2	2a, 2c	0
EPM 2	O157:H7	Human	Stx2	2a, 2c	97
EPM 94	O157:H7	Animal	Stx2	2c	95
Raph/4	O165:H⁻	Human	Stx2	2a, 2c	0
EPM O3	O172:NM	Animal	Stx2	2a	44
EPM O22	ONT:H16	Animal	Stx2	2b	46
EPM 59	ONT:H16	Animal	Stx2	2d	97
EPM 81	ONT:H38	Animal	Stx2	2a	27
BA 1189	ONT:H49	Human	Stx2	2a, 2d	24
BA 1132	ONT:H49	Human	Stx2	2a, 2c, 2d	7
EPM 79	O22:H16	Animal	Stx1/2	1a, 2c, 2d	53
BA 3003	O48:H7	Human	Stx1/2	1a, 2a	0
EPM O36	O75:H8	Animal	Stx1/2	1c, 2b	0
EPM 4	O93:H19	Human	Stx1/2	1a, 2d	78
EPM 53	O98:H17	Animal	Stx1/2	1a, 2a, 2c	100

Table 1. *Cont.*

Strain	Serotype	Source	Stx	Subtype	Neutralization Rate (%)
EPM 55	O98:H17	Animal	Stx1/2	1a, 2a, 2c	85
EPM 9	O103:H2	Human	Stx1/2	1a, 2c	46
EPM 66	O105:H18	Animal	Stx1/2	1a, 2a, 2b	7
EPM O55	O146:H21	Animal	Stx1/2	1a, 2a, 2b	24
3104-88	O157:H7	Human	Stx1/2	1a, 2a	27
C7-88	O157:H7	Human	Stx1/2	1a, 2NT	15
EDL 933	O157:H7	Food	Stx1/2	1a, 2a	80
EPM 45	O181:H1	Animal	Stx1/2	1a, 2a	80
18 (ICB)	ND	ND	Stx1/2	1NT, 2NT	87
Purified Stx2	-	-	Stx2	2a	84

ND—not determined; NT—not typeable.

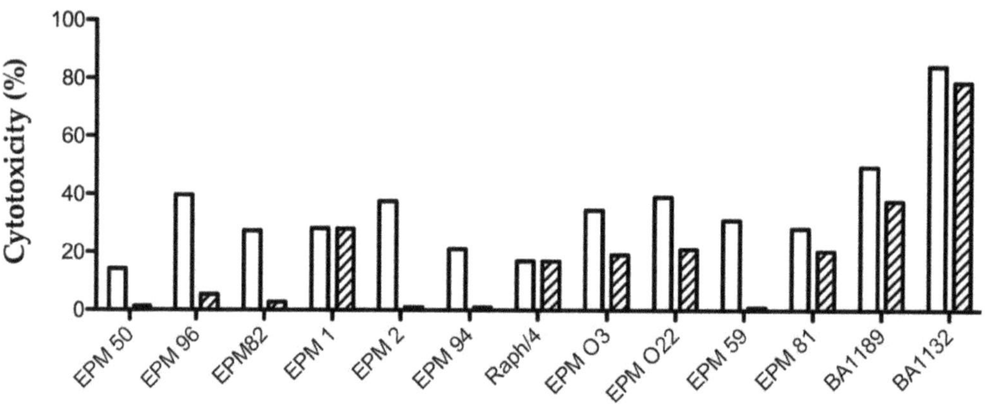

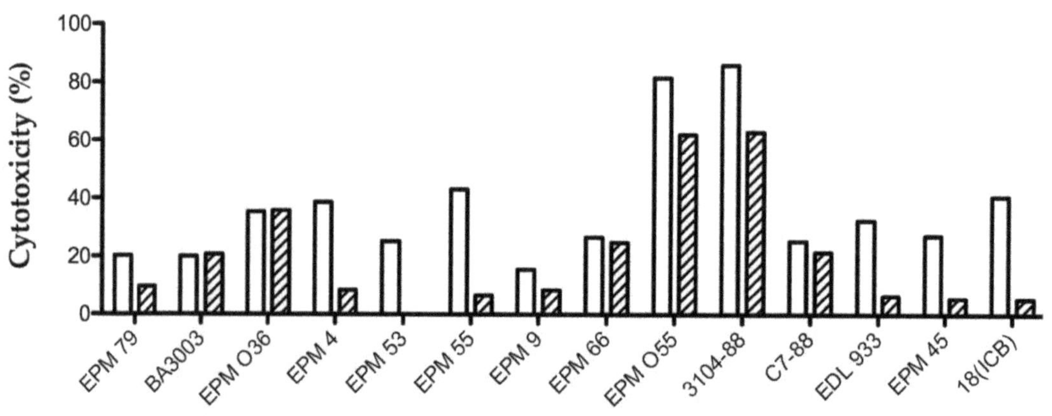

Figure 2. Percentage of cytotoxicity of STEC supernatants from strains bearing *stx2* or *stx1/2* genes; in absence (white bars) and in presence of FabF8:Stx2 (flared bars). Data represents biological duplicates of three independent experiments.

2.3. *FabF8:Stx2 Protects Cell Viability of Human Glomerular Endothelial Cells (HGEC) from Stx2 Effects*

Considering the deleterious effects of Stx2 on the HGEC viability, we evaluate the FabF8:Stx2 ability to neutralize Stx2 cytotoxicity on HGEC. In a dose-dependent manner, the FabF8:Stx2 significantly neutralized the cytotoxic effects caused by 0.5 ng/mL Stx2 in HGEC ($p < 0.05$, $n = 3$) in both tested treatment conditions (pre-incubation or co-incubation). After the Stx2 treatment, the HGEC viability percentage was 41.0 ± 1.2%. The highest protection of HGEC viability was observed with 10 µg/mL of FabF8:Stx2 and no significant differences were found between pre-incubation and co-incubation, since cell viability percentage with 10 µg/mL FabF8:Stx2 was 89.6 ± 5.0% and 81 ± 1.7% for each condition, respectively (Figure 3). To calculate the percentage of FabF8:Stx2 protection, we first calculate the maximum prevention possible to obtain in the HGEC viability by subtracting the percentage of viability after Stx2 treatment to the viability of controls. Then, we calculated the additional % of HGEC viability obtained with FabF8:Stx2 by subtracting the viability % of Stx2 treated cells to pre-incubation and co-incubation. Finally, with these results, the percentage of protection with FabF8:Stx2 at both experimental conditions was calculated with the following formula: [(% of cell viability Pre/Co treated with FabF8:Stx2—% of cell viability in Stx2 treated cells)/(% of control cell viability—% of cell viability in Stx2 treated cells)] × 100. The protection obtained with 10 µg/mL FabF8:Stx2 was 83.0 ± 5.1% at the pre-incubation condition, and 67.5 ± 1.7% at the co-incubation condition, without statistical differences (Table 2).

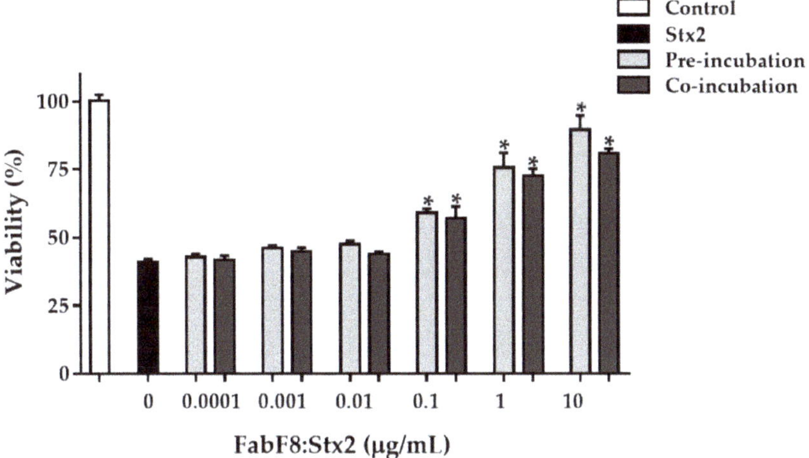

Figure 3. FabF8:Stx2 protects human renal endothelial cells (HGEC) against Stx2 cytotoxicity. HGEC were pre-treated with different concentrations of FabF8:Stx2 (1 h at 37 °C), and Stx2 (0.5 ng/mL) was then added, or cells were co-treated with FabF8:Stx2 (0.0001 µg/mL to 10 µg/mL) and Stx2 (0.5 ng/mL) simultaneously. Results are expressed as means ± SD of three experiments, pre/co-incubation vs. Stx2, * $p < 0.05$.

Table 2. Percentage of FabF8:Stx2 protection against Stx2 in HGEC cells.

FabF8:Stx2 (g/mL)	Stx2 Cytotoxicity Prevention (%)	
	Pre-Incubation	Co-Incubation
0	0	0
0.1	30.5 ± 1.5	27.1 ± 4.2
1	58.5 ± 5.4	53.5 ± 2.6
10	83.0 ± 5.1	67.5 ± 1.7

2.4. FabF8:Stx2 Antibodies Prevent Detachment and Swelling Caused by Stx2 in HGEC

Morphology of HGEC treated with Stx2 in the presence of FabF8:Stx2 antibodies was evaluated. This recombinant antibody fragment significantly prevented HGEC detachment and intracellular edema caused by 0.5 ng/mL Stx2 (Figure 4A). The prevention obtained on cell detachment with pre-incubation and co-incubation conditions was 62.0 ± 4.0 and 45.0 ± 3.0 ($p < 0.05$, $n = 3$), respectively. In addition, cell detachment protection was significantly greater with pre-incubation than co-incubation (Figure 4B). However, when the cell area was analyzed, practically a total protection from swelling was achieved with FabF8:Stx2 with both experimental conditions assayed, pre-incubation: 95.0 ± 6.0% and co-incubation: 90.0 ± 5.5%, $p < 0.05$, $n = 3$ (Figure 4C).

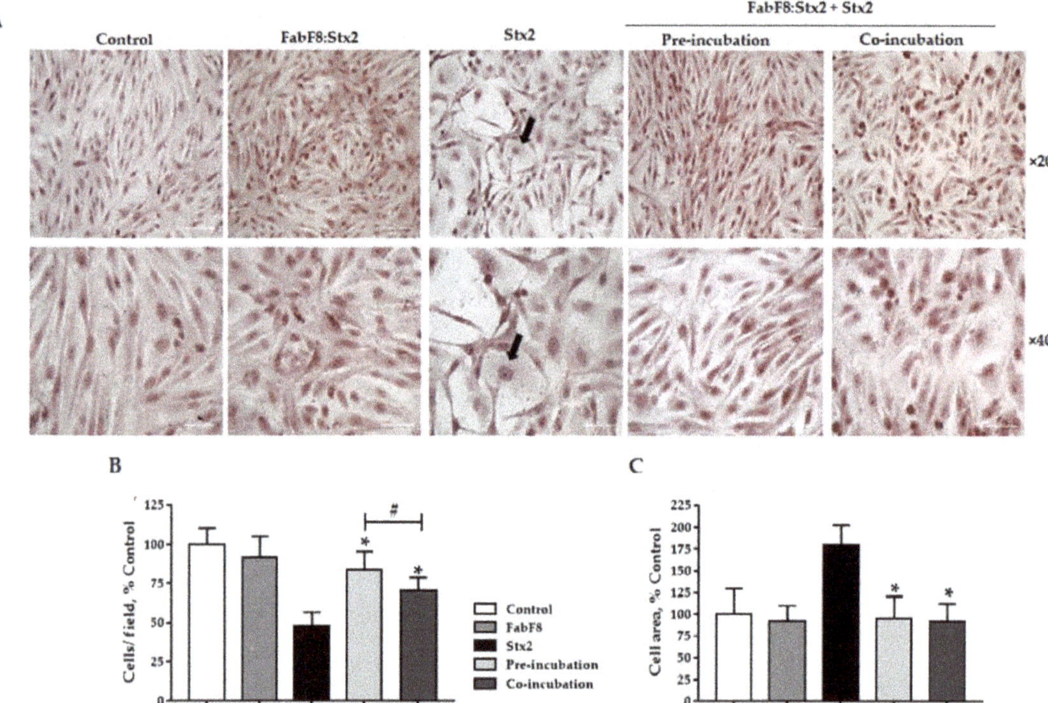

Figure 4. FabF8:Stx2 protects human glomerular endothelial cells (HGEC) from Stx2-induced morphological alterations (**A**) Cell morphology was evaluated and the number of HGEC (**B**) were analyzed by light microscopy (×200 and ×400). HGEC areas (**C**) were measured using Image J software. The black arrows indicate intracellular edema. Results are expressed as means ± SD of three experiments. One hundred percent represents the values of controls. Stx2 vs. Ctrl, * $p < 0.05$. Pre/co-incubation vs. Stx2, # $p < 0.05$.

2.5. FabF8:Stx2 Antibodies Avoid Apoptosis Induced by Stx2 in HGEC

Apoptosis is the principal cell death mechanism triggered by Stx2. We previously showed that this toxin-induced necrosis and apoptosis in HGEC [28]. Following, we evaluated the ability of FabF8:Stx2 antibodies to avoid necrosis and apoptosis by analyzing HGEC stained with acridine orange/ethidium bromide by fluorescence microscopy (Figure 5A). The FabF8:Stx2 (1 µg/mL), significantly decreased the apoptosis caused by 0.5 ng/mL Stx2 in both experimental conditions (pre-incubation: 3.3 ± 0.9% and co-incubation: 11.6 ± 1.4 vs. Stx2: 45.0 ± 2.0%, $p < 0.05$, $n = 3$). Furthermore, FabF8:Stx2 at pre-incubation conditions was more effective than co-incubation to prevent apoptosis (93.0 ± 0.90 % vs. 75.0 ± 1.4%, respectively, $p < 0.05$, $n = 3$) (Figure 5B). No significant differences were found for necrosis (Figure 5C).

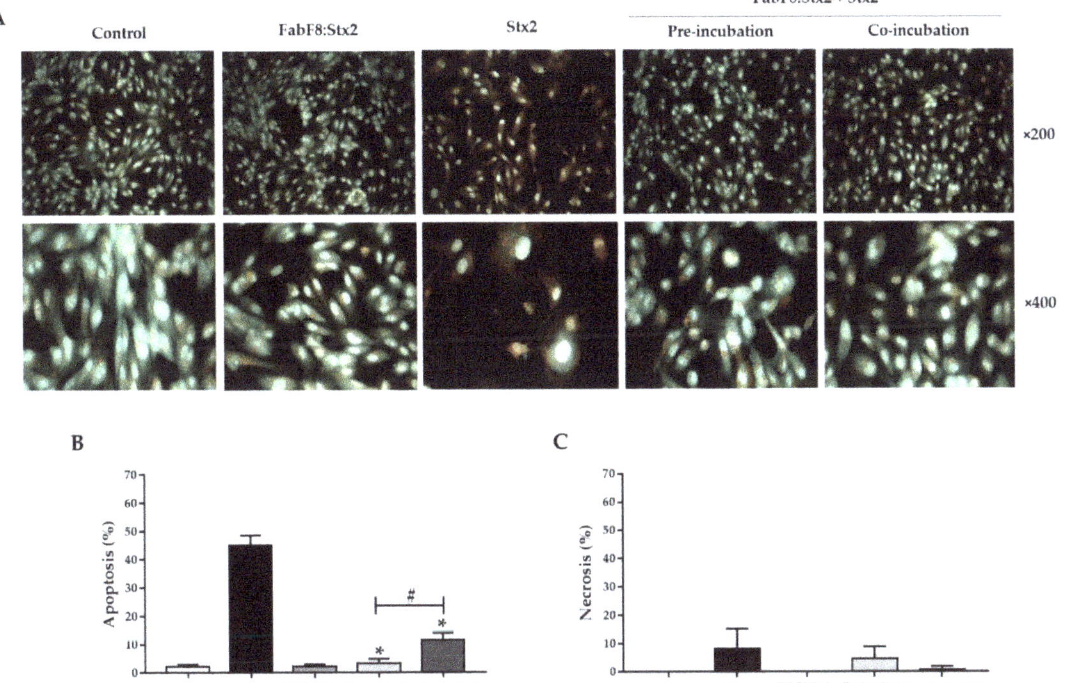

Figure 5. FabF8:Stx2 prevents apoptosis induced by Stx2 in human glomerular endothelial cells (HGEC). The percentage of necrotic and apoptotic cells after 72 h of treatments was established morphologically by fluorescence microscopy after staining with acridine orange/ethidium (×200 and ×400). A representative experiment is shown in panel (**A**). Results are expressed as means ± SD of three experiments. Apoptosis (**B**), * $p < 0.05$ for pre/co-incubation vs. Stx2. # $p < 0.05$ for pre-incubation vs. co-incubation. Necrosis (**C**), ns.

3. Discussion

Shiga toxin (Stx) is central to the development of hemolytic uremic syndrome (HUS). The supportive treatment is the current default procedure for STEC-infected patients, also, the administration of some antibiotic classes could increase the Stx production or release, which could lead to a risk of catastrophic consequences with HUS development, making this treatment option highly controversial [6,10]. Therefore, it is mandatory to develop either an effective treatment or a prevention method for the deleterious effects of Stx intoxication [19]. The STEC prevention is focused on individual and industry levels, such as hygiene procedures, meat processing protocols, and slaughterhouse maintenance, for example. Regarding therapy, it is focused mainly on inpatient supportive care, even though some strategies are in development aiming at different stages of infection, such as bacterial growth control without increasing Stx secretion, toxin trafficking interference and cellular response to the toxin. Moreover, it is worth mentioning the challenges regarding a therapeutic approach against Stx-induced symptoms, especially for clinical trials, such as the low incidence of STEC infections and HUS, the lack of highly specific diagnostic screening, and the narrow therapeutic window (onset of disease 3 days after infection, HUS development one week after the first symptom), which is also hypothetical [29]. Thus, despite great achievements towards a therapeutic tool against Stx, a specific treatment remains elusive.

Specific antibodies against Stx as a tool either to prevent or treat the HUS disease process is a promising approach [20]. Indeed, some other recombinant antibodies have also shown neutralizing ability against Stx2 in vitro or in vivo. Such as, the family Stx2B-binding VHHs which were constructed with one anti-Stx2B VHH, and two copies were fused to one anti-human albumin VHH, neutralizing Stx2 in vitro [24]. Another VHH also protected mice against Stx2 intoxication, but it was not humanized [30]. Concerning scFv, the one described by Maa et al. [31] and Luz et al. [25], neutralizes the cytotoxic ability of Stx2 in vivo and in vitro, respectively, however, none of them is a human antibody, produced in a bacterial system, which impairs their use as therapeutic agents and costs of production.

The recombinant human Fab and F(ab')2 fragments characterized by Akiyoshi et al. [32] showed neutralizing capacity, however, the production was dependable on mammalian cells (CHO), having a high cost for obtaining as with hybridoma technology. Therefore, the library F [27] was employed to select specific Fab high binders against Stx toxins. Two phage clones showed high affinity and binding ability against Stx2. The FabC11:Stx2 was the first to be characterized and showed cross-reactivity with Stx1 besides being able to prevent Stx2 toxicity to human kidney cells and in mice [21,22].

Herein, the other Fab selected against Stx2 (FabF8:Stx2) was characterized and employed in different cell assays. The variety of toxin subtypes that could be expressed by a diverse set of STEC serotypes able to express one or more toxin types at the same time is a major challenge for antibody-based Shiga toxin neutralizers, once to be universally effective, should neutralize multiple Stx1 and Stx2 subtypes [14]. In the present study, using the gold standard Vero cell neutralization assay, we observed that FabF8:Stx2 neutralized the cytotoxicity of 23 of 27 supernatants from Stx2 or Stx1/Stx2-producing STEC strains. These strains belong to different serotypes and present diverse *stx* subtypes, it is worth mentioning that no differences were found with neutralization percentage and its *stx* subtype, even though the FabF8:Stx2 was generated against a Stx2a toxin, some strains harboring *stx2a* gene were not neutralized whereas two non-Stx2a producing were neutralized. This kind of investigation is not commonly employed, usually, most neutralization assays are tested against one type and/or one subtype of the toxin, therefore in this work, for the first time, we showed how one recombinant monoclonal antibody neutralizes different Stx combinations obtained from the STEC isolates culture.

The human microvascular endothelial cells are an excellent cell model for in vitro therapeutic studies once it can express 50-fold higher Gb_3 levels compared to the endothelial cells of large vessels [33]. In this sense, previously, we developed human glomerular endothelial cells (HGEC) primary cultures and demonstrated the decrease of cell viability by apoptosis and endothelial injury like that documented in kidney biopsies of HUS patients after incubation with Stx2 [28]. In this work, additionally, we assayed FabF8:Stx2 antibodies on HGEC exposed to Stx2 and we were able to corroborate their great effectiveness on the protection of Stx2 cytotoxicity on HGEC, in about 80–90% at the pre-incubation condition. These results were coincident with the high capacity of these antibodies to prevent HGEC apoptosis in about 75–90% under pre-incubation and co-incubation conditions. Furthermore, at pre-incubation, cell detachment was avoided in approximately 60–65% and swelling, in about 90–95%.

Previously, we demonstrated that by using 10 μg/mL FabC11:Stx1/Stx2 we observed lower protection of the HGEC viability (54.0% and 52.0%, for pre-incubation and co-incubation, respectively) compared to the same concentration of FabF8:Stx2 when cells were exposed to a 1 CD_{50} of Stx2, therefore, preventing Stx2 toxicity on human kidney cells [26]. Additionally, FabC11:Stx1/Stx2 cell detachment protection was also lower than FabF8:Stx2, which showed protection levels of 43.5% under pre-incubation and 36% under co-incubation conditions. With respect to swelling, although we demonstrated good prevention, pre-incubation: 97.0% and co-incubation: 81.0%, it is noteworthy that this protection had been obtained with a higher concentration of FabC11:Stx1/Stx2 (10 μg/mL) compared with FabF8:Stx2 (1 μg/mL). Our results conclusively demonstrate the efficacy of

FabF8:Stx2 antibodies to avoid the cytotoxic effects of Stx2 on human renal microvascular endothelial cells, one of the principal target cells for this toxin, raising the possibility of the development of a new therapeutic molecule against Stx2 toxicity.

4. Conclusions

The present work showed encouraging results about the effectiveness of FabF8:Stx2 antibodies to neutralize the cytotoxic effects of both purified Stx2 and Stx2 or Stx1/Stx2 produced by STEC strains. Thus, they could be a promising therapeutic strategy to prevent kidney damage and the subsequent development of HUS. Future studies will be focused on analyzing the efficacy of FabF8:Stx2 in in vivo models.

5. Materials and Methods

5.1. Bacterial Strains, Plasmids, and Antigen

The bacterial Phage-resistant *Escherichia coli* Omnimax (Invitrogen) was used for Phage Display assays. For Fab cloning and expression, *Escherichia coli* DH5a (Thermo Fisher Scientific) and BL21(DE3) pLysS (Novagen) were used, respectively. The plasmid vectors used were phagemid HP153 and pFabHis-MBP [25]. The bacterial strains used in this study were strains previously defined as STEC by gene presence and Stx1 or Stx2 production [34], including the prototype EDL933 (O157:H7) [35] (Table 1). The antigens Stx2 and Stx1 were commercially available and acquired from Phoenix Laboratory, Tufts Medical Center, Boston, MA, USA.

5.2. Antibody Generation and Characterization

The FabF8:Stx2 was selected by phage display, using a human synthetic antibody library (library F), which displays on the M13 bacteriophage surface a Fab antibody fragment [27]. Selection and panning were performed as described by Sidhu and Fellouse [36] using the protocol of selection against immobilized antigens. In summary, it was used 5 μg/mL toxin (100 μL/well) in phosphate-buffered saline (PBS) to coat a microplate (Maxisorp, Nunc) in the first round and 2.5 μg/mL toxin (100 μL/well) in PBS for the followed selection rounds. The same was performed with the negative protein control (MPB). The coated plate was incubated at room temperature for 2 h or 18 h at 4 °C, followed by blocking for 1 h with 200 μL/well PBS-0.2% BSA. Next, the phage library solution in PBT buffer (PBS-0.2% BSA-0.05% Tween-20) was added to the negative control wells (100 μL/well) and the plate was incubated at room temperature for 2 h with gentle shaking. The content of the control wells containing the phage library solution was removed and placed into the toxin-coated wells. Next, the non-binded phages were removed by washing them 10 times with PT buffer (PBS-0.05% Tween-20). Toxin-bound phages were eluted by adding 100 μL/well of 100 mM HCl and incubating at room temperature for 5 min. To neutralize the pH of the eluent, it was transferred to a new 1.5-mL microfuge tube containing 1.0 M Tris-HCl, pH 11. Half of the eluted phage solution was added to 10 volumes of actively growing *E. coli* omnimax (OD600 < 1.0) in 2YT/tet medium, which was then incubated at 37 °C for 20 min with shaking at 200 rpm before M13-K07 helper phage were added (10^{10} infectious units (IU)/mL) and the whole culture was incubated at 37 °C with shaking for an additional hour. The culture was then transferred to a 30 mL 2YT/carb/kan medium, and cells grew at 37 °C overnight before phage was harvested for the next round of panning. A serial dilution on LB/carb plates was performed to determine the number of phages eluted and four panning rounds were performed. The phage selected was sequenced and forwarded to cloning and production. The cloning of the Fab expression vector, Fab fragment expression, and purification was performed as previously reported by Luz et al. [25].

FabF8:Stx2 Characterization

Surface Plasmon Resonance—The antibody affinity was determined by surface plasmon resonance (BIAcore T200, Cytiva, Little Chalfont, UK) following the manufacturer's recommendations. The experiments used HBS-EP buffer, pH 7.4, containing 10 mM HEPES, 150 mM NaCl, 3 mM EDTA, and 0.05% Tween 20 as the running buffer. Briefly, Stx2 (purchased from Tufts University School of Medicine, Boston, MA, USA) at 5 μg/mL in 10 mM sodium acetate buffer, pH 5.5 was immobilized (152 RU) on CM5 sensor chips activated by mixing equal amounts of *N*-ethyl-*N'*-(dimethyl aminopropyl) carbodiimide (EDC) and *N*-hydroxysuccinimide (NHS) following the standard immobilization protocol. The sample preparation was in HBS-EP buffer (0–7.4 μM, twofold dilutions) and the kinetic study was performed by a multicycle model at 25 °C and a flow rate of 30 μL/min (contact of 120 s and dissociation of 600 s). The sensor chip was regenerated between cycles by a 15 μL pulse of 100 mM glycine containing 2 mM $MgCl_2$, pH 2. The kinetic affinity constant (KD) was calculated using BIAevaluation version 3.0, using the Langmuir 1:1 binding model. Stx2 monoclonal antibody was employed as a control [36]. The experiments were performed in duplicate.

EC_{50} definition—Half-maximal effective concentration (EC_{50}) was performed as described by Luz et al. [25] by coating a 384-well plate (Maxisorp) with 2 μg/well of antigen and incubating overnight at 4 °C with gentle shaking, followed by blocking step with 0.2% PB buffer for 1 h at room temperature with gentle shaking. A log 3 serial Fab/scFv dilutions, starting with 20 μg/mL, were performed in PBT, and incubated for 30 min at room temperature with gentle shaking. The assay development was performed after 30 min incubation with gentle shaking using HRP antibody/anti-Flag conjugated to peroxidase (1:5000) in PBT followed by addition of TMB (1:1) and stop with 1 M H_3PO_4. Several washes with PBT were performed between each incubation. The plate was read with a 450 nm filter. Specificity and absence of cross-reactivity were performed by ELISA as described by Luz et al. [25] using a 96-well plate (Maxisorp) coated with different concentration Stx2 or Stx1 purified toxins (5 μg/mL) incubated 18 h at 4 °C with gentle shaking, followed by blocking with 0.2% PB buffer for 1 h at room temperature with gentle shaking. The EC_{50} concentration of FabF8:Stx2 was added to the plate and incubated for 30 min at room temperature with gentle shaking, followed by 8 times washing with PT. Next, it was added (100 μL/well) into the wells, HRP antibody/anti-Flag conjugated to peroxidase (1:5000) in PBT, which was then incubated for 30 min at room temperature with gentle shaking. Again, the plate was washed 8 times with PT. The reaction was developed by adding 30 μL/well of TMB (1:1) and stopped by adding 30 μL/well of 1 M H_3PO_4, and the plate was read with a 450 nm filter.

5.3. Vero Cell Antibody Neutralization Assay

The certified Vero cell lineage was purchased from Instituto Adolfo Lutz (São Paulo, SP, Brazil). The STEC bacterial supernatant was obtained as described by Shiga et al. [37]. Vero cells (1×10^5 cells/mL) were grown in 96-well plates in Dulbecco's medium (DMEM) supplemented with 10% FBS and 30 μg/mL gentamicin, at 37 °C in a 5% CO_2 atmosphere, for 24 h. The FabF8:Stx2 neutralizing ability was determined by pre incubating for 2 h the Stx2 toxin at the CD_{50}, (i.e., 0.5 μg/mL) defined by Rocha et al. [38] or bacterial supernatants (diluted 1:50) with the same volume of an EC_{50} concentration of Fab diluted in DMEM supplemented with 2% of FBS at 37 °C for 72 h with 5% CO_2. After incubation, the viable cells were accessed by 3-(4,5-dimethylthiazol-2-yl)-2,5-diphenyltetrazolium bromide (MTT) (Sigma-Aldrich, St Louis, MO, USA) as described by the manufacturer's instructions. These assays were performed three times in duplicate.

5.4. Primary Culture

The human glomerular endothelial cells (HGEC) were obtained as previously described by Amaral et al. [28] from kidneys of human pediatric patients, under proper consent and ethical approval (N°: 035 LUP1S0/19). HGEC were cultivated in M199 media, supplemented with 20% fetal calf serum (FCS), 3.2 mM L-glutamine, 100 U/mL penicillin/streptomycin (GIBCO, Waltham, MA, USA), and 25 μg/mL endothelial cell growth supplement (ECGS, Sigma, St. Louis, MO, USA). All the experiments were performed with

HGEC between 2–7 passages and were previously characterized for positive expression of von Willebrand factor and platelet/endothelial cell adhesion molecule 1 (PECAM-1). Moreover, experiments were carried out at growth-arrested conditions using M199 medium supplemented with 10% FCS without ECGS [28].

5.5. Stx2 Neutralization Assay in HGEC

Neutralization assays were developed according to two procedures: pre-incubation and co-incubation. The pre-incubation was performed by pre-treat HGEC primary cultures with different FabF8:Stx2 concentrations (1 h at 37 °C) before exposure of cells to Stx2 for 72 h. On the other hand, for co-incubation, cells were treated simultaneously with FabF8:Stx2 and Stx2 for 72 h. FabF8:Stx2 concentrations used in the experiments ranged from 0.0001 to 10 μg/mL and Stx2 was assayed at the dilution required to kill 50% of cells (1 CD_{50} = 0.5 ng/mL). The molar ratios used ranged from 0.07:5 to 7000:5 for FabF8:Stx2. Finally, the ability of FabF8:Stx2 to neutralize Stx2 was analyzed by HGEC cell viability as is described below.

5.6. Neutral Red Viability Assay

The neutral red uptake assay was used to analyze the HGEC cell viability as previously described [28]. In summary, HGEC cells were grown with a complete medium, until confluence, in 96-well plates. After 72 h of treatments, freshly diluted neutral red (Sigma Aldrich, St. Louis, MO, USA) was added to cells to a final concentration of 10 mg/mL, followed by an additional incubation of 1 h at 37 °C in 5% CO_2. Then, cells were washed and fixed with 1% $CaCl_2$/1% formaldehyde, followed by lysis with 1% acetic acid in 50% ethanol. Absorbance at 540 nm was measured in an automated plate spectrophotometer. Results were expressed as viability percentage, in which 100% represents cells incubated under identical conditions but without treatment. The 100% of HGEC viability protection was considered when Stx2 cytotoxic effects were totally reversed.

5.7. Cell Morphology Analysis

HGEC cells were grown on gelatinized glass coverslips (12 mm) and treated as it was described above. For cell morphology analysis, FabF8:Stx2 were used at 1 μg/mL. Following, cells were fixed with 96% v/v alcohol for 2 h at room temperature and stained with hematoxylin/eosin (H&E). Subsequently, HGEC were analyzed by light microscopy (×200 and ×400, Zeiss Axiophot, Zeiss, Heidelberg, Germany). The percentage of cells/field was obtained from photographs of 10 randomly selected fields. Cells were then counted and averaged, and the percentage of cells per field was estimated by considering the average number of controls as 100% (percentage of cells/field = (number of treated cells × 100)/number of control cells). Furthermore, the percentage of cell area was calculated from the same photographs. For that, the cell area was analyzed in each cell by using the Image J software (NIH) according to the manual instructions. The cell area average was calculated for each condition and the cell area of controls was considered as 100% (percentage of cell area/field = (cell area of treated cells × 100)/cell area of control cells [28]. Results were expressed as means ± standard deviation of the mean (SD). The percentages of protection from cell detachment and intracellular edema were calculated considering 100% prevention when these alterations were totally reversed.

5.8. Necrosis and Apoptosis Analysis

HGEC were grown on gelatinized glass coverslips (12 mm) and then treated as it was described for item 5.7. After treatments, the percentage of necrotic and apoptotic cells were analyzed morphologically by fluorescence microscopy. For that, cells were stained with acridine orange/ethidium bromide (1:1, v/v) at a final concentration of 100 μg/mL [28]. In the analysis, it was considered that live cells have normal nuclei when presented with green chromatin and organized structures. Apoptotic cells contain fragmented or condensed chromatin (green or orange). Finally, necrotic cells have similar normal nuclei staining as live cells, but with the chromatin in orange instead of green. The percentage of apoptotic and necrotic cells/field was obtained from photographs of 10 randomly selected fields. Cells were then counted, and the percentage of necrotic and apoptotic cells was estimated by considering the total number of cells/field as 100% (percentage of necrotic or apoptotic cells/field = (number of necrotic or apoptotic cells $\times$ 100)/total number of cells). Results were expressed as means $\pm$ standard deviation of the mean (SD). Percentages of prevention from necrosis and apoptosis were calculated by considering 100% of protection when these alterations were totally reversed.

5.9. Data Analysis

The data are presented as mean $\pm$ SD. ANOVA was used to calculate differences between groups and Tukey's multiple comparisons test was used as a posteriori. All Statistical analysis was performed using Graph Pad Prism Software 5.0 (San Diego, CA, USA).

Supplementary Materials: The following are available online at https://www.mdpi.com/article/10.3390/toxins13110825/s1, Figure S1: Binding of Fab:F8 to Stx2 measured by Biacore.

Author Contributions: Conceptualization, D.L., M.M.A. and R.M.F.P.; Formal analysis, D.L., F.D.G., R.L.F., B.S.M., B.E.C.G., W.Q., F.S. and G.C.; Funding acquisition, D.L., C.I., S.S.S., M.M.A. and R.M.F.P., S.S.S.; Investigation, D.L., F.D.G. and F.S.; Methodology, D.L., F.D.G., R.L.F., B.S.M., B.E.C.G., W.Q., A.P., F.S., G.C. and M.M.A.; Project administration, M.M.A. and R.M.F.P.; Resources, D.L., A.M.M., C.I., S.S.S., M.M.A. and R.M.F.P.; Supervision, C.I., S.S.S., M.M.A. and R.M.F.P.; Validation, D.L.; Visualization, F.D.G., R.L.F., B.S.M., B.E.C.G., W.Q., A.M.M., A.P., F.S., C.I., G.C. and S.S.S.; writing—original draft, D.L., M.M.A. and R.M.F.P.; Writing—review and editing, D.L., F.D.G., B.E.C.G., W.Q., F.S., C.I., G.C., S.S.S., M.M.A. and R.M.F.P. All authors have read and agreed to the published version of the manuscript.

Funding: This work was supported by São Paulo Research Foundation (FAPESP–2015/17178-2;—2018/13895-0 to R.M.F.P and 2013/03160-9, 2019/24276-1 to D.L) and by the "National Agency for Promotion of Science and Technology" (grant number ANPCYT-PICT 2017-0617 to María Marta Amaral), the "University of Buenos Aires" (grant number UBACYT-20020170200154BA to María Marta Amaral and UBACYT-20020170100600BA to Cristina Ibarra) and the "National Scientific and Technical Research Council" grant number CONICET: PUE 0041 to María Marta Amaral and Cristina Ibarra). RLF, an undergraduate student was a recipient of a fellowship from FAPESP (2018/24659-5) and Fundação Butantan, and currently, she is the recipient of a fellowship from the Brazilian National Council (PIBIC-CNPq). RMFP and AMM received a fellowship from National Council of Scientific and Technological Development (CNPq 303969/2017-2 and 307045/2020-0, respectively).

Institutional Review Board Statement: The study was conducted according to the guidelines of the Declaration of Helsinki and approved by the Human Ethics Committee of the University of Buenos Aires (18 August 2018) and by the Bioethics Committee of National Hospital "Alejandro Posadas", Buenos Aires, Argentina (n°: 035 LUP1S0/19 (13), 10 June 2019).

Informed Consent Statement: Human glomerular endothelial cells (HGEC) were isolated from kidney fragments removed from normal areas from different pediatric patients with segmental uropathies or tumors in one pole and normal creatinine that were undergoing nephrectomies performed at National Hospital "Alejandro Posadas", Buenos Aires, Argentina. Written informed consent was obtained from the next of kin, caretakers, or guardians on the behalf of the minors/children participants involved in our study.

Data Availability Statement: The data presented in this study are available on request from the corresponding authors. The data are not available in the repository of the Butantan Institute (https://repositorio.butantan.gov.br (accessed on 16 November 2021)).

Conflicts of Interest: The authors declare no conflict of interest.

References

1. Rangel, J.M.; Sparling, P.H.; Crowe, C.; Griffin, P.M.; Swerdlow, D.L. Epidemiology of *Escherichia coli* O157:H7 Outbreaks, United States, 1982–2002. *Emerg. Infect. Dis.* **2005**, *11*, 603–609. [CrossRef] [PubMed]
2. Frank, C.; Werber, D.; Cramer, J.P.; Askar, M.; Faber, M.; an der Heiden, M.; Bernard, H.; Fruth, A.; Prager, R.; Spode, A.; et al. Epidemic Profile of Shiga-Toxin–Producing *Escherichia coli* O104:H4 Outbreak in Germany. *N. Engl. J. Med.* **2011**, *365*, 1771–1780. [CrossRef] [PubMed]
3. Terajima, J.; Iyoda, S.; Ohnishi, M.; Watanabe, H. Shiga Toxin (Verotoxin)-Producing *Escherichia coli* in Japan. *Microbiol. Spectr.* **2014**, *2*, 2. [CrossRef]
4. Ministerio de Salud de la Nación Argentina. Boletín Integrado de Vigilancia. 2021. Available online: https://bancos.salud.gob.ar/recurso/boletin-integrado-de-vigilancia-n560-se-302021 (accessed on 13 October 2021).
5. Pianciola, L.; Rivas, M. Genotypic Features of Clinical and Bovine *Escherichia coli* O157 Strains Isolated in Countries with Different Associated-Disease Incidences. *Microorganisms* **2018**, *6*, 36. [CrossRef]
6. Alconcher, L.F.; Rivas, M.; Lucarelli, L.I.; Galavotti, J.; Rizzo, M. Shiga toxin-producing *Escherichia coli* in household members of children with hemolytic uremic syndrome. *Eur. J. Clin. Microbiol. Infect. Dis.* **2019**, *39*, 427–432. [CrossRef]
7. Repetto, H.A. Epidemic hemolytic-uremic syndrome in children. *Kidney Int.* **1997**, *52*, 1708–1719. [CrossRef] [PubMed]
8. Repetto, H.A.; Rodríguez de Córdoba, E.; Arrizurieta, E.; Rivas, M.; y Ibarra, C. Microangiopatía trombótica y Síndrome Hemolítico Urémico. In *Nefrología Clínica*, 3rd ed.; Editorial Médica Panamericana: Buenos Aires, Argentina, 2014; Chapter 25; pp. 352–363.
9. Spinale, J.M.; Ruebner, R.L.; Copelovitch, L.; Kaplan, B.S. Long-term outcomes of Shiga toxin hemolytic uremic syndrome. *Pediatr. Nephrol.* **2013**, *28*, 2097–2105. [CrossRef] [PubMed]
10. Fuller, C.A.; Pellino, C.A.; Flagler, M.J.; Strasser, J.E.; Weiss, A.A. Shiga toxin subtypes display dramatic differences in potency. *Infect Immun.* **2011**, *79*, 1329–1337. [CrossRef] [PubMed]
11. Kavaliauskiene, S.; Lingelem, A.B.D.; Skotland, T.; Sandvig, K. Protection against Shiga Toxins. *Toxins* **2017**, *9*, 44. [CrossRef]
12. Sandvig, K.; Garred, O.; Prydz, K.; Kozlov, J.V.; Hansen, S.H.; Deurs, B.V. Retrograde transport of endocytosed Shiga toxin to the endoplasmatic reticulum. *Nature* **1992**, *358*, 510–512. [CrossRef]
13. Hall, G.; Kurosawa, S.; Stearns-Kurosawa, D.J. Shiga Toxin Therapeutics: Beyond Neutralization. *Toxins* **2017**, *9*, 291. [CrossRef] [PubMed]
14. Kakoullis, L.; Papachristodoulou, E.; Chra, P.; Panos, G. Shiga toxin-induced haemolytic uraemic syndrome and the role of antibiotics: A global overview. *J. Infect.* **2019**, *79*, 75–94. [CrossRef]
15. Cody, E.M.; Dixon, B.P. Hemolytic Uremic Syndrome. *Pediatr. Clin. North Am.* **2019**, *66*, 235–246. [CrossRef] [PubMed]
16. Melton-Celsa, A.R.; O'Brien, A.D. New Therapeutic Developments against Shiga Toxin-Producing *Escherichia coli*. *Microbiol. Spectr.* **2014**, *2*, 5. [CrossRef]
17. Mühlen, S.; Dersch, P. Treatment Strategies for Infections with Shiga Toxin-Producing *Escherichia coli*. *Front. Cell. Infect. Microbiol.* **2020**, *10*, 169. [CrossRef]
18. Dowling, T.C.; Chavaillaz, P.A.; Young, D.G.; Melton-Celsa, A.; O'Brien, A.; Thuning-Roberson, C.; Edelman, R.; Tacket, C.O. Phase 1 safety and pharmacokinetic study of chimeric murine-human monoclonal antibody c alpha Stx2 administered intravenously to healthy adult volunteers. *Antimicrob. Agents Chemother.* **2005**, *49*, 1808–1812. [CrossRef]
19. Bitzan, M.; Poole, R.; Mehran, M.; Sicard, E.; Brockus, C.; Thuning-Roberson, C.; Rivière, M. Safety and Pharmacokinetics of Chimeric Anti-Shiga Toxin 1 and Anti-Shiga Toxin 2 Monoclonal Antibodies in Healthy Volunteers. *Antimicrob. Agents Chemother.* **2009**, *53*, 3081–3087. [CrossRef]
20. López, E.L.; Contrini, M.M.; Glatstein, E.; Ayala, S.G.; Santoro, R.; Allende, D.; Ezcurra, G.; Teplitz, E.; Koyama, T.; Matsumoto, Y.; et al. Safety and pharmacokinetics of urtoxazumab, a humanized monoclonal antibody, against Shiga-like toxin 2 in healthy adults and in pediatric patients infected with Shiga-like toxin-producing *Escherichia coli*. *Antimicrob. Agents Chemother.* **2010**, *54*, 239–243. [CrossRef]
21. Taylor, C.M.; Bitzan, M.; Reymond, D. A Phase II study assessing monoclonal antibodies against Shiga toxin 1 and 2 in Shiga toxin-producing *E. coli*-infected children. Shigatec, Boston, USA, October 21, 2011. *Pediatr. Nephrol.* **2011**, *26*, 1595–1596.
22. Holliger, P.; Hudson, P.J. Engineered antibody fragments and the rise of single domains. *Nat. Biotechnol.* **2005**, *23*, 1126–1136. [CrossRef]
23. Nilvebrant, J.; Sidhu, S.S. Construction of Synthetic Antibody Phage-Display Libraries. In *Methods in Molecular Biology*; Springer: Berlin/Heidelberg, Germany, 2018; Volume 1701, pp. 45–60.
24. Mejías, M.P.; Hiriart, Y.; Lauché, C.; Fernández-Brando, R.J.; Pardo, R.; Bruballa, A.; Ramos, M.V.; Goldbaum, F.A.; Palermo, M.S.; Zylberman, V. Development of camelid single chain antibodies against Shiga toxin type 2 (Stx2) with therapeutic potential against Hemolytic Uremic Syndrome (HUS). *Sci. Rep.* **2016**, *6*, 24913. [CrossRef]
25. Luz, D.; Chen, G.; Maranhão, A.; Rocha, L.B.; Sidhu, S.; Piazza, R.M.F. Development and Characterization of Recombinant Antibody Fragments That Recognize and Neutralize In Vitro Stx2 Toxin from Shiga Toxin-Producing *Escherichia coli*. *PLoS ONE* **2015**, *10*, e0120481. [CrossRef]
26. Luz, D.; Amaral, M.M.; Sacerdoti, F.; Bernal, A.M.; Quintilio, W.; Moro, A.M.; Palermo, M.S.; Ibarra, C.; Piazza, R.M.F. Human Recombinant Fab Fragment Neutralizes Shiga Toxin Type 2 Cytotoxic Effects in vitro and in vivo. *Toxins* **2018**, *10*, 508. [CrossRef]

27. Persson, H.; Ye, W.; Wernimont, A.; Adams, J.J.; Koide, A.; Koide, S.; Lam, R.; Sidhu, S.S. CDR-H3 diversity is not required for antigen recognition by synthetic antibodies. *J. Mol. Biol.* **2012**, *425*, 803–811. [CrossRef]
28. Amaral, M.M.; Sacerdoti, F.; Jancic, C.; Repetto, H.A.; Paton, A.W.; Paton, J.C.; Ibarra, C. Action of Shiga Toxin Type-2 and Subtilase Cytotoxin on Human Microvascular Endothelial Cells. *PLoS ONE* **2013**, *8*, e70431. [CrossRef]
29. Bitzan, M. Treatment options for HUS secondary to *Escherichia coli* O157:H7. *Kidney Int.* **2009**, *75*, S62–S66. [CrossRef] [PubMed]
30. Tremblay, J.M.; Mukherjee, J.; Leysath, C.E.; Debatis, M.; Ofori, K.; Baldwin, K.; Boucher, C.; Peters, R.; Beamer, G.; Sheoran, A.; et al. A Single VHH-Based Toxin-Neutralizing Agent and an Effector Antibody Protect Mice against Challenge with Shiga Toxins 1 and 2. *Infect. Immun.* **2013**, *81*, 4592–4603. [CrossRef] [PubMed]
31. Ma, Y.; Mao, X.; Li, J.; Li, H.; Feng, Y.; Chen, H.; Luo, P.; Gu, J.; Yu, S.; Zeng, H.; et al. Engineering an anti-Stx2 antibody to control severe infections of EHEC O157:H7. *Immunol. Lett.* **2008**, *121*, 110–115. [CrossRef] [PubMed]
32. Akiyoshi, D.E.; Sheoran, A.S.; Rich, C.M.; Richard, L.; Chapman Bonofiglio, S.; Tzipori, S. Evaluation of Fab and F(ab′) 2 Fragments and Isotype Variants of a Recombinant Human Monoclonal Antibody against Shiga Toxin 2. *Infect. Immun.* **2010**, *78*, 1376–1382. [CrossRef]
33. Obrig, T.G. Escherichia coli Shiga Toxin Mechanisms of Action in Renal Disease. *Toxins* **2010**, *2*, 2769–2794. [CrossRef]
34. Silva, M.A.; Santos, A.R.R.; Rocha, L.B.; Caetano, B.A.; Mitsunari, T.; Santos, L.I.; Polatto, J.M.; Horton, D.S.P.Q.; Guth, B.E.C.; Dos Santos, L.F.; et al. Development and Validation of Shiga Toxin-Producing *Escherichia coli* Immunodiagnostic Assay. *Microorganisms* **2019**, *7*, 276. [CrossRef]
35. Perna, N.T.; Plunkett, G., 3rd; Burland, V.; Mau, B.; Glasner, J.D.; Rose, D.J.; Mayhew, G.F.; Evans, P.S.; Gregor, J.; Kirkpatrick, H.A.; et al. Genome sequence of enterohaemorrhagic *Escherichia coli* O157:H7. *Nature* **2001**, *409*, 529–533. [CrossRef] [PubMed]
36. Sidhu, S.S.; Fellouse, F.A. Synthetic therapeutic antibodies. *Nat. Chem. Biol.* **2006**, *2*, 682–688. [CrossRef]
37. Shiga, E.; Guth, B.; Piazza, R.; Luz, D. Comparative analysis of rapid agglutination latex test using single-chain antibody fragments (scFv) versus the gold standard Vero cell assay for Shiga toxin (Stx) detection. *J. Microbiol. Methods* **2020**, *175*, 105965. [CrossRef] [PubMed]
38. Rocha, L.B.; Luz, D.; Moraes, C.T.P.; Caravelli, A.; Fernandes, I.; Guth, B.E.C.; Horton, D.S.P.Q.; Piazza, R.M.F. Interaction between Shiga Toxin and Monoclonal Antibodies: Binding Characteristics and in Vitro Neutralizing Abilities. *Toxins* **2012**, *4*, 729–747. [CrossRef] [PubMed]

Article

Individual Variability in *Bothrops atrox* Snakes Collected from Different Habitats in the Brazilian Amazon: New Findings on Venom Composition and Functionality

Leijane F. Sousa [1], Matthew L. Holding [2], Tiago H. M. Del-Rei [1], Marisa M. T. Rocha [3], Rosa H. V. Mourão [4], Hipócrates M. Chalkidis [5], Benedito Prezoto [6], H. Lisle Gibbs [7] and Ana M. Moura-da-Silva [1,*]

1 Laboratório de Imunopatologia, Instituto Butantan, Av. Vital Brazil, 1500, São Paulo 05503-900, SP, Brazil; leijianes@gmail.com (L.F.S.); tiago.moretto@esib.butantan.gov.br (T.H.M.D.-R.)
2 Life Sciences Institute, University of Michigan, Ann Arbor, MI 48104, USA; matthewholding28@gmail.com
3 Laboratório de Herpetologia, Instituto Butantan, Av. Vital Brazil, 1500, São Paulo 05503-900, SP, Brazil; marisa.rocha@butantan.gov.br
4 Laboratório de Bioprospecção e Biologia Experimental, Universidade Federal do Oeste do Pará—UFOPA, Rua Vera Paz, s/n, Santarém 68035-110, PA, Brazil; mouraorhv@yahoo.com
5 Laboratório de Pesquisas Zoológicas, Unama Centro Universitário da Amazônia, Santarém 68010-200, PA, Brazil; chalkidis@hotmail.com
6 Laboratório de Farmacologia, Instituto Butantan, Av. Vital Brazil, 1500, São Paulo 05503-900, SP, Brazil; benedito.prezoto@butantan.gov.br
7 Department of Evolution, Ecology, and Organismal Biology, Ohio State University, Columbus, OH 43210-1242, USA; gibbs.128@osu.edu
* Correspondence: ana.moura@butantan.gov.br; Tel.: +55-11-2627-9779

Citation: Sousa, L.F.; Holding, M.L.; Del-Rei, T.H.M.; Rocha, M.M.T.; Mourão, R.H.V.; Chalkidis, H.M.; Prezoto, B.; Gibbs, H.L.; Moura-da-Silva, A.M. Individual Variability in *Bothrops atrox* Snakes Collected from Different Habitats in the Brazilian Amazon: New Findings on Venom Composition and Functionality. *Toxins* **2021**, *13*, 814. https://doi.org/10.3390/toxins13110814

Received: 13 September 2021
Accepted: 10 November 2021
Published: 18 November 2021

Abstract: Differences in snake venom composition occur across all taxonomic levels and it has been argued that this variation represents an adaptation that has evolved to facilitate the capture and digestion of prey and evasion of predators. *Bothrops atrox* is a terrestrial pitviper that is distributed across the Amazon region, where it occupies different habitats. Using statistical analyses and functional assays that incorporate individual variation, we analyzed the individual venom variability in *B. atrox* snakes from four different habitats (forest, pasture, degraded area, and floodplain) in and around the Amazon River in Brazil. We observed venom differentiation between spatially distinct *B. atrox* individuals from the different habitats, with venom variation due to both common (high abundance) and rare (low abundance) proteins. Moreover, differences in the composition of the venoms resulted in individual variability in functionality and heterogeneity in the lethality to mammals and birds, particularly among the floodplain snakes. Taken together, the data obtained from individual venoms of *B. atrox* snakes, captured in different habitats from the Brazilian Amazon, support the hypothesis that the differential distribution of protein isoforms results in functional distinctiveness and the ability of snakes with different venoms to have variable toxic effects on different prey.

Keywords: *Bothrops atrox*; individual variability; venom heterogeneity

Key Contribution: This study provides new and important insights about functional differences in venom, which could influence the expression of the snake venom phenotypes in specimens inhabiting different habitats, including nearby geographic regions.

1. Introduction

Snake venoms are complex mixtures of toxic components belonging to multipleprotein families [1], each of which expresses several isoforms that are present in the venoms in different proportions [2]. The concentration of each isoform present in the venoms is highly variable and modulates venom function [3] and, as a consequence, the complexity of the venoms is associated with differences in their toxicity to a wide range of prey [4]. In addition, differences in venom composition are argued to be an adaptation that has evolved to facilitate the capture and digestion of prey and evasion of predators and is observed across all taxonomic levels, particularly in species with wide distributions [5,6].

Bothrops atrox (common lancehead) is a terrestrial pitviper widely distributed across the Amazon, from tropical lowlands to the rainforest of northern South America east of the Andes [7], where it occupies different types of habitats. Variability is widely reported in the composition of *B. atrox* venom and has been associated with ontogeny [8–10], geographical distribution [11–14], and environmental characteristics [15]. This species is responsible for most of the human envenomations in the Amazon region [16]. In envenomed patients, the intraspecific variability in *B. atrox* venom composition may hamper the patients' prognosis as venom isoforms are involved in distinct clinical symptoms [17]. In some cases, functionally relevant isoforms present in high levels in venoms of a particular group of snakes may show lower reactivity with antivenoms and reduce the effectiveness of the treatment of some victims of snakebite [15,18].

Genetic differences among populations may correspond to the expression of different venom isoforms with distinct or similar functions [2,14]. However, venom variability also occurs among specimens from the same geographical areas [2] and may occur during the life span of individuals as ontogenetic variation related to an increase in body size, allowing capture and digestion of larger prey [19,20].

Large rivers in the Amazon basin contain ecologically diverse habitats and these have been associated with diversification in specific groups of vertebrates, such as birds [21]. In the west of Pará State, Brazil, the Amazon River is wide and has distinct habitats, such as upland forests on either bank and floodplain habitats. These represent distinct habitats that could generate diversification in venom phenotypes. For the last few years, our studies have been focusing on the variability in venom composition in this particular region of the Brazilian Amazon. Snakes have been collected at forest, pasture, and floodplain areas on both banks of the Amazon river. Using genome-scale RADseq data, we showed an interesting pattern of gene dispersal, suggesting a role for the Amazon River as a driver of in situ divergence both by impeding (but not preventing) gene flow and through parapatric differentiation along an ecological gradient [22]. The transcriptomes of the venom glands from the snakes collected in the northern or southern banks of the river denoted the same pattern of transcripts regarding the major toxin groups, but with the expression of different alleles or paralogs in snakes from the northern or the southern banks [2]. Using the comprehensive transcriptomic annotations described above, we compared the phenotype of pooled venoms from *B. atrox* snakes collected forest, pasture, or floodplain habitats by comparing the major toxic activities with data from free-label proteomics of the whole pools of venoms and the identification of relevant isoforms separated from these pools by RP-HPLC chromatographies, also by the proteomics of each fraction [15]. We observed two predominant phenotypes: pooled venoms from the forest, pasture, and degraded areas were more hemorrhagic, while the venom pool from snakes collected at the floodplain was more procoagulant. However, these analyses used pooled venom samples and so the level and significance of venom variation at the level of the individual snake are not understood. This information could help clarify the relative importance of habitat compared to individual variation as modulators of the venom variability observed in our previous report [15].

Here, using individual venom samples extracted from the snakes collected at the same habitats, and the proteomics information obtained previously, we investigated the hypothesis that the heterogeneity in venom composition could be higher in more unsettled environments. We found that the variability in the composition and functional activities is higher within individual venoms from snakes collected at the floodplain habitat, an extremely dynamic environment subjected to drastic seasonal changes, thus supporting our initial assumption. Moreover, our findings provide a deeper view of the main toxins and biological activities related to the individual venom variability within these *B. atrox* groups and suggest that the functional diversity of the venoms appears to be relevant to the ability of these snakes to persist in highly variable unsettled environments, such as the floodplain habitat.

2. Results

2.1. Individual Variation of Venom Composition Is Associated with Habitats

Chromatographic analyses by RP-HPLC were our first approach to investigate the individual variability of *B. atrox* venoms among the specimens collected in each habitat: forest, pasture, degraded area, and floodplain. Overall, similar profiles can be seen in all chromatograms, although the relative heights or areas of peaks varied greatly between venoms even among snakes collected at the same habitat (Figure 1 and Supplementary Figure S1, for individual chromatograms). In all venom chromatographies, the highest peaks were eluted after 80 min, while peaks eluted between 50 and 70 min showed the most variable area percentages, being higher in floodplain venoms and lower in pasture venoms. According to our previous study [15], Snake Venom Metalloproteinases (SVMPs) are the main components eluted after 80 min, while Phospholipases A_2 (PLA_2s), C-type Lectin-like proteins (CTLs), and Snake Venom Serine proteinases (SVSPs) are the prevalent protein families in the fractions eluted in the intermediate steps (50–70 min).

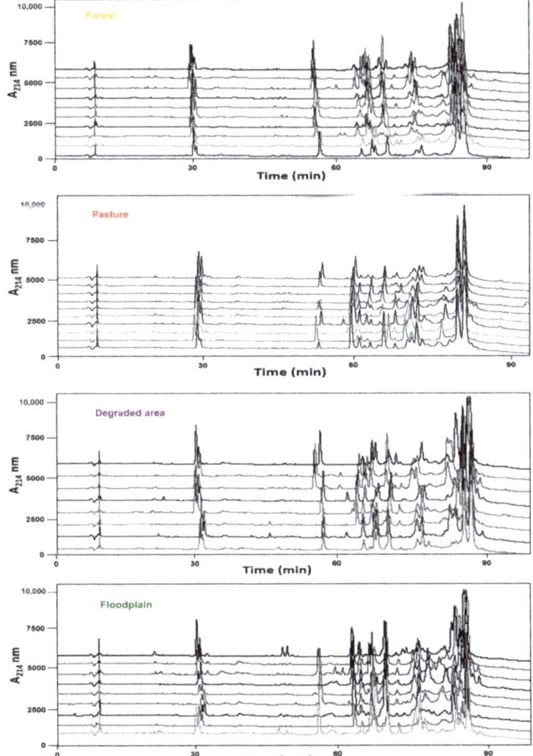

Figure 1. Chromatographic profiles of *B. atrox* venoms from different habitats west of Pará State, Brazilian Amazon. Individual venom samples (5 mg) of *B. atrox* snakes collected at the forest, pasture, degraded area, or floodplain were applied to a Vydac C-18 column. Mobile phases used were 0.1% TFA in water (solution A) or 0.1% TFA in acetonitrile (solution B). Proteins were gradient-eluted at 2 mL/min (5% B for 5 min, 5–15% B over 10 min, 15–45% B over 60 min, 45–70% B over 10 min, 70–100% over 5 min, and 100% B over 10 min). Separation was monitored at 214 nm.

2.2. Population-Level Differentiation between Spatially Disparate *B. atrox* Groups

First, we identified the population-level differentiation among venoms of snakes collected at the different habitats, based on the individual venom composition of each group of snakes. The PERMANOVA analysis (Table 1) showed that the environment explained 22.5% of the total variation in venom peak abundances in the sampled snakes (df = 3.33, F = 3.2, $p < 0.0001$). Further exploration of these differences with compositionally robust PCA (Figure 2, score plots) showed that PC1 largely differentiates floodplain from pasture venoms. Forest and degraded venoms occupying an intermediate space on the PC1 axis. Components with the most prominent loadings are fraction 10 and the fractions eluted after 85 min. The highest negative values of PC1, responsible for clustering the pasture venoms, were fraction 23 (PC1 = −0.7308, PC2 = −0.6035) and fraction 21 (PC1 = −0.5028, PC2 = −0.6410). These fractions were characterized in our previous study [15] as PIII-class and PI-class SVMP isoforms, respectively. The highest positive values of PC1, responsible for clustering the floodplain venoms, were fraction 20 (PC1 = 0.2958, PC2 = −0.3732), fraction 24 (PC1 = 0.1241, PC2 = 0.2006), and fraction 10 (PC1 = 0.2284, PC2 = −0.0353), which in our previous study have been characterized as different isoforms of the PIII-class and PI-class SVMPs, with minor proportions of CTL isoforms, and an SVSP, respectively (Figure 2, loading plots).

Table 1. Posthoc pairwise PERMANOVA analyses were used to test for significant venom compositional differentiation among all pairs of populations.

	Forest	Pasture	Floodplain	Degraded Area
Forest		<0.0001	<0.0001	0.34
Pasture	19.4		<0.0001	<0.0001
Floodplain	16.0	23.2		0.01
Degraded area	6.4	17.4	12.2	

Values shown are *p*-values (above diagonal) and R2 values (below the diagonal) for each comparison.

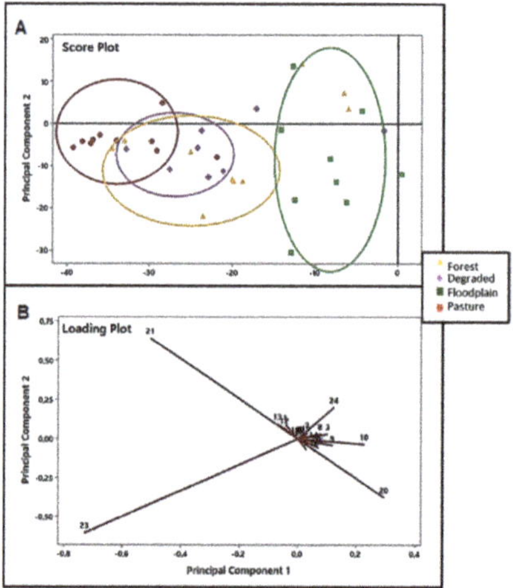

Figure 2. Principal component analysis based on peak areas of *B. atrox* individual venoms according to their chromatographic profiles by RP-HPLC, using a C-18 column. Score (**A**) and loading (**B**) plots in 2D graphs of the principal components axes (PC1 = 36% and PC2 = 22.5%) of venoms from *B. atrox* snakes captured in the forest (▲), pasture (●), floodplain (■), and degraded (◆) habitats.

In addition, posthoc comparisons of individual population pairs using PERMANOVA showed that the spatially proximate degraded and forest habitats were the only locations between which venom composition did not significantly differ. All other population pairs differed significantly, with the greatest degree of differentiation between the spatially disparate floodplain and pasture populations.

Next, peak-by-peak comparisons showed that several chromatographic peaks contributed to population-level differentiation among venoms. After FDR-correction, nine peaks showed significant variation: Peaks 1, 3, 5, 7, 8, 12, 20, 21, and 23 (Figure 3).

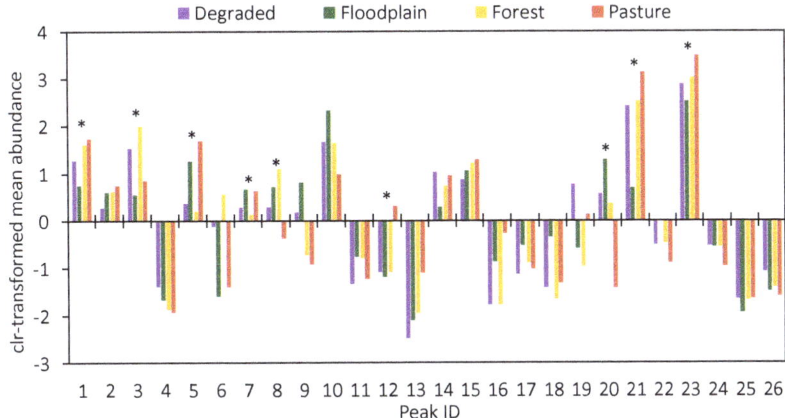

Figure 3. Variation in the chromatographic fractions in *B. atrox* snakes from different habitats. The centered log-ratio mean abundance for each reversed-phase high-performance liquid chromatography peak was plotted. Negative numbers correspond to low-abundance peaks, whereas positive numbers correspond to high abundance peaks. (*) Asterisks indicate peaks that show significant population-level variation. X-axis labels correspond to RP-HPLC peak numbers (1 to 26).

In terms of comparisons between habitats, venoms from the forest or recently degraded area were similar in relative peak abundances. Pasture venom was distinct with a higher abundance of Peaks 5 and 7, which correspond to the acidic PLA_2s, Peak 12, with CTL as the major toxin, and Peaks 21 and 23, which include the SVMPs. Significantly lower abundances were observed in pasture venoms for Peaks 8 and 20, which have as major toxins PLA_2s and PI-SVMPs, respectively. Peak 20 was practically absent in pasture venoms (Figure 1). Floodplain venoms were distinct in peaks related to SVMPs and PLA_2s: Peaks 1, 21, and 23, which contain SVMPs and disintegrins, were present at a lower abundance, while Peak 20, which contains mostly PI-SVMPs, is at a comparatively higher abundance. In peaks containing PLA_2s, Peak 3, which contains K-49 basic PLA_2s, is less abundant, while Peak 7, which contains acidic PLA_2s, and Peak 8, rich in D-49 basic PLA_2s, were proportionally higher in the venoms from the floodplain.

2.3. Venom Differentiation Is Not Limited to Rare (Low Abundance) Proteins

We also investigated whether the venom variability in *B. atrox* groups would be associated with the degree to which the venom proteins were expressed in the different habitats. RP-HPLC peaks for each habitat were classified into two sets: low or high abundance, based on the clr-transformed mean of each peak and a nonparametric MANOVA analysis, as previously described [23], where rare proteins = the mean of an individual peak < the geometric mean, and abundant proteins = mean of an individual peak > the geometric mean (Figure 4). Proteins showing deviations off the middle line are those in which the expression levels are different between the compared populations whereas those that are on the line are similar.

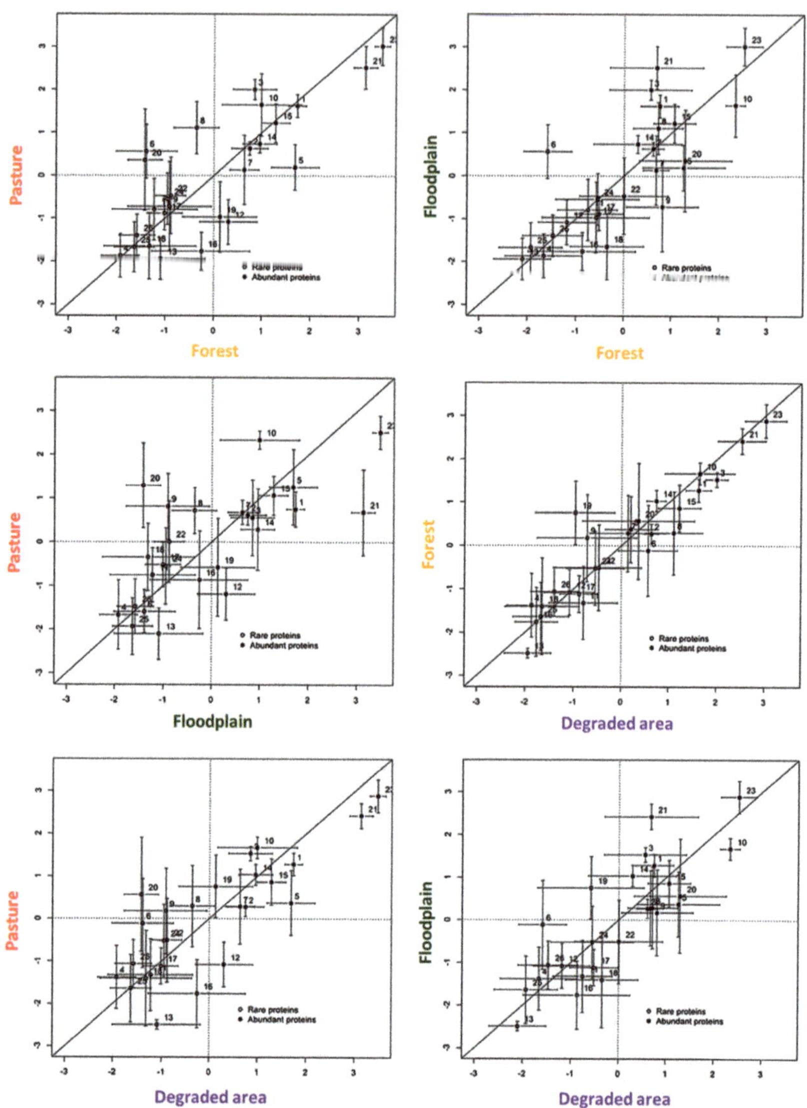

Figure 4. Comparisons of the degree of differentiation in low vs. high abundance proteins based on chromatographic peak areas from *B. atrox* individual venoms. The clr mean was plotted for each RP-HPLC peak across the different habitat (x-axis) and (y-axis) populations for *B. atrox* snakes. Bars indicate the SE, a solid line indicates a perfect agreement, dashed lines indicate the origin (i.e., the geometric mean), and proteins less than these values were considered low-expression proteins.

In our PERMANOVA analyses, only in the spatially proximate degraded and forest habitats the venom composition did not significantly differ (Table 1). Accordingly, minor differences in both rare and abundant proteins were observed between these populations, which showed only slight differences in the rare proteins. However, in the populations from the other habitats, the venom differentiation was not constrained to rare proteins. On the contrary, in some cases, these proteins showed more similarity across populations from different habitats (for example floodplain vs. forest; pasture vs. forest) than proteins classified as abundant. In fact, except for peak 23 (rich in PIII-class SVMPs), abundant proteins showed differences across all habitats. Interestingly, the abundant proteins would be closely related to the functional activity of the venom toward their prey. As shown before [15], the conserved fraction 23 contains Batroxrhagin, a multifunctional PIII-class SVMP extremely conserved in different samples of *B. atrox* venom [2]. However, in Figure 4, we show that other abundant toxins are differentially expressed between the snakes from different habitats: fraction 21, which predominantly contains a PI-class SVMP, had lower expression in floodplain venoms; fraction 10, which contains mostly SVSPs, had higher expression in floodplain venoms; and fraction 20, which predominantly contains a PI-class SVMP and CTLs, was relatively over-expressed in venoms from pasture compared to the other areas.

2.4. Specific-Level Differentiation among Venoms of Snakes Collected in the Same Environment

Once the variability among the groups was defined, we quantified intrapopulation variability in venom composition, among specimens collected in close proximity in the same areas. As shown in Figure 1, HPLC chromatograms show variable profiles within venoms from snakes collected in the same environment (for more details see Supplementary Figure S1). Consistent with this pattern, we found great heterogeneity in the percentage area of each peak in the groups (Table 2).

In every habitat, individual venoms showed important differences in both their percent area and the presence/absence of some peaks. Venoms from pasture and floodplain snakes were again the most distinct among the individuals from the same group (Table 2; Supplementary Figure S1), with venoms from the floodplain being the most heterogeneous. From Table 2 we emphasize three regions: In the PLA_2-eluting fractions (Peaks 3 and 5), Peak 3 was absent in several chromatograms, but its absence is offset by the homogeneous increase of Peak 5 in pasture specimens. Venoms from floodplain specimens showed the most heterogeneous distribution of both PLA_2 fractions. CTL fractions (Peaks 15–17) are expressed at low but homogeneous levels in pasture venoms with increased variation in venoms from the forest and degraded area, reaching the highest heterogeneity in venoms from the floodplain. A notable observation is the high level of variation in the expression of the peaks containing SVMPs (peaks from 21 to 23) in floodplain specimens. These are the most abundant and homogeneous fractions in venoms from the other areas, but in venoms from the floodplain, fraction 21 is very low or even absent in venoms of five specimens while fraction 22, almost not present in venoms from other areas, is detectable in high levels in venoms from four specimens from the floodplain, indicating a high variety of SVMP isoforms venoms from snakes in this environment. It is important to note that the higher variability observed in venoms from floodplain snakes is not due to the two distant floodplain spots of snake collection. The differences appointed here can be observed within the snakes collected at Santarém (ATXV 5, 7, 8, 9, and 16) or Oriximiná (ATXV 10, 11, 12, and 13), which are approximately 300 km apart.

2.5. Differences in the Composition of Individual Venoms Resulting in Functional Variability

The most variable components among the venoms included SVMPs, SVSPs, and PLA_2s. Our next step was to evaluate some of the main biological activities related to these protein families. With the goal of reducing the number of experimental animals for toxicity tests for ethical reasons and due to the limited amounts of some individual venom samples, venoms from only 16 specimens were evaluated per functional test, comprising four from each habitat.

Table 2. Comparison of the chromatographic peak areas (%) from *B. atrox* individual venoms according to the habitat.

Area	Snake	Peaks (%Area)																									
		1	2	3	4	5	6	7	8	9	10	11	12	13	14	15	16	17	18	19	20	21	22	23	24	25	26
Forest	F 24	6.15	1.72	7.11	0.00	0.00	0.00	0.20	4.81	1.68	5.38	0.36	0.00	0.78	1.44	6.30	0.00	0.92	0.00	0.24	0.00	23.64	0.00	32.79	0.00	0.00	0.00
	F 25	4.54	2.15	7.84	0.00	0.00	0.00	0.59	5.04	3.83	9.77	0.64	0.00	0.00	4.40	3.47	0.00	0.59	0.00	1.97	10.29	12.28	0.00	29.10	0.00	0.00	0.70
	F 26	4.68	2.19	8.12	1.02	0.00	0.00	9.06	1.24	6.49	3.98	1.20	0.00	0.00	4.53	7.61	0.31	0.75	0.00	0.30	9.38	19.40	0.00	5.57	7.04	0.45	1.17
	F 28	5.20	2.89	6.02	0.00	0.00	0.00	0.60	3.15	2.93	8.83	0.30	0.00	1.79	6.51	0.00	0.00	0.39	0.28	0.69	16.06	13.16	0.00	27.72	0.00	0.22	0.00
	F 29	5.65	2.11	12.00	0.00	0.00	0.00	0.53	5.62	9.84	8.93	1.04	0.00	0.00	2.38	6.73	0.68	0.43	0.22	2.25	0.13	12.96	10.32	8.23	3.88	0.92	0.00
	F 30	5.96	2.70	9.88	0.00	0.00	0.00	0.28	4.70	3.55	6.26	0.33	0.00	1.76	1.19	0.79	0.00	0.67	0.16	2.05	14.50	12.61	0.00	28.62	0.00	0.00	0.00
	F 31	6.63	1.81	6.69	0.00	0.00	0.00	0.54	3.78	0.94	9.28	0.63	0.00	1.29	4.85	0.00	0.00	0.72	1.56	0.72	0.00	23.17	0.00	35.78	0.00	0.00	0.00
	F 33	6.14	1.65	9.98	0.00	0.97	0.91	5.08	0.24	7.66	0.00	0.80	0.00	0.00	0.00	3.48	1.18	0.00	0.71	0.00	1.75	1.12	14.45	36.50	0.00	0.00	0.79
	F 34	3.65	2.18	4.78	0.00	0.00	0.00	0.68	5.03	6.06	12.76	1.45	2.07	2.91	0.26	0.00	0.00	0.00	1.19	0.00	2.03	17.28	0.00	28.86	0.00	1.80	0.41
	F 35	10.21	1.96	12.83	0.00	0.00	0.00	0.40	6.78	3.14	11.52	0.43	2.42	4.41	0.00	0.00	0.00	0.00	0.36	0.33	0.00	23.20	0.00	7.16	9.56	0.00	0.00
Degraded	D 3	7.64	2.21	8.76	0.00	0.00	0.00	0.32	5.15	4.74	5.70	0.31	0.00	0.00	5.41	0.53	0.00	0.53	0.26	2.72	21.16	13.46	0.00	7.37	7.63	0.25	0.00
	D 4	3.41	3.05	7.00	0.00	5.29	0.23	6.75	0.00	0.00	11.66	0.00	2.11	3.09	0.00	0.00	0.00	0.86	3.66	4.26	0.00	19.24	0.00	23.98	0.00	0.00	0.32
	D 6	4.75	1.74	8.46	0.00	4.23	4.57	4.36	0.00	0.00	5.44	0.95	4.77	3.13	0.16	0.00	0.00	1.00	0.20	11.15	0.12	17.05	0.00	15.42	4.01	2.20	1.89
	D 7	3.17	1.34	6.38	1.54	5.45	3.28	0.15	1.35	2.74	5.95	0.83	0.00	3.80	2.92	0.00	0.00	0.00	0.86	7.06	9.32	13.16	0.00	27.29	0.00	1.18	0.60
	D 8	6.45	1.61	5.38	0.00	0.63	1.97	0.10	2.05	1.70	8.46	0.05	0.00	1.97	5.75	0.63	0.00	0.60	0.00	0.33	11.89	14.75	0.00	30.02	0.00	0.00	0.00
	D 9	4.63	1.55	4.02	0.00	6.16	3.26	0.00	4.83	2.43	5.80	1.25	0.00	0.00	4.20	4.39	0.00	0.00	1.91	3.06	2.39	15.32	0.00	32.33	0.00	0.00	0.35
	D 10	4.37	1.39	5.08	0.92	0.00	0.00	0.18	3.51	4.67	6.30	0.34	4.76	0.00	0.00	3.81	0.19	0.31	0.41	5.16	5.34	18.50	0.00	26.55	0.00	0.00	0.89
	D 11	4.32	1.19	4.37	0.00	0.00	0.00	0.60	3.84	4.18	7.49	0.52	0.00	3.03	4.87	1.89	0.00	0.31	0.11	0.66	0.66	22.48	0.00	33.95	0.00	0.00	0.49
Floodplain	V 5	9.11	1.50	0.00	0.00	5.18	6.62	0.48	4.93	0.32	17.68	1.26	1.69	0.76	1.10	0.38	0.83	0.52	0.60	1.64	4.91	12.09	4.04	7.13	7.37	0.00	0.00
	V 7	4.76	4.03	0.00	0.00	9.91	1.50	0.53	3.01	3.32	13.12	1.30	0.00	1.01	4.47	0.00	0.00	2.26	0.96	2.34	0.00	26.46	4.98	6.25	6.75	0.08	0.00
	V 8	2.22	3.91	4.29	1.07	4.86	5.23	0.22	4.24	6.47	8.46	1.06	0.00	5.44	6.96	0.00	1.48	1.37	2.60	0.21	2.28	13.92	0.00	15.77	0.00	1.56	1.09
	V 9	1.50	2.82	0.00	0.00	17.31	3.19	0.18	1.10	0.44	9.17	0.40	0.00	2.79	3.01	0.00	15.41	0.00	0.00	5.59	3.74	0.00	2.97	23.94	0.00	1.01	1.72
	V 10	3.12	2.60	9.77	0.17	0.00	0.00	0.25	4.08	12.31	13.50	1.81	0.00	0.00	6.10	7.34	2.72	0.49	0.83	0.20	4.34	1.62	0.52	20.18	0.00	0.17	0.71
	V 11	1.71	1.65	7.82	0.00	10.61	2.80	0.59	1.39	7.26	18.15	1.12	0.19	0.71	1.46	5.46	0.32	0.57	0.00	1.46	11.89	0.75	0.00	21.92	0.00	0.00	0.10
	V 12	1.23	2.03	17.20	0.00	0.70	0.12	0.33	10.21	1.32	11.49	0.00	0.28	2.23	3.46	0.00	1.38	0.06	0.00	7.69	10.56	8.53	0.00	17.43	0.00	0.00	0.00
	V 13	3.02	3.99	0.00	0.00	14.00	0.00	0.88	0.73	5.78	20.93	0.00	0.00	0.00	16.52	2.29	2.20	1.00	0.00	2.42	8.56	0.94	0.00	11.99	0.00	0.00	0.43
	V 16	4.05	1.63	0.00	0.00	6.12	0.00	0.00	1.63	1.86	12.40	0.81	0.00	0.00	0.00	4.81	0.00	0.92	0.56	0.80	25.43	0.00	4.01	33.16	0.00	0.00	0.00
Pasture	O 1	5.50	4.14	4.57	0.20	6.92	0.88	0.00	0.88	0.59	4.49	1.54	1.50	2.21	2.13	0.00	0.00	0.00	0.00	0.82	0.00	25.56	0.00	35.50	0.00	0.00	0.00
	O 2	5.19	4.28	4.19	0.00	6.25	0.92	0.00	0.68	0.48	4.11	1.13	1.30	2.39	2.50	0.00	0.00	0.00	0.00	0.52	0.00	26.23	0.00	37.75	0.00	0.00	0.00
	O 3	8.11	1.81	0.00	0.00	2.04	2.48	0.20	0.00	4.87	7.69	0.49	2.13	2.68	3.42	0.00	0.00	0.00	0.00	1.77	0.00	26.34	0.00	35.31	0.00	0.00	0.00
	O 5	7.88	1.76	0.00	0.00	2.06	2.92	0.00	0.00	3.54	6.20	0.58	0.00	3.08	3.52	0.00	0.00	0.50	0.00	3.31	0.00	26.73	0.00	36.31	0.00	0.00	0.00
	O 6	5.89	2.46	1.99	0.00	9.85	2.06	0.00	0.53	0.00	0.45	3.86	2.22	2.09	5.84	0.00	0.00	0.81	2.31	1.42	0.24	23.13	0.00	31.52	0.00	0.42	0.69
	O 7	5.47	1.98	4.40	0.00	5.84	1.85	0.00	0.23	0.00	5.12	1.05	2.86	1.38	4.13	0.00	0.00	0.79	1.80	0.56	0.00	26.66	0.00	33.27	0.00	0.00	0.15
	O 9	4.89	2.44	2.84	1.14	17.82	3.51	0.00	2.30	0.91	0.00	3.46	0.00	6.70	6.16	0.00	1.34	0.27	0.21	5.08	0.29	11.00	2.16	24.22	0.00	1.12	0.00
	O 15	5.32	1.27	2.11	0.00	7.27	0.00	0.00	0.00	0.00	4.20	1.18	4.71	16.76	2.29	1.29	0.00	0.40	0.00	2.19	0.20	25.50	0.00	21.71	0.00	0.18	0.00
	O 16	4.79	2.18	5.99	0.17	1.17	0.96	0.08	2.53	0.00	8.06	0.00	7.32	1.98	5.93	0.00	0.00	0.41	0.00	5.71	0.00	20.37	0.00	30.20	0.00	0.14	0.39
	O 19	4.84	0.57	1.23	0.00	12.43	0.00	0.40	1.15	0.00	9.93	0.00	1.53	6.13	1.60	0.29	0.00	0.00	0.00	2.30	1.25	19.40	0.00	30.30	0.00	0.32	0.23

Gradual scales in green or red represent values above or below the average of peak area, respectively.

As shown in Figure 5, individual variation was observed in the catalytic activities of the major enzymatic components (SVMPs, SVSPs, and PLA_2s) among the venom samples from each habitat (Figure 5A,C,E), except in the venoms from the floodplain, in which the SVMP catalytic activity was similar and low. There were significant differences among the venoms collected at the same habitat in hemorrhagic (Figure 5B) and myotoxic (Figure 5F) activities. The four venoms from the floodplain snakes (V5, V8, V13, and V16) induced hemorrhagic spots comparable to those induced by snake venoms from the other habitats, with an emphasis on the V8 snake, whose venom had the highest hemorrhagic activity in the floodplain group. For myotoxic activity, the greatest variation was found among the venoms from floodplain snakes: this group was the only one in which all tested venoms showed statistically significant differences in activity (Figure 5F).

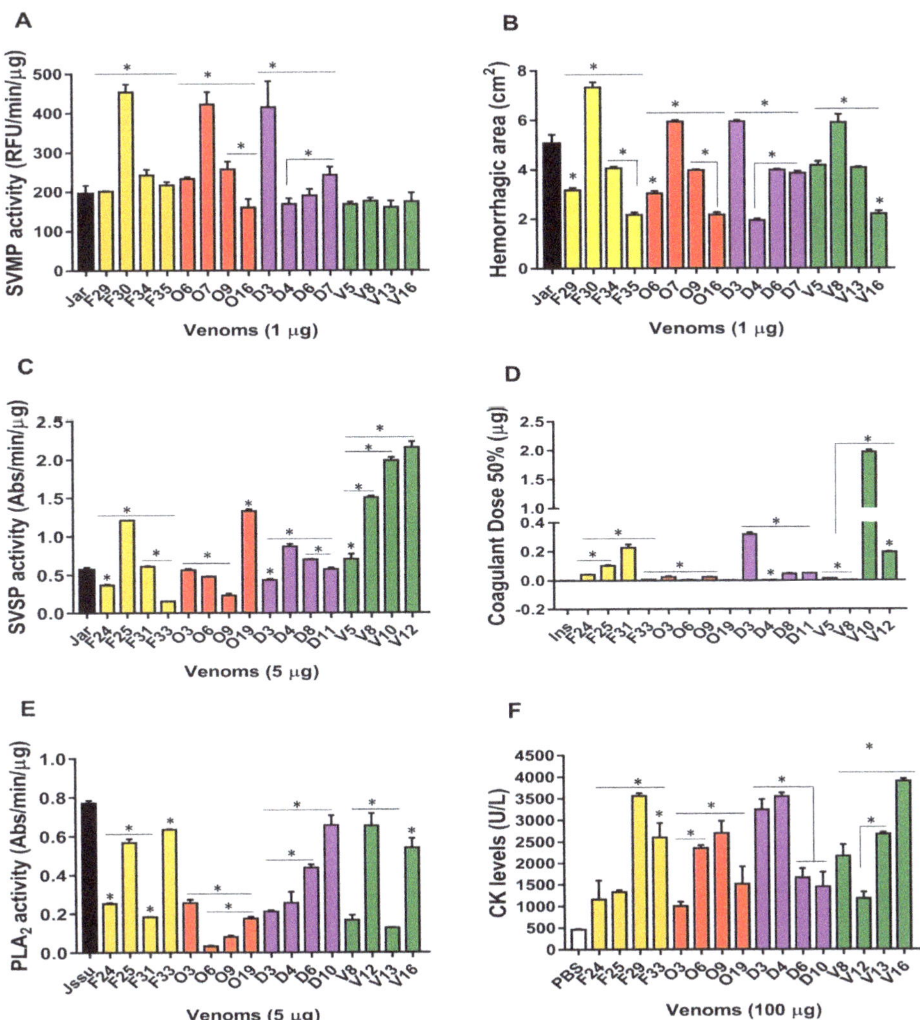

Figure 5. Functional assays: (**A**) SVMP catalytic activity: evaluated by hydrolysis of FRET substrate (Abz-AGLA-EDDnp), expressed as RFU/min/µg of venom. (**B**) Hemorrhagic activity: evaluated by the size of the lesions observed 3 h after venom injection (10 µg) into the dorsal skin of mice, and expressed in cm^2. (**C**) SVSPs catalytic activity: evaluated by hydrolysis of the chromogenic synthetic substrate (L-BAPNA), and expressed in Abs/min/µg of venom. (**D**) Pro-coagulant activity: evaluated the clotting times measured by thromboelastography of recalcified plasma from chickens, and the results

were expressed in terms of Coagulation Dose 50% (CD_{50}). **(E)** PLA_2s activity: evaluated by hydrolysis of the chromogenic substrate (NOBA), and the results were expressed in Abs/min/µg of venom. **(F)** Myotoxic activity: evaluated by the creatine kinase activity in mice serum 3 h after venom injection, and the results expressed in U/L. The data shown represent the mean + SD of three independent experiments. Controls: PBS and venom pools from *Bothrops jararaca*—Jar; *Bothrops jararacussu*—Jssu; *Bothrops insularis*—Ins. (*) Asterisks indicate significant variations among venoms from a same habitat.

The greatest within-group difference was found in procoagulant activity among the floodplain snakes, in which the DC_{50} values varied from 0.0035 to 1.965 µg for V8 and V10 snakes, respectively (Figure 5D). No other group showed such a large variation in DC_{50} values. However, the wide range of activity in this group was mostly due to the very high DC_{50} value of only one specimen (V10). On the other hand, pasture venoms were similar in procoagulant activity with only a small amount of variation observed in the venoms from the forest and degraded areas.

2.6. Individual Heterogeneity in Venom-Induced Lethality

Finally, to link functional variation to toxicity to specific prey, we decided to investigate the ability of the venoms from each habitat to kill rodents (mice) or birds (chicks). We also quantify individual variation in venom-induced lethality within the same habitat. Venoms from five specimens from each habitat were injected into groups of six mice or six chicks, and the time of death of the animals was monitored over 48 h (Figure 6).

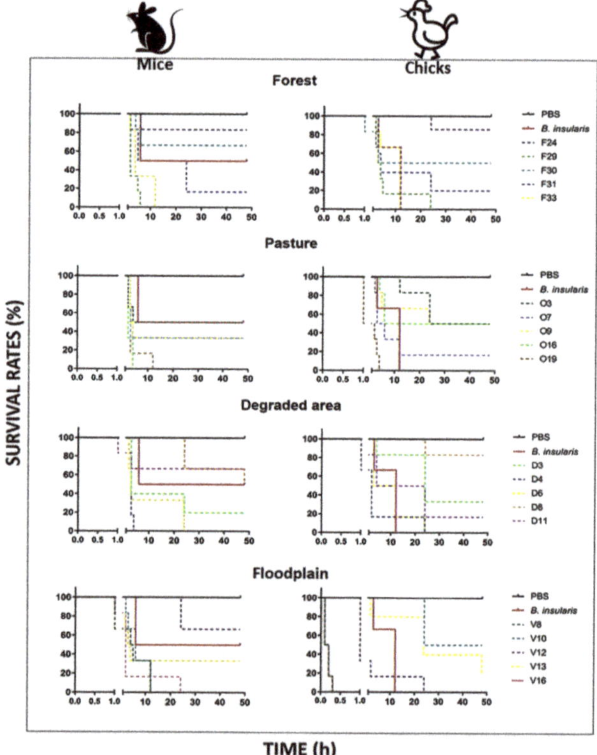

Figure 6. Time of death induced by individual *B. atrox* venoms in avian and murine models. Single doses of selected venom (200 µg) were injected i.p. into groups of six animals (mice or chicks). Next, the deaths were followed up every hour until 6 h after, then at 12, 24, and 48 h after venom injection. The data are representative of two independent experiments.

We observed a great deal of variation in the lethal activity of venoms from individual venoms from all groups analyzed. In venoms from the forest, degraded area, and pasture areas, mortality of experimental animals was observed from 1 to 2 h after venom injection, while in a number of cases, several animals survived through 48 h of observation. Similar results were observed for both the mice and birds, which matches the generalist diet preferences of ***B. atrox***. There were some quite interesting patterns of toxicity in floodplain venoms. These venoms had similar activity to venoms from the forest, pasture, and degraded area when injected in mice. However, some venoms from the floodplain habitat (V8 and V16) induced deaths only a few minutes after its inoculation in chicks, a pattern that was not observed in mice. In addition, one venom (V12) was noticeably more lethal to chicks, killing half the animals in the first two hours and all animals of the group within 24 h, whereas in mice, the same venom induced the death of only two animals, 12 and 24 h after inoculation. However, not all floodplain venoms were highly lethal to birds. The V10 snake venom was much more lethal to rodents than to birds, killing all mice within 24 h, and only two chicks until the end of the experiment. The venom of snake V13 killed the same number of chicks and mice (only 2), although at different time intervals. Similar to the variations observed in the venom composition, the differential lethality to chicks observed with floodplain venoms was not due to the two distant floodplain spots of snake collection. Venoms that preferentially killed chicks were from snakes collected at Santarém (ATXV 8 and 16) or Oriximiná (ATXV 12 and 13).

3. Discussion

In this study, we analyzed the influence of different environments on the individual variability of venom samples obtained from *B. atrox* snakes captured in four different habitats of the Brazilian Amazon: forest, pasture, degraded area, and floodplain. Our analyses revealed clear differences in both venom composition and its biological activities among snakes from different habitats. However, major differences were also observed among venom samples collected from snakes in the floodplain habitat, a dynamic environment, subject to periods of annual drought and floods.

In a previous report involving *B. atrox* snakes, the variability in venom composition was attributed mainly to low-abundance proteins that would be a genetic reservoir for quick adaptive changes [2], while the most abundant isoforms from each toxin family were considered as "core toxins", conserved in venoms of most specimens and responsible for the major activities of the venom [2]. A similar observation was reported by Margres and collaborators [23], showing that, in *Crotalus adamanteus*, *Sistrurus miliarius*, and *Agkistrodon piscivorus* snakes, the differences in the expression levels were present mostly in the low-expression proteins. However, in the present study, both rare (low expression) and abundant (high expression) proteins show contribute to venom variation within and between snakes from different habitats.

For example, our data also showed that Peak 23 was composed mostly of Batroxrhagin, a P-III class SVMP from *B. atrox* venom isolated by our group [24], was widely conserved in all the 37 venoms analyzed, within and across all groups; this confirms a previous observation [2] that Batroxrhagin would act as a "core toxin" in *B. atrox* venoms, acting on important physiological targets of diverse prey types. In contrast, other abundant venom proteins showed less conservation and could represent more specific or even "adaptive" variation in these proteins. Clear examples of abundant fractions differentially expressed between the groups are fraction 21, which contains predominantly a PI-class SVMP and had lower expression in floodplain venoms; fraction 10, which contains mostly SVSPs and had higher expression in floodplain venoms; and fraction 20, which contains predominantly a PI-class SVMP and CTLs and is higher expressed in venoms from pasture compared to the other areas. Other fractions rich in SVSPs and CLTs were also were variable. These proteins participate to the hemorrhagic and coagulant activity of *B. atrox* venom and the observed differences in expression of the isoforms may alter procoagulant activity and to promote differential lethality in rodents and birds, suggesting an important role of the environment in selecting for venom variation.

Individual variation in the abundance of SVMPs, SVSPs, and PLA_2s could have arisen as a result of environmental differences between habitats. In our previous study [15], we compared the functional profiles of pools of venoms from *B. atrox* snakes from the same habitats and observed a less hemorrhagic and more procoagulant phenotype in the pool of venoms from floodplain snakes. The pool of venoms from the floodplain snakes was also significantly faster to induce clotting in several types of plasmas, including avian plasma [25]. Here, we used similar experimental approaches to investigate the individual variability within each group of snakes, introducing modifications in the tests of DC_{50}, which were performed in the presence of calcium, to ensure the detection of coagulotoxins, dependent or not on this cofactor. In addition, our tests were performed with avian plasma as previously reported [26,27], which allowed us to construct better dose–response curves to assess the procoagulant effects of the individual venoms.

Venoms from *B. atrox* specimens from the floodplain showed low SVMP activity and higher SVSP catalytic activities. However, the hemorrhagic and procoagulant phenotypes observed in the pool of venoms were not consistent across all individual venoms collected from this environment. Some individual venoms such as from V5 and V8 snakes, showed both a potent procoagulant action on avian plasma and high hemorrhagic activities in mice. Moreover, the higher DC_{50} value observed, which corresponds to the less procoagulant venom, was precisely from a floodplain snake venom (V10), demonstrating that not all snakes from the floodplain have venoms that induce potent clotting of avian plasma, but the range of clotting activity was the highest in this environment.

Bernardoni and collaborators [3] analyzed *Bothrops neuwiedi* venom and found SVMPs isoforms that are functionally different and capable of affecting different targets in the hemostatic systems of birds, rodents, and humans. They suggested that some SVMPs are less selective, guaranteeing the action of the venom on different targets, while other isoforms are more selective, modulating the action of the venom for specific prey. More recently, we also investigated the effects of these SVMP isoforms on amphibian plasma (*Rinella marina*), demonstrating the coagulotoxic effects of these toxins on different types of animal plasmas [18]. In the present study, some fractions rich in SVMPs had a very variable distribution in the *B. atrox* venoms of all habitats and may be involved in the variability observed in the coagulotoxic activity observed in these individual venoms. In contrast, the major SVMP eluted in Peak 23 was very conserved in the venoms of all 37 specimens analyzed, confirming the previous assumption [2] that Batroxhagin is the core SVMP of *B. atrox* venom. On the other hand, the strong hemorrhagic action observed in some floodplain venoms (V5, V8, and V13), is compatible with a prominent increase of the fractions represent by Peaks 21 and 22 that shows height/area comparable to the ones observed in venoms from the other habitats. These peaks contain Atroxlisin-Ia, a PI-class SVMP that induces hemorrhage comparable to the P-III class enzymes [28].

During envenomation, various components present in snake venoms can act synergistically to cause the prey's organism to collapse, which usually results in rapid death. The speed to kill/immobilize prey is crucial for food acquisition of terrestrial viperids during hunting [29,30]. Previous works also report that lethality profiles of snake venoms can be variable in different animal models, such as mammals, birds, reptiles, and amphibians [5,31]. For this reason, we included the functional characterization of the individual venoms in two different animal models, birds (chicks) and mammals (mice). We found substantial differences in venom activities in both, the number of test animals dying, and the time of deaths in the two models. Forest venoms showed similar lethality profiles in rodents and birds across individual venom samples, while venoms from pasture and degraded areas were slightly more lethal to rodents. Of particular note, venoms from floodplain snakes were highly variable according to their ability to prey on mammals or birds. For example, the venoms of the snakes V8 and V16 were more toxic to chicks, inducing deaths within few minutes in birds, even more quickly than the venom of *B. insularis*, a snake with a diet specialized in birds [32]. The venom of the snake V10 however was more lethal to rodents. Interestingly, the venoms with the highest procoagulant activity are more lethal to birds while V10 venom, the weakest procoagulant, killed predominantly mice. The limitations imposed by the small available quantities of venom prevented using the same individual venoms in all the functional tests performed or to carry out dose–response analyses, hindering a more direct assessment of the relationship between lethality and the other toxic activities evaluated.

Nevertheless, our findings strongly suggest that the procoagulant activity of the floodplain venoms contributes significantly to their lethal effects in rodents and birds. The most striking and differential characteristics of the floodplain venoms are its potent effects on coagulation in different types of plasma. The venom of the snake V10 (from floodplain habitat) was more efficient in killing rodents than birds, and this same venom had the highest EC_{50} value on avian plasma. A possible explanation for the remarkable action of some floodplain venoms on hemostasis and velocity to induce death in birds could be directly linked to the floodplain environment, where these snakes were captured. Floodplains are periodically flooded by the lateral overflow of waters rich in sediments from the Amazon River. The floodplain regions alternate cyclical periods of drought and flood, remaining flooded for a few months during the year, this seasonal variation being driven by flood and precipitation seasons [33]. Balancing selection pressures could result in greater venom heterogeneity, as evidenced by the differential lethality to rodents or birds observed in some floodplain venoms, while the snake venoms from the other habitats presented a more homogeneous lethality pattern.

In line with our findings, Smiley-Walters et al. [31] showed that variation in the venom of *Sistrurus miliarius* at the population level has an adaptive role in terms of toxicity to prey. Testing adaptive hypotheses requires careful analysis of phenotypic characters whose variation has clear functional consequences. Venoms directly affect the ability of an individual snake to immobilize and kill its prey [31,34–37]. For *B. atrox*, a snake with a generalist diet that includes arthropods, frogs, lizards, birds, and small mammals [38,39], the functional diversity of the venoms could represent a local adaptation of individual venoms in each population to different sets of prey. Within an ecological context, the nature of the interactions between species, including prey–predator relationships, may change among populations [40,41], especially for species with a wide geographical distribution. of the functional versatility and diversity of *B. atrox* venoms may explain its wide distribution throughout all Amazon regions [7], reflectinghow this species adapts to prey communities in different habitats.

4. Conclusions

Individual venom variation of *B. atrox* snakes, captured in different habitats within the Brazilian Amazon, support the idea that dynamic environments may select for more variable venoms. Such differential distribution of protein isoforms leads to differential function and toxic effects on different prey. The fact that the greater heterogeneity in terms of composition/toxicity of the venom was found in snakes from the floodplain habitat suggests that balancing selection for expression of different isoforms could be enacted by the drastic seasonal changes (drought/flood) present in this type of environment. Variable selective pressures in different habitats may likely exert localized impacts on the venom phenotype in a species with a wide geographic distribution, such as *B. atrox*.

5. Material and Methods

5.1. Snakes and Venoms

Thirty-seven male and female adult specimens of *B. atrox* snakes, with sizes ranging from 71.2 to 124.5 cm, were captured in four environments at Santarem and Oriximiná, in the western region of the state of Pará, Brazil (Supplementary Figure S2) under ICMBio/SISBio license 32098-1 and SISGEN number A78BD88: (1) Pasture: ten adult snakes were collected from a pasture area in the municipality of Oriximiná (ATXO 1, 2, 3, 5, 6,

7, 9, 15, 16, 19) on the north shore of the Amazon River. This site was historically Terra Firme forest characterized by large trees that were cleared for pasture ~20 years earlier. (2) Floodplain (ATXV): five adults were collected from a seasonally flooded island in the main course of the Amazon river near Santarem (at Urucurituba: ATXV 5, 7, 8, and 9; at Tapará: ATXV 16) or Oriximiná (ATXV 10, 11, 12, 13). These areas are typical of floodplain habitats subject to periodic flooding by the Amazon River and formed by the deposition of sediments that have led to the formation of many islands. The typical vegetation consists of grasses that grow on highly fertile alluvial soils. (3) Forest: ten snakes were captured at the Floresta Nacional do Tapajós, a protected area located in the municipality of Belterra (ATXF 24, 25, 26, 28, 29, 30, 31, 33, 34, and 35) next to the Tapajos River about 50 km south of the Amazon River. This site also represents the upland Terra Firme forest. (4) Degraded: eight snakes were collected at a recently degraded (ATXD 3, 4, 6, 7, 8, 9, and 10) area contiguous to the forest area at Floresta Nacional do Tapajós, which was cleared for pasture. After capture, the snakes were transferred to the Herpetarium of Laboratório de Pesquisas Zoológicas, Universidade da Amazônia (UNAMA), in Santarém, PA. Venom samples were collected using manual extraction techniques, freeze-dried, and kept at −20 °C until use. Animal care and procedures used in the handling of snakes were undertaken according to the guidelines and permits (CEUAIB 1244/14, Instituto Butantan, São Paulo, Brazil).

5.2. Chromatographic Analysis

The *B. atrox* venom samples were individually fractionated by reverse-phase high-performance liquid chromatography (RF-HPLC) in the Vydac C18 column as previously described [15]. The fractions had their toxin composition predicted according to peak shape and retention time comparing to a previous standard chromatography of *B. atrox* venom from snakes collected in the same areas from which components present in each fraction were identified and quantified by free-label mass spectrometry [15].

5.3. Functional Assays

5.3.1. Enzymatic Assays on Synthetic Substrates

SVMP, SVSP, and PLA_2 enzymatic activities of individual venoms from *B. atrox* were assayed using synthetic substrates according to the procedures previously standardized in our lab [42]. Briefly, the substrate used for SVMPs was Abz-AGLA-EDDnp substrate (Peptide International, Gardner, MA, USA), and the results are expressed as Relative Fluorescence Units-RFU/min/µg. The PLA_2 activity was assayed using the substrate 4-nitro-3-[octanoyloxy] benzoic acid (Enzo Life Sciences, New York, NY, USA) and activity were determined according to the absorbance at 425 nm and expressed as Absorbance/min/µg of venom. For SVSP catalytic activity, the chromogenic synthetic substrate benzoyl-arginyl-p-nitroanilide (L-BAPNA) (Sigma-Aldrich®, St. Louis, MO, USA) was used and the hydrolysis of the substrate was expressed as the increase in Absorbance/min/µg of venom. The results represent the mean ± SD of three independent experiments, undertaken in triplicate.

5.3.2. Procoagulant Activity

The procoagulant activity was evaluated in recalcified chicken (White leghorn) plasma, obtained under license CEUAIB n° 13,710-14, Instituto Butantan, by thromboelastometry using a four-channel ROTEM® system (Pentapharm, Munich, Germany), as previously described [26]. The plasma was prepared as 3.2% citrated stock and stored at −80 °C until needed in 1 mL aliquots. For the experiments, an aliquot was defrosted by placing it into a 37 °C water bath for 10 min. Venom samples and controls were dissolved in PBS (60 µL) and added in the specific cups with 20 µL of $CaCl_2$ (0.02 M) and 260 µL of citrated plasma (final volume = 340 µL), pipette-mixed, and the clotting time was immediately recorded. PBS and ellagic acid were used as negative and positive controls, respectively. Under such conditions, recalcified chicken plasma treated with PBS only is unclottable for at least 30 min (CT value = 1800 s). The results are shown as Coagulant Dose 50% (DC_{50}), defined as the amount of venom (µg) that reduces the CT parameter to 900 s.

5.3.3. Venom Activities Using Animal Models

Hemorrhagic and myotoxic activities were carried out as previously described [43] using male Swiss mice (18–20 g) under the approval of the Butantan Institute Ethics Committee on Animal Use (Protocol Number: 13710-14). Briefly, for hemorrhagic activity, samples containing 10 μg of each venom pool, diluted in 50 μL of PBS, were injected intradermally into the dorsal skin of mice. After 3 h, the animals were euthanized in a CO_2 chamber, the dorsal skin was removed, and the hemorrhagic halos were measured. Groups of 5 animals were tested, and control group animals were injected with PBS only or *B. jararaca* venom (10 μg). Results represent the mean ± SE of the area of hemorrhagic spots (cm^2) from mice tested in at least 2 independent experiments. The myotoxic activity was assayed using 50 μg of venom pools in 30 μL of PBS, injected intramuscularly into the gastrocnemius muscle in Swiss mice. After 3 h, the animals were bled via ophthalmic plexus and the sera were assayed for creatine-kinase activity with a commercial kit CK-UV (Bioclin), according to the manufacturer's instructions. Groups of 5 animals were tested, and animals from the control groups were injected with PBS only or *B. jararacussu* venom (50 μg). Results represent the mean ± SE of the CK activity in mice sera (U/mL) from mice tested in at least 2 independent experiments.

The lethality induced by the venoms was tested in murine (Swiss mice 18–20 g body weight) and avian (three days old Bovan chicks 20–30 g body weight) models with protocols approved by the Animal Ethical Use Committees of the Instituto Butantan (CEUAIB n° 13710-14). Groups of 6 mice or chicks were injected intraperitoneally with a single dose (200 μg) of venom samples, diluted in 500 μL of PBS. Following injection, the times of death for each group were recorded hourly until the sixth hour, and then at 12, 24, and 48 h after inoculation of the samples. After 48 h, the surviving animals were euthanized, using a CO_2 chamber for the mice and injection of sodium pentobarbital overdose for the chicks. The data obtained were graphed on a survival curve plot. The tests were undertaken in two independent experiments.

5.4. Statistical Analysis

We tested for significant compositional differentiation using a hierarchical multivariate approach that is robust to the compositional nature of the data [44]. First, we brought the relative abundance data out of the compositional simplex space using the isometric log-ratio (ilr) transformation, which also avoids the zero-sum constraint of the centered-log ratio (clr) transformations but at the cost of one dimension of the data. The ilr-transformed peak data were then subjected to permutational multivariate analysis of variance based on distances using the adonis function from the vegan R package [45], with all peak dimensions as a response and environment as a predictor. Significance and R^2 were estimated using 10,000 permutations of the raw data. We then visualized the venom phenotype of each snake in multivariate space using robust principal components analysis (PCA) as implemented in the pcaCoDA function from the robCompositions R package [46].

After detection of a significant global association between environment and venom composition, we conducted pairwise posthoc PERMANOVA, as described above, between the venom compositions of all pairs of environments to determine which environments differed significantly. Lastly, we preserved the full dimensionality of the data but removed the data from the simplex using the clr-transformation, to determine which peaks specifically varied across environments. To do this, we use the lm function from R stats, with the clr-transformed abundance of an individual peak as the response and environment of origin as the predictor. We repeated this test for all 26 peaks and used the false-discovery rate (fdr) approach to correct the *p*-values for multiple tests.

For the functional analyses (in vitro and in vivo assays), firstly, the data were evaluated for a normal distribution (Shapiro–Wilk), and then differences among the means were evaluated by one-way ANOVA, followed by a Tukey post-test (for multiple comparisons). Data that did not meet the normality criteria were analyzed using a non-parametric test (Kruskal–Wallis). Results represent the mean and standard deviation or standard error, as appropriate, and the level of significance was set at $p \leq 0.05$. Data were analyzed using the GraphPad Prism statistical program (version 7.00 for Windows, GraphPad Software, San Diego, CA, USA).

Supplementary Materials: The following are available online at https://www.mdpi.com/article/10.3390/toxinstoxins13110814/s1, Figure S1: Chromatographic profiles of *B. atrox* venoms of different habitats from west of Pará State, Brazilian Amazon; Figure S2: Characteristics and localization of areas of snake collection.

Author Contributions: Conceptualization, L.F.S. and A.M.M.-d.-S.; formal analysis, L.F.S., M.L.H., R.H.V.M., B.P., H.L.G. and A.M.M.-d.-S.; funding acquisition, A.M.M.-d.-S.; investigation, L.F.S., T.H.M.D.-R., M.M.T.R., R.H.V.M. and B.P.; methodology, L.F.S., M.L.H., T.H.M.D.-R., M.M.T.R., B.P. and H.L.G.; project administration, A.M.M.-d.-S.; resources, H.M.C.; software, M.L.H.; supervision, A.M.M.-d.-S.; writing—original draft, L.F.S.; writing—review and editing, L.F.S., M.L.H., R.H.V.M., H.L.G. and A.M.M.-d.-S. All authors have read and agreed to the published version of the manuscript.

Funding: This research was funded by Coordenação de Aperfeiçoamento de Pessoal nível superior (CAPES 063/2010-Toxinology-AUXPE 1209/2011); Fundação de Amparo à Pesquisa de São Paulo (FAPESP 2016/50127-5; 2014/13124-2; 2017/15170-0) and Conselho Nacional de Desenvolvimento Científico e Tecnológico (CNPq) (grant number 303958/2018-9). Additional support was provided by Fundação de Amparo à Pesquisa do Estado do Amazonas-FAPEAM (PRO-ESTADO) to AMMS. as Manaus Visiting Professor. LFS and THMDR were students in the Science Graduate Program—Toxinology (Instituto Butantan).

Institutional Review Board Statement: This study was conducted according to the guidelines of the Declaration of Helsinki. The snakes were captured under ICMBio/SISBio license 32098-1 and SISGEN number A78BD88 and CEUAIB 1244/14, and its handling was undertaken according to the guidelines and permits (CEUAIB 1244/14, Instituto Butantan). The procedures involving mice, chickens, and chicks were approved by the Ethics Committee on Animal Use of the Instituto Butantan (Protocol Number: 13710-14).

Conflicts of Interest: The authors declare no conflict of interest. The funders had no role in the design of the study; in the collection, analyses, or interpretation of data; in the writing of the manuscript, or in the decision to publish the results.

References

1. Calvete, J.J.; Juárez, P.; Sanz, L. Snake venomics. Strategy and applications. *J. Mass Spectrom.* **2007**, *42*, 1405–1414. [CrossRef]
2. Amazonas, D.R.; Portes-Junior, J.A.; Nishiyama-Jr, M.Y.; Nicolau, C.A.; Chalkidis, H.M.; Mourão, R.H.V.; Grazziotin, F.G.; Rokyta, D.R.; Gibbs, H.L.; Valente, R.H.; et al. Molecular mechanisms underlying intraspecific variation in snake venom. *J. Proteom.* **2018**, *181*, 60–72. [CrossRef] [PubMed]
3. Bernardoni, J.L.; Sousa, L.F.; Wermelinger, L.S.; Lopes, A.S.; Prezoto, B.C.; Serrano, S.M.T.; Zingali, R.B.; Moura-da-Silva, A.M. Functional Variability of Snake Venom Metalloproteinases: Adaptive Advantages in Targeting Different Prey and Implications for Human Envenomation. *PLoS ONE* **2014**, *9*, e109651. [CrossRef] [PubMed]
4. Holding, M.L.; Strickland, J.L.; Rautsaw, R.M.; Hofmann, E.P.; Mason, A.J.; Hogan, M.P.; Nystrom, G.S.; Ellsworth, S.A.; Colston, T.J.; Borja, M.; et al. Phylogenetically diverse diets favor more complex venoms in North American pitvipers. *Proc. Natl. Acad. Sci. USA* **2021**, *118*, e2015579118. [CrossRef] [PubMed]
5. Gibbs, H.L.; Mackessy, S.P. Functional basis of a molecular adaptation: Prey-specific toxic effects of venom from Sistrurus rattlesnakes. *Toxicon* **2009**, *53*, 672–679. [CrossRef] [PubMed]
6. Casewell, N.R.; Wüster, W.; Vonk, F.J.; Harrison, R.A.; Fry, B.G. Complex cocktails: The evolutionary novelty of venoms. *Trends Ecol. Evol.* **2013**, *28*, 219–229. [CrossRef]
7. Campbell, J.A.; Lamar, W.W. *The Venomous Reptiles of the Western Hemisphere*; Cornell University Press: Ithaca, NY, USA, 2004.
8. López-Lozano, J.L.; de Sousa, M.V.; Ricart, C.A.; Chávez-Olortegui, C.; Flores Sanchez, E.; Muniz, E.G.; Bührnheim, P.F.; Morhy, L. Ontogenetic variation of metalloproteinases and plasma coagulant activity in venoms of wild Bothrops atrox specimens from Amazonian rain forest. *Toxicon* **2002**, *40*, 997–1006. [CrossRef]
9. Saldarriaga, M.M.; Otero, R.; Núñez, V.; Toro, M.F.; Díaz, A.; Gutiérrez, J.M. Ontogenetic variability of Bothrops atrox and Bothrops asper snake venoms from Colombia. *Toxicon* **2003**, *42*, 405–411. [CrossRef]
10. Guércio, R.A.; Shevchenko, A.; López-Lozano, J.L.; Paba, J.; Sousa, M.V.; Ricart, C.A. Ontogenetic variations in the venom proteome of the Amazonian snake Bothrops atrox. *Proteom. Sci.* **2006**, *4*, 11. [CrossRef]

11. Salazar, A.M.; Rodriguez-Acosta, A.; Girón, M.E.; Aguilar, I.; Guerrero, B. A comparative analysis of the clotting and fibrinolytic activities of the snake venom (*Bothrops atrox*) from different geographical areas in Venezuela. *Thromb. Res.* **2007**, *120*, 95–104. [CrossRef]
12. Núñez, V.; Cid, P.; Sanz, L.; De La Torre, P.; Angulo, Y.; Lomonte, B.; Gutiérrez, J.M.; Calvete, J.J. Snake venomics and antivenomics of Bothrops atrox venoms from Colombia and the Amazon regions of Brazil, Perú and Ecuador suggest the occurrence of geographic variation of venom phenotype by a trend towards paedomorphism. *J. Proteom.* **2009**, *73*, 57–78. [CrossRef] [PubMed]
13. Calvete, J.J.; Sanz, L.; Perez, A.; Borges, A.; Vargas, A.M.; Lomonte, B.; Angulo, Y.; Maria Gutierrez, J.; Chalkidis, H.M.; Mourao, R.H.V.; et al. Snake population venomics and antivenomics of Bothrops atrox: Paedomorphism along its transamazonian dispersal and implications of geographic venom variability on snakebite management. *J. Proteom.* **2011**, *74*, 510–527. [CrossRef]
14. Moretto Del-Rei, T.H.; Sousa, L.F.; Rocha, M.M.T.; Freitas-de-Sousa, L.A.; Travaglia-Cardoso, S.R.; Grego, K.; Sant'Anna, S.S.; Chalkidis, H.M.; Moura-da-Silva, A.M. Functional variability of Bothrops atrox venoms from three distinct areas across the Brazilian Amazon and consequences for human envenomings. *Toxicon* **2019**, *164*, 61–70. [CrossRef] [PubMed]
15. Sousa, L.F.; Portes-Junior, J.A.; Nicolau, C.A.; Bernardoni, J.L.; Nishiyama, M.Y., Jr.; Amazonas, D.R.; Freitas-de-Sousa, L.A.; Mourao, R.H.V.; Chalkidis, H.M.; Valente, R.H.; et al. Functional proteomic analyses of Bothrops atrox venom reveals phenotypes associated with habitat variation in the Amazon. *J. Proteom.* **2017**, *159*, 32–46. [CrossRef]
16. Monteiro, W.M.; Contreras-Bernal, J.C.; Bisneto, P.F.; Sachett, J.; Mendonça da Silva, I.; Lacerda, M.; Guimarães da Costa, A.; Val, F.; Brasileiro, L.; Sartim, M.A.; et al. *Bothrops atrox*, the most important snake involved in human envenomings in the amazon: How venomics contributes to the knowledge of snake biology and clinical toxinology. *Toxicon X* **2020**, *6*, 100037. [CrossRef]
17. Moura-da-Silva, A.M.; Contreras-Bernal, J.C.; Cirilo Gimenes, S.N.; Freitas-de-Sousa, L.A.; Portes-Junior, J.A.; da Silva Peixoto, P.; Kei Iwai, L.; Mourão de Moura, V.; Ferreira Bisneto, P.; Lacerda, M.; et al. The relationship between clinics and the venom of the causative Amazon pit viper (*Bothrops atrox*). *PLoS Negl. Trop. Dis.* **2020**, *14*, e0008299. [CrossRef]
18. Sousa, L.F.; Bernardoni, J.L.; Zdenek, C.N.; Dobson, J.; Coimbra, F.; Gillett, A.; Lopes-Ferreira, M.; Moura-da-Silva, A.M.; Fry, B.G. Differential coagulotoxicity of metalloprotease isoforms from Bothrops neuwiedi snake venom and consequent variations in antivenom efficacy. *Toxicol. Lett.* **2020**, *333*, 211–221. [CrossRef]
19. Freitas-de-Sousa, L.A.; Nachtigall, P.G.; Portes-Junior, J.A.; Holding, M.L.; Nystrom, G.S.; Ellsworth, S.A.; Guimarães, N.C.; Tioyama, E.; Ortiz, F.; Silva, B.R.; et al. Size matters: An evaluation of the molecular basis of ontogenetic modifications in the composition of Bothrops jararacussu snake venom. *Toxins* **2020**, *12*, 791. [CrossRef]
20. Schonour, R.B.; Huff, E.M.; Holding, M.L.; Claunch, N.M.; Ellsworth, S.A.; Hogan, M.P.; Wray, K.; McGivern, J.; Margres, M.J.; Colston, T.J.; et al. Gradual and Discrete Ontogenetic Shifts in Rattlesnake Venom Composition and Assessment of Hormonal and Ecological Correlates. *Toxins* **2020**, *12*, 659. [CrossRef]
21. Harvey, M.G.; Aleixo, A.; Ribas, C.C.; Brumfield, R.T. Habitat Association Predicts Genetic Diversity and Population Divergence in Amazonian Birds. *Am. Nat.* **2017**, *190*, 631–648. [CrossRef] [PubMed]
22. Gibbs, H.L.; Sovic, M.; Amazonas, D.; Chalkidis, H.; Salazar-Valenzuela, D.; Moura-Da-Silva, A.M. Recent lineage diversification in a venomous snake through dispersal across the Amazon River. *Biol. J. Linnean Soc.* **2018**, *123*, 651–665. [CrossRef]
23. Margres, M.J.; Wray, K.P.; Seavy, M.; McGivern, J.J.; Herrera, N.D.; Rokyta, D.R. Expression Differentiation Is Constrained to Low-Expression Proteins over Ecological Timescales. *Genetics* **2016**, *202*, 273–283. [CrossRef]
24. Freitas-de-Sousa, L.A.; Amazonas, D.R.; Sousa, L.F.; Sant'Anna, S.S.; Nishiyama, M.Y., Jr.; Serrano, S.M.T.; Junqueira-de-Azevedo, I.L.M.; Chalkidis, H.M.; Moura-da-Silva, A.M.; Mourao, R.H.V. Comparison of venoms from wild and long-term captive Bothrops atrox snakes and characterization of Batroxrhagin, the predominant class PIII metalloproteinase from the venom of this species. *Biochimie* **2015**, *118*, 60–70. [CrossRef] [PubMed]
25. Sousa, L.F.; Zdenek, C.N.; Dobson, J.S.; Op den Brouw, B.; Coimbra, F.; Gillett, A.; Del-Rei, T.H.M.; Chalkidis, H.M.; Sant'Anna, S.; Teixeira-da-Rocha, M.M.; et al. Coagulotoxicity of *Bothrops* (Lancehead Pit-Vipers) Venoms from Brazil: Differential Biochemistry and Antivenom Efficacy Resulting from Prey-Driven Venom Variation. *Toxins* **2018**, *10*, 411. [CrossRef] [PubMed]
26. Oguiura, N.; Kapronezai, J.; Ribeiro, T.; Rocha, M.M.; Medeiros, C.R.; Marcelino, J.R.; Prezoto, B.C. An alternative micromethod to access the procoagulant activity of Bothrops jararaca venom and the efficacy of antivenom. *Toxicon* **2014**, *90*, 148–154. [CrossRef]
27. Prezoto, B.C.; Tanaka-Azevedo, A.M.; Marcelino, J.R.; Tashima, A.K.; Nishiduka, E.S.; Kapronezai, J.; Mota, J.O.; Rocha, M.M.T.; Serino-Silva, C.; Oguiura, N. A functional and thromboelastometric-based micromethod for assessing crotoxin anticoagulant activity and antiserum relative potency against Crotalus durissus terrificus venom. *Toxicon* **2018**, *148*, 26–32. [CrossRef]
28. Freitas-de-Sousa, L.A.; Colombini, M.; Lopes-Ferreira, M.; Serrano, S.M.T.; Moura-da-Silva, A.M. Insights into the Mechanisms Involved in Strong Hemorrhage and Dermonecrosis Induced by Atroxlysin-Ia, a PI-Class Snake Venom Metalloproteinase. *Toxins* **2017**, *9*, 239. [CrossRef]
29. Greene, H.W. Snakes. In *The Evolution of Mistery in Nature*; University of Califórnia Press: Berkeley, CA, USA, 1997.
30. Saviola, A.J.; Gandara, A.J.; Bryson, R.W.; Mackessy, S.P. Venom phenotypes of the Rock Rattlesnake (*Crotalus lepidus*) and the Ridge-nosed Rattlesnake (*Crotalus willardi*) from México and the United States. *Toxicon* **2017**, *138*, 119–129. [CrossRef] [PubMed]
31. Smiley-Walters, S.A.; Farrell, T.M.; Gibbs, H.L. Evaluating local adaptation of a complex phenotype: Reciprocal tests of pigmy rattlesnake venoms on treefrog prey. *Oecologia* **2017**, *184*, 739–748. [CrossRef]
32. Zelanis, A.; de Souza Ventura, J.; Chudzinski-Tavassi, A.M.; de Fátima Domingues Furtado, M. Variability in expression of Bothrops insularis snake venom proteases: An ontogenetic approach. *Comp. Biochem. Physiol. C Toxicol. Pharmacol.* **2007**, *145*, 601–609. [CrossRef]

33. McGrath, D.; Castro, F.; Futemma, C.; Amaral, B.; Calabria, J. Fisheries and resource management on the Lower Amazon floodplain. *Human Ecol.* **1993**, *21*, 167–195. [CrossRef]
34. Thomas, R.G.; Pough, F.H. The effect of rattlesnake venom on digestion of prey. *Toxicon* **1979**, *17*, 221–228. [CrossRef]
35. Torres-Bonilla, K.A.; Schezaro-Ramos, R.; Floriano, R.S.; Rodrigues-Simioni, L.; Bernal-Bautista, M.H.; Alice da Cruz-Höfling, M. Biological activities of Leptodeira annulata (banded cat-eyed snake) venom on vertebrate neuromuscular preparations. *Toxicon* **2016**, *119*, 345–351. [CrossRef]
36. Zimmerman, K.D.; Gates, G.R.; Heatwole, H. Effects of venom of the olive sea snake, Aipysurus laevis, on the behaviour and ventilation of three species of prey fish. *Toxicon* **1990**, *28*, 1469–1478. [CrossRef]
37. Richards, D.P.; Barlow, A.; Wüster, W. Venom lethality and diet: Differential responses of natural prey and model organisms to the venom of the saw-scaled vipers (Echis). *Toxicon* **2012**, *59*, 110–116. [CrossRef]
38. Martins, M.; Gordo, M. *Bothrops atrox* (common lancehead). *Diet Herpetol. Rev.* **1993**, *24*, 2.
39. Martins, M.; Marques, O.A.V.; Sazima, I.; Schuett, G.; Höggren, M.; Green, H.W. Ecological and phylogenetic correlates of feeding habits in Neotropicalpitvipers of the genus Bothrops. In *Biology of the Pit Vipers*; Schuett, G.W., Höggren, M., Douglas, M.E., Greene, H.W., Eds.; Eagle Mountain Publishing: Tyler, TX, USA, 2002; pp. 307–328.
40. Harrison, R.A.; Moura-da-Silva, A.M.; Laing, G.D.; Wu, Y.; Richards, A.; Broadhead, A.; Bianco, A.E.; Theakston, R.D.G. Antibody from mice immunized with DNA encoding the carboxyl-disintegrin and cysteine-rich domain (JD9) of the haemorrhagic metalloprotease, Jararhagin, inhibits the main lethal component of viper venom. *Clin. Exp. Immunol.* **2000**, *121*, 358–363. [CrossRef]
41. Callaway, R.M.; Brooker, R.W.; Choler, P.; Kikvidze, Z.; Lortie, C.J.; Michalet, R.; Paolini, L.; Pugnaire, F.I.; Newingham, B.; Aschehoug, E.T.; et al. Positive interactions among alpine plants increase with stress. *Nature* **2002**, *417*, 844–848. [CrossRef]
42. Knittel, P.S.; Long, P.F.; Brammall, L.; Marques, A.C.; Almeida, M.T.; Padilla, G.; Moura-da-Silva, A.M. Characterising the enzymatic profile of crude tentacle extracts from the South Atlantic jellyfish Olindias sambaquiensis (Cnidaria: Hydrozoa). *Toxicon* **2016**, *119*, 1–7. [CrossRef] [PubMed]
43. Sousa, L.F.; Nicolau, C.A.; Peixoto, P.S.; Bernardoni, J.L.; Oliveira, S.S.; Portes-Junior, J.A.; Mourao, R.H.V.; Lima-dos-Santos, I.; Sano-Martins, I.S.; Chalkidis, H.M.; et al. Comparison of Phylogeny, Venom Composition and Neutralization by Antivenom in Diverse Species of *Bothrops* Complex. *PLoS Negl. Trop. Dis.* **2013**, *7*. [CrossRef] [PubMed]
44. Margres, M.J.; McGivern, J.J.; Seavy, M.; Wray, K.P.; Facente, J.; Rokyta, D.R. Contrasting modes and tempos of venom expression evolution in two snake species. *Genetics* **2015**, *199*, 165–176. [CrossRef] [PubMed]
45. Oksanen, J.; Kindt, R.; O'Hara, B. Vegan: Community Ecology Package. R Package Version 1.17-0. Available online: http://cran.r-project.org/web/packages/vegan/ (accessed on 12 September 2021).
46. Templ, M.; Hron, K.; Filzmoser, P. robCompositions: An R-package for Robust Statistical Analysis of Compositional Data. In *Compositiona Data Analysis: Theory and Applications*; John Wiley & Sons: Hoboken, NJ, USA, 2011; pp. 341–355.

Article

Modulation of Adhesion Molecules Expression by Different Metalloproteases Isolated from *Bothrops* Snakes

Bianca C. Zychar [1,*], Patrícia B. Clissa [2], Eneas Carvalho [3], Adilson S. Alves [4], Cristiani Baldo [5], Eliana L. Faquim-Mauro [2] and Luís Roberto C. Gonçalves [1,*]

1 Laboratory of Pathophysiology, Butantan Institute, São Paulo 05503-900, Brazil
2 Laboratory of Immunopathology, Butantan Institute, São Paulo 05503-900, Brazil; patricia.clissa@butantan.gov.br (P.B.C.); eliana.faquim@butantan.gov.br (E.L.F.-M.)
3 Laboratory of Bacteriology, Butantan Institute, São Paulo 05503-900, Brazil; eneas.carvalho@butantan.gov.br
4 Department. of Physiology and Biophysics, Institute of Biomedical Sciences, University of São Paulo, São Paulo 05503-900, Brazil; adilson@icb.usp.br
5 Department of Biochemistry and Biotechnology, State University of Londrina, Paraná 86051-990, Brazil; critianibaldo@uel.br
* Correspondence: bianca.zychar@butantan.gov.br (B.C.Z.); luis.goncalves@butantan.gov.br (L.R.C.G.)

Abstract: Snake venom metalloproteinases (SVMP) are involved in local inflammatory reactions observed after snakebites. Based on domain composition, they are classified as PI (pro-domain + proteolytic domain), PII (PI + disintegrin-like domains), or PIII (PII + cysteine-rich domains). Here, we studied the role of different SVMPs domains in inducing the expression of adhesion molecules at the microcirculation of the cremaster muscle of mice. We used Jararhagin (Jar)—a PIII SVMP with intense hemorrhagic activity, and Jar-C—a Jar devoid of the catalytic domain, with no hemorrhagic activity, both isolated from *B. jararaca* venom and BnP-1—a weakly hemorrhagic P1 SVMP from *B. neuwiedi* venom. Toxins (0.5 µg) or PBS (100 µL) were injected into the scrotum of mice, and 2, 4, or 24 h later, the protein and gene expression of CD54 and CD31 in the endothelium, and integrins (CD11a and CD11b), expressed in leukocytes were evaluated. Toxins induced significant increases in CD54, CD11a, and CD11b at the initial time and a time-related increase in CD31 expression. In conclusion, our results suggest that, despite differences in hemorrhagic activities and domain composition of the SVMPs used in this study, they behave similarly to the induction of expression of adhesion molecules that promote leukocyte recruitment.

Keywords: *Bothrops*; metalloproteases; inflammation; microcirculation; adhesion molecules; leukocyte-endothelium interactions

Key Contribution: We show that SVMPs with different domain compositions can induce the expression of adhesion molecules into the microvasculature of mouse cremaster muscle; independent of their hemorrhagic activity, which can contribute to the knowledge of local inflammatory reactions observed in *Bothrops* snakebites.

Citation: Zychar, B.C.; Clissa, P.B.; Carvalho, E.; Alves, A.S.; Baldo, C.; Faquim-Mauro, E.L.; Gonçalves, L.R.C. Modulation of Adhesion Molecules Expression by Different Metalloproteases Isolated from *Bothrops* Snakes. *Toxins* **2021**, *13*, 803. https://doi.org/10.3390/toxins13110803

Received: 18 October 2021
Accepted: 9 November 2021
Published: 15 November 2021

Publisher's Note: MDPI stays neutral with regard to jurisdictional claims in published maps and institutional affiliations.

1. Introduction

Snakebite accidents in humans are a severe global health problem that mainly affects the poor and economically active population, causing significant social issues [1,2]. The World Health Organization (WHO) estimates that approximately 2.7 million snakebite accidents occur annually worldwide. As a result of these accidents, an estimated 81,000 to 138,000 deaths occur, and 400,000 survivors experience permanent sequelae [3,4].

In Brazil, an average of 29,000 snakebites occur annually, resulting in approximately 120 deaths/year [5], in addition to underreported cases. Of these 29,000 envenomations, approximately 90% are caused by snakes of the *Bothrops* genus [6].

In *Bothrops* snakebites, severe tissue loss at the site of the bite is observed as a result of hemorrhage and an exacerbated local inflammatory response induced by the venom.

Among the components present in these venoms, metalloproteases are considered to be some of the main factors responsible for local inflammation and necrosis [7–9].

Some of the most abundant proteins found in *Bothrops* venoms are snake venom metalloproteinases (SVMPs), zinc-dependent proteinases that belong to the reprolysin subfamily [10]. Analysis of the gene expression in the venom gland of *Bothrops jararaca* (*Bj*) snakes showed that more than 50% of transcribed genes are SVMPs [10]. SVMPs are classified as PI to PIII, according to the presence or absence of disintegrin and cysteine-rich domains together with a typical metalloproteinase domain, at least in the precursor molecule form [11].

Several hemorrhagic metalloproteases have been isolated from *Bothrops jararaca* venom (*Bj*V) [12–14]. One of the best characterized proteins is jararhagin (Jar), a protein with a molecular weight of 52 kDa that contains PI, metalloproteinase, ECD-disintegrin (ECD: Glu-Cys-Asp) and cysteine-rich domains [14], which are characteristics of an SVMP of the PIII class. In general, the toxins belonging to this class are highly hemorrhagic, and this activity depends on the metalloproteinase domain. However, ECD-disintegrin and cysteine-rich domains have also been shown to be important for the biological functions of these toxins [15–17].

Two forms of Jar are present in *Bj*V: the molecule with the three domains, as described above, and jararhagin-C (Jar-C), a nonhemorrhagic molecule containing only the ECD-disintegrin and cysteine-rich domains produced by the proteolytic cleavage of Jar [18]. The absence of the catalytic site in this toxin does not prevent the inflammatory activity of the toxin. Jar-C triggers the local release of cytokines and induces alterations in leukocyte-endothelium interactions, similar to the inflammatory response caused by Jar [9,19].

PI-class SVMPs contain only a catalytic domain, have a molecular mass ranging from 20 to 30 kDa, possess fibrin(ogeno)lytic activity, and mostly present weak hemorrhagic activity. Metalloproteases isolated from *Bothrops neuwiedi* venoms (*Bn*V), such as neuwiedase (PI-class SVMP), degrade fibrinogen, fibrin, type I collagen, fibronectin, and laminin and induce inflammatory reactions [20].

Two other PI-SVMPs analogous to neuwiedase were isolated from *Bn*V: BnP1 and BnP2. However, only BnP1 exerts biological effects similar to those of Jar on the hydrolysis of coagulation factors, the degradation of extracellular matrix components, and the apoptosis induction in endothelial cells. Nevertheless, Jar possesses strong hemorrhagic activity, while BnP1 presents weak hemorrhagic activity [21].

The inflammatory response starts with sequential and orchestrated phenomena in the vascular endothelium, resulting in the loss of vasomotor reactivity and cell recruitment. Endothelial dysfunction causes an imbalance in microcirculatory homeostasis, activating the immune response that triggers the production and release of inflammatory mediators, such as tumor necrosis factor-alpha (TNF-α), interleukin-1 (IL-1) and interleukin-6 (IL-6) and the subsequent release of vasoactive substances, alteration of blood flow and increased vascular permeability and leukocyte migration to inflamed tissues after *Bothrops* envenomation [19,22].

In addition, the release of these chemotactic agents mediates a sequence of adhesive contacts between leukocytes and endothelial cells [23]. These adhesive contacts, as well as the exit of these cells from postcapillary venules, are mediated by the expression of adhesion molecules [24]. Leukocyte transmigration to tissue is composed of a complex series of events depending on the time and expression levels of various adhesion molecules through a precise mechanism with additive/cooperative interaction potential that is mediated by the opening of junctions of adjacent endothelial cells (via para-cellular) or through the cell body (transcellular) [25,26].

Cell adhesion molecules are glycoproteins expressed on the cell surface that mediate contact between two cells or between cells and the endothelium. Three families of adhesion molecules have been identified: (1) selectins, which are predominantly responsible for the initial contact of leukocytes with the endothelial vasculature in the leukocyte recruitment and "rolling" stages [27]; (2) integrins, which are composed of the α subunit, also known as CD11, and β subunit and mediate the firm adhesion of leukocytes to the endothelium, especially integrins of the β2 or CD18 family [28]; and (3) CAMs (cell adhesion molecules), proteins belonging to the immunoglobulin superfamily that are involved in adhesion and migration between leukocytes and endothelial cells [27,28].

Few studies have examined the participation of different domains of metalloproteases in inflammatory reactions, and even fewer have assessed their roles in modulating the expression of adhesion molecules during the process of leukocyte recruitment and cell migration after exposure to these toxins. An understanding of the mechanisms that contribute to the inflammatory response caused by *Bothrops* envenomation may help resolve the local reaction observed. Thus, this study investigated the effects of different SVMP domains on modulating the expression of adhesion molecules on leukocytes and the microvasculature of mice exposed to three different toxins isolated from *Bothrops* venoms: Jar and Jar-C, which are a PIII-SVMP and disintegrin-like protein isolated from *Bj*V, respectively, and BnP1, a PI-SVMP with weak hemorrhagic action isolated from *Bn*V.

2. Results

2.1. Expression of the ICAM-1 and PECAM-1 Proteins in the Cremaster Muscle

The microcirculation of the cremaster muscle in animals injected with the toxins Jar, Jar-C, and BnP1 presented positive labeling for both adhesion molecules compared to the microcirculation of the control group. When the levels of these proteins were quantified using ELISA, these three toxins induced a similar pattern of protein expression based on the level of fluorescence and time course of appearance (Figure 1A–E).

Toxins induced an increase ICAM-1 expression in the early stages (2 h and 4 h) compared with the control group (Figure 1A,B); however, at 24 h postinjection, the group that received BnP1 did not exhibit a morphological difference from the control group (Figure 1C). Quantification using the immunoassay confirmed the increase in the expression of ICAM 1 in groups injected with the toxins 2 and 4 h after the injection compared to the control group. A statistically significant difference was observed in animals injected with Jar between 2 and 4 h, and this pattern was exclusively induced by Jar (Figure 1D). Twenty-four hours after the injection of those three toxins, the concentration of ICAM-1 returned to a value similar to its initial level and did not differ from the control group (Figure 1D).

PECAM-1 expression in the microvasculature of cremaster muscle was significantly increased 4 h and 24 h after the injection of the three toxins when compared to the control group, as evidenced by the results of immunofluorescence assays (Figure 1B,C). The quantification of PECAM-1 levels using ELISA revealed constitutive expression in the control group, and an evident increase in the expression of this molecule was induced by the three toxins over time. This difference was greater at 24 h after toxin injection (Figure 1E). Notably, no nonspecific immunofluorescence labeling for the primary antibody was observed, since staining was also conducted using samples from in naive animals (data not shown).

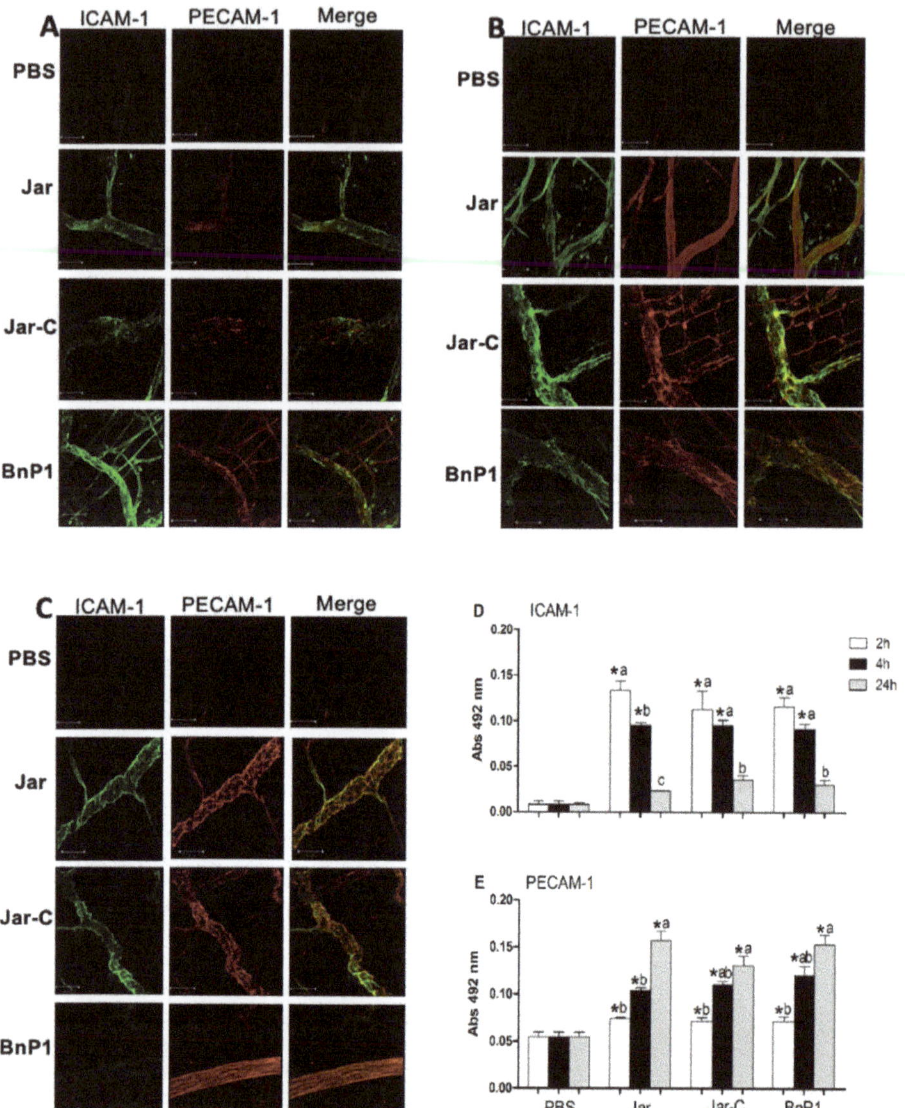

Figure 1. Expression of ICAM-1 and PECAM-1 in the cremaster muscle at different time points. The panels present photomicrographs of immunofluorescence labeling analyzed using confocal microscopy at 2 h (**A**), 4 h (**B**) and 24 h (**C**) after the subcutaneous injection of PBS, Jar, Jar-C or BnP1 (0.5 μg/100 μL) in the mouse scrotum. Bar, 50 μm. Quantification of the protein levels of ICAM-1 (**D**) or PECAM-1 (**E**) in homogenates of cremaster muscle determined using ELISA. Absorbance was measured at 492 nm. The results are presented as the means ± S.E.M. ($n = 5$). * $p < 0.05$ compared to the PBS group at any time point. Different letters (a, b, c) in the same group represent significant differences (* $p < 0.05$).

2.2. Expression of mRNAs Encoding the Adhesion Molecules ICAM-1 and PECAM-1 in the Cremaster Muscle of Mice

The expression of the mRNAs encoding ICAM-1 and PECAM-1 present on the vascular endothelium of the cremaster muscle of mice was evaluated in samples obtained at 30 min,

2, and 6 h after the injection of Jar, Jar-C or BnP1 using real-time PCR and compared with the nontreated control group (PBS). According to our results, the three toxins increased the expression of the mRNA encoding the cell adhesion molecule ICAM-1 at 30 min after toxin injection (Figure 2A).

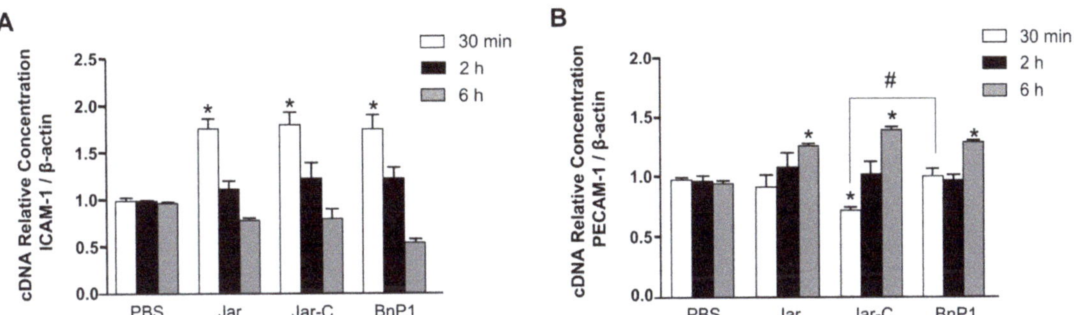

Figure 2. Expression of mRNAs encoding the adhesion molecules ICAM-1 (**A**) and PECAM-1 (**B**) in cremaster muscle. Jar, Jar-C, BnP1 (0.5 μg/100 μL) or PBS (100 μL) was injected, and the muscle was isolated for total RNA extraction. Gene expression was quantified using real-time PCR. Graph bars show the relative expression of each mRNA compared with the PBS group after normalization to the housekeeping gene β-actin. The results are presented as the means ± S.E.M. of three independent experiments ($n = 5$) analyzed in triplicate. * $p < 0.05$ compared to samples treated with the control, # $p < 0.05$ compared to the BnP1 group.

The PECAM-1 mRNA expression pattern differed from that observed for ICAM-1. PECAM-1 was expressed later (6 h) (Figure 2B). At the early time points of 30 min and 2 h, the toxin-injected groups showed no significant differences from the control group. The profile induced by Jar-C showed a slight but significant difference compared to the PBS and BnP1 groups, as PECAM-1 expression decreased in the first 30 min (Figure 2B).

2.3. Expression of CD11a and CD11b on the Surface of Leukocytes Present in the Peritoneal Exudates from Mice Injected with Different Toxins

The expression of the CD11a and CD11b adhesion molecules on the surface of leukocytes present in the mouse peritoneal exudates was investigated using flow cytometry 4 h after the injection of Jar, Jar-C or BnP1 and compared to the control group (Figures 3 and 4). Leukocytes from the PBS-treated group showed basal expression of CD11a (Figure 3A,E), and the number of cells expressing this molecule increased in the groups of mice injected with the three toxins (Figure 3B–E).

The expression of CD11b was not detected in peritoneal exudate cells from the control group. However, two other cell populations showed differences in the expression of the CD11b molecule, i.e., one population exhibited intermediate expression of CD11b and was designated CD11b+ cells, and the other exhibited high expression of this molecule and was designated CD11b++ cells (Figure 4A,E,F). In groups injected with the toxins, an increase in the CD11b+ population was observed compared to that observed in the exudate from the group injected with PBS (Figure 4A–E), but no differences in the CD11b++ population were observed (Figure 4F).

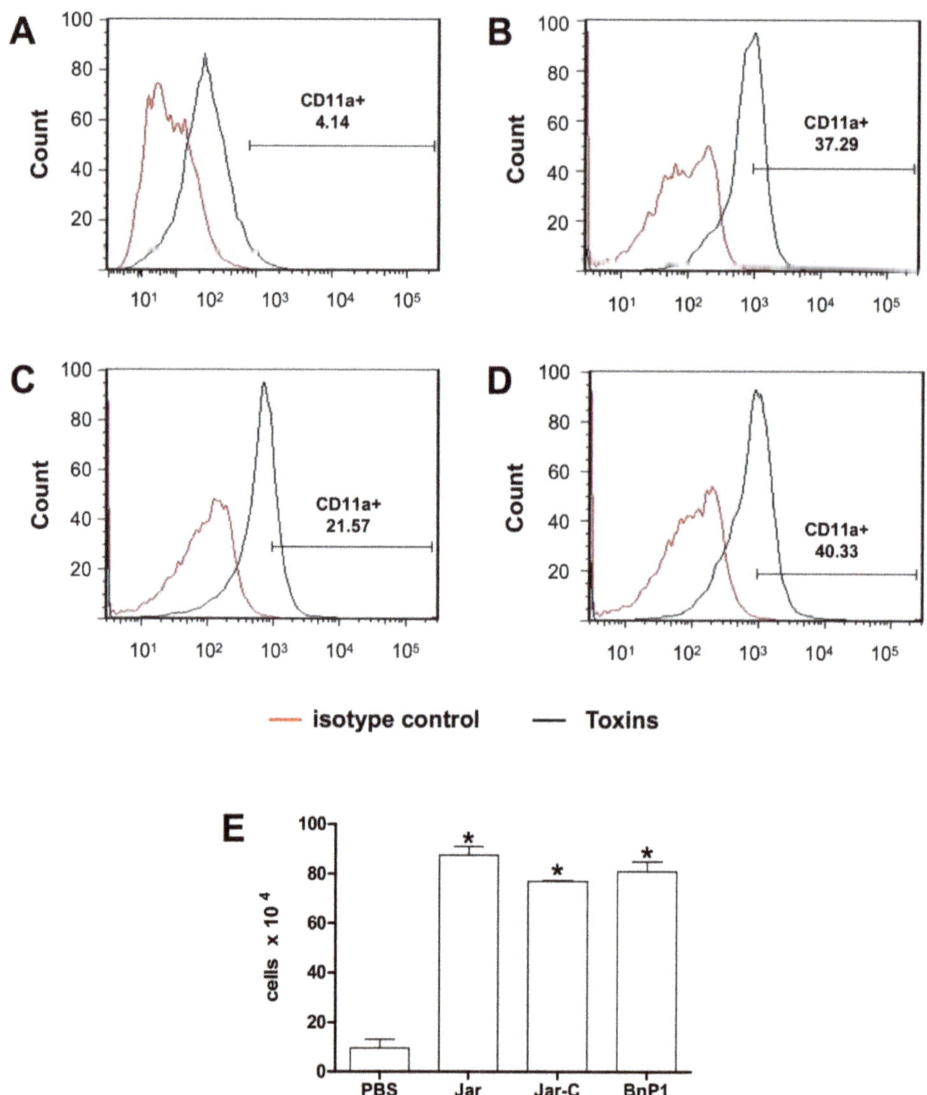

Figure 3. Expression of CD11a on the surface of leukocytes present in the peritoneal exudate of mice injected with different toxins. Peritoneal exudate cell suspensions were obtained within 4 h after the injection of PBS (**A**), Jar (**B**), Jar-C (**C**) or BnP1 (**D**) (2 μg/300 μL). Cells were incubated with anti-CD11a-FITC antibodies or isotype control-FITC. All incubations with anti-CD11a-FITC were performed on duplicate samples and analyzed using flow cytometry. Histograms are representative of one experiment. The bar graph shows the average numbers of CD11a-positive cells in each experimental group ± SD from three independent experiments (**E**). * $p < 0.05$ compared to the PBS group.

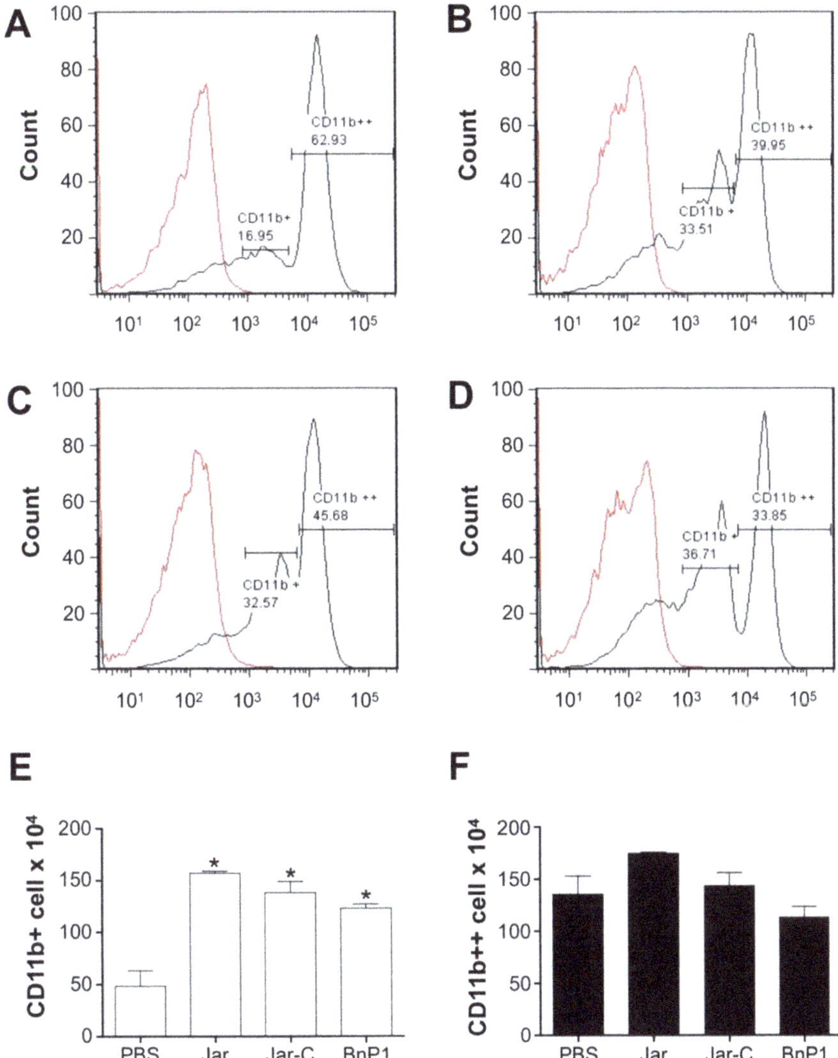

Figure 4. Expression of CD11b on the surface of leukocytes present in the peritoneal exudate of mice injected with different toxins. Peritoneal exudate cell suspensions were obtained within 4 h after the injection of PBS (**A**), Jar (**B**), Jar-C (**C**) or BnP1 (**D**) (2 μg/300 μL). Cells were incubated with anti-CD11b-FITC antibodies or isotype control-FITC. All incubations with anti-CD11b-FITC were performed on duplicate samples and analyzed using flow cytometry. Histograms are representative of one experiment. The bar graph shows the average numbers of positively stained cells designated as CD11b+ (**E**) or CD11b++ (**F**) in each experimental group ± SD from three independent experiments. * $p < 0.05$ compared to the PBS group.

3. Discussion

As shown in our study, three SVMP toxins, Jar, Jar-C, and BnP1, isolated from *Bothrops* venoms, induce the expression of adhesion molecules on the microvasculature of murine cremaster muscle. Each of these toxins shows essential differences in the compositions of their structural domains. In general, PIII SVMPs (Jar) are more hemorrhagic than PI SVMPs (BnP1), although both degrade extracellular matrix molecules [29,30]. The stronger hemorrhagic activity of PIII SVMPs has been attributed to the presence of a disintegrin-like domain adjacent to the catalytic domain. This disintegrin domain favors the anchoring of the molecule to the basement membrane, amplifying the hemorrhagic potential [30]. Some SVMPs of *Bothrops* venoms degrade components of the basement membrane, increasing vascular permeability, and induce the expression of adhesion molecules [30,31]. Although Jar-C does not contain the catalytic domain, the ECD-disintegrin and cysteine-rich domains have also been shown to be essential for the inflammatory function of these toxins [9,16].

Unlike the hemorrhagic action of the SVMPs, the toxins Jar, Jar-C, and BnP1 exerted similar effects on inducing the expression of adhesion molecules responsible for changes in the leukocyte-endothelium interaction, regardless of the composition of their domains. Our results reveal the responses of endothelial and leukocyte cells to the presence of toxins through the expression of the key molecules ICAM-1, CD11a, and CD11b in the early stages of cellular migration and the expression of PECAM-1 in the late stages studied (Figures 1–4).

Leukocyte recruitment is regulated by factors that alter microvessels to promote firm adhesion and cell migration to the site of infection or tissue injury, inducing inflammation and subsequent tissue repair [25]. A cascade of events occurs to achieve this effect, including endothelial cell activation, the release of inflammatory mediators and the expression of adhesion molecules. These processes are often amplified by pathogen-associated molecular patterns (PAMPs) [32], damage-associated molecular patterns (DAMPs) [33] or venom-associated molecular patterns (VAMPs) [34] that function as alarmins stimulating the innate immune response alone or through interactions with cytokines and chemicals.

Bothrops venoms induce inflammatory responses and activate signaling pathways that culminate in the transcription of inflammatory genes such as cytokines and eicosanoids [35,36], inducing endothelial activation to promote capture, rolling, firm adhesion and cell migration [37,38].

Previous studies by our group have identified the critical role for SVMPs in the proinflammatory cytokine action induced by *Bothrops jararaca* and *Philodryas patagoniensis* venoms; both venoms contain a large amount of metalloproteases [8,39]. Additionally, we recently reported that Jar, Jar-C, and BnP1 induce cell adhesion and migration in postcapillary venules in the cremaster muscle of mice, as observed using intravital microscopy [9]. These data corroborate the increase in the expression of adhesion molecules observed in the present study.

Other PIII SMVPs, namely HF3 isolated from *Bothrops jararaca* venom and patagonfibrase isolated from the Dipsadidae snake *Philodryas patagoniensis*, also induce changes in leukocyte-endothelial interactions [39,40].

The recruitment of leukocytes to the injured tissues is a characteristic event of the inflammatory response, and the leukocyte-endothelium interaction is promoted by the expression of cell adhesion glycoproteins on the surface of leukocytes and endothelial cells. This event occurs mainly when integrins on leukocytes bind to adhesion molecules on the endothelium [23–25,41].

Increases in the expression of ICAM-2, PECAM-1, and Jam-A have been described in the literature after the injection of TNF-α and IL-1β, and the interaction between these adhesion molecules facilitates the migration of leukocytes to extravascular tissue [25,41]. Notably, the proinflammatory role of Jar might be attributed to its ability to process the tumor necrosis factor (TNF-α) precursor [7], and it is still capable of inducing the mRNA expression of proinflammatory cytokines such as TNF-α, IL-6, and IL-1β [42]. Additionally, once the leukocyte adheres to ICAM-1, this binding increases its avidity through interactions with integrins, facilitating the transmigration process on the leukocyte surface [43]. This observation might explain the increased expression of ICAM-1, CD11a and CD11b observed in the first hours evaluated after the injection of the different toxins (Figures 1, 2A, 3 and 4).

In general, leukocytes are recruited during an inflammatory insult, and this mobility is essential for the defense of the organism. Neutrophils are more frequently observed and are quickly recruited and activated [41], although other circulating leukocytes also participate in this process, depending on the inflammatory stimulus. Based on our results, both LFA-1 (CD11a/CD18) and Mac-1 (CD11b/CD18) are expressed after the injection of toxins, suggesting that both adhesion molecules participate in this leukocyte-endothelium interaction. However, the differences between these molecules should be highlighted. Both LFA-1 and MAC-1 are members of the β2 integrin family but are expressed in different cell types. MAC-1 is present on neutrophils and monocytes [25,42], identifying the two types of cell populations present in the peritoneal exudate from mice 4 h after the injection of the different toxins (Figure 4), whereas LFA-1 is expressed on all effector leukocytes (Figure 3).

Phillipson et al. [44] described that the adherence of neutrophils to the activated endothelium is mainly mediated by LFA-1, but intraluminal crawling, which follows adherence and spreading, depends on MAC-1. Sumagin et al. [45] observed an increase in CD11b expression in monocytes and a decrease in this expression in neutrophils in postcapillary venules of the cremaster muscle of mice without inflammatory stimulation. Some authors have also shown that the MAC-1 blockade prevents leukocyte recruitment [46,47]. The balance and regulation of this signaling via integrins contribute to leukocyte transmigration guided by endothelial cells.

PECAM-1 is the central adhesion molecule related to cellular transmigration. This protein is expressed at high levels in endothelial cells. It accumulates in intercellular contacts and is associated with the junctional regulation of endothelial complexes such as β-catenin, VE-cadherin, and adherent endothelial junctions, signaling remodeling of the cytoskeleton during the transmigration process [44,48]. We observed a significant increase in the expression of the PECAM-1 mRNA (Figure 2B) and protein (Figure 1E) in the cremaster muscle of mice injected with all toxins at the later time points compared with the ICAM-1 molecule that is responsible for firm adhesion, which was expressed in the first studied periods (Figures 1D and 2A).

These molecules have also been shown to play important roles in other studies examining *Bothrops* venoms. L-selectin, ICAM-1, and PECAM-1 molecules participate in the leukocyte recruitment induced by *Bothrops asper* venom, or BaP1, a P1 SVMP [31,35]. This inflammatory effect was related to the actions of inflammatory cytokines and leukotrienes released by the venom that act on endothelial cells and leukocytes [49].

The results obtained in the present study confirm the importance of metalloproteases in inducing the local inflammatory response observed in previous studies [49–51], particularly in microcirculatory changes [8,9]. Nevertheless, our results reveal the initial participation of the molecules ICAM-1, CD11a, and CD11b and the late expression of PECAM-1. This expression follows the pattern of cell adhesion and migration in vivo and is compatible with acute inflammation and consistent with results observed in other studies [9,38]. The participation of other adhesion molecules in the observed changes should not be excluded.

4. Conclusions

In the present study, we compared different SVMPs: Jar, a PIII with high hemorrhagic activity, Jar-C, which is devoid of hemorrhagic activity, and BnP1, a PI SVMP with low hemorrhagic activity. Despite the differences in the composition and hemorrhagic actions of these toxins, they similarly modulate the expression of adhesion molecules that promote leukocyte recruitment. In addition, their expression followed a sequential pattern, consistent with the recruitment and adhesion processes that were observed in previous in vivo studies performed using intravital microscopy. Therefore, we suggest a possible mechanism underlying the leukocyte-endothelium interaction after exposure to snake venom toxins. A similar pattern was observed even in other inflammatory pathologies. These data contribute to the understanding of local reactions observed after envenomation induced by *Bothrops* snakebites. The same approach can be used to evaluate if other toxins, such as PLA_2 and LAAO present in other *Bothrops* venoms, contribute to the inflammatory process observed in this envenoming.

5. Materials and Methods

5.1. Animals

Male Swiss mice weighing 20–25 g supplied by the housing facility of the Instituto Butantan were used. The animals were maintained for two days in a 12:12 h light: dark cycle and received water and food ad libitum, before experiments. All experimental procedures were conducted according to the ethical parameters proposed by the International Society of Toxinology and the Brazilian College of Experimental Animals and were approved by the Ethical Committee for the Use of Animals of Butantan Institute (n° 466/08).

5.2. Toxins

Venoms of *Bothrops jararaca* and *B. neuwiedi* were obtained from snakes housed in captivity at the Butantan Institute Serpentarium (Laboratório de Herpetologia). Jar and Jar-C were isolated from *B. jararaca* snake venom, and BnP1 was isolated from *B. neuwiedi* snake venom, as previously described [13,19,21]. The purity of toxins was tested using 12.5% SDS–PAGE under reducing conditions [52], and all samples were subjected to treatment with Triton X-114 to remove eventual remaining LPS contamination [53]. After Triton X-114 treatment, LPS absence was confirmed by the Limulus Amebocyte Lysate test (LAL, Charles River). Jar and BnP1 used for the LAL assay were previously heated at 70 °C to inactivate the proteolytic activity, and the biological activities of toxins after Triton X-114 treatment were tested by determining the hemorrhagic activity (Jar and BnP1) or by performing a platelet aggregation inhibition assay (Jar-C). All the toxins used in the experiments were LPS-free. The doses of toxins used were based in a previous study [9].

5.3. Expression of Adhesion Molecules

5.3.1. Evaluation of Protein Expression in Endothelial Cells of the Cremaster Muscle Using Immunohistochemistry

Adhesion molecules expression was assessed 2, 4, or 24 h after the subcutaneous administration of toxins (0.5 μg/100 μL) or 100 μL of phosphate-buffered saline (PBS, as a negative control) in the scrotum (n = 5/group). For the immunohistochemical study, animals were anesthetized and transcardiacally perfused with PBS followed by a fixative solution comprising 4% paraformaldehyde (PFA) dissolved in 0.1 M phosphate buffer (PB, pH 7.4). After perfusion, the cremaster muscle was collected, postfixed with 4% PFA for 6 h, and then transferred to cryoprotectant solution of 30% sucrose in PB for 48 h. The cremaster muscle was incubated with anti-ICAM-1 and anti-PECAM-1 primary antibodies (1:100, Biosciences Product) diluted in 0.3% Triton X-100 for 48 h at room temperature with constant agitation. The tissue was washed with PBS (3 times/10 min) and incubated with the secondary antibody (FITC-conjugated AffiniPure goat anti-mouse IgG and TRICT-conjugated AffiniPure, Jackson ImmunoResearch Laboratories; diluted 1:50 in 0.3% Triton X-100) for 2 h at room temperature. After three washes with PBS, the tissues were mounted on slides with Vectashield (Vector Labs, Burlingame, CA, USA) [30]. The tissues were stored at 4 °C, and the fluorescence was measured using a Zeiss confocal microscope (Zeiss LSM Image Browser program version 4.0.0.157).

5.3.2. Evaluation of Protein Expression in Endothelial Cells of the Cremaster Muscle Using Indirect ELISA Assays

Adhesion molecules expression was also evaluated using indirect ELISA assays [54] after the same treatment described for the immunohistochemical assay. The cremaster muscle was removed, placed in 200 μL of 1 M Trisma solution (pH 7.5) containing protease inhibitors (EDTA 1 mM, PMSF 2 mM and aprotinin 0.3 μM), and disrupted with an ultrasonic homogenizer. The homogenate was centrifuged at 12,000× g for 20 min at 4 °C, and the protein present in the supernatant was quantified using the Bradford method. ELISA microplate wells were coated overnight at 4 °C with 50 μg of protein diluted in 100 mM carbonate-bicarbonate buffer (pH 9.6). The plates were washed three times with PBS-Tween (0.05%) and blocked with 1% bovine serum albumin (BSA) for 1 h at 37 °C.

The plates were washed again and incubated overnight at 4 °C with anti-ICAM-1 and anti-PECAM-1 antibodies (1:500, Biosciences Product). Then, after a new series of three washes, the plates were incubated for 2 h at 37 °C with anti-IgG conjugated with peroxidase (1:2000 —Jackson ImmunoResearch Laboratories). The antibodies were diluted in 1% BSA/PBS-Tween 0.05%. The reaction was visualized using 0.7 mg of OPD (o-phenylenediamine) diluted in 100 mM phosphate-citrate buffer (pH 5.5) containing 0.05% H_2O_2, stopped after 10 min by adding 50 µL/well of 8 N H_2SO_4, and quantified by reading the absorbance at 492 nm using an ELISA plate reader (Multiskan EX Thermo Scientific). All ELISA experiments were performed in triplicate in three independent experiments.

5.4. Real-Time PCR Analysis of Gene Expression in Endothelial Cells of the Cremaster Muscle

The expression of adhesion molecule transcripts was measured 30 min, 2 h, and 6 h after the subcutaneous administration of toxins (0.5 µg/100 µL) or PBS (100 mL) as a negative control into the scrotum. Total RNA was extracted from the dissected cremaster muscle in TRIzol solution (Invitrogen) using an ultrasonic homogenizer. Five micrograms of total RNA were reverse transcribed into cDNAs using Superscript III RT (Invitrogen). Quantitative real-time PCR was performed in a Line Gene K Thermal Cycler (Hangzhou Bioer Technology Co., Hangzhou, China) using a Platinum Syber Green qPCR Supermix-UDG kit (Invitrogen) with the following thermal cycling protocol: 15 min at 95 °C followed by 40 cycles of 15 s at 95 °C, 30 s at 58 °C, and 30 s at 72 °C. All procedures were performed according to the manufacturer's instructions. The primer sequences for mouse β-actin (FW: 5′CCCAGGCATTGCTGACAGG3′; RV: 5′TGGAAGGTGGACAGTGAGGC3′) were designed using sequence alignments obtained from NIH/NCBI based on the published RNA sequence. The primer sequences for mouse ICAM-1 (FW: 5′AAGGAGATCACATTCACGGT3′; RV: 5′GCCTCGGAGACATTAGAGAA3′) and PECAM-1 (FW: 5′AGAGACGGTCTTGTCGCAGTA3′; RV: 5′CGCACACCTGGATCGG3′) were obtained from the literature [55]. The relative mRNA expression level was determined using the $2^{-\Delta\Delta}$Ct (cycle threshold) method, and data were normalized to β-actin mRNA expression levels. All real-time PCR experiments were performed in triplicate using samples from three independent animals.

5.5. Integrin Expression in Peritoneal Leucocytes

The expression of CD11a and CD11b integrins in leukocytes that migrated to the peritoneal cavity in response to toxin injection was evaluated using flow cytometry. For this experiment, the mice were injected (i.p.) with toxins (2.0 µg/300 µL) or sterile PBS and sacrificed in a CO_2 chamber 4 h later. Their peritoneal cavities were washed with 5.0 mL of PBS. The peritoneal fluid was collected and centrifuged (1000× g/10 min at 4 °C), and the precipitate was resuspended in red blood cell lysis buffer and centrifuged (1000× g/10 min at 4 °C). The pellet was resuspended in 1 mL of PBS. The total leukocyte count was determined using an automatic veterinary hematology analyzer (Mindray BC 2800 VET). The peritoneal cells from different animals were pooled to obtain a concentration of 10^6 cells/mL. Cells were incubated with anti-mouse FcγRII/RIII at a concentration of 1 µg/10^6 cells (BD System) for 30 min at 4 °C followed by washes with PBS containing 1% serum albumin (BSA) and centrifugation (300× g/10 min). The expression of different molecules was analyzed by incubating the cells with 1:200 dilutions of Fluorescein (FITC)-conjugated anti-mouse CD11a and CD11b antibodies for 30 min at 4 °C. All cell suspensions were also incubated with the respective fluorescein-labeled isotype control monoclonal antibodies (e-Bioscience). The cells were washed and resuspended in PBS containing 0.1% paraformaldehyde. The flow cytometry analyses (10^4 events per data acquisition file) were performed with FACSCalibur using Cell Quest software (Becton Dickinson). All flow cytometry experiments were performed in triplicate in three independent experiments.

5.6. Statistical Analysis

The results are presented as the means ± standard errors of the different mice/group for each experiment and were analyzed using one-way ANOVA followed by Tukey's test. Differences in results were considered statistically significant when $p < 0.05$.

Author Contributions: Conceptualization: B.C.Z., P.B.C., L.R.C.G.; Investigation: B.C.Z., P.B.C., E.C., E.L.F.-M., A.S.A.; Resources: A.S.A., C.B.; Supervision and Funding acquisition: L.R.C.G.; Writing original draft: B.C.Z., P.B.C., L.R.C.G.; Writing revision/editing: B.C.Z., P.B.C., E.C., A.S.A., C.B., E.L.F.-M., L.R.C.G. All authors have read and agreed to the published version of the manuscript.

Funding: This research was funded by Fundação Butantan and Fundação de Amparo à Pesquisa do Estado de São Paulo—FAPESP (grant n° 2008/00108-8 and 2010/09002-8), and LRCG and ELFM were research fellows of Conselho Nacional de Desenvolvimento Científico e Tecnológico—CNPq (process 308083/2013-0, 309392/2015-2, and 312096/2018-6).

Institutional Review Board Statement: All experimental procedures were conducted according to the ethical parameters proposed by the International Society of Toxinology and the Brazilian College of Experimental Animals and were approved by the Ethical Committee for the Use of Animals of Butantan Institute (n° 466/08), approval date (09/04/2008).

Informed Consent Statement: Not applicable.

Data Availability Statement: The data presented in this study are available on request from the corresponding authors.

Acknowledgments: The authors also thank Luiz R. G. Britto for assistance with his laboratory facilities and for providing helpful suggestions.

Conflicts of Interest: The authors have no conflict of interest regarding this work to declare.

References

1. Gutiérrez, J.M.; Calvete, J.J.; Habib, A.G.; Harrison, R.A.; Williams, D.J.; Warrell, D.A. Snakebite envenoming. *Nat. Rev. Dis. Prim.* **2017**, *3*, 17063. [CrossRef] [PubMed]
2. Harrison, R.A.; Hargreaves, A.; Wagstaff, S.C.; Faragher, B.; Lalloo, D.G. Snake envenoming: A disease of poverty. *PLoS Negl. Trop Dis.* **2009**, *3*, e569. [CrossRef] [PubMed]
3. World Health Organization (WHO). *Seventy-First World Health Assembly Resolution WHA71.5 on Addressing the Burden of Snakebite Envenoming*; WHO: Geneva, Switzerland, 2018; Volume 1, pp. 24–26. Available online: http://apps.who.int/gb/ebwha/pdf_files/EB142/B142_R4-en.pdf (accessed on 17 October 2021).
4. Kasturiratne, A.; Wickremasinghe, A.R.; De Silva, N.; Gunawardena, N.K.; Pathmeswaran, A.; Premaratna, R.; Savioli, L.; Lalloo, D.; De Silva, H.J. The global burden of snakebite: A literature analysis and modelling based on regional estimates of envenoming and deaths. *PLoS Med.* **2008**, *5*, e218. [CrossRef] [PubMed]
5. Ministério da Saúde do Brasil. *Manual de Diagnóstico e Tratamento de Acidentes por Animais Peçonhentos*; FUNASA: Brasilia, Brazil, 2001; p. 112.
6. Warrell, D.A. Snake bite. *Lancet* **2010**, *375*, 77–88. Available online: https://linkinghub.elsevier.com/retrieve/pii/S0140673609617542 (accessed on 17 October 2021). [CrossRef]
7. Moura-da Silva, A.M.; Laing, G.D.; Paine, M.J.; Dennison, J.M.; Politi, V.; Cramptom, J.M.; Theakston, R.D. Processing of pro-tumor necrosis factor-alpha by venom metalloproteinase: A hypothesis explaining local tissue damage follwing snakebite. *Eur. J. Immunol.* **1996**, *26*, 2000–2005. [CrossRef]
8. Zychar, B.C.; Dale, C.S.; Demarchi, D.S.; Gonçalves, L.R.C. Contribution of metalloproteases, serine proteases and phospholipases A2 to the inflammatory reaction induced by *Bothrops jararaca* crude venom in mice. *Toxicon* **2010**, *55*, 227–234. [CrossRef] [PubMed]
9. Zychar, B.C.; Clissa, P.B.; Carvalho, E.; Baldo, C.; Gonçalves, L.R.C. Leukocyte recruitment induced by snake venom metalloproteinases: Role of the catalytic domain. *Biochem. Biophys. Res. Commun.* **2020**, *521*, 402–407. [CrossRef]
10. Cidade, D.A.; Simão, T.A.; Dávila, A.; Wagner, G.; Junqueira-De-Azevedo, I.D.L.; Ho, P.L.; Bon, C.; Zingali, R.B.; Albano, R.M. *Bothrops jararaca* venom gland transcriptome: Analysis of the gene expression pattern. *Toxicon* **2006**, *48*, 437–461. [CrossRef]
11. Fox, J.W.; Serrano, S.M.T. Insights into and speculations about snake venom metalloproteinase (SVMP) synthesis, folding and disulfide bond formation and their contribution to venom complexity. *FEBS J.* **2008**, *275*, 3016–3030. [CrossRef]
12. Mandelbaum, F.R.; Reichel, A.P.; Assakura, M.T. Isolation and characterization of a proteolytic enzyme from the venom of the snake *Bothrops jararaca* (Jararaca). *Toxicon* **1982**, *20*, 955–972. [CrossRef]
13. Paine, M.J.; Desmond, H.P.; Theakston, R.G.D.; Crampton, J.M. Purification, cloning and molecular characterization of high molecular weight hemorrhagic metalloproteinase, jararhagin, from *Bothrops jararaca* venom. *J. Biol. Chem.* **1992**, *267*, 22869–22876. [CrossRef]

14. Maruyama, M.; Tanigawa, M.; Sugiki, M.; Yoshida, E.; Mihara, H. Purification and characterization of low molecular weight fibrinolytic/hemorrhagic enzymes from snake (*Bothrops jararaca*) venom. *Enzyme Protein.* **1993**, *47*, 124–135. [CrossRef] [PubMed]
15. Serrano, S.M.; Kim, J.; Wang, D.; Dragulev, B.; Shannon, J.D.; Mann, H.H.; Veit, G.; Wagener, R.; Koch, M.; Fox, J.W. The cysteine-rich domain of snake venom metalloproteinases is a ligand for von Willebrand factor a domains: Role in substrate targeting. *J. Biol. Chem.* **2006**, *281*, 39746–39756. [CrossRef]
16. Moura-Da-Silva, A.M.; Butera, D.; Tanjoni, I. Importance of snake venom metalloproteinase in cell biology: Effects on platelets, inflammatory and endothelial cell. *Curr. Pharm. Des.* **2007**, *13*, 2893–2905. [CrossRef] [PubMed]
17. Tanjoni, I.; Evangelista, K.; Della Casa, M.S.; Butera, D.; Magalhães, G.S.; Baldo, C.; Clissa, P.B.; Fernandes, I.; Eble, J.; Moura-Da-Silva, A.M. Different regions of the class PIII snake venom metalloproteinase jararhagin are involved in binding to α2β1 integrin and collagen. *Toxicon* **2010**, *55*, 1093–1099. [CrossRef]
18. Moura-da Silva, A.M.; Della-Casa, M.S.; David, A.S.; Assakura, M.T.; Butera, D.; Lebrun, I.; Shannon, J.D.; Serrano, S.M.T.; Fox, J.W. Evidence of heterogeneous forms of the snake venom metalloproteinase jararhagin: A factor contributing to snake venom variability. *Arch. Bioch. Biophys.* **2003**, *409*, 395. [CrossRef]
19. Clissa, P.B.; Lopes-Ferreira, M.; Della-Casa, M.S.; Farsky, S.H.P.; Moura-da Silva, A.M. Importance of jararhagin disintegrin-like and cysteine-rich domains in the early events of local inflammatory response. *Toxicon* **2006**, *47*, 591–596. [CrossRef] [PubMed]
20. Rodrigues, V.D.M.; Soares, A.; Guerra-Sá, R.; Rodrigues, V.; Fontes, M.; Giglio, J.R. Structural and functional characterization of neuwidase, anon-hemorrhagic fibrin (ogen)olyitc metalloprotease from *Bothrops neuwiedi* snake venom. *Arch. Bioch. Biophys.* **2000**, *381*, 213–224. [CrossRef]
21. Baldo, C.; Tanjoni, I.; León, I.; Batista, I.; Della-Casa, M.; Clissa, P.; Weinlich, R.; Lopes-Ferreira, M.; Lebrun, I.; Amarante-Mendes, G.P.; et al. BnP1, a novel P-I metalloproteinase from *Bothrops neuwiedi* venom: Biological effects benchmarking relatively to jararhagin, a P-III SVMP. *Toxicon* **2008**, *51*, 54–65. [CrossRef] [PubMed]
22. Hermant, B.; Bibert, S.; Concord, E.; Dublet, B.; Weidenhaupt, M.; Vernet, T.; Gulino-Debrac, D. Identification of proteases involved in the proteolysis of vascular endothelium cadherin during neutrophil transmigration. *J. Biol. Chem.* **2003**, *18*, 14002–14012. [CrossRef]
23. Nolte, D.; Kuebler, W.M.; Muller, W.A.; Wolff, K.D.; Messmer, K. Attenuation of leukocyte sequestration by selective blockade of PECAM-1 or VCAM-1 in murine endotoxemia. *Eur. Surg. Res.* **2004**, *26*, 331–337. [CrossRef] [PubMed]
24. Dejana, E.; Orsenigo, F.; Lampugnani, M.G. The role of adherens junctions and VE-cadherin in the control of vascular permeability. *J. Cell Biol.* **2008**, *121*, 2115–2122. [CrossRef] [PubMed]
25. Ley, K.; Laudanna, C.; Cybulsky, M.I.; Nourshargh, S. Getting to the site of inflammation: The leukocyte adhesion cascade updated. *Nat. Rev. Immunol.* **2007**, *7*, 678–689. [CrossRef] [PubMed]
26. Carman, C.V.; Springer, T.A. A Transmigratory cup in leukocyte diapedesis both through individual vascular endothelial cells and between them. *J. Cell Biol.* **2004**, *167*, 377–388. [CrossRef] [PubMed]
27. Butcher, E.C. Leukocyte-endothelial cell recognition: Three (or more) steps to specificity and diversity. *Cell* **1991**, *67*, 1033–1036. [CrossRef]
28. Rose, D.M.; Alon, R.; Ginsberg, M.H. Integrin modulation and signaling in leukocyte adhesion and migration. *Immunol. Rev.* **2007**, *218*, 126–134. [CrossRef]
29. Baldo, C.; Jamora, C.; Yamanouye, N.; Zorn, T.M.; Moura-da-Silva, A.M. Mechanisms of vascular damage by hemorrhagic snake venom metalloproteinases: Tissue distribution and in situ hydrolysis. *PLoS Negl. Trop. Dis.* **2010**, *4*, e727. [CrossRef]
30. Herrera, C.; Voisin, M.B.; Escalante, T.; Rucavado, A.; Nourshargh, S.; Gutierrez, J.M. Effects of PI and PIII Snake Venom Haemorrhagic Metalloproteinases on the Microvasculature: A Confocal Microscopy Study on the Mouse Cremaster Muscle. *PLoS ONE* **2016**, *11*, e0168643. [CrossRef]
31. Fernandes, C.M.; Zamuner, S.R.; Zuliani, J.P.; Rucavado, A.; Gutiérrez, J.M.; Teixeira, C.F.P. Inflammatory effects of BaP1 a metalloproteinase isolated from *Bothrops asper* snake venom: Leukocyte recruitment and release of cytokines. *Toxicon* **2006**, *47*, 549–559. [CrossRef] [PubMed]
32. Paudel, Y.N.; Angelopoulou, E.; Piperi, C.; Balasubramaniam, V.R.; Othman, I.; Shaikh, M.F. Enlightening the role of high mobility group box 1 (HMGB1) in inflammation: Updates on receptor signalling. *Eur. J. Pharmacol.* **2019**, *858*, 172487. [CrossRef]
33. Harris, H.E.; Andersson, U.; Pisetsky, D.S. HMGB1: A multifunctional alarmin driving autoimmune and inflammatory disease. *Nat. Rev. Rheumatol.* **2012**, *8*, 195–202. [CrossRef]
34. Moreira, V.; Teixeira, C.; Borges da Silva, H.; D'Império Lima, M.R.; Dos-Santos, M.C. The role of TLR_2 in the acute inflammatory response induced by *Bothrops atrox* snake venom. *Toxicon* **2016**, *118*, 121–128. [CrossRef]
35. Zamunér, R.S.; Zuliani, J.P.; Fernandes, C.M.; Gutiérrez, J.M.; Teixeira, F.P.C. Inflammation induced by *Bothrops* asper venom: Release of proinflammatory cytokines and eicosanoids, and role of adhesion molecules in leukocyte infiltration. *Toxicon* **2005**, *46*, 806–813. [CrossRef] [PubMed]
36. Gutierrez, J.M.; Rucavado, A.; Escalante, T.; Herrera, C.; Fernandez, J.; Lomonte, B.; Fox, J.W. Unresolved issues in the understanding of the pathogenesis of local tissue damage induced by snake venoms. *Toxicon* **2018**, *148*, 123–131. [CrossRef] [PubMed]
37. Farsky, S.H.; Goncalves, L.R.C.; Cury, Y. Characterization of local tissue damage evoked by *Bothrops jararaca* venom in the rat connective tissue microcirculation: An intravital microscopic study. *Toxicon* **1999**, *37*, 1079–1083. [CrossRef]
38. Zychar, B.C.; Castro, N.C., Jr.; Marcelino, J.R.; Gonçalves, L.R.C. Phenol used as a preservative in *Bothrops* antivenom induces impairment leukocyte-endothelium alteraction. *Toxicon* **2008**, *51*, 1151–1157. [CrossRef]

39. Peichoto, M.C.; Zychar, B.C.; Tavares, F.L.; Gonçalves, L.R.C.; Acosta, O.; Santoro, M.L. Inflammatory effects of patagonfibrase, a metalloproteinase from *Philodryas patagoniensis* (Patagonia Green Racer; Dipsadidae) venom. *Exp. Biol. Med.* **2011**, *236*, 1166–1172. [CrossRef] [PubMed]
40. Menezes, M.C.; Paes Leme, A.F.; Melo, R.L.; Silva, C.A.; Della Casa, M.; Bruni, F.M.; Lima, C.; Lopes-Ferreira, M.; Camargo, A.C.M.; Fox, J.W.; et al. Activation of leukocyte rolling by the cysteine-rich domain and the hyper-variable region of HF3, a snake venom hemorrhagic metalloproteinase. *FEBS Lett.* **2008**, *582*, 3915–3921. [CrossRef] [PubMed]
41. Ley, K.; Hoffman, H.M.; Kubes, P.; Cassatella, M.A.; Zychlinsky, A.; Hedrick, C.C.; Catz, S.D. Neutrophils: New insights and open questions. *Sci. Immunol.* **2018**, *3*, eaat4579. [CrossRef]
42. Clissa, P.B.; Laing, G.D.; Theakston, R.D.; Mota, I.; Taylor, M.J.; Moura-da-Silva, A.M. The effect of jararhagin, a metalloproteinase from *Bothrops jararaca* venom, on pro-inflammatory cytokines released by murine peritoneal adherent cells. *Toxicon* **2001**, *39*, 1567–1573. [CrossRef]
43. Muller, W.A. Mechanisms of leukocyte transendothelial migration. *Annu. Rev. Pathol. Mech. Dis.* **2011**, *6*, 323–344. [CrossRef]
44. Phillipson, M.; Heit, B.; Colarusso, P.; Liu, L.; Ballantyne, C.M.; Kubes, P. Intraluminal crawling of neutrophils to emigration sites: A molecularly distinct process from adhesion in the recruitment cascade. *J. Exp. Med.* **2006**, *203*, 2569–2575. [CrossRef]
45. Sumagin, R.; Prizant, H.; Lomakina, E.; Waugh, R.E.; Sarelius, I.H. LFA-1 and Mac-1 Define characteristically different intraluminal crawling and emigration patterns for monocytes and neutrophils in situ. *J. Immun.* **2010**, *185*, 7057–7066. [CrossRef] [PubMed]
46. Sisco, M.; Chao, J.D.; Kim, I.; Mogford, J.E.; Mayadas, T.N.; Mustoe, T.A. Delayed wound healing in Mac-1-deficient mice is associated with normal monocyte recruitment. *Wound Repair. Regen.* **2007**, *15*, 566–571. [CrossRef] [PubMed]
47. Wolf, D.; Anto-Michel, N.; Blankenbach, H.; Wiedemann, A.; Buscher, K.; Hohmann, J.D.; Lim, B.; Bäuml, M.; Marki, A.; Mauler, M.; et al. A ligand-specific blockade of the integrin Mac-1 selectively targets pathologic inflammation while maintaining protective host-defense. *Nat. Commun.* **2018**, *9*, 525. [CrossRef]
48. Nourshargh, S.; Hordijk, P.L.; Sixt, M. Breaching multiple barriers: Leukocyte motility through venular walls and the interstitium. *Nat. Rev.* **2010**, *11*, 366–378. [CrossRef]
49. Zamuner, S.R.; Teixeira, C.F.P. Cell adhesion molecules involved in the leukocyte recruitment induced by venom of the snake *Bothrops jararaca*. *Mediators Inflamm.* **2002**, *11*, 351–357. [CrossRef] [PubMed]
50. Dale, C.S.; Goncalves, L.R.C.; Juliano, L.; Juliano, M.A.; Moura-da-Silva, A.M.; Giorgi, R. The C-terminus of murine S100A9 inhibits hyperalgesia and edema induced by jararhagin. *Peptides* **2004**, *25*, 81–89. [CrossRef]
51. Stroka, A.; Donato, J.L.; Bon, C.; Hyslop, S.; De Araujo, A.L. Purification and characterization of a hemorrhagic metalloproteinase from *Bothrops lanceolatus* (Fer-de-lance) snake venom. *Toxicon* **2005**, *45*, 411–420. [CrossRef] [PubMed]
52. Laemmli, U.K. Cleavage of structural proteins during the assembly of the head of bacteriophage T4. *Nature* **1970**, *227*, 680–685. [CrossRef]
53. Aida, Y.; Pabst, M.J. Removal of endotoxin from protein solutions by phase separation using Triton X-114. *J. Immunol. Methods* **1990**, *132*, 191–195. [CrossRef]
54. De Castro, K.L.P.; Lopes-de-Souza, L.; de Oliveira, D.; Machado-de-Ávila, R.A.; Paiva, A.L.B.; de Freitas, C.F.; Ho, P.L.; Chávez-Olórtegui, C.; Duarte, C.G. A Combined Strategy to Improve the Development of a Coral Antivenom against *Micrurus* spp. *Front. Immunol.* **2019**, *10*, 2422. [CrossRef] [PubMed]
55. Guri, A.J.; Hotencillas, R.; Bassaganya-Riera, J. Abscisic acid ameliorates experimental IBD by downregulating cellular adhesion molecule expression and suppressing immune cell infiltration. *Clin. Nutr.* **2010**, *286*, 2504–2516. [CrossRef] [PubMed]

Article

Observation of *Bothrops atrox* Snake Envenoming Blister Formation from Five Patients: Pathophysiological Insights

Sarah N. C. Gimenes [1], Jacqueline A. G. Sachett [2,3], Mônica Colombini [1], Luciana A. Freitas-de-Sousa [1], Hiochelson N. S. Ibiapina [2,4], Allyson G. Costa [2,5], Monique F. Santana [2,4,5], Jeong-Jin Park [6], Nicholas E. Sherman [6], Luiz C. L. Ferreira [4], Fan H. Wen [7], Wuelton M. Monteiro [2,4], Ana M. Moura-da-Silva [1,4,*] and Jay W. Fox [6,*]

1 Laboratório de Imunopatologia, Instituto Butantan, São Paulo 05503-900, SP, Brazil; sarah.gimenes@univie.ac.at (S.N.C.G.); monica.colombini@butantan.gov.br (M.C.); luciana.sousa@butantan.gov.br (L.A.F.-d.-S.)
2 Escola Superior de Ciências da Saúde, Universidade do Estado do Amazonas, Manaus 69050-030, AM, Brazil; jsachett@uea.edu.br (J.A.G.S.); naajibe@gmail.com (H.N.S.I.); allyson.gui.costa@gmail.com (A.G.C.); monique.freire20@gmail.com (M.F.S.); wueltonmm@gmail.com (W.M.M.)
3 Departamento de Ensino e Pesquisa, Fundação de Dermatologia Alfredo da Matta, Manaus 69065-130, AM, Brazil
4 Departamento de Ensino e Pesquisa, Fundação de Medicina Tropical Dr. Heitor Vieira Dourado, Manaus 69040-000, AM, Brazil; ferreira.luiz@gmail.com
5 Departamento de Ensino e Pesquisa, Fundação de Hematologia e Hemoterapia do Amazonas, Manaus 69040-010, AM, Brazil
6 School of Medicine, University of Virginia, Charlottesville, VA 22903, USA; jp2ht@virginia.edu (J.-J.P.); nes3f@virginia.edu (N.E.S.)
7 Núcleo de Produção de Soros, Instituto Butantan, São Paulo 05503-900, SP, Brazil; fan.hui@butantan.gov.br
* Correspondence: ana.moura@butantan.gov.br (A.M.M.-d.-S.); jwf8x@virginia.edu (J.W.F.)

Citation: Gimenes, S.N.C.; Sachett, J.A.G.; Colombini, M.; Freitas-de-Sousa, L.A.; Ibiapina, H.N.S.; Costa, A.G.; Santana, M.F.; Park, J.-J.; Sherman, N.E.; Ferreira, L.C.L.; et al. Observation of *Bothrops atrox* Snake Envenoming Blister Formation from Five Patients: Pathophysiological Insights. *Toxins* **2021**, *13*, 800. https://doi.org/10.3390/toxins13110800

Received: 7 October 2021
Accepted: 9 November 2021
Published: 13 November 2021

Publisher's Note: MDPI stays neutral with regard to jurisdictional claims in published maps and institutional affiliations.

Abstract: In the Brazilian Amazon, *Bothrops atrox* snakebites are frequent, and patients develop tissue damage with blisters sometimes observed in the proximity of the wound. Antivenoms do not seem to impact blister formation, raising questions regarding the mechanisms underlying blister formation. Here, we launched a clinical and laboratory-based study including five patients who followed and were treated by the standard clinical protocols. Blister fluids were collected for proteomic analyses and molecular assessment of the presence of venom and antivenom. Although this was a small patient sample, there appeared to be a correlation between the time of blister appearance (shorter) and the amount of venom present in the serum (higher). Of particular interest was the biochemical identification of both venom and antivenom in all blister fluids. From the proteomic analysis of the blister fluids, all were observed to be a rich source of damage-associated molecular patterns (DAMPs), immunomodulators, and matrix metalloproteinase-9 (MMP-9), suggesting that the mechanisms by which blisters are formed includes the toxins very early in envenomation and continue even after antivenom treatment, due to the pro-inflammatory molecules generated by the toxins in the first moments after envenomings, indicating the need for local treatments with anti-inflammatory drugs plus toxin inhibitors to prevent the severity of the wounds.

Keywords: *Bothrops atrox*; blister; local damage; DAMPs; snake venom; antivenom; snakebite

Key Contribution: Blisters are sometimes observed around the wound of patients bitten by the *Bothrops atrox* snake. Blisters are associated with a poorer prognosis in local envenomations by pit vipers as they increase the chance of infection and extensive necrosis of the local wounds, and antivenoms have usually little impact on preventing blister formation. In this study, using clinical and laboratory-based observations, we increased the knowledge about the mechanisms underlying blister formation, which is initiated by the toxins very early in envenomation and continues even after antivenom treatment, due to the pro-inflammatory molecules generated by the toxins in the first moments after envenomings. This indicates the need for local treatments with anti-inflammatory drugs plus toxin inhibitors to prevent the severity of the wounds.

1. Introduction

Snakebite envenoming (SBE) is classified as a neglected tropical disease that mainly affects rural areas in developing countries [1]. SBE victims display a wide array of symptoms and pathophysiology, largely depending on the type of snake that causes the wound, as well as intrinsic factors, such as the site of the bite, amount of venom injected, the health status of the victim, and the time elapsed before antivenom administration [2]. In Brazil, the northern states, which include the Amazon Forest, have the highest SBE prevalence. *Bothrops atrox* is the species responsible for the majority of SBE in the region, causing a significant detrimental economic and public health impact on the rural communities [3,4]. In this region, the typical local symptoms observed in *B. atrox* envenomation patients include edema, erythema, pain, and effects of tissue damage at the site of the bite, such as necrosis, inflammation and occasionally blistering in the proximity of the wound [5–7].

Blister formation following SBE, particularly in the case of *B. atrox,* occurs well after the bite [8,9]. Generally, blisters are related to a poor local prognosis as they increase the chance of infection and necrosis [10–13]. What is unknown is why there is such a significant delay in the formation of blisters following envenoming and why it appears antivenom treatment does not seem to prevent the appearance of blisters. This leads to a further question of what role venom components play in the pathophysiology of blister formation. It would seem unlikely that the venom components play a direct role in causing a blister given the delay in blister appearance, but perhaps it is possible they play an indirect role, whereby the actions of venom on the host ultimately lead to the production of agents or the activation of pathways that are known to produce blisters in other pathological conditions.

Some studies have attributed blister formation to snake venom metalloproteinases (SVMPs) [14]. SVMPs represent a class of zinc-dependent enzymes with molecular masses ranging from 20 kDa to 110 kDa [15]. This family of venom proteinases has been shown to give rise to the signature of the local and systemic hemorrhage associated with viperid SBEs, as well as playing a role in the recruitment of an immune infiltrate and the local production of cytokines and chemokines [16–18]. These activities are primarily the result of the proteolytic abilities of the SVMPs to degrade a variety of extracellular matrix (ECM) components, including laminin, fibronectin, nidogen, and collagen, giving rise to both structural and functional contributions to the pathophysiology of SBE [19,20]. These toxins have also been implicated in inflammatory reactions associated with envenomation at the onset of local tissue damage [18]. Furthermore, it is relevant that viperid venom in general and its SVMP components, in particular, can produce a wound exudate rich in damage-associated molecular patterns (DAMPs), which contribute to chemokine and cytokine production and tissue permeability [12,13,21]. As such, it is possible that this is a contributing factor by which venom and/or SVMPs may contribute to blister formation.

This report presents clinical and laboratory data derived from five *B. atrox*-envenomed patients focusing on blister production in these patients. These data are discussed to provide insight into blister formation in humans and how this may be important in other features of post-acute envenomation disabilities, and how this information may be considered for application in the clinical care of snakebite patients.

2. Results

2.1. Clinical Observations

The clinical descriptions of the patients included in this study are presented in Table 1. The severity of the envenomation was determined from the patient clinical data, which were primarily focused on the extent of local edema throughout the bitten limb and systemic manifestations of the patient [22]. Moderate envenomation is characterized by significant pain, apparent edema that goes beyond the envenomed anatomical site, and sometimes the presence of blistering. Severe envenoming is classified by the presence of intense pain and extensive edema, involving the entire envenomed limb, often accompanied by blistering, secondary infection, severe bleeding, hypotension, shock, and acute renal failure [22]. Based on these observations, three patients in the cohort (P1, P2, and P3) were clinically classified as suffering from "moderate" envenoming and two patients (P4 and P5) from "severe" envenoming. All patients, except one, were envenomed in the foot and there was a wide range of times recorded for the period of envenomation to hospitalization (2 h–11 h). Notably, all patients received antivenom at the time of admission. The times from envenoming until blister formation ranged from 63 h to 137 h. Surprisingly, a review of the clinical data did not show any correlation between the time of envenoming, hospitalization, antivenom administration, or envenoming severity with the time until the formation of blisters. However, this could simply be a reflection of the small sample size in this preliminary study.

Table 1. Clinical data of patients.

Clinical Data/Patient		**Patient 1**	**Patient 2**	**Patient 3**	**Patient 4**	**Patient 5**
Envenomation severity		Moderate	Moderate	Moderate	Severe	Severe
Gender		Male	Female	Male	Male	Male
Bite site		Foot	Foot	Foot	Leg	Foot
Tourniquet use		No	No	No	Yes	No
Time until hospital admission		5 h	2 h	6 h	11 h	4 h
Time from hospital admission to blister appearance		58 h	135 h	77 h	111 h	80 h
Time from envenomation to blister appearance		63 h	137 h	83 h	122 h	84 h
Blister site *		Foot	Foot	Foot	Leg	Foot
Pain	Time 0	10	8	7	10	10
	Time 24	4	7	5	0	5
(0 to 10 scale) **	Time 48	0	5	4	0	0
	Time 72	0	5	3	0	7
	Time 144	0	4	2	8	0
Edema ***	Time 0	Moderate	Moderate	Mild	Moderate	Moderate
	Time 24	Moderate	Moderate	Moderate	Moderate	Severe
	Time 48	Moderate	Moderate	Moderate	Moderate	Severe
	Time 72	Moderate	Moderate	Moderate	Moderate	Moderate
	Time 144	Moderate	Mild	Mild	Moderate	Moderate
Ecchymosis		No	After 48 h	No	No	No
Necrosis		No	No	No	No	No
Infection		No	Yes	Yes	Yes	Yes
Lactic Dehydrogenase #	Time 0	464	234	715	935	264
	Time 24	335	221	501	305	255
	Time 48	308	115	281	286	344
	Time 72	313	283	302	328	255
	Time 144	285	263	289	230	289
C-reactive protein ##	Time 0	6,5	6,5	48	96	6,5
	Time 24	96	96	48	96	48
	Time 48	96	96	192	24	48
	Time 72	24	96	96	48	96
	Time 144	48	96	96	12	48

* In all patients, blisters were hemorrhagic and appeared in the perilesional area, next to the bite. ** Pain (0–10) scale of 0 to 10 with 10 as highest; *** Edema was classified according to its extension, in segments as recommended by the Brazilian Ministry of Health [22]: mild = 1 to 2 segments, moderate = 3 to 4 segments, and severe = 5 affected segments. # Lactic Dehydrogenase normal value 190 UI/l; ## C-reactive protein normal value 0.8 mg/dL.

The clinical data also provided insight into the systemic and local symptoms experienced in this patient group. Firstly, from a review of the systemic markers, an increased inflammatory state after envenomation was observed in all patients, as evidenced by increased levels of C-reactive protein (CRP). All patients experienced increased CRP levels from time 0 (immediately before antivenom treatment), which confirms that all patients had an acute inflammatory condition following envenoming (Table 1).

From the analysis of the local symptoms, all five patients reported intense pain on the day of hospital admission into the following day. All were noted to have the presence of moderate to severe edema during the time of hospitalization. Interestingly, in this particular patient cohort, blister formation was not evident until after a minimum of 58 h from the time of hospital admission (Table 1). Furthermore, at admission, all patients presented elevated levels of lactic dehydrogenase (LDH) (>190 UI/I), a marker indicative of tissue damage, suggesting local damage was well underway at the time of admission and subsequent antivenom administration.

An important aspect common to the northern region of Brazil, and observed in some of these patients as well, is the long distances and difficulties in medical transport, hence the time from the accident to the patient being seen at the hospital is often longer than six hours [2]. The delays in patient care, along with the use of substances from traditional medicine and inappropriate practices, such as tourniquets, may aggravate the conditions at the bite site, leading to a high frequency of local complications resulting from the envenomings [9]. It is important to emphasize that one of our patients reported the use of tourniquets. Pardal and colleagues (2004) showed that tourniquet use is a common practice in *Bothrops* accidents [23]. Our data showed that the patient (Patient 4) who used a tourniquet before arrival at the hospital showed more severe inflammatory clinical parameters (Table 1). This observation highlights the importance of avoiding alternative methods of envenomation treatment, which may aggravate the patient's clinical condition [22].

In general, it is considered that antivenom administration is relatively modest in its effectiveness at preventing local tissue damage from SBEs. The reason for the lower local efficacy of antivenom has been attributed to the low likelihood of the antivenom reaching the tissues in time to neutralize the critical venom components involved in local damage. In this study, all patients received antivenom therapy well after envenomation and all had delayed blister formation. This suggests that the mechanism through which blisters are formed, regardless of their delayed appearance, happens very early in envenomation.

2.2. *Laboratory Characterization for the Presence of Venom and Antivenom in Patient Serum and Blister Fluid*

Previously, we have shown the presence of venom proteins in human blister fluid resulting from snakebites [12], and therefore we investigated whether venom proteins and antivenom were present in this cohort of patients' blisters, as well as the relative amount of venom present in their serum. We used enzyme-linked immunosorbent assay (ELISA) to evaluate the venom protein concentration present in the serum of patients taken at the time of admission, before antivenom administration, and in their blisters that developed at much later times (Figure 1A). Not unexpectedly, the patients had varying venom concentrations in their serum, no doubt reflecting many different factors. Interestingly, patients 1, 3, and 5 who showed higher levels of serum concentration of venom were also the patients who had the shorter time intervals between envenomation and blister formation suggesting a higher venom amount injected at the time of the bite and thus potentially a more rapid and possibly higher production of the agents involved in subsequent blister formation. Another interesting observation from these data is that Patients 2 and 5 had higher venom concentrations in their blister fluid compared with their serum, suggesting that venom proteins were preferentially concentrated in the tissues adjacent to the wound.

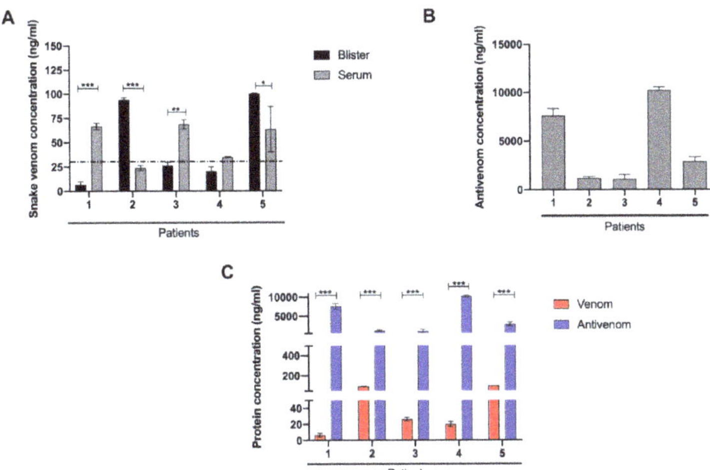

Figure 1. Quantitative analysis of snake venom and antivenom by enzyme-linked immunosorbent assay (ELISA). (**A**) Venom concentration in the serum at the time of patient admission and in the blister fluid. (**B**) Antivenom concentration in the blister fluid. (**C**) Comparative analysis of *B. atrox* venom and antivenom in the blister fluid. Results are expressed as the mean ± sd of three independent readings. *—$p < 0.1$; **—$p < 0.05$, ***—$p < 0.01$.

Figure 1B shows the concentrations of antivenom measured in the contents of the blisters. The presence of antivenom was observed in blister fluids of all patients; however, the relative amounts of antivenom in the blister fluids were variable with Patients 1 and 4, being significantly higher than Patients 2, 3, and 5. Finally, we compared the protein concentrations of both venom and antivenom in the blister fluid (Figure 1C). In all cases, there appeared to be a higher protein concentration of antivenom compared with the venom in each blister, suggesting that the venom components are likely to be complexed with antivenom, with a significant amount of uncomplexed antivenom remaining in the blister. Hence, it is probable that most of the venom components found in the blister did not play a significant direct role in situ for blister formation, but acted to initiate a multi-factorial process well before the blister formation.

To determine whether uncomplexed antivenom was present in the blister fluids as well as to possibly get a sense of what venom components they could detect, we performed Western blots of the fluid against *B. atrox* venom. As seen in Figure 2, all patients' fluids collected from the blisters contained antibodies able to react with *B. atrox* venom components. A closer examination of the Western blots suggested the presence of antibodies in Patients 1, 2, and 4, which could recognize higher molecular mass targets consistent with PIII-class SVMPs (~50 kDa). Interestingly, none of the blister fluids had antibodies that recognized the targets in the region of the PI-class SVMPs (~20 kDa). One potential explanation for this is that the antibody population in circulation represents the antibodies that remain after neutralizing venom components early in antivenom infusion. Thus, it may be that much of the PI-SVMP neutralizing antibodies were depleted from the antivenom before fluid filling these blisters, which appeared well after envenomation and antivenom treatment. However, we also have to consider that the *Bothrops* antivenom presents a higher reactivity to PIII-class SVMPs, recognizing preferentially epitopes located at the Disintegrin-like/Cysteine Rich Domains [24]. Regardless, it is fascinating that both venom, likely in complex with antivenom, and uncomplexed antivenom were detected in the blister fluid. The role this may play, if any, in blister formation will be discussed below.

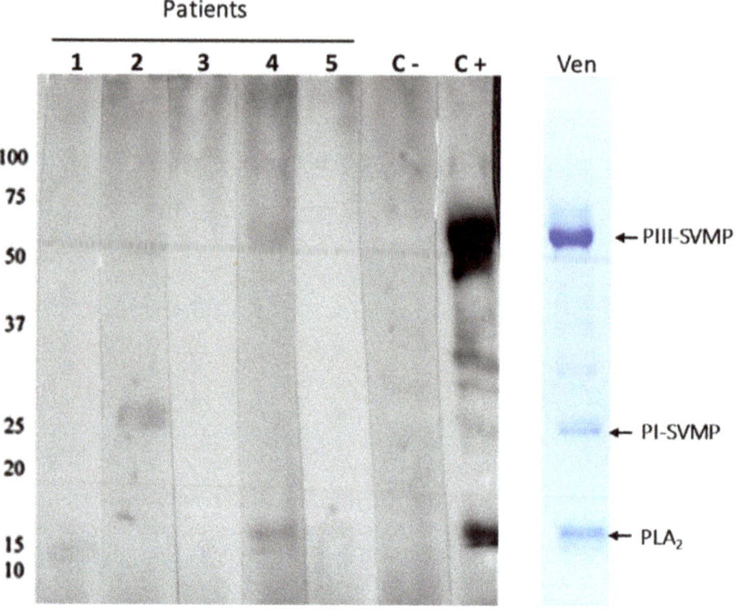

Figure 2. Recognition of *B. atrox* venom antigens by the antivenom present in the Blister fluids. *B. atrox* venom was electrophoresed under reducing conditions in 12% SDS-PAGE gels (Ven). Proteins contained in the gels were transferred to nitrocellulose membranes, which were incubated with the patients' blister fluids (1–5), and the capacity of the antivenom present in the blister content to bind to venom proteins was detected by chemiluminescence using peroxidase-labeled anti-horse antibody and substrate. The images were captured after 15 s of membrane exposure. The positive control (C+) was the commercial antivenom similar to the one used to treat the patients, and the negative control (C−) was the plasma from a volunteer who never received antivenom as treatment. The numbers at the left indicate the migration of molecular mass markers and at the right, the bands corresponding to PIII-class snake venom metalloproteinases (SVMPs), PI-class SVMPs, and Phospholipases A_2 (PLA_2).

2.3. Proteomic Characterization of Patient Blister Fluids

A total of 647 proteins were identified in the blister fluid from the five patients (Supplementary Table S1). Cluster analysis of these data did not provide any correlation to the clinical nor laboratory observations (data not shown). Proteins found in all of the blister fluids were compared to the control serum and included DAMPs, immunomodulators, complement, and ECM proteins (Table 2). Most of the complement proteins were of a lower abundance in the blister fluid than in the control serum, which is not surprising given that it is likely much of the complement system was activated and consumed by envenomation [25]. Interestingly, as observed in our previous studies [12,13,21], there was an increase in a variety of DAMPs in the fluids compared with the control serum, most notably the S100 protein family. In addition, several patients' fluids showed an increase in ECM components, such as the collagens. Another observation was the increase in matrix metalloproteinase-9 (MMP-9) in four of the five patients' blister fluids (Table 2). As MMP-9 has been implicated in blister formation, this was to be expected [26,27].

Table 2. Most abundant extracellular matrix proteins, immunomodulators, and damage-associated molecular patterns (DAMPs) identified in blister fluids.

Identified Proteins	Ac. Number	Patient 1	Patient 2	Patient 3	Patient 4	Patient 5	Control Serum
Complement C3	P01024	106	128	69	264	101	122
Complement factor B	P00751	2	4	4	4	5	56
Complement C4-B	P0C0L5	0	0	3	28	2	37
Complement component C7	P10643	0	0	0	0	1	25
Complement component C9	P02748	0	0	0	0	1	21
Complement factor H	Q03591	0	0	5	0	5	20
Complement factor I	P05156	0	0	1	0	0	16
Complement C1s	P09871	0	1	2	3	1	16
Complement component C8	P07357	0	0	2	0	1	15
Complement C1r	Q9NZP8	1	0	0	4	0	13
Complement factor H-related protein 1	Q03591	0	0	0	0	1	2
Complement C2	Q8SQ75	4	5	0	22	2	0
Fibrinogen alpha chain	P02671	32	37	11	30	14	13
Fibrinogen gamma chain	P02679	1	3	9	7	12	0
Fibrinogen beta chain	P02675	0	2	12	6	11	0
Serum amyloid A-4	P35542	0	0	2	1	2	9
Proteoglycan 4	Q92954	0	2	0	0	0	2
Heat shock 70 kDa protein 1B	P0DMV9	6	25	18	6	37	0
Heat shock protein HSP 90-alpha	P07900	0	0	4	0	3	0
Heat shock protein HSP 90-beta	P08238	0	0	1	0	2	0
Basement membrane-specific heparan sulfate proteoglycan core protein	O05793	0	0	0	1	0	0
Putative histone H2B type 2-C	Q6DN03	0	0	4	2	0	0
Histone H4	P35059	0	8	30	15	0	0
Myosin regulatory light chain sqh	P40423	0	0	1	0	1	0
Myosin-9	P35579	0	6	50	6	84	0
Protein S100-A4	P26447	11	10	6	4	4	0
Protein S100-A6	P06703	3	4	2	0	5	0
Protein S100-A8	P05109	2	5	6	8	23	0
Protein S100-A9	P06702	0	10	19	11	44	0
Protein S100-A12	P80511	0	6	6	6	17	0
Protein S100-A11	P31949	2	3	7	4	8	0
Protein S100-P	P25815	3	5	3	3	8	0
Annexin A1	P04083	0	0	0	1	4	0
Annexin A3	P12429	0	5	0	4	56	0
Annexin A5	Q5R1W0	0	0	0	0	19	0
Annexin A6	P08133	0	0	0	0	4	0
Alpha-2-HS-glycoprotein	P02765	30	15	5	25	3	39
Vitronectin	P04004	2	6	2	4	6	32
Fibronectin	P04937	0	1	4	0	2	17
Lumican	P51884	7	0	1	14	0	10
Proteoglycan 4	Q92954	0	2	0	0	0	2
EGF-containing fibulin-like extracellular matrix protein 1	O35568	0	0	1	1	0	2

Table 2. *Cont.*

Identified Proteins	Ac. Number	Patient 1	Patient 2	Patient 3	Patient 4	Patient 5	Control Serum
Collagen alpha-1(III) chain	P02461	0	0	0	1	0	0
Collagen alpha-1(I) chain	P02452	1	0	2	12	3	0
Collagen alpha-1(VI) chain	P12109	0	0	0	4	0	0
Collagen alpha-1(XXVII) chain	Q5QNQ9	0	0	1	0	0	0
Collagen alpha-3(VI) chain	P12111	0	0	1	0	0	0
Matrix metalloproteinase-9	P14780	0	1	6	1	13	0
Olfactomedin-4	Q6UX06	0	0	0	4	21	0

The numbers in each cell are the abundance, according to the normalized spectra count. The red boxes represent values twice higher; blue boxes represent values twice lower; white boxes represent values without significant difference compared to the normal human serum.

Following the molecular analysis of the endogenous factors, which likely contribute to local damage, we identified the presence of immunomodulators and DAMPs. These were generally observed to be at a higher abundance in the blister fluid than the normal serum (Table 2), reflecting the fact that DAMPs and immunomodulators are well-described markers associated with a pro-inflammatory effect. The presence of immunomodulators and DAMPs in blister fluid suggests that the local microenvironment could at some level be contributing to the blistering phenomenon.

Thus, from the analyses of the local effects, it suggests that blistering is a molecular process that may involve some snake venom toxins, from which SVMPs are good candidates. SVMPs are responsible for ECM protein degradation and the subsequent release of endogenous pro-inflammatory molecules, which increase vessel permeability and contribute to tissue damage and blister formation [21,28,29]. In this context, in the blister contents we found some products associated with ECM degradation, DAMPs, and immunomodulators triggered by snake venom toxins. Rucavado and colleagues (2016) investigated the presence of proinflammatory molecules in the exudate generated from the action of snake venoms in experimental models early after venom injection. Proteomics studies showed the presence of cytokines and DAMPs released after 1 h of snake venom inoculation. In addition, the injection of these same exudates into mice skin increased vascular permeability. These findings suggest that DAMPs and cytokines present in the exudate may interfere with hemostasis through inflammatory pathways, such as Toll-like receptors (TLRs) [21].

The molecular mechanism for tissue damage and blister formation in SBEs by activating the cascade of immunomodulators and TLRs can also be related to endogenous proteinases (MMPs). Endogenous proteinases can be released by infiltrating inflammatory cells triggering proteolytic and inflammatory cascades, which indicates a crucial role of a metalloprotease in wounds [27,30]. In bullous pemphigoid disease, an increasing perivascular inflammatory infiltration and activity of endogenous metalloproteinases, especially MMP-9 enhancing blisters, has been observed [26,27]. The presence of MMP-9 in the blister content from our patients suggests a combined action of endogenous and exogenous proteinases in pathophysiological events. MMPs can be up-regulated by the activation of Toll-like receptor pathways (TLR-4 and TLR-2) [31]. This leads to the speculation that MMP-9 in the blister may act on local ECM proteins and contribute to the blistering effect.

3. Discussion

In general, there is a paucity of studies on snake venom-induced blistering, although when observed in patients they can be quite dramatic, even though they do not significantly contribute to venom morbidity. In one study using an animal model of toxin-induced blister formation, a PI SVMP, BaP1, from *Bothrops asper* was shown to cause blister formation within an hour of injection of 80 μg of the toxin into mouse muscle [14]. In the wound exudate, endogenous MMPs and the toxin were identified along with several extracellular matrix fragments. This suggested to those authors that the SVMP and the endogenous MMPs likely were involved in blister formation via proteolytic degradation of matrix proteins involved in epidermal/dermal association. Another investigation of blister formation utilizing the same toxin, but in a mouse ear model, also produced a rapid formation of blisters [32]. The authors attributed blister formation to the direct proteolytic action of the SVMP on known substrates at the dermal–epidermal junction, such as collagen type IV and laminin. In spite of the use of these elegant models, it is difficult to expect that they fully reflect the situation of blister formation in envenomed humans. In the mouse studies, the authors used a relatively high concentration of the toxin, which could be why blisters formed so rapidly in those models and thus could be attributed to a direct effect of the toxin on key proteins involved with dermal/epidermal association. Furthermore, mouse skin is quite different from human skin, primarily in that human skin is significantly thicker and firm, whereas mouse skin is thinner and loose, being comprised of only two or three cell layers in the epidermis compared to five to ten in humans [33]. Therefore, it seems unlikely that the mechanism of blister formation by the direct action of SVMPs at the dermal–epidermal junction is the same in humans, and thus this warrants different considerations.

In this study, we identified the presence of immunomodulators and DAMPs in blister fluid. These were generally observed to be at a higher abundance in the blister fluid in comparison with normal serum, reflecting the fact that DAMPs and immunomodulators are typically associated with a pro-inflammatory effect, contributing, at some level, to blistering phenomenon. Another important observation was that both antivenom and venom reach and remain in the envenomation region for a long period. However, the presence of antivenom in the blister did not prevent local tissue damage or blister formation. These observations highlight the immediacy of venom action at the site of envenomation activating endogenous pathways that promote tissue damage even after antivenom administration. In this study, it was evident that antivenom reaches the injured limb, and the activation of such endogenous pro-inflammatory pathways explains the general considerations that antivenom administration is relatively modest in its effectiveness in preventing local tissue damage in patients. Blister formation could be an independent process of the local protective effect of antivenom.

In this context, our results suggest that the pathophysiology of blister formation is more likely related to the generation of proinflammatory molecules (DAMPs and immunomodulators) by the toxins before antivenom administration. Although blisters do not play a significant role in morbidity, our investigation also illustrates how the initial action of the venom can contribute to pathophysiology that occurs at a relatively long period after envenoming, regardless of treatment with antivenom. The mechanism by which blisters are formed may be extended to other effects, leading to tissue damage in these patients, which may also be induced before antivenom administration. Finally, as has been long recognized, there remains an urgent need for therapeutic approaches that can rapidly be deployed at the envenomation site to attenuate the toxic activities that occur virtually immediately and have long-lasting effects, both locally and systemically. In this regard, our data strongly support the need for local treatments with anti-inflammatory drugs plus toxin inhibitors to prevent the severity of the wounds.

4. Materials and Methods

4.1. Ethical Statement and Clinical Data Collection

Patients included in the study (5) were hospitalized due to snakebite, at the Fundação de Medicina Tropical Dr. Heitor Vieira Dourado (FMT-HVD), a reference hospital for snakebite treatment in Manaus, State of Amazonas, Brazil. Eligible patients were those diagnosed with clinical signs of envenomation by *Bothrops* snakes, which developed blisterings at the bitten limb during the time of hospital treatment. This study was approved by the Ethics Review Board of FMT-HVD (approval number CAAE 19380913.6.0000.5016/2013). All participants signed a consent form after an explanation of the study aims. Upon admission, epidemiological and clinical information were collected using a standardized questionnaire. Envenomation was classified as mild, moderate, or severe, according to the Brazilian Ministry of Health guidelines [22]. The presence of pain, local bleeding, ecchymosis, necrosis, and systemic bleeding were also recorded. Patients were treated according to the Brazilian Ministry of Health protocols [22]. To collect the blister fluids, the area was cleaned using 2% chlorhexidine, and the fluid was collected using a sterile needle (13 × 4.5 mm) and a 1mL syringe. Subsequently, blister fluid was aliquoted and transferred to a sterile microtube and frozen at −80 °C until analysis.

4.2. ELISA Quantification of Antivenom and Venom in Blister Contents

For antivenom detection, a 96-well plate was coated for 18 h at room temperature with 10 µg/mL of *B. atrox* venom in PBS (100 µL/well). The plates were then washed with PBS containing 0.05% TWEEN 20 solution and then blocked for 2 h at 37 °C with 300 µL/well of phosphate-buffered saline (PBS) containing 2% bovine serum albumin (BSA). After blocking, the plates were washed and incubated for 2 h at 37 °C with the blister fluid 1000 times diluted in PBS containing 1% BSA and 0.05% TWEEN 20 or different dilutions of the standard curve, prepared with a known concentration of IgG (Fab$'_2$ fragment) isolated from the same antivenom used for patient treatment. After a new washing step, peroxidase-labeled anti-horse IgG antibodies (SIGMA) were placed in each well, 2000 times diluted, and incubated at 37 °C for one hour. Antigen/antibody binding was detected by the addition of ortho-phenylenediamine. The antivenom concentration was calculated by linear regression of the standard curve. The same methodology was used for the detection of venom in the blister fluids, in which the plates were coated with Bothrops antivenom in a PBS buffer. After washing, the plates were incubated for 4 h at 37 °C with blister fluid (diluted 1:5 in PBS containing 1% BSA and 0.05% TWEEN 20) and the standard curve was prepared with *B. atrox* venom at known concentrations. The reaction was detected by rabbit serum anti-*B. atrox* venom followed by the addition of peroxidase labelled sheep anti-rabbit IgG (SIGMA) and ortho-phenylenediamine. The plates were read using a wavelength of 492 nm, and the venom concentration was calculated by linear regression of the standard curve. Experiments were carried out in triplicate and the results were expressed as mean ± S.E.M.

4.3. Western Blot of Antivenom in Blister Contents

The reactivity of blister contents with venom proteins was evaluated by Western blot. Briefly, *B. atrox* venom was electrophoresed under reducing conditions in 12% dodecyl sulfate-polyacrylamide gel electrophoresis (SDS-PAGE) gels. Proteins contained in the gels were transferred to nitrocellulose membranes, which were blocked for 2 h at 37 °C with 500 µL of PBS containing 2% BSA (SIGMA-USA). Then, each membrane was incubated overnight at 4 °C with blister fluids from each patient, diluted at 1:2 in PBS containing 1% BSA. Following incubation, the membranes were washed with PBS containing 0.05% TWEEN. The capacity of the antivenom present in the blister content to bind to snake venom proteins immobilized on the nitrocellulose membranes was detected by chemiluminescence using peroxidase-labeled anti-horse antibody and ®Super Signal West Pico substrate from Thermo Scientific (Waltham, MA, USA). The images were captured after 15 s of membrane exposure. The positive control was Butantan commercial antivenom, similar to the one used to treat the patients, and the negative control was the plasma from a volunteer who never received antivenom as treatment.

4.4. Proteomic Analyses

The proteins (~10 ug) contained in the patient blisters and the serum of normal volunteers were submitted to reduction with 10 mM dithiothreitol (DTT) in 0.1 M ammonium bicarbonate and alkylation with 50 mM iodoacetamide in 0.1 M ammonium bicarbonate (both room temperature for 0.5 h). The samples were then digested overnight at 37 °C with 1 µg trypsin in 50 mM ammonium bicarbonate. The samples were acidified with acetic acid to stop digestion and were then spun down. The supernatant was evaporated to 20 µL for liquid chromatography–mass spectrometry (LC–MS) analysis.

The LC–MS system utilized for proteomic analyses was a Thermo Electron Q Exactive HF mass spectrometer system with an Easy Spray ion source connected to a Thermo 75 µm × 15 cm C18 Easy Spray column (through pre-column). Samples of 1 µg were injected and the peptides eluted from the column by an acetonitrile/0.1 M acetic acid gradient at a flow rate of 0.3 µL/min over 2.0 h. The nanospray ion source was operated at 1.9 kV. The analysis produces approximately 25,000 MS/MS spectra of ions ranging in abundance over several orders of magnitude. The data were then analyzed by database searching using the Sequest search algorithm against Uniprot Human. An analysis of the spectra generated was performed using carbamidomethylation on cysteine as a fixed modification, with oxidation of methionine as a variable modification. For the analysis of the results and validation of peptide and protein identifications, data obtained were exported to Scaffold (version 4.3.2, Proteome Software Inc., Portland, OR, USA).

Supplementary Materials: The following are available online at https://www.mdpi.com/article/10.3390/toxins13110800/s1, Supplementary Table S1: Protein identification in blister fluids of *B. atrox* snakebite patients and serum of control volunteers by LC-MS.

Author Contributions: Conceptualization, S.N.C.G., W.M.M., A.M.M.-d.-S. and J.W.F.; methodology, S.N.C.G., M.C., L.A.F.-d.-S., J.-J.P. and N.E.S.; formal analysis, J.A.G.S., A.G.C., L.C.L.F. and F.H.W.; resources, J.A.G.S., H.N.S.I. and M.F.S.; writing—original draft preparation, S.N.C.G.; writing—review and editing, A.M.M.-d.-S., J.W.F. and W.M.M.; supervision, A.M.M.-d.-S. and J.W.F.; funding acquisition, A.M.M.-d.-S. and J.W.F. All authors have read and agreed to the published version of the manuscript.

Funding: This research was funded by Fundação de Amparo à Pesquisa de São Paulo (FAPESP 2016/50127-5; 2017/24546-3; 2018/13108-8; 2019/08208-6); Conselho Nacional de Desenvolvimento Científico e Tecnológico (CNPq) (grant number 303958/2018-9), and Fundação de Amparo à Pesquisa do Estado do Amazonas-FAPEAM (PRO-ESTADO; FAPEAM PPSUS 287/2013); Office of Research Core Administration, University of Virginia. A.M.M.-d.-S. and W.M.M. are CNPq productivity fellows.

Institutional Review Board Statement: Ethical approval for human information collection was obtained from the Fundação de Medicina Tropical Doutor Heitor Vieira Dourado (approval number CAAE 19380913.6.0000.5016/2013 approval date 13 December 2013).

Informed Consent Statement: Informed consent was obtained from all subjects involved in the study.

Conflicts of Interest: The authors declare no conflict of interest. The funders had no role in the design of the study; in the collection, analyses, or interpretation of data; in the writing of the manuscript; or in the decision to publish the results.

References

1. WHO. *Rabies and Envenomings. A Neglected Public Health Issue*; World Health Organization: Geneva, Switzerland, 2007; ISBN 978-92-4-156348-2.
2. Feitosa, E.L.; Sampaio, V.S.; Salinas, J.L.; Queiroz, A.M.; da Silva, I.M.; Gomes, A.A.; Sachett, J.; Siqueira, A.M.; Ferreira, L.C.; Dos Santos, M.C.; et al. Older Age and Time to Medical Assistance Are Associated with Severity and Mortality of Snakebites in the Brazilian Amazon: A Case-Control Study. *PLoS ONE* **2015**, *10*, e0132237. [CrossRef]
3. Magalhães, S.F.V.; Peixoto, H.M.; Moura, N.; Monteiro, W.M.; de Oliveira, M.R.F. Snakebite envenomation in the Brazilian Amazon: A descriptive study. *Trans. R. Soc. Trop. Med. Hyg.* **2019**, *113*, 143–151. [CrossRef]
4. Feitosa, E.S.; Sampaio, V.; Sachett, J.; Castro, D.B.; Noronha, M.; Lozano, J.L.; Muniz, E.; Ferreira, L.C.; Lacerda, M.V.; Monteiro, W.M. Snakebites as a largely neglected problem in the Brazilian Amazon: Highlights of the epidemiological trends in the State of Amazonas. *Rev. Soc. Bras. Med. Trop.* **2015**, *48* (Suppl. 1), 34–41. [CrossRef]
5. Alves, E.C.; Sachett, J.A.G.; Sampaio, V.S.; Sousa, J.D.B.; Oliveira, S.S.; Nascimento, E.F.D.; Santos, A.D.S.; da Silva, I.M.; da Silva, A.M.M.; Wen, F.H.; et al. Predicting acute renal failure in Bothrops snakebite patients in a tertiary reference center, Western Brazilian Amazon. *PLoS ONE* **2018**, *13*, e0202361. [CrossRef]
6. Monteiro, W.M.; Contreras-Bernal, J.C.; Bisneto, P.F.; Sachett, J.; Mendonça da Silva, I.; Lacerda, M.; Guimarães da Costa, A.; Val, F.; Brasileiro, L.; Sartim, M.A.; et al. *Bothrops atrox*, the most important snake involved in human envenomings in the amazon: How venomics contributes to the knowledge of snake biology and clinical toxinology. *Toxicon X* **2020**, *6*, 100037. [CrossRef]

7. Oliveira, S.S.; Alves, E.C.; Santos, A.S.; Pereira, J.P.T.; Sarraff, L.K.S.; Nascimento, E.F.; de-Brito-Sousa, J.D.; Sampaio, V.S.; Lacerda, M.V.G.; Sachett, J.A.G.; et al. Factors Associated with Systemic Bleeding in a Tertiary Hospital in the Brazilian Amazon. *Toxins* **2019**, *11*, 22. [CrossRef]
8. Moura-da-Silva, A.M.; Contreras-Bernal, J.C.; Cirilo Gimenes, S.N.; Freitas-de-Sousa, L.A.; Portes-Junior, J.A.; da Silva Peixoto, P.; Kei Iwai, L.; Mourão de Moura, V.; Ferreira Bisneto, P.; Lacerda, M.; et al. The relationship between clinics and the venom of the causative Amazon pit viper (Bothrops atrox). *PLoS Negl. Trop. Dis.* **2020**, *14*, e0008299. [CrossRef]
9. Wen, F.H.; Monteiro, W.M.; Moura da Silva, A.M.; Tambourgi, D.V.; da Silva, I.M.; Sampaio, V.S.; dos Santos, M.C.; Sachett, J.; Ferreira, L.C.L.; Kalil, J.; et al. Snakebites and Scorpion Stings in the Brazilian Amazon: Identifying Research Priorities for a Largely Neglected Problem. *PLoS Negl. Trop. Dis.* **2015**, *9*, e0003701. [CrossRef]
10. Escalante, T.; Rucavado, A.; Pinto, A.F.M.; Terra, R.M.S.; Gutierrez, J.M.; Fox, J.W. Wound Exudate as a Proteomic Window to Reveal Different Mechanisms of Tissue Damage by Snake Venom Toxins. *J. Proteome Res.* **2009**, *8*, 5120–5131. [CrossRef] [PubMed]
11. Eming, S.A.; Koch, M.; Krieger, A.; Brachvogel, B.; Kreft, S.; Bruckner-Tuderman, L.; Krieg, T.; Shannon, J.D.; Fox, J.W. Differential proteomic analysis distinguishes tissue repair biomarker signatures in wound exudates obtained from normal healing and chronic wounds. *J. Proteome Res.* **2010**, *9*, 4758–4766. [CrossRef]
12. Macêdo, J.K.A.; Joseph, J.K.; Menon, J.; Escalante, T.; Rucavado, A.; Gutiérrez, J.M.; Fox, J.W. Proteomic Analysis of Human Blister Fluids Following Envenomation by Three Snake Species in India: Differential Markers for Venom Mechanisms of Action. *Toxins* **2019**, *11*, 246. [CrossRef]
13. Rucavado, A.; Escalante, T.; Kalogeropoulos, K.; Camacho, E.; Gutiérrez, J.M.; Fox, J.W. Analysis of wound exudates reveals differences in the patterns of tissue damage and inflammation induced by the venoms of Daboia russelii and Bothrops asper in mice. *Toxicon* **2020**, *186*, 94–104. [CrossRef] [PubMed]
14. Rucavado, A.; Núñez, J.; Gutiérrez, J.M. Blister formation and skin damage induced by BaP1, a haemorrhagic metalloproteinase from the venom of the snake Bothrops asper. *Int. J. Exp. Pathol.* **1998**, *79*, 245–254. [PubMed]
15. Fox, J.W.; Serrano, S.M.T. Structural considerations of the snake venom metalloproteinases, key members of the M12 reprolysin family of metalloproteinases. *Toxicon* **2005**, *45*, 969–985. [CrossRef] [PubMed]
16. Clissa, P.B.; Laing, G.D.; Theakston, R.D.G.; Mota, I.; Taylor, M.J.; Moura-da-Silva, A.M. The effect of jararhagin, a metalloproteinase from Bothrops jararaca venom, on pro-inflammatory cytokines released by murine peritoneal adherent cells. *Toxicon* **2001**, *39*, 1567–1573. [CrossRef]
17. Rucavado, A.; Escalante, T.; Teixeira, C.F.P.; Fernandes, C.M.; Diaz, C.; Gutierrez, J.M. Increments in cytokines and matrix metalloproteinases in skeletal muscle after injection of tissue-damaging toxins from the venom of the snake Bothrops asper. *Mediat. Inflamm.* **2002**, *11*, 121–128. [CrossRef]
18. Moura-da-Silva, A.M.; Butera, D.; Tanjoni, I. Importance of snake venom metalloproteinases in cell biology: Effects on platelets, inflammatory and endothelial cells. *Curr. Pharm. Des.* **2007**, *13*, 2893–2905. [CrossRef]
19. Baramova, E.N.; Shannon, J.D.; Bjarnason, J.B.; Fox, J.W. Degradation of extracellular matrix proteins by hemorrhagic metalloproteinases. *Arch. Biochem. Biophys.* **1989**, *275*, 63–71. [CrossRef]
20. Gutierrez, J.M.; Rucavado, A.; Chaves, F.; Diaz, C.; Escalante, T. Experimental pathology of local tissue damage induced by Bothrops asper snake venom. *Toxicon* **2009**, *54*, 958–975. [CrossRef]
21. Rucavado, A.; Nicolau, C.A.; Escalante, T.; Kim, J.; Herrera, C.; Gutiérrez, J.M.; Fox, J.W. Viperid Envenomation Wound Exudate Contributes to Increased Vascular Permeability via a DAMPs/TLR-4 Mediated Pathway. *Toxins* **2016**, *8*, 349. [CrossRef]
22. Brasil. Ministério da Saúde. Secretaria de Vigilância em Saúde. Coordenação-Geral de Desenvolvimento da Epidemiologia em Serviços. Guia de Vigilância em Saúde. Chapter 11. p. 69. Available online: http://bvsms.saude.gov.br/bvs/publicacoes/guia_vigilancia_saude_4ed.pdf (accessed on 8 September 2021).
23. Pardal, P.P.O.; Souza, S.M.; Monteiro, M.R.C.C.; Fan, H.W.; Cardoso, J.L.C.; Franca, F.O.S.; Tomy, S.C.; Sano-Martins, I.S.; Sousa-e-Silva, M.C.C.; Colombini, M.; et al. Clinical trial of two antivenoms for the treatment of Bothrops and Lachesis bites in the north eastern Amazon region of Brazil. *Trans. R. Soc. Trop. Med. Hyg.* **2004**, *98*, 28–42. [CrossRef]
24. Sousa, L.F.; Nicolau, C.A.; Peixoto, P.S.; Bernardoni, J.L.; Oliveira, S.S.; Portes-Junior, J.A.; Mourao, R.H.V.; Lima-dos-Santos, I.; Sano-Martins, I.S.; Chalkidis, H.M.; et al. Comparison of Phylogeny, Venom Composition and Neutralization by Antivenom in Diverse Species of *Bothrops* Complex. *PLoS Negl. Trop. Dis.* **2013**, *7*, e2442. [CrossRef]
25. Stone, S.F.; Isbister, G.K.; Shahmy, S.; Mohamed, F.; Abeysinghe, C.; Karunathilake, H.; Ariaratnam, A.; Jacoby-Alner, T.E.; Cotterell, C.L.; Brown, S.G. Immune response to snake envenoming and treatment with antivenom; complement activation, cytokine production and mast cell degranulation. *PLoS Negl. Trop. Dis.* **2013**, *7*, e2326. [CrossRef]
26. Riani, M.; Le Jan, S.; Plée, J.; Durlach, A.; Le Naour, R.; Haegeman, G.; Bernard, P.; Antonicelli, F. Bullous pemphigoid outcome is associated with CXCL10-induced matrix metalloproteinase 9 secretion from monocytes and neutrophils but not lymphocytes. *J. Allergy Clin. Immunol.* **2017**, *139*, 863–872. [CrossRef]
27. Hiroyasu, S.; Turner, C.T.; Richardson, K.C.; Granville, D.J. Proteases in Pemphigoid Diseases. *Front. Immunol.* **2019**, *10*, 1454. [CrossRef]
28. Gutiérrez, J.M.; Calvete, J.J.; Habib, A.G.; Harrison, R.A.; Williams, D.J.; Warrell, D.A. Snakebite envenoming. *Nat. Rev. Dis Primers* **2017**, *3*, 17063. [CrossRef]
29. Bickler, P.E. Amplification of Snake Venom Toxicity by Endogenous Signaling Pathways. *Toxins* **2020**, *12*, 68. [CrossRef] [PubMed]

30. Niimi, Y.; Pawankar, R.; Kawana, S. Increased expression of matrix metalloproteinase-2, matrix metalloproteinase-9 and matrix metalloproteinase-13 in lesional skin of bullous pemphigoid. *Int. Arch. Allergy Immunol.* **2006**, *139*, 104–113. [CrossRef] [PubMed]
31. Voelcker, V.; Gebhardt, C.; Averbeck, M.; Saalbach, A.; Wolf, V.; Weih, F.; Sleeman, J.; Anderegg, U.; Simon, J. Hyaluronan fragments induce cytokine and metalloprotease upregulation in human melanoma cells in part by signalling via TLR4. *Exp. Dermatol.* **2008**, *17*, 100–107. [CrossRef] [PubMed]
32. Jimenez, N.; Escalante, T.; Gutierrez, J.M.; Rucavado, A. Skin pathology induced by snake venom metalloproteinase: Acute damage, revascularization, and re-epithelization in a mouse ear model. *J. Investig. Dermatol.* **2008**, *128*, 2421–2428. [CrossRef]
33. Zomer, H.D.; Trentin, A.G. Skin wound healing in humans and mice: Challenges in translational research. *J. Dermatol. Sci.* **2018**, *90*, 3–12. [CrossRef] [PubMed]

MDPI

Article

Systemic Effects of Hemorrhagic Snake Venom Metalloproteinases: Untargeted Peptidomics to Explore the Pathodegradome of Plasma Proteins

Luciana Bertholim [1], Alison F. A. Chaves [1], Ana K. Oliveira [1,†], Milene C. Menezes [1], Amanda F. Asega [1], Alexandre K. Tashima [2], Andre Zelanis [3] and Solange M. T. Serrano [1,*]

1 Laboratório de Toxinologia Aplicada, Center of Toxins, Immune-Response and Cell Signalig, CeTICS, Instituto Butantan, São Paulo 05503-900, SP, Brazil; luciana.nasciben@gmail.com (L.B.); alison.chaves@butantan.gov.br (A.F.A.C.); ankaoliv@gmail.com (A.K.O.); menezes.milene@gmail.com (M.C.M.); amanda.asega@gmail.com (A.F.A.)
2 Department of Biochemistry, Escola Paulista de Medicina, Federal University of Sao Paulo, Sao Paulo 04023-901, SP, Brazil; aktashima@unifesp.br
3 Functional Proteomics Laboratory, Department of Science and Technology, Federal University of São Paulo (UNIFESP), 330 Talim St., São José dos Campos 12231-280, SP, Brazil; andre.zelanis@unifesp.br
* Correspondence: solange.serrano@butantan.gov.br
† Present address: Department of Pathology, University of Virginia School of Medicine, Charlottesville, VA 22908, USA.

Citation: Bertholim, L.; Chaves, A.F.A.; Oliveira, A.K.; Menezes, M.C.; Asega, A.F.; Tashima, A.K.; Zelanis, A.; Serrano, S.M.T. Systemic Effects of Hemorrhagic Snake Venom Metalloproteinases: Untargeted Peptidomics to Explore the Pathodegradome of Plasma Proteins. *Toxins* **2021**, *13*, 764. https://doi.org/10.3390/toxins13110764

Received: 19 August 2021
Accepted: 6 October 2021
Published: 28 October 2021

Abstract: Hemorrhage induced by snake venom metalloproteinases (SVMPs) is a complex phenomenon that involves capillary disruption and blood extravasation. HF3 (hemorrhagic factor 3) is an extremely hemorrhagic SVMP of *Bothrops jararaca* venom. Studies using proteomic approaches revealed targets of HF3 among intracellular and extracellular proteins. However, the role of the cleavage of plasma proteins in the context of the hemorrhage remains not fully understood. The main goal of this study was to analyze the degradome of HF3 in human plasma. For this purpose, approaches for the depletion of the most abundant proteins, and for the enrichment of low abundant proteins of human plasma, were used to minimize the dynamic range of protein concentration, in order to assess the proteolytic activity of HF3 on a wide spectrum of proteins, and to detect the degradation products using mass spectrometry-based untargeted peptidomics. The results revealed the hydrolysis products generated by HF3 and allowed the identification of cleavage sites. A total of 61 plasma proteins were identified as cleaved by HF3. Some of these proteins corroborate previous studies, and others are new HF3 targets, including proteins of the coagulation cascade, of the complement system, proteins acting on the modulation of inflammation, and plasma proteinase inhibitors. Overall, the data indicate that HF3 escapes inhibition and sculpts the plasma proteome by degrading key proteins and generating peptides that may act synergistically in the hemorrhagic process.

Keywords: *Bothrops jararaca*; HF3; human plasma; proteolysis; snake venom metalloproteinase

Key Contribution: Application of untargeted peptidomics to identify the degradome of human plasma proteins incubated with a potent hemorrhagic SVMP, HF3 of *B. jararaca*. Sixty-one proteins were identified as cleaved by HF3; in four types of plasma preparations; and are involved in the systemic effects triggered by the SVMP.

1. Introduction

Among the drastic consequences of viperid snakebite envenomation, manifestations of local tissue damage, such as hemorrhage and myonecrosis, may result in permanent tissue damage and sequelae [1,2]. In cases of severe envenomation, bleeding in organs distant from the site of bite, such as the heart, lungs, kidneys, and brain, may also occur [3–5]. Coagulopathy, including procoagulant, blood-clotting, fibrinolytic, and anticoagulant effects, is another cause of morbidity and mortality upon viperid snakebite accidents [6,7]. Local and systemic effects of viperid envenomation involve the synergistic effects of snake venom metalloproteinases (SVMPs) on plasma proteins, connective tissue, platelets, and blood vessels. SVMPs are zinc-dependent enzymes classified in the M12B subfamily of metallopeptidases, in which the P-III class protein precursors are comprised pro-, catalytic, disintegrin-like, and cysteine-rich domains [8]. SVMPs target specific capillary basement components, cell surface proteins, and extracellular matrix and plasma proteins, thereby promoting capillary rupture and content extravasation, and resulting in hemostasis disturbance and hemorrhage [8–11].

HF3 is a very potent P-III class SVMP of *Bothrops jararaca* venom that induces local hemorrhage with minimum hemorrhagic doses of 15 ng on rabbit skin, and 160 ng on mouse skin [12,13]. The precursor of HF3 is composed of 606 amino acid residues, including five putative *N*-glycosylation sites. The calculated molecular mass of the mature form of HF3 is 46 kDa, whereas on SDS-PAGE, it shows a mobility corresponding to a protein of ~70 kDa, indicating that it is heavily glycosylated [13,14]. HF3 was shown to degrade proteins of the plasma and extracellular matrix, including fibrinogen, fibronectin, vitronectin, von Willebrand factor, collagens IV and VI, laminin, matrigel, antithrombin III, complement components C3 and C4, prothrombin, and plasminogen in vitro [9,15]. Moreover, HF3 showed degradation or limited proteolysis of the proteoglycans aggrecan, brevican, biglycan, decorin, glypican-1, lumican, mimecan, and syndecan-1 [15]. Proteins extracted from the hemorrhagic dorsal skin of mice injected with HF3 were submitted to SDS-PAGE and immunostained with specific anti-proteoglycan antibodies, resulting in the demonstration of in vivo cleavage of biglycan, decorin, glypican-1, lumican, and syndecan-1 [15]. Interestingly, HF3 cleaved the platelet derived growth factor receptor (PDGFR; alpha and beta), and PDGF, in vitro, and both receptor forms were also detected as degraded in vivo in the hemorrhagic process generated by HF3 in the mouse skin [15]. Moreover, the proteolytic activity of HF3 is not affected by plasma proteinase inhibitors, including α2-macroglobulin, which is cleaved by HF3 [16]. The disintegrin-like and cysteine-rich domains of HF3 play a role in its activities upon cells, as reported on their ability to inhibit collagen-induced platelet aggregation, to activate macrophage phagocytosis mediated by αMβ2 integrin, and to induce inflammation by increasing leukocyte rolling in the microcirculation [14,17,18].

The potent hemorrhage generated by HF3 on the mouse skin was analyzed using proteomic approaches, which corroborated the hydrolysis of intracellular, extracellular, and plasma proteins, including some proteoglycans [19]. Indeed, the cleavage of proteoglycans suggested a critical role of the destabilization of the mouse skin integrity in the hemorrhagic process generated by HF3, along with the release of pro-inflammatory fragments acting in the imbalance of tissue homeostasis [15,19]. Despite the strong evidence of the role of proteolysis in the local hemorrhage promoted by SVMPs, the full substrate repertoire of HF3 is unknown. Recently, we reported positional proteomic studies of HF3 cleavage sites in mouse embryonic fibroblast secreted proteins using terminal amine isotopic labeling of substrates (TAILS), which revealed a number of substrates, including proteins of the extracellular matrix and focal adhesions, and the cysteine protease inhibitor cystatin-C [20]. Proteomic identification of cleavage site specificity (PICS) was also used for identifying cleavage sites and sequence preferences in peptides upon incubation with HF3. Two studies using tryptic libraries of proteins from human plasma [21] and from THP-1 monocytic cells [20] revealed a clear preference for leucine at P1′ position and the influence of amino acid sequences adjacent to the scissile bond in the substrate specificity of HF3, similarly to other metalloproteinases from viperid venoms [22].

The aim of this study was to gain new insights into the mechanisms of hemorrhage production by HF3 by expanding the analysis of the substrate repertoire of this SVMP on plasma proteins. To this end, approaches for the depletion of the most abundant proteins and for the enrichment of low abundant proteins of the human plasma were used to minimize the dynamic range of protein concentration. In order to assess the proteolytic activity of HF3 on a wide spectrum of proteins, we used untargeted peptidomics to detect the degradation products by mass spectrometry.

2. Results

The aim of this study was to evaluate the proteolytic activity of HF3 on human plasma proteins. In order to overcome the large protein dynamic range and complexity of human plasma [23,24], four different types of human plasma preparations were compared in in vitro incubations with HF3: whole plasma [P(W)]; plasma depleted of albumin [P(Alb-D)]; plasma depleted of 20 most abundant proteins [P(20-MAP-D)]; and plasma enriched of low-abundance proteins [P(LAP-E)]. Further, the resulting degradation products present in the peptide fraction were analyzed by LC-MS/MS and a database search using tools of Mascot in conjunction with the Trans-Proteomic Pipeline (TPP) and PEAKS Studio. An overview of the study is provided in Figure 1, as well as the nomenclature used for the different plasma fractionation methods.

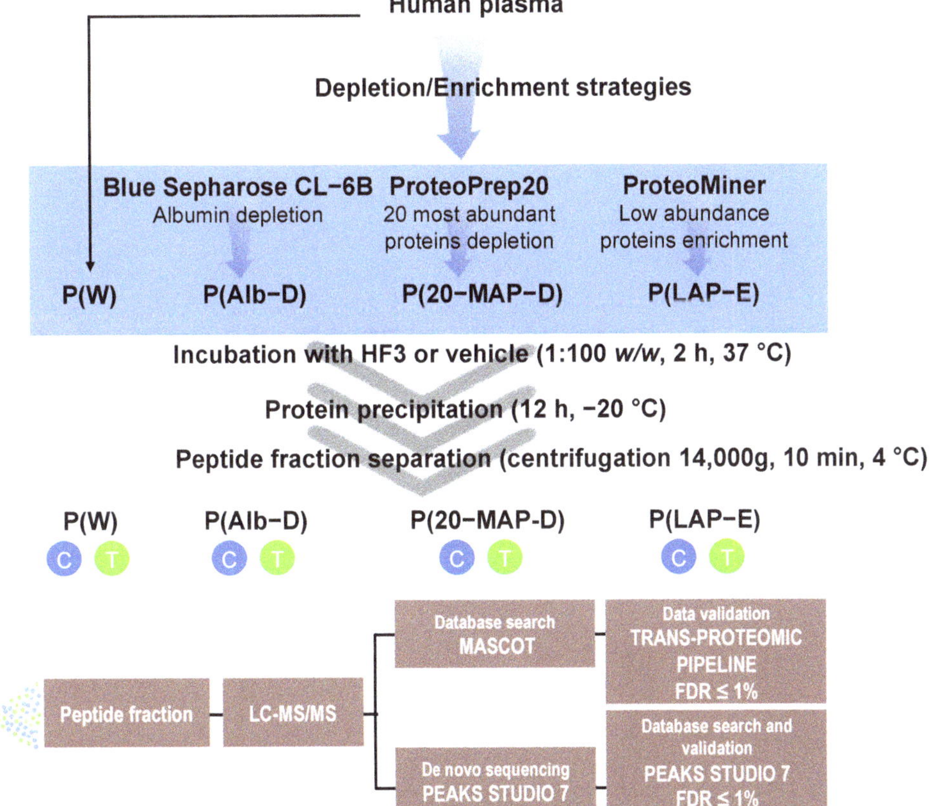

Figure 1. Experimental workflow for the analysis of the proteolytic activity of HF3 on human plasma proteins. Human plasma was depleted of (i) albumin, or (ii) 20 most abundant proteins, or (iii) enriched of low abundant proteins. Whole plasma [P(W)], albumin-depleted plasma [P(Alb-D)], plasma depleted of the 20 most abundant proteins [P(20-MAP-D)], and plasma enriched of low abundant proteins [P(LAP-E)] were individually incubated with HF3, or vehicle, for 2 h at 37 °C. Peptide fractions from control (C) or HF3-treated (T) samples were obtained and analyzed by LC-MS/MS, and peptides were identified using Mascot in conjunction with the TPP and PEAKS Studio.

Three individual experiments of incubation of plasma proteins with HF3 were carried out with each of the four types of plasma samples, P(W), P(Alb-D), P(20-MAP-D), and P(LAP-E), and the following criteria were used in the untargeted peptidomic analysis to accept a protein as being cleaved by HF3: (i) peptides identified in the control samples were disregarded in the analysis, therefore only proteins that showed peptides identified exclusively in the sample of plasma treated with HF3 were considered; (ii) only proteins identified in at least two experiments were considered.

2.1. Electrophoretic Profiles of Plasma Samples Incubated with HF3

As shown in Figure 2, the three methods used for the decomplexation of the plasma proteome to achieve adequate sampling of proteins resulted in distinct electrophoretic profiles, containing proteins of 15−250 kDa. The incubation of these plasma samples with HF3 resulted in protein bands that decreased in intensity in the treated plasma in comparison to the control plasma, indicating proteins that may have been degraded by HF3. The bands that increased in intensity in the treated plasma indicated HF3 proteolysis products derived from proteins of higher molecular mass. In the case of P(W) incubated with HF3, it was possible to observe an increase in the intensity of bands in the region above 225 kDa, and a slight variation in the mobility of an intense band in the region of 52 kDa. P(Alb-D) treated with HF3 showed a decrease in the intensity of bands of ~140 kDa and above, and also a small alteration in the mobility of the ~52 kDa band. On the other hand, the electrophoretic profile of P(20-MAP-D) treated with HF3 showed only a slight decrease in the intensity of the bands in the region between 80 kDa and 120 kDa. However, upon incubation with HF3, P(LAP-E) showed a clear decrease in the intensity of the bands in of ~55 kDa, 40 kDa, and 25 kDa, and an increase in the intensity of the bands of 34 kDa and 22 kDa.

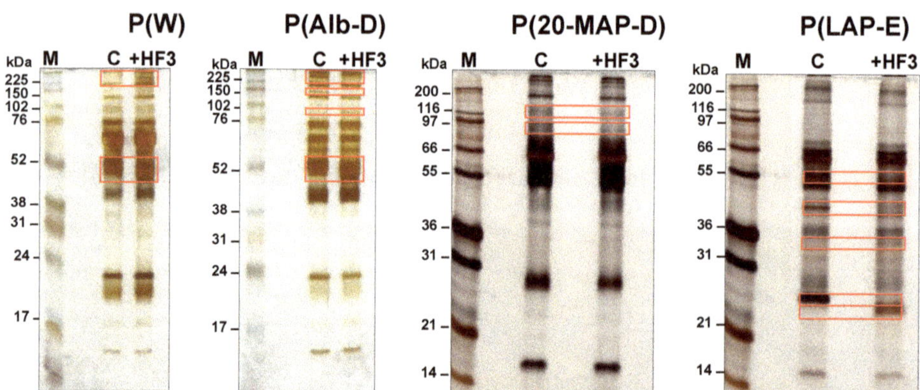

Figure 2. SDS-PAGE profile (12% SDS-polyacrylamide gel) of plasma samples incubated with HF3. M: molecular mass markers; C: control plasma; +HF3: plasma incubated with HF3. Red rectangles indicate regions showing different staining intensities between the control and HF3-treated plasma. Proteins were stained with silver.

Overall, all electrophoretic profiles indicated differences between control and treated plasma samples, evidencing the proteolytic effect of HF3 on plasma proteins in vitro. Interestingly, the electrophoretic profile of P(LAP-E) showed the most significant differences between control and treated samples, possibly due to the fact that the method applied for the enrichment of low abundant proteins of the plasma proteome was more efficient in minimizing the dynamic range of protein concentration, thus enabling a better visualization of the degraded proteins.

2.2. Peptidomic Profiling of Plasma Samples Incubated with HF3

The peptide fractions of control and treated samples of P(W), P(Alb-D), P(20-MAP-D), and P(LAP-E) were subjected to LC-MS/MS, and analyses of peptide identification and spectra counting were carried out by: (i) a Mascot database search and validation of results using the PeptideProphet and ProteinProphet tools of the TPP platform (the results of this approach were designated as 'Mascot/TPP'); and (ii) de novo sequencing, a database search, and validation of results by Peaks Studio 7 program (the results of this approach were designated as 'PEAKS') (Supplementary Tables S1–S8).

Figure 3 shows Venn diagrams comparing the total numbers of plasma proteins identified as cleaved by HF3 using both identification approaches. The analysis of whole plasma [P(W)] incubated with HF3 showed 36 proteins as cleaved by HF3 using both identification approaches (Table 1), whereas five proteins were identified exclusively by PEAKS. In the plasma depleted of albumin [P(Alb-D)], we identified 21 proteins cleaved by HF3 using both identification approaches (Table 2), including three proteins exclusively identified by Mascot/TPP, and two proteins exclusively by PEAKS. In the plasma depleted of the 20 most abundant proteins [P(20-MAP-D)], 30 proteins were identified as cleaved by HF3 using both identification approaches (Table 3), with one protein identified exclusively by Mascot/TPP, and five by PEAKS. The incubation of HF3 with the plasma enriched of low-abundance proteins [P(LAP-E)] revealed 38 substrates by both approaches (Table 4), and 16 proteins identified only by PEAKS.

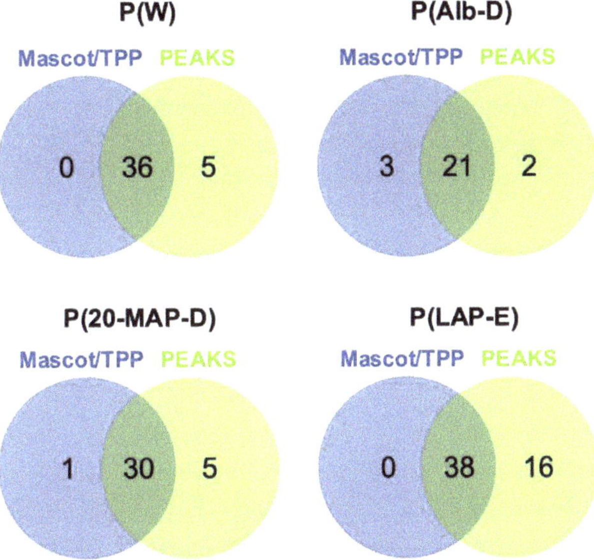

Figure 3. Summary of numbers of plasma proteins identified as degraded by HF3, by LC-MS/MS analysis of the peptide fraction, using four different plasma preparations. Comparison of results obtained using Mascot/TPP and PEAKS Studio 7 for peptide identification.

Table 1. HF3 substrates revealed by the analysis of P(W) peptide fraction by LC-MS/MS. Only proteins that showed peptides identified exclusively in the sample of plasma treated with HF3 were considered. The boldface indicates the ten most degraded proteins according to the number of identified spectra.

		Number of Spectra * (exp. 1/exp. 2/exp. 3)	
Identification	**Protein**	**Mascot/TPP**	**PEAKS**
ACTB_HUMAN	Actin cytoplasmic 1	0/0/0	7/3/0
A1AT_HUMAN	Alpha-1-antitrypsin	1/0/1	6/10/5
A2AP_HUMAN	Alpha-2-antiplasmin	12/13/13	20/24/18
FETUA_HUMAN	**Alpha-2-HS-glycoprotein**	**75/94/92**	**270/437/391**
A2MG_HUMAN	Alpha-2-macroglobulin	2/0/2	9/4/8
APOA1_HUMAN	Apolipoprotein A-I	9/25/14	43/58/50
APOA2_HUMAN	**Apolipoprotein A-II**	**13/30/19**	**63/70/60**
APOA4_HUMAN	Apolipoprotein A-IV	5/6/7	23/21/16
APOC1_HUMAN	Apolipoprotein C-I	0/0/0	4/5/0
APOC2_HUMAN	Apolipoprotein C-II	7/17/11	30/41/30
APOC3_HUMAN	Apolipoprotein C-III	9/19/13	34/43/30
APOE_HUMAN	Apolipoprotein E	3/4/3	10/9/10
APOF_HUMAN	Apolipoprotein F	2/5/3	7/11/12
APOL1_HUMAN	Apolipoprotein L1	4/9/5	36/38/35
CLUS_HUMAN	**Clusterin**	**17/42/29**	**89/108/90**
FA5_HUMAN	Coagulation factor V	0/0/0	3/0/3
C1R_HUMAN	Complement C1r subcomponent	1/2/2	18/15/15
C1S_HUMAN	Complement C1s subcomponent	3/3/2	26/20/16
CO3_HUMAN	Complement C3	9/18/16	20/26/21
CO4A_HUMAN	**Complement C4-A**	**19/30/24**	**61/76/67**
CO9_HUMAN	Complement component C9	0/2/1	3/4/0
CFAH_HUMAN	Complement factor H	1/2/0	3/2/0
FIBA_HUMAN	**Fibrinogen alpha chain**	**126/221/163**	**369/602/462**
FIBB_HUMAN	**Fibrinogen beta chain**	**16/52/32**	**39/53/45**
FINC_HUMAN	Fibronectin	7/12/6	29/29/22
GELS_HUMAN	Gelsolin	5/8/7	28/21/25
HRG_HUMAN	Histidine-rich glycoprotein	10/10/8	24/55/39
H2BFS_HUMAN	Histone H2B type F-S	0/0/0	4/1/0
IGHG1_HUMAN	Ig gamma-1 chain C region	0/5/1	6/8/7
IGHM_HUMAN	Ig mu chain C region	2/5/2	6/6/6
ITIH1_HUMAN	Inter-alpha-trypsin inhibitor heavy chain H1	6/5/6	17/17/19
ITIH2_HUMAN	**Inter-alpha-trypsin inhibitor heavy chain H2**	**51/92/79**	**204/230/224**
ITIH3_HUMAN	Inter-alpha-trypsin inhibitor heavy chain H3	1/2/1	6/4/5
ITIH4_HUMAN	**Inter-alpha-trypsin inhibitor heavy chain H4**	**24/37/28**	**93/103/94**
KNG1_HUMAN	Kininogen-1	8/8/10	33/21/30
PEDF_HUMAN	Pigment epithelium-derived factor	3/4/3	5/5/4
PZP_HUMAN	Pregnancy zone protein	0/1/2	5/5/5

Table 1. *Cont.*

Identification	Protein	Number of Spectra * (exp. 1/exp. 2/exp. 3)	
		Mascot/TPP	PEAKS
THRB_HUMAN	**Prothrombin**	**17/25/20**	**63/62/65**
ALBU_HUMAN	Serum albumin	0/0/0	0/3/2
TTHY_HUMAN	Transthyretin	6/17/17	43/43/43
VTNC_HUMAN	**Vitronectin**	**18/31/24**	**74/78/71**

* Only spectra identified exclusively in the HF3-treated samples were considered.

Table 2. HF3 substrates revealed by the analysis of P(Alb-D) peptide fraction by LC-MS/MS. Only proteins that showed peptides identified exclusively in the sample of plasma treated with HF3 were considered. The boldface indicates the ten most degraded proteins according to the number of identified spectra.

Identification	Protein	Number of Spectra * (exp. 1/exp. 2/exp. 3)	
		Mascot/TPP	PEAKS
ACTG_HUMAN	Actin cytoplasmic 1	0/0/0	1/0/1
A1AT_HUMAN	**Alpha-1-antitrypsin**	**1/10/11**	**16/24/25**
A2AP_HUMAN	**Alpha-2-antiplasmin**	**10/12/11**	**17/18/20**
FETUA_HUMAN	**Alpha-2-HS-glycoprotein**	**84/109/117**	**219/226/230**
APOA1_HUMAN	**Apolipoprotein A-I**	**13/0/1**	**33/8/6**
APOA2_HUMAN	Apolipoprotein A-II	5/3/1	17/13/11
APOA4_HUMAN	Apolipoprotein A-IV	8/5/7	13/13/15
APOC2_HUMAN	Apolipoprotein C-II	4/3/2	11/8/8
APOC3_HUMAN	**Apolipoprotein C-III**	**6/12/10**	**21/25/27**
APOE_HUMAN	Apolipoprotein E	3/2/2	2/4/4
APOL1_HUMAN	Apolipoprotein L1	0/4/1	4/8/10
CERU_HUMAN	Ceruloplasmin	0/2/3	2/6/9
CLUS_HUMAN	**Clusterin**	**1/2/2**	**5/16/12**
CO3_HUMAN	Complement C3	0/2/4	0/2/8
FIBA_HUMAN	**Fibrinogen alpha chain**	**0/3/14**	**2/13/41**
FIBB_HUMAN	**Fibrinogen beta chain**	**7/14/20**	**13/20/28**
FINC_HUMAN	Fibronectin	0/1/2	0/0/0
GPX3_HUMAN	Glutathione peroxidase 3	0/0/0	5/7/0
IGHG1_HUMAN	Ig gamma-1 chain C region	2/1/0	0/0/0
ITIH2_HUMAN	**Inter-alpha-trypsin inhibitor heavy chain H2**	**2/7/18**	**7/26/42**
ITIH4_HUMAN	Inter-alpha-trypsin inhibitor heavy chain H4	0/4/4	0/8/7
MGAP_HUMAN	Isoform 3 of MAX gene-associated protein	4/2/0	0/0/0
KNG1_HUMAN	Kininogen-1	1/2/1	2/8/6
THRB_HUMAN	Prothrombin	1/2/5	2/6/12
PON1_HUMAN	**Serum paraoxonase/arylesterase 1**	**7/15/21**	**27/36/52**
TTHY_HUMAN	**Transthyretin**	**27/33/34**	**44/57/68**

* Only spectra identified exclusively in the HF3-treated samples were considered.

Table 3. HF3 substrates revealed by the analysis of P(20-MAP-D) peptide fraction by LC-MS/MS. Only proteins that showed peptides identified exclusively in the sample of plasma treated with HF3 were considered. The boldface indicates the ten most degraded proteins according to the number of identified spectra.

		Number of Spectra * (exp. 1/exp. 2/exp. 3)	
Identification	**Protein**	**Mascot/TPP**	**PEAKS**
ACTB_HUMAN	Actin cytoplasmic 1	6/0/2	18/0/2
A2AP_HUMAN	**Alpha-2-antiplasmin**	**70/28/20**	**81/39/30**
FETUA_HUMAN	**Alpha-2-HS-glycoprotein**	**270/150/134**	**454/232/245**
A2MG_HUMAN	Alpha-2-macroglobulin	5/4/3	7/6/4
APOA1_HUMAN	**Apolipoprotein A-I**	**69/39/36**	**91/58/62**
APOA2_HUMAN	**Apolipoprotein A-II**	**39/19/21**	**56/30/37**
APOA4_HUMAN	**Apolipoprotein A-IV**	**28/48/15**	**56/71/31**
APOC1_HUMAN	Apolipoprotein C-I	0/0/0	2/2/0
APOC2_HUMAN	Apolipoprotein C-II	39/4/10	46/6/18
APOC3_HUMAN	Apolipoprotein C-III	55/9/10	68/14/17
APOE_HUMAN	Apolipoprotein E	4/9/2	8/11/7
APOF_HUMAN	Apolipoprotein F	17/7/2	26/4/0
APOL1_HUMAN	Apolipoprotein L1	8/4/7	8/3/19
CBPB2_HUMAN	Carboxypeptidase B2	0/0/0	1/0/4
CLUS_HUMAN	Clusterin	5/2/8	10/4/30
F13A_HUMAN	Coagulation factor XIII A chain	5/0/1	5/0/1
C1R_HUMAN	Complement C1r subcomponent	5/0/2	0/0/0
CO3_HUMAN	Complement C3	5/2/3	5/2/6
CO4A_HUMAN	Complement C4-A	6/0/10	9/0/15
ECM1_HUMAN	Extracellular matrix protein 1	0/0/0	4/3/0
FIBA_HUMAN	**Fibrinogen alpha chain**	**159/50/83**	**162/0/122**
FIBB_HUMAN	**Fibrinogen beta chain**	**31/38/14**	**34/44/17**
FINC_HUMAN	Fibronectin	4/1/8	6/5/22
GELS_HUMAN	**Gelsolin**	**49/18/14**	**68/23/42**
HRG_HUMAN	Histidine-rich glycoprotein	2/6/14	5/7/29
IGHG1_HUMAN	Ig gamma-1 chain C region	4/0/9	4/0/12
ITIH2_HUMAN	**Inter-alpha-trypsin inhibitor heavy chain H2**	**20/18/34**	**31/23/69**
ITIH3_HUMAN	Inter-alpha-trypsin inhibitor heavy chain H3	0/0/0	2/0/9
ITIH4_HUMAN	**Inter-alpha-trypsin inhibitor heavy chain H4**	**131/44/38**	**174/56/80**
FIBG_HUMAN	Isoform Gamma-A of Fibrinogen gamma chain	2/19/0	3/19/0
KNG1_HUMAN	Kininogen-1	7/1/3	17/3/9
THRB_HUMAN	Prothrombin	1/2/23	1/2/61
SAA4_HUMAN	Serum amyloid A-4 protein	0/0/0	7/1/0
TETN_HUMAN	Tetranectin	5/2/5	5/2/6
TTHY_HUMAN	Transthyretin	9/2/10	10/2/15
VTNC_HUMAN	Vitronectin	33/9/11	43/12/15

* Only spectra identified exclusively in the HF3-treated samples were considered.

Table 4. HF3 substrates revealed by the analysis of P(LAP-E) peptide fraction by LC-MS/MS. Only proteins that showed peptides identified exclusively in the sample of plasma treated with HF3 were considered. The boldface indicates the ten most degraded proteins according to the number of identified spectra.

		Number of Spectra * (exp 1/exp 2/exp 3)	
Identification	**Protein**	**Mascot/TPP**	**PEAKS**
A1AT_HUMAN	Alpha-1-antitrypsin	3/15/1	5/18/3
A2AP_HUMAN	Alpha-2-antiplasmin	1/2/0	2/4/0
FETUA_HUMAN	**Alpha-2-HS-glycoprotein**	**32/31/30**	**51/60/49**
APOA1_HUMAN	**Apolipoprotein A-I**	**80/127/17**	**171/258/37**
APOA2_HUMAN	**Apolipoprotein A-II**	**37/69/20**	**56/132/26**
APOA4_HUMAN	**Apolipoprotein A-IV**	**52/86/11**	**124/201/20**
APOC1_HUMAN	Apolipoprotein C-I	11/17/0	34/45/2
APOC2_HUMAN	Apolipoprotein C-II	22/21/5	27/32/9
APOC3_HUMAN	**Apolipoprotein C-III**	**19/51/5**	**43/88/11**
APOE_HUMAN	**Apolipoprotein E**	**29/9/4**	**57/30/13**
APOF_HUMAN	Apolipoprotein F	1/2/0	1/2/0
APOL1_HUMAN	Apolipoprotein L1	0/0/0	0/3/2
C4BPA_HUMAN	C4b-binding protein alpha chain	0/0/0	3/5/0
CBPB2_HUMAN	Carboxypeptidase B2	1/0/1	1/0/1
CLUS_HUMAN	Clusterin	1/18/0	7/49/0
CO1A2_HUMAN	Collagen alpha-2(I) chain	0/0/0	1/0/1
COL11_HUMAN	Collectin-11	0/2/1	3/5/2
C1QB_HUMAN	Complement C1q subcomponent subunit B	0/0/0	1/0/1
C1S_HUMAN	Complement C1s subcomponent	0/0/0	1/3/0
CO3_HUMAN	Complement C3	8/9/7	13/22/21
CO4A_HUMAN	Complement C4-A	4/14/4	17/30/11
CFAH_HUMAN	Complement factor H	0/3/2	0/7/2
FHR1_HUMAN	Complement factor H-related protein 1	0/0/0	3/8/0
FIBA_HUMAN	**Fibrinogen alpha chain**	**82/141/40**	**149/241/63**
FIBB_HUMAN	**Fibrinogen beta chain**	**160/29/9**	**84/50/7**
FINC_HUMAN	Fibronectin	4/7/6	4/10/15
FCN2_HUMAN	Ficolin-2	1/2/0	2/4/0
LG3BP_HUMAN	Galectin-3-binding protein	0/0/0	2/2/0
GELS_HUMAN	Gelsolin	3/2/1	5/6/2
GPX3_HUMAN	Glutathione peroxidase 3	1/0/1	1/2/1
HRG_HUMAN	Histidine-rich glycoprotein	0/0/0	1/5/0
HABP2_HUMAN	Hyaluronan-binding protein 2	2/3/0	2/6/0
IGHG1_HUMAN	Ig gamma-1 chain C region	0/2/4	0/6/10
IGKC_HUMAN	Ig kappa chain C region	1/9/0	3/14/1
KV309_HUMAN	Ig kappa chain V-III region VG (Fragment)	0/0/0	1/6/0
KV402_HUMAN	Ig kappa chain V-IV region Len	0/0/0	1/2/0

Table 4. *Cont.*

		Number of Spectra * (exp 1/exp 2/exp 3)	
Identification	**Protein**	**Mascot/TPP**	**PEAKS**
LV101_HUMAN	Ig lambda chain V-I region V	0/0/0	1/3/2
LV403_HUMAN	Ig lambda chain V-IV region Hil	0/0/0	0/3/1
LAC2_HUMAN	Ig lambda-2 chain C regions	3/4/0	8/12/9
IGLL5_HUMAN	Immunoglobulin lambda-like polypeptide 5	0/0/0	0/8/12
ITIH1_HUMAN	Inter-alpha-trypsin inhibitor heavy chain H1	2/4/0	3/8/0
ITIH2_HUMAN	**Inter-alpha-trypsin inhibitor heavy chain H2**	**23/24/2**	**35/42/3**
ITIH4_HUMAN	Inter-alpha-trypsin inhibitor heavy chain H4	10/7/2	21/20/8
FIBG_HUMAN	Isoform Gamma-A of fibrinogen gamma chain	8/9/0	19/25/0
KNG1_HUMAN	Kininogen-1	2/3/0	6/6/3
PLMN_HUMAN	Plasminogen	0/0/0	1/6/0
PRAP1_HUMAN	Proline-rich acidic protein 1	0/0/0	4/1/2
THRB_HUMAN	Prothrombin	7/20/5	21/43/12
ALBU_HUMAN	Serum albumin	1/10/1	2/25/6
SAA4_HUMAN	Serum amyloid A-4 protein	4/3/0	14/11/1
TTHY_HUMAN	Transthyretin	2/5/0	7/13/3
PROC_HUMAN	Vitamin K-dependent protein C	0/0/0	0/1/1
PROS_HUMAN	Vitamin K-dependent protein S	2/1/0	4/5/1
VTNC_HUMAN	**Vitronectin**	**19/43/13**	**41/85/20**

* Only spectra identified exclusively in the HF3-treated samples were considered.

Regarding the different methods of depletion of abundant plasma proteins and enrichment of low-abundant proteins, the four-way Venn diagrams displayed in Figure 4 show that the analysis of the peptide fraction of P(LAP-E) provided a higher number of identified HF3 substrates, with 11 substrates indicated by the approach of Mascot/TPP, and 19 substrates indicated by PEAKS. Unexpectedly, the analysis of P(W) incubated with HF3 also provided results on HF3 substrates which were identified exclusively by this method (six substrates by the Mascot/TPP identification approach, and seven substrates by PEAKS), however, it is worth mentioning that the initial amount of proteins used in the incubation of P(W) with HF3 was higher (200 μg) compared to 50 μg for P(Alb-D), P(20-MAP-D), and P(LAP-E). This result can be attributed to the fact that there was less manipulation of the whole plasma sample compared to other approaches, which involved at least one more protein depletion/enrichment step, performed separately (biological replicates), before incubation with the proteinase.

In general, in each method of preparation of human plasma for incubation with HF3, most proteins considered as a target of HF3 for proteolysis were identified by both bioinformatics approaches used, reinforcing the results (Figure 4). Interestingly, both approaches resulted in a higher number of substrates identified in the P(W) and P(LAP-E) plasma samples. Moreover, although the PEAKS approach revealed a higher number of HF3 substrates (70), 53 proteins were identified by both bioinformatics approaches.

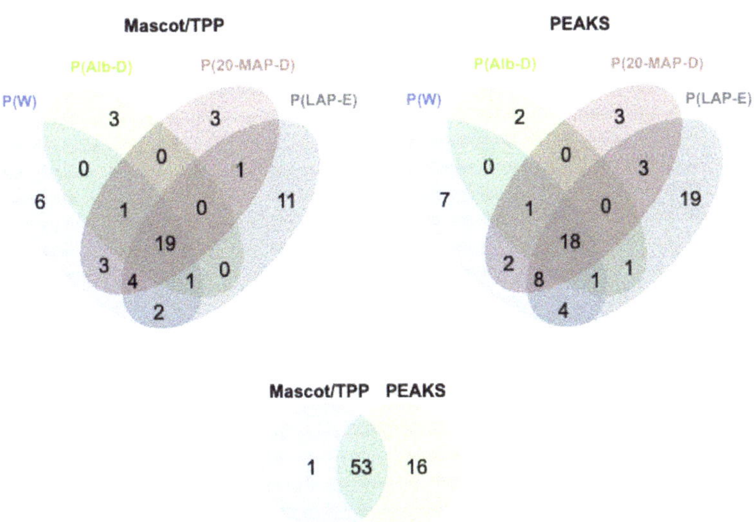

Figure 4. Summary of proteins identified as HF3 substrates. Upper panels: Four-way Venn diagrams of substrate identification using four samples of human plasma, and two methods of protein identification. Lower panel: Summary of proteins identified by each method of protein identification.

2.3. Proteins Degraded by HF3 in the Human Plasma

Table 1 shows the list of 41 proteins identified as substrates of HF3 in the P(W) sample. The ten most degraded substrates are proteins involved in functions in the coagulation cascade, complement system, protein transport, and proteinase inhibition.

The identification of substrates of HF3 in the plasma depleted of albumin resulted in only 26 proteins (Table 2). The ten most degraded substrates are proteins involved in functions in the coagulation cascade, complement system, lipid transport, hormone transport, and proteinase inhibition.

In the plasma depleted of the 20 most abundant proteins, 36 proteins were identified as degraded by HF3 (Table 3), whereas in the plasma submitted to enrichment of the low-abundant proteins, 54 proteins were detected as cleaved by HF3 (Table 4). Despite the higher number of substrates identified in the latter, the ten most degraded proteins in both types of plasma preparations were nearly the same. Considering the bioinformatics approaches applied for peptide identification, overall, when using PEAKS, higher numbers of spectra were identified for all degraded proteins.

Overall, a total of 61 proteins (Table 5) were detected as cleaved by HF3, including 18 that were identified in all types of plasma preparations: alpha-2-antiplasmin; alpha-2-HS-glycoprotein; apolipoproteins A-I, A-II, A-IV, C-II, C-III, E, and L1; clusterin; complement C3; fibrinogen alpha and beta chains; inter-alpha-trypsin inhibitor heavy chains H2 and H4; kininogen-1; prothrombin; and transthyretin.

Table 5. 61 proteins identified as cleaved by HF3 in human plasma.

Actin Cytoplasmic-1	Fibrinogen
Alpha-1-antitrypsin	Fibronectin
Alpha-2-antiplasmin	Ficolin-2
Alpha-2-HS-glycoprotein	Galectin-3-binding protein
Apha-2-macroglobulin	Gelsolin
Apolipoprotein A-I	Glutathione peroxidase 3
Apolipoprotein A-II	Histidine rich glycoprotein
Apolipoprotein A-IV	Histone H2B type F-S
Apolipoprotein C-I	Hyaluronan-binding protein 2
Apolipoprotein C-II	Ig gamma
Apolipoprotein C-IIII	Ig kappa
Apolipoprotein E	Ig lambda
Apolipoprotein F	Ig mu chain C region
Apolipoprotein L1	Ig lambda-like polypeptide 5
C4b-binding protein alpha chain	Inter-alpha-trypsin inhibitor H1−H4 chains
Carboxypeptidade B2	Isoform 3 of MAX gene-associated protein
Ceruloplasmin	Kininogen
Clusterin	Pigment ephitelium-derived factor
Coagulation factor V	Plasminogen
Coagulation factor XIII A chain	Pregnancy zone protein
Collagen alpha-2(I) chain	Proline-rich acidic protein 1
Collectin 11	Prothrombin
Complement C1q subcomponent subunit B	Serum albumin
Complement C1r subcomponent	Serum amyloid A-4 protein
Complement C1s subcomponent	Serum paraoxonase/arylesterase 1
Complement component C3	Tetranectin
Complement component C4-A	Transthyretin
Complement component C9	Vitamin K-dependent protein C
Complement factor H	Vitamin K-dependent protein S
Complement factor H-related protein	Vitronectin
Extracellular matrix protein 1	

We further investigated the newly identified proteins as HF3 substrates in the human plasma. For this analysis, we selected some proteins that are commercially available and that were detected as cleaved by HF3 using LC-MS/MS analysis. The proteins were incubated at a 1:10 (*w/w*) enzyme-to-substrate ratio with HF3 for 2 h and subjected to SDS-PAGE (Figure 5). Apolipoprotein A-IV was completely degraded by HF3. Apolipoprotein E was almost completely degraded by HF3 to generate a fragment of ~15 kDa. After incubation with HF3, clusterin showed a slight reduction of molecular mass, indicating that HF3 may have promoted its limited proteolysis. HF3 also promoted the limited proteolysis of α-2-antiplasmin, resulting in fragments of ~68 kDa and 55 kDa. The 120 kDa band of high molecular weight kininogen was almost completely degraded by HF3, resulting in a stable fragment of ~68 kDa. In the case of transthyretin, its dimer and monomer bands remained unchanged after incubation with HF3.

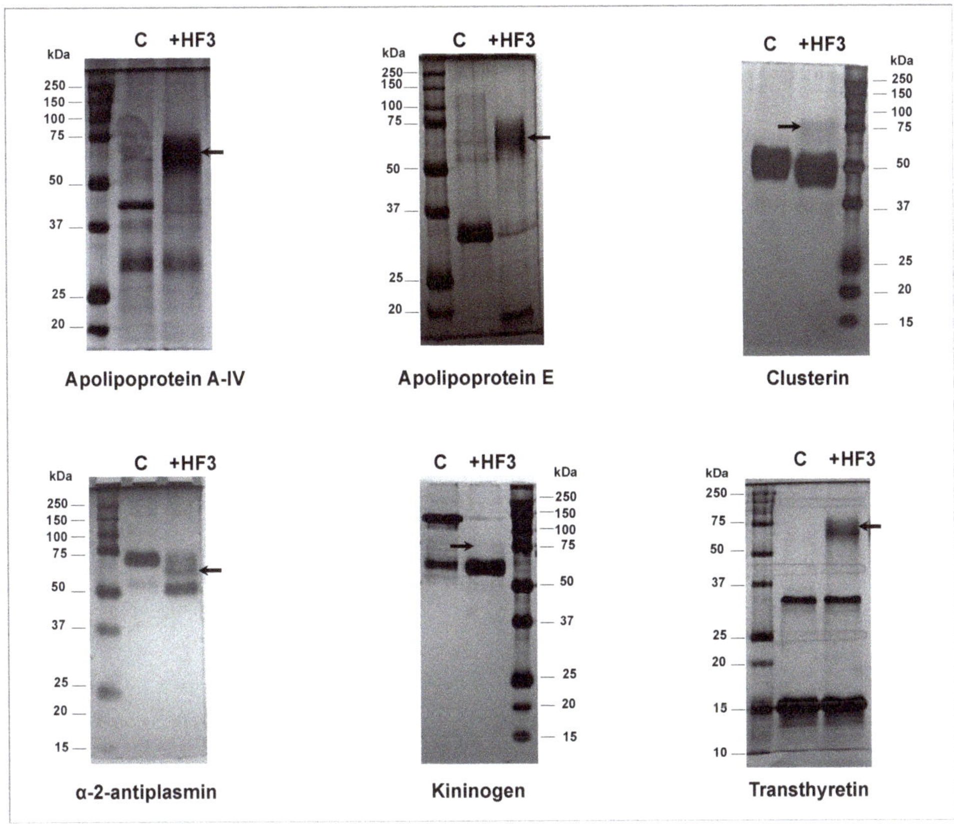

Figure 5. Proteolytic activity of HF3 upon isolated plasma proteins. Apolipoprotein A-IV, apolipoprotein E, clusterin, α-2-antiplasmin, kininogen, and transthyretin were incubated at a 1:10 (*w/w*) enzyme-to-substrate ratio with HF3 for 2 h, as described in Materials and Methods, and subjected to SDS-PAGE. Arrows indicate the band of HF3. Numbers on the right and left indicate molecular mass marker mobility. Proteins were silver stained.

Figure 6 shows the protein-protein interaction network of 61 proteins cleaved by HF3 in the human plasma, visualized using STRING analysis [25], evidencing that most of them (53) have connected molecular functions related mainly to the activation and control of the coagulation and complement systems.

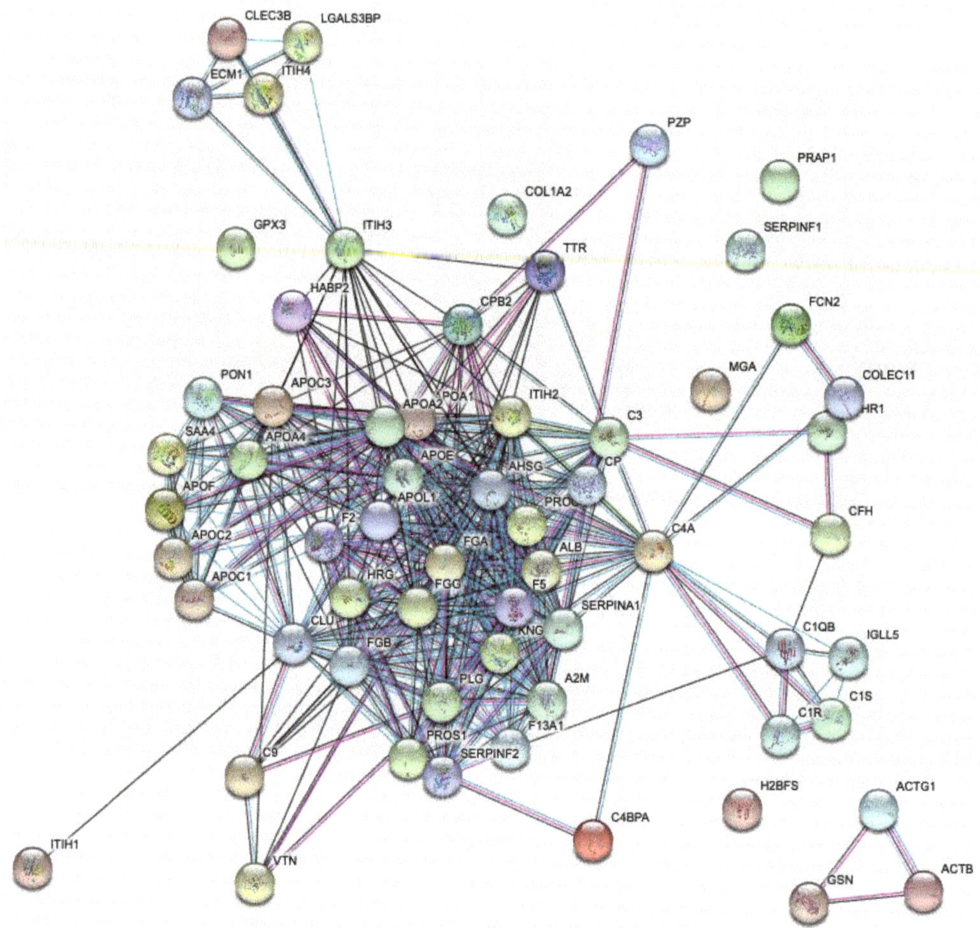

Figure 6. Protein-protein interactions of proteins identified as degraded in human plasma by HF3 according to the STRING database (61 proteins; 401 edges (expected 27); PPI enrichment p-value < 1.0^{-16}). The connecting lines between protein nodes indicate protein-protein interactions.

2.4. *Mapping the Primary Specificity of HF3 on Plasma Proteins*

A large number of peptides derived from the activity of HF3 on plasma proteins were identified in the experiments performed with P(W), P(Alb-D), P(20-MAP-D), and P(LAP-E), and these were used for the evaluation of the proteins that were hydrolyzed by HF3, as well as to carry out the mapping of amino acid sequences adjacent to the cleavage sites, which are preferential for the proteinase (primary specificity). For this purpose, we used the amino acid sequences of the peptides, corresponding to the positions P1′–P6′ [26], identified in the samples treated with HF3, and originating from the proteins considered as substrate, whereas the complementary amino acid sequences (P6-P1) adjacent to the cleavage sites of the respective peptides were obtained using a tool available online at http://clipserve.clip.ubc.ca/pics; accessed on 1 April 2014 [27], which also generated the graphical representation (heat map) of the proteinase cleavage consensus sequence. The heat maps representing the occurrence of amino acids at positions P6-P6′ are described in Figure 7, corresponding to the identification results obtained by the Mascot/TPP and

PEAKS approaches. It was interesting to verify that, although the number of cleavage sites identified by the PEAKS approach was higher than that of Mascot/TPP, the heat maps generated with data from both approaches from data from all plasma samples were very similar and evidenced a clear preference of HF3 for Leu at the P1′ position.

The analysis of hydrolysis products of plasma proteins incubated with HF3 showed that fibrinogen was cleaved in the alpha, beta, and gamma chains (Supplementary Material Tables S1–S8). These data corroborate previous studies that described extensive cleavage of the alpha chain, followed by more limited beta chain cleavage [9] by HF3, however, proteolysis of the fibrinogen gamma chain by HF3 was unknown. Here, peptides from the C-terminal region of the fibrinogen gamma chain were identified as a product of HF3 activity in P(20-MAP-D) and P(LAP-E). Studies have shown that the C-terminal region of the fibrinogen gamma chain plays important roles in platelet interaction [28] and fibrin stabilization [29], and that the C-terminal peptide also plays a role in the inhibition of platelet aggregation [30]. Thus, for the validation of peptides generated by cleavage of fibrinogen by HF3 in plasma, it was incubated with isolated fibrinogen, followed by analysis of the resulting peptide fraction by LC-MS/MS. This analysis revealed the peptides generated by the enzymatic activity of HF3, and indirectly, the cleavage points of HF3 in the protein (Supplementary Material Table S9). Figure 8 shows the location of these peptides in the fibrinogen sequence, corresponding to 73 peptides of the alpha chain and 15 peptides of the beta chain. Eight fibrinogen gamma chain peptides were also identified, including two located in the C-terminal region, thus confirming the data obtained by the incubation of HF3 with plasma.

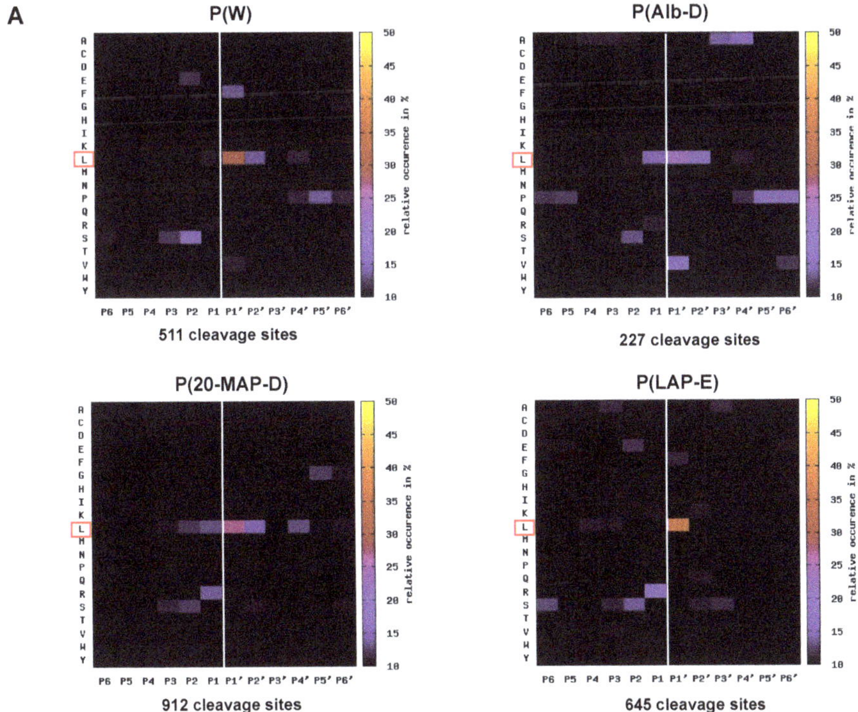

Figure 7. *Cont.*

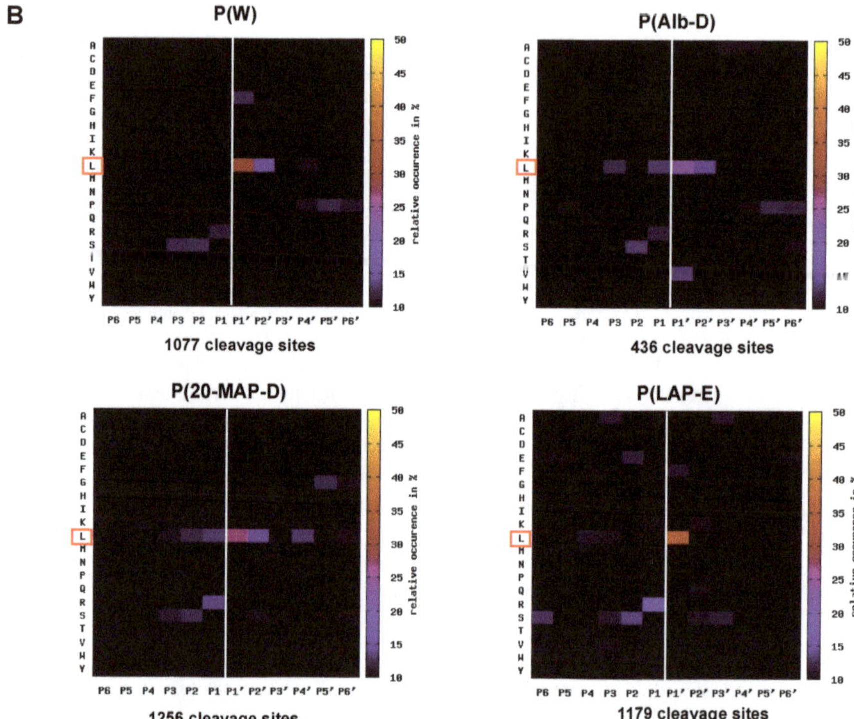

Figure 7. Substrate specificity assessed by the identification of peptides generated by the proteolytic activity of HF3 on human plasma proteins. (**A**) Heat maps showing the relative occurrence (in %) of each amino acid residue and the fold-change over the natural abundance of amino acids, identified using Mascot/TPP. (**B**) Heat maps showing the relative occurrence (in %) of each amino acid residue and the fold-change over the natural abundance of amino acids, identified using PEAKS. Only peptides derived from proteins identified as HF3 substrates were considered, and peptides identified both in control and treated samples were excluded. Heat maps were created at http://clipserve.clip.ubc.ca/pics (accessed on 1 April 2014).

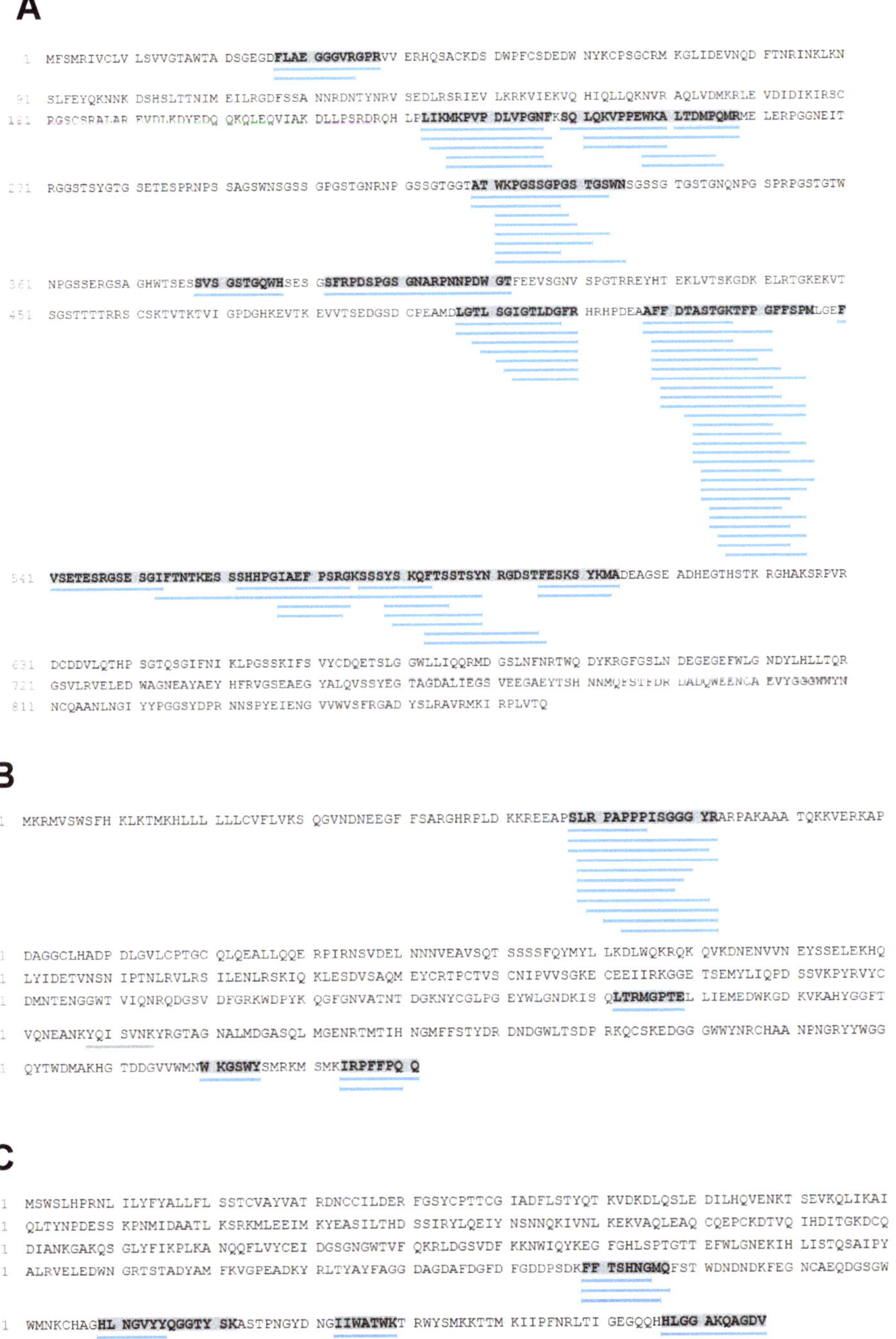

Figure 8. Amino acid sequence of human fibrinogen alpha (**A**), beta (**B**), and gamma (**C**) chains, indicating the identified cleavage products (blue bars) generated by the incubation with HF3. Graphical view generated with PEAKS Studio 7.

3. Discussion

3.1. Human Plasma Preparations for Degradomic Analysis

The application of methods for the depletion of the most abundant proteins in human plasma, and for the enrichment of low-abundant proteins, resulted in samples with significantly different electrophoretic profiles, which allowed us to evaluate the proteolytic activity of HF3 on a wide spectrum of proteins in concentrations that allowed the detection of degradation products by mass spectrometry. Regarding the electrophoretic profile, P(LAP-E) showed the clearest differences in comparison to whole plasma, a fact that can be directly attributed to the principle of the ProteoMiner technology, which, by equalizing the proteome of a sample, significantly alters the concentration of proteins poorly represented in a proteome. Millioni et al., 2011, compared techniques for depleting the 20 most abundant proteins with the ProteoPrep 20 column and the enrichment of less abundant proteins with ProteoMiner, for the analysis of the plasma proteome [31]. They reported that the methods were complementary regarding the identified proteins, however, a higher number of proteins was identified in the plasma proteome depleted of the 20 most abundant proteins. Hakimi et al., 2014, tested the efficacy and reproducibility of these methods for plasma protein depletion/enrichment, and by comparing 18 randomly chosen proteins, they found that both techniques were reproducible [32]. However, the protein enrichment approach with ProteoMiner provided a greater number of identified proteins. It is worth emphasizing that a large part of the studies that employ techniques for the depletion of the most abundant proteins in the plasma or enrichment of low-abundant proteins aim to characterize the plasma proteome itself, and more often, the search for biomarkers [33–37]. Studies aimed at analyzing the repertoire of substrates of a proteinase on a given proteome are less frequent, and generally employ methods for the depletion of abundant proteins after the action of the proteinase, in a step prior to one-dimensional or two-dimensional electrophoresis [18,19,38], or do not employ protein depletion/enrichment techniques [39–41]. In the present study, plasma protein depletion/enrichment techniques were used in a step prior to incubation with the proteinase. Thus, it was possible to evaluate the proteolytic activity of HF3 through the identification of peptides from the cleaved proteins, for the characterization of the HF3 degradome in the human plasma in vitro.

3.2. Analysis of the Plasma Peptide Fraction

Considering that the peptides identified by mass spectrometry after the incubation of plasma with HF3, which were absent in their respective control, are probably products arising from the cleavage of plasma proteins by HF3, the structure of these peptides indicates the substrates degraded by this proteinase. To elucidate the HF3 cleavage products by the analysis of the peptide fraction, we used two bioinformatics approaches with restrictive criteria for peptide identification. In addition, to be considered as HF3 substrate, proteins whose peptides were identified in the peptide fraction were again subjected to restrictive criteria for comparing samples treated with HF3 and control: (i) the peptides identified in the control samples were subtracted from the list of peptides identified in the HF3-treated samples, as they were potentially not generated by HF3, thus, only peptides identified exclusively in the treated sample remained in the analysis; (ii) proteins were considered cleaved by HF3 if their peptides (hydrolysis products) were present in at least two of the three experiments performed. In view of these criteria, we identified that 70 protein entries present in the UniProt database (considering individual oligomeric protein chains as different entries) were identified as HF3 substrates, and of these, 61 are unique proteins. Some of these proteins corroborate data described in the literature and others are considered new substrates of this metalloproteinase.

In general, for each method of preparation of human plasma, a large portion of the proteins considered as a target of HF3 were identified by both bioinformatics approaches used. The PEAKS approach provided a greater number of proteins identified in P(W), P(20-MAP-D), and P(LAP-E), whereas the Mascot/TPP approach provided a greater number of proteins identified as substrate in the P(Alb-D) approach. Zhang et al., 2012, compared, for data from LC-MS/MS analysis of trypsin digestion of *Pseudomonas aeruginosa*, the number of peptide-to-spectrum match (PSM) identified by PEAKS and Mascot, using a database search [42]. The study showed that PEAKS provided 30% more PSMs when compared to Mascot. In the present study, higher numbers of spectral counts and identified peptides were detected in the analysis of the peptide fraction of the plasma incubated with HF3 by PEAKS, nevertheless, most substrates were identified by both bioinformatics approaches.

The direct incubation of some of the new substrate candidates with HF3 and the visualization of cleavage products by SDS-PAGE confirmed its proteolytic activity upon human plasma proteins, as detected by LC-MS/MS analysis. Furthermore, it revealed different outcomes, i.e., limited proteolysis or degradation, resulting from the hydrolysis of plasma proteins by HF3. In the case of transthyretin, however, as most peptides identified as cleavage products derived from the incubation of HF3 with plasma were derived from its N- and C-terminal regions, its direct incubation with HF3 did not show any change in the bands corresponding to its dimer and monomer forms, as visualized by SDS-PAGE.

3.3. Analysis of the Primary Specificity of HF3 on Plasma Proteins

The use of peptides generated by the cleavage of plasma proteins by HF3 proved to be an excellent tool to evaluate its primary specificity, since a large number of cleavage sites were analyzed (2295 identified by Mascot/TPP and 3948 by PEAKS) and showed the preference of HF3 for Leu in the P1′ position. In agreement with these results, the analysis of the primary specificity of HF3, using data resulting from cleavage events in native proteins present in the secretome of a mouse embryonic fibroblast analyzed by TAILS technology, showed a predominance of Leu at P1′ position [20]. Interestingly, the analysis of the activity of HF3 on human plasma peptide substrates by the PICS approach showed a similar pattern, in which Leu had a higher occurrence at P1′ position, followed by hydrophobic residues in P2′ [21]. The preference for Leu at the P1′ subsite is a common feature of metalloproteinases, such as matrix metalloproteases (MMPs) [43]. In the study by Kleifeld et al., 2010, the primary specificity of matrix metalloproteinase 2 (MMP-2) was evaluated by the TAILS method, showing a strong preference for Leu at the P1′ position (more than 45% occurrence) and Pro at the P3 position [44]. Using the same approach, Prudova et al., 2010, found that MMP-2 had a strong preference for Leu in position P1′ and Pro in P3, whereas MMP-9 exhibited a more relaxed preference for Leu in P1′ and a very strong preference for Pro in P3, reflecting the differences in the S1′subsites of these proteinases [45]. MMP-10 also revealed its preference for Leu at the P1′ position and Pro at the P3 position [46]. Unlike what has been shown in these studies, the results of the analysis on the primary specificity of HF3 on plasma proteins did not indicate a significant preference for any amino acid at positions other than P1′, indicating that amino acids present at positions adjacent to the cleavage site do not have a relevant role in determining the hydrolysis of peptide bonds in proteins by HF3.

3.4. Plasma Proteins Cleaved by HF3

Fibrinogen peptides (alpha and beta chains), vitronectin, and fibronectin were considered as cleaved by HF3, corroborating previous studies by our group that showed that these proteins are hydrolyzed by HF3 in vitro [9]. The results obtained in the present study also confirm data from Paes Leme et al., 2012 [19], who showed that fibronectin and fibrinogen were degraded in the plasma of mice injected in the dorsal skin with HF3. Although fibrinogen alpha and beta chains have already been described as HF3 substrates, in this study, the fibrinogen gamma chain was also detected as a target of this metalloproteinase. Peptides from the fibrinogen gamma chain were identified as a product of HF3 activity on P(20-MAP-D) and P(LAP-E). In addition, incubation of HF3 with isolated fibrinogen, followed by analysis of the resulting peptide fraction by LC-MS/MS, confirmed the presence of these peptides. These analyses revealed the release of peptides from the C-terminal portion of the fibrinogen gamma chain by HF3. In addition to the important role of this region in the stabilization of fibrin by cross-linking [29], studies have shown that the C-terminal region of the gamma chain is involved in its interaction with the αIIbβ3 integrin in platelets [28,47,48]. Farrell et al., 1992, also found that a recombinant fibrinogen, which contained an interruption in the C-terminal region of the gamma chain, lost much of its ability to mediate platelet aggregation [49]. In addition, the C-terminal gamma-chain fibrinogen peptide (HHLGGAKQAGDV) demonstrated the ability to inhibit platelet aggregation and fibrinogen binding to rabbit platelets, through its direct interaction with GPIIb-IIIa [30]. Thus, it is possible that the cleavage of the fibrinogen gamma chain by HF3, detected upon incubation with plasma or isolated fibrinogen, results in the release of the C-terminal peptide, potentially affecting platelet aggregation.

In the same work by Paes Leme et al., 2012, alpha-1-antitrypsin, a serine proteinase inhibitor, was found in lower concentration in the skin injected with HF3 in comparison with the control skin. In the present study, we found that HF3 was able to cleave this protein in vitro, in agreement with its lower abundance in the skin of mice treated with HF3 [19]. Alpha-1-antitrypsin has anti-apoptotic effects and can act as a negative inflammatory regulator [50], thus, considering that HF3 and other metalloproteinases from venoms have pro-inflammatory activity, the degradation of alpha-1-antitrypsin could facilitate the development of inflammation. Peptides derived from alpha-2-macroglobulin and transthyretin were detected in the peptide fraction of plasma treated with HF3, indicating their proteolysis. These proteins were, however, found in higher abundance in the skin of mice injected with HF3, according to Paes Leme et al., 2012 [19], probably due to the increase of blood in the hemorrhagic skin. However, in agreement with the present study, we showed that alpha-2-macroglobulin was not able to inactivate the proteolytic activity of HF3 in vitro, and instead, was partially hydrolyzed by HF3 [16].

Plasminogen, a component of the fibrinolytic system, was identified as a substrate for HF3, but its limited cleavage was only detected in P(LAP-E), with a low number of identified peptides. Plasminogen is a single-chain glycoprotein that circulates in plasma in zymogen form and, when activated by urokinase (u-PA) or tissue plasminogen activator (t-PA) by cleavage of the Arg580-Val581 bond, it is converted to plasmin, and thus is able to cleave fibrin [51]. Angiostatin is an antiangiogenic protein that has been identified as the primary factor controlling the dormant state of cells in a secondary metastatic tumor, inhibiting angiogenesis and resulting in decreased blood flow and reduction in tumor size [52]. Angiostatin is a 38 kDa internal fragment of plasminogen [53], and angiostatin-like molecules can be generated in a variety of ways, including processing of plasminogen by various matrix metalloproteinases [54]. Ho et al., 2002, showed that metalloproteinases from *Bothrops* venoms, when incubated with plasminogen, were capable of generating a product of 38 kDa, and whose N-terminal sequencing evidenced the cleavage in the Ser460-Val461 bond, indicating the generation of an angiostatin-like protein [55]. Here, peptides identified in the peptide fraction of plasma treated with HF3 indicated cleavages in plasminogen close to the regions of initiation and termination of angiostatin. Furthermore, in a recent study, the incubation of HF3 with the isolated plasminogen did generate a product with a molecular mass close to 38 kDa [15], indicating that the limited proteolysis of plasminogen by HF3 indeed generates a protein similar to angiostatin. As angiostatin modulates the rate of plasminogen activation through non-competitive inhibition of the tissue-type plasminogen activator, the cleavage of plasminogen by HF3 and the generation of angiostatin would result in the decrease of the concentration of plasminogen in plasma, and as a consequence, its activity in the fibrinolytic system would be compromised. The impairment in the generation of plasmin, through the decrease of the concentration of plasminogen, would significantly affect the negative feedback of the coagulation cascade and, in turn, would also contribute to the fibrinogen consumption during the coagulopathies triggered by snake envenomation.

Prothrombin (coagulation factor II) was found in higher abundance in the plasma of mice injected with HF3 in the dorsal skin, in comparison to the plasma of mice treated with the control solution [19], by the analysis of plasma proteins subjected to in solution digestion with trypsin, and spectra count by LC-MS/MS. In the present study, prothrombin was identified as cleaved by HF3 by analyzing the plasma peptide fraction. The cleavage of prothrombin has also been shown by incubating the isolated protein with HF3, followed by SDS-PAGE, which revealed products of ~ 28, 30, 35, and 50 kDa [15]. The presence of many peptides from prothrombin in the peptide fraction of plasma treated with HF3 indicates that this protein is rather degraded and not activated by the proteinase, and as a consequence, prothrombin would be unavailable to participate in the blood coagulation process, where it is activated by factor Xa and converted into thrombin. The increase in prothrombin in the plasma of mice injected with HF3, verified in the study by Paes Leme et al., 2012 [19], could be explained as an attempt by the organism (mouse) to counteract, not only the hemorrhagic process, but also the cleavage of the protein itself by HF3, providing more protein to the coagulation cascade.

Alpha-2-antiplasmin is a serine proteinase inhibitor (serpin) that acts to protect fibrin clots from plasmin-mediated cleavage [56]. The plasmin inhibition by the serpin occurs primarily by the binding of the Gln41 residue from its N-terminal region to the Lys342 residue of the fibrin alpha chain, mediated by factor XIIIa. Then, the C-terminal domain of alpha-2-antiplasmin interacts with plasmin, so that its Arg403 aligns and forms a covalent bond with the Ser residue present in the active site of plasmin, forming an inactive enzyme-inhibitor complex [57]. Alpha-2-antiplasmin can be found in plasma either in its mature form (452 amino acids) or in its propeptide-containing form (464 amino acids) [58]. Pro-alpha-2-antiplasmin has inhibitory activity on plasmin, however, its ability to cross-link fibrin is reduced by approximately one third compared to the mature protein [59]. The peptide products generated from alpha-2-antiplasmin by HF3 indicated that both the mature protein and pro-protein forms were cleaved. However, no peptide corresponding to the exact sequence of the pro-domain was identified (M-E-P-L-G-R-Q-L-T-S-G-P), indicating that HF3 would not be able to activate the pro-protein. Furthermore, we detected that HF3 was able to cleave peptides from the N-terminal region of alpha-2-antiplasmin, containing the Gln41 residue involved in the crosslinking to fibrin. Moreover, the C-terminal region was also degraded by the HF3, but the sequence containing the Arg403 residue did not undergo hydrolysis. In view of these findings, we can suggest that HF3 would prevent the binding of alpha-2-antiplasmin to fibrin, however, the effect of HF3 on its inhibitory activity needs further investigation.

In addition to prothrombin, coagulation factor XIII was identified as a substrate of HF3. However, only two peptides (^{29}TVELQGVVPR^{38} and ^{464}LIVTKQIGGDGMMDITDT^{481}) were identified as generated by HF3 from factor XIII (chain A) in P(20-MAP-D), indicating limited proteolysis. The analysis of the activity of HF3 upon isolated factor XIII by SDS-PAGE revealed only a weak protein band of ~28 kDa, which could correspond to the region corresponding to residues 482−732 of the A chain of Factor XIII [15]. Factor XIII is a transglutaminase that circulates in the human plasma as a heterotetramer of two A and two B chains [60]. Activated Factor XIII (FXIIIa) is responsible for stabilizing fibrin by introducing covalent cross-links [61]. FXIIIa is also responsible for cross-linking alpha-2-antiplasmin and fibrin. The activation of FXIII is catalyzed by thrombin that cleaves the Arg38-Gly39 bonds of the A chain of the A2B2 tetramer, leading to chain dissociation and exposure of the cysteine residue of the catalytic site [62]. The identification of the peptide ^{29}TVELQGVVPR^{38}, which contains the thrombin cleavage site, is an indication of a possible activation of Factor XIII by HF3, however, as other FXIII A chain peptides were identified, it is not possible to state whether the effect of HF3 on this protein is degradation or activation.

Kininogen was identified as a substrate for HF3 in all plasma samples used in this study, and by both bioinformatics methods for protein identification. Peptides derived from both high molecular weight and low molecular weight kininogens have been identified and their extensive proteolysis has been indicated. High molecular weight kininogen participates in the coagulation cascade and the kallikrein-kinin system, whereas low molecular weight kininogen participates only in the latter [63]. These proteins are identical throughout their heavy chains, in the region containing bradykinin, and in the twelve amino acids in the N-terminal region of their light chains. Their heavy chains confer them the function of cysteine proteinase inhibitors [64]. The peptides bradykinin or lysyl-bradykinin are released from both kininogen forms when cleaved by kallikreins. In addition to its vasodilatory function, bradykinin (R-P-P-G-F-S-P-F-R) is an inflammation-mediating peptide [65]. The clotting activity of high molecular weight kininogen resides in its light chain [66], the region through which it binds to prekallikrein and factor XI. Furthermore, the same light chain contains a region rich in histidine, responsible for its binding to negatively charged surfaces [67]. The peptides identified as hydrolysis products generated by HF3 are located in the heavy chain of both kininogens, which would compromise their cysteine proteinase inhibitor function. In addition, the light chain of the high molecular weight kininogen was also cleaved by HF3, which would compromise its role in the coagulation cascade. Finally, we found that the region containing the peptide corresponding to bradykinin is cleaved by HF3, but no peptide corresponding to the exact bradykinin sequence was identified, so further investigations are needed to evaluate whether proteolysis of this region would have any effect on the kallikrein-kinin system.

Ceruloplasmin was also identified as a substrate for HF3, although a low number of peptides was detected as a hydrolysis product, and only in the peptide fraction of P(Alb-D) after treatment with HF3. This protein has a high degree of homology with the factor V (A chain) of the coagulation cascade [68]. Ceruloplasmin carries more than 95% of the copper present in plasma, and its function is related to the regulation of copper and iron homeostasis, in addition to its role in angiogenesis and antioxidant activity [69]. Ceruloplasmin is also an acute phase protein and its plasma level increases in response to inflammation [70,71]. Walker and Fay (1990) reported its ability to bind to protein C, thus protecting factors Va and VIII from inactivation catalyzed by activated protein C. Thus, its hydrolysis by HF3 could be involved in the hemorrhagic process [72].

The complement system, which is part of the innate immune system, is composed of more than 30 plasma proteins and cell surface receptors that react with each other in a range of functions, including direct cell lysis and enhancement of B and T cell responses, and induce a series of inflammatory responses that contribute to fighting infection [73–75]. Some proteins are proteinases that are activated in a proteolytic cleavage cascade, similar to the blood coagulation cascade [76]. The activation of the complement system by snake venom components and its role in the envenomation has been shown with venoms from the *Elapidae* and *Viperidae* families [77,78]. Among other potential new targets for HF3, this study identified some components of the complement cascade: C1r; C1s; C3; C4 and C9; C4b binding protein (alpha chain); complement factor H; and factor H-related protein 1. Cleavage of components C3 and C4 was confirmed by incubation of HF3 with these isolated proteins [15], indicating that they would not be available in plasma to participate in the complement cascade activation process, hence HF3 would play a role as a modulator of this cascade of proteinases.

Peptides from clusterin (apoliprotein J) were identified in the peptide fraction of P(W), P(Alb-D), P(20-MAP-D), and P(LAP-E) after incubation with HF3, indicating an extensive degradation of this protein. Clusterin is a chaperone protein with anti-inflammatory and cytoprotective activity, which inhibits MMP-9, MMP-2, MMP-3, and MMP-7 [79], so its proteolysis by HF3 could prevent its anti-inflammatory activity. Further, it is possible to suggest that, opposite to what happens with MMPs, clusterin would not be an inhibitor of HF3.

The results of the present study also showed potential new HF3 targets, such as alpha-2-HS-glycoprotein, or fetuin-A, which was shown as extensively degraded by HF3. This protein functions as an important component of several physiological or pathological mechanisms, including vascular calcification, bone metabolism, insulin resistance, keratinocyte migration, and breast cancer tumor cell proliferative signaling [80]. In response to acute inflammation, alpha-2-HS-glycoprotein appears to act as an anti-inflammatory modulator. Alpha-2-HS-glycoprotein has been shown to suppress the excessive release of tumor necrosis factor from activated macrophages [80–82]. Hence, the proteolysis of alpha-2-HS-glycoprotein by HF3 would impair its anti-inflammatory activity. Meprins are zinc-dependent, astacin-like metalloproteinases that play a pivotal role in inflammation by activating cytokines, and the potential role of endogenous meprin inhibitor has also been attributed to alpha-2-HS-glycoprotein [80,83,84]. Guerranti et al., 2010 [38], reported the proteolysis of alpha-2-HS-glycoprotein after incubation of human plasma with *Echis carinatus* venom, however, the component of the venom responsible for its degradation has not been investigated. To our knowledge, there are no reports on the cleavage of alpha-2-HS-glycoprotein by other venom metalloproteinases, and its degradation by HF3 would potentially have a pro-inflammatory effect.

In the study by Paes Leme et al., 2012 [19], it was found that apolipoprotein A-II, the second major component of HDL particles, was clearly degraded in the hemorrhagic process generated by HF3. Here we show that not only apolipoprotein A-II, but also other proteins of the same family (apolipoprotein AI, A-IV, CI, C-II, C-III, E, F, and L1), were detected as substrates of HF3. Apolipoproteins bind and transport lipids, and members of classes A, C, and E play an important role in lipoprotein metabolism and in inflammatory processes [85]. Previous studies have shown that apolipoproteins AI and E are cleaved by matrix metalloproteinase 14 (MMP-14) [39], whereas apolipoproteins C-II and A-IV are cleaved by matrix metalloproteinases 7 and 14 [86–88]. The cleavage of apolipoproteins A-I and A-II by metalloproteinases from the venom of *Cerastes cerastes* was observed by El-Asmar and Swaney (1988) [89], and the cleavage of apolipoprotein A-I by proteinases in the venom of *Echis carinatus* was also reported by Guerranti et al., 2010 [38]. However, the role of proteolysis of these apolipoproteins in the context of envenomation is not well established, and a hypothesis that can be suggested is that the degradation products of these proteins could play a role in modulating the pro-inflammatory activity of venom metalloproteinases. Whether this modulation would be characterized by the increase or by the limitation of the pro-inflammatory effects is a matter of future investigation.

Gelsolin was identified as a substrate for HF3 in P(W), P(20-MAP-D), and P(LAP-E), and was extensively degraded. In another study by our group on the degradation of proteins secreted by fibroblasts, the cleavage of gelsolin by HF3 was also identified by the TAILS approach [20]. Plasma gelsolin, together with the vitamin D-binding protein, forms part of the actin scavenger system [90,91]. Excessive release of intracellular actin or decreased activity of the scavenger system is associated with pathological conditions such as small blood vessel obstruction, clots, endothelial damage, hepatic necrosis, and septic shock [92]. Studies using degradomics approaches showed that gelsolin is a substrate of matrix metalloproteinases, and MMP-3 is able to cleave it more efficiently, followed by MMP-2, MMP-1, and MMP-14 [93]. Thus, as in the case of MMPs, gelsolin cleavage by HF3 could affect the actin scavenging system and promote pathological conditions induced by an excess of unremoved extracellular actin.

Many peptides from the H1, H2, H3, and H4 heavy chains of inter-alpha trypsin inhibitor family proteins were identified in the peptide fraction of plasma after incubation with HF3, indicating extensive cleavage of these polypeptide chains. Proteins of the inter-alpha trypsin inhibitor family are composed of a common light chain (the chondroitin sulfate proteoglycan bikunin), and various heavy chains [94,95]. Bikunin shows inhibitory activity on a wide spectrum of proteinases, including some of pathological importance, such as trypsin, chymotrypsin, plasmin, elastase, and cathepsin. Bikunin is also recognized as an anti-inflammatory mediator [96]. The binding of the heavy chains of the inter-alpha trypsin inhibitor to hyaluronan may be involved in the stabilization and integrity of the extracellular matrix, and provides anti-inflammatory properties [97–99]. Catanese and Kress (1985) reported on the in vitro cleavage of the inter-alpha trypsin inhibitor by metalloproteinases from viperid, colubrid, and elapid venoms, generating products still capable of inhibiting trypsin [100]. The degradation of H1−H4 chains of inter-alpha trypsin inhibitor by HF3 would likely not impair the ability of its light chain to inhibit serine proteinases, but would affect its anti-inflammatory activity.

The histidine-rich glycoprotein was also identified as cleaved by HF3, however, this cleavage was very specific and generated peptides within the sequence HSHGPPLPQGPP-PLLPM of its proline-rich region [101]. Histidine-rich glycoprotein is found in plasma, leukocytes, platelet α-granules, and megakaryocytes. It contains various binding domains, and interacts with the heme group, plasminogen, heparin, fibrinogen, thrombospondin, immunoglobulin G, complement components, and Zn^{2+} ions [102]. Histidine-rich glycoprotein can also interact with cell-associated molecules, including Fcγ receptors, heparan sulfate, and phospholipids [102]. In the context of this study, the most interesting aspect of the histidine-rich glycoprotein lies in its ability to modulate components of the coagulation cascade. It can bind to heparin released by mast cells, preventing it from inhibiting the procoagulant activity of monocytes at the site of inflammation and thrombosis [101,103]. Some studies have demonstrated the profibrinolytic effect of histidine-rich glycoprotein, by its ability to bind to plasminogen, stimulating its cleavage by plasmin [104]. Other studies, on the other hand, have investigated its action as an antifibrinolytic agent, as the histidine-rich glycoprotein binding to plasminogen could interfere with the interaction of plasminogen with fibrin clots, thus inhibiting plasmin-mediated fibrinolysis [101,105]. Thus, the cleavage of histidine-rich glycoprotein by HF3 would have a variety of implications in the context of hemorrhage, mostly related to its role in regulating hemostasis.

Other proteins were detected as cleaved by HF3, but a low number of peptides were found as hydrolysis products, and mostly in only one type of plasma preparation: carboxypeptidase B2; collectin-11; ficolin-2; galectin-3 binding protein; glutathione peroxidase 3; histone H2B type FS; hyaluronan binding protein; MAX gene-associated protein isoform 3; pigmented epithelial-derived factor; pregnancy zone protein; proline-rich acidic proteins; amyloid A protein -4 serum; immunoglobulins; tetranectin; and serum paraoxonase/arylesterase. These proteins do not have a defined role in the generation of hemorrhage, and thus the implications of their hydrolysis by HF3 need further investigation.

4. Conclusions

The human plasma depletion methods used in this study provided heterogeneous samples with respect to the range of dynamic protein concentration, which were compatible with the peptidomic analysis, and generated complementary results for the elucidation of the HF3 degradome. The two bioinformatics approaches used for the analysis of the peptide fraction gave robustness to the set of obtained results, since most proteins identified as HF3 substrates were detected by both. The determination of the primary specificity of HF3 on protein substrates showed that Leu at P1′ is a major determinant of HF3 primary specificity, which agrees with previous studies using peptides, and reinforces the importance of this residue at P1′, regardless of the substrate structure.

As a result of this study, knowledge about the HF3 substrate repertoire in human plasma has been expanded in terms of number, as well as protein classes and functions (Figure 9). Taken together, the results illustrate the proteolytic signature of human plasma in the context of HF3-induced hemorrhage. By acting on distinct substrates, which are part of a highly connected biological circuit, the proteolytic signaling triggered by HF3 may not be fully anticipated by the results of in vitro incubation with single substrates. Actually, the activated/impaired biological pathways involved in the hemorrhagic and pro-inflammatory effects of SVMPs are the result of complex signaling circuits, which are significantly affected by limited proteolysis and protein degradation. In this regard, it was interesting to note that the hydrolysis of some proteins by HF3 seems to lead to antagonistic results, such as the hydrolysis of fibrinogen and plasminogen, which play roles in different steps of blood coagulation and fibrinolysis. In general, the characterization of HF3 substrate degradome in the human plasma suggests that it acts in a dysregulated manner, refractory to plasma inhibitors, causing an imbalance in hemostasis.

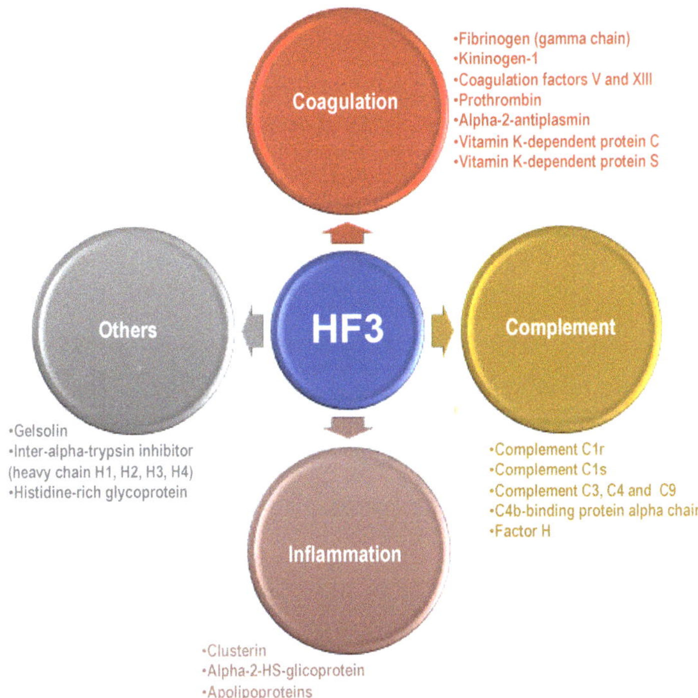

Figure 9. Schematic representation of some of the new substrates of HF3 revealed in this study, according to biological function.

The hydrolysis of human plasma proteins by an uncontrolled, exogeneous metalloproteinase has a direct impact in the plasma proteome, and it can be hypothesized that some hydrolysis products could also play synergistic roles in the pro-inflammatory and hemorrhagic processes generated by HF3.

5. Materials and Methods

5.1. HF3

HF3 (Uniprot entry Q98UF9) was purified as described previously (Oliveira et al., 2009) from *B. jararaca* venom provided by the Laboratory of Herpetology of Butantan Institute (São Paulo, Brazil), and identified by trypsin digestion and mass spectrometric (LC–MS/MS) analysis.

5.2. Analytical Procedures

Protein and peptide contents were quantified using, respectively, Bradford assay (Sigma-Aldrich, St. Louis, MO, USA) and micro-BCA assay (Pierce) kits according to the manufacturers' recommendations. SDS–PAGE was carried out according to Laemmli (1970) [106]. Silver staining was carried out according to Mortz et al., 2001 [107].

5.3. Human Plasma

Procedures using human blood in this study were approved by ethics committee of Instituto Federal de Educação, Ciência e Tecnologia de São Paulo, Brazil, and registered under CAAE 19892213.7.0000.5473. Human blood was collected from three volunteers who declared they had not used painkillers for at least 10 days. In order to obtain human plasma, blood was collected and anticoagulated with 0.1 vol. of 3.8% sodium citrate and centrifuged at 1000× *g* for 10 min at 4 °C. Whole plasma samples (n = 3) were designated as P(W).

5.4. Albumin-Depleted Human Plasma

To deplete serum albumin from human plasma, we employed Blue Sepharose CL-6B resin affinity chromatography (Sigma-Aldrich, St. Louis, MO, USA; [108,109]). Three independent experiments were carried out with plasma samples from three individuals. The procedure was performed using a microtube with a filter, according to the manufacturer's instructions. In the upper part of the filter microtube, 0.5 mL of resin was added, and steps for resin conditioning, binding and elution, and resin reconstitution were carried out by fractioning the total volume required in 0.5 mL aliquots, added to the top of the microtube, followed by mild manual agitation and centrifugation for 1 min at 150× *g*, with disposal or storage of the eluted solution conducted according to the stage. The following solutions were added: (i) conditioning: 2.5 mL of ultrapure water and 4.0 mL of binding buffer (20 mM HEPES, pH 7.4), with disposal of the eluted solution; (ii) binding: the lower outlet duct of the micro tube containing the resin was closed, and 40 µL of plasma diluted in 310 µL of binding buffer was loaded. The tube was kept at room temperature for 30 min under agitation. After centrifugation, the eluted material, corresponding to the plasma depleted of albumin, was stored. For the elution of proteins bound to the resin (eluate), 4 mL of 2 M NaCl were added to the resin, and the eluted solution was stored. For resin reconstitution, 1 mL of 6 M guanidine hydrochloride and 4 mL of binding buffer were applied to the resin, and the eluted solution was discarded. Albumin-depleted plasma samples were designated as P(Alb-D).

5.5. Human Plasma Depleted of 20 Most Abundant Proteins

Depletion of the 20 most abundant proteins from human plasma was performed with the ProteoPrep 20 Plasma Immunodepletion Kit (Sigma-Aldrich, St. Louis, MO, USA), according to the manufacturer's instructions. Three independent experiments were carried out with plasma samples from three individuals. Briefly, samples of 8 µL of human plasma were diluted to 100 µL with phosphate-buffered saline (PBS) and loaded to a micro-spin column, previously equilibrated with PBS. After incubation for 20 min at room temperature, the non-bound protein fraction was recovered by centrifugation at 2000 rpm for 1 min and the flow-through, containing plasma depleted of the 20 most abundant proteins P(20-MAP-D), was collected in a clean tube. The remaining unbound proteins were washed twice by adding 100 µL of PBS, centrifuging, and collecting the wash in the same tube. The depletion procedure was repeated five times, and flow-throughs were pooled and then concentrated using a Centricon YM-3 filter (Millipore). After each depletion procedure, the micro-spin column was regenerated with 2 mL of 0.1 M Glycine-HCl, pH 2.5, and TWEEN 20 in order to elute bound proteins, and stored at 5 °C in 5 mL of PBS with the addition of 10 µL of ProteoPrep Preservative Concentrate. Human plasma depleted of 20 most abundant proteins was designated as P(20-MAP-D).

5.6. Human Plasma Enriched in Low-Abundance Proteins

The enrichment of low-abundant proteins in human plasma was performed with the ProteoMiner Protein Enrichment Kit (Bio-Rad) [37] according to the manufacturer's instructions. Three independent experiments were carried out with plasma samples from three individuals. In brief, 200 µL of human plasma was loaded onto the column previously conditioned with PBS, pH 7.4, and incubated for 2 h at room temperature. After centrifugation, the flow-through fraction was stored for further analysis. The fraction that bound to the resin, containing the low-abundant proteins, was eluted with 8 M urea containing 2% CHAPS. For desalinization, the proteins in solution were precipitated by the addition of eight volumes of cold acetone and one volume of cold methanol, and stored for 12 h at −20 °C. After centrifugation for 10 min at 14,000× *g* at 4 °C, the precipitate was washed with cold methanol and resuspended in 60 µL of 100 mM NaOH and 340 µL of 50 mM HEPES, pH 7.5. Plasma enriched of low-abundant proteins was designated as P(LAP-E).

5.7. Proteolytic Activity of HF3 on Human Plasma

For each of the four types of human plasma preparations, three biological replicates were performed. For each experiment, P(Alb-D), P(20-MAP-D), and P(LAP-E); 50 µg were separately incubated with 0.5 µg HF3 (1:100; *w/w*) in 250 mM ammonium acetate containing $CaCl_2$ 1 mM for 2 h at 37 °C. P(W) (200 µg) was incubated with 2 µg HF3 under the same conditions. Samples of P(Alb-D), P(20-MAP-D), P(LAP-E), and P(W) were incubated without HF3, as a control, under identical conditions. Reactions were stopped by adding eight volumes of cold acetone and one volume of cold methanol, and incubated for 12 h at −20 °C. Peptide fractions (supernatant) were obtained by centrifugation at 14,000 g for 10 min at −4 °C, and subsequently dried using a SpeedVac concentrator. The protein fraction (precipitate) was separated and stored at −20 °C until use.

5.8. LC-MS/MS Analysis of the Plasma Peptide Fraction

Previously to LC-MS/MS analysis, plasma peptide fractions containing hydrolysis products resulting from the proteolytic activity of HF3 were subjected to removal of traces of detergent using Macro Spin Columns (Harvard Apparatus). Samples were then desalted with Sep-Pak Light C18 (Waters) cartridges, vacuum dried, and resuspended in 20 µL of 0.1% formic acid. Aliquots of 10 µL were separated by RP-HPLC on an EASY-nLC II (Thermo Scientific, Waltham, MA, USA) using a column (75 µm i.d. × 10 cm) packed with 5 µm C18 beads (Phenomenex), and coupled to an LTQ-Orbitrap Velos mass spectrometer (Thermo Fisher Scientific, Waltham, MA, USA). The gradient was 5–40% acetonitrile in 0.1 M formic acid over 90 min, at a flow rate of 300 nL/min. The mass spectrometer was operated in data dependent mode, in which one full MS scan was acquired in the m/z range of 400–2000 at 60,000 resolution, followed by MS/MS acquisition using high-energy collision dissociation of the six most intense ions from the MS scan, at 15,000 resolution. A dynamic peak exclusion was applied to avoid the same m/z of being selected for the next 25 s, using a ± 1 Da mass tolerance window around the precursor ion mass.

5.9. Mass Spectrometry Data Analysis

Two strategies were employed for the identification of peptides originated from HF3 cleavage of plasma proteins. In the first, we used the Mascot program, and validated the results using the PeptideProphet and ProteinProphet tools of the Trans-Proteomic Pipeline (TPP) platform (Mascot/TPP). In the second, de novo sequencing, searching the database, and validating the results were carried out using the Peaks Studio 7 program (Peaks).

Mascot/TPP: Acquired MS/MS raw data were converted to the mgf and mzXML format using MS-Convert. Database searches were performed against the human UniProtKB/Swiss-Prot protein database (available at http://www.uniprot.org; accessed on 3 February 2014) using Mascot 2.4.1 (Matrix Science) with the following parameters: no enzyme specificity indicated; 10 ppm precursor tolerance; 20 mmu fragment ion tolerance; variable Met oxidation (+15.9949 Da); variable *N* terminal acetylation (+42.0105 Da); and variable Asn and Gln deamidation (+0.9840 Da). Mascot search results were further processed using the Trans-Proteomic Pipeline (TPP, version 4.6) (Keller and Shteynberg, 2011). Peptides were included in the analysis if they were identified at a false discovery rate (FDR) of ≤1% at peptide level (PeptideProphet), and (ii) proteins at an FDR of ≤1% at protein level (ProteinProphet).

Peaks: Acquired MS/MS raw data were imported into Peaks Studio 7 software (Ma et al., 2003). De novo analysis was performed with the following parameters: no enzyme specificity indicated; 10 ppm precursor tolerance; 0.02 Da fragment ion tolerance; variable Met oxidation (+15.9949 Da); variable *N* terminal acetylation (+42.0105 Da); and variable Asn and Gln deamidation (+0.9840 Da). After de novo sequencing, a database search (Peaks DB) was performed against the human UniProtKB/Swiss-Prot protein database (available at http://www.uniprot.org; accessed on 10 February 2014), using the same parameters. Peptides were included in the analysis if they were identified at a Peaks DB FDR of ≤1%.

5.10. Identification of Peptides Generated the Incubation of Fibrinogen with HF3

Fibrinogen (50 μg) was incubated with HF3 (0.5 μg) in 0.025 M Tris-HCl, pH 7.5, 5 mM $CaCl_2$ for 2 h at 37 °C. The reaction was stopped by adding eight volumes of cold acetone and one volume of cold methanol, and stored for 12 h at −20 °C. After centrifugation at 14,000× *g* for 10 min at 4 °C, the supernatant, corresponding to the peptide fraction, was dried in a SpeedVac concentrator, subjected to desalination procedures using Sep-Pak Light C18 cartridges, vacuum dried, and resuspended in 20 μL of 0.1% formic acid for analysis by LC-MS/MS, as described above, and a database search using the Peaks Studio 7 program.

5.11. Validation of HF3 Substrates in Human Plasma by Incubation with HF3

Apolipoprotein A-IV, clusterin, α-2-antiplasmin, kininogen, and transthyretin (2 μg; Sigma-Aldrich, St. Louis, MO, USA) were incubated with HF3 (200 ng) (1:10 (*w/w*) enzyme-to-substrate ratio) in 0.05 M Tris-HCl, pH 8.0, 1.0 mM $CaCl_2$ for 2 h at 37 °C. Apoliprotein E (4 μg; Sigma-Aldrich, St. Louis, MO, USA) was incubated with HF3 (400 ng) (1:10 (*w/w*) enzyme-to-substrate ratio) in the same buffer. A sample of each protein was incubated without enzymes under identical conditions. Reactions were stopped by adding a Laemmli sample buffer, and subjected to SDS-PAGE.

Supplementary Materials: The following are available online at https://www.mdpi.com/article/10.3390/toxins13110764/s1, Supplementary Tables S1–S8: Peptides identified by LC-MS/MS analysis of the peptide fraction of human plasma after incubation with HF3. Supplementary Table S9: Peptides identified by LC-MS/MS analysis of the peptide fraction of human fibrinogen after incubation with HF3.

Author Contributions: Conceptualization, S.M.T.S.; experimental design, A.Z., L.B. and S.M.T.S.; data collection: A.F.A.C., A.K.O., A.Z., L.B. and M.C.M.; data analysis, A.F.A., A.F.A.C., A.K.T., A.Z., L.B. and S.M.T.S.; manuscript writing, S.M.T.S.; funding acquisition, S.M.T.S.; manuscript revising, all authors. All authors have read and agreed to the published version of the manuscript.

Funding: This research was funded by São Paulo Research Foundation (FAPESP 2011/16623-1; 2011/08514-8; 2013/07467-1; 2020/12317-2) and the National Council for Scientific and Technological Development (CNPq 308133/2015-3 to SMTS).

Institutional Review Board Statement: Not applicable.

Informed Consent Statement: Not applicable.

Data Availability Statement: All mass spectrometry peptidomics data have been deposited to the ProteomeXchange Consortium (http://proteomecentral.proteomexchange.org), via the PRIDE partner repository [110] with the dataset identifier: PXD027997.

Conflicts of Interest: The authors declare no conflict of interest.

References

1. Gutiérrez, J.M.; Escalante, T.; Rucavado, A.; Herrera, C. Hemorrhage caused by snake venom metalloproteinases: A journey of discovery and understanding. *Toxins* **2016**, *8*, 93. [CrossRef] [PubMed]
2. Kasturiratne, A.; Wickremasinghe, A.R.; De Silva, N.; Gunawardena, N.K.; Pathmeswaran, A.; Premaratna, R.; Savioli, L.; Lalloo, D.G.; De Silva, H.J. The global burden of snakebite: A literature analysis and modelling based on regional estimates of envenoming and deaths. *PLoS Med.* **2008**, *5*, e218. [CrossRef]
3. Ownby, C.L.; Bjarnason, J.; Tu, A.T. Hemorrhagic toxins from rattlesnake (Crotalus atrox) venom. Pathogenesis of hemorrhage induced by three purified toxins. *Am. J. Pathol.* **1978**, *93*, 201–218.
4. Ohsaka, A.; Ikezawa, H.; Kondo, H.; Kondo, S.; Uchida, N. Haemorrhagic activities of habu snake venom, and their relations to lethal toxicity, proteolytic activities and other pathological activities. *Br. J. Exp. Pathol.* **1960**, *41*, 478–486.
5. Bjarnason, J.B.; Fox, J.W. Hemorrhagic metalloproteinases from snake venoms. *Pharmacol. Ther.* **1994**, *62*, 325–372. [CrossRef]
6. White, J. Snake venoms and coagulopathy. *Toxicon* **2005**, *45*, 951–967. [CrossRef]
7. Isbister, G. Procoagulant snake toxins: Laboratory studies, diagnosis, and understanding snakebite coagulopathy. *Semin. Thromb. Hemost.* **2009**, *35*, 93–103. [CrossRef] [PubMed]

8. Fox, J.W.; Serrano, S.M.T. Insights into and speculations about snake venom metalloproteinase (SVMP) synthesis, folding and disulfide bond formation and their contribution to venom complexity. *FEBS J.* **2008**, *275*, 3016–3030. [CrossRef]
9. Oliveira, A.K.; Leme, A.P.; Asega, A.F.; Camargo, A.C.M.; Fox, J.W.; Serrano, S.M.T. New insights into the structural elements involved in the skin haemorrhage induced by snake venom metalloproteinases. *Thromb. Haemost.* **2010**, *104*, 485–497. [CrossRef] [PubMed]
10. Escalante, T.; Rucavado, A.; Fox, J.W.; Gutiérrez, J.M. Key events in microvascular damage induced by snake venom hemorrhagic metalloproteinases. *J. Proteom.* **2011**, *74*, 1781–1794. [CrossRef]
11. Kini, R.M.; Koh, C.Y. Metalloproteases Affecting blood coagulation, fibrinolysis and platelet aggregation from snake venoms: Definition and nomenclature of interaction sites. *Toxins* **2016**, *8*, 284. [CrossRef]
12. Assakura, M.T.; Reichl, A.P.; Mandelbaum, F.R. Comparison of immunological, biochemical and biophysical properties of three hemorrhagic factors isolated from the venom of Bothrops jararaca (jararaca). *Toxicon* **1986**, *24*, 943–946. [CrossRef]
13. Oliveira, A.K.; Leme, A.F.P.; Assakura, M.T.; Menezes, M.C.; Zelanis, A.; Tashima, A.K.; Lopes-Ferreira, M.; Lima, C.; Camargo, A.C.; Fox, J.W.; et al. Simplified procedures for the isolation of HF3, bothropasin, disintegrin-like/cysteine-rich protein and a novel P-I metalloproteinase from Bothrops jararaca venom. *Toxicon* **2009**, *53*, 797–801. [CrossRef]
14. Silva, C.A.; Zuliani, J.P.; Assakura, M.T.; Mentele, R.; Camargo, A.C.; Teixeira, C.F.; Serrano, S.M. Activation of αMβ2-mediated phagocytosis by HF3, a P-III class metalloproteinase isolated from the venom of Bothrops jararaca. *Biochem. Biophys. Res. Commun.* **2004**, *322*, 950–956. [CrossRef] [PubMed]
15. Asega, A.F.; Menezes, M.C.; Trevisan-Silva, D.; Carvalho, D.C.; Bertholim, L.; Oliveira, A.K.; Zelanis, A.; Serrano, S.M.T. Cleavage of proteoglycans, plasma proteins and the platelet-derived growth factor receptor in the hemorrhagic process induced by snake venom metalloproteinases. *Sci. Rep.* **2020**, *10*, 12912. [CrossRef] [PubMed]
16. Asega, A.F.; Oliveira, A.K.; Menezes, M.C.; Neves-Ferreira, A.G.C.; Serrano, S.M. Interaction of Bothrops jararaca venom metalloproteinases with protein inhibitors. *Toxicon* **2014**, *80*, 1–8. [CrossRef] [PubMed]
17. Menezes, M.C.; de Oliveira, A.K.; Melo, R.L.; Lopes-Ferreira, M.; Rioli, V.; Balan, A.; Leme, A.F.P.; Serrano, S.M. Disintegrin-like/cysteine-rich domains of the reprolysin HF3: Site-directed mutagenesis reveals essential role of specific residues. *Biochimie* **2011**, *93*, 345–351. [CrossRef]
18. Menezes, M.C.; Imbert, L.; Kitano, E.S.; Vernet, T.; Serrano, S.M.T. Recombinant expression of the precursor of the hemorrhagic metalloproteinase HF3 and its non-catalytic domains using a cell-free synthesis system. *Amino Acids* **2016**, *48*, 2205–2214. [CrossRef] [PubMed]
19. Leme, A.P.; Sherman, N.; Smalley, D.M.; Sizukusa, L.O.; Oliveira, A.K.; Menezes, M.C.; Fox, J.W.; Serrano, S.M.T. Hemorrhagic activity of HF3, a snake venom metalloproteinase: Insights from the proteomic analysis of mouse skin and blood plasma. *J. Proteome Res.* **2012**, *11*, 279–291. [CrossRef]
20. Zelanis, A.; Oliveira, A.K.; Prudova, A.; Huesgen, P.F.; Tashima, A.K.; Kizhakkedathu, J.; Overall, C.M.; Serrano, S.M.T. Deep profiling of the cleavage specificity and human substrates of snake venom metalloprotease HF3 by proteomic identification of cleavage site specificity (PICS) using proteome derived peptide libraries and terminal amine isotopic labeling of substrates. *J. Proteome Res.* **2019**, *18*, 3419–3428. [CrossRef]
21. Bertholim, L.; Zelanis, A.; Oliveira, A.K.; Serrano, S.M.T. Proteome-derived peptide library for the elucidation of the cleavage specificity of HF3, a snake venom metalloproteinase. *Amino Acids* **2016**, *48*, 1331–1335. [CrossRef]
22. Leme, A.P.; Escalante, T.; Pereira, J.G.; Oliveira, A.K.; Sanchez, E.O.F.; Gutiérrez, J.M.; Serrano, S.M.T.; Fox, J.W. High resolution analysis of snake venom metalloproteinase (SVMP) peptide bond cleavage specificity using proteome based peptide libraries and mass spectrometry. *J. Proteom.* **2011**, *74*, 401–410. [CrossRef] [PubMed]
23. Anderson, N.L.; Anderson, N.G. The human plasma proteome. *Mol. Cell. Proteom.* **2002**, *1*, 845–867. [CrossRef]
24. Geyer, P.E.; Kulak, N.A.; Pichler, G.; Holdt, L.M.; Teupser, D.; Mann, M. Plasma Proteome profiling to assess human health and disease. *Cell Syst.* **2016**, *2*, 185–195. [CrossRef]
25. Szklarczyk, D.; Morris, J.H.; Cook, H.; Kuhn, M.; Wyder, S.; Simonovic, M.; Santos, A.; Doncheva, N.T.; Roth, A.; Bork, P.; et al. The STRING database in 2017: Quality-controlled protein–protein association networks, made broadly accessible. *Nucleic Acids Res.* **2017**, *45*, D362–D368. [CrossRef]
26. Schechter, I.; Berger, A. On the size of the active site in proteases. I. Papain. *Biochem. Biophys. Res. Commun.* **1967**, *27*, 157–162. [CrossRef]
27. Schilling, O.; Huesgen, P.; Barré, O.; Keller, U.A.D.; Overall, C.M. Characterization of the prime and non-prime active site specificities of proteases by proteome-derived peptide libraries and tandem mass spectrometry. *Nat. Protoc.* **2011**, *6*, 111–120. [CrossRef] [PubMed]
28. Kloczewiak, M.; Timmons, S.; Lukas, T.J.; Hawiger, J. Platelet receptor recognition site on human fibrinogen. Synthesis and structure-function relationship of peptides corresponding to the carboxy-terminal segment of the gamma chain. *Biochemistry* **1984**, *23*, 1767–1774. [CrossRef]
29. Chen, R.; Doolittle, R.F. Isolation, Characterization, and location of a donor-acceptor unit from cross-linked fibrin. *Proc. Natl. Acad. Sci. USA* **1970**, *66*, 472–479. [CrossRef]
30. Packham, M.A.; Taylor, D.M.; Yeo, E.L.; Gemmell, C.H.; Patil, S.; Lam, S.C.-T.; Rand, M.L. The fibrinogen γ chain dodecapeptide inhibits agonist-induced aggregation of rabbit platelets and fibrinogen binding to rabbit glycoprotein IIb-IIIa. *Thromb. Haemost.* **1999**, *82*, 1680–1686. [CrossRef]

31. Millioni, R.; Tolin, S.; Puricelli, L.; Sbrignadello, S.; Fadini, G.P.; Tessari, P.; Arrigoni, G. High Abundance proteins depletion vs low abundance proteins enrichment: Comparison of methods to reduce the plasma proteome complexity. *PLoS ONE* **2011**, *6*, e19603. [CrossRef] [PubMed]
32. Hakimi, A.; Auluck, J.; Jones, G.D.D.; Ng, L.L.; Jones, D.J.L. Assessment of reproducibility in depletion and enrichment workflows for plasma proteomics using label-free quantitative data-independent LC-MS. *Proteomics* **2014**, *14*, 4–13. [CrossRef]
33. Polaskova, V.; Kapur, A.; Khan, A.; Molloy, M.; Baker, M.S. High-abundance protein depletion: Comparison of methods for human plasma biomarker discovery. *Electrophoresis* **2010**, *31*, 471–482. [CrossRef]
34. Govorukhina, N.; Keizer-Gunnink, A.; van der Zee, A.; de Jong, S.; de Bruijn, H.; Bischoff, R. Sample preparation of human serum for the analysis of tumor markers. *J. Chromatogr. A* **2003**, *1009*, 171–178. [CrossRef]
35. Gong, Y.; Li, X.; Yang, B.; Ying, W.; Li, D.; Zhang, Y.; Dai, S.; Cai, Y.; Wang, J.; He, A.F.; et al. Different immunoaffinity fractionation strategies to characterize the human plasma proteome. *J. Proteome Res.* **2006**, *5*, 1379–1387. [CrossRef] [PubMed]
36. Quero, C.; Colomé, N.; Prieto, M.R.; Carrascal, M.; Posada, M.; Gelpí, E.; Abian, J. Determination of protein markers in human serum: Analysis of protein expression in toxic oil syndrome studies. *Proteomics* **2004**, *4*, 303–315. [CrossRef] [PubMed]
37. Boschetti, E.; Righetti, P.G. The ProteoMiner in the proteomic arena: A non-depleting tool for discovering low-abundance species. *J. Proteom.* **2008**, *71*, 255–264. [CrossRef]
38. Guerranti, R.; Cortelazzo, A.; Hope-Onyekwere, N.S.; Furlani, E.; Cerutti, H.; Puglia, M.; Bini, L.; Leoncini, R. In vitro effects of Echis carinatus venom on the human plasma proteome. *Proteomics* **2010**, *10*, 3712–3722. [CrossRef] [PubMed]
39. Hwang, I.K.; Park, S.M.; Kim, S.Y.; Lee, S.-T. A proteomic approach to identify substrates of matrix metalloproteinase-14 in human plasma. *Biochim. Biophys. Acta Proteins Proteom.* **2004**, *1702*, 79–87. [CrossRef]
40. Escalante, T.; Rucavado, A.; Pinto, A.F.M.; Terra, R.M.S.; Gutiérrez, J.M.; Fox, J.W. Wound exudate as a proteomic window to reveal different mechanisms of tissue damage by snake venom toxins. *J. Proteome Res.* **2009**, *8*, 5120–5131. [CrossRef]
41. Menezes, M.C.; Kitano, E.S.; Bauer, V.C.; Oliveira, A.K.; Cararo-Lopes, E.; Nishiyama, M.Y.; Zelanis, A.; Serrano, S.M. Early response of C2C12 myotubes to a sub-cytotoxic dose of hemorrhagic metalloproteinase HF3 from Bothrops jararaca venom. *J. Proteom.* **2019**, *198*, 163–176. [CrossRef]
42. Zhang, J.; Xin, L.; Shan, B.; Chen, W.; Xie, M.; Yuen, D.; Zhang, W.; Zhang, Z.; Lajoie, G.A.; Ma, B. PEAKS DB: De novo sequencing assisted database search for sensitive and accurate peptide identification. *Mol. Cell. Proteomics* **2012**, *11*, M111.010587. [CrossRef] [PubMed]
43. Eckhard, U.; Huesgen, P.F.; Schilling, O.; Bellac, C.L.; Butler, G.S.; Cox, J.H.; Dufour, A.; Goebeler, V.; Kappelhoff, R.; Keller, U.A.D.; et al. Active site specificity profiling of the matrix metalloproteinase family: Proteomic identification of 4300 cleavage sites by nine MMPs explored with structural and synthetic peptide cleavage analyses. *Matrix Biol.* **2016**, *49*, 37–60. [CrossRef]
44. Kleifeld, O.; Doucet, A.; Keller, U.A.D.; Prudova, A.; Schilling, O.; Kainthan, R.K.; Starr, A.E.; Foster, L.J.; Kizhakkedathu, J.N.; Overall, C.M. Isotopic labeling of terminal amines in complex samples identifies protein N-termini and protease cleavage products. *Nat. Biotechnol.* **2010**, *28*, 281–288. [CrossRef] [PubMed]
45. Prudova, A.; Keller, U.A.D.; Butler, G.S.; Overall, C.M. Multiplex N-terminome analysis of MMP-2 and MMP-9 substrate degradomes by iTRAQ-TAILS quantitative proteomics. *Mol. Cell. Proteom.* **2010**, *9*, 894–911. [CrossRef]
46. Schlage, P.; Egli, F.E.; Nanni, P.; Wang, L.W.; Kizhakkedathu, J.N.; Apte, S.; Keller, U.A.D. Time-resolved analysis of the matrix metalloproteinase 10 substrate degradome. *Mol. Cell. Proteom.* **2014**, *13*, 580–593. [CrossRef]
47. Kloczewiak, M.; Timmons, S.; Bednarek, M.A.; Sakon, M.; Hawiger, J. Platelet receptor recognition domain on the .gamma. chain of human fibrinogen and its synthetic peptide analogues. *Biochemistry* **1989**, *28*, 2915–2919. [CrossRef]
48. Springer, T.A.; Zhu, J.; Xiao, T. Structural basis for distinctive recognition of fibrinogen γC peptide by the platelet integrin αIIbβ. *J. Cell Biol.* **2008**, *182*, 791–800. [CrossRef] [PubMed]
49. Farrell, D.H.; Thiagarajan, P.; Chung, D.W.; Davie, E.W. Role of fibrinogen alpha and gamma chain sites in platelet aggregation. *Proc. Natl. Acad. Sci. USA* **1992**, *89*, 10729–10732. [CrossRef]
50. Stockley, R.A. The multiple facets of alpha-1-antitrypsin. *Ann. Transl. Med.* **2015**, *3*, 130–137. [CrossRef]
51. Robbins, K.C.; Summaria, L.; Hsieh, B.; Shah, R.J. The peptide chains of human plasmin. Mechanism of activation of human plasminogen to plasmin. *J. Biol. Chem.* **1967**, *242*, 2333–2342. [CrossRef]
52. O'Reilly, M.S. Angiostatin: A novel angiogenesis inhibitor that mediates the suppression of metastases by a lewis lung carcinoma. *Cell* **1994**, *79*, 315–328. [CrossRef]
53. Cao, Y.; Xue, L. Angiostatin. *Semin. Thromb. Hemost.* **2004**, *30*, 83–93. [CrossRef]
54. Cornelius, A.L.; Nehring, L.C.; Harding, E.; Bolanowski, M.; Welgus, H.G.; Kobayashi, D.K.; Pierce, R.A.; Shapiro, S.D. Matrix metalloproteinases generate angiostatin: Effects on neovascularization. *J. Immunol.* **1998**, *161*, 6845–6852.
55. Ho, P.L.; Serrano, S.M.T.; Chudzinski-Tavassi, A.M.; Moura-Da-Silva, A.M.; Mentele, R.; Caldas, C.; Oliva, M.L.V.; Batista, I.D.F.C.; de Oliveira, M.L.S. Angiostatin-like molecules are generated by snake venom metalloproteinases. *Biochem. Biophys. Res. Commun.* **2002**, *294*, 879–885. [CrossRef]
56. Moroi, M.; Aoki, N. Isolation and characterization of alpha2-plasmin inhibitor from human plasma. A novel proteinase inhibitor which inhibits activator-induced clot lysis. *J. Biol. Chem.* **1976**, *251*, 5956–5965. [CrossRef]
57. Lee, K.N.; Lee, C.S.; Tae, W.-C.; Jackson, K.W.; Christiansen, V.J.; Mckee, P.A. Crosslinking of α2-Antiplasmin to Fibrin. *Ann. N. Y. Acad. Sci.* **2006**, *936*, 335–339. [CrossRef]

58. Koyama, T.; Koike, Y.; Toyota, S.; Miyagi, F.; Suzuki, N.; Aoki, N. Different NH2-terminal form with 12 additional residues of α2-plasmin inhibitor from human plasma and culture media of HEP G2 cells. *Biochem. Biophys. Res. Commun.* **1994**, *200*, 417–422. [CrossRef]
59. Sumi, Y.; Ichikawa, Y.; Nakamura, Y.; Miura, O.; Aoki, N. Expression and characterization of Pro α2-Plasmin Inhibitor. *J. Biochem.* **1989**, *106*, 703–707. [CrossRef] [PubMed]
60. Muszbek, L.; Bereczky, Z.; Bagoly, Z.; Komáromi, I.; Katona, É. Factor XIII: A coagulation factor with multiple plasmatic and cellular functions. *Physiol. Rev.* **2011**, *91*, 931–972. [CrossRef]
61. Pisano, J.J.; Finlayson, J.S.; Peyton, M.P. Cross-link in fibrin polymerized by Factor XIII: Egr-(ggr-Glutamyl) lysine. *Science* **1968**, *160*, 892–893. [CrossRef] [PubMed]
62. Komáromi, I.; Bagoly, Z.; Muszbek, L. Factor XIII: Novel structural and functional aspects. *J. Thromb. Haemost.* **2011**, *9*, 9–20. [CrossRef] [PubMed]
63. Kitamura, N.; Kitagawa, H.; Fukushima, D.; Takagaki, Y.; Miyata, T.; Nakanishi, S. Structural organization of the human kininogen gene and a model for its evolution. *J. Biol. Chem.* **1985**, *260*, 8610–8617. [CrossRef]
64. Salvesen, G.; Parkes, C.; Abrahamson, M.; Grubb, A.; Barrett, A.J. Human low-Mr kininogen contains three copies of a cystatin sequence that are divergent in structure and in inhibitory activity for cysteine proteinases. *Biochem. J.* **1986**, *234*, 429–434. [CrossRef] [PubMed]
65. Kaplan, A.P.; Silverberg, M. The coagulation-kinin pathway of human plasma. *Blood* **1987**, *70*, 1–15. [CrossRef]
66. Thompson, E.R.; Mandle, R.; Kaplan, A.P. Characterization of human high molecular weight kininogen. Procoagulant activity associated with the light chain of kinin-free high molecular weight kininogen. *J. Exp. Med.* **1978**, *147*, 488–499. [CrossRef] [PubMed]
67. Kato, H.; Sugo, T.; Ikari, N.; Hashimoto, N.; Maruyama, I.; Han, Y.N.; Iwanaga, S.; Fujii, S. Role of bovine high-molecular-weight (HMW) kininogen in contact-mediated activation of bovine Factor XII. *Adv. Exp. Med. Biol.* **1979**, *120*, 19–37.
68. Zaitsev, V.N.; Zaitseva, I.; Papiz, M.; Lindley, P.F. An X-ray crystallographic study of the binding sites of the azide inhibitor and organic substrates to ceruloplasmin, a multi-copper oxidase in the plasma. *JBIC J. Biol. Inorg. Chem.* **1999**, *4*, 579–587. [CrossRef] [PubMed]
69. Ehrenwald, E.; Fox, P. Isolation of nonlabile human ceruloplasmin by chromatographic removal of a plasma metalloproteinase. *Arch. Biochem. Biophys.* **1994**, *309*, 392–395. [CrossRef] [PubMed]
70. Gitlin, J.D. Transcriptional regulation of ceruloplasmin gene expression during inflammation. *J. Biol. Chem.* **1988**, *263*, 6281–6287. [CrossRef]
71. Fleming, R.E.; Whitman, I.P.; Gitlin, J.D. Induction of ceruloplasmin gene expression in rat lung during inflammation and hyperoxia. *Am. J. Physiol. Cell. Mol. Physiol.* **1991**, *260*, L68–L74. [CrossRef]
72. Walker, F.J.; Fay, P.J. Characterization of an interaction between protein C and ceruloplasmin. *J. Biol. Chem.* **1990**, *265*, 1834–1836. [CrossRef]
73. Reid, K.B.M.; Porter, R.R. The proteolytic activation systems of complement. *Annu. Rev. Biochem.* **1981**, *50*, 433–464. [CrossRef]
74. Fearon, D.T.; Carroll, M.C. Regulation of B lymphocyte responses to foreign and self-antigens by the CD19/CD21 complex. *Annu. Rev. Immunol.* **2000**, *18*, 393–422. [CrossRef] [PubMed]
75. Carroll, M.C. The complement system in regulation of adaptive immunity. *Nat. Immunol.* **2004**, *5*, 981–986. [CrossRef]
76. Arlaud, G.J.; Volanakis, J.E.; Thielens, N.M.; Narayana, S.V.; Rossi, V.; Xu, Y. The Atypical serine proteases of the complement system. *Adv. Immunol.* **1998**, *69*, 249–307. [CrossRef]
77. Vogel, C.-W.; Müller-Eberhard, H.J. The cobra complement system: I. The alternative pathway of activation. *Dev. Comp. Immunol.* **1985**, *9*, 311–325. [CrossRef]
78. Tambourgi, D.; van den Berg, C.W. Animal venoms/toxins and the complement system. *Mol. Immunol.* **2014**, *61*, 153–162. [CrossRef]
79. Jeong, S.; Ledee, D.; Gordon, G.M.; Itakura, T.; Patel, N.; Martin, A.; Fini, M.E. Interaction of Clusterin and matrix metalloproteinase-9 and its implication for epithelial homeostasis and inflammation. *Am. J. Pathol.* **2012**, *180*, 2028–2039. [CrossRef]
80. Mori, K.; Emoto, M.; Inaba, M. Fetuin-A: A multifunctional protein. *Recent Pat. Endocr. Metab. Immune Drug Discov.* **2011**, *5*, 124–146. [CrossRef]
81. Wang, H.; Zhang, M.; Soda, K.; Sama, A.; Tracey, K.J. Fetuin protects the fetus from TNF. *Lancet* **1997**, *350*, 861–862. [CrossRef]
82. Ombrellino, M.; Wang, H.; Yang, H.; Zhang, M.; Vishnubhakat, J.; Frazier, A.; Scher, L.A.; Friedman, S.G.; Tracey, K.J. Fetuin, a negative acute phase protein, attenuates TNF synthesis and the innate inflammatory response to carrageenan. *Shock* **2001**, *15*, 181–185. [CrossRef] [PubMed]
83. Hedrich, J.; Lottaz, D.; Meyer, K.; Yiallouros, I.; Jahnen-Dechent, W.; Stöcker, W.; Becker-Pauly, C. Fetuin-A and cystatin C are endogenous inhibitors of human meprin metalloproteases. *Biochemistry* **2010**, *49*, 8599–8607. [CrossRef] [PubMed]
84. Dholey, Y.; Chaudhuri, A.; Chakraborty, S.S. An integrated in silico approach to understand protein–protein interactions: Human meprin-β with fetuin-A. *J. Biomol. Struct. Dyn.* **2020**, *38*, 2080–2092. [CrossRef] [PubMed]
85. Gisterå, A.; Hansson, G.K. The immunology of atherosclerosis. *Nat. Rev. Nephrol.* **2017**, *13*, 2368–2380. [CrossRef]
86. Kim, M.-S.; Pinto, S.M.; Getnet, D.; Nirujogi, R.S.; Manda, S.S.; Chaerkady, R.; Madugundu, A.K.; Kelkar, D.S.; Isserlin, R.; Jain, S.; et al. A draft map of the human proteome. *Nature* **2014**, *509*, 575–581. [CrossRef]

87. Kim, S.Y.; Park, S.M.; Lee, S.-T. Apolipoprotein C-II is a novel substrate for matrix metalloproteinases. *Biochem. Biophys. Res. Commun.* **2006**, *339*, 47–54. [CrossRef]
88. Park, J.Y.; Park, J.H.; Jang, W.; Hwang, I.-K.; Kim, I.J.; Kim, H.-J.; Cho, K.-H.; Lee, S.-T. Apolipoprotein A-IV is a novel substrate for matrix metalloproteinases. *J. Biochem.* **2012**, *151*, 291–298. [CrossRef]
89. El-Asmar, M.; Swaney, J. Proteolysis in vitro of low and high density lipoproteins in human plasma by Cerastes cerastes (Egyptian sand viper) venom. *Toxicon* **1988**, *26*, 809–816. [CrossRef]
90. Lind, S.E.; Smith, D.B.; Janmey, P.A.; Stossel, T.P. Role of plasma gelsolin and the vitamin D-binding protein in clearing actin from the circulation. *J. Clin. Investig.* **1986**, *78*, 736–742. [CrossRef]
91. Haddad, J.G.; Harper, K.D.; Guoth, M.; Pietra, G.G.; Sanger, J.W. Angiopathic consequences of saturating the plasma scavenger system for actin. *Proc. Natl. Acad. Sci. USA* **1990**, *87*, 1381–1385. [CrossRef] [PubMed]
92. Epstein, F.H.; Lee, W.M.; Galbraith, R.M. The extracellular actin-scavenger system and actin toxicity. *N. Engl. J. Med.* **1992**, *326*, 1335–1341. [CrossRef] [PubMed]
93. Park, S.-M.; Hwang, I.K.; Kim, S.Y.; Lee, S.-J.; Park, K.-S.; Lee, S.-T. Characterization of plasma gelsolin as a substrate for matrix metalloproteinases. *Proteomics* **2006**, *6*, 1192–1199. [CrossRef]
94. Zhuo, L.; Hascall, V.C.; Kimata, K. Inter-α-trypsin Inhibitor, a covalent protein-glycosaminoglycan-protein complex. *J. Biol. Chem.* **2004**, *279*, 38079–38082. [CrossRef]
95. Lord, M.S.; Melrose, J.; Day, A.J.; Whitelock, J.M. The inter-α-trypsin inhibitor family: Versatile molecules in biology and pathology. *J. Histochem. Cytochem.* **2020**, *68*, 907–927. [CrossRef]
96. Fries, E.; Blom, A. Bikunin—Not just a plasma proteinase inhibitor. *Int. J. Biochem. Cell Biol.* **2000**, *32*, 125–137. [CrossRef]
97. Bost, F.; Diarra-Mehrpour, M.; Martin, J.-P. Inter-alpha-trypsin inhibitor proteoglycan family. A group of proteins binding and stabilizing the extracellular matrix. *Eur. J. Biochem.* **1998**, *252*, 339–346. [CrossRef] [PubMed]
98. Zhuo, L.; Kimata, K. Structure and function of inter-α-trypsin inhibitor heavy chains. *Connect. Tissue Res.* **2008**, *49*, 311–320. [CrossRef]
99. He, H.; Zhang, S.; Tighe, S.; Son, J.; Tseng, S.C.G. Immobilized heavy chain-hyaluronic acid polarizes lipopolysaccharide-activated macrophages toward M2 phenotype. *J. Biol. Chem.* **2013**, *288*, 25792–25803. [CrossRef]
100. Catanese, J.J.; Kress, L.F. Enzymatic digestion of human plasma inter-α-trypsin inhibitor by snake venom metalloproteinases. *Comp. Biochem. Physiol. Part B Comp. Biochem.* **1985**, *80*, 507–512. [CrossRef]
101. Jones, A.L.; Hulett, M.; Parish, C.R. Histidine-rich glycoprotein: A novel adaptor protein in plasma that modulates the immune, vascular and coagulation systems. *Immunol. Cell Biol.* **2005**, *83*, 106–118. [CrossRef]
102. Poon, I.; Patel, K.K.; Davis, D.A.S.; Parish, C.; Hulett, M. Histidine-rich glycoprotein: The Swiss Army knife of mammalian plasma. *Blood* **2011**, *117*, 2093–2101. [CrossRef] [PubMed]
103. Leung, L.; Saigo, K.; Grant, D. Heparin binds to human monocytes and modulates their procoagulant activities and secretory phenotypes. Effects of histidine-rich glycoprotein. *Blood* **1989**, *73*, 177–184. [CrossRef]
104. Borza, D.-B.; Morgan, W.T. Acceleration of plasminogen activation by tissue plasminogen activator on surface-bound histidine-proline-rich glycoprotein. *J. Biol. Chem.* **1997**, *272*, 5718–5726. [CrossRef] [PubMed]
105. Lijnen, H.; Hoylaerts, M.; Collen, D. Isolation and characterization of a human plasma protein with affinity for the lysine binding sites in plasminogen. Role in the regulation of fibrinolysis and identification as histidine-rich glycoprotein. *J. Biol. Chem.* **1980**, *255*, 10214–10222. [CrossRef]
106. Laemmli, U.K. Cleavage of Structural Proteins during the Assembly of the Head of Bacteriophage T4. *Nature* **1970**, *227*, 680–685. [CrossRef] [PubMed]
107. Mortz, E.; Krogh, T.N.; Vorum, H.; Görg, A. Improved silver staining protocols for high sensitivity protein identification using matrix-assisted laser desorption/ionization-time of flight analysis. *Proteomics* **2001**, *1*, 1359–1363. [CrossRef]
108. Travis, J.; Pannell, R. Selective removal of albumin from plasma by affinity chromatography. *Clin. Chim. Acta* **1973**, *49*, 49–52. [CrossRef]
109. Subramanian, S.; Ross, P.D. Dye-ligand affinity chromatography: The interaction of cibacron blue F3GA® with Proteins and Enzyme. *Crit. Rev. Biochem.* **1984**, *16*, 169–205. [CrossRef]
110. Vizcaíno, J.A.; Deutsch, E.W.; Wang, R.; Csordas, A.; Reisinger, F.; Ríos, D.; Dianes, J.A.; Sun, Z.; Farrah, T.; Bandeira, N.; et al. ProteomeXchange provides globally coordinated proteomics data submission and dissemination. *Nat. Biotechnol.* **2014**, *32*, 223–226. [CrossRef]

MDPI

Article

Bothrops Jararaca Snake Venom Modulates Key Cancer-Related Proteins in Breast Tumor Cell Lines

Carolina Yukiko Kisaki [1], Stephanie Santos Suehiro Arcos [1], Fabio Montoni [1], Wellington da Silva Santos [1], Hamida Macêdo Calacina [1], Ismael Feitosa Lima [1], Daniela Cajado-Carvalho [1], Emer Suavinho Ferro [2], Milton Yutaka Nishiyama-Jr [1,]* and Leo Kei Iwai [1,]*

1 Laboratory of Applied Toxinology (LETA) and Center of Toxins, Immune-Response and Cell Signaling (CeTICS), Butantan Institute, São Paulo 05503-900, Brazil; carolkisaki@hotmail.com (C.Y.K.); stephanie.arcos@usp.br (S.S.S.A.); fabio.montoni@esib.butantan.gov.br (F.M.); wellington.silva@esib.butantan.gov.br (W.d.S.S.); hamidamacedo33@gmail.com (H.M.C.); ismael.lima@butantan.gov.br (I.F.L.); daniela.carvalho@butantan.gov.br (D.C.-C.)

2 Department of Pharmacology, Biomedical Sciences Institute (ICB), University of São Paulo (USP), São Paulo 05508-000, Brazil; eferro@usp.br

* Correspondence: milton.nishiyama@butantan.gov.br (M.Y.N.-J.); leo.iwai@butantan.gov.br (L.K.I.)

Abstract: Cancer is characterized by the development of abnormal cells that divide in an uncontrolled way and may spread into other tissues where they may infiltrate and destroy normal body tissue. Several previous reports have described biochemical anti-tumorigenic properties of crude snake venom or its components, including their capability of inhibiting cell proliferation and promoting cell death. However, to the best of our knowledge, there is no work describing cancer cell proteomic changes following treatment with snake venoms. In this work we describe the quantitative changes in proteomics of MCF7 and MDA-MB-231 breast tumor cell lines following treatment with *Bothrops jararaca* snake venom, as well as the functional implications of the proteomic changes. Cell lines were treated with sub-toxic doses at either 0.63 μg/mL (low) or 2.5 μg/mL (high) of *B. jararaca* venom for 24 h, conditions that cause no cell death per se. Proteomics analysis was conducted on a nano-scale liquid chromatography coupled on-line with mass spectrometry (nLC-MS/MS). More than 1000 proteins were identified and evaluated from each cell line treated with either the low or high dose of the snake venom. Protein profiling upon venom treatment showed differential expression of several proteins related to cancer cell metabolism, immune response, and inflammation. Among the identified proteins we highlight histone H3, SNX3, HEL-S-156an, MTCH2, RPS, MCC2, IGF2BP1, and GSTM3. These data suggest that sub-toxic doses of *B. jararaca* venom have potential to modulate cancer-development related protein targets in cancer cells. This work illustrates a novel biochemical strategy to identify therapeutic targets against cancer cell growth and survival.

Keywords: mass spectrometry; proteome; snake venom; *Bothrops jararaca*; breast cancer

Key Contribution: We describe that *B. jararaca* snake venom modulates specific protein pathways related to cancer cell growth and invasion, which could be a useful approach to identify novel therapeutic targets for cancer treatment.

Citation: Kisaki, C.Y.; Arcos, S.S.S.; Montoni, F.; da Silva Santos, W.; Calacina, H.M.; Lima, I.F.; Cajado-Carvalho, D.; Ferro, E.S.; Nishiyama-Jr, M.Y.; Iwai, L.K. Bothrops Jararaca Snake Venom Modulates Key Cancer-Related Proteins in Breast Tumor Cell Lines. *Toxins* **2021**, *13*, 519. https://doi.org/10.3390/toxins13080519

Received: 5 July 2021
Accepted: 19 July 2021
Published: 25 July 2021

Publisher's Note: MDPI stays neutral with regard to jurisdictional claims in published maps and institutional affiliations.

1. Introduction

Ophidian accidents constitute a serious public health problem in Brazil, with an average of 29,000 cases and 125 deaths reported every year (Brazilian Ministry of Health, 2019) [1]. Approximately 80% of ophidian accidents are caused by snakes of the Viperidae family, more specifically of the *Bothrops* genus [2]. Among them, about 25% lead to death or sequels capable of generating temporary or permanent incapacity for work and customary activities. Venom from the *Bothrops jararaca* (*B. jararaca*) snake is a complex mixture composed of proteins, peptides, amino acids, nucleotides, lipids, and carbohydrates that present a range of different actions when they are isolated or together [3–7], leading to hemotoxic, cardiotoxic, cytotoxic, or neurotoxic effects [8,9]. Several previous reports have defined the proteomics composition of *Bothrops* venoms [5,10–16]. These studies have shown that *Bothrops* venoms are composed of various classes of toxin, including metalloproteinases, serine proteinases, phospholipases A2, and C-type lectins, the most abundant components participating in the local and systemic envenomation effects.

The venom of *B. jararaca* engenders three main activities: proteolytic, coagulant, and hemolytic. The proteolytic activity causes degradation of extracellular matrix proteins, plasma, and cell surface [17], which represents an important factor for the clinical characterization of a bothropic accident [2,18]. In addition, venom can cause local tissue lesion, myonecrosis, edema, cardiovascular alterations, hypovolemic shock, coagulation alteration and renal alterations, resulting from the combined action of the enzymatic and toxic activity of the venom [19]. About 90 to 95% of the dry weight of the *B. jararaca* venom is composed of a complex mixture of proteins, mainly metalloproteinases, serine proteinases, phospholipases (PLA2), and L-amino acid oxidases. The metalloproteinases comprise most of the venom composition [20,21]. They are proteolytic enzymes associated with fibrinolysis and coagulation, and they are involved in cell migration and tissue repair, besides being related to pathological effects such as cancer [22,23]. In terms of therapeutic interventions, protease inhibitors have been shown to inhibit homeostasis and thrombosis by acting on the coagulation cascade [24]. The third major component of the venom, PLA2, is an enzyme capable of hydrolyzing the ester bonds at the sn2 position of glycerolphospholipids, releasing arachidonic acid, important for the biosynthesis of many mediators involved in inflammation, such as prostaglandins, thromboxanes, and leukotrienes [25]. Finally, the L-amino acid oxidases (LAAOs), which make up about 1 to 9% of the venom composition [26], are flavoenzymes belonging to the class of oxidoreductases, which produce alpha-keto acid, hydrogen peroxide, and ammonia [27,28]. However, when there is a high production of hydrogen peroxide, it has been found that L-amino acid oxidases can induce apoptosis in mammalian endothelial cells [29].

Snake venom constituents have been isolated and studied for their therapeutic potential in the treatment of various diseases. One example is Eptifibatide, marketed as Integrilin, derived from the *Echis carinatus* snake venom and produced by Millennium Pharmaceuticals and Schering-Plow. It is used as an antiplatelet drug [30]. Another example is the angiotensin I converting enzyme inhibitor Captopril produced by Bristol-Myers Squibb whose active component was derived from *B. jararaca* venom. It is used for the treatment against hypertension and renal insufficiency [31,32]. In addition to the potential use of the derivatives of snake venom toxins in the treatment of non-malignant diseases [33], several studies have described anti-tumorigenic characteristics of snake venom, stating that snake venom may be capable of inhibiting cell proliferation and promoting cell death by different means: inducing apoptosis in cancer cells by increasing the influx of Ca^{2+}, inducing the release of cytochrome C, decreasing or increasing the expression of proteins that control the cell cycle, and causing damage to cell membranes [26,34–36]. With the goal of searching novel therapy against cancer, studies have characterized the proteins, peptides or enzymes derived from snake venom to identify components that are capable of interfering with the transport of substances or signal transduction across the membrane or disrupting the cell membrane [35,37].

With the rapid advances of nano-scale liquid chromatography (nLC) and mass spectrometry (MS) technologies in the last two decades, nLC-MS/MS-based proteomics analysis has been widely applied as a powerful tool for biomarker discovery to improve cancer therapy [38,39]. A number of studies have characterized the biochemical and physiological action of venom or isolated venom derivatives on cell lines or tissues [40–43]. In addition, several works have shown proteomic changes of cancer cell lines upon drug treatment that suggest molecular mechanisms of drug action, including diverse effects on proteasome regulation, metabolic processes, and oxidative stress [44,45].

However, to the best of our knowledge, there has been no report that describes the effects of *B. jararaca* snake venom treatment on breast cancer-related cell proteome. In the present study, nLC-MS/MS was used to characterize the effects of sub-toxic doses of *B. jararaca* snake venom on two different breast cancer cell lines MCF7 and MDA-MB-231. MCF7 and MDA-MB-231 are non-metastatic and metastatic tumor cell lines, respectively. They are characterized by a high degree of glycolytic efficiency that promotes the interaction between the tumor cell and the extracellular matrix [46]. Although both cell lines are from breast origin, they are molecularly distinct. MCF7 are estrogen and progesterone receptors positive and HER2 negative, while MDA-MB-231 are triple negative (estrogen receptor, progesterone receptor, and HER2 negatives) and prone to cytotoxic agents because of their impaired DNA repairing capability which is in part due to mutation in the p53 gene [47,48]. Proteomic changes observed herein upon treatment with *B. jararaca* snake venom in these cell lines highlight proteins and cell pathways that could be targeted in cancer therapy.

2. Results

2.1. The Cytotoxicity of B. jararaca Snake Venom in MCF7 and MDA-MB-231 Cells

The *B. jararaca* venom cytotoxicity assay on MCF7 and MDA-MB-231 cell lines was monitored using the WST-1 reagent. This analysis showed that although cell viability was similar between both MCF7 and MDA-MB-231 cell lines, they had different venom resistance profiles where MDA-MB-231 cells showed to be more resistant to the venom when compared to the MCF7 cells. Although both cell lines started to die at doses higher than 2.5 μg/mL, at the 5.0 μg/mL of venom, only about half of the MDA-MB-231 cells have died while most of the MCF7 have died at this venom concentration (Figure 1). Lethal concentration 50 (LC50) was determined as 4.50 μg/mL for MCF7 and 4.76 μg/mL for MDA-MB-231 cell line. Interestingly, the treatment of cells with the low dose at 0.5 μg/mL of venom also killed more MCF7 cells compared to the MDA-MB-231 cells when compared to the venom doses of 0.63 μg/mL and 1.25 μg/mL (Figure 1).

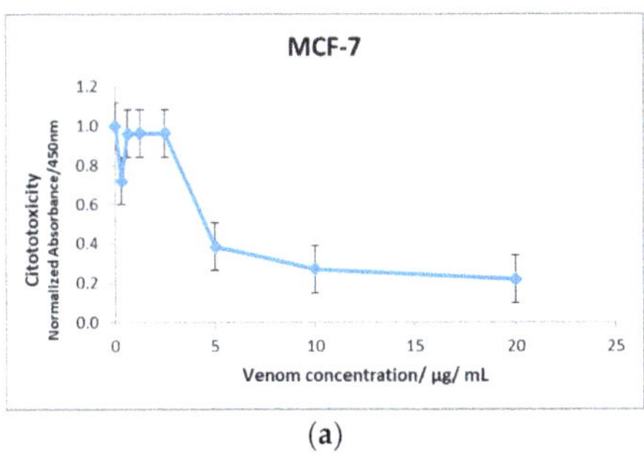

(a)

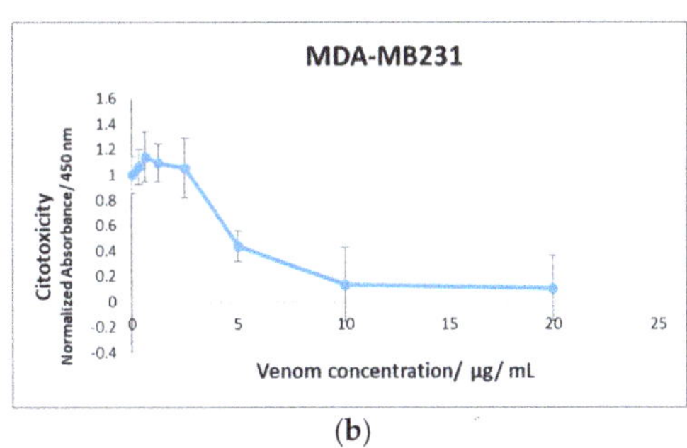

(b)

Figure 1. Cytotoxicity assay of (**a**) MCF7 and (**b**) MDA-MB-231 cell lines treated with *B. jararaca* snake venom ranging from 0 to 20 μg/mL for 24 h. Experiment was performed using the WST-1 reagent kit.

2.2. Optical Microscopy Analysis of MCF7 and MDA-MB-231 Cells under B. jararaca Venom Treatment

Optical microscopy analysis at 10× magnification of MCF7 and MDA-MB-231 cell lines treated with concentrations higher than 2.5 μg/mL of *B. jararaca* snake venom showed cellular morphological changes such as cell shrinkage and cell birefringence change (Figure S1). At the 20 μg/mL of venom treatment all MCF7 cells detached from the plate, whereas the MDA-MB-231 cell line continued to show morphological death-like changes, but the cells did not detach from the plate (Figure S1).

Based on the cytotoxicity assays and visualization of cell morphology changes through the microscope images, two working concentrations, representing a low dose of 0.63 μg/mL and a high sub-toxic dose of 2.5 μg/mL of venom, were selected for further proteomics analysis.

2.3. *Mass Spectrometry-Based Proteomics of MCF7 and MDA-MB-231 Cells Treated with B. jararaca Venom*

Protein identification was performed analyzing the raw data using the MaxQuant software against the *Homo sapiens* database downloaded from Uniprot. Analysis of the MCF7 cell line treated with different venom concentrations allowed the identification of 789 proteins from which 704 proteins were identified in the control group (non-venom treatment), 725 proteins were identified in cells treated with 0.63 µg/mL of venom, and 713 proteins were identified in the cells treated with 2.5 µg/mL of venom. Comparative analysis of the identified proteins showed that among all of the 789 proteins identified, 637 proteins were in common to all conditions, 657 proteins were in common between the control and the 0.63 µg/mL venom treatment, 656 proteins were in common between control and the 2.5 µg/mL venom treatment, and 677 proteins were in common between the 0.63 µg/mL and 2.5 µg/mL venom treatments (Figure 2a, Table S1a). In addition, we also identified exclusive proteins in each condition: 28 proteins in the control group, 28 proteins in 0.63 µg/mL venom treatment, and 17 proteins exclusive in the 2.5 µg/mL venom treatment (Figure 2a, Table S1a).

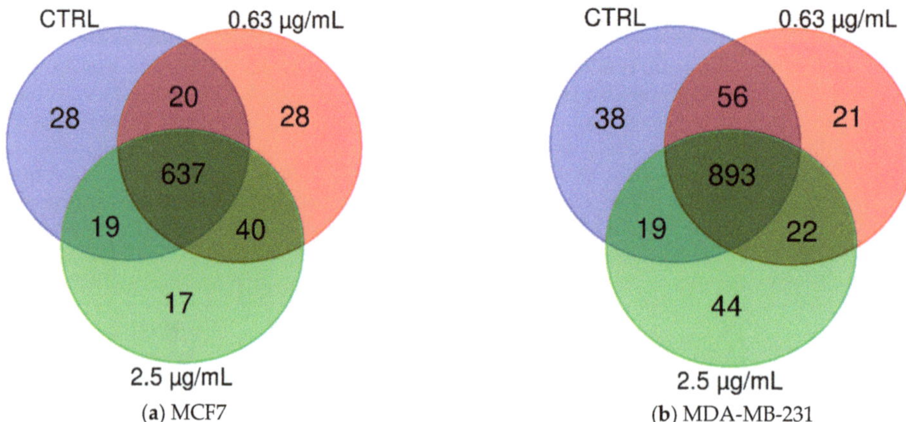

Figure 2. Diagrammatic representation of the comparative analysis of identified proteins under the three conditions: no venom control group, low venom dose at 0.63 µg/mL and high venom sub-toxic dose of 2.5 µg/mL of *B. jararaca* venom. (**a**) MCF7 and (**b**) MDA-MB-231.

In the MDA-MB-231 cell line, mass spectrometry-based proteomics analysis allowed us to identify a total of 1093 proteins from which 1006 proteins in the control group, 992 proteins in the cells treated with 0.63 µg/mL of venom, and 978 proteins in the cells treated with 2.5 µg/mL of venom. Comparative analysis of the identified proteins showed 893 proteins identified in common to all conditions, 949 proteins in common between the control group and 0.63 µg/mL venom treatment, 912 proteins were in common between control group and 2.5 µg/mL venom treatment, and 915 proteins were in common between the 0.63 µg/mL and 2.5 µg/mL venom treatments (Figure 2a, Table S1a). We also observed 38 exclusive proteins in the control group, 21 proteins in 0.63 µg/mL venom treatment, and 44 proteins exclusive in the 2.5 µg/mL venom treatment (Figure 2b, Table S1b).

2.4. *Semi-Quantitative Proteomics Analysis: MCF7 and MDA-MB-231 Cell Line Protein Abundance Variation*

In general, we observed a higher number of proteins whose abundance had changed more than 1.5× or less than 0.67× in MCF7 cell lines when compared to the MDA-MB-231 cell lines. The semi-quantitative analysis of the MCF7 cell line (Table S2a,) treated with 2.5 µg/mL venom allowed us to identify 137 proteins, whose abundance changed over

1.5× (fold change FC ≥ 1.5, marked in light red) when compared to the control group, from which 55 proteins presented FC ≥ 2.0 (marked in red). We highlight 12 highly abundant proteins with FC ≥ 3.0, marked in dark red: Sorting nexin-3 (SNX3), Purine nucleoside phosphorylase (HEL-S-156an), Peroxisome proliferator activated receptor interacting complex protein (PRIC295), Small nuclear ribonucleoprotein component (SNRP116), Eukaryotic translation initiation factor 4B (EIF4B), Methylcrotonoyl-CoA carboxylase beta chain (MCCC2), 26S proteasome non-ATPase regulatory subunit 5 (PSMD5), Heterogeneous nuclear ribonucleoprotein R (HNRNPR), Full-length cDNA clone CS0DJ015YJ12 of T cells (PSME2), Isoleucyl-tRNA synthetase (IARS), Large proline-rich protein BAG6 (BAG6), and Glutathione S-transferase (GSTM3). In addition, we identified 23 proteins with FC ≤ 0.67 (marked in light green) from which five proteins presented FC ≤ 0.5 (marked in dark green): Histone H4 (HIST1H4J), ATP synthase subunit d, mitochondrial (ATP5PD), Voltage-dependent anion-selective channel protein 2 (VDAC2), 4a-hydroxytetrahydrobiopterin dehydratase (PCBD), and Histone H3 (H3F3B). At the low 0.63 μg/mL venom treatment, we identified 25 proteins with FC ≥ 1.5 (marked in light red) from which only two proteins presented FC ≥ 2.0 (marked in red): Anterior gradient 2 homolog (AGR2) and Leucine-rich PPR-motif containing protein (LRPPRC); and 19 proteins with FC ≤ 0.67 (marked in light green) from which two proteins with FC ≤ 0.5 (marked in dark green): PCBD and 40S ribosomal protein S29 (RPS29). The description of the highlighted proteins is shown in Table 1.

Semi-quantitative proteomic analysis of the MDA-MB-231 cell line (Table S2b) treated with 2.5 μg/mL allowed the identification of 34 proteins whose abundance changed (FC) ≥1.5 over the control (marked in light red), nine proteins with FC ≥ 2 (marked in red) from which we highlight three proteins with FC ≥ 3 (marked in dark red): Histone H3.2 (H3C15/HIST2H3), 14 kDa phosphohistidine phosphatase (HEL-S-132P), and Mitochondrial carrier homolog 2 (MTCH2). Moreover, 41 proteins presented FC ≤ 0.67 (light green) from which we highlight four proteins with FC ≤ 0.5 (marked in dark green): DnaJ homolog subfamily A member 1 (DNAJA1), Insulin-like growth factor 2 mRNA-binding protein 1 (IGF2BP1), Cysteine-rich angiogenic inducer 61 (CYR61), and Thrombospondin-1 (THBS1). At the lower 0.63 μg/mL venom treatment, 16 proteins presented FC ≥ 1.5 (marked in light red) from which we highlight H3C15/HIST2H3 with FC = 3.8 and MTCH2 with FC = 2.3 (marked in red); and 28 proteins with FC ≤ 0.67 (marked in light green) from which 12 proteins with FC ≤ 0.5 (marked in dark green): 60S ribosomal protein L37 (RPL37), D-3-phosphoglycerate dehydrogenase (HEL-S-113), ATPase inhibitor, mitochondrial (ATP5IF1), Non-histone chromosomal protein HMG-14 (HMGN1), RCC2 protein (RCC2), Serine/threonine-protein phosphatase PP1-gamma catalytic subunit (PPP1CC), D-dopachrome decarboxylase (DDT), Ran GTPase-activating protein 1 (RANGAP1), dCTP pyrophosphatase 1 (DCTPP1), IGF2BP1, THBS1, and CYR61 (Table 1).

Table 1. Description of the highlighted proteins and their association with cancer.

Protein	Protein Name	Protein Description Related to Cancer	Cancer Type Association	References
AMOT	Angiomotin	Plays a central role in tight junction maintenance. Appears to regulate endothelial cell migration and tube formation. May also play a role in the assembly of endothelial cell-cell junctions. Plays a critical role in angiogenesis, proliferation and migration and invasion of cancer cells	BL, BR, CE, CL, CR, EN, HN, KD, LE, LI, LA, LS, OV, PR, ST	[49–51]
ATP5PD	ATP synthase peripheral stalk subunit D	Mitochondrial ATP synthase catalyzes ATP synthesis, utilizing an electrochemical gradient of protons across the inner membrane during oxidative phosphorylation. Linked to failure of therapy, disease progression, and poor survival in patients with cancer. High expression of ATP5PD has been observed in several types of cancer	BR, CE, CL, EN, GL, HN, LI, LU, LY, ME, OV, PA, PR, RE, SK, TE, TY	[52]
ATR	Serine/threonine-protein kinase	Plays important roles for cell survival and is considered a major mediator of DNA response in human cells, preventing cells with damaged or incompletely replicated DNA from entering mitosis when cells are damaged by radiotherapy or chemotherapy during cancer treatment	BL, BR, CE, CR, EN, GB, HN, KD, LU, LA, LS, OV, ST, TY	[53,54]
CYR61	Cysteine-rich heparin-binding protein 61	Plays an important role in cell proliferation, survival, chemotaxis, angiogenesis, adhesion, and migration of different types of cells. Participate in key different cellular events during vascular development, angiogenesis, wound healing and the development and progression of various types of cancers	BN, BR, CR, EN, GA, GB, GL, LI, LU, OV, PA, PR, ST, TE, UR	[55,56]
GSTM3	The glutathione S-transferase Mu 3	Part of the GSTs enzymes that have functions such as immunological system evasion and inhibition of apoptosis. Involved in prostaglandin and leukotriene synthesis and metabolization of both endogenous compounds and xenobiotics such as chemotherapeutic drugs, insecticides, carcinogens, and oxidative stress byproducts	BL, BR, CR, EN, LE, LA, LS, OV, PA, ST, TY, UR	[57,58]
H3F3B/H3C15	H3 histone family member 3B	Core component of nucleosome. Histones play a central role in transcription regulation, DNA repair, DNA replication and chromosomal stability and are related to different types of cancer. H3F3B mutation has been described to lead to some human cancers	BL, BN, BR, CE, CH, CR, EN, GB, HN, LU, OV, UR	[59–61]
HEL-S-156an PNP	Purine nucleoside phosphorylase	Catalyze the phosphorolysis of purine nucleosides. Mutations which result in nucleoside phosphorylase deficiency result in defective T-cell (cell-mediated) immunity but can also affect B-cell immunity and antibody responses. High expression of PNP has been observed in several types of cancer	BR, CL, CR, GA, GL, KD, LI, LU, LA, LY, ME, OV, PR, TY	[62–64]

Table 1. *Cont.*

Protein	Protein Name	Protein Description Related to Cancer	Cancer Type Association	References
HIST1H4J	Histone H4	Core component of nucleosome. Histones play a central role in transcription regulation, DNA repair, DNA replication and chromosomal stability. Post-translational alterations of histones have been shown to affect the activation and repression of oncogenes and tumor suppressor genes	BR, CE, CR, GL, HN, LI, LU, ME, OV, PA, PR, SK, ST, TE, TY	[65–68]
IGF2BP1	Insulin like growth factor 2 MRNA binding protein 1	RNA-binding factor that recruits target transcripts to cytoplasmic protein–RNA complexes (mRNPs). IGF2BP1 has an oncogenic role, characterized by changes in actin dynamics, migration, invasion, proliferation, and self-renewal. Play a role in resistance to drugs	BR, CE, CR, EN, GB, HN, LI, LU, LA, LS, ME, OV, PR, ST, TE, UR	[69–71]
KRT1	Keratin	May regulate the activity of kinases such as PKC and SRC via binding to integrin beta-1 (ITB1) and to the receptor of activated protein C kinase 1 (RACK1). High expression of KRT1 protein has been observed in several types of cancer and is correlated with advanced melanoma tumor stage and infiltration of immune cells	BR, CE, CR, EN, GB, HN, KD, LI, LS, LA, OV, SK, ST, UR	[72,73]
LAP3	Leucine aminopeptidase 3	Cytosolic metallopeptidase that catalyzes the removal of unsubstituted N-terminal hydrophobic amino acids from various peptides. Involved in the metabolism of glutathione and in the degradation of glutathione S-conjugates, which may play a role in the control of the cell redox status. Related to protein renewal. Have a potential for determining the prognosis for breast cancer	BL, BR, CR, EN, HN, KD, LI, LA, LS, PA, ST, TY	[74–77]
MCCC2	Methylcrotonoyl-CoA carboxylase beta chain, mitochondrial	Enzyme that catalyzes the conversion of 3-methylcrotonyl-CoA to 3-methylglutaconyl-CoA, a critical step for leucine and isovaleric acid catabolism. Overexpression of MCCC2 is associated with tumor stage, node, metastasis, lymph node metastasis and predicts unfavorable prognosis. Additionally, involved in the development and formation of some tumors, such as breast cancer	BL, BR, CR, EN, HN, KD, LI, LA, OV, PA, PR, ST, TY	[78]
MTCH2	Mitochondrial carrier homolog 2	Member of the SLC25 family of nuclear-encoded transporters that are localized in the inner mitochondrial membrane. Members of this superfamily are involved in many metabolic pathways and cell functions. Associated with metastasis and tumor cell survival. Indirect involvement in the expression of miR-135b mRNA, which is one of the proteins responsible for tumorigenicity	BL, BR, CR, EN, HN, KD, LI, LA, ME, OV, PA, PR, ST, TY	[79–81]

Table 1. *Cont.*

Protein	Protein Name	Protein Description Related to Cancer	Cancer Type Association	References
PCBD	4a-hydroxytetrahydrobiopterin dehydratase	Involved in tetrahydrobiopterin biosynthesis. Regulates various aspects of cell morphogenesis and differentiation as a cofactor for the homeobox transcription factor. Several types of cancer show expression or alteration in the homeobox genes. PCBD degradation increases cell survival and proliferation, and inhibits tumor cell differentiation	BR, CL, CR, EN, LE, LI, LU, OV, PA, PR, RE, SK	[82–84]
PRIC295	Peroxisome proliferator-activated receptor-a (PPARα)-interacting cofactor	Functions as a transcriptional coactivator for nuclear receptors. Enhances the activation of PPARα and PPARγ and plays a key role in lipid metabolism and energy combustion regulating the genes for fatty acid oxidation. Observed to be significantly enhanced in chemotherapy recurrence when compared to chemotherapy treatment in ovarian cancer patients	BL, BR, CR, EN, GB, HN, KD, LE, LA, LS, ME, OV, PA, PR, ST, TY	[85–87]
PSMD5	The 26S proteasome non-ATPase regulatory subunit 5	Acts as a chaperone during the assembly of the 26S proteasome. Expression reduced in several types of cancer including intestinal and colorectal tumors	BL, BR, CR, EN, HN, PR, ST, TY	[88]
PSME2	The proteasome activator complex subunit 2	Implicated in immunoproteasome assembly and required for efficient antigen processing. Member of the PSME family that regulates proteasome function. Elevated expression of PSME have also been associated with several types of cancer	BR, CR, EN, HN, LC, LA, LS, ME, PR, ST	[89–91]
RPS29	Ribosomal protein S29	Belongs to the universal ribosomal protein uS14 family. Related to have tumor suppressor activity for ras-transformed NIH3T3 cells. High expression RPS29 mRNA levels observed in adenomas	BR, CE, CR, EN, HN, KD, LA, OV, ST	[92,93]
SNRP116/EFTUD2	Small nuclear ribonucleoprotein component	Required for pre-mRNA splicing as a component of the spliceosome, including pre-catalytic, catalytic, and post-catalytic spliceosomal complexes. Knockout of EFTUD2 suppressed the development and tumor progression due to impaired activation of NF-kB signaling in macrophages	BL, BR, CR, EN, GB, HN, KD, LA, LS, ME, OV, PR, ST, TE, TY, UR	[94,95]
SNX3	The sorting nexin 3	Phosphoinositide-binding protein required for multivesicular body formation. Plays a role in protein transport between cellular compartments. The knockdown of SNX3 is associated with degradation of the EGF receptor which is related to resistance to chemotherapy and radiotherapy	BR, CR, EN, GL, HN, KD, LU, OV, RE, TY	[96,97]
THBS1	Thrombospondin 1	Adhesive glycoprotein that mediates cell-to-cell and cell-to-matrix interactions. Influences angiogenesis modulation by regulating adhesion, invasion, metastasis, migration, proliferation, and apoptosis and has been implicated in numerous types of cancers	BR, CR, EN, ES, HN, KD, LA, LS, LY, ME, OV, PA, PR, SK, ST, TE, TY	[98–100]

Table 1. *Cont.*

Protein	Protein Name	Protein Description Related to Cancer	Cancer Type Association	References
TUFM	Tu translation elongation factor, mitochondrial	Promotes the GTP-dependent binding of aminoacyl-tRNA to the A-site of ribosomes during protein biosynthesis. Plays important roles in the regulation of autophagy and innate immunity. TUFM is highly expressed in several types of cancers	BR, CR, EN, ES, GA, GS, HN, LI, LU, LA, ME, PA, PR, RE, SK, ST, TE	[101–103]
UQCRC1	The ubiquinol-cytochrome C reductase core protein 1	Component of the ubiquinol-cytochrome c reductase complex, which is part of the mitochondrial respiratory chain. High expression was observed in several types of cancer. Negative expression correlated significantly with clinical and pathological parameters including tumor stage, vascular invasion, and lymph node metastasis, suggesting that the reduction of this protein is associated with tumor progression	BR, CR, EN, GB, GL, HN, LI, LA, LY, ME, OV, PA, PR, RE, ST, TE, TY, UR	[104–106]
VDAC1 and VDAC2	The voltage-dependent anion selective channel 1 and 2	Forms a channel through the mitochondrial outer membrane and plasma membrane allowing diffusion of small hydrophilic molecules. In the plasma membrane it is involved in cell volume regulation and apoptosis. The abnormal expression or mal functioning of VDACs has been reported in multiple tumors and it has been considered as a biomarker capable of predicting treatment failure and breast cancer recurrence	BL, BR, CR, EN, GB, HN, LI, LA, LS, ME, OV, PA, PR, RE, ST, TY, UR	[107–110]

Note: This description may include information from UniProtKB, PhosphoSitePlus (v.6.5.9.3), GeneCards (Human Gene Database), and Human Protein Atlas [111–113]. Bladder: BL, bone (osteosarcoma): BN, breast: BR, cervical: CE, chondroblastoma: CH colon: CL, colorectal: CR, endometrial: EN, esophageal: ES, gastric: GA, gastrointestinal stromal tumor: GS glioblastoma: GB, glioma: GL, head and neck: HN, kidney: KD, laryngeal carcinoma: LC, leukemia: LE, liver: LI, lung: LU, lung adenocarcinoma: LA, lung squamous: LS, lymphoma: LY, melanoma: ME, ovarian: OV, pancreatic: PA, prostate: PR, renal: RE, skin: SK, stomach: ST, testis: TE, thyroid: TY, urothelial: UR.

2.5. Hierarchical Clustering Analysis

Hierarchical clustering analysis of differentially expressed proteins with FC $\geq$ 1.5 identified in both MCF7 and MDA-MB-231 cell lines characterized those proteins in seven major clusters (Figure 3, Table S3). Clusters 1–4 had proteins with fold change (FC) increased at 2.5 µg/mL of venom treatment in MCF7 cells and presented no change in MDA-MB-231 cell lines. Proteins in these clusters include HEL-S-156 and PRIC295 in cluster 1 and PSMD5 and PSME2 in cluster 2. Cluster 5 identified proteins that decreased in FC at 2.5 µg/mL of venom treatment in both cell lines including HIST1H4J, VDAC1, and VDAC2. Cluster 6 identified proteins that increased FC at 2.5 µg/mL of venom treatment in MDA-MB-231 cell line including H3F3B, LAP3, and KRT1, and cluster 7 identified proteins that did not change when cells were treated with low and high venom treatment, but they presented higher FC change in MDA-MB-231 when compared to MCF7 cell line such as RPS29 (Table 1).

2.6. Principal Component Analysis

PCA was applied to the differentially expressed proteins identified from both MCF7 and MDA-MB-231 cell lines based on the log2 FC of cells treated with low and high *B. jararaca* venom compared to the PBS treatment control group (Figure 4). The projection into the component space shows a distinct coordinated activity of proteins between the cell lines conditions. Orthogonal vectors show highly positive correlation between both venom concentrations in MDA-MB-231 cell line which may represent similar cell line responses to the different venom concentrations, but they present a highly negative correlation to both venom concentrations in MCF7 and respective set of expressed proteins. In the MCF7 cell line, however, we observe a negative correlation between the 0.63 µg/mL and 2.5 µg/mL venom treatment, indicating a differential response upon low and high dose venom treatment. PCA also shows differential correlation between both cell lines where the first two components showed 40% variation in the PC1 and 26.7% variation in the PC2 between the MCF7 and MDA-MB-231 cell lines. In addition, we observed clusters of proteins positively correlating with both low and high venom treatment in MDA-MB-231 cell line such as LAP3, H3F3B, and KRT1. On the other hand, we also observed proteins such as H2AC20 and TUFM correlating with low venom treatment, and proteins such as HEL-S-156an correlating with high venom treatment on MCF7 cell line (Table 1).

2.7. Gene Ontology Functional Analysis

The most enriched protein families and functional categories were analyzed based on highly abundant proteins with FC $\geq$ 1.5 for each cell line and treatment at low and high *B. jararaca venom* conditions (Figures S2–S5). The functional ontology classification analysis of these sets of proteins showed that both MCF7 and MDA-MB-231 venom-treated cell lines showed similar enriched categories. In addition, the most prominent enrichment was identified for treated cells with the sub-toxic dose of 2.5 µg/mL of venom. The molecular function enrichment analysis in both the MCF7 and MDA-MB-231 cell lineages showed an enrichment of proteins related to binding, structural molecule activity, and catalytic activity. In addition, the MCF7 cells had enriched, albeit in a lower amount, proteins related to function and transcriptional regulatory activity and carrier activity (Figure S2). The functional classification analysis related to biological processes showed enriched proteins related to the metabolic process and the cellular component organization or biogenesis. Moreover, the analysis of proteins identified in the MCF7 cell line presented proteins related to the cellular process, localization, biological regulation, stimulus response, developmental process, multicellular organismal process, and the immune system process (Figure S3). The analysis of protein distribution by cellular components showed an enrichment related to the "cell", protein complex, and organelle (Figure S4), and the enrichment analysis of protein family classification showed an enrichment of cytoskeleton proteins, ligase, nucleic acid binding, signaling molecule, modulating enzyme, calcium binding protein, and hydrase (Figure S5).

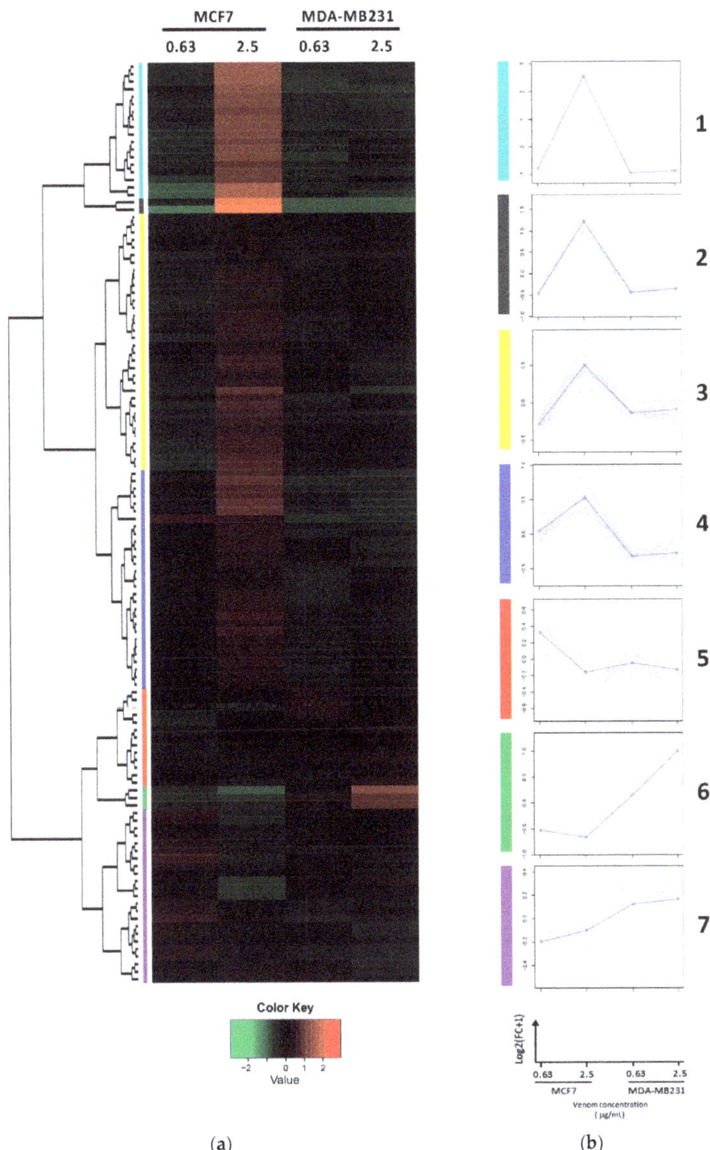

(a) (b)

Figure 3. Hierarchical clustering of differentially expressed proteins detected in both MCF7 and MDA-MB-231 cells treated with low (0.63 µg/mL) and high (2.5 µg/mL) *B. jararaca* venom for 24 h. (**a**) Heatmap representation of the hierarchical clustering of proteins detected in both cell lines with quantification in at least two replicates showing the changes in protein abundance. The protein fold change is log2 transformed and normalized with mean-centering scale. (**b**) Protein Clusters extracted from the hierarchical clustering. X axis: Cell types treated with different *B. jararaca* venom concentrations (MCF7 0.63 µg/mL; MCF7 2.5 µg/mL, MDA-MB-231 0.63 µg/mL, MDA-MB-231 2.5 µg/mL); Y axis: mean-centered log2 Fold Change. Grey lines: individual proteins; Black line: average expression values per cluster.

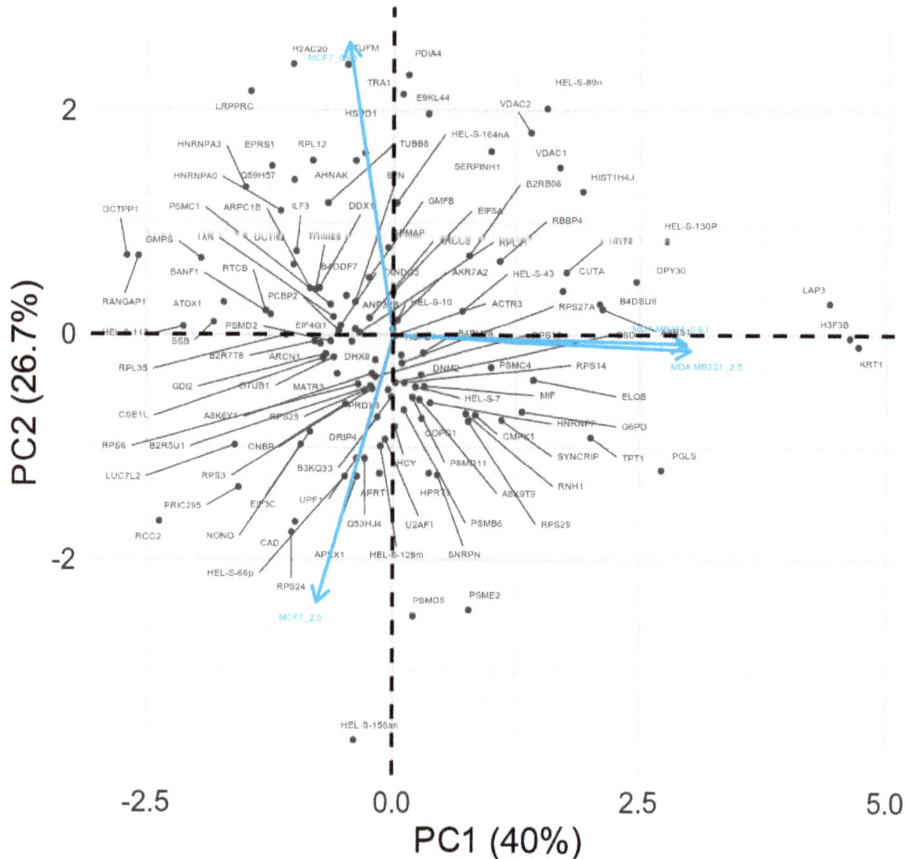

Figure 4. Comparison of protein log2 Fold Change profiles across treated cell lines. Principal component analysis in a 2D graph represented by the first two components PC1 and PC2 explains 67.7% of the protein variability among the different conditions. Vectors that are closer are highly correlated. Vectors representing the conditions which are orthogonal or well-spaced in terms of the observed proteome indicate that those proteins can be closely related to each specific cell line condition.

2.8. Protein–Protein Interaction Analysis

STRING protein–protein interaction network analysis tool was used to evaluate protein–protein interactions identified among proteins with FC $\geq$ 1.5 from each cell line treated with *B. jararaca* venom. According to Doncheva and colleagues [114], STRING indicates interactions according to co-expression analyzes and evolutionary signals in all genomes based on data described in the literature between genes or proteins using functional classification systems such as Gene Ontology, KEGG (Kyoto Encyclopedia of Genes and Genomes), and Reactome.

Analysis of protein–protein interactions (PPIs) of proteins whose abundance increased more than 1.5× in the MCF7 cell line treated with 2.5 μg/mL of venom showed a high interconnection of proteins related to metabolic process (in red) and metabolism pathways (in blue) (Figure 5a). Interestingly, among the proteins identified in the MCF7, we observed clusters of highly connected proteins related to proteasome pathway (in green) and mRNA splicing (in yellow). We highlight the proteins PSMD2 and PSMD11 (26S proteasome non-ATPase regulatory subunit 2 and 11, respectively), PSME1 and PSME2 (Proteasome activator complex subunit 1 and 2, respectively), PSMC1, PSMC4, and PSMC5

(26S proteasome regulatory subunit 4, 6B, and 8, respectively), TXNL (Thioredoxin-like protein 1), and USP14 (Ubiquitin carboxyl-terminal hydrolase 14), which are all members of the proteasome pathway (in green) and proteasome complex, and, together with MST4 (Serine/threonine-protein kinase 26) and PSMD5 (26S proteasome non-ATPase regulatory subunit 5) they are all related to apoptosis (Figure 5a). PPI analysis of proteins identified in the MDA-MB-231 in the same conditions showed less interaction among the proteins identified with FC $\geq$ 1.5 (Figure 5b).

PPI analysis of the MCF7 cell lines with FC $\geq$ 1.5 treated with 0.63 µg/mL and FC $\leq$ 0.67 when cells were treated with 0.63 µg/mL and 2.5 µg/mL of venom are shown in Figure S6. Additionally, PPI analysis of proteins identified in the MDA-MB-231 at 0.63 µg/mL and 2.5 µg/mL venom treatment presenting FC $\leq$ 0.67 and FC $\geq$ 2.5 are shown in Figure S7.

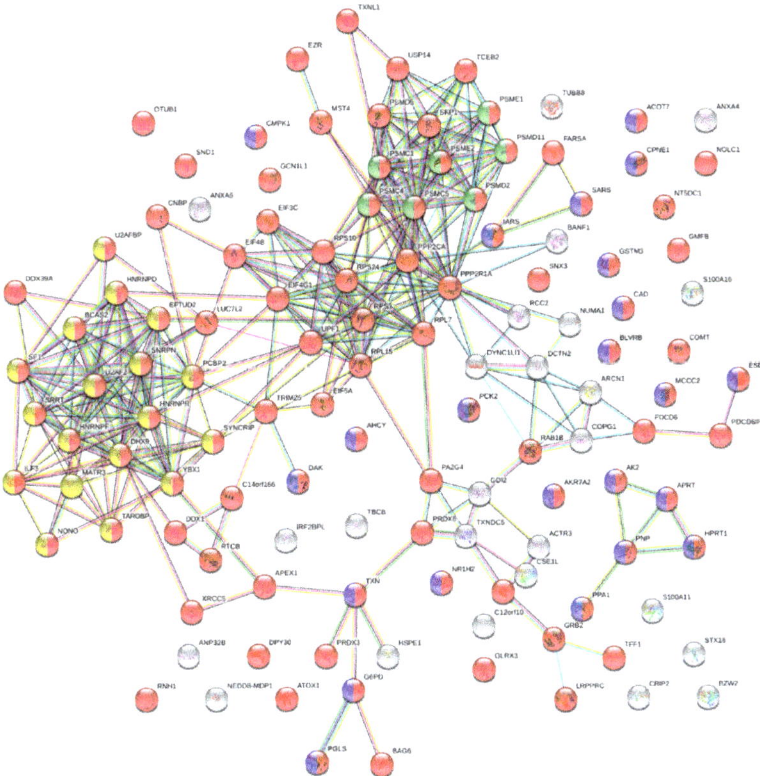

Red: Biological Process of Metabolism. Blue: Metabolism Pathways (Reactome). Yellow: mRNA Splicing local network cluster (STRING). Green: Proteasome pathway (KEGG). Number of nodes: 125, Number of edges: 327, Average node degree: 5.23, Avg. local clustering coefficient: 0.47, Expected number of edges: 120, PPI enrichment p-value: $< 1.0 \times 10^{-16}$.

(a)

Figure 5. *Cont.*

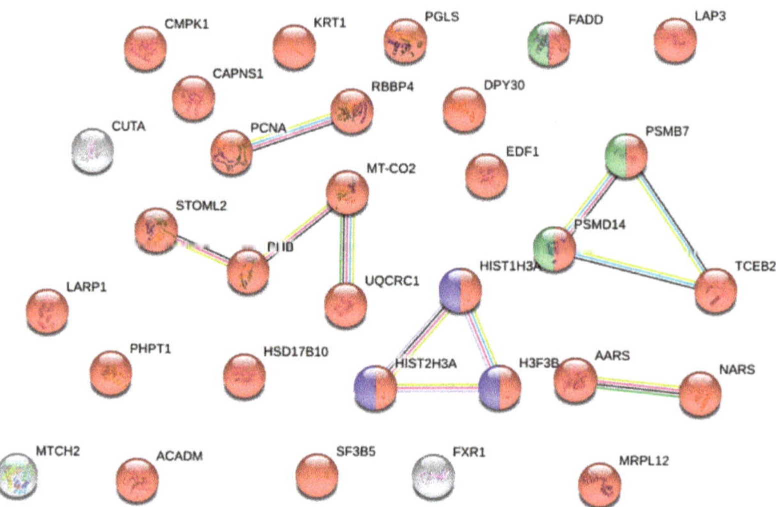

Red: Biological Process of Metabolism. Blue: Histone H3 complex (SMART, Simple Modular Architecture Research Tool). Green: apoptosis (Reactome pathways). Number of nodes: 31, Number of edges: 11, Average node degree: 0.71, Avg. local clustering coefficient: 0.387, Expected number of edges: 5, PPI enrichment p-value: 0.0192

(**b**)

Figure 5. Protein–protein interaction of proteins identified in (**a**) MCF7 and (**b**) MDA-MB-231 cell lines presenting FC $\geq$ 1.5 at 2.5 g/mL *B. jararaca* venom treatment.

2.9. Exclusive Proteins

We further analyzed proteins exclusively identified in one or two of the three conditions in both MCF7 cell line (Table S4a, Figure S8) and MDA-MB-231 cell line (Table S4b, Figure S8). Proteins identified only on PBS treated cells (control group), suggesting that the expression of the proteins was inhibited with the addition of venom, allowed us to identify 28 proteins in the MCF7 cell line from which we highlight proteins: eukaryotic peptide chain release factor subunit 1 (ETF1), serine/arginine repetitive matrix protein 2 (SRRM2), PHD finger-like domain containing protein 5A (PHF5A), and lamina-associated polypeptide 2 (TMPO). The analysis of the MDA-MB-231 cell line allowed us to identify 41 proteins exclusively expressed at the control group from which we highlight Plasminogen activator inhibitor 1 (SERPINE1), MHC class I antigen (HLA-C), Transcription factor BTF3 (BTF3L4), Cytochrome c oxidase subunit 6C (COX6C), Myb-binding protein 1A (MYBBP1A), and MYCBP protein (MYCBP) (Figure S8a, Table S4).

Of the proteins identified exclusively when cells were treated with 0.63 µg/mL of venom there were 28 proteins from MCF7 cell line, from which we highlight Cytochrome b-c1 complex subunit 1, mitochondrial (UQCRC1), Latexin (LXN), Vesicle-associated membrane protein-associated protein B/C (VAPB), Cysteine and glycine-rich protein 1 (CSRP1), and DDB1- and CUL4-associated factor 7 (DCAF7). From MDA-MB-231 cell line, 20 proteins were identified from which we highlight Eukaryotic translation initiation factor 2 subunit 2 (EIF2S2), low molecular weight phosphotyrosine protein phosphatase (ACP1), Transcription factor A, mitochondrial (TFAM), and Rae1 protein homolog (RAE1) (Table 1 and Table S4, Figure S8b).

Proteins identified exclusively at the high dose treatment of 2.5 µg/mL of venom included 17 proteins detected in the MCF7 cell line, from which we highlight 26S proteasome non-ATPase regulatory subunit 6 (PSMD6) and 60S ribosomal protein L36 (RPL36).

Similarly, we identified 44 proteins in MDA-MB-231 cell line from which we highlight protein Ubiquitin carboxyl-terminal hydrolase (HEL-117) (Figure S8c, Table S4).

We also identified proteins that were expressed exclusively when cells were treated with low and high doses of *B. jararaca* venom. From the MCF7 cell line, we identified 40 proteins from which we highlight: Serine/threonine-protein kinase ATR (ATR) and Jupiter microtubule-associated homolog 1 (JPT1). From the MDA-MB-231 cell line, we identified 22 proteins from which we highlight Angiomotin (AMOT) and Small nuclear ribonucleoprotein G (SNRPG) (Table 1, Tables S3 and S4, Figure S8d).

Among the proteins that were identified in the control group and at the 2.5 μg/mL venom treatment, we highlight: Small nuclear ribonucleoprotein Sm D1 (SNRPD1) in MCF7, and Epididymis secretory protein Li 71 (Hel-S-71) and Isoform Far upstream element-binding protein 3 (FUBP3) in MDA-MB-231 cell line (Figure S8e, Table S4). Of proteins that were identified in the control group and low 0.63 μg/mL dose of venom, we highlight Serine/arginine-rich splicing factor 10 (SRSF10) in MCF7 cell line, and Histone H2B type 2-E (HIST2H2BE), 60S ribosomal protein L34 and L37a (RPL34, RPL37A) in MDA-MB-231 cell line (Figure S8f, Tables S3 and S4).

3. Discussion

The present study successfully identified cancer-related proteins that undergo significant changes upon *B. jararaca* venom treatment of MCF7 and MDA-MB-231 cells, including SNX3, HEL-S-156an, UQCRC1, RPL36, and ATR identified in MCF7 cell line and H3C15/HIST2H3, HEL-S-132P, MTCH2, TFAM, KCTD12, RPL34, and RPL37A identified in MDA-MB-231 cell line, and histone H3F3B and LAP3 in both cell lines.

With the discovery of rattlesnake venom's antitumor activity in 1931 by Essex and Priestley [115] and later the angiotensin-converting enzyme inhibitor drug Captopril in 1981, developed originally from *B. jararaca* snake venom [116–118], several groups have been focusing on to the therapeutic potentials of bioactive compounds in snake venoms by isolating and characterizing components of the venom and analyzing their pharmacological potential that may lead to the development of more efficacious drugs [119,120].

Snake venoms have a complex mixture of proteins that can account for 95% of the total dried weight. Metalloproteases, serine proteases, LAAOs, and PLA2s are some of the most abundant proteins present in bothropic venoms, and the toxic action of these proteins are responsible for myotoxic effects, disruption in the coagulation cascade through thrombin-like proteins (serine proteases), imbalance in the blood homeostasis by the action of some metalloproteinases, apoptosis induction, and changes in the protein expression in oxidative stress and energy metabolism by the increase of Ca^{2+} influx. Several studies have revealed that these enzymes are critical regulators of cancer pathologies in several types of cancers including breast cancer [121–127]. Moreover, it has been previously shown that *B. jararaca* venom may have an important antitumor effect on Ehrilich ascites tumor cells in vivo and in vitro [36].

Aiming at the drug development, several groups have used proteomics approach to observe up- or down-regulated effects of proteins resulting from disease activity or side effects of the treatments [44,45,128–131]. Moreover, the quest for the identification of biological markers or biomarkers using venom that may help in the early detection of pathologies such as cancer, and the quest for products able to evaluate the metastatic potential and propose new forms of treatment to tumors, remains an important goal of research groups and pharmaceutical companies.

Proteomics studies using tumor cell lines has been performed to systematically characterize protein complexes that may be important in disease or drug development, comparing the protein levels of cells in normal versus pathological situations [132]. Studies have shown that venoms from several snakes such as *Calloselasma rhodostoma*, *Macrovipera lebetina* and *Bothrops mattogrossensis* significantly inhibit cell viability, either with crude or fractionated venoms and in different tumor cell lines such as LS174T (colorectal adenocarcinoma), HCT116 (colorectal carcinoma), HT29 (colorectal adenocarcinoma), HEL92.1.7 (erythroleukemia), and SK-BR-3 (breast adenocarcinoma) cell lines [133,134].

To evaluate the cytotoxicity of *B. jararaca* venom, we tested two different invasive breast cancer cell lines with different phenotypic and genotypic differences, MCF7 and MDA-MB-231 cell lines. They were treated with different venom concentrations ranging from 0.32 μg/mL to 20 μg/mL. When treated with over 10 μg/mL of venom both MCF7 and MDA-MB-231 cell lines showed rounded and irregular morphologies and the MCF7 cell line detached from the plate. These results are similar to the results observed by Bernardes-Oliveira et al. (2016) when testing 50 μg/mL of venom from *B. jararaca* and *Bothrops erythromelas* for 48 h on SiHa HPV-16 and HeLa cells, both derived from cervical tumors. They observed that the cells became more rounded, with gradual size reduction and detachment from the cell monolayer [135]. In addition to the morphological analyses, we evaluated cell viability through mitochondrial cell activity using WST-1 assay for cell proliferation and viability. We observed that cell death of both MCF7 and MDA-MB-231 cell lines started with 2.5 μg/mL of *B. jararaca* venom treatment. Based on both morphological and cell viability analyses, we selected the concentrations of 0.63 μg/mL (low sub-toxic dose) and 2.5 μg/mL (high sub-toxic dose) for further proteomic analyses.

Semi-quantitative proteomics analysis of the MDA-MB-231 cell line treated with the low 0.63 μg/mL and the sub-lethal dose of 2.5 μg/mL of *B. jararaca* venom presented less proteins with FC $\geq$ 1.5 when compared to the MCF7 cell line. We identified 137 proteins with FC $\geq$ 1.5 at 2.5 μg/mL venom treatment in MCF7 while 34 proteins in MDA-MB-231 at 2.5 μg/mL venom treatment. At the low 0.63 μg/mL venom treatment, 25 proteins in MCF7 presented FC $\geq$ 1.5 and 16 proteins in MDA-MB-231. However, MDA-MB-231 showed more proteins with FC $\leq$ 0.67 compared to MCF7 cell lines. MDA-MB-231 showed 28 proteins presenting FC $\leq$ 0.67 at 0.63 μg/mL venom treatment and 19 proteins in MCF7; and 41 proteins with FC $\leq$ 0.5 at 2.5 μg/mL venom treatment in MDA-MB-231 cell lines compared to 23 proteins in MCF7 cell line. Among these proteins, our data showed that *B. jararaca* venom was able to modulate, up or down, the abundance of several proteins playing essential roles in tumorigenicity. For example, the venom treatment increased the abundance of SNX3, HEL-S-156, MCCC2, and GSTM3 in the MCF7 cell line, and H3C15/HIST2H3 and MTCH2 in the MDA-MB-231 cell line whose overexpression has been related to increase in tumorigenicity (Table 1). Furthermore, among the proteins that decreased the abundance presenting FC $\leq$ 0.67, we highlight PCBD, PSMD5, RPS29, H3F3B, VDAC1 and VDAC2, ATP5PD, and HIST1H4J in MCF7 cell line and IGF2BP1, THBS1, and CYR61 in MDA-MB-231 cell line. These proteins have also been described to be related to several mechanisms and pathways in the onset of different types of cancer (Table 1). Additionally, cytotoxicity analysis performed in our laboratory (data not shown) evaluated the effects of *B. jararaca* venom on non-tumor HEK293 and HUVEC cell lines, which showed relatively less sensitivity to cell-death induction by snake venoms compared to MCF7 and MDA-MB-231 cancer cells. These data suggest that cancer cells could be more sensitive to snake venom components than non-cancer cells in certain situations.

Proteomic analysis also revealed exclusive proteins, i.e., proteins that were only identified in one or two of the three conditions analyzed. Among these proteins, we highlight the UQCRC1, identified in the MCF7 cell line only when these cells were treated with 0.63 μg/mL venom. High expression of this protein was observed in 74% of cases of breast cancer and 34% of ovarian cancer [104] (Table 1). Among proteins identified exclusively in the MCF7 cell line we highlight ATR (serine/threonine-protein kinase ATR) which plays important roles for cell survival and is considered a major mediator of DNA response in human cells, preventing cells with damaged or incompletely replicated DNA from entering mitosis when cells are damaged by radiotherapy or chemotherapy during cancer treatment [53]. In the MDA-MB-231 cell line, we highlight AMOT (angiomotin) protein identified only when cells were treated with either low or high concentration venom treatment and is known to play a critical role in angiogenesis, proliferation, and migration of cancer cells and is also known to promote proliferation and invasion of several types of tumor cells, including breast, prostate, colon, cervical, and liver cancer (Table 1).

Semi-quantitative proteomics analysis of proteins with FC $\geq$ 1.5 after addition of venom in both MCF7 and MDA-MB-231 cell lines also identified enrichment of proteins related to protein metabolism pathways as well as cellular stress response pathway. The protein metabolism pathway ranges from protein synthesis to post-translational protein modification and degradation [136]. It is likely that the set of modifications undergone in the protein metabolism pathway undermine the tumorigenic activity of the cell line MCF7 and MDA-MB-231. Therefore, the activation of the apoptotic pathways using snake venom may be a potential treatment approach aiming at tumor degradation preferentially over the normal cells. The ribosome pathway enrichment was also observed in both cell lines. The lack of ribosomes in the cell impairs cell growth even under optimal conditions for it to happen [137]. Protein enrichment in ribosome pathway suggests increased protein synthesis, which may be related to damage repair mechanisms and cellular responses to external stimuli. Alterations in ribosomal pathways are related to cellular alterations and susceptibility to cancer. However, there is still no consensus if deregulation of ribosomes alteration is a cancer consequence or cause [138]. In addition, downregulation of cell maintenance-related proteins may also be related to tumorigenicity, thus opening the door for debate on the role of ribosomes in tumor cell lines [139].

The increase of terms involved in cellular stress response pathway, the metabolic pathway, the proteasome pathway, and the spliceosome pathway in MCF7 cell line suggests that the cells activated the metabolic pathways in response to stress, mRNA transcription and spliceosome activation, and ultimately activation of the proteasome. The proteasome is a cell apparatus that has several cellular functions, such as cell cycle regulation, differentiation, signal transduction pathway, antigen processing for immune responses, stress signaling, inflammatory responses and apoptosis, the latter being mediated by ubiquitination [140–142]. Since the inhibition of proteasome is related to cancer onset (i.e., p53 stability), proteolytic activity of the proteasome and its role in both biology and cancer treatment suggest an effective way to treat cancer [143,144].

In the MCF7 cell line treated with *B. jararaca* snake venom, a higher number of proteins changed in abundance when compared to the MDA-MB-231 cell line. Some of our observations showed that *B. jararaca* venom was able to up or down modulate the abundance of proteins that plays essential roles in tumorigenicity, conversely to what has been frequently described in the literature to decrease or increase cell tumorigenicity [113,145]. For example, the venom treatment was able to increase abundance of SNX3, HEL-S-156, MCCC2, and GSTM3 in the MCF7 cell line and H3C15/HIST2H3 and MTCH2 in the MDA-MB-231 cell line whose overexpression is related to increase in tumorigenicity. Although we cannot explain the mechanism on how the increase in abundance of these already overexpressed proteins is contributing to cell death of MCF7 and MDA-MB-231 cells, we have clues based on a recent published manuscript by Dias and colleagues (2019). The authors demonstrated that overloading the cell stress pathway in Y1 adrenocortical mouse tumor cell line under the effect of FGF-2 growth factor disrupted the cell homeostasis and sensitized the Y1 tumor cells to stress-oriented therapeutic inhibitors. Dias concludes that further stimulation of the same signaling pathways may further increase mobilization and dependence on stress response pathways in tumor cells, improving the efficacy and selectivity of therapeutic interventions [146]. This data suggests that *B. jararaca* snake venom or some of its content may be used to down-modulate already known overexpressed proteins in cancer or overexpress oncogenic proteins that lead cells to stress, by disrupting the tumor homeostasis leading cells to death. This is particularly interesting for cancer types that are more resistant to the chemotherapeutic agents, such as the MCF7 cells. In this case, *B. jararaca* snake venom would work as an adjuvant treatment of cancer, making cells more sensitive to the cytotoxic agents.

Moreover, due to the complexity of *B. jararaca* venom and the complex response of different cell lines, this snake venom is also a promising candidate in the prognostic aid of different tumors, assisting in the assessment of tumor level, as highlighted for the several overexpressed proteins. For example, targeted-knockdown of some of these proteins may result in tumor growth inhibition. In addition, *B. jararaca* venom has induced overexpression of proteins related to pathways that impairs antitumor activity such as the protein metabolism pathways, proteasome, and cellular stress response pathways.

It is possible that many responses observed here were due to specific or multiple venom components. Nevertheless, the current approach of applying nLC-MS/MS proteomics to cancer cells treated with crude snake venom is a promising strategy for the identification of proteins with potential application in cancer cell therapy. Overall, this study was able to identify several cancer-related proteins that undergo significant changes upon venom treatment of MCF7 and MDA-MB-231 cells. Future studies should address specific mechanisms by which the snake venom and some of its content may contribute to MCF7 and MDA-MB-231 cell death and survival.

4. Conclusions

Quantitative proteomic analysis of breast cancer cell lines allowed us to identify several proteins whose abundance (FC) increased more than 1.5×, and proteins that abundance decreased less than 0.67×, after 24 h treatment with *B. jararaca* at either low sub-toxic dose of 0.63 μg/mL or high sub-toxic dose of 2.5 μg/mL. Most of these proteins identified suggest that the treatment with the venom may activate mitochondrial apoptotic pathways leading cells to death. In addition, several of the identified proteins play important roles related to cell proliferation, invasion, metastasis, apoptosis, and stress response. Therefore, these data show that *B. jararaca* venom or some of its toxin or components can inhibit tumor cell proliferation and survival and can potentially be used to identify novel targets for cancer therapy.

5. Materials and Methods

5.1. Cell Culture and Maintenance

MCF7 and MDA-MB-231 breast cancer cell lines were previously acquired from ATCC (Manassas, VA, USA) and maintained at our laboratory cell bank. Cells were thawed and cultured at 37 °C in 5% CO_2 in RPMI 1640 culture medium (Gibco, Life Technologies, Grand Island, NY, USA) supplemented with 10% inactivated fetal bovine serum (Cultilab, Campinas, Brazil) and 25 mg/mL ampicillin and 100 mg/mL streptomycin antibiotics. Cells were split when the confluence reached 80%.

5.2. Bothrops jararaca Venom

The venom of *B. jararaca* (lot 01/09-2) used in this study was a pool made from the extraction of 697 snakes collected at various Brazilian locations (in the states of São Paulo, Paraná, and Santa Catarina). Extracted venom were pooled and lyophilized at the department of herpetology at Butantan Institute under the coordination of Dr. Marisa Maria Teixeira da Rocha, and assigned for use and testing at the Laboratory of Applied Toxinology at Butantan Institute. The methods and use of venom in this work were approved by Butantan Institute Ethics Committee under the certification CEUAIB #9766150719 (2019).

5.3. Cell Viability Assay

The cytotoxicity assays were carried out by treating the cells with different concentrations of venom in the range from 0.1 μM to 20 μM. The cytotoxicity tests were performed using the colorimetric method of cell viability analysis using the WST-1 cell proliferation reagent kit (Roche, Mannheim, Germany) according to manufacturer's instructions. Briefly, 10^4 cells were plated in each well of a flat bottom 96-well microtiter plate. Cells were cultured at 37 °C in 5% CO_2 and on the day prior to analysis (70–80% confluency), culture medium from each well was exchanged to 100 μL of medium in the absence or presence of venom in different concentrations. After 24 h, 10 μL of the WST-1 reagent previously dissolved into an Electro Coupling Solution (ECS) were added to each well and incubated for 4 h at 37 °C in 5% CO_2. The absorbance at 450 nm of each well was measured using a microplate reader FlexStation 3 spectrophotometer (Molecular Devices, San Jose, CA, USA). The determination of the absorbance and the quantification of viable cells were calculated using SoftMax Pro (version 5.1, Molecular Devices) and Microsoft Excel software.

5.4. Cell Treatment with *B. jararaca* Venom

In a six-well plate, 2×10^5 cells were plated and after the cells had reached a confluence of 70–80% (2–3 days), different concentrations of *B. jararaca* venom were added to the culture medium, starting from the cell cytotoxicity threshold up to a 1000-fold dilution for 24 h. For mass spectrometry analysis, cells were washed with ice-cold PBS and lysed with 1 mL ice-cold 8 M urea supplemented with cOmplete™ protease and phosphatase inhibitors cocktail (Merck Millipore, Burlington, MA, USA). Fifty microliters of the cell lysate were separated for protein quantification using BCA (Bicinchoninic acid assay, Thermo Pierce, Walthan, MA, USA) and the lysate were stocked at −80 °C until further preparation of the sample for mass spectrometry analysis. Both cytotoxicity and proteomics experiments were performed in three biological replicates.

5.5. Sample Preparation for Proteomic Analysis

To the cell lysates, four microliters of 10 mM dithiothreitol (DTT) were added to reduce the disulfide bonds and incubated at 56 °C for 1 h. After incubation with DTT, 40 μL of iodoacetamide (IAA) was added for the alkylation of the cysteines for 1 h in the dark at room temperature. After incubation, 2.5 mL of ammonium acetate and 20 μL of trypsin (Sigma Aldrich, St. Louis, MO, USA) were added enough to have an enzyme:protein ratio of 1:50 and incubated at 37 °C overnight. The reaction was stopped by adding 30 μL of 100% acetic acid. Samples volume were reduced in a SpeedVac (Hetovac VR-1, Heto Lab Equipment, Allerød, Denmark) until the volume has lowered to 50 μL and were further desalted on in-house manufactured stop-and-go extraction tips (Stage-Tip) with three SDB-XC (styrene-divinylbenzene, Empore, 3M, Royersford, PA, USA) membranes as previously described [147]. The stage-tips were initially conditioned with 100% methanol followed by 0.1% formic acid. Samples were applied to the stage-tips and the peptides bound to the membranes were eluted with 50% acetonitrile, 0.1% formic acid. The eluate was lyophilized for further analysis in the mass spectrometer.

5.6. Mass Spectrometry Analysis

Dried samples were resuspended in 0.1% formic acid and analyzed on LTQ-Orbitrap Velos mass spectrometer (Thermo Scientific, Bremen, Germany) coupled to an EASY II nano liquid chromatographer (Thermo Scientific) using the shotgun approach. The spectrometer was equipped with a nanospray source connected to an in-house prepared analytical column (10 cm × ID 75 μm × OD 360 μm) packaged with 7 cm of 5 μm C18 resin (Jupiter, Phenomenex, Torrance, CA, USA). Precolumn (7 cm × ID 75 μm × OD 360 μm) was also prepared in-house packed with 5 cm of 10 μm C18 resin (Acqua, Phenomenex). The LC-MS/MS analyses were carried out by injecting 5 μg of the peptide extract and the peptides eluted from the column with a gradient of 5–40% in 100 min of solvent B (acetonitrile 90%, formic acid 0.1%) at a flow rate of 200 nL/min. The nanospray source was operated at 1.8 kV. The peptide mixture was analyzed by the acquisition of spectra in the full scan mode at a resolution of 30,000 for the determination of molecular masses (MS) of up to 10,000 Da. The 10 most intense peaks were automatically selected via data dependent acquisition (DDA) for the subsequent acquisition of spectra of the ions product to MS/MS for amino acid sequence determination at a resolution of 7500, a maximum injection time of 30 ms, a range of 200 to 2000 m/z, and a dynamic exclusion of 70 s. Protein identification was performed using Mascot (Matrix Science, version 2.4.0, Boston, MA, USA), MaxQuant (version 1.4.1.2, www.maxquant.net accessed on 1 October 2019), and Peaks Studio (version 10, Bioinformatic Solutions Inc, Toronto, Canada) against the *Homo sapiens* database downloaded from the Uniprot in March 2019 (www.uniprot.org

accessed on 1 October 2019). As search parameters, oxidation of methionine was set as a variable modification and carbamidomethylation of cysteine was set as a fixed modification. Searches were performed with mass error tolerance of 10 ppm for MS and 0.3 Da for MS/MS, and trypsin was selected as the proteolytic enzyme used in the digestion of proteins, and up to two miscleavages were allowed. Mass spectrometry-based proteomic analysis of MCF7 and MDA-MB-231 cell lysates, treated or not treated with low or high doses of *B. jararaca* venom were performed in triplicate. All raw data files for these analyses were uploaded and are available at: http://massive.ucsd.edu/MSV000084138/ accessed on 5 July 2021.

5.7. Proteome Functional and Enrichment Analysis

The entire proteome was analyzed to classify protein profiles by biological processes, molecular function, cellular components, and cellular pathways using Gene Ontology (geneontology.org accessed on 1 October 2019) and PantherDB (pantherdb.org accessed on 1 October 2019). The proteins presenting different profiles in the venom conditions were also analyzed using KEGG (www.genome.jp/kegg accessed on 1 October 2019) canonical pathways to determine the protein group or pathways that have undergone changes with different concentrations of the venom treatment. The PANTHER Classification System (Protein ANalysis THrough Evolutionary Relationships) is a research tool that covers evolutionary and functional information on protein family genes and has been widely used to understand protein evolution and its functional classification [148,149]. We conducted the functional classification for the molecular function, cell component, biological process, and protein family classification profiles [150,151] of the different proteins in each cell line and treatment at low and high *B. jararaca* venom that presented FC $\geq$ 1.5 in order to identify an enrichment pattern of protein families and pathways affected by the venom in the cell lines.

5.8. Semi-Quantitative Proteomics Analysis

Semi-quantitative protein analysis was performed using the Label Free Quantification MaxLFQ algorithm from MaxQuant software with an FDR rate of $\leq$1% to compare relative abundance of proteins in each of the cell lines [152]. Data generated by MaxQuant were entered into the Perseus program to further perform statistical and bioinformatics analyzes [153]. Proteins identified in the contaminant database and the decoy database were removed. As a protein identification criterion, it was considered that only peptides identified with the posterior error probability (PEP) $\leq$ 0.01 in at least one biological replicate, the minimum identification of eight ions belonging to the b and y ion series in the MS/MS spectra, and the occurrence of at least one unique peptide. We considered the intensity values of the LFQ that are normalized by the Maxquant software based on the sum of the intensity of all peptides of all identified proteins. LFQ data was considered for calculation, when the intensity data were present in at least two out of three replicates. Protein abundance or fold change (FC) analysis was performed using Microsoft Excel software. In addition, proteins that had a zero value in two of the three conditions were analyzed separately. The breast cancer cell lines treated with the whole *B. jararaca* venom for two different doses were compared according to the differentially expressed proteins with FC $\geq$ 1.5 by hierarchical clustering using as variables the average log2 fold change for the three replicates for each treatment, normalized with mean-centering. The Clustering analysis was performed using R statistical software version 3.6.3 (http://www.R-project.org accessed on 1 October 2019). The set of protein dissimilarities were computed using the "Euclidean" distance with the function "dist" to the hierarchical clustering based on the package and function "hclust". There was employed the agglomerative method with "ward.D2". The fold change and the protein–protein interactions provided by information from the String database [154] was used to explore the biological interactions of proteins identified as differentially abundant between control and cells treated with *B. jararaca* venom using the *Homo sapiens* reference genome.

5.9. Principal Component Analysis

Principal component analysis (PCA) was applied to find which linear combinations of the differentially expressed proteins with ≥1.5 would explain most of the variability for the different cell lineage conditions. The PCA analysis was performed using the R-statistics packages FactoMineR (accessed on 1 November 2020) [155] and Factoextra (http://www.sthda.com/english/rpkgs/factoextra accessed on 1 November 2020) for graphical visualization.

Supplementary Materials: The following are available online at https://www.mdpi.com/article/10.3390/toxins13080519/s1, Figure S1: Optical microscopy analysis of breast cancer cell lines treated with different concentrations of *B. jararaca* snake venom. MCF7 (a) and MDA-MB-231 (b) cell lines (10× magnification), Figure S2: Functional classification of the proteins presenting fold change (FC) ≥ 1.5 according to Molecular Function GO enrichment analysis. (a) MCF7 cell line. (b) MDA-MB-231 cell line, Figure S3: Functional classification of the proteins presenting FC ≥ 1.5 according to the Biological Process GO enrichment analysis. (a) MCF7 cell line. (b) MDA-MB-231 cell line, Figure S4: Functional classification of the proteins presenting FC ≥ 1.5 according to the Cellular Component GO enrichment analysis. (a) MCF7 cell line. (b) MDA-MB-231 cell line, Figure S5: Functional classification of the proteins presenting FC ≥ 1.5 according to Protein Family Classification GO enrichment analysis. (a) MCF7 cell line. (b) MDA-MB-231 cell line, Figure S6a: Protein-protein interaction of proteins identified in MCF7 cell line presenting FC ≤ 0.67 in cells treated with 2.5 µg/mL of *B. jararaca* venom compared to control, Figure S6b: Protein-protein interaction of proteins identified in MCF7 cell line presenting FC ≥ 1.5 in cells treated with 0.67 µg/mL of *B. jararaca* venom compared to control, Figure S6c: Protein-protein interaction of proteins identified in MCF7 cell line presenting FC ≤ 0.67 when in cells treated with 0.63 µg/mL of *B. jararaca* venom compared to control, Supplemental Figure S7a: Protein-protein interaction of proteins identified in MDA-MB-231 cell line presenting FC ≤ 0.67 in cells treated with 2.5 µg/mL of *B. jararaca* venom compared to control, Figure S7b: Protein-protein interaction of proteins identified in MDA-MB-231 cell line presenting FC ≤ 0.67 in cells treated with 0.63 µg/mL of *B. jararaca* venom compared to control, Figure S7c: Protein-protein interaction of proteins identified in MDA-MB-231 cell line presenting FC ≥ 2.5 in cells treated with 0.63 µg/mL of *B. jararaca* venom compared to control, Figure S8: Intensity plots of proteins identified exclusively in one or two of the conditions tested (Control, 0.63 µg/mL or 2.5 µg/mL of *B. jararaca* venom) in both MCF7 and MDA-MB-231 cell lines (Values presented in Table S4). (a) Proteins identified exclusively in the control group. (b) Proteins identified exclusively in the 0.63 µg/mL of venom treatment. (c) Proteins identified exclusively in the 2.5 µg/mL of venom treatment. (d) Proteins identified exclusively in the 0.63 and 2.5 µg/mL of venom treatments. (e) Proteins identified exclusively in the control group and 2.5 µg/mL of venom treatments. (f) Proteins identified exclusively in the control group and 0.63 µg/mL of venom treatment, Table S1a: Mass spectrometry-based proteomics analysis of MCF7 cell line treated with PBS (CTRL), 0.63 µg/mL and 2.5 µg/mL of *B. jararaca* snake venom, Table S2a: Mass spectrometry-based proteomics analysis of MCF7 cell lines treated with PBS (CTRL), 0.63 mg/mL and 2.5 µg/mL of *B. jararaca* snake venom. Fold change based on the control LFQ intensity, Table S1b: Mass spectrometry-based proteomics analysis of MDA-MB-231 cell line treated with PBS (CTRL), 0.63 µg/mL, and 2.5 µg/mL of *B. jararaca* snake venom, Table S2b: Mass spectrometry-based proteomics analysis of MDA-MB-231 cell lines treated with PBS (CTRL), 0.63 µg/mL and 2.5 µg/mL of *B. jararaca* snake venom. Fold change based on the control LFQ intensity, Table S3: Protein list of each of the seven clusters corresponding to Figure 3, Table S4a: MCF7 proteins identified exclusively at one or two of the conditions tested: PBS (CTRL), 0.63 µg/mL and 2.5 µg/mL of *B. jararaca* snake venom, Table S4b: MDA-MB-231 proteins identified exclusively at one or two of the conditions tested: PBS (CTRL), 0.63 µg/mL and 2.5 µg/mL of *B. jararaca* snake venom.

Author Contributions: Conceptualization, L.K.I.; methodology, L.K.I.; validation, C.Y.K. and S.S.S.A.; formal analysis, C.Y.K., S.S.S.A., F.M., W.d.S.S., H.M.C., D.C.-C., E.S.F., M.Y.N.-J. and L.K.I., investigation, C.Y.K., S.S.S.A., I.F.L., and D.C.-C.; resources, L.K.I.; data curation, M.Y.N.-J.; writing—original draft preparation, C.Y.K., S.S.S.A., E.S.F., M.Y.N.-J. and L.K.I.; writing—review and editing, C.Y.K., E.S.F., M.Y.N.-J. and L.K.I.; supervision, E.S.F., M.Y.N.-J. and L.K.I.; project administration, L.K.I.; funding acquisition, E.S.F. and L.K.I. All authors have read and agreed to the published version of the manuscript.

Funding: This work was supported by grants 2013/07467-1, 2016/04000-3, and 2017/17943-6 from the São Paulo Research Foundation (FAPESP). C.Y.K. was supported by FAPESP master's degree program fellowship 2017/06496-9. E.S.F. was supported by Erna and Jakob Michael "Visiting Professorship", Department of Biological Regulation, Weizmann Institute of Science, Rehovot, Israel and grant 302809/2016-3 from the Brazilian National Council for Scientific and Technological Development, CNPq, Brazil. F.M., W.d.S.S. and S.S.S.A. were supported by the Coordination for the Improvement of Higher Education Personnel (CAPES, Brazil) and CNPq institutional fellowship #88882.442313/2019-01 (Butantan Institute, Brazil) for W.d.S.S. and # 131408/2019-4 for S.S.S.A. (University of São Paulo, Brazil).

Institutional Review Board Statement: The methods and use of venom in this work were approved by Butantan Institute Ethics Committee under the certification CEUAIB #9766150719 (2019).

Informed Consent Statement: Not Applicable.

Data Availability Statement: All generated raw files for these analyses were uploaded and are available at: http://massive.ucsd.edu/MSV000084138/ (acccessed on 5 July 2021).

Acknowledgments: We thank Solange M.T. Serrano, Hugo A. Armelin and Zhibin Chen for useful discussion.

Conflicts of Interest: The authors declare that there is no conflict of interest.

References

1. Brazilian Ministry of Health. Accidents at Work by Venomous Animals among Workers at the Countryside, Forest, and Water in Brazil from 2007 to 2017 (Port). 2019; Volume 50. Available online: https://portalarquivos2.saude.gov.br/images/pdf/2019/marco/29/2018-059.pdf (accessed on 1 March 2021).
2. Ribeiro, L.A.; Jorge, M.T. Bites by snakes in the genus Bothrops: A series of 3139 cases. *Rev. Soc. Bras. Med. Trop.* **1997**, *30*, 475–480. [CrossRef]
3. Tanjoni, I.; Weinlich, R.; Della-Casa, M.S.; Clissa, P.B.; Saldanha-Gama, R.F.; de Freitas, M.S.; Barja-Fidalgo, C.; Amarante-Mendes, G.P.; Moura-da-Silva, A.M. Jararhagin, a snake venom metalloproteinase, induces a specialized form of apoptosis (anoikis) selective to endothelial cells. *Apoptosis* **2005**, *10*, 851–861. [CrossRef] [PubMed]
4. Calvete, J.J.; Juarez, P.; Sanz, L. Snake venomics. Strategy and applications. *J. Mass Spectrom.* **2007**, *42*, 1405–1414. [CrossRef] [PubMed]
5. Fox, J.W.; Serrano, S.M. Exploring snake venom proteomes: Multifaceted analyses for complex toxin mixtures. *Proteomics* **2008**, *8*, 909–920. [CrossRef] [PubMed]
6. Doley, R.; Kini, R.M. Protein complexes in snake venom. *Cell. Mol. Life Sci.* **2009**, *66*, 2851–2871. [CrossRef]
7. Serrano, S.M.; Shannon, J.D.; Wang, D.; Camargo, A.C.; Fox, J.W. A multifaceted analysis of viperid snake venoms by two-dimensional gel electrophoresis: An approach to understanding venom proteomics. *Proteomics* **2005**, *5*, 501–510. [CrossRef]
8. Jin, H.; Varner, J. Integrins: Roles in cancer development and as treatment targets. *Br. J. Cancer* **2004**, *90*, 561–565. [CrossRef]
9. White, J. Snake venoms and coagulopathy. *Toxicon* **2005**, *45*, 951–967. [CrossRef]
10. Gay, C.; Sanz, L.; Calvete, J.J.; Pla, D. Snake Venomics and Antivenomics of Bothrops diporus, a Medically Important Pitviper in Northeastern Argentina. *Toxins* **2015**, *8*, 9. [CrossRef]
11. Sanz, L.; Perez, A.; Quesada-Bernat, S.; Diniz-Sousa, R.; Calderon, L.A.; Soares, A.M.; Calvete, J.J.; Caldeira, C.A.S. Venomics and antivenomics of the poorly studied Brazil's lancehead, Bothrops brazili (Hoge, 1954), from the Brazilian State of Para. *J. Venom. Anim. Toxins Incl. Trop. Dis.* **2020**, *26*, e20190103. [CrossRef]
12. Tashima, A.K.; Zelanis, A.; Kitano, E.S.; Ianzer, D.; Melo, R.L.; Rioli, V.; Sant'anna, S.S.; Schenberg, A.C.; Camargo, A.C.; Serrano, S.M. Peptidomics of three Bothrops snake venoms: Insights into the molecular diversification of proteomes and peptidomes. *Mol. Cell. Proteom.* **2012**, *11*, 1245–1262. [CrossRef]
13. Gren, E.C.K.; Kitano, E.S.; Andrade-Silva, D.; Iwai, L.K.; Reis, M.S.; Menezes, M.C.; Serrano, S.M.T. Comparative analysis of the high molecular mass subproteomes of eight Bothrops snake venoms. *Comp. Biochem. Physiol. Part D Genom. Proteom.* **2019**, *30*, 113–121. [CrossRef]

14. Zelanis, A.; Tashima, A.K.; Pinto, A.F.; Paes Leme, A.F.; Stuginski, D.R.; Furtado, M.F.; Sherman, N.E.; Ho, P.L.; Fox, J.W.; Serrano, S.M. Bothrops jararaca venom proteome rearrangement upon neonate to adult transition. *Proteomics* **2011**, *11*, 4218–4228. [CrossRef]
15. Farias, I.B.; Morais-Zani, K.; Serino-Silva, C.; Sant'Anna, S.S.; Rocha, M.; Grego, K.F.; Andrade-Silva, D.; Serrano, S.M.T.; Tanaka-Azevedo, A.M. Functional and proteomic comparison of Bothrops jararaca venom from captive specimens and the Brazilian Bothropic Reference Venom. *J. Proteom.* **2018**, *174*, 36–46. [CrossRef]
16. Augusto-de-Oliveira, C.; Stuginski, D.R.; Kitano, E.S.; Andrade-Silva, D.; Liberato, T.; Fukushima, I.; Serrano, S.M.; Zelanis, A. Dynamic Rearrangement in Snake Venom Gland Proteome: Insights into Bothrops jararaca Intraspecific Venom Variation. *J. Proteome Res.* **2016**, *15*, 3752–3762. [CrossRef]
17. Laing, G.D.; Clissa, P.B.; Theakston, R.D.; Moura-da-Silva, A.M.; Taylor, M.J. Inflammatory pathogenesis of snake venom metalloproteinase-induced skin necrosis. *Eur. J. Immunol.* **2003**, *33*, 3458–3463. [CrossRef]
18. Azevedo-Marques, M.M.; Cupo, P.; Hering, S.E. Acidentes Por Animais PeÇonhentos: Serpentes PeÇonhentas. *Medicina* **2003**, *36*. [CrossRef]
19. Bjarnason, J.B.; Fox, J.W. Hemorrhagic metalloproteinases from snake venoms. *Pharmacol. Ther.* **1994**, *62*, 325–372. [CrossRef]
20. Cidade, D.A.; Simao, T.A.; Davila, A.M.; Wagner, G.; Junqueira-de-Azevedo, I.L.; Ho, P.L.; Bon, C.; Zingali, R.B.; Albano, R.M. Bothrops jararaca venom gland transcriptome: Analysis of the gene expression pattern. *Toxicon* **2006**, *48*, 437–461. [CrossRef]
21. Pereira, L.M.; Messias, E.A.; Sorroche, B.P.; Oliveira, A.D.N.; Arantes, L.; de Carvalho, A.C.; Tanaka-Azevedo, A.M.; Grego, K.F.; Carvalho, A.L.; Melendez, M.E. In-depth transcriptome reveals the potential biotechnological application of Bothrops jararaca venom gland. *J. Venom. Anim. Toxins Incl. Trop. Dis.* **2020**, *26*, e20190058. [CrossRef]
22. Serrano, S.M.; Maroun, R.C. Snake venom serine proteinases: Sequence homology vs. substrate specificity, a paradox to be solved. *Toxicon* **2005**, *45*, 1115–1132. [CrossRef]
23. Fox, J.W.; Serrano, S.M. Insights into and speculations about snake venom metalloproteinase (SVMP) synthesis, folding and disulfide bond formation and their contribution to venom complexity. *FEBS J.* **2008**, *275*, 3016–3030. [CrossRef]
24. Matsui, T.; Fujimura, Y.; Titani, K. Snake venom proteases affecting hemostasis and thrombosis. *Biochim. Biophys. Acta (BBA)-Protein Struct. Mol. Enzym.* **2000**, *1477*, 146–156. [CrossRef]
25. Da Silva Cunha, K.C.; Fuly, A.L.; de Araujo, E.G. A phospholipase A(2) isolated from Lachesis muta snake venom increases the survival of retinal ganglion cells in vitro. *Toxicon* **2011**, *57*, 580–585. [CrossRef]
26. Jain, D.; Kumar, S. Snake venom: A potent anticancer agent. *Asian Pac. J. Cancer Prev.* **2012**, *13*, 4855–4860. [CrossRef] [PubMed]
27. Tan, N.; Fung, S.Y. Snake Venom L-Amino Acid Oxidases potential biomedical applications. *Mal. J. Biochem. Mol. Biol.* **2008**, *16*, 1–10.
28. Tan, K.K.; Bay, B.H.; Gopalakrishnakone, P. L-amino acid oxidase from snake venom and its anticancer potential. *Toxicon* **2018**, *144*, 7–13. [CrossRef]
29. Pawelek, P.D.; Cheah, J.; Coulombe, R.; Macheroux, P.; Ghisla, S.; Vrielink, A. The structure of L-amino acid oxidase reveals the substrate trajectory into an enantiomerically conserved active site. *EMBO J.* **2000**, *19*, 4204–4215. [CrossRef]
30. Phillips, D.R.; Scarborough, R.M. Clinical Pharmacology of Eptifibatide. *Am. J. Cardiol.* **1997**, *80*, 11B–20B. [CrossRef]
31. Ferreira, S.H. A Bradykinin-Potentiating Factor (Bpf) Present in the Venom of Bothrops Jararca. *Br. J. Pharmacol. Chemother.* **1965**, *24*, 163–169. [CrossRef]
32. Camargo, A.C.; Ianzer, D.; Guerreiro, J.R.; Serrano, S.M. Bradykinin-potentiating peptides: Beyond captopril. *Toxicon* **2012**, *59*, 516–523. [CrossRef]
33. Fox, J.W.; Serrano, S.M. Approaching the golden age of natural product pharmaceuticals from venom libraries: An overview of toxins and toxin-derivatives currently involved in therapeutic or diagnostic applications. *Curr. Pharm. Des.* **2007**, *13*, 2927–2934. [CrossRef]
34. Marsh, N.; Williams, V. Practical applications of snake venom toxins in haemostasis. *Toxicon* **2005**, *45*, 1171–1181. [CrossRef]
35. Vyas, V.K.; Brahmbhatt, K.; Bhatt, H.; Parmar, U. Therapeutic potential of snake venom in cancer therapy: Current perspectives. *Asian Pac. J. Trop. Biomed.* **2013**, *3*, 156–162. [CrossRef]
36. da Silva, R.J.; da Silva, M.G.; Vilela, L.C.; Fecchio, D. Antitumor effect of Bothrops jararaca venom. *Mediat. Inflamm.* **2002**, *11*, 99–104. [CrossRef]
37. Yamazaki, Y.; Morita, T. Snake venom components affecting blood coagulation and the vascular system: Structural similarities and marked diversity. *Curr. Pharm. Des.* **2007**, *13*, 2872–2886. [CrossRef] [PubMed]
38. Dias, M.H.; Kitano, E.S.; Zelanis, A.; Iwai, L.K. Proteomics and drug discovery in cancer. *Drug Discov. Today* **2016**, *21*, 264–277. [CrossRef] [PubMed]
39. Hanahan, D.; Weinberg, R.A. Hallmarks of cancer: The next generation. *Cell* **2011**, *144*, 646–674. [CrossRef] [PubMed]
40. Mora, R.; Valverde, B.; Diaz, C.; Lomonte, B.; Gutierrez, J.M. A Lys49 phospholipase A(2) homologue from Bothrops asper snake venom induces proliferation, apoptosis and necrosis in a lymphoblastoid cell line. *Toxicon* **2005**, *45*, 651–660. [CrossRef] [PubMed]
41. Bateman, E.; Venning, M.; Mirtschin, P.; Woods, A. The effects of selected Australian snake venoms on tumour-associated microvascular endothelial cells (TAMECs) in vitro. *J. Venom. Res.* **2013**, *4*, 21–30.
42. Moreira, V.; Lomonte, B.; Vinolo, M.A.; Curi, R.; Gutierrez, J.M.; Teixeira, C. An Asp49 phospholipase A2 from snake venom induces cyclooxygenase-2 expression and prostaglandin E2 production via activation of NF-kappaB, p38MAPK, and PKC in macrophages. *Mediat. Inflamm.* **2014**, *2014*, 105879. [CrossRef]

43. Leiguez, E.; Giannotti, K.C.; Moreira, V.; Matsubara, M.H.; Gutierrez, J.M.; Lomonte, B.; Rodriguez, J.P.; Balsinde, J.; Teixeira, C. Critical role of TLR2 and MyD88 for functional response of macrophages to a group IIA-secreted phospholipase A2 from snake venom. *PLoS ONE* **2014**, *9*, e93741. [CrossRef]
44. Frejno, M.; Meng, C.; Ruprecht, B.; Oellerich, T.; Scheich, S.; Kleigrewe, K.; Drecoll, E.; Samaras, P.; Hogrebe, A.; Helm, D.; et al. Proteome activity landscapes of tumor cell lines determine drug responses. *Nat. Commun.* **2020**, *11*, 3639. [CrossRef] [PubMed]
45. Ruprecht, B.; Di Bernardo, J.; Wang, Z.; Mo, X.; Ursu, O.; Christopher, M.; Fernandez, R.B.; Zheng, L.; Dill, B.D.; Wang, H.; et al. A mass spectrometry-based proteome map of drug action in lung cancer cell lines. *Nat. Chem. Biol.* **2020**, *16*, 1111–1119. [CrossRef] [PubMed]
46. Furtado, C.M.; Marcondes, M.C.; Sola-Penna, M.; de Souza, M.L.; Zancan, P. Clotrimazole preferentially inhibits human breast cancer cell proliferation, viability and glycolysis. *PLoS ONE* **2012**, *7*, e30462. [CrossRef] [PubMed]
47. Hui, L.; Zheng, Y.; Yan, Y.; Bargonetti, J.; Foster, D.A. Mutant p53 in MDA-MB-231 breast cancer cells is stabilized by elevated phospholipase D activity and contributes to survival signals generated by phospholipase D. *Oncogene* **2006**, *25*, 7305–7310. [CrossRef]
48. Neve, R.M.; Chin, K.; Fridlyand, J.; Yeh, J.; Baehner, F.L.; Fevr, T.; Clark, L.; Bayani, N.; Coppe, J.P.; Tong, F.; et al. A collection of breast cancer cell lines for the study of functionally distinct cancer subtypes. *Cancer Cell* **2006**, *10*, 515–527. [CrossRef]
49. Lv, M.; Shen, Y.; Yang, J.; Li, S.; Wang, B.; Chen, Z.; Li, P.; Liu, P.; Yang, J. Angiomotin Family Members: Oncogenes or Tumor Suppressors? *Int. J. Biol. Sci.* **2017**, *13*, 772–781. [CrossRef]
50. Zhang, H.; Fan, Q. MicroRNA-205 inhibits the proliferation and invasion of breast cancer by regulating AMOT expression. *Oncol. Rep.* **2015**, *34*, 2163–2170. [CrossRef]
51. Couderc, C.; Boin, A.; Fuhrmann, L.; Vincent-Salomon, A.; Mandati, V.; Kieffer, Y.; Mechta-Grigoriou, F.; Del Maestro, L.; Chavrier, P.; Vallerand, D.; et al. AMOTL1 Promotes Breast Cancer Progression and Is Antagonized by Merlin. *Neoplasia* **2016**, *18*, 10–24. [CrossRef] [PubMed]
52. Song, K.H.; Kim, J.H.; Lee, Y.H.; Bae, H.C.; Lee, H.J.; Woo, S.R.; Oh, S.J.; Lee, K.M.; Yee, C.; Kim, B.W.; et al. Mitochondrial reprogramming via ATP5H loss promotes multimodal cancer therapy resistance. *J. Clin. Investig.* **2018**, *128*, 4098–4114. [CrossRef] [PubMed]
53. Qiu, Z.; Oleinick, N.L.; Zhang, J. ATR/CHK1 inhibitors and cancer therapy. *Radiother. Oncol.* **2018**, *126*, 450–464. [CrossRef] [PubMed]
54. Karnitz, L.M.; Zou, L. Molecular Pathways: Targeting ATR in Cancer Therapy. *Clin. Cancer Res.* **2015**, *21*, 4780–4785. [CrossRef] [PubMed]
55. O'Kelly, J.; Chung, A.; Lemp, N.; Chumakova, K.; Yin, D.; Wang, H.J.; Said, J.; Gui, D.; Miller, C.W.; Karlan, B.Y.; et al. Functional domains of CCN1 (Cyr61) regulate breast cancer progression. *Int. J. Oncol.* **2008**, *33*, 59–67. [CrossRef] [PubMed]
56. Huang, Y.T.; Lan, Q.; Lorusso, G.; Duffey, N.; Ruegg, C. The matricellular protein CYR61 promotes breast cancer lung metastasis by facilitating tumor cell extravasation and suppressing anoikis. *Oncotarget* **2017**, *8*, 9200–9215. [CrossRef]
57. Hayes, J.D.; Flanagan, J.U.; Jowsey, I.R. Glutathione transferases. *Annu. Rev. Pharmacol. Toxicol.* **2005**, *45*, 51–88. [CrossRef] [PubMed]
58. Meding, S.; Balluff, B.; Elsner, M.; Schone, C.; Rauser, S.; Nitsche, U.; Maak, M.; Schafer, A.; Hauck, S.M.; Ueffing, M.; et al. Tissue-based proteomics reveals FXYD3, S100A11 and GSTM3 as novel markers for regional lymph node metastasis in colon cancer. *J. Pathol.* **2012**, *228*, 459–470. [CrossRef] [PubMed]
59. Saucedo-Cuevas, L.P.; Ruppen, I.; Ximenez-Embun, P.; Domingo, S.; Gayarre, J.; Munoz, J.; Silva, J.M.; Garcia, M.J.; Benitez, J. CUL4A contributes to the biology of basal-like breast tumors through modulation of cell growth and antitumor immune response. *Oncotarget* **2014**, *5*, 2330–2343. [CrossRef]
60. Lan, F.; Shi, Y. Histone H3.3 and cancer: A potential reader connection. *Proc. Natl. Acad. Sci. USA* **2015**, *112*, 6814–6819. [CrossRef]
61. Ayoubi, H.A.; Mahjoubi, F.; Mirzaei, R. Investigation of the human H3.3B (H3F3B) gene expression as a novel marker in patients with colorectal cancer. *J. Gastrointest. Oncol.* **2017**, *8*, 64–69. [CrossRef]
62. Roberts, E.L.; Newton, R.P.; Axford, A.T. Plasma purine nucleoside phosphorylase in cancer patients. *Clin. Chim. Acta* **2004**, *344*, 109–114. [CrossRef]
63. Vareed, S.K.; Bhat, V.B.; Thompson, C.; Vasu, V.T.; Fermin, D.; Choi, H.; Creighton, C.J.; Gayatri, S.; Lan, L.; Putluri, N.; et al. Metabolites of purine nucleoside phosphorylase (NP) in serum have the potential to delineate pancreatic adenocarcinoma. *PLoS ONE* **2011**, *6*, e17177. [CrossRef]
64. Dummer, R.; Duvic, M.; Scarisbrick, J.; Olsen, E.A.; Rozati, S.; Eggmann, N.; Goldinger, S.M.; Hutchinson, K.; Geskin, L.; Illidge, T.M.; et al. Final results of a multicenter phase II study of the purine nucleoside phosphorylase (PNP) inhibitor forodesine in patients with advanced cutaneous T-cell lymphomas (CTCL) (Mycosis fungoides and Sezary syndrome). *Ann. Oncol.* **2014**, *25*, 1807–1812. [CrossRef]
65. Tan, E.; Besant, P.G.; Zu, X.L.; Turck, C.W.; Bogoyevitch, M.A.; Lim, S.G.; Attwood, P.V.; Yeoh, G.C. Histone H4 histidine kinase displays the expression pattern of a liver oncodevelopmental marker. *Carcinogenesis* **2004**, *25*, 2083–2088. [CrossRef]
66. Besant, P.G.; Attwood, P.V. Histone H4 histidine phosphorylation: Kinases, phosphatases, liver regeneration and cancer. *Biochem. Soc. Trans.* **2012**, *40*, 290–293. [CrossRef]

67. Long, M.; Sun, X.; Shi, W.; Yanru, A.; Leung, S.T.C.; Ding, D.; Cheema, M.S.; MacPherson, N.; Nelson, C.J.; Ausio, J.; et al. A novel histone H4 variant H4G regulates rDNA transcription in breast cancer. *Nucleic. Acids Res.* **2019**, *47*, 8399–8409. [CrossRef] [PubMed]
68. Nacev, B.A.; Feng, L.; Bagert, J.D.; Lemiesz, A.E.; Gao, J.; Soshnev, A.A.; Kundra, R.; Schultz, N.; Muir, T.W.; Allis, C.D. The expanding landscape of 'oncohistone' mutations in human cancers. *Nature* **2019**, *567*, 473–478. [CrossRef] [PubMed]
69. Bell, J.L.; Wachter, K.; Muhleck, B.; Pazaitis, N.; Kohn, M.; Lederer, M.; Huttelmaier, S. Insulin-like growth factor 2 mRNA-binding proteins (IGF2BPs): Post-transcriptional drivers of cancer progression? *Cell. Mol. Life Sci.* **2013**, *70*, 2657–2675. [CrossRef]
70. Fakhraldeen, S.A.; Clark, R.J.; Roopra, A.; Chin, E.N.; Huang, W.; Castorino, J.; Wisinski, K.B.; Kim, T.; Spiegelman, V.S.; Alexander, C.M. Two Isoforms of the RNA Binding Protein, Coding Region Determinant-binding Protein (CRD-BP/IGF2BP1), Are Expressed in Breast Epithelium and Support Clonogenic Growth of Breast Tumor Cells. *J. Biol. Chem.* **2015**, *290*, 13386–13400. [CrossRef]
71. Kim, T.; Havighurst, T.; Kim, K.; Albertini, M.; Xu, Y.G.; Spiegelman, V.S. Targeting insulin-like growth factor 2 mRNA-binding protein 1 (IGF2BP1) in metastatic melanoma to increase efficacy of BRAF(V600E) inhibitors. *Mol. Carcinog.* **2018**, *57*, 678–683. [CrossRef] [PubMed]
72. Han, W.; Hu, C.; Fan, Z.J.; Shen, G.L. Transcript levels of keratin 1/5/6/14/15/16/17 as potential prognostic indicators in melanoma patients. *Sci. Rep.* **2021**, *11*, 1023. [CrossRef]
73. Palko, E.; Poliska, S.; Sziklai, I.; Penyige, A. Analysis of KRT1, KRT10, KRT19, TP53 and MMP9 expression in pediatric and adult cholesteatoma. *PLoS ONE* **2018**, *13*, e0200840. [CrossRef]
74. Hitzerd, S.M.; Verbrugge, S.E.; Ossenkoppele, G.; Jansen, G.; Peters, G.J. Positioning of aminopeptidase inhibitors in next generation cancer therapy. *Amino Acids* **2014**, *46*, 793–808. [CrossRef]
75. Tian, S.Y.; Chen, S.H.; Shao, B.F.; Cai, H.Y.; Zhou, Y.; Zhou, Y.L.; Xu, A.B. Expression of leucine aminopeptidase 3 (LAP3) correlates with prognosis and malignant development of human hepatocellular carcinoma (HCC). *Int. J. Clin. Exp. Pathol.* **2014**, *7*, 3752–3762.
76. He, X.; Huang, Q.; Qiu, X.; Liu, X.; Sun, G.; Guo, J.; Ding, Z.; Yang, L.; Ban, N.; Tao, T.; et al. LAP3 promotes glioma progression by regulating proliferation, migration and invasion of glioma cells. *Int. J. Biol. Macromol.* **2015**, *72*, 1081–1089. [CrossRef] [PubMed]
77. Zhang, S.; Yang, X.; Shi, H.; Li, M.; Xue, Q.; Ren, H.; Yao, L.; Chen, X.; Zhang, J.; Wang, H. Overexpression of leucine aminopeptidase 3 contributes to malignant development of human esophageal squamous cell carcinoma. *J. Mol. Histol.* **2014**, *45*, 283–292. [CrossRef]
78. Liu, Y.; Yuan, Z.; Song, C. Methylcrotonoyl-CoA carboxylase 2 overexpression predicts an unfavorable prognosis and promotes cell proliferation in breast cancer. *Biomark. Med.* **2019**, *13*, 427–436. [CrossRef] [PubMed]
79. Arco, A.D.; Satrustegui, J. New mitochondrial carriers: An overview. *Cell. Mol. Life Sci.* **2005**, *62*, 2204–2227. [CrossRef] [PubMed]
80. Hao, J.M.; Chen, J.Z.; Sui, H.M.; Si-Ma, X.Q.; Li, G.Q.; Liu, C.; Li, J.L.; Ding, Y.Q.; Li, J.M. A five-gene signature as a potential predictor of metastasis and survival in colorectal cancer. *J. Pathol.* **2010**, *220*, 475–489. [CrossRef] [PubMed]
81. Arigoni, M.; Barutello, G.; Riccardo, F.; Ercole, E.; Cantarella, D.; Orso, F.; Conti, L.; Lanzardo, S.; Taverna, D.; Merighi, I.; et al. miR-135b coordinates progression of ErbB2-driven mammary carcinomas through suppression of MID1 and MTCH2. *Am. J. Pathol.* **2013**, *182*, 2058–2070. [CrossRef]
82. De Palma, G.; Sallustio, F.; Curci, C.; Galleggiante, V.; Rutigliano, M.; Serino, G.; Ditonno, P.; Battaglia, M.; Schena, F.P. The Three-Gene Signature in Urinary Extracellular Vesicles from Patients with Clear Cell Renal Cell Carcinoma. *J. Cancer* **2016**, *7*, 1960–1967. [CrossRef]
83. Nunes, F.D.; de Almeida, F.C.; Tucci, R.; de Sousa, S.C. Homeobox genes: A molecular link between development and cancer. *Pesquisa Odontol. Bras.* **2003**, *17*, 94–98. [CrossRef]
84. Samuel, S.; Naora, H. Homeobox gene expression in cancer: Insights from developmental regulation and deregulation. *Eur. J. Cancer* **2005**, *41*, 2428–2437. [CrossRef]
85. Pyper, S.R.; Viswakarma, N.; Jia, Y.; Zhu, Y.J.; Fondell, J.D.; Reddy, J.K. PRIC295, a Nuclear Receptor Coactivator, Identified from PPARalpha-Interacting Cofactor Complex. *PPAR Res.* **2010**, *2010*, 173907. [CrossRef]
86. Viswakarma, N.; Jia, Y.; Bai, L.; Gao, Q.; Lin, B.; Zhang, X.; Misra, P.; Rana, A.; Jain, S.; Gonzalez, F.J.; et al. The Med1 subunit of the mediator complex induces liver cell proliferation and is phosphorylated by AMP kinase. *J. Biol. Chem.* **2013**, *288*, 27898–27911. [CrossRef]
87. Ahmed, N.; Escalona, R.; Leung, D.; Chan, E.; Kannourakis, G. Tumour microenvironment and metabolic plasticity in cancer and cancer stem cells: Perspectives on metabolic and immune regulatory signatures in chemoresistant ovarian cancer stem cells. *Semin. Cancer Biol.* **2018**, *53*, 265–281. [CrossRef]
88. Levin, A.; Minis, A.; Lalazar, G.; Rodriguez, J.; Steller, H. PSMD5 Inactivation Promotes 26S Proteasome Assembly during Colorectal Tumor Progression. *Cancer Res.* **2018**, *78*, 3458–3468. [CrossRef] [PubMed]
89. Chai, F.L.Y.; Bi, J.; Chen, L.Z.F.; Cui, Y.; Bian, X.; Jiang, J. High expression of REGγ is associated with metastasis and poor prognosis of patients with breast cancer. *Int. J. Clin. Exp. Pathol.* **2014**, *7*, 7834. [PubMed]
90. Chai, F.; Liang, Y.; Bi, J.; Chen, L.; Zhang, F.; Cui, Y.; Jiang, J. REGgamma regulates ERalpha degradation via ubiquitin-proteasome pathway in breast cancer. *Biochem. Biophys. Res. Commun.* **2015**, *456*, 534–540. [CrossRef]
91. Wang, X.; Tu, S.; Tan, J.; Tian, T.; Ran, L.; Rodier, J.F.; Ren, G. REG gamma: A potential marker in breast cancer and effect on cell cycle and proliferation of breast cancer cell. *Med. Oncol.* **2011**, *28*, 31–41. [CrossRef] [PubMed]

92. Coppock, D.; Kopman, C.; Gudas, J.; Cina-Poppe, D.A. Regulation of the quiescence-induced genes: Quiescin Q6, decorin, and ribosomal protein S29. *Biochem. Biophys. Res. Commun.* **2000**, *269*, 604–610. [CrossRef]
93. Notterman, D.A.; Alon, U.; Sierk, A.J.; Levine, A.J. Transcriptional gene expression profiles of colorectal adenoma, adenocarcinoma, and normal tissue examined by oligonucleotide arrays. *Cancer Res.* **2001**, *61*, 7.
94. Lv, Z.; Wang, Z.; Luo, L.; Chen, Y.; Han, G.; Wang, R.; Xiao, H.; Li, X.; Hou, C.; Feng, J.; et al. Spliceosome protein Eftud2 promotes colitis-associated tumorigenesis by modulating inflammatory response of macrophage. *Mucosal Immunol.* **2019**, *12*, 1164–1173. [CrossRef]
95. Sato, N.; Maeda, M.; Sugiyama, M.; Ito, S.; Hyodo, T.; Masuda, A.; Tsunoda, N.; Kokuryo, T.; Hamaguchi, M.; Nagino, M.; et al. Inhibition of SNW1 association with spliceosomal proteins promotes apoptosis in breast cancer cells. *Cancer Med.* **2015**, *4*, 268–277. [CrossRef] [PubMed]
96. Pons, V.; Luyet, P.P.; Morel, E.; Abrami, L.; van der Goot, F.G.; Parton, R.G.; Gruenberg, J. Hrs and SNX3 functions in sorting and membrane invagination within multivesicular bodies. *PLoS Biol.* **2008**, *6*, e214. [CrossRef]
97. Mendelsohn, J. The epidermal growth factor receptor as a target for cancer therapy. *Endocr. Relat. Cancer* **2001**, *8*, 3–9. [CrossRef]
98. John, A.S.; Rothman, V.L.; Tuszynski, G.P. Thrombospondin-1 (TSP-1) Stimulates Expression of Integrin alpha6 in Human Breast Carcinoma Cells: A Downstream Modulator of TSP-1-Induced Cellular Adhesion. *J. Oncol.* **2010**, *2010*, 645376. [CrossRef] [PubMed]
99. Jayachandran, A.; Anaka, M.; Prithviraj, P.; Hudson, C.; McKeown, S.J.; Lo, P.H.; Vella, L.J.; Goding, C.R.; Cebon, J.; Behren, A. Thrombospondin 1 promotes an aggressive phenotype through epithelial-to-mesenchymal transition in human melanoma. *Oncotarget* **2014**, *5*, 5782–5797. [CrossRef]
100. Tzeng, H.T.; Tsai, C.H.; Yen, Y.T.; Cheng, H.C.; Chen, Y.C.; Pu, S.W.; Wang, Y.S.; Shan, Y.S.; Tseng, Y.L.; Su, W.C.; et al. Dysregulation of Rab37-Mediated Cross-talk between Cancer Cells and Endothelial Cells via Thrombospondin-1 Promotes Tumor Neovasculature and Metastasis. *Clin. Cancer Res.* **2017**, *23*, 2335–2345. [CrossRef]
101. Shi, H.; Hayes, M.; Kirana, C.; Miller, R.; Keating, J.; Macartney-Coxson, D.; Stubbs, R. TUFM is a potential new prognostic indicator for colorectal carcinoma. *Pathology* **2012**, *44*, 506–512. [CrossRef]
102. Weng, X.; Zheng, S.; Shui, H.; Lin, G.; Zhou, Y. TUFM-knockdown inhibits the migration and proliferation of gastrointestinal stromal tumor cells. *Oncol. Lett.* **2020**, *20*, 250. [CrossRef]
103. He, K.; Guo, X.; Liu, Y.; Li, J.; Hu, Y.; Wang, D.; Song, J. TUFM downregulation induces epithelial-mesenchymal transition and invasion in lung cancer cells via a mechanism involving AMPK-GSK3beta signaling. *Cell. Mol. Life Sci.* **2016**, *73*, 2105–2121. [CrossRef] [PubMed]
104. Kulawiec, M.; Arnouk, H.; Desouki, M.M.; Kazim, L.; Still, I.; Singh, K.K. Proteomic analysis of mitochondria-to-nucleus retrograde response in human cancer. *Cancer Biol. Ther.* **2006**, *5*, 967–975. [CrossRef] [PubMed]
105. Li, W.; Wubulikasimu, G.; Zhao, X.; Wang, C.; Liu, R.; Wang, L.; Zhu, X.; Chen, Z. UQCRC1 downregulation is correlated with lymph node metastasis and poor prognosis in CRC. *Eur. J. Surg. Oncol.* **2019**, *45*, 1005–1010. [CrossRef]
106. Wang, Q.; Li, M.; Gan, Y.; Jiang, S.; Qiao, J.; Zhang, W.; Fan, Y.; Shen, Y.; Song, Y.; Meng, Z.; et al. Mitochondrial Protein UQCRC1 is Oncogenic and a Potential Therapeutic Target for Pancreatic Cancer. *Theranostics* **2020**, *10*, 2141–2157. [CrossRef] [PubMed]
107. Mazure, N.M. VDAC in cancer. *Biochim. Biophys. Acta Bioenerg.* **2017**, *1858*, 665–673. [CrossRef]
108. Sotgia, F.; Fiorillo, M.; Lisanti, M.P. Mitochondrial markers predict recurrence, metastasis and tamoxifen-resistance in breast cancer patients: Early detection of treatment failure with companion diagnostics. *Oncotarget* **2017**, *8*, 68730–68745. [CrossRef]
109. Camara, A.K.S.; Zhou, Y.; Wen, P.C.; Tajkhorshid, E.; Kwok, W.M. Mitochondrial VDAC1: A Key Gatekeeper as Potential Therapeutic Target. *Front. Physiol.* **2017**, *8*, 460. [CrossRef]
110. Magri, A.; Reina, S.; De Pinto, V. VDAC1 as Pharmacological Target in Cancer and Neurodegeneration: Focus on Its Role in Apoptosis. *Front. Chem.* **2018**, *6*, 108. [CrossRef] [PubMed]
111. Thul, P.J.; Akesson, L.; Wiking, M.; Mahdessian, D.; Geladaki, A.; Ait Blal, H.; Alm, T.; Asplund, A.; Bjork, L.; Breckels, L.M.; et al. A subcellular map of the human proteome. *Science* **2017**, *356*. [CrossRef]
112. Uhlen, M.; Fagerberg, L.; Hallstrom, B.M.; Lindskog, C.; Oksvold, P.; Mardinoglu, A.; Sivertsson, A.; Kampf, C.; Sjostedt, E.; Asplund, A.; et al. Tissue-based map of the human proteome. *Science* **2015**, *347*, 1260419. [CrossRef] [PubMed]
113. Uhlen, M.; Zhang, C.; Lee, S.; Sjostedt, E.; Fagerberg, L.; Bidkhori, G.; Benfeitas, R.; Arif, M.; Liu, Z.; Edfors, F.; et al. A pathology atlas of the human cancer transcriptome. *Science* **2017**, *357*. [CrossRef]
114. Doncheva, N.T.; Morris, J.H.; Gorodkin, J.; Jensen, L.J. Cytoscape StringApp: Network Analysis and Visualization of Proteomics Data. *J. Proteome Res.* **2019**, *18*, 623–632. [CrossRef]
115. Essex, H.E.; Priestley, J.T. Effect of Rattlesnake Venom on Flexner-Jobling's Carcinoma in the White Rat (Mus Norvegicus Albinus.). *Exp. Biol. Med.* **1931**, *28*, 550–551. [CrossRef]
116. Ferreira, S.H.; Greene, L.H.; Alabaster, V.A.; Bakhle, Y.S.; Vane, J.R. Activity of various fractions of bradykinin potentiating factor against angiotensin I converting enzyme. *Nature* **1970**, *225*, 379–380. [CrossRef] [PubMed]
117. Cushman, D.W.; Ondetti, M.A. History of the design of captopril and related inhibitors of angiotensin converting enzyme. *Hypertension* **1991**, *17*, 589–592. [CrossRef] [PubMed]
118. Smith, C.G.; Vane, J.R. The discovery of captopril. *FASEB J.* **2003**, *17*, 788–789. [CrossRef] [PubMed]
119. Ma, R.; Mahadevappa, R.; Kwok, H.F. Venom-based peptide therapy: Insights into anti-cancer mechanism. *Oncotarget* **2017**, *8*, 100908–100930. [CrossRef]

120. Li, L.; Huang, J.; Lin, Y. Snake Venoms in Cancer Therapy: Past, Present and Future. *Toxins* **2018**, *10*, 346. [CrossRef] [PubMed]
121. Sukocheva, O.; Menschikowski, M.; Hagelgans, A.; Yarla, N.S.; Siegert, G.; Reddanna, P.; Bishayee, A. Current insights into functions of phospholipase A2 receptor in normal and cancer cells: More questions than answers. *Semin. Cancer Biol.* **2019**, *56*, 116–127. [CrossRef]
122. Peng, Z.; Chang, Y.; Fan, J.; Ji, W.; Su, C. Phospholipase A2 superfamily in cancer. *Cancer Lett.* **2021**, *497*, 165–177. [CrossRef]
123. Cathcart, J.; Pulkoski-Gross, A.; Cao, J. Targeting Matrix Metalloproteinases in Cancer: Bringing New Life to Old Ideas. *Genes Dis.* **2015**, *2*, 26–34. [CrossRef]
124. Di Cara, G.; Marabeti, M.R.; Musso, R.; Riili, I.; Cancemi, P.; Pucci Minafra, I. New Insights into the Occurrence of Matrix Metalloproteases-2 and -9 in a Cohort of Breast Cancer Patients and Proteomic Correlations. *Cells* **2018**, *7*, 89. [CrossRef] [PubMed]
125. Quintero-Fabian, S.; Arreola, R.; Becerril-Villanueva, E.; Torres-Romero, J.C.; Arana-Argaez, V.; Lara-Riegos, J.; Ramirez-Camacho, M.A.; Alvarez-Sanchez, M.E. Role of Matrix Metalloproteinases in Angiogenesis and Cancer. *Front. Oncol.* **2019**, *9*, 1370. [CrossRef]
126. Bezerra, P.H.A.; Ferreira, I.M.; Franceschi, B.T.; Bianchini, F.; Ambrosio, L.; Cintra, A.C.O.; Sampaio, S.V.; de Castro, F.A.; Torqueti, M.R. BthTX-I from Bothrops jararacussu induces apoptosis in human breast cancer cell lines and decreases cancer stem cell subpopulation. *J. Venom. Anim. Toxins Incl. Trop. Dis.* **2019**, *25*, e20190010. [CrossRef] [PubMed]
127. Murray, A.S.; Hyland, T.E.; Sala-Hamrick, K.E.; Mackinder, J.R.; Martin, C.E.; Tanabe, L.M.; Varela, F.A.; List, K. The cell-surface anchored serine protease TMPRSS13 promotes breast cancer progression and resistance to chemotherapy. *Oncogene* **2020**, *39*, 6421–6436. [CrossRef] [PubMed]
128. Krasny, L.; Huang, P.H. Data-independent acquisition mass spectrometry (DIA-MS) for proteomic applications in oncology. *Mol. Omics* **2021**, *17*, 29–42. [CrossRef] [PubMed]
129. Zhao, W.; Li, J.; Chen, M.M.; Luo, Y.; Ju, Z.; Nesser, N.K.; Johnson-Camacho, K.; Boniface, C.T.; Lawrence, Y.; Pande, N.T.; et al. Large-Scale Characterization of Drug Responses of Clinically Relevant Proteins in Cancer Cell Lines. *Cancer Cell* **2020**, *38*, 829–843.e4. [CrossRef]
130. Klaeger, S.; Heinzlmeir, S.; Wilhelm, M.; Polzer, H.; Vick, B.; Koenig, P.A.; Reinecke, M.; Ruprecht, B.; Petzoldt, S.; Meng, C.; et al. The target landscape of clinical kinase drugs. *Science* **2017**, *358*. [CrossRef]
131. Qi, Z.; Zhang, L.; Chen, Y.; Ma, X.; Gao, X.; Du, J.; Zhang, F.; Cheng, X.; Cui, W. Biological variations of seven tumor markers. *Clin. Chim. Acta* **2015**, *450*, 233–236. [CrossRef]
132. Lamond, A.I.; Mann, M. Cell biology and the genome projects a concerted strategy for characterizing multiprotein complexes by using mass spectrometry. *Trends Cell Biol.* **1997**, *7*, 139–142. [CrossRef]
133. Tavares, C.; Maciel, T.; Burin, S.; Ambrosio, L.; Ghisla, S.; Sampaio, S.; Castro, F. l-Amino acid oxidase isolated from Calloselasma rhodostoma snake venom induces cytotoxicity and apoptosis in JAK2V617F-positive cell lines. *Rev. Bras. Hematol. Hemoter.* **2016**, *38*, 128–134. [CrossRef]
134. Zakraoui, O.; Marcinkiewicz, C.; Aloui, Z.; Othman, H.; Grepin, R.; Haoues, M.; Essafi, M.; Srairi-Abid, N.; Gasmi, A.; Karoui, H.; et al. Lebein, a snake venom disintegrin, suppresses human colon cancer cells proliferation and tumor-induced angiogenesis through cell cycle arrest, apoptosis induction and inhibition of VEGF expression. *Mol. Carcinog.* **2017**, *56*, 18–35. [CrossRef] [PubMed]
135. Bernardes-Oliveira, E.; Gomes, D.L.; Martelli Palomino, G.; Juvenal Silva Farias, K.; da Silva, W.D.; Rocha, H.A.; Goncalves, A.K.; Fernandes-Pedrosa, M.F.; Crispim, J.C. Bothrops jararaca and Bothrops erythromelas Snake Venoms Promote Cell Cycle Arrest and Induce Apoptosis via the Mitochondrial Depolarization of Cervical Cancer Cells. *Evid. Based Complement. Altern. Med.* **2016**, *2016*, 1574971. [CrossRef]
136. Knorre, D.G.; Kudryashova, N.V.; Godovikova, T.S. Chemical and functional aspects of posttranslational modification of proteins. *Acta Nat.* **2009**, *1*, 23.
137. Freed, E.F.; Bleichert, F.; Dutca, L.M.; Baserga, S.J. When ribosomes go bad: Diseases of ribosome biogenesis. *Mol. Biosyst* **2010**, *6*, 481–493. [CrossRef]
138. Ruggero, D.; Pandolfi, P.P. Does the ribosome translate cancer? *Nat. Rev. Cancer* **2003**, *3*, 179–192. [CrossRef]
139. Bilanges, B.; Stokoe, D. Mechanisms of translational deregulation in human tumors and therapeutic intervention strategies. *Oncogene* **2007**, *26*, 5973–5990. [CrossRef] [PubMed]
140. Hershko, A. Ubiquitin: Roles in protein modification and breakdown. *Cell* **1983**, *34*, 11–12. [CrossRef]
141. Hershko, A.; Ciechanover, A. The ubiquitin system. *Annu. Rev. Biochem.* **1998**, *67*, 425–479. [CrossRef]
142. Jankowska, E.; Stoj, J.; Karpowicz, P.; Osmulski, P.A.; Gaczynska, M. The proteasome in health and disease. *Curr. Pharm. Des.* **2013**, *19*, 19.
143. Pajonk, F.; McBride, W.H. The Proteasome in Cancer Biology and Treatment. *Radiat. Res.* **2001**, *156*, 447–459. [CrossRef]
144. Teicher, B.A.; Ara, G.; Herbst, R.; Palombella, V.J.; Adams, J. The proteasome inhibitor PS-341 in cancer therapy. *Clin. Cancer Res.* **1999**, *5*, 2638–2645. [PubMed]
145. Thul, P.J.; Lindskog, C. The human protein atlas: A spatial map of the human proteome. *Protein Sci.* **2018**, *27*, 233–244. [CrossRef]
146. Dias, M.H.; Fonseca, C.S.; Zeidler, J.D.; Albuquerque, L.L.; da Silva, M.S.; Cararo-Lopes, E.; Reis, M.S.; Noel, V.; Dos Santos, E.O.; Prior, I.A.; et al. Fibroblast Growth Factor 2 lethally sensitizes cancer cells to stress-targeted therapeutic inhibitors. *Mol. Oncol.* **2019**, *13*, 290–306. [CrossRef] [PubMed]

147. Montoni, F.; Andreotti, D.Z.; Eichler, R.; Santos, W.D.S.; Kisaki, C.Y.; Arcos, S.S.S.; Lima, I.F.; Soares, M.A.M.; Nishiyama-Jr, M.Y.; Nava-Rodrigues, D.; et al. The impact of rattlesnake venom on mice cerebellum proteomics points to synaptic inhibition and tissue damage. *J. Proteom.* **2020**, *221*, 103779. [CrossRef]
148. Mi, H.; Muruganujan, A.; Huang, X.; Ebert, D.; Mills, C.; Guo, X.; Thomas, P.D. Protocol Update for large-scale genome and gene function analysis with the PANTHER classification system (v.14.0). *Nat. Protoc.* **2019**, *14*, 703–721. [CrossRef] [PubMed]
149. Mi, H.; Poudel, S.; Muruganujan, A.; Casagrande, J.T.; Thomas, P.D. PANTHER version 10: Expanded protein families and functions, and analysis tools. *Nucleic Acids Res.* **2016**, *44*, D336–D342. [CrossRef]
150. Thomas, P.D.; Campbell, M.J.; Kejariwal, A.; Mi, H.; Karlak, B.; Daverman, R.; Diemer, K.; Muruganujan, A.; Narechania, A. PANTHER: A library of protein families and subfamilies indexed by function. *Genome Res.* **2003**, *13*, 2129–2141. [CrossRef]
151. Thomas, P.D.; Kejariwal, A.; Campbell, M.J.; Mi, H.; Diemer, K.; Guo, N.; Ladunga, I.; Ulitsky-Lazareva, B.; Muruganujan, A.; Rabkin, S.; et al. PANTHER: A browsable database of gene products organized by biological function, using curated protein family and subfamily classification. *Nucleic Acids Res.* **2003**, *31*, 334–341. [CrossRef]
152. Cox, J.; Hein, M.Y.; Luber, C.A.; Paron, I.; Nagaraj, N.; Mann, M. Accurate proteome-wide label-free quantification by delayed normalization and maximal peptide ratio extraction, termed MaxLFQ. *Mol. Cell Proteom.* **2014**, *13*, 2513–2526. [CrossRef] [PubMed]
153. Tyanova, S.; Temu, T.; Sinitcyn, P.; Carlson, A.; Hein, M.Y.; Geiger, T.; Mann, M.; Cox, J. The Perseus computational platform for comprehensive analysis of (prote)omics data. *Nat. Methods* **2016**, *13*, 731–740. [CrossRef] [PubMed]
154. Szklarczyk, D.; Franceschini, A.; Wyder, S.; Forslund, K.; Heller, D.; Huerta-Cepas, J.; Simonovic, M.; Roth, A.; Santos, A.; Tsafou, K.P.; et al. STRING v10: Protein-protein interaction networks, integrated over the tree of life. *Nucleic Acids Res.* **2015**, *43*, D447–D452. [CrossRef]
155. Lê, S.; Josse, J.; Husson, F. FactoMineR: AnRPackage for Multivariate Analysis. *J. Stat. Softw.* **2008**, *25*. [CrossRef]

Review

Fish Cytolysins in All Their Complexity

Fabiana V. Campos [1], Helena B. Fiorotti [1,2], Juliana B. Coitinho [1] and Suely G. Figueiredo [1,*]

1 Centro de Ciências da Saúde, Departamento de Ciências Fisiológicas, Universidade Federal do Espírito Santo (UFES), Av. Marechal Campos 1468, Vitória 29040-090, Brazil; fabiana.v.campos@ufes.br (F.V.C.); helena.fiorotti@esib.butantan.gov.br (H.B.F.); juliana.b.goncalves@ufes.br (J.B.C.)
2 Laboratório de Bioquímica, Instituto Butantan, Av. Vital Brasil 1500, São Paulo 05503-900, Brazil
* Correspondence: suely.figueiredo@ufes.br; Tel.: +55-27-3335-7349; Fax: +55-27-3335-7342

Abstract: The majority of the effects observed upon envenomation by scorpaenoid fish species can be reproduced by the cytolysins present in their venoms. Fish cytolysins are multifunctional proteins that elicit lethal, cytolytic, cardiovascular, inflammatory, nociceptive, and neuromuscular activities, representing a novel class of protein toxins. These large proteins (MW 150–320 kDa) are composed by two different subunits, termed α and β, with about 700 amino acid residues each, being usually active in oligomeric form. There is a high degree of similarity between the primary sequences of cytolysins from different fish species. This suggests these molecules share similar mechanisms of action, which, at least regarding the cytolytic activity, has been proved to involve pore formation. Although the remaining components of fish venoms have interesting biological activities, fish cytolysins stand out because of their multifunctional nature and their ability to reproduce the main events of envenomation on their own. Considerable knowledge about fish cytolysins has been accumulated over the years, although there remains much to be unveiled. In this review, we compiled and compared the current information on the biochemical aspects and pharmacological activities of fish cytolysins, going over their structures, activities, mechanisms of action, and perspectives for the future.

Keywords: fish venoms; cytolysins; multifunctionality; pore formation

Key Contribution: In this review, we discussed the current knowledge on structural and physio-pharmacological properties of fish cytolysins, the multifunctional toxins responsible for the major symptoms observed upon envenomation by fish. In addition, we attempted to provide some insights into unknown aspects of their multi-mechanistic mode of action, which may be useful for future studies on these interesting molecules.

Citation: Campos, F.V.; Fiorotti, H.B.; Coitinho, J.B.; Figueiredo, S.G. Fish Cytolysins in All Their Complexity. *Toxins* **2021**, *13*, 877. https://doi.org/10.3390/toxins13120877

Received: 12 November 2021
Accepted: 4 December 2021
Published: 9 December 2021

1. Introduction

Hundreds of venomous ray-finned fish species (Actinopterygii) are distributed in tropical and temperate marine and freshwater environments worldwide, being commonly associated with human injuries. Although this significant biodiversity translates into a promising source of bioactive compounds, the study of fish venoms is rather underrepresented in the literature [1]. Nevertheless, the number of accidents caused by venomous fish is by no means small, and some can have serious consequences and even be fatal [2,3].

Venomous fish species belonging to the order Scorpaeniformes, which includes the families Synanceiidae (e.g., stonefish and devil stinger), Scorpaenidae (e.g., scorpionfish and lionfish), and Tetrarogidae (e.g., waspfish) [4] have been frequently associated with envenomation events [3]. Weeverfish from the family Trachinidae, order Perciformes, are also worthy of note [5] (Figure 1).

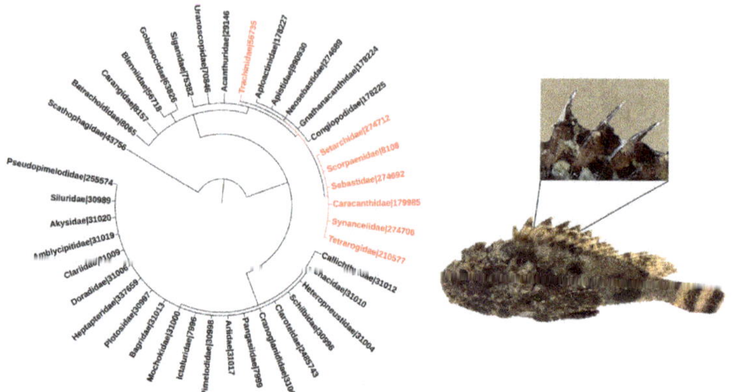

Figure 1. Phylogeny of venomous ray-finned fish families. The families included in the tree were listed based on the minimum estimate of venomous species according to [1]. Families belonging to the order Scorpaeniformes are highlighted in red, and so is the family Trachinidae from the order Perciformes. Numbers refer to NCBI taxonomic IDs. The phylogenetic tree was generated using the Interactive Tree of Life (iTOL–https://itol.embl.de; last accessed on 3 November 2021) tool [6]. The images adjacent to the phylogenetic tree depict a specimen of the scorpionfish *Scorpaena plumieri*, a member of the Scorpaenidae family, zooming in its dorsal venomous spines.

The venom apparatus of these fish is similar, and so are the biological activities associated with their venoms. The apparatus invariably consists of mucous-covered spines that can be located in different regions of the fish, being more common in the dorsal region (Figure 1), although several species have pectoral fins modified as a venom injection system. The secretory system is found in the anterolateral cavities of the spines, and may comprise well-defined glands or a set of more primitive, specialized secretory cells. Envenomation occurs when the victim threads or touches the spines, tearing the mucous sheath by mechanical pressure and causing the venom to be released into the wounds [7–9].

Excruciating pain is the most prevalent symptom of envenomation by fish, although local inflammation and a number of systemic responses, of which the most severe are cardiovascular and respiratory disturbances, are also observed [3,9,10]. Although envenomation by fish is rarely fatal to humans, under certain experimental conditions fish venoms are invariably lethal to animals [11–15].

Fish use their venoms mostly for defensive purposes, and the molecular composition of these venoms—combined with the evolution of strategies, such as mimicry and aposematism—reflects this fact [16]. Several bioactive compounds, including enzymes, such as hyaluronidases and proteases, large proteinaceous toxins, lectins, and peptides, among others, have been identified in fish venoms [17,18]. However, these venoms are somewhat unique in the sense that the biological activities underlying the major symptoms of envenomation are all induced by members of a high-molecular weighted protein family present in these venoms.

When first described, these extremely labile toxic components were referred to as "lethal factors" [12,19]. Physio-pharmacological studies on these lethal factors confirmed their multifunctional nature. These toxins reproduce the pain, inflammation, neurotoxic, myotoxic, and cardiotoxic effects induced by the crude venoms (for review, see [20,21]). In spite of this multifunctionality, the lethal activity associated with these toxins is attributed mainly to cardiac collapse and respiratory arrest [15,19,22].

In addition to the aforementioned effects, the multifunctional lethal factors also show strong hemolytic activity in vitro—a common feature in scorpaenoid fish venoms [13,19,23–28]; therefore, they are more accurately called membrane-damaging toxins or cytolysins [25]. A few cytolysins have been successfully identified in fish venoms and had some of their structural and functional features described (Table 1).

Table 1. Venomous fish and their cytolysins.

Family	Species	Cytolysin	Activity Described
Synanceiidae			
Estuarine stonefish	*Synanceia horrida*	SNTX *-Stonustoxin	Lethal [12] Hemolytic [12] Cardiovascular [22,29,30] Neuromuscular [31] Edematogenic [12]
	Synanceia trachynis	TLY *-Trachynilysin	Lethal [25] Hemolytic [32] Cardiovascular [33] Neuromuscular [32,34,35]
Reef stonefish	*Synanceia verrucosa*	VTX *-Verrucotoxin neoVTX *-Neoverrucotoxin	Lethal [19] Hemolytic [19] Cardiovascular [36–38] Lethal [39] Hemolytic [39]
Devil stinger	*Inimicus japonicus*	IjTx **-*I. japonicus* toxin	Hemolytic [40]
Scorpaenidae			
Scorpionfish	*Scorpaena plumieri*	Sp-CTx *-*S. plumieri* cytolytic toxin	Lethal [15] Hemolytic [41,42] ardiovascular [15,41,42] Edematogenic [43] Nociceptive [43]
	Scorpaena guttata	Unnamed toxin **	Lethal [44]
	Sebastapistes strongia	SsTx ***-*S. strongia* toxin [45]	-
	Scorpaenopsis oxycephala	SoTx ***-*S. oxycephala* toxin [45]	-
	Dendrochirus zebra	DzTx ***-*D. zebra* toxin [45]	-
	Sebasticus marmoratus	SmTx ***-*S. marmoratus* toxin [45]	-
Lionfish	*Pterois volitans*	PvTx ***-*P. volitans* toxin [14]	-
	Pterois antennata	PaTx ***-*P. antennata* toxin [14]	-
	Pterois lunulata	PlTx **-*P. lunulata* toxin	Hemolytic [40]
Trachinidae			
Greater weeverfish	*Trachinus draco*	Dracotoxin *	Lethal [13] Hemolytic [13]
Lesser weeverfish	*Echiichthys vipera* #	Trachinine *	Lethal [11]
Tetrarogidae			
Waspfish	*Hypodytes rubriprinnis*	HrTx **-*H. rubriprinnis* toxin	Hemolytic [40]

(*) purified protein; (**) semi purified lethal/hemolytic large protein; (***) identified at the cDNA level only; (#) also known as *Trachinus vipera*.

The wide spectrum of biological activities displayed by fish cytolysins implies that they play a major role in the envenomation process, which makes them subjects worthy of investigation. In this review, we will present a detailed overview of the biochemical and functional features of the known fish cytolysins. These toxins will be compared as to their structural and functional differences and similarities, and what has been put forward so far regarding their mechanisms of action will be discussed.

2. The Search for the Lethal Factor: Identification and/or Purification of Fish Cytolysins

Fish venoms can be collected through different methods, which may vary depending on the presence of defined glands or more primitive secretory systems. For those with defined glands—e.g., stonefish—the venom can be collected through puncture of the gland after removal of the tegument [46]. For those with primitive secretory cells—e.g., scorpionfish—the venom is usually obtained through the batch method, in which the spines are stripped and immersed in buffered solution, or the aspiration method, in which the spines are stripped and the venom aspirated from the grooves [44]. More recent methods, such as the sponge-in-a-tube method [47], in which a microtube containing a sponge is pressed against the spine to rupture the tegument and collect the venom, have the advantage of not requiring the removal of the spines and the sacrifice of specimens.

The search for a lethal factor in fish venoms started with the study of stonefish species, the most venomous of them all. The nature of the molecule responsible for the lethal activity associated with the venom of *Synanceia horrida* was first glimpsed in 1961, when it was found that only one of the seven protein fractions obtained through starch gel electrophoresis of the venom contained the heat-labile lethal material [46]. This fraction was almost twice as lethal as the crude venom when injected intravenously (i.v.) into mice tails (LD_{50}: 18 versus 30 μg of nitrogen/kg). Ten years later, the lethal activity of the venom of the scorpionfish *Scorpaena guttata* was associated to a semi-purified instable fraction, which was almost three times as lethal as the crude venom (LD_{50}: 0.9 versus 2.8 mg/kg, i.v.) [44]. The extreme lability of these fractions hindered the establishment of isolation processes, representing a major bottleneck for their research, which remained stagnant for many years.

By the end of the 20th century, when some of the difficulties imposed by the aforesaid lability of fish cytolysins were overcome through the establishment of proper purification and storage conditions [13,25,41,43], the biochemical and pharmacological characterization of these toxins was greatly accelerated. In addition, the combination of different methods, including liquid chromatography, cDNA cloning, automated Edman degradation, mass spectrometry (MS), and X-ray crystallography resulted in significant advances in the biochemical characterization of these molecules.

Cytolysins have been successfully purified from fish venoms through chromatographic steps using gel filtration, ion exchange, hydrophobic interaction, or adsorption, combined or not with fractionation by saline precipitation. Hemolytic and/or lethal activities were usually employed as a way of tracking the success of the purification process of these proteins [11–13,19,32,39,41].

From the venoms of the stonefish *S. horrida* and *Synanceia trachynis*, were purified stonustoxin (SNTX) [12] and trachynilysin (TLY) [32], respectively. The native molecular weights of SNTX and TLY were estimated in 148 kDa and 158 kDa, respectively. Both are heterodimeric proteins composed of two distinct subunits, named α and β, with masses of 71 kDa and 79 kDa for SNTX, and 76 kDa and 83 kDa for TLY. SNTX accounted for 9% of the protein content of the crude venom and was 22 times more lethal to mice (LD_{50} 0.017 μg/g; i.v.) [12,25,32].

Two different cytolysins, verrucotoxin (VTX) and neoverrucotoxin (neoVTX), were isolated from the venom of the reef stonefish *Synanceia verrucosa* [19,39]. VTX is a 320-kDa glycoprotein that in native form is organized in a tetrameric scaffold comprising two 83-kDa α and two 78-kDa β subunits [19]. It was twice as potent as the crude venom, being immediately lethal to mice at less than 60 ng/g (i.v.). NeoVTX, on the other hand, is a dimeric 166-KDa protein that comprises two distinct subunits (75-kDa α and 80-kDa β) and lacks carbohydrate moieties, showing structural features comparable to those of SNTX and TLY, while considerably differing from VTX [39]. It was also lethal to mice, with an LD_{50} of 47 μg/kg (i.v.).

The venom of the spotted scorpionfish *Scorpaena plumieri*, although considerably less harmful to humans than that of stonefish, also contains a lethal factor, named *S. plumieri* cytolytic toxin (Sp-CTx), which was first purified with very low yields (1%) [41]. Sp-CTx was then shown to be 12.3-fold more hemolytic (EC_{50} 56 ng/mL) than the crude venom. A molecular mass of 121 kDa was estimated for the native Sp-CTx, while a mass of 65,251 Da *m/z* was revealed by its mass spectrum, pointing to it having a dimeric nature with subunits of very similar molecular masses, much like SNTX and TLY [41]. However, unlike TLY and SNTX, Sp-CTx is a glycoprotein, displaying typical N- and O- linked glycoconjugate residues [41,43]. Three years later, the same group optimized the purification protocol of Sp-CTx and this new method increased the final yield by 13-fold when compared to the previous one [42]. The dimeric nature of Sp-CTx was then confirmed by cross-linking studies using bis-(sulfosuccinimidyl) suberate (BS3). Although lethality was not directly assessed in either of these previous reports, Sp-CTx was proved to be lethal when experiments conducted on anesthetized rats revealed that the animals eventually died upon receiving 70 μg/kg (i.v.) of the toxin [15].

In addition to the aforementioned toxins—which are the best chemically and functionally characterized fish cytolysins—a few such molecules have been purified from the venoms of other fish species.

From the lesser (*Echiichthys vipera*—also known as *Trachinus vipera*) and greater (*Trachinus draco*) weeverfish venoms were purified the cytolysins trachinine [11] and dracotoxin [13], respectively. Trachinine was purified by preparative electrophoresis and shown to be a 324-kDa molecule of tetrameric nature (81-kDa subunits), similar to VTX. It showed an LD_{100} of < 100 μg/kg in mice (i.v.) [11]. Dracotoxin, unlike all the other known fish cytolysins, is a monomeric protein with MW estimated in 105 kDa both by SDS-PAGE and gel filtration. It was purified with high yields (36%) and it was lethal to mice (minimum lethal dose of 11 μg/g; i.v.) and ~20-fold more hemolytic than the crude venom of *T. draco* [13].

Semi-purified hemolytic fractions were obtained from lionfish (*Pterois lunulata*), devil stinger (*Inimicus japonicus*), and waspfish (*Hypodytes rubripinnis*) species. Based on the data obtained through gel filtration, immunoblotting, and cDNA cloning, these toxins were determined to be 160-kDa heterodimers composed of 80-kDa α and β subunits [40].

In spite of some variation as to the oligomeric functional state assumed by fish cytolysins purified so far, they are all high-molecular-mass proteins that appear to be active when assembled into oligomers, which could explain their extreme lability. Regarding the quaternary scaffold of these toxins, it is worthy of note that those described as dimeric had their native masses estimated by gel filtration chromatography, while, incidentally, those deemed tetrameric had their masses estimated by gel electrophoresis using the method described by [48].

Lethal and hemolytic activities have also been described in the venoms of the lionfish species *Pterois antennata* and *Pterois volitans* [14], although the actual molecules responsible for these activities have not yet been purified. Nevertheless, both venoms contain toxins identified through cDNA cloning and immunoblotting that share high sequence identity with stonefish cytolysins [14]. The venoms of the scorpaenoid fish species *Sebastapistes strongia*, *Scorpaenopsis oxycephala*, *Sebasticus marmoratus*, and *Dendrochirus zebra* also contain toxins, whose sequences were deduced from cDNA and genomic DNA data, that showed similarity with known fish cytolysins [45].

Finally, several studies have reported antigenic cross-reactivity between fish venoms. The antivenom raised against the venom of *S. trachynis* (SFAV)—the only commercially available fish antivenom—is effective in neutralizing not only the in vivo and in vitro effects of *S. trachynis* venom, but also those of other fish venoms [24,49–51]. In addition, it has been shown that the cytolysins isolated so far cross-react with SFAV and their effects can be neutralized by this antivenom [41,51], suggesting a close similarity between these molecules.

3. Analyzing the Structural Features of Fish Cytolysins

Before comparing and discussing the biological activities of different fish cytolysins, we will describe what is known so far regarding their structures. This overview should provide a useful background for a better understanding of such activities, particularly regarding the hemolytic activity associated with these toxins.

The complete amino acid sequences of several fish cytolysins were determined by cDNA cloning and/or genomic sequencing [13,14,39,40,45,52–54] (Figure 2).

Chain α

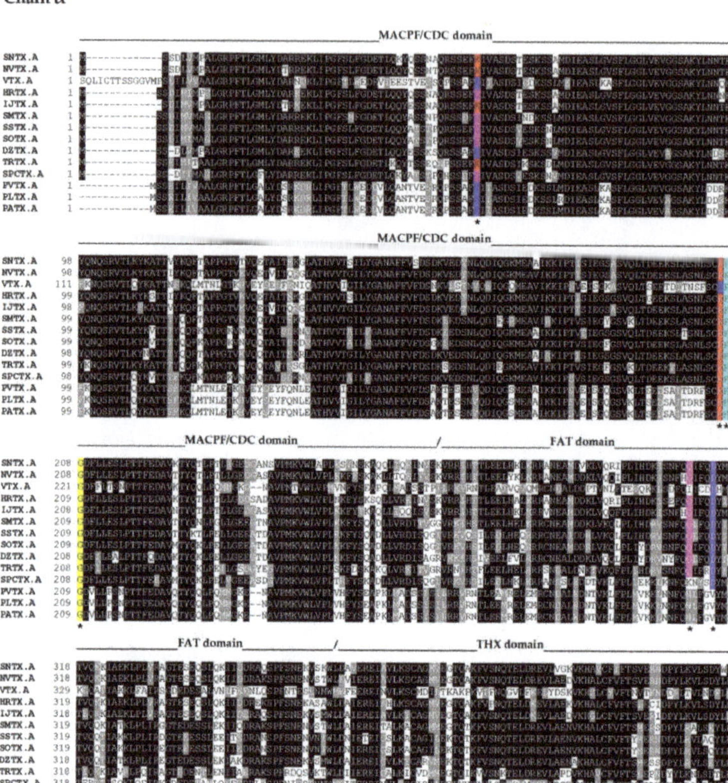

Figure 2. *Cont.*

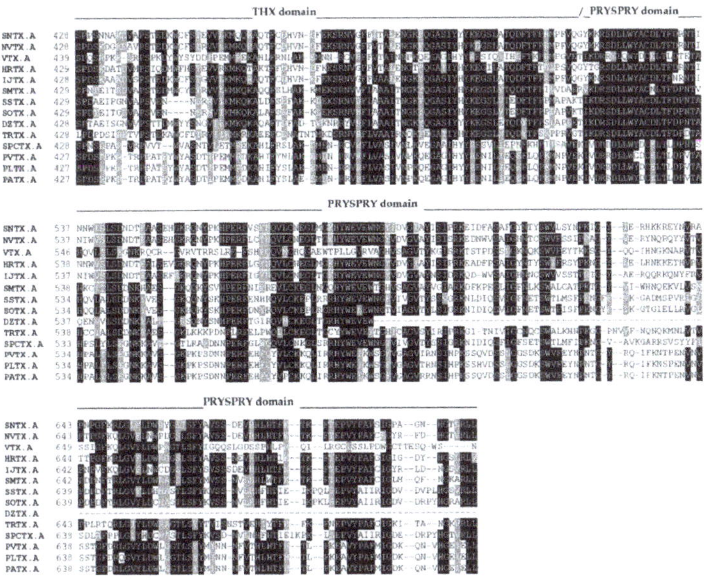

Chain β

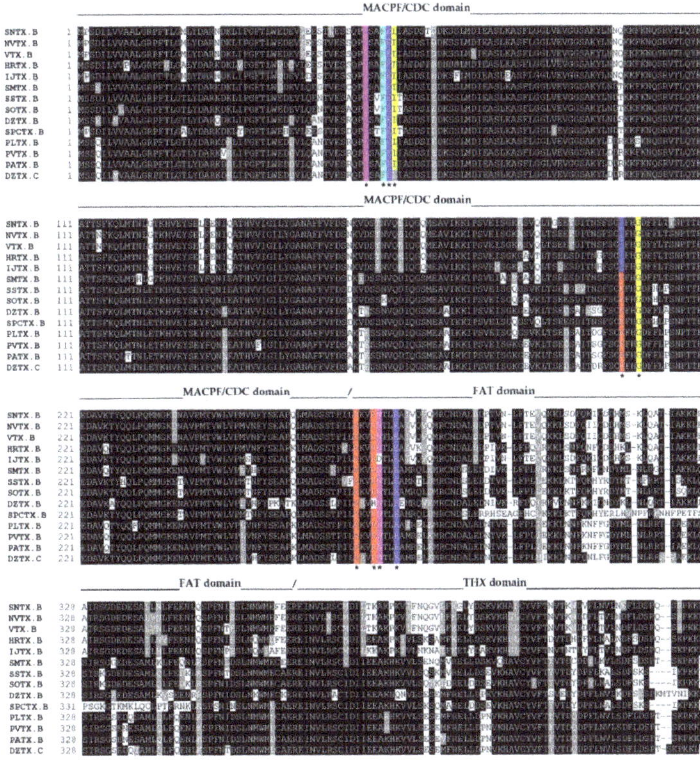

Figure 2. *Cont.*

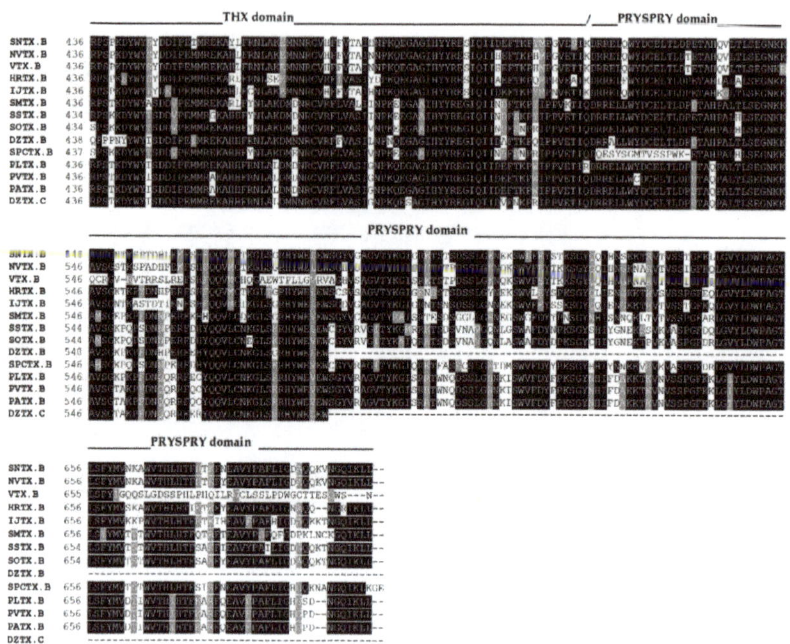

Figure 2. Primary sequence alignment of α and β chains of the cytolysins SNTX [A (Q98989); B (Q91453)] from *S. horrida*, VTX [A (CAA69254.1-partial); B (Q98993)] and neoVTX [A (A0ZSK3), B (A0ZSK4)] from *S. verrucosa*, HrTx [A (BAM74459.1); B (BAM74460.1)] from *H. rubriprinnis*, IjTx [A (BAM74457.1); B (BAM74458.1)] from *I. japonicus*, SmTx [A (AIC84040.1); B (AIC84041.1)] from *S. marmoratus*, SsTx [A (AIC84036.1); B (AIC84037.1)] from *S. strongia*, SoTx [A (AIC84038.1); B (AIC84039.1)] from *S. oxycephala*, DzTx [A (AIC84042.1–partial); B (AIC84043.1–partial)] from *D. zebra*, trachinine [A (AHY22717.1)] from *E. vipera*, Sp-CTx [A (A0A2P1BRQ0), B (A0A2P1BRP3)] from *S. plumieri*, PlTx [A (BAM74455.1); B (BAM74456.1)] from *P. lunulata*, PvTx [A (BAK18814.1); B (BAK18815.1)] from *P. volitans*, and PaTx [A (BAK18812.1); B (BAK18813.1)] from *P. antennata*. Alignments were performed on Muscle server. The sequence of dracotoxin from *T. draco* was not found. The species *S. trachynis* has been reclassified as a synonym of *S. horrida*, so that the sequence of TLY is reported as an alternative name for SNTX, under the same access code. Trachinine is referred to as echiitoxin on NCBI and only the α chain was found. Amino acid residues mentioned at the text are colored as yellow (nonpolar side chain); pink (polar side chain); blue (negatively charged side chain); red (positively charged side chain); and cyan (aromatic side chain) and marked with *. Domain marking was performed based on SNTX structure [55].

These sequences, along with others totaling one hundred sequences (cover > 83%), were acquired by submitting the primary sequences of the α and β subunits of SNTX to the BLASTP algorithm [56] on the NCBI webserver (https://blast.ncbi.nlm.nih.gov/Blast.cgi; last accessed on 6 October 2021), employing the non-redundant protein sequences (nr) database from fish (taxid 7898).

Fish cytolysins share high primary sequence similarity, with identity ranging from 45 to 94% and ~20% of residues present at conserved positions (Figure 2), suggesting a strong structural correlation between these molecules. However, only SNTX had its three-dimensional structure determined [55]. The crystallographic model (3.1 Å resolution) of SNTX (Figure 3A) showed that the α and β chains form a stable dimer with an extensive parallel interface along its longitudinal axis, maintained through polar interactions, such as hydrogen bonds and electrostatic interactions.

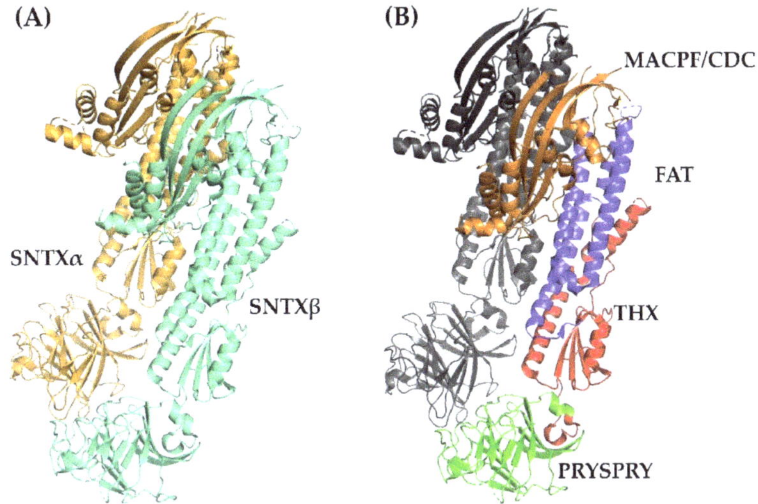

Figure 3. Crystallographic model of stonustoxin (SNTX). (**A**)—SNTX heterodimer showing the α (SNTXA-orange) and β (SNTXB-green) chains. (**B**)—SNTX domains: N-terminal MACPF/CDC domain (residues 1–265-orange); FAT domain (residues 266-385-blue); THX domain (residues 386-517-red), and C-terminal PRISPY domain (residues 518-703-green). Figure produced on Pymol using the model 4wvm deposited in the Protein Data Bank [55].

Fold recognition searches [55] showed that SNTX belongs to a branch of the "perforin-like" superfamily and revealed the presence of four domains (Figures 2 and 3B) in each chain: (i) N-terminal MACPF/CDC domain (residues 1–265), homologous to the Membrane Attack Complex -Perforin/Cholesterol-Dependent Cytolysin domain. This domain is found in several species and represents a superfamily of pore-forming toxins; (ii) FAT domain (residues 266–385), with high structural similarity to the focal adhesion-targeting domain of human kinase-1. This domain is also found in several proteins and plays a role in the assembly of signaling complexes; (iii) THX domain (residues 386–517), with high structural similarity with thioredoxin-3 (THX-3) from *Saccharomyces cerevisiae*; and (iv) C-terminal domain (residues 518–703), which is similar to the PRYSPRY (PRY SPla and the RYanodine Receptor) domain, member of the tripartite motif family (TRIM) that participates in immune recognition in intracellular bacteria and viruses.

The sequences of the other fish cytolysins were analyzed using the Conserved Domains tool of the NCBI server (https://www.ncbi.nlm.nih.gov/Structure/cdd/wrpsb.cgi; last accessed on 6 October 2021) (Figure 2). Much like SNTX, they all showed the presence of the C-terminal SPRY (residues from 500 to 700, approximately) and the THX (residues from 380 to 500, approximately) domains on each subunit.

The similarity between these cytolysins and SNTX allows us to infer that some of the features described for the latter are shared among them all. For instance, the THX domain—which is usually associated with redox reactions—most likely plays a structural role, as the key cysteines of the THX catalytic motif are not conserved in these molecules. In addition, there appears to be electrostatic complementarity between the α and β chains through highly conserved charged residues (D314 and Q310 in SNTX-α; R269, N273, R272, and E276 in SNTX-β), which most likely participate in the polar interactions that take place in the interface between the chains. These interactions can be disrupted under the conditions used in purification and mass estimation processes, which may also account for the differences observed in the quaternary arrangements of fish cytolysins. Finally, four cysteine residues are conserved in all cytolysins analyzed, pointing to a possible role in heterodimer stability through disulfide bridges. However, the formation of these bonds was not observed in the three-dimensional structure of SNTX [55].

4. The Ubiquitous Hemolytic Activity of Fish Cytolysins: From the Definition of the Pore-Forming Mechanism to its Structural Confirmation

The cytolytic activity of almost all fish cytolysins isolated so far has been extensively characterized against erythrocytes, hence their being referred to as hemolysins in some instances. In spite of some divergence in the literature as to the potency and the species—specificity of the hemolytic activity of fish cytolysins, which can by at least partly explained by different experimental conditions—they are all most potent against rabbit erythrocytes [13,24,25], although the reason for such marked sensitivity remains unknown. In addition, a weak hemolytic activity against rat, mice, and cattle erythrocytes has been reported for some fish cytolysins, while those from human, sheep, pig, and chicken appear to be resistant to the lytic activity of these toxins [12,13,25].

There is also some evidence as to the lytic activity of fish cytolysins against other cell types. For instance, the prolonged exposure of neuro-glioblastoma NG108-15 cells to TLY led to an irreversible increase in membrane permeability [57]. In addition, SNTX lysed the membranes of platelets in rabbit blood in a dose-dependent way [58].

Cytolytic activity is often associated with the action of enzymes such as proteases and phospholipases and, in some cases, of C-type lectins such as CEL-III from the sea cucumber *Cucumaria echinata* [59,60]. However, fish venoms lack phospholipase activity [24,26,27,61–63] and, although enzymatic activity such as proteolytic and hyaluronidase has been described in these venoms, no such activity was associated with the purified lethal/hemolytic factors [12]. As to C-type lectins, although such molecules have been described in fish venoms [64], they are not associated with the hemolytic activity induced by these venoms [65]. Therefore, fish cytolysins were very soon believed to lyse erythrocytes through a direct, non-enzymatic mechanism [26].

The formation of non-selective transmembrane pores was soon shown to be the mechanism by which fish cytolysins destroy cells [42,57,66]. The pore-forming activities of SNTX and Sp-CTx were demonstrated through osmotic assays, in which the kinetics of the hemolysis was evaluated in the presence of osmotic protectants of different sizes [42,66]. While the smaller compounds raffinose and saccharose failed to prevent hemolysis, polyethylene glycols (PEGs) of ~1000–2000 reduced the rate of hemolysis induced by the toxins and PEGs > 3000 conferred almost full protection against it [42,66]. The pores formed by SNTX in the membrane of rabbit erythrocytes were estimated to be 2.5–3.2 nm in diameter [66].

The search for what mediates the interaction between fish cytolysins and the cell membrane began as soon as the first such toxins were purified. For instance, the hemolytic activity of the *T. draco* venom was found to be preceded by the binding of its hemolytic component to a protein receptor (glycophorin) on the surface of erythrocytes [13]. From then on, it became clear that cationic amino acid residues present in fish cytolysins are essential for their interaction with anionic, neutral or zwitterionic lipids in the membrane of erythrocytes, eventually leading to hemolysis.

In 1997, it was observed that SNTX no longer induced hemolysis when positively charged lysine and arginine residues in its surface were chemically modified, although the toxin's secondary structure remained unaffected [66]. The modification of cationic residues also inhibited the lethal activity associated with SNTX [67]. These data are in agreement with structural motifs observed in these toxins, for instance, an amphiphilic, ~20-residues long α-helix flanked by regions rich in basic residues was predicted in both α- and β-subunits of SNTX, and it is believed to be the cationic site crucial for the toxin's hemolytic activity [52].

As expected, the hemolytic activity of SNTX and neoVTX against rats and rabbit erythrocytes, respectively, was competitively inhibited by negatively charged lipids, most potently by cardiolipin and less so by phosphatidylserine and the gangliosides GM1 and GM2 [39,66]. Although cardiolipin was the most potent inhibitor, it is not present in the erythrocyte membrane, thus the hemolytic activity must be actually triggered by the electrostatic interaction between the toxins' cationic residues and anionic phospholipids such as phosphatidylserine, which is abundant in the membrane of erythrocytes [39]. It stands to reason that the species-specificity of the hemolytic activity of fish cytolysins could be related to the density of these lipids in the erythrocyte membrane.

The electrostatic interaction between anionic lipids and toxins is influenced by the number of negative charges and the pKa of acidic groups in the lipids, as indicated by the different inhibitory potencies of different lipids [66]. This conclusion is further supported by the fact that the maximal hemolytic activity of Sp-CTx is achieved between pH 8 and 9 [43]. Although the neutral lipids phosphatidylcholine, phosphatidylethanolamine, sphingomyelin, and cholesterol did not affect the hemolytic activity of SNTX and neoVTX [39,66], that which was induced by Sp-CTx in rabbit erythrocytes was inhibited by phosphatidylethanolamine, phosphatidylglycerol, and cholesterol [43]. Similar results regarding the role of differently charged lipids on the hemolytic activity were found for the venomous extract of the lionfish *P. volitans* [68], which is yet another indication that this venom contains cytolysins homologous to those identified in other fish species.

The presence of the MACPF/CDC domain in fish cytolysins [53,55] and the inhibitory effect of cholesterol on the hemolytic activity of Sp-CTx [43] suggest that this lipid is also required for the interaction between these toxins and the erythrocyte membrane, much like the MACPF/CDC proteins found in bacteria.

The hemolytic activity of fish cytolysins is influenced by factors other than the cationic residues present in their surfaces. That of Sp-CTx, for instance, is calcium dependent, being abolished by zinc ions [43]. Furthermore, five of the 15 cysteine residues and 10 of the 18 cysteine residues of SNTX and neoVTX, respectively, are free, and these free thiol groups, much like the tryptophan residues in the surface of SNTX, also play an important role in the hemolytic activity induced by these toxins [39,52,69].

All this evidence pointing to a pore-formation mechanism being responsible for the hemolytic activity of fish cytolysins was corroborated by the three-dimensional structure of SNTX. Pore-formation was shown to be the result of an interaction between the α- and β-chains along their longitudinal axis through charge complementarity between the MACPF/CDC domains. These domains—which are formed by twisted four-stranded antiparallel β-sheets with two adjacent bundles of α-helices (transmembrane helices—TMH)—interact on the heterodimer to form a soluble early pre-pore. After that, the TMH from each chain unwind to form a large, continuous β-sheet that comprises the β-barrel of the pore (Figure 4).

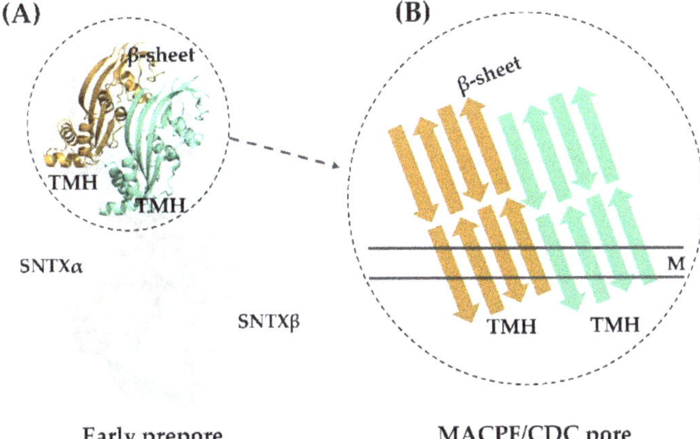

Figure 4. Transmembrane pore formation mechanism. The SNTX heterodimer represents an early and soluble phase of the pore formation mechanism (**A**). After the interaction between chains α and β, the MACPF/CDC domain helices undergo a conformational change to form the continuous β-sheet of the transmembrane pore (**B**). (**A**) was produced on Pymol using the model 4wvm deposited in the Protein Data Bank [55] and (**B**) was adapted from [55]. TMH—transmembrane helices; M—membrane.

This arrangement is in agreement with the complementarity between MACPF/CDC monomers described for other perforins, which directs initial oligomerization events (pre-pore assembly). In SNTX, this pre-pore contains 20 SNTX subunits (or 10 SNTX-α/β heterodimers) that line up along a horizontal plane [55].

Furthermore, the PRYSPRY domain was shown to be responsible for the initial interaction of SNTX with the cell surface. Consistent with this function, this domain is located in the solvent-exposed face of SNTX, in a position analogous to the Ig and C2 domains of CDC and perforins, which mediate protein–protein and protein–lipid interactions in the recognition of pathogens by TRIM immune proteins [55].

The high sequence similarity and the presence of conserved domains in the other fish cytolysins identified so far allows us to predict that these proteins perform their cytolytic functions through the same mechanism. For instance, the residues G208 (α-chain) and G209 (β-chain) of SNTX, reported as necessary for pre-pore formation, are present in all cytolysins. So are the residues F206 (α-chain) and F54, I56, and S52 (β-chain), related to the formation of hydrogen bonds between the β4-strand of SNTX-α and the β1-strand of SNTX-β, which are also involved in pre-pore formation (Figure 3).

Moreover, the pairs K205 (α-chain)-E55 (β-chain) and K54 (α-chain)-E206 (β-chain) present in SNTX's β4 (α-chain) and β1 (β-chain) strands form ionic pairs analogous to a partially closed zipper, which contributes to pore formation. These ionic pairs are conserved in almost all cytolysins, in spite of exchanges as to which residue belongs to each chain.

In addition to the structural evidence, the formation of large ring-shaped pores was visualized in rabbit erythrocytes by transmission electron microscopy (outer dimension: 257 ± 5.7 Å (mean ± SEM); lumen i.d.: 117 ± 4.5 Å) [55], in agreement with what had been predicted for the pores formed by SNTX and Sp-CTx [42,66].

5. Pharmacological Activities Associated to the Multifunctional Fish Cytolysins

5.1. Cardiovascular Activity

The most potent and best-studied pharmacological activity associated with the venoms of Scorpaeniformes is that on the cardiovascular system [9,23,28]. The profound cardiovascular alterations induced by these venoms are attributed to the action of cytolysins, being closely related to the lethality of these toxins [15,19,22].

The molecular mechanisms underlying the cardiovascular effects induced in vivo and in vitro by cytolysins from the venoms of stonefish and scorpionfish have been fairly well documented. Despite some variability, which could simply reflect the fact that different studies employed different doses and experimental models, it is a consensus that these toxins cause drastic changes in blood pressure, which are the result of their effects on the blood vessels and heart.

When the purification of VTX from the venom of *S. verrucosa* was first described, it was observed that the injection of this cytolysin (~16 μg/kg, i.v.) into anesthetized rats induced an immediate hypotensive effect that evolved to cardiac arrest [19]. A drastic, irrecoverable fall in the blood pressure of anesthetized rats has also been described for SNTX (20 μg/kg, i.v.) and Sp-CTx (70 μg/kg, i.v.), although a biphasic response characterized by an increase followed by a decrease on arterial pressure was observed in these instances [15,22].

Consistent with the hypotension observed in vivo, a significant vasorelaxation has been associated with the action of fish cytolysins on vascular preparations [22,29,30,41]. In addition, a biphasic response on rat aortic rings pre-contracted with phenylephrine, characterized by an endothelium- and dose-dependent relaxation followed by a contractile phase, was associated with both SNTX and Sp-CTx, albeit with different intensities. Regardless, the fact that both toxins induced relaxation of isolated aortic rings in the picomolar range (68 pM and 1 pM for SNTX and Sp-CTx, respectively), attests to the high hypotensive potency of these molecules [22,29,41].

The vasorelaxant phase of the effect induced by SNTX and Sp-CTx was irreversibly abolished by 25 and 10 μM of the nitric oxide (NO) synthesis inhibitor N^G-nitro-L-arginine methyl ester (L-NAME), respectively [22,41]. In addition, the relaxant response associated with SNTX was reversibly inhibited by the guanylate cyclase inhibitors methylene blue (10 μM) and hemoglobin (5 mM) [22]. It was also reduced by the H_2S inhibitors DL-propargylglycine (PAG; 1 mM) and β-cyano-L-alanine (BCA; 1 mM), being potentiated when these inhibitors were individually combined with L-NAME (1 μM) [30]. The K^+ channel blocker tetraethylammonium (TEA; 3 mM) and the substance P (SP) receptor antagonist N-acetyl-L-tryptophan-3,5-bis(trifluoromethyl)-benzyl ester (NATB; 0.5 mM) also reduced the SNTX-induced vasorelaxation, which was completely blocked by the iNOS inhibitor AMT-HCl (0.5 mM) [29].

Based on these results, it has been proposed that SNTX acts by binding directly to the SP receptor in the endothelial cell or by promoting SP release. The consequent increase in intracellular [Ca^{2+}] stimulates iNOS and increases the production of NO, which will be released by the endothelial cell and absorbed by the smooth muscle cell. There, it increases the levels of cGMP, which will then positively modulate K^+ channels, hyperpolarizing the membrane and causing relaxation [29]. The SNTX-induced relaxation also involves the synergistic action of H_2S and NO [30]. The high sequence identity and structural similarities between fish cytolysins, as well as the antigenic cross-reactivity displayed by these toxins [14,40,41], suggest that the aforementioned mechanism of action can be extended to cytolysins other than SNTX.

The effect of fish cytolysins on the heart itself has also been explored to a certain extent. In 1997, the cardiotoxicity of the freshly purified but unstable VTX (1 μg/mL) and an impure but more stable fraction named p-VTX (0.001–1 μg/mL) was assessed [36]. Both exerted negative inotropic and chronotropic effects on the peak contraction and accelerated the relaxation phase in frog atrial fibers [36]. The negative inotropic effect induced by p-VTX was reduced in higher external [Ca^{2+}], while the chronotropic effect remained unaffected in these conditions. Moreover, p-VTX hyperpolarized the fibers and decreased the duration of both the plateau and the hyperpolarizing phase of action potentials recorded in these fibers. Both muscular and electrical activities of the atrial fibers was counteracted by glibenclamide, an ATP-sensitive K^+ channel (K_{ATP}) blocker. The nature of these responses suggests p-VTX acts by inhibiting Ca^{2+} influx into the fibers, possibly by negatively modulating Ca^{2+} channels, and by positively modulating K_{ATP} channels.

On the other hand, studies conducted in guinea pig ventricular myocytes, employing electrophysiology and various pharmacological strategies, showed that, unlike what had been observed in atrial fibers, VTX inhibits K_{ATP} currents via the activation of the muscarinic M3 receptor-PKC pathway and increases L-type Ca^{2+} currents by activating the β-adrenergic-cAMP-PKA pathway [37,38]. As pointed out by [37], this discrepancy regarding the effects of VTX on frog atrial fibers and guinea pig ventricular cells could be simply the consequence of different species having different cardiac regulatory mechanisms. Accordingly, TLY (1 μg/mL) acted much like VTX in frog atrial fibers, hyperpolarizing the membrane, shortening the action potentials, and inducing a negative inotropic effect, in addition to a contracture due to acetylcholine release. However, this negative inotropic effect associated with TLY, although Ca^{2+}-dependent, was not blocked by Cd^{2+}, indicating it does not involve voltage-dependent Ca^{2+} channels [33].

In addition to its effects in vivo and on isolated vessels, Sp-CTx had its cardiotoxic activity assessed in rat isolated hearts and papillary muscles [15]. The toxin (10^{-9} to 5×10^{-6} M) induced a transient, concentration-dependent positive inotropic effect, characterized by a drastic increase in left ventricular pressure. It also induced vasoconstriction on the coronary bed by increasing vascular perfusion pressure. The biphasic pattern described in anesthetized rats and rat aortic rings [15,41] was not observed in isolated hearts.

In isolated papillary muscle, Sp-CTx (0.1 μM) increased myocardial contractility through a pathway involving the activation of β-adrenoreceptors, as the pre-treatment with the β-blocker propranolol and the catecholamine releasing agent tyramine inhibited this response [15]. Post-rest contraction experiments suggested that the cardiac response induced by Sp-CTx involves an increase in sarcolemmal Ca^{2+} influx, which was corroborated by the reversible 18% increase in L-type Ca^{2+} currents induced by 1 nM of this toxin in isolated ventricular myocytes [15]. Taken together, these results point to Sp-CTx acting by increasing Ca^{2+} influx into cardiac cells via modulation of L-type Ca^{2+} channels through the activation of the β-adrenergic signaling pathway, much like what had been described for VTX in guinea pig ventricular myocytes [38].

It has been suggested that the pore-forming mechanism underlying the hemolytic activity of fish cytolysins might also account for some of the other pharmacological activities associated with these toxins [22,55,57,66]. However, the aforementioned studies showing that SNTX, TLY, VTX, and Sp-CTx mediate their cardiovascular effects through complex pathways involving the release of vasorelaxant mediators and the modulation of membrane receptors and/or ion channels [15,33,36–38], point to pore-formation not being the underlying cause of these responses. It is also important to stress that, although seemingly conflicting depending on the experimental conditions, these pathways could actually contribute in a synergistic way to potentiate the cardiovascular toxicity attributed to these cytolysins.

5.2. Neuromuscular Activity

In addition to the cardiovascular effects induced by fish venoms, symptoms observed after envenomation, such as weakness, muscle spasms, and even paralysis, are an evident sign of neurotoxicity. Indeed, under experimental conditions, the crude venoms of *S. verrucosa, S. horrida (trachynis), P. volitans*, and *T. draco* have had their neurotoxic effects demonstrated [13,70–72].

At least for the estuarine stonefish venom, the neuromuscular activity could be associated directly to its cytolysins—SNTX and TLY. In 1994, the effects of SNTX (8 to 50 μg/mL) on the neuromuscular function were investigated in mouse and chick skeletal muscles in vitro and rat skeletal muscle in vivo [31]. SNTX produced an irreversible, concentration- and time-dependent block of nerve- and muscle-evoked twitches of the mouse hemidiaphragm. Similar effects were observed in the chick biventer cervicis muscle, in which nerve-evoked twitches and the contractures induced by acetylcholine (200 μM), carbachol (8 μM), and KCl (40 mM) were blocked by SNTX (22 and 44 μg/mL). The ability of SNTX to block muscle-evoked twitches, as well as the marked blockade of this effect by dantrolene sodium (6 μM)—a muscle relaxant that inhibits Ca^{2+} release from intracellular stores—but not by the nicotinic receptor antagonist tubocurarine (15 μM), suggested that SNTX acts directly on the muscle to inhibit neuromuscular function, inducing contractures via a Ca^{2+}-dependent rather than an acetylcholine-dependent mechanism. In addition, SNTX did not block conduction in the frog sciatic nerve, which also points to the conclusion that the toxin affects neuromuscular function via myotoxicity [31].

On the other hand, TLY—which had its neuromuscular activities rather well explored—was reported to act presynaptically by causing the release and depletion of neurotransmitters from nerve terminals [32], as had been previously shown for the crude venom itself [71]. This cytolysin (2 to 20 μg/mL) irreversibly increased the frequency of miniature end plate potentials in frog cutaneous pectoris preparations, indicating it stimulated the quantal release of acetylcholine [32]. A similar outcome was found in *Torpedo marmorata* neuromuscular junction, in which TLY (30 to 60 nM) increased the spontaneous quantal neurotransmitter release in an irreversible way [34].

This neurosecretory role attributed to TLY was supported by the observation that botulinum toxin—a neurotoxin that disturbs exocytosis—inhibits the release of catecholamines induced by this cytolysin (60 nM) in bovine adrenal chromaffin cells [35]. However, unlike in motor nerve endings, where TLY depleted the number of small vesicles almost completely (86%) without affecting that of large-dense core vesicles [32], in chromaffin cells, this cytolysin mediated a soluble N-ethylmaleimide-sensitive fusion protein attachment protein receptor (SNARE)-dependent release of catecholamines from large core vesicles [35].

As expected, this neurosecretory effect was Ca^{2+}-dependent, although voltage-gated Ca^{2+} channels do not seem to participate in the process, as the specific inhibition of various Ca^{2+} channel isoforms failed to affect this response in both neuromuscular junction and chromaffin cells [34,35]. Nevertheless, fluorescence-based evidence showed that TLY does promote a transient influx of Ca^{2+} into chromaffin cells in some other way, and its effect on neurosecretion was in fact blocked by the non-specific inhibitor La^{3+}. In addition, the toxin induced a more sustained, localized increase in intracellular Ca^{2+} levels through release from caffeine-sensitive intracellular stores [35].

Further evidence as to how TLY affects membrane permeability came to light in 2002, when the voltage-clamp technique was used to investigate the mechanism of action underlying the neurosecretory effects induced by this cytolysin [57]. The perfusion of neuroblastoma/glioma cells (NG108-15) with TLY (12 nM) for ~1 min increases membrane conductance, by means of an inward cationic current that was inhibited by anti-TLY antibodies and La^{3+}. A lack of ionic selectivity was confirmed by the current reversing around 0 mV. The rather complex nature of the macroscopic current induced by TLY, which comprises both distinct steps and low fluctuations, and the unusually high single channel conductance associated to the current steps, point to the target of TLY being other than pre-existing channels. In addition, TLY-induced currents were also recorded in membrane patches free from channels. Taken together, these results, and the fact that, once bound, the toxin cannot be washed out, indicate pore-formation as the most likely mechanism underlying the increase in conductance promoted by TLY in neuroblastoma/glioma cells [57]. The fact that the membrane must be exposed to a certain amount of toxin for some time in order to suffer permeability changes supports this conclusion, being in fact a requirement predicted by the pore-forming model proposed for SNTX [55,66].

5.3. Pain-Inducing and Inflammatory Activities

Immediate pain is the most distressing feature of envenomation by fish venoms, being disproportionate to the resulting wound in the absence of notable secondary tissue injury. In fact, counteracting pain is usually the main focus of treatment when the envenomation does not progress to systemic complications. Some experimental evidence points to the participation of cytolysins in the pain as well as in the local inflammatory process that rapidly develops on the site of injury caused by fish venoms.

The first association between fish cytolysins and these local effects was made when SNTX was first isolated in 1991 [12]. It was then observed that the toxin (0.15–0.4 mg, intraplantar injection—i.pl.) induced a dose-dependent edema (a minimum edema dose of 0.15 μg) within 1 h of injection into mouse hind paws, which lasted for more than 24 h and could not be blocked by pre-treatment with the antihistamine diphenhydramine (50 mg/kg, intraperitoneal injection—i.p.). No reference was made then as to the possible nociceptive effects induced by SNTX.

On the other hand, nocitoxin, a mildly hemolytic toxin isolated from the venom of the bullrout *Nothestes robusta*, was deemed to be the pain-causing protein in this venom, as only the fraction containing it could reproduce the algesic effect induced by the crude venom in human subjects [27]. There is, however, some doubt as to its falling into the fish cytolysin category, as it did not cross-react with the stonefish antivenom.

The nociceptive and inflammatory activities of Sp-CTx have been relatively well described when compared to those of other fish cytolysins. The first indication that Sp-CTx was at least partly responsible for the pain and swelling induced by the venom of *S. plumieri* came from the fact that these responses were blocked by the stonefish antivenom, which, as previously mentioned, reacted against a protein with the biochemical characteristics of this toxin [51]. The confirmation came in 2018, when [43] showed that the injection of Sp-CTx into mouse hind paws (0.1–3 μg, i.pl.) caused a dose-dependent nociceptive response and an intense, sustained, and also dose-dependent edema response that reached its peak within 60 min and persisted for up to 4 days.

Further insights into the mechanism underlying the local inflammatory reaction induced by fish cytolysins were obtained when these same authors reported that the swelling induced by Sp-CTx was reduced by pre-treatment with the bradykinin B2 receptor antagonist HOE-140 (100 nm/kg, i.p.) or the serine-proteases inhibitor aprotinin (8 mg/kg, i.p.) [43]. These findings indicate that the kallikrein-kinin system participates in this response, as had been already proposed when the inflammatory activity of the crude venom of *S. plumieri* was investigated a few years earlier [73]. However, these drugs did not prevent the edema response completely [43], suggesting that additional pathways might contribute to the onset of the inflammatory response. Moreover, the nociceptive activity induced by Sp-CTx could not be blocked by either drug, which indicates that it does not share the same mechanism described for the inflammatory activity [43].

That is, in fact, one of the biggest gaps in the study of fish venoms, and, by proxy, of fish cytolysins: the mechanisms behind the pain-inducing activity associated with these molecules remain largely unknown. It has been suggested that pore formation might be the underlying cause [43,55], but there is also evidence against this hypothesis. For instance, while investigating the pain caused by the lionfish *P. volitans* venom, [74] found that the heat-labile, protein components responsible for this activity appear to act specifically on neuropeptidergic nociceptors, although the identity of these components has yet to be determined.

6. Conclusions

Fish cytolysins stand out among other animal toxins because they alone can reproduce the main in vitro and in vivo effects induced by the venoms in which they are contained. These multifunctional toxins account for the extreme pain and inflammatory events experienced by the victims of envenomation by fish and affect the neuromuscular and cardiovascular systems in a significant way, potentially leading to death. Not to mention the species-specific hemolytic activity that, although seemingly irrelevant where envenomation is concerned, characterizes this type of toxin.

Although intriguing, this multifunctionality also adds to the complexity surrounding the mechanisms of action underlying the different activities displayed by these toxins (Figure 5). At this point, it is established that fish cytolysins destroy cells by forming non-selective pores in the membrane, but how much this pore-forming ability actually contributes to the pharmacological activities induced by these molecules is a matter for discussion.

Ample evidence points to fish cytolysins affecting the cardiovascular and neuromuscular systems through the modulation of signaling pathways that might vary in different tissues and species. However, that does not exclude the possibility that these toxins might also be able to form pores in the membrane of the cells that compose these systems. Based on the data gathered so far, we propose it to be a function of dose, time of exposure, and—naturally—the presence of proper recognition sites in the aforesaid membranes. Future studies on fish cytolysins should take this possibility into account.

All in all, much has been done regarding the investigation of biochemical and pharmacological features of fish cytolysins, considering how very labile and complex these molecules are. Nevertheless, as exposed in this review, there are still considerable gaps and contradictions, especially where what we believe to be their multi-mechanistic mode of action is concerned. The major role played by fish cytolysins in the envenomation process, added to the many questions raised by their multifunctionality, fully justifies the quest for a better understanding of how these molecules act.

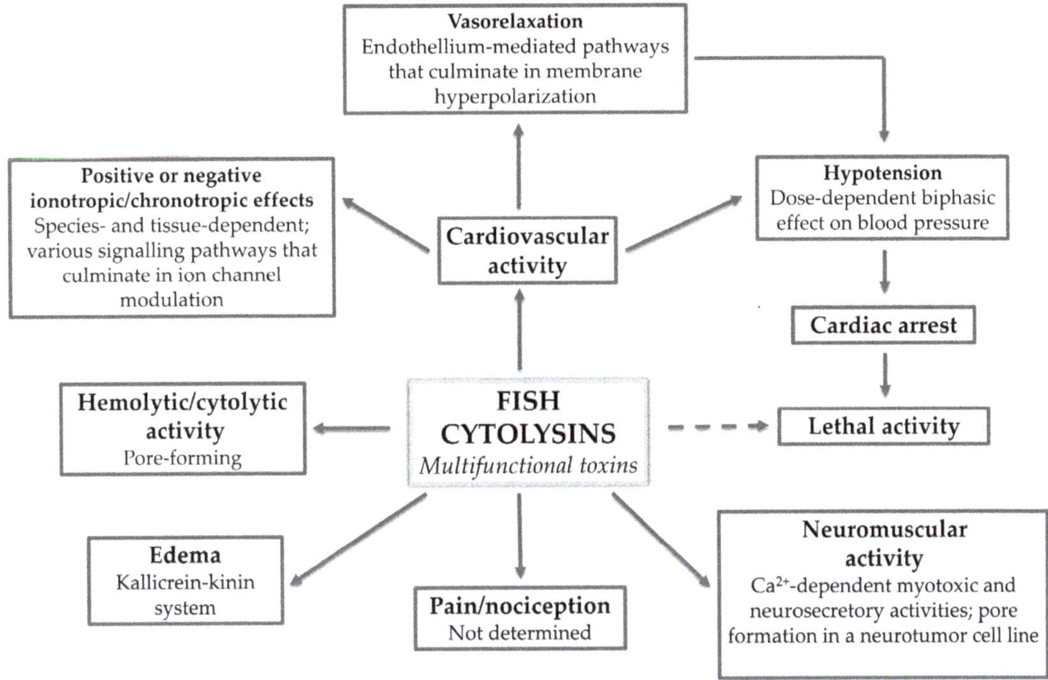

Figure 5. Fish cytolysins: multifunctional toxins. The various pharmacological activities displayed by multifunctional fish cytolysins and what has been determined regarding the mechanisms of action underlying each activity.

Author Contributions: Conceptualization, F.V.C., J.B.C. and S.G.F.; investigation, F.V.C., H.B.F., J.B.C. and S.G.F.; resources, J.B.C. and S.G.F.; writing—original draft preparation, F.V.C., H.B.F., J.B.C. and S.G.F.; writing—review and editing, F.V.C. and S.G.F.; supervision, J.B.C. and S.G.F.; funding acquisition, S.G.F. All authors have read and agreed to the published version of the manuscript.

Funding: This research was funded by CAPES, postdoctoral fellowship to FVC; Instituto Butantan, Ph.D. fellowship to HBF; and INCTTox (573790/2008-6) and CAPES Toxinologia (063/10-23038006280/20+11-07), grants to SGF.

Institutional Review Board Statement: Not applicable.

Informed Consent Statement: Not applicable.

Data Availability Statement: Not applicable.

Conflicts of Interest: The authors declare no conflict of interest.

References

1. Smith, W.L.; Stern, J.H.; Girard, M.G.; Davis, M.P. Evolution of Venomous Cartilaginous and Ray-Finned Fishes. *Integr. Comp. Biol.* **2016**, *56*, 950–961. [CrossRef] [PubMed]
2. Haddad, V., Jr. *Atlas de Animais Aquáticos Perigosos do Brasil: Guia Médico de Identificação e Tratamento (Atlas of Dangerous Aquatic Animals of Brazil: A Medical Guide of Identification and Treatment)*; Roca: São Paulo, Brazil, 2000.
3. Haddad, V., Jr.; Martins, I.A.; Makyama, H.M. Injuries caused by scorpionfishes (*Scorpaena plumieri* Bloch, 1789 and *Scorpaena brasiliensis* Cuvier, 1829) in the Southwestern Atlantic Ocean (Brazilian coast): Epidemiologic, clinic and therapeutic aspects of 23 stings in humans. *Toxicon* **2003**, *42*, 79–83. [CrossRef]
4. Smith, W.L.; Wheeler, W.C. Venom Evolution Widespread in Fishes: A Phylogenetic Road Map for the Bioprospecting of Piscine Venoms. *J. Hered.* **2006**, *97*, 206–217. [CrossRef] [PubMed]
5. Gorman, L.M.; Judge, S.J.; Fezai, M.; Jemaà, M.; Harris, J.B.; Caldwell, G.S. The venoms of the lesser (*Echiichthys vipera*) and greater (*Trachinus draco*) weever fish—A review. *Toxicon* **2020**, *6*, 100025. [CrossRef] [PubMed]

6. Letunic, I.; Bork, P. Interactive Tree of Life (iTOL) v4: Recent updates and new developments. *Nucleic Acids Res.* **2019**, *47*, W256–W259. [CrossRef] [PubMed]
7. Halstead, B.W. Injurious effects from the sting of the scorpionfish Scorpaena guttata with report of a case. *Calif. Med.* **1951**, *74*, 395–396. [PubMed]
8. Roche, E.T.; Halstead, B.W. *The Venom Apparatus of California Rockfishes (Family Scorpaenidae)*; Fish Bulletin of the Department of Fish and Game of the State of California; State of California, Department of Fish and Game: Sacramento, CA, USA, 1972; Volume 156, pp. 1–49.
9. Church, J.E.; Hodgson, W.C. The pharmacological activity of fish venoms. *Toxicon* **2002**, *40*, 1083–1093. [CrossRef]
10. Reckziegel, G.C.; Dourado, F.S.; Garrone-Neto, D.; Haddad, V., Jr. Injuries caused by aquatic animals in Brazil: An analysis of the data present in the information system for notifiable diseases. *Rev. Soc. Bras. Med. Trop.* **2015**, *48*, 460–467. [CrossRef] [PubMed]
11. Perriere, C.; Goudey-Perriere, F.; Petek, F. Purification of a lethal fraction from the venom of the weever fish, *Trachinus vipera* C.V. *Toxicon* **1988**, *26*, 1222–1227. [CrossRef]
12. Poh, C.; Yuen, R.; Khoo, H.; Chung, M.; Gwee, M.; Gopalakrishnakone, P. Purification and partial characterization of stonustoxin (lethal factor) from Synanceja horrida venom. *Comp. Biochem. Physiol. Part B Comp. Biochem.* **1991**, *99*, 793–798. [CrossRef]
13. Chhatwal, I.; Dreyer, F. Biological properties of a crude venom extract from the greater weever fish Trachinus draco. *Toxicon* **1992**, *30*, 77–85. [CrossRef]
14. Kiriake, A.; Shiomi, K. Some properties and cDNA cloning of proteinaceous toxins from two species of lionfish (*Pterois antennata* and *Pterois volitans*). *Toxicon* **2011**, *58*, 494–501. [CrossRef] [PubMed]
15. Gomes, H.L.; Menezes, T.; Malacarne, P.F.; Roman-Campos, D.; Gondim, A.N.; Cruz, J.; Vassallo, D.; Figueiredo, S.G. Cardiovascular effects of Sp-CTx, a cytolysin from the scorpionfish (*Scorpaena plumieri*) venom. *Toxicon* **2016**, *118*, 141–148. [CrossRef] [PubMed]
16. Harris, R.J.; Jenner, R.A. Evolutionary Ecology of Fish Venom: Adaptations and Consequences of Evolving a Venom System. *Toxins* **2019**, *11*, 60. [CrossRef]
17. Borges, M.H.; Andrich, F.; Lemos, P.H.; Soares, T.G.; Menezes, T.N.; Campos, F.V.; Neves, L.X.; Castro-Borges, W.; Figueiredo, S.G. Combined proteomic and functional analysis reveals rich sources of protein diversity in skin mucus and venom from the Scorpaena plumieri fish. *J. Proteom.* **2018**, *187*, 200–211. [CrossRef] [PubMed]
18. Ziegman, R.; Undheim, E.A.B.; Baillie, G.; Jones, A.; Alewood, P.F. Investigation of the estuarine stonefish (*Synanceia horrida*) venom composition. *J. Proteom.* **2019**, *201*, 12–26. [CrossRef] [PubMed]
19. Garnier, P.; Goudey-Perrière, F.; Breton, P.; Dewulf, C.; Petek, F.; Perrière, C. Enzymatic properties of the stonefish (*Synanceia verrucosa* Bloch and Schneider, 1801) venom and purification of a lethal, hypotensive and cytolytic factor. *Toxicon* **1995**, *33*, 143–155. [CrossRef]
20. Figueiredo, S.G.; Andrich, F.; Lima, C.; Lopes-Ferreira, M.; Haddad Jr., V. Venomous fish: A brief overview. In *Animal Toxins: State of the Art. Perspectives on Health and Biotechnology*; De Lima, M.E., Pimenta, A.M.C., Martin-Eauclaire, M.F., Zingali, R., Rochat, H., Eds.; UFMG: Belo Horizonte, Brazil, 2009; pp. 73–95.
21. Ziegman, R.; Alewood, P. Bioactive Components in Fish Venoms. *Toxins* **2015**, *7*, 1497–1531. [CrossRef] [PubMed]
22. Low, K.S.; Gwee, M.; Yuen, R.; Gopalakrishnakone, P.; Khoo, H. Stonustoxin: A highly potent endothelium-dependent vasorelaxant in the rat. *Toxicon* **1993**, *31*, 1471–1478. [CrossRef]
23. Carlson, R.W.; Schaeffer, R.C., Jr.; La Grange, R.G.; Roberts, C.M.; Russell, F.E. Some pharmacological properties of the venom of the scorpionfish Scorpaena guttata. *Toxicon* **1971**, *9*, 379–391. [CrossRef]
24. Shiomi, K.; Hosaka, M.; Fujita, S.; Yamanaka, H.; Kikuchi, T. Venoms from six species of marine fish: Lethal and hemolytic activities and their neutralization by commercial stonefish antivenom. *Mar. Biol.* **1989**, *103*, 285–289. [CrossRef]
25. Kreger, A.S. Detection of a cytolytic toxin in the venom of the stonefish (*Synanceia trachynis*). *Toxicon* **1991**, *29*, 733–743. [CrossRef]
26. Khoo, H.E.; Yuen, R.; Poh, C.H.; Tan, C.H. Biological activities ofSynanceja horrida (stonefish) venom. *Nat. Toxins* **1992**, *1*, 54–60. [CrossRef] [PubMed]
27. Hahn, S.; O'Connor, J. An investigation of the biological activity of bullrout (*Notesthes robusta*) venom. *Toxicon* **2000**, *38*, 79–89. [CrossRef]
28. Carrijo, L.C.; Andrich, F.; de Lima, M.E.; Cordeiro, M.N.; Richardson, M.; Figueiredo, S.G. Biological properties of the venom from the scorpionfish (*Scorpaena plumieri*) and purification of a gelatinolytic protease. *Toxicon* **2005**, *45*, 843–850. [CrossRef] [PubMed]
29. Sung, J.M.L.; Low, K.S.Y.; Khoo, H.E. Characterization of the mechanism underlying stonustoxin-mediated relaxant response in the rat aorta in vitro. *Biochem. Pharmacol.* **2002**, *63*, 1113–1118. [CrossRef]
30. Liew, H.; Khoo, H.; Moore, P.; Bhatia, M.; Lu, J.; Moochhala, S. Synergism between hydrogen sulfide (H2S) and nitric oxide (NO) in vasorelaxation induced by stonustoxin (SNTX), a lethal and hypotensive protein factor isolated from stonefish *Synanceja horrida* venom. *Life Sci.* **2007**, *80*, 1664–1668. [CrossRef] [PubMed]
31. Low, K.S.; Gwee, M.C.; Yuen, R.; Gopalakrishnakone, P.; Khoo, H. Stonustoxin: Effects on neuromuscular function in vitro and in vivo. *Toxicon* **1994**, *32*, 573–581. [CrossRef]
32. Colasante, C.; Meunier, F.A.; Kreger, A.S.; Molgó, J. Selective Depletion of Clear Synaptic Vesicles and Enhanced Quantal Transmitter Release at Frog Motor Nerve Endings Produced by Trachynilysin, a Protein Toxin Isolated from Stonefish (*Synanceia trachynis*) Venom. *Eur. J. Neurosci.* **1996**, *8*, 2149–2156. [CrossRef]

33. Sauviat, M.-P.; Meunier, F.A.; Kreger, A.; Molgó, J. Effects of trachynilysin, a protein isolated from stonefish (*Synanceia trachynis*) venom, on frog atrial heart muscle. *Toxicon* **2000**, *38*, 945–959. [CrossRef]
34. Ouanounou, G.; Mattei, C.; Meunier, F.A.; Kreger, A.S.; Molgó, J. Trachynilysin, a protein neurotoxin isolated from stonefish (*Synanceia trachynis*) venom, increases spontaneous quantal acetylcholine release from Torpedo marmorata neuromuscular junctions. *Cybium* **2000**, *24*, 149–156.
35. Meunier, F.A.; Mattei, C.; Chameau, P.; Lawrence, G.; Colasante, C.; Kreger, A.S.; Dolly, J.O.; Molgó, J. Trachynilysin mediates SNARE-dependent release of catecholamines from chromaffin cells via external and stored Ca^{2+}. *J. Cell Sci.* **2000**, *113*, 1119–1125. [CrossRef]
36. Garnier, P.; Sauviat, M.-P.; Goudey-Perriere, F.; Perriere, C. Cardiotoxicity of verrucotoxin, a protein isolated from the venom of Synanceia verrucosa. *Toxicon* **1997**, *35*, 47–55. [CrossRef]
37. Wang, J.-W.; Yazawa, K.; Hao, L.-Y.; Onoue, Y.; Kameyama, M. Verrucotoxin inhibits KATP channels in cardiac myocytes through a muscarinic M3 receptor-PKC pathway. *Eur. J. Pharmacol.* **2007**, *563*, 172–179. [CrossRef] [PubMed]
38. Yazawa, K.; Wang, J.-W.; Hao, L.-Y.; Onoue, Y.; Kameyama, M. Verrucotoxin, a stonefish venom, modulates calcium channel activity in guinea-pig ventricular myocytes. *Br. J. Pharmacol.* **2007**, *151*, 1198–1203. [CrossRef]
39. Ueda, A.; Suzuki, M.; Honma, T.; Nagai, H.; Nagashima, Y.; Shiomi, K. Purification, properties and cDNA cloning of neoverrucotoxin (neoVTX), a hemolytic lethal factor from the stonefish Synanceia verrucosa venom. *Biochim. Biophys. Acta BBA-Gen. Subj.* **2006**, *1760*, 1713–1722. [CrossRef] [PubMed]
40. Kiriake, A.; Suzuki, Y.; Nagashima, Y.; Shiomi, K. Proteinaceous toxins from three species of scorpaeniform fish (lionfish Pterois lunulata, devil stinger Inimicus japonicus and waspfish Hypodytes rubripinnis): Close similarity in properties and primary structures to stonefish toxins. *Toxicon* **2013**, *70*, 184–193. [CrossRef] [PubMed]
41. Andrich, F.; Carnielli, J.; Cassoli, J.; Lautner, R.; Santos, R.; Pimenta, A.; de Lima, M.; Figueiredo, S. A potent vasoactive cytolysin isolated from Scorpaena plumieri scorpionfish venom. *Toxicon* **2010**, *56*, 487–496. [CrossRef] [PubMed]
42. Gomes, H.L.; Andrich, F.; Fortes-Dias, C.L.; Perales, J.; Teixeira-Ferreira, A.; Vassallo, D.V.; Cruz, J.S.; Figueiredo, S.G. Molecular and biochemical characterization of a cytolysin from the Scorpaena plumieri (scorpionfish) venom: Evidence of pore formation on erythrocyte cell membrane. *Toxicon* **2013**, *74*, 92–100. [CrossRef] [PubMed]
43. Malacarne, P.F.; Menezes, T.N.; Martins, C.W.; Naumann, G.B.; Gomes, H.L.; Pires, R.G.; Figueiredo, S.G.; Campos, F.V. Advances in the characterization of the Scorpaena plumieri cytolytic toxin (Sp-CTx). *Toxicon* **2018**, *150*, 220–227. [CrossRef]
44. Schaeffer, R.C.; Carlson, R.W.; Russell, F.E. Some chemical properties of the venom of the scorpionfish Scorpaena guttata. *Toxicon* **1971**, *9*, 69–78. [CrossRef]
45. Chuang, P.-S.; Shiao, J.-C. Toxin gene determination and evolution in scorpaenoid fish. *Toxicon* **2014**, *88*, 21–33. [CrossRef]
46. Saunders, P.R.; Tökés, L. Purification and properties of the lethal fraction of the venom of the stonefish Synanceja horrida (Linnaeus). *Biochim. Biophys. Acta BBA-Bioenerg.* **1961**, *52*, 527–532. [CrossRef]
47. Frederico, A.; Américo, D.; Stéphane, B.; Beatriz, R.; Francisco, V.; Joana, R.; José, M.; Luisa, M.; Pedro, F.; Lino, C.J.; et al. A simple and practical technique for fish venom extraction—Protein content analysis for future biotechnological applications. *Front. Mar. Sci.* **2016**, *3*. [CrossRef]
48. Hedrick, J.L.; Smith, A.J. Size and charge isomer separation and estimation of molecular weights of proteins by disc gel electrophoresis. *Arch. Biochem. Biophys.* **1968**, *126*, 155–164. [CrossRef]
49. Church, J.E.; Hodgson, W.C. Stonefish (*Synanceia* spp.) antivenom neutralizes the in vitro and in vivo cardiovascular activity of soldierfish (*Gymnapistes marmoratus*) venom. *Toxicon* **2001**, *39*, 319–324. [CrossRef]
50. Church, J.E.; Hodgson, W.C. Stonefish (*Synanceia trachynis*) Antivenom: In Vitro Efficacy and Clinical Use. *J. Toxicol. Toxin Rev.* **2003**, *22*, 69–76. [CrossRef]
51. Gomes, H.L.; Menezes, T.N.; Carnielli, J.B.; Andrich, F.; Evangelista, K.S.; Chávez-Olórtegui, C.; Vassallo, D.V.; Figueiredo, S.G. Stonefish antivenom neutralises the inflammatory and cardiovascular effects induced by scorpionfish Scorpaena plumieri venom. *Toxicon* **2011**, *57*, 992–999. [CrossRef] [PubMed]
52. Ghadessy, F.J.; Chen, D.; Kini, M.; Chung, M.C.M.; Jeyaseelan, K.; Khoo, H.E.; Yuen, R. Stonustoxin Is a Novel Lethal Factor from Stonefish (*Synanceja horrida*) Venom. *J. Biol. Chem.* **1996**, *271*, 25575–25581. [CrossRef] [PubMed]
53. Costa, F.L.S.; De Lima, M.E.; Figueiredo, S.G.; Ferreira, R.S.; Prates, N.S.; Sakamoto, T.; Salas, C.E. Sequence analysis of the cDNA encoding for SpCTx: A lethal factor from scorpionfish venom (*Scorpaena plumieri*). *J. Venom. Anim. Toxins Incl. Trop. Dis.* **2018**, *24*, 24. [CrossRef]
54. Garnier, P.; Ducancel, F.; Ogawa, T.; Boulain, J.C.; Goudey-Perrière, F.; Perrière, C.; Ménez, A. Complete amino-acid sequence of the beta-subunit of VTX from venom of the stonefish (*Synanceia verrucosa*) as identified from cDNA cloning experiments. *Biochim. Biophys. Acta* **1997**, *1337*, 1–5. [CrossRef]
55. Ellisdon, A.; Reboul, C.; Panjikar, S.; Huynh, K.; Oellig, C.A.; Winter, K.L.; Dunstone, M.A.; Hodgson, W.; Seymour, J.; Dearden, P.; et al. Stonefish toxin defines an ancient branch of the perforin-like superfamily. *Proc. Natl. Acad. Sci. USA* **2015**, *112*, 15360–15365. [CrossRef]
56. Altschul, S.F.; Madden, T.L.; Schäffer, A.A.; Zhang, J.; Zhang, Z.; Miller, W.; Lipman, D.J. Gapped BLAST and PSI-BLAST: A new generation of protein database search programs. *Nucleic Acids Res.* **1997**, *25*, 3389–3402. [CrossRef] [PubMed]

57. Ouanounou, G.; Malo, M.; Stinnakre, J.; Kreger, A.S.; Molgó, J. Trachynilysin, a Neurosecretory Protein Isolated from Stonefish (*Synanceia trachynis*) Venom, Forms Nonselective Pores in the Membrane of NG108-15 Cells. *J. Biol. Chem.* **2002**, *277*, 39119–39127. [CrossRef] [PubMed]
58. Khoo, H.; Hon, W.; Lee, S.; Yuen, R. Effects of stonustoxin (lethal factor from *Synanceja horrida* venom) on platelet aggregation. *Toxicon* **1995**, *33*, 1033–1041. [CrossRef]
59. Hatakeyama, T.; Nagatomo, H.; Yamasaki, N. Interaction of the Hemolytic Lectin CEL-III from the Marine Invertebrate Cucumaria echinata with the Erythrocyte Membrane. *J. Biol. Chem.* **1995**, *270*, 3560–3564. [CrossRef]
60. Oda, T.; Tsuru, M.; Hatakeyama, T.; Nagatomo, H.; Muramatsu, T.; Yamasaki, N. Temperature- and pH-dependent cytotoxic effect of the hemolytic lectin CEL-III from the marine invertebrate Cucumaria echinata on various cell lines. *J. Biochem.* **1997**, *121*, 560–567. [CrossRef]
61. Hopkins, B.J.; Hodgson, W.C.; Sutherland, S.K. Pharmacological studies of stonefish (*Synanceja trachynis*) venom. *Toxicon* **1994**, *32*, 1197–1210. [CrossRef]
62. Lopes-Ferreira, M.; Barbaro, K.; Cardoso, D.; Moura-Da-Silva, A.M.; Mota, I. Thalassophryne nattereri fish venom: Biological and biochemical characterization and serum neutralization of its toxic activities. *Toxicon* **1998**, *36*, 405–410. [CrossRef]
63. Hopkins, B.J.; Hodgson, W.C. Cardiovascular studies on venom fromthe soldierfish (*Gymnapistes marmoratus*). *Toxicon* **1998**, *36*, 973–983. [CrossRef]
64. Andrich, F.; Richardson, M.; Naumann, G.; Cordeiro, M.; Santos, A.; Santos, D.; Oliveira, J.; De Lima, M.; Figueiredo, S. Identification of C-type isolectins in the venom of the scorpionfish Scorpaena plumieri. *Toxicon* **2015**, *95*, 67–71. [CrossRef] [PubMed]
65. Campos, F.V.; Menezes, T.N.; Malacarne, P.F.; Costa, F.L.S.; Naumann, G.B.; Gomes, H.L.; Figueiredo, S.G. A review on the Scorpaena plumieri fish venom and its bioactive compounds. *J. Venom. Anim. Toxins Incl. Trop. Dis.* **2016**, *22*, 35. [CrossRef] [PubMed]
66. Chen, D.; Kini, M.; Yuen, R.; Khoo, H.E. Haemolytic activity of stonustoxin from stonefish (*Synanceja horrida*) venom: Pore formation and the role of cationic amino acid residues. *Biochem. J.* **1997**, *325*, 685–691. [CrossRef] [PubMed]
67. Khoo, H.E.; Chen, D.S.; Yuen, R. The role of cationic amino acid residues in the lethal activity of stonustoxin from stonefish (*Synanceja horrida*) venom. *Biochem. Mol. Biol. Int.* **1998**, *44*, 643–646. [PubMed]
68. Sáenz, A.; Ortiz, N.; Lomonte, B.; Rucavado, A.; Díaz, C. Comparison of biochemical and cytotoxic activities of extracts obtained from dorsal spines and caudal fin of adult and juvenile non-native Caribbean lionfish (Pterois volitans/miles). *Toxicon* **2017**, *137*, 158–167. [CrossRef] [PubMed]
69. Yew, W.S.; Khoo, H.E. The role of tryptophan residues in the hemolytic activity of stonustoxin, a lethal factor from stonefish (*Synanceja horrida*) venom. *Biochimie* **2000**, *82*, 251–257. [CrossRef]
70. Breton, P.; Delamanche, I.; Bouee, J.; Goudey-Perriere, F.; Perriere, C. Toxicon. Verrucotoxin and neurotoxic effects of stonefish (*Synanceia verrucosa*) venom. *Toxicon* **1999**, *37*, 1213.
71. Kreger, A.; Molgó, J.; Comella, J.; Hansson, B.; Thesleff, S. Effects of stonefish (Synanceia trachynis) venom on murine and frog neuromuscular junctions. *Toxicon* **1993**, *31*, 307–317. [CrossRef]
72. Cohen, A.S.; Olek, A.J. An extract of lionfish (Pterois volitans) spine tissue contains acetylcholine and a toxin that affects neuromuscular transmission. *Toxicon* **1989**, *27*, 1367–1376. [CrossRef]
73. Menezes, T.; Carnielli, J.B.; Gomes, H.L.; Pereira, F.E.; Lemos, E.M.; Bissoli, N.S.; Lopes-Ferreira, M.; Andrich, F.; Figueiredo, S.G. Local inflammatory response induced by scorpionfish Scorpaena plumieri venom in mice. *Toxicon* **2012**, *60*, 4–11. [CrossRef]
74. Mouchbahani-Constance, S.; Lesperance, L.S.; Petitjean, H.; Davidova, A.; Macpherson, A.; Prescott, S.A.; Sharif-Naeini, R. Lionfish venom elicits pain predominantly through the activation of nonpeptidergic nociceptors. *Pain* **2018**, *159*, 2255–2266. [CrossRef] [PubMed]

 **toxins** **MDPI**

Review

Inflammatory Effects of *Bothrops* Phospholipases A$_2$: Mechanisms Involved in Biosynthesis of Lipid Mediators and Lipid Accumulation

Vanessa Moreira [1,†], Elbio Leiguez [2,†], Priscila Motta Janovits [2], Rodrigo Maia-Marques [2], Cristina Maria Fernandes [2] and Catarina Teixeira [2,*]

1 Departamento de Farmacologia, Escola Paulista de Medicina, Universidade Federal de Sao Paulo, Sao Paulo 04044-020, Brazil; vmoreira@unifesp.br
2 Laboratório de Farmacologia, Instituto Butantan, Sao Paulo 05503-900, Brazil; leiguez@uni9.pro.br (E.L.); priscilajanovits@gmail.com (P.M.J.); rodrigo.marques@esib.butantan.gov.br (R.M.-M.); cristina.fernandes@butantan.gov.br (C.M.F.)
* Correspondence: catarina.teixeira@butantan.gov.br
† These authors contributed with equal importance.

Abstract: Phospholipases A$_2$s (PLA$_2$s) constitute one of the major protein groups present in the venoms of viperid and crotalid snakes. Snake venom PLA$_2$s (svPLA$_2$s) exhibit a remarkable functional diversity, as they have been described to induce a myriad of toxic effects. Local inflammation is an important characteristic of snakebite envenomation inflicted by viperid and crotalid species and diverse svPLA$_2$s have been studied for their proinflammatory properties. Moreover, based on their molecular, structural, and functional properties, the viperid svPLA$_2$s are classified into the group IIA secreted PLA$_2$s, which encompasses mammalian inflammatory sPLA$_2$s. Thus, research on svPLA$_2$s has attained paramount importance for better understanding the role of this class of enzymes in snake envenomation and the participation of GIIA sPLA$_2$s in pathophysiological conditions and for the development of new therapeutic agents. In this review, we highlight studies that have identified the inflammatory activities of svPLA$_2$s, in particular, those from *Bothrops* genus snakes, which are major medically important snakes in Latin America, and we describe recent advances in our collective understanding of the mechanisms underlying their inflammatory effects. We also discuss studies that dissect the action of these venom enzymes in inflammatory cells focusing on molecular mechanisms and signaling pathways involved in the biosynthesis of lipid mediators and lipid accumulation in immunocompetent cells.

Keywords: *Bothrops* phospholipases A$_2$; inflammation; lipid mediators; signaling pathways

Key Contribution: This review provides an overview and recent advances in the understanding of inflammatory mechanisms triggered by svPLA$_2$s with a focus on their actions on lipid mediator biosynthetic pathways and lipid accumulation in immunocompetent cells.

Citation: Moreira, V.; Leiguez, E.; Janovits, P.M.; Maia-Marques, R.; Fernandes, C.M.; Teixeira, C. Inflammatory Effects of *Bothrops* Phospholipases A$_2$: Mechanisms Involved in Biosynthesis of Lipid Mediators and Lipid Accumulation. *Toxins* **2021**, *13*, 868. https://doi.org/10.3390/toxins13120868

Received: 3 November 2021
Accepted: 30 November 2021
Published: 4 December 2021

1. Introduction

Bothrops spp. snakes are responsible for the majority of snakebites in Latin America. Envenomation by these snakes induces severe pathological alterations at the site of venom injection, characterized by an intense local inflammatory reaction associated with myonecrosis, pain, and hemorrhage, potentially leading to permanent tissue damage and disability [1–6]. The local inflammatory response to bothropic envenomation involves a set of events including an increase in vascular permeability, edema formation, hyperalgesia, the activation and infiltration of immunocompetent cells, and a complex network of inflammatory mediators driving the inflammatory response [2]. These events are caused by the activation of critical host defense mechanisms of the victims by direct and indirect actions of the toxins present in the snake venom. Proteomic studies of *Bothrops* snake venoms revealed that the phospholipases A_2 (PLA_2s) are ubiquitous components of these venoms and play an important role in the pathophysiology of envenoming by these snakes [7–11]. During envenomation, in addition to aiding prey digestion, these toxins have been described to display myotoxic, cytotoxic, hemolytic, hypotensive, anticoagulant, platelet aggregation inhibition/activation, and proinflammatory effects [12–14]. Regarding inflammation, in addition to exerting direct effects on cell membranes, these lipolytic enzymes can recruit mammalian PLA_2s analogs of similar activity and trigger endogenous signaling systems that display and amplify the cell injury and host defense mechanisms triggered by the whole venom. This amplification is responsible for many acutely important consequences of *Bothrops* envenoming. Yet, as discussed below, studies with in vitro and in vivo models aiming to understand the inflammatory action of isolated venom PLA_2s can contribute to the knowledge of the local inflammatory mechanisms induced by *Bothrops* snake venoms and those from the Viperidae family [3,15–17]. These studies might lead to the discovery of new therapeutic targets for a more efficient treatment of envenoming by viperid snakes, since the currently available antivenoms have low effectiveness to neutralize the local events promoted by their venoms [18]. Finally, due to the structural and functional similarities to mammalian group (G) IIA PLA_2, the *Bothrops* PLA_2s can constitute useful tools for studies on the roles of human GIIA PLA_2s in inflammatory diseases. In this regard, the effectiveness of varespladib, an inhibitor of svPLA_2s [19] in attenuating inflammatory events caused by Viperidae snake venoms has been demonstrated in mice experimental model [20].

1.1. Phospholipases A_2

Phospholipases A_2 constitute a group of enzymes with diverse biological functions, ranging from homeostasis and membrane remodeling to the generation of metabolites and second messengers involved in biological processes and signal transduction. These enzymes hydrolyze the acyl ester bond at the sn-2 position of phospholipids, generating free fatty acids, such as arachidonic acid (AA) and oleic acid, and lysophospholipids, such as lyso-PAF [21,22]. Currently, PLA_2s are classified into close families, including secretory PLA_2 (sPLA_2), cytosolic phospholipase (cPLA_2), and Ca^{2+}-independent PLA_2 (iPLA_2); platelet activating factor acetylhydrolase (PAF); and lysosomal PLA_2 (LPLA_2), PLA/acyltransferase (PLAAT), α/β hydrolase (ABHD), adipose-PLA_2 (AdPLA), and glycosylphosphatidylinositol (GPI)-specific PLA_2 [23–25]. This classification of PLA_2s is based on the cell location, amino acid sequence, molecular weight, presence of intramolecular disulfide bridges, calcium requirement, and catalytic activity, including the hydrolysis at the sn-2 position of glycerophospholipids [22,26]. The sPLA_2 family is largely distributed in nature, and its components are classified into I, II, III, V, IX, X, XI, XII, XIII, and XIV groups. They are present in high concentrations in snake, bee, and wasp venoms and in several mammalian organs, cells, and pancreatic juice [21,27]. Among them, we highlight group I PLA_2s, comprising the secreted enzymes found in the snake venoms of the Elapidae and Hydrophiidae families and in the pancreas secretion of mammals. Meanwhile, group II consists of sPLA_2s found in snake venoms of the Viperidae family and in mammalian tissues and are expressed in inflammatory processes [13,28].

For a long time, the pathophysiological activities of sPLA_2s were related exclusively to their enzymatic activity, capable of providing a substrate for the synthesis of second messengers and inflammatory mediators. However, alternative mechanisms of action have been associated with the ability of sPLA_2 to interact with receptors or even specific domains present on cell membrane surfaces, such as heparan sulfate proteoglycans, which have already been described to be important for triggering the activation of other sPLA_2s in the target cells [29–32]. The identification of sPLA_2s binding proteins was initially achieved by Lambeau et al. (1989) [33], who used the sPLA_2 from the *Oxyuranus scutelatus* snake venom, called OS2. Due to its prevalence in brain tissue, this protein was denominated an N (neuronal) type receptor. Then, through screening in various tissues and cells, for studies of other OS2 binding proteins, a second type of PLA_2 receptor was described, present in rabbit skeletal muscles, called an M (muscle) type receptor [34]. The presence of this latter receptor was identified in several tissues and in neutrophils, monocytes, and human alveolar macrophages [35,36], but not in murine peritoneal macrophages [37]. The M-type phospholipase A_2 receptor has a high degree of homology (~30%) to mannose receptors, a member of the lectin receptor family, constitutively expressed in macrophages [38,39]. Mannose receptors are involved in phagocytosis, antigen processing [29,38,40], and the production of inflammatory cytokines by macrophages [40–42]. Additionally, this receptor has been demonstrated to play a role in cell proliferation and AA release, via MAPKs (protein kinase activated by the mitogens family), induced by the group IB $sPLA_2s$ [43–45]. Furthermore, it has been revealed that group IIA $sPLA_2s$ can bind to mannose receptors and promote the release of IL-6 by human alveolar macrophages [46].

Investigations into the biological role of mammalian group IIA PLA_2, also known as inflammatory PLA_2, in the development of several pathologies of inflammatory and immunological origin have been described. Several studies have revealed that group IIA $sPLA_2s$ are present in high levels in rheumatoid arthritis [47–50], acute pancreatitis [51–53], septic shock [54,55], Crohn's disease and ulcerative colitis [56–58], respiratory distress syndrome [59–61], bronchial asthma and allergic rhinitis [59,62], atherosclerosis [63,64], autoimmune diseases [65], and cancer [66–69]. These observations imply that both local and systemic inflammation are associated with the release of $sPLA_2s$ in vivo, thus raising the unclear question of the role of these PLA_2s in inflammatory reactions. Additionally, it was found that proinflammatory cytokines, such as interleukin (IL)-1β, IL-6, and tumor necrosis factor alpha (TNF-α), induce, in a variety of tissues, the gene transcription of $sPLA_2s$ and the subsequent increase in their secretion, thus, supporting the hypothesis of the involvement of $sPLA_2s$ in inflammation [70–73]. In addition, $sPLA_2s$ activate intracellular signaling events in cells that participate in inflammatory processes, caused by the generation of second messengers, and the phosphorylation of kinases as of MAPK [74–77]. Thus, $sPLA_2s$ represent an important target for investigations regarding the mechanisms of inflammatory events.

1.2. *Inflammation—General Concepts and Signaling Pathways*

Inflammation is a response of body tissues to noxious conditions for restoring homeostasis, setting the stage for the healing and reconstitution of injured tissue. The acute inflammatory response to injury involves functional alterations of microvessels that occur early after injury and develop at varying rates. The major features of these alterations include transient vasoconstriction, followed by vasodilation, then, leakage of protein-rich fluid from the microcirculation leading to edema formation, and movement of phagocytic leukocytes into the site of injury followed by local pain [78]. Immunocompetent cells, such as neutrophils and monocytes, found in blood circulation are capable of rapidly infiltrating tissues; macrophages and dendritic cells reside within tissue and play key roles in tissue surveillance and antigen presentation [79,80]. The vascular and cellular reactions are triggered and highly regulated by chemical factors, called inflammatory mediators, which include cytokines, chemokines, vasoactive amines, and eicosanoids and are produced by plasma components or are released in close proximity to the injury by endothelial cells, tissue-resident leukocytes such as mast cells, and macrophages at the early stages, followed by infiltrated leukocytes. The effects of inflammatory mediators involve the engagement of specific receptors, which then display signaling pathways responsible for the immune response [81–83]. Parallel to changes in blood flow, the margination of leukocytes begins, and leukocytes adhere to the microvascular endothelium through rolling and firm adhesion, and then moving through the vascular wall into the interstitial tissue. The various steps in the leukocyte migration are regulated by different subsets of cell-adhesion molecules expressed by both leukocyte and endothelial cells [84,85]. Thereafter, the phagocytosis of offending agents by migrated leukocytes occurs, followed by a release of lysosomal enzymes and an increase in the oxidative metabolism in leukocytes, known as respiratory burst, resulting in the production of microbicidal agents, such as superoxide anion (O_2^-) and hydrogen peroxide (H_2O_2) [86,87]. Four major classes of receptor-mediated phagocytosis exist: receptors of complement that recognize complement-coated particles; Fcg receptors, which are constitutively active for phagocytosis of IgG-coated particles; mannose receptors that recognize mannose and fucose on the surface of pathogens; and β-glucan receptors that recognize β-glucans-bearing ligands [87]. Furthermore, receptors for the Fc portion of immunoglobulin G and mannose/fucose residues lead to the release of proinflammatory mediators and reactive oxygen [88].

Despite the stereotyped features associated with an inflammatory response, the signal pathways involved in this response are determined by the nature of the inflammatory trigger, the sensors that detect them, the inflammatory mediators released, and the tissue affected. A number of surface and cytosolic receptors expressed in innate immune cells can sense pathogen-associated molecular patterns (PAMPs), damage-associated molecular patterns (DAMPs), or venom-associated molecular patterns (VAMPs) with high sensitivity and specificity. The recognition of these molecules is achieved by interacting with pattern recognition receptors (PRRs) [89]. These sensing receptors include Toll-like receptors (TLRs) [90,91], C-type lectin receptors (CLRS), RIG-I-like receptors (RLRs), and nucleotide-binding domain leucine-rich repeat (NLRs) or nucleotide-binding and oligomerisation domain (NOD)-like receptors [92,93]. Among them, TLRs (TLR1-TLR10) are highly conserved transmembrane proteins that play an important role in recognizing microbial pathogens and endogenous damage molecules, thereby triggering the generation of signals for the production of proinflammatory proteins and cytokines, via cooperation of adaptor proteins (MyD88, TIRAP, TRIF, and TRAM) [94,95]. The recognition of specific molecular patterns by NLRs induces the oligomerisation of proteins in the cytosol, generating platforms called inflammasomes [96]. The inflammasome is a high molecular weight protein complex that elicits the activation of inflammatory caspases and the processing of pro-interleukin-1β (pro-IL-1β) and pro-IL-18, generating the mature biologically active cytokines and a rapid inflammatory form of cell death termed pyroptosis [97]. Several distinct inflammasomes have been identified, including NLR and the pyrin domain containing receptor 1 (NLRP1), NLRP3, and NLR; the caspase recruitment domain containing receptor 4 (NLRC4); and the AIM2-like receptors (ALR) family [98,99]. These receptors positively regulate genes related to inflammatory mediators, including cytokines and key enzymes in the biosynthetic pathways of lipid mediators known as eicosanoids [100].

The early induction of most inflammatory transcripts depends on networks of transcription factors whose activation is coupled to pathways of signal transduction. The nuclear factor-kappa B (NF-κB) is a major and the best-studied transcription factor of inflammatory response [101,102]. The binding of activated NF-κB to the nuclear promoter region of diverse inflammatory factors leads to the transcriptional activation and expression of inflammatory mediators and enzymes. Currently, the major signaling pathways involved in inflammation are mediated by the cascade phosphorylation of protein kinases, such as the mitogen activated kinases (MAPKs) encompassing ERK1/2, SAPK, c-Jun NH2-terminal or JNK and p38MAPKs, phosphatidylinositol 3 kinase (PI3K), protein kinase C (PKC), and protein tyrosine kinase (PTK). These kinases mediate distinct intracellular signaling pathways associated with cytokines production, cytokine receptors, growth factors, mobilization of intracellular Ca^{2+}, and regulate a variety of functions of immunocompetent cells, including cell migration, phagocytosis, degranulation, respiratory burst, and programmed cell death [103–112].

An effective acute inflammatory response results in the removal of noxious factors followed by the resolution and repair stages [113]. The shift in inflammatory markers, including the switch from proinflammatory mediators to anti-inflammatory, resolution-inducing mediators (lipoxins, maresins, protectins, and resolvins), is vital for the change from inflammation to resolution [114]. This switch drives the transition from neutrophil to monocyte recruitment into the affected sites, resulting in the clearance of dead cells and other debris, assisted by the lymphatic system, and the initiation of tissue repair at the damaged site [115]. However, if the acute events are not properly controlled, the inflammatory response becomes detrimental to the host. Yet, if the acute response does not succeed in neutralizing the injurious stimulus, the resolution phase might not be appropriately induced, and a chronic inflammatory state may ensue, leading to several inflammatory-mediated diseases [116,117].

2. Inflammatory Effects of sPLA$_2$ from *Bothrops* spp. Venoms

2.1. *Bothrops svPLA$_2$s Induce Inflammatory Events and Activate Defense Functions in Leukocytes*

Phospholipases A$_2$ of GIIA are major components of *Bothrops* spp. snake venoms and play important roles in the pathophysiology of envenoming by these snakes, including the inflammatory response. Although these enzymes conserve a chemistry and catalytic structure, the natural evolution of viperid venoms introduced alterations in their primary amino acid residues, generating various other biological and toxicological effects [118]. In general, the GIIA sPLA$_2$s found in viperid snake venoms are classified as sPLA$_2$s, known as 'classic', containing an amino acid aspartate at position 49 (Asp49) and catalyzing the hydrolysis of the ester bond at position sn-2 of glycerophospholipids in a Ca^{2+} dependent manner. Meanwhile, the other type of sPLA$_2$s is described as 'variant' and contains a lysine at the same position 49 (Lys49), with or without low catalytic activity [119]. Such a substitution affects the ability of these proteins to bind to Ca^{2+}, an essential cofactor for the stabilization of tetrahedral intermediate, which occurs in the catalytic reaction performed by the Asp49-sPLA$_2$s [120]. Despite the lack of enzymatic activity, sPLA$_2$s-Lys49 homologues maintain their damaging capacity in membranes through a mechanism that is not completely understood and independent of Ca^{2+} [12,121,122].

It has long been demonstrated that viperid sPLA$_2$s are potent inductors of inflammation. Although they present differences in their catalytic activity, both viperid Asp49 and Lys49 PLA$_2$ homologues are capable of inducing local inflammation in diverse experimental models [123–125]. As such, this group of enzymes is considered to be a major component responsible for the severe local edema in envenomings by *Bothrops* spp. The inflammatory response to venom PLA$_2$s is characterized by edema and the marked infiltration of leukocytes into the site of toxin injection. Studies on the mechanism of local edema induced by viperid sPLA$_2$s (svPLA$_2$s) have demonstrated an early increase in vascular permeability and a local release of inflammatory mediators, which act synergistically to cause the initiation and development of the inflammatory events. Among these mediators are vasoactive amines, including histamine, serotonin, and substance P, as well as vasodilating prostaglandins. Yet, in vivo studies employing a pharmacological approach have demonstrated that antagonists of serotonin and H1 receptors of histamine reduced the progression of edema induced by both catalytically active and inactive variants of sPLA$_2$ isolated from *B. asper* [126], *B. neuwiedii* [127], *B. jararacussu* [128,129], or *B. insularis* [130]. In support of these reports, the release of histamine and serotonin by mast cells was observed following stimulation with bothropic sPLA$_2$s from *B. jararacussu* [128,131]. Consistent with this evidence, the contribution of mast cells to edema formation induced by viperid PLA$_2$s was further observed in in vitro experimental models demonstrating the ability of sPLA$_2$s isolated from *B. pirajai*, *B. jararacussu*, and *B. atrox* snake venoms to degranulate mast cells [125,128,131,132]. It is well known that upon activation, mast cells secrete and synthesize an array of inflammatory mediators, which trigger the earliest events of inflammation [133,134]. Moreover, the contribution of the catalytic activity for the edematogenic effect of the enzymatic active Asp49 from bothropic PLA$_2$s was suggested by studies, revealing that the chemical modification of this sPLA$_2$ by p-bromophenacyl bromide inhibited edema formation induced by these viperid PLA$_2$s [126–128]. In addition, the role of lipid mediators, such as PAF and eicosanoids, for hyperalgesia induced by catalytic active venom PLA$_2$s was highlighted by studies using a pharmacological approach [135]. These authors suggested that the enzymatic hydrolysis of membrane phospholipids played a role in these events by directly releasing the precursors of lipid mediators, such as lyso-PAF and AA.

As mentioned previously, leukocytes are central components of inflammation. An important cellular component exists in the inflammatory response to *Bothrops* $sPLA_2s$. As such, the stimulatory effect of piratoxin-I, bothropstoxin-I, and -II from *B. pirajai* and *B. jararacussu*, respectively, on neutrophil chemotaxis was demonstrated in an in vitro experimental model [136]. This effect was revealed to involve the interaction of these $sPLA_2s$ with surface heparan binding sites of neutrophils, followed by the release of chemotactic mediator leukotriene B_4 (LTB_4) and PAF, and is independent of enzyme activity. Furthermore, the ability of these venom PLA_2s to recruit an endogenous PLA_2 through the activation of GTP-binding protein and PKC was added to the mechanisms by which they cause neutrophil migration [137]. Moreover, studies conducted using in vivo experimental models have demonstrated the ability of *Bothrops* $sPLA_2s$ to induce a marked influx of polymorphonuclear and mononuclear cells into the site of their injection, as demonstrated for both catalytic active and non-catalytic venom PLA_2s, such as MT-III and MT-II from *B. asper* snake venom [123,138]. A similar effect was reported by other authors, investigating various $sPLA_2s$ isolated from different *Bothrops* spp. snake venoms, such as Bothropstoxin (BthTX)-I and BthTX-II; *B. jararacussu* [131], BnSP-7, a catalytically inactive PLA_2 from *B. pauloensis* [139], BatroxPLA_2 from *B. atrox* [140] and BJ-PLA_2-I from *B. jararaca* [141] in in vivo experimental models. The $sPLA_2$-induced leukocyte migration was linked to the upregulation of adhesion molecules, such as l-selectin, LFA-1, and CD18, which in turn was associated with the release of inflammatory cytokines IL-1β, IL-6, and TNF-α with chemotactic activity by resident leukocytes, primarily macrophages [123]. Cytokines, chemokines, and leukotriene B_4 are among the major mediators regulating the expression of adhesion molecules and chemotaxis of leukocytes [142–144]. Consistent with this information, increased serum levels of IL-6, IL-1, and TNF-α induced by Bbil-TX from *B. bilineata* snake venom were observed in a mouse experimental model [145]. In addition, there are reports that two Lys49 PLA_2s isolated from *B. mattogrossensis* (BmaTX-I and BmaTX-II) venom were able to induce the release of IL-1β by murine neutrophils in culture [146] and that BatroxPLA_2, an acidic $sPLA_2$ from *B. atrox* venom, induced the release of IL-6, PGE_2, and LTB_4 from murine macrophages in culture [140]. In this context, the involvement of inflammasomes in the production of IL-1β induced by *Bothrops* $sPLA_2s$ was recently investigated. The participation of NLP3 inflammasome via the activation of caspase 1 in the production of IL-1β induced by BthTX-I, a Lys49-PLA_2 from *B. jararacussu* venom, injected into mouse gastrocnemius muscle was reported [147]. In addition, the participation of inflammasomes in BthTX-I-induced production of IL-1β was demonstrated in peritoneal macrophages. This effect was demonstrated to be dependent on caspase 1/11, ASC, and NLRP3 and was associated with the release of ATP and activation of P2X7 receptors [148]. Despite the importance of cytokines, chemokines, and eicosanoids in orchestrating the events of inflammation and the potent proinflammatory effects triggered by viperid $sPLA_2s$, including those from *Bothrops* genus, a complete picture of the inflammatory mediators released by immunocompetent cells upon stimulus by *Bothrops* $sPLA_2s$ has yet to be further investigated. Moreover, the mechanisms involved in the production and release of these mediators and the possible crosstalk between them remain to be better clarified. Regarding the mechanisms involved in the biosynthesis of lipid mediators induced by *Bothrops* $sPLA_2s$, the progress made is presented in this review as a separate item (Section 2.2).

It is well recognized that the activation of innate effector functions, such as phagocytosis, and the production of microbicidal substances in leukocytes are critical for host defense and tissue repair. Regarding phagocytosis, studies have demonstrated the activity of *Bothrops* $sPLA_2s$ to induce phagocytosis following the activation of distinct receptors in immune-competent cells. In this sense, it was demonstrated that MT-II and MT-III, isolated from *B. asper* snake venom, can directly stimulate phagocytosis by macrophages in culture. MT-II significantly increased phagocytosis mediated by all classes of receptors, whereas MT-III increased phagocytosis via only mannose and beta-glucan receptors. This suggests that although the catalytic activity of *Bothrops* $sPLA_2s$ is not an essential requirement for enhancing macrophage phagocytosis, it may drive the class of phagocytosis receptors involved in this process. Molecular regions distinct from the catalytic network are likely involved in this effect [138]. In addition, the signaling pathways mediating zymosan phagocytosis, induced by both MT-II and MT-III, were investigated, with a focus on lipid second messengers. This study demonstrated that whereas the effect of MT-III, catalytically active, was dependent on the activation of endogenous $iPLA_2$, the effect of MT-II was dependent on both endogenous $iPLA_2$ and $cPLA_2$. Likewise, COX-2 and 5-LO-derived metabolites in addition to PAF were involved in the signaling events required for phagocytosis induced by both venom $sPLA_2$s [138]. In line with these data, BaltTX-I, devoid of catalytic activity and isolated from *B. alternatus* snake venom, was reported to activate the phagocytosis of serum-opsonized zymosan by murine macrophages, indicating the involvement of complement receptors. In addition, the participation of PKC was demonstrated. Nonetheless, BaltTX-II, a catalytically active $sPLA_2$ isolated from the same venom did not stimulate phagocytosis in macrophages, lending support to previous findings that the catalytic activity of *Bothrops* $sPLA_2$s is not essential for the stimulation of phagocytosis via complement receptor [149]. In addition, the $sPLA_2$s isolated from Panamanian *B. asper* snake venom, pMTX-III (catalytically active Asp49) and pMTX-II and -IV, two enzymatically inactive Lys49 isoforms, were described to induce phagocytosis via mannose receptor and superoxide production in macrophages [150]. The mechanisms underlying the differences between the catalytic and non-catalytic active *Bothrops* PLA_2s, regarding the activation of phagocytosis in macrophages and the participation of distinct receptors in their effects, require further clarification.

Concomitantly with phagocytosis, there is an increase in the oxidative metabolism, also referred to as respiratory burst, in leukocytes. In this context, the literature reveals that viperid $sPLA_2$s can trigger the respiratory burst in immunocompetent cells. In the first study describing the ability of *Bothrops* $sPLA_2$ to induce the release of microbicidal agents, the authors demonstrated that MT-II and MT-III, isolated from *B. asper* snake venom, induced the release of H_2O_2 by macrophages, with MT-III being the more potent stimulator [151]. In agreement with this evidence, it has been demonstrated that BaltTX-I and BaltTX-II from *B. alternatus* snake venom induced superoxide production by macrophages in culture in a process mediated by PKC [149]. In addition, other authors have revealed that the three $sPLA_2$s from *B. atrox* venom, namely BaTX-I, a Lys49 variant devoid of catalytic activity; BaTX-II, a catalytically active Asp49; and $BaPLA_2$, an acidic Asp49 $sPLA_2$ induced the release of the superoxide anion by the J774A.1 lineage macrophages in culture [152]. BaTX-I was the only $sPLA_2$ able to stimulate complement receptor-mediated phagocytosis, but all studied $sPLA_2$s could increase the macrophage lysosomal volume [152]. These data demonstrate the ability of *Bothrops* PLA_2s to trigger the respiratory burst, which is an essential process for the elimination of harmful agents. Although the structural determinants of such an effect were not investigated, it is likely that neither the enzymatic activity nor the basic or acidic characteristic of PLA_2 is essential for the activation of the respiratory burst.

An additional defensive strategy important for host defense is the neutrophil extracellular trap, or 'NET'. The formation of NET (NETosis) occurs through the release of nuclear DNA, forming a sticky 'net' of extracellular fibers that can halt the dissemination of pathogens and toxins [153,154]. Despite its importance in the inflammatory response, little attention has been paid to the involvement of this defense mechanism in the effects of viperid $sPLA_2$s. Yet, a report indicates that BaTX-II, an Asp49 PLA_2 isolated from *B. atrox* snake venom, can activate human neutrophils in culture to produce hydrogen peroxide via the PI3K signaling pathway. Furthermore, this $sPLA_2$ stimulated neutrophils to secrete MPO, NETs, and inflammatory mediators, including IL-1β, IL-8, and LTB_4 [155]. Therefore, the activation of neutrophilic functions, including toxin trapping and inactivation, is likewise involved in the inflammatory response to *Bothrops* $sPLA_2$s. Further studies are necessary to amplify the knowledge regarding the participation of NETs in inflammation induced by *Bothrops* spp. $sPLA_2$s. Interestingly, in contrast to the reported ability of *Bothrops* $sPLA_2$s to activate distinct inflammatory functions in leukocytes, a report revealed that CB (Crotoxin B), a catalytically active $sPLA_2$ isolated from *Crotalus durissus terrificus*, which is a subunit of crotoxin complex [156,157], could, per se, display inhibitory effects in macrophage functions, including spreading and phagocytosis [158]. Such an inhibitory effect suggests an anti-inflammatory activity for this particular viperid $sPLA_2$ [159]. In agreement with this idea, CB was reported to reduce the release of inflammatory cytokines, including IL-6 and TNF-α, and increase the release of PGE_2 and lipoxin A4, both immunomodulatory lipid mediators, in dendritic cells [160]. A summary of the inflammatory activities of $svPLA_2s$ is illustrated in Figure 1. In Table 1, the $svPLA_2s$-induced inflammatory responses are summarized according to the amino acid residue at position 49 and basic and acidic characteristics.

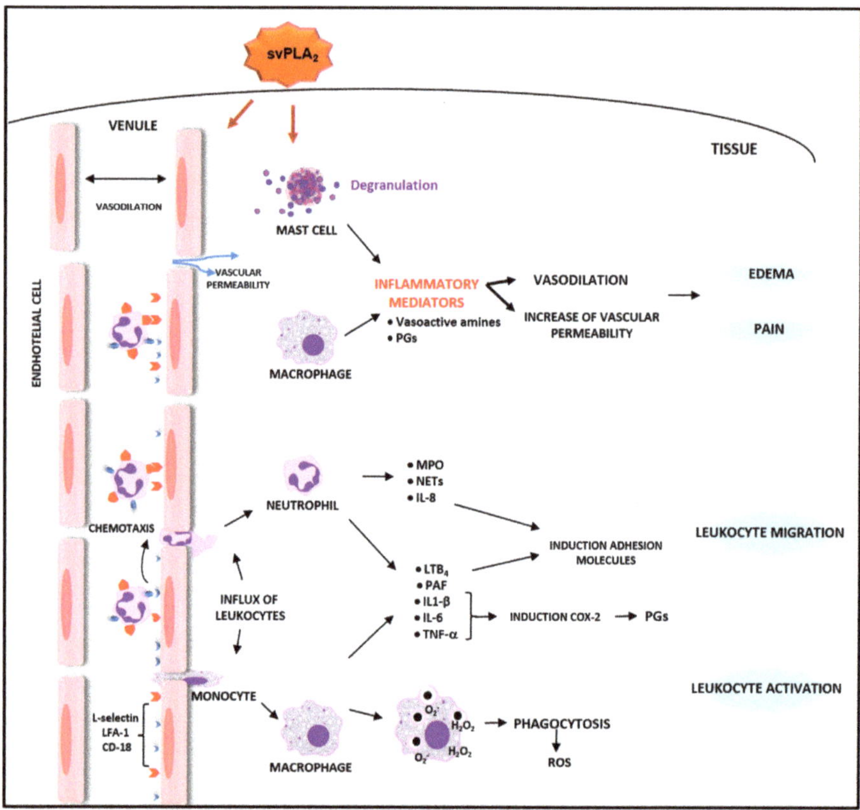

Figure 1. Scheme of inflammatory activities of $svPLA_2s$. The $svPLA_2s$ induce inflammatory events, characterized by activation of innate immune cells and endothelial cells and release of several inflammatory mediators that interfere in the vascular dynamic. $svPLA_2s$ induce mast cells degranulation and activation of resident macrophages with release of inflammatory mediators such as prostaglandins (PGs), histamine, serotonin, and substance P, which lead to vasodilation, increase of vascular permeability, culminating in edema formation and pain. In addition, $svPLA_2s$ activate phagocytosis by macrophages and increase the local production of oxygen reactive species (ROS). Furthermore, $svPLA_2s$, along with vascular alterations and produced inflammatory mediators, increase the expression of adhesion molecules such as LFA, CD-18 and L-selectin. These adhesion molecules, in turn, promote chemotaxis and leukocyte migration. The $svPLA_2s$ induce production of myeloperoxidase (MPO) and release of NETs by neutrophils. Both neutrophils and macrophages release proinflammatory mediators such as platelet-activating factor (PAF), IL-8, LTB_4, IL-1β, IL-6, and TNF-α. These last three mediators are involved in the upregulation of COX-2 isoform, and release of PGs, thus amplifying the inflammatory response induced by $svPLA_2s$.

Table 1. Inflammatory activities of basic and acidic svPLA$_2$s isoforms.

PLA$_2$	Origin	Basic or Acid	Type of PLA$_2$ Variant	Inflammatory Activity/ Experimental Model	Refs.
Piratoxin-I	*B. pirajai*	Basic	Lys49	Increase in vascular permeability (in vivo) Mast cell degranulation, neutrophil chemotaxis (in vitro)	[125,136]
P-1	*B. neuwiedii*	Acidic	nd	Edema (in vivo)	[126]
P-2	*B. neuwiedii*	Acidic	nd	Edema (in vivo)	[126]
SIIISPIIA	*B. jararacussu*	Acidic	Asp49	Edema (in vivo)	[129]
SIIISPIIB	*B. jararacussu*	Acidic	Asp49	Edema (in vivo)	[129]
SIIISPIIIA	*B. jararacussu*	Acidic	Asp49	Edema (in vivo)	[129]
SIIISPIIIB	*B. jararacussu*	Acidic	Asp49	Edema (in vivo)	[129]
BintTX-I	*B. insularis*	Acidic	Asp49	Edema (in vivo)	[130]
Bothropstoxin-I (BthTX-I)	*B. jararacussu*	Basic	Lys49	Edema, leukocyte migration, mast cell degranulation (in vivo) Neutrophil chemotaxis, activation of inflammasome (in vitro)	[128,131,136,147,148]
Bothropstoxin-II (BthTX-II)	*B. jararacussu*	Basic	Asp49	Edema, leukocyte migration, mast cell degranulation (in vivo) Neutrophil chemotaxis (in vitro)	[128,131,136]
Myotoxin-II (MT-II)	*B. asper*	Basic	Lys49	Increase in vascular permeability, leukocyte migration, release of mediators, hyperalgesia, eicosanoid production, COX-2 expression (in vivo) Phagocytosis, H_2O_2 production COX-2 expression, lipid droplet formation (in vitro)	[123,126,135,138,161–167]
Myotoxin-III (MT-III)	*B. asper*	Basic	Asp49	Increase in vascular permeability, leukocyte migration, release of mediators, hyperalgesia, eicosanoid production; COX-2 expression (in vivo) Phagocytosis, H_2O_2 production, COX-2 expression, lipid droplet formation, preadipocyte activation (in vitro)	[123,126,135,138,162–166,168–170]
BnSP-7	*B. pauloensis*	Basic	Lys49	Edema (in vivo)	[139]
BatroxPLA$_2$	*B. atrox*	Acidic	Asp49	Leukocyte chemotaxis, mediators release (in vivo) Mast cell degranulation (in vitro)	[140]
BJ-PLA$_2$-I	*B. jararaca*	Acidic	Asp49	Leukocyte migration, mediators release (in vivo)	[141]
Bbil-TX	*B. bilineata*	Basic	nd	Neutrophil migration, mediators release (in vivo)	[145]
BmaTX-I	*B. mattogrossensis*	Basic	Lys49	Mediator release (in vitro)	[146]
BmaTX-II	*B. mattogrossensis*	Basic	Lys49	Mediator release (in vitro)	[146]
BaltTX-I	*B. alternatus*	Basic	Lys49	Phagocytosis, superoxide production (in vitro)	[149]
BaltTX-II	*B. alternatus*	Basic	Asp49	Superoxide production (in vitro)	[149]
pMTX-II	*B. asper*	Basic	Lys49	Phagocytosis, superoxide production (in vitro)	[150]
pMTX-III	*B. asper*	Basic	Asp49	Phagocytosis, superoxide production (in vitro)	[150]
pMTX-IV	*B. asper*	Basic	Lys49	Phagocytosis, superoxide production (in vitro)	[150]
BaTX-I	*B. atrox*	Basic	Lys49	Superoxide production, lipid droplet formation (in vitro)	[155]

Table 1. *Cont.*

PLA_2	Origin	Basic or Acid	Type of PLA_2 Variant	Inflammatory Activity/ Experimental Model	Refs.
BaTX-II	*B. atrox*	Basic	Asp49	Superoxide and H_2O_2 production, MPO release, NET formation, lipid droplet formation (in vitro)	[155]
$BaPLA_2$	*B. atrox*	Acidic	Asp49	Superoxide production, lipid droplet formation (in vitro)	[140,155]
$BaPLA_2I$	*B. atrox*	Basic	nd	Mast cell degranulation, edema (in vivo)	[132]
$BaPLA_2III$	*B. atrox*	Neutral	nd	Mast cell Degranulation, edema (in vivo)	[132]

nd, not described.

2.2. *Influence of Bothrops svPLA$_2$s on Pathways of Arachidonic Acid Metabolism*

It is well established that sPLA$_2$s play key modulatory roles in numerous cellular processes in physiological and pathological conditions by regulating the release of AA from membrane phospholipids [27,171]. It has long been recognized that the AA-derived lipid mediators are potent mediators of inflammation [83]. The AA is rapidly metabolized by several enzyme complexes, including cyclooxygenases (COX), lipoxygenases (LOX), and cytochrome P450 (CYP450). These enzymatic pathways promote the synthesis of oxygenated and bioactive products, generically called eicosanoids, which include prostaglandins (PG), leukotrienes (LT), hydroperoxyeicosatetraenoic acids (HPETEs), hydroxyeicosatetraenoic acids (HETEs), epoxides (EETs), and lipoxins (LX) [172–178]. A summary of the cascades involved in biosynthesis of eicosanoids is shown in Figure 2.

It is important to emphasize that COX-1 is a constitutive isoform present in most tissues and is responsible for generating PGs for diverse physiological functions [179–182]. In contrast, COX-2 is upregulated by inflammatory cytokines and growth factors [183,184] and is constitutively expressed in some tissues [185,186].

Regarding the production of inflammatory eicosanoids by *Bothrops* svPLA$_2$, studies have demonstrated that the intraperitoneal injection of MT-III [187] and MT-II [162] in mice induced an early and transient release of PGD_2, followed by a rapid and sustained release of PGE_2. Likewise, in mice injected with BatroxPLA$_2$ [140], from *B. atrox* snake venom, and BJ-PLA$_2$-I from *B. jararaca* [141], an early release of PGE_2 was observed. The in vivo experimental models of previous studies have revealed that *B. asper* sPLA$_2$s induce the release of other eicosanoids, such as thromboxane A_2 (TXA_2) and LTB_4 [123]. Moreover, an Asp49 svPLA$_2$ from *B. atrox* venom [140] stimulated the production of LTB_4, lipoxin, and PGE_2.

PGE_2 and PGD_2 are important modulators of vasodilation, and PGE_2 can potentiate an increase in vascular permeability, promoted by mediators of this phenomenon, with a consequent formation of edema [188,189]. Studies using pharmacological treatment with non-steroidal anti-inflammatory compounds were crucial in demonstrating the participation of these COXs-derived lipid mediators on edema [126,190] and hyperalgesia [135], induced by *B. asper* sPLA$_2$s. In addition, studies demonstrating that MT-III and MT-II upregulated COX-2 protein expression in peritoneal leukocytes without altering the constitutive expression of COX-1 evidenced the ability of these venom PLA$_2$s to influence downstream cyclooxygenase isozymes and suggested this as a mechanism by which these sPLA$_2$s induced the production of prostaglandins [162,163]. Moreover, these findings suggested that the catalytic activity of these bothropic PLA$_2$s did not contribute to the induction of PG biosynthesis, since MT-II, devoid of catalytic activity, caused the same effect.

In this regard, studies have demonstrated that the IκB phosphorylation inhibitor TPCK effectively prevented both MT-II- or MT-III-induced COX-2 expression, suggesting that the activation of NF-κB was critical for the induction of COX-2 expression by these bothropic svPLA$_2$s. The involvement of NF-κB as the mechanism underlying this venom sPLA$_2$s-induced upregulation of COX-2 expression was further confirmed by results that revealed the inhibition of the NF-κB nuclear translocation site, markedly reduced svPLA$_2$s-induced COX-2 expression and, as a consequence, reduced PGE$_2$ production by macrophages in culture [162,164].

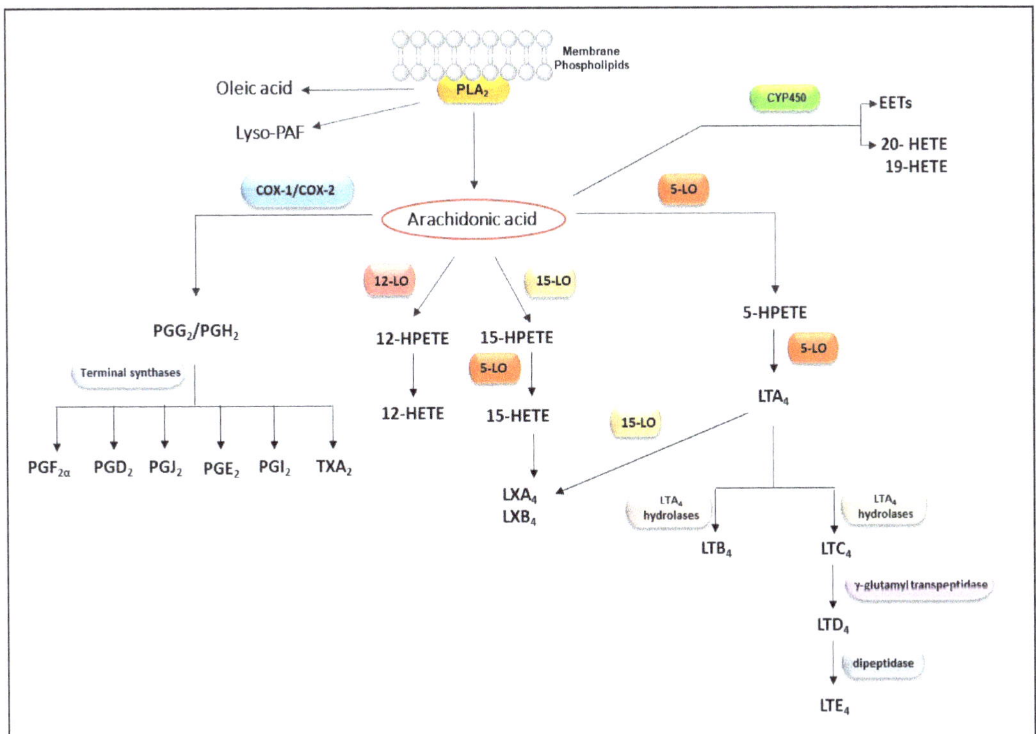

Figure 2. Scheme of arachidonic acid metabolism by several enzymatic pathways leading to production of bioactive lipid mediators. Abbreviations: (PLA$_2$) phospholipase A$_2$, (Lyso-PAF) lysophospholipid-platelet-activating factor, (COX-1) cyclooxygenase-1, (COX-2) cyclooxygenase-2, (5-LO) 5-lipoxygenase, (15-LO) 15-lipoxygenase, (12-LO) 12-lipoxygenase, cytochrome P450 (CYP450), (PGG$_2$) prostaglandin G$_2$, (PGH$_2$) prostaglandin H$_2$, (TXA$_2$) thromboxane A$_2$, (PGE$_2$) prostaglandin E$_2$, (PGD$_2$) prostaglandin D$_2$, (PGJ$_2$) prostaglandin J$_2$, (PGF$_{2\alpha}$) prostaglandin F$_2$ alpha, (PGI$_2$) prostacyclin, (15-HPETE) 15-hydroperoxyeicosatetraenoic, (15-HETE) 15-hydroxyeicosatetraenoic acid, (12-HPETE) 12-hydroperoxyeicosatetraenoic, (12-HETE) 12-hydroxyeicosatetraenoic acid, (5-HPETE) 5-hydroperoxyeicosatetraenoic, (5-HETE) 5-hydroxyeicosatetraenoic acid, (LTA$_4$) leukotriene A$_4$, (LTB$_4$) leukotriene B$_4$, (LTC4) leukotriene C$_4$, (LTD$_4$) leukotriene D$_4$, (LTE$_4$) leukotriene E$_4$, (LXA$_4$) lipoxin A$_4$, (LXB$_4$), lipoxin B$_4$, (19-HETE) 19-hydroxyeicosatetraenoic acid, (20-HETE) 20-hydroxyeicosatetraenoic acid, (EETs) epoxyeicosatrienoic acids.

Studies employing mouse resident peritoneal macrophages or neutrophils in culture revealed that viperids sPLA$_2$s induced a marked release of PGE$_2$ in cell supernatants, accompanied by the release of AA [162,165,166,191]. These data support the results in vivo and serve as evidence that immune innate leukocytes, such as resident macrophages and neutrophils, are important sources of PGs under in vivo stimuli by sPLA$_2$ from *Bothrops* spp. snake venoms. Interestingly, studies have demonstrated that the incubation of resident peritoneal macrophages with MT-II or MT-III significantly increased the concentration of AA [162,164]. Although the release of AA induced by the Asp49 sPLA$_2$ was approximately 20 times greater than that induced by MT-II, it demonstrated that the catalytic activity of viperid sPLA$_2$s was not an essential requirement for inducing COX-2 expression and PGE$_2$ production. According to Kini and Evans (1995) [192], in mechanisms independent on catalytic activity, as in the case of MT-II, the interaction of sPLA$_2$s to acceptor regions can cause the biological effect or interfere with the interaction of target proteins with their physiological ligands. Furthermore, some effects may result from combinations of both enzymatic and non-enzymatic mechanisms [192], leading to the activation of several signaling pathways; this should be considered when interpreting the effects of group IIA Lys49 svPLA$_2$s. In this context, since crosstalk among sPLA$_2$, cPLA$_2$, and iPLA$_2$s has been demonstrated to occur in several physiological and inflammatory conditions [193–196], the contribution of prey/victim cPLA$_2$ and/or iPLA$_2$ to the increased production of PGs and upregulation of COX-2 protein expression induced by svPLA$_2$ variants was evaluated in diverse in vitro experimental models [162,165,187]. Thus, the pharmacological treatment of cells with the cPLA$_2$ inhibitor but not the iPLA$_2$ inhibitor decreased the release of AA and the production of PGE$_2$ and PGD$_2$ induced by svPLA$_2$. In contrast, these pretreatments did not modify the MT-III-induced COX-2 expression but reduced the COX-2 expression induced by MT-II. These results demonstrate that cPLA$_2$ is required for distinct actions of MT-II in the PG biosynthetic pathway in macrophages [162,187]. This is consistent with the reported functional cooperation between intracellular PLA$_2$s and GIIA sPLA$_2$ for PG biosynthetic responses in several other cell systems [171,197–199]. The role of cPLA$_2$ as a key enzyme in supplying AA for COX-2-dependent PGs production is well established [171,200,201]. Taken together, the available data demonstrate that the Asp49 svPLA$_2$s are functionally coupled with cPLA$_2$, since prior activation of cPLA$_2$ is required for MT-III to act with downstream enzymes for PG biosynthesis in macrophages and neutrophils [165,187]. Interestingly, the association of Lys49 sPLA$_2$ with cPLA$_2$, in addition to being important for the supply of AA for the production of PG, appears to modulate the transcription and protein expression of COX-2 inflammatory isoform. The mechanisms involved in the coupling between the venom GIIA sPLA$_2$s and mammalian cPLA$_2$ have yet to be investigated. One possibility is that GIIA svPLA$_2$s activate cPLA$_2$ by distinct signaling cascades that mimic the transducing mechanism conveyed by physiological activators of cPLA$_2$, such as MAPKs, since this enzyme family is likewise important for the activation of NF-κB [202]. In line with this concept, the Asp49 svPLA$_2$, MT-III, in addition to Lys49 PLA$_2$, MT-II, from *B. asper* snake venom, were revealed to stimulate the phosphorylation of protein kinases, including the MAPKs, such as p38MAPK and ERK1/2 [167,169,171]. Moreover, other protein kinases, including PI3K, PKC, and PTK, were reported to be phosphorylated in macrophages stimulated by both MT-III and MT-II [162,164]. In this regard, studies have demonstrated that MT-III (Asp49 sPLA$_2$) stimulates PKC and p38MAPK pathways to positively modulate PGE$_2$ production and COX-2 expression via NF-κB, while MT-II (Lys49 PLA$_2$) displays similar effects by activating PKC, ERK1/2, and PTK in murine peritoneal macrophages [167,169,171]. Since PTK is involved in the activation of MAPKs, which, in turn, are essential for cPLA$_2$ activation, this signaling protein might be involved in the activation of cPLA$_2$ by MT-II [202–204]. Furthermore, another pathway implicated in the release of PGE$_2$ and the expression of COX-2 induced by MT-III was demonstrated to be independent of NF-κB activation. This pathway involved the activation of ERK1/2 by the 12-HETE pathway, the main product of 12-LO [166]. The involvement of another transcription factor in these MT-III-induced effects was suggested by the authors. Together, these findings reveal the variety and complexity of the mechanisms involved in the effects of svPLA$_2$ leading to the generation of lipid mediators. The signaling molecules and pathways acting in an innate immune cell (macrophage) upon stimulus either by Asp49 or Lys49 svPLA$_2$ are summarized in Figure 3.

Although the participation of the M-type PLA$_2$ receptor or another type of interaction of *Bothrops* sPLA$_2$ with membrane sites for the stimulation of signal transduction pathways has not yet been demonstrated, a study revealed for the first time the involvement of the TLRs in the inflammatory response induced by MT-III (Asp49-PLA$_2$) from *B. asper* in macrophages [205]. The involvement of TLR2 and MyD88 adapter molecules was demonstrated to be critical in producing PGs, COX-2 protein expression, and cytokines IL-1β and IL-10 induced by this svPLA$_2$. An indirect mechanism for the activation of

TLRs through the release of DAMPs was suggested by the authors, since the analysis of the fatty acids released by the hydrolysis of membrane phospholipids by MT-III revealed high levels of oleic and palmitic acids. In this context, it is known that arachidonic, oleic, and palmitic acids produced by membrane cleavage by $sPLA_2$s are important bioactive mediators involved in the induction and release of COX-2 and PGE_2, respectively, through the activation of intracellular signaling mechanisms in several cell types [206–209].

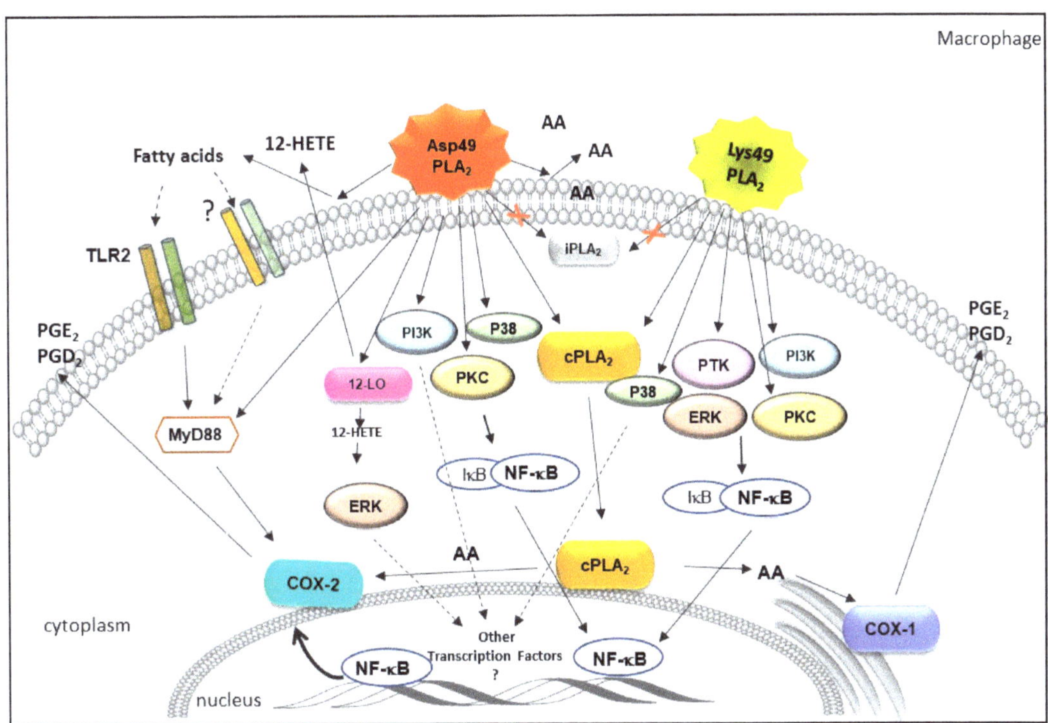

Figure 3. Schematic representation of signaling pathways stimulated by Asp49 and Lys49 PLA_2s from *B. asper* snake venom to produce prostaglandins in macrophages. Asp49 PLA_2 induces AA and fatty acids release from macrophage membrane. Free fatty acids can act as DAMPs and activate TLR2 or other TLRs (still unknown), via activation of adapter protein MyD88, leading to COX-2 protein expression and release of PGs, Asp49PLA_2 upregulates the 12-LO pathway, culminating the release of 12-HETE. 12-HETE, in turn, activates the ERK1/2 pathway, leading to COX-2 protein expression and PG release, independent on NF-κB translocation. Asp49PLA_2 also activates the signaling protein PI3K leading to COX-2 expression and production of PGs independent on NF-κB activation. Asp49 PLA_2 also activates PKC and p38 MAPK pathways promoting COX-2 expression and production of PGs via NF-κB activation. In addition, Asp49PLA_2 provides AA for activation of COX-1 activity which is followed by production of proinflammatory PGs. Asp49 and Lys49 PLA_2s both produce PGs by pathways independent on $iPLA_2$. Although both $sPLA_2$ from bothropic venom produce PGs via crosstalk with $cPLA_2$, only Lys49PLA_2 induces COX-2 expression dependent on this cytosolic PLA_2. Lys49PLA_2 activates signaling pathways mediated by p38 MAPK, PTK, PKC, and ERK1/2. All these kinase pathways, except for p38 MAPK, are involved in NF-κB activation and COX-2 protein expression and PG production. Full arrows indicate actions already studied and demonstrated. Dotted arrows indicate hypothesized or unknown effects.

2.3. *Bothrops svPLA$_{2s}$ Trigger Lipid Accumulation in Immunocompetent Cells*

As mentioned, the high enzymatic activity exhibited by ophidian $sPLA_2$ provides a microenvironment rich in free fatty acids that exert stimulating effects in immunocompetent cells, leading to the biosynthesis of lipid mediators. Moreover, excess free fatty acids exert cytotoxic effects that trigger the activation of cellular mechanisms capable of converting free fatty acids into metabolites of lower toxicity, known as neutral lipids (triacylglycerol and cholesterol—energetic body reserve) [210]. In recent decades, numerous studies have demonstrated the activation of intracellular metabolic pathways responsible for the metabolization of free fatty acids in neutral lipids and the consequent formation of dynamic organelles called lipid droplets (LDs) [211,212]. These organelles are composed of a hydrophobic neutral lipid core surrounded by a phospholipid monolayer membrane, which contains numerous proteins related to cellular activation in addition to structural proteins, such as perilipin 2 (PLIN2), which plays an important role in LD assembly and the formation of foam macrophage [213], marker cells in metabolic diseases, such as atherosclerosis and obesity [214]. LDs are commonly present in adipocytes due to the already-established role of these cells in the supply of energy in mammalian organisms [215]. In addition to the relevance regarding lipid homeostasis, the direct relationship between increased LD formation and inflammatory processes was evidenced by numerous studies [216–218]. In this sense, mammalian group IIA $sPLA_2s$ have been identified as potential plasma biomarkers for diseases related to lipid imbalance, such as atherosclerotic cardiovascular disease and obesity [219]. Alongside a marked inflammatory reaction, these metabolic diseases are characterized by lipid accumulation in immunocompetent cells [220,221]. In line with the ability of $svPLA_2s$ to elicit an inflammatory response characterized by a high level of inflammatory mediators and free fatty acids, it was demonstrated that MT-III (Asp49 PLA_2), isolated from *B. asper* venom, induced LD formation enriched by PLIN2 protein in mice peritoneal macrophages [170]. This effect was likewise observed in rat vascular smooth muscle cells isolated from the thoracic aorta stimulated by MT-III [168]. Moreover, the ability of MT-II, a Lys49 PLA_2 homologue devoid of catalytic activity from *B. asper* venom, to directly activate macrophages to form LDs was reported [167]. This effect was reproduced by a synthetic peptide corresponding to the C-terminal sequence 115–129 of MT-II, evidencing the critical role of C-terminus for the MT-II-induced effect [167]. Similarly, BaTX-I, a catalytically inactive Lys49 variant; BaTX-II, a catalytically active Asp49; and $BaPLA_2$, an acidic catalytically active Asp49 PLA_2, isolated from *B. atrox* snake venom, increased the number of LDs on murine macrophages cell line J774A.1 [152]. The formation of LDs upon stimulus by MT-III was likewise demonstrated in human monocytes of peripheral blood [222].

Considering the above information, LD formation induced by $svPLA_2s$ in phagocytes proved to be inherent to the action of $sPLA_2s$ regardless of the catalytic activity. The capability of *Bothrops* $svPLA_2s$ to induce the formation of LDs is related to the activation of PRRs of an innate immune response, kinase proteins, and intracellular PLA_2 signaling pathways involved in cellular metabolism, proliferation, and differentiation [170,205]. Hence, by using gene knockout mice cells and a pharmacological approach, the participation of TLR2, MYD 88 adaptor protein, and CD36 in LD formation in macrophages in culture has been reported [170,205]. In addition, the upregulation of CD36 receptors was observed in these cells. Considering the participation of the MYD 88 adaptor molecule, the involvement of other TLRs has yet to be investigated. In line with the reported Asp49 $svPLA_2$ action on PRRs, the upregulation of both SRA-1, from the scavenger receptor family, and LOX-1, an LDL receptor, was demonstrated in mouse aortic smooth muscle cells (VSMCs) [168]. These findings demonstrate that the inflammatory response elicited by Asp49 $svPLA_2$ likewise involved the upregulation of PRRs associated with lipid uptake in immunocompetent cells. This fact indicates that $svPLA_2s$ can be useful tools in studies aiming to understand the diseases associated with lipid imbalance.

In addition to providing the synthesis of mediators, TLR2 activation elicited by MT-III action was related to cytoskeleton activation [205], a critical step in the transport of structural proteins into LDs, such as PLIN2. Cytoskeletal activation involves the activation of kinase proteins [223]. Consistent with this information, the participation of kinase proteins in LD formation induced by both Asp49 and Lys49 $svPLA_2s$ from *B. asper* venom has been demonstrated through a pharmacological approach and the detection of phosphorylated kinase proteins. Yet, MT-II- and MT-III-induced LD biogenesis is dependent on the activation of PKC, PI3K, p38MAPK, and ERK1/2 signaling pathways in mice peritoneal macrophages [224]. It is well known that PKC regulates a variety of processes associated with lipid droplet biology, such as adipocyte differentiation [225], magnolol-induced lipolysis [226], cholesterol-induced targeting of caveolin to lipid droplets [227], and the expression of the PAT family [228]. Hence, the activation of the PKC signaling pathway in macrophages stimulated by the *Bothrops* PLA_2s MT-II and MT-III in peritoneal macrophages may be implicated in an increase in PLIN2 protein expression, since LD formation induced by both $svPLA_2$s has always been accompanied by an increase in PLIN2 protein expression in macrophages.

PI3K/AKT is a classical pathway involved in insulin resistance, cell growth, and lipid metabolism associated with the inhibition of cholesterol efflux leading to LD formation [229]. In this sense, the participation of the PIK3 signaling pathway in MT-III-induced LDs in macrophages and vascular smooth muscle cells stimulated with MT-III has been demonstrated [167,224]. In the case of vascular smooth muscle cells, the activation of PI3K was related to the uptake of fatty acids to LDs by macropinocytosis [167]. Furthermore, the findings that MT-III increases phagocytic activity and upregulates macrophage markers in VSMCs reinforce the importance of this class of enzymes as inducers of factors implicated in the formation of foamy cells in both mononuclear phagocytic cells and VSMCs, which are key elements in the development of metabolic diseases.

The MAPK signaling pathway has been revealed to mediate the activation of intracellular PLA_2s in physiologic and inflammatory contexts [230,231]. It has been demonstrated that the ERK1/2 signaling pathway is implicated in LD formation via the activation of phospholipase D (PLD) and phosphorylation of dynein [232,233]. Consistent with this information, the critical role of ERK1/2 for LD formation induced by MT-II and MT-III in macrophages and VSMCs was reported [167,224]. Regarding p38 MAPK, the literature evidenced its importance in the development of atherosclerosis [234] and the apoptosis of foam macrophages. Macrophage death is a feature of atherosclerotic plaque linked to necrosis and plaque destabilization [235]. Interestingly, although MT-III has been demonstrated to activate apoptotic pathways, including the p38 MAPK signaling pathway, DGAT, ACAT, $cPLA_2$, and LD formation in macrophages, no change in cell viability was observed. Further studies may clarify this lack of apoptotic effect.

It is known that the MAP kinases signaling pathway is implicated in intracellular PLA_2 activation, including the Ca^{2+}-dependent cytosolic group IVA PLA_2 ($cPLA_{2\alpha}$) and the Ca^{2+}-independent group VIA PLA_2 ($iPLA_2$) involved in both physiological and pathophysiological conditions [236,237]. The biogenesis of LDs in CHO-K1 cells submitted to an enriched environment of fatty acids demonstrated ERK, p38, and JNK signaling pathway activation, with JNK cascade being responsible for $cPLA_{2\alpha}$ phosphorylation in this event. Of note, $cPLA_2$ was likewise implicated in LD biogenesis stimulated by MT-III associated with the activation of ERK and the p38 MAPK signaling pathway [224]. Considering the ability of $cPLA_2$ to mobilize cell membrane fatty acids [238], the aforementioned activation should amplify the action of MT-III and provide a greater substrate for the metabolization and formation of LDs. Moreover, the biogenesis of LDs induced by the $svPLA_2$s MT-III and MT-II is dependent on the activation of $iPLA_2$ signaling pathways [167,224]. This signaling pathway was associated with the processing of fatty acids into triacylglycerol, a relevant component in the constitution of LDs [214]. Hence, the crosstalk already evidenced between intracellular PLA_2 and $svPLA_2$ to elicit inflammatory conditions [169] might contribute to elucidating mechanisms related to the formation of LDs.

The peroxisome proliferator-activated receptors (PPARs) are transcription factors belonging to the family of nuclear receptors that regulate glucose homeostasis, inflammation, and lipid metabolism. Three proteins, encoded by distinct genes, have been identified: PPAR-α, PPAR-β/δ, and PPAR-γ, which control gene expression by binding to PPREs in the promoters [239]. The activation of PPARs is a tightly regulated process implicated in the control of lipid homeostasis, which involves the biogenesis of LDs and protein expression involved in lipid uptake, including PRRs and structural protein PLIN2, and enzymes implicated in neutral lipid synthesis (triacylglycerol and cholesterol) [240]. PPARs have been demonstrated to increase in foam macrophages [241]. In this sense, it was revealed that MT-III induced the upregulation of the transcription factors PPAR-γ and PPAR-β/δ, in addition to the translocation of these factors to the nucleus of mouse peritoneal macrophages. The pharmacological blockage of the PPAR-β/δ transcription factor abolished the increase in PLIN2 and CD36 protein expression induced by MT-III. Moreover, the PPAR-γ blockage caused a reduction in LD formation and abolished CD36 receptor protein expression induced by MT-III. Since an increased expression of CD36 and PLIN2 is related to macrophage differentiation into foam cells [242,243], these findings suggest that MT-III induces foam cell formation by this route. In addition, MT-III caused an increase in the levels of triacylglycerol and cholesterol due to the uptake of free fatty acids. These effects were mediated by DGAT and ACAT enzymes, which are involved in the synthesis of triacylglycerol and cholesterol, respectively [244]. In agreement with this study, a significant increase in triacylglycerol and cholesterol levels was observed in human monocytes under MT-III stimulation [222]. This effect was dependent on fatty acid reacylation. Moreover, the fatty acid composition of triacylglycerol and cholesterol induced by MT-III was compatible with fatty acids released by the enzymatic action of this $svPLA_2$ on cell membranes. According to the above information, the mechanisms triggered by MT-III, in both mice peritoneal macrophages and human monocytes, align with macrophages differentiation into foamy cells, a cell type characteristic of inflammatory metabolic diseases, such as atherosclerosis [245]. Similarly, MT-III could stimulate LD formation in VSCMs. This lipid accumulation was likewise mediated by the activation of transcription factors PPAR-γ and PPAR-β/δ and DGAT and ACAT enzymes. Moreover, it is noteworthy that VSCMs under stimulation of MT-III exhibited an increase in the protein levels of PRRs, SRA-1 (scavenger receptor type 1), and LOX-1 (lectin-like oxidized low-density lipoprotein receptor-1). Interestingly, the blockage of these receptors did not alter the formation of LDs induced by MT-III, but the upregulation of LOX-1 was associated with an increased uptake of acetylated-low density lipoprotein (acLDL) in VSMCs stimulated by this $svPLA_2$. This higher uptake of acLDL by VSMCs identifies new pathways involved in the accumulation of lipids triggered by a $sPLA_2$ that is not directly linked to the reacylation of free fatty acids. In addition, lipid accumulation induced by MT-III in VSCMs was related to the expression of ATP-binding cassette transporters ABCA1 and ABCG1, responsible for the efflux of cholesterol of macrophage-derived foam cells [242]. Although the signaling pathway by which MT-III induces an increased expression of the factors implicated in lipid homeostasis has not been fully elucidated, these studies have broadened the knowledge about the actions of $svPLA_2$s on the formation of LDs and the synthesis of lipid mediators and provided new insights into the actions of group IIA $sPLA_2$s in diseases related to lipid imbalance. The pathways and factors involved in lipid accumulation in an innate immune cell (macrophage) upon stimulus by $svPLA_2$s are summarized in Figure 4.

Another aspect related to the metabolism of free fatty acids is the biosynthesis of lipid mediators [246–248]. On the one hand, it is known that the synthesis of eicosanoids is closely related to the triggering of the inflammatory process induced by $svPLA_2$s, regardless of the catalytic activity on the cell membranes [13]. In this sense, the immunofluorescence approach has demonstrated that LDs stimulated by $svPLA_2$s synthesize PGE_2 [167,249]. On the other hand, some eicosanoids are implicated in the resolution of inflammatory processes, such as PGJ_2. This mediator was co-located in LDs in macrophage peritoneal mice stimulated by MT-II [161], indicating, for the first time, that the LDs not only are related to the production of inflammatory mediators but also might play a role in regulating this process.

Adipose tissue is the principal organ responsible for balancing energy metabolism in the mammalian body. An imbalance in adipose tissue functions is linked to the triggering of the inflammatory process observed in metabolic diseases, including obesity [250,251]. Recently, it has been demonstrated that MT-III activated proinflammatory mechanisms in 3T3-L1 preadipocytes, including the biosynthesis of PGE_2 and PGI_2, lipid mediators implicated in preadipocytes differentiation into mature adipocytes, and IL-6 and MCP-1 [169]. In these cells, PGE_2 production induced by MT-III was dependent on $cPLA_2$ activation, the upregulation of COX-2 and mPGES-1, and the engagement of the PGE_2 EP4 receptor. In addition, the release of IL-6 and MCP-1 was dependent on EP4 or EP3 activation, respectively. These data indicate that the production of PGE_2 is critical for the activation of proinflammatory pathways associated with cytokine production in preadipocytes stimulated by this $svPLA_2$. Furthermore, MT-III upregulated the gene expression of the adipokines leptin and adiponectin in preadipocytes [169]. These mediators have been described as regulating appetite and satiety, glucose and lipid metabolism, inflammation, and immune functions [252–255]. Although the mechanisms related to the release of adipokines and the ability of MT-III to induce lipid accumulation in adipocytes have not yet been investigated, these data offer new directions for investigating the actions triggered by $svPLA_2$s and mammalian GIIA $sPLA_2$s.

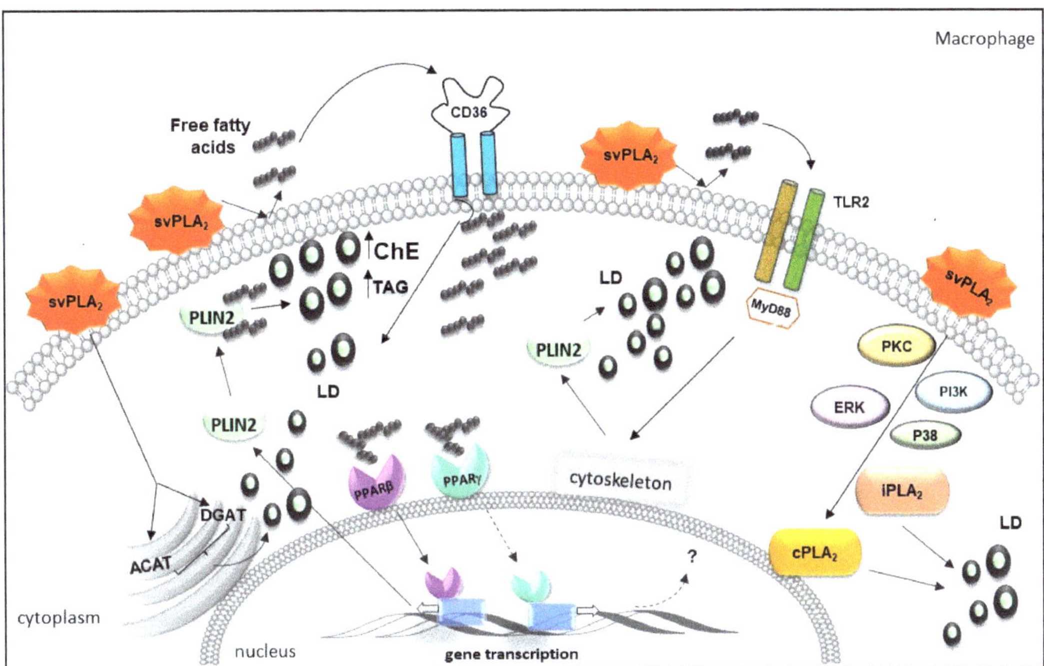

Figure 4. Schematic representation of the mechanisms and factors involved in lipid accumulation induced by $svPLA_2$s, in macrophages. $svPLA_2$ acts on membrane phospholipids generating free fatty acids, which are ligands of and may activate the TLR2, CD36, and cytoplasmic transcription receptors and factors PPARs. $svPLA_2$ induces the activation of transcription factors PPAR-γ and PPAR-δ/β and increases protein expression of PPARs and CD36. PPAR-γ, PPAR-β/δ, TLR2/MyD88, and CD36 receptors, as well as DGAT and ACAT enzymes are involved in the lipid droplets formation stimulated by MT-III. PPAR-β/δ, but not PPAR-γ, is implicated in upregulation of PLIN2 protein expression, induced by MT-III. Moreover, TLR2 and the Myd88 adaptor molecules participate in the recruitment of the PLIN2 protein via cytoskeleton activation stimulated by MT-III. In addition, LD formation induced by $svPLA_2$ is dependent on activation of PKC, PI3K, p38MAP, ERK, $cPLA_2$, and $iPLA_2$ signaling pathways.

3. Conclusions

The existing literature demonstrates that the $svPLA_2$s trigger a cascade of inflammatory events including edema formation, leukocyte recruitment into tissues, release of a complex network of inflammatory mediators, and increased oxidative stress in experimental animal models that mimic the inflammatory responses elicited by viperid snake venoms, especially those from the *Bothrops* genus, in the victims. The catalytic activity of the svPLA$_2$s is not strictly required by these proteins for the triggering of all the inflammatory responses, since the catalytically inactive Lys49 PLA$_2$ variants can display inflammatory events that are qualitatively similar to those of Asp49 PLA$_2$s. In addition to cell migration, the svPLA$_2$s can activate distinct functions of immunocompetent cells that include phagocytosis, the respiratory burst, NET formation, production of cytokines, chemokines and multiple reactive cleavage products such as lysophospholipids, polyunsaturated fatty acids and eicosanoids, as well as formation of LDs. The highly complex network of mediators, particularly lipid mediators, modulates a variety of inflammatory events triggered by this class of snake venom toxins. The effects triggered by svPLA$_2$s in inflammatory cells that lead to generation of lipid mediators has been associated with the activation of distinct signaling pathways of inflammatory kinase proteins by mechanisms dependent and independent of NF-kB. Moreover, the inflammatory response elicited by svPLA$_2$s in leukocytes also involves upregulation of PRRs of innate immune response, the crosstalk between the svPLA$_2$ and intracellular PLA$_2$s, and upregulation of factors implicated in lipid homeostasis. Although much has been learned regarding the inflammatory actions of svPLA$_2$s, many knowledge gaps still exist and need to be addressed. There is still considerable work to be done before we fully understand the complex interactions that occur among svPLA$_2$s and immunocompetent cells and tissues that lead to inflammation. The cell acceptors and/or receptors involved in the actions of svPLA$_2$s in these cells and the signaling pathways elicited and how they interact with each other remain to be clarified. In addition, the actual types and subtypes of receptors activated by the principal mediators produced by svPLA$_2$s and the mechanisms involved in coupling between the svPLA$_2$s and endogenous PLA$_2$s have yet to be investigated. Recently, the stimulatory activity of a svPLA$_2$ on adipose tissue cells leading to increased biosynthesis of PGE$_2$ and other inflammatory mediators including adipokines was demonstrated. This information offers new directions for investigating the actions triggered by svPLA$_2$s and mammalian GIIA sPLA$_2$ and gives insights into the potential role of the adipocytes as target cells for viperid snake venoms. Finally, a deeper and comprehensive understanding of the mechanisms underlying the inflammatory actions of svPLA$_2$s will give new insights into (i) the actions of group IIA sPLA$_2$s in diseases related to lipid imbalance and inflammation and (ii) a better understanding of the pathophysiology of *Bothrops* envenomation. Within this frame, the acquired knowledge might pave the way for the development of novel therapeutic approaches aimed at counteracting the prominent inflammation caused by *Bothrops* snakebite envenoming.

Author Contributions: Conceptualization, C.T., V.M., E.L. and C.M.F.; writing—original draft, C.T., V.M., E.L., C.M.F., P.M.J. and R.M.-M.; writing—review and editing, C.T., V.M., E.L., C.M.F., P.M.J. and R.M.-M.; visualization, V.M. and R.M.-M.; supervision, C.T. All authors have read and agreed to the published version of the manuscript.

Funding: This research was funded by Fundação Butantan, São Paulo, Brazil. E.L. is a recipient of a postdoctural fellowship from Coordenação de Aperfeicçoamento de Pessoal de Nível Superior (CAPES). C.T. is a recipient of a Conselho Nacional de Desenvolvimento Científico e Tecnológico fellowship (PQ-CNPq), grant number 310930/201.7. R.M.M. is recipient of a doctoral fellowship from Fundação Butantan, grant FB 001/0708/003.434/2021.

Institutional Review Board Statement: Not applicable.

Informed Consent Statement: Not applicable.

Data Availability Statement: Not applicable.

Conflicts of Interest: The authors declare no conflict of interest.

References

1. Teixeira, C.F.P.; Cury, Y.; Oga, S.; Jancar, S. Hyperalgesia induced by *Bothrops jararaca* venom in rats: Role of eicosanoids and platelet activating factor (PAF). *Toxicon* **1994**, *32*, 419–426. [CrossRef]
2. Teixeira, C.; Cury, Y.; Moreira, V.; Picolo, G.; Chaves, F. Inflammation induced by *Bothrops asper* venom. *Toxicon* **2009**, *54*, 67–76. [CrossRef]
3. Mamede, C.C.N.; de Sousa Simamoto, B.B.; da Cunha Pereira, D.F.; de Oliveira Costa, J.; Ribeiro, M.S.M.; de Oliveira, F. Edema, hyperalgesia and myonecrosis induced by Brazilian bothropic venoms: Overview of the last decade. *Toxicon* **2020**, *187*, 10–18. [CrossRef] [PubMed]
4. Gutiérrez, J.M.; Rucavado, A.; Chaves, F.; Díaz, C.; Escalante, T. Experimental pathology of local tissue damage induced by *Bothrops asper* snake venom. *Toxicon* **2009**, *54*, 958–975. [CrossRef] [PubMed]
5. *Ministério da Saúde Guia de Vigilância Epidemiológica*; Ministério da Saúde: Brasília, Brazil, 2019; ISBN 9788533416321.
6. Echeverría, S.; Leiguez, E.; Guijas, C.; do Nascimento, N.G.; Acosta, O.; Teixeira, C.; Leiva, L.C.; Rodríguez, J.P. Evaluation of pro-inflammatory events induced by Bothrops alternatus snake venom. *Chem. Biol. Interact.* **2018**, *281*, 24–31. [CrossRef] [PubMed]
7. Jorge, R.J.B.; Monteiro, H.S.A.; Gonçalves-Machado, L.; Guarnieri, M.C.; Ximenes, R.M.; Borges-Nojosa, D.M.; Luna, K.P.D.O.; Zingali, R.B.; Corrêa-Netto, C.; Gutiérrez, J.M.; et al. Venomics and antivenomics of *Bothrops erythromelas* from five geographic populations within the *Caatinga ecoregion* of northeastern Brazil. *J. Proteom.* **2015**, *114*, 93–114. [CrossRef] [PubMed]
8. Nicolau, C.A.; Carvalho, P.C.; Junqueira-de-Azevedo, I.L.M.; Teixeira-Ferreira, A.; Junqueira, M.; Perales, J.; Neves-Ferreira, A.G.C.; Valente, R.H. An in-depth snake venom proteopeptidome characterization: Benchmarking *Bothrops jararaca*. *J. Proteom.* **2017**, *151*, 214–231. [CrossRef]
9. Sousa, L.F.; Portes-Junior, J.A.; Nicolau, C.A.; Bernardoni, J.L.; Nishiyama, M.Y., Jr.; Amazonas, D.R.; Freitas-de-Sousa, L.A.; Mourão, R.H.; Chalkidis, H.M.; Valente, R.H.; et al. Functional proteomic analyses of Bothrops atrox venom reveals phenotypes associated with habitat variation in the Amazon. *J. Proteom.* **2017**, *159*, 32–46. [CrossRef] [PubMed]
10. De Farias, I.B.; de Morais-Zani, K.; Serino-Silva, C.; Sant'Anna, S.S.; da Rocha, M.M.; Grego, K.F.; Andrade-Silva, D.; Serrano, S.M.T.; Tanaka-Azevedo, A.M. Functional and proteomic comparison of *Bothrops jararaca* venom from captive specimens and the Brazilian Bothropic Reference Venom. *J. Proteom.* **2018**, *174*, 36–46. [CrossRef] [PubMed]
11. Mora-Obando, D.; Salazar-Valenzuela, D.; Pla, D.; Lomonte, B.; Guerrero-Vargas, J.A.; Ayerbe, S.; Gibbs, H.L.; Calvete, J.J. Venom variation in *Bothrops asper* lineages from North-Western South America. *J. Proteom.* **2020**, *229*, 103945. [CrossRef] [PubMed]
12. Gutiérrez, J.M.; Lomonte, B. Phospholipases A2: Unveiling the secrets of a functionally versatile group of snake venom toxins. *Toxicon* **2013**, *62*, 27–39. [CrossRef] [PubMed]
13. Bickler, P.E. Amplification of Snake Venom Toxicity by Endogenous Signaling Pathways. *Toxins* **2020**, *12*, 68. [CrossRef] [PubMed]
14. Hiu, J.J.; Yap, M.K.K. Cytotoxicity of snake venom enzymatic toxins: Phospholipase A2 and L-amino acid oxidase. *Biochem. Soc. Trans.* **2020**, *48*, 719–731. [CrossRef] [PubMed]
15. Gutiérrez, J.M.; Chaves, F.; Bolaños, R.; Cerdas, L.; Rojas, E.; Arroyo, O.; Portilla, E. Neutralizacion de los efectos locales del veneno de *Bothrops asper* por un antiveneno polivalente. *Toxicon* **1981**, *19*, 493–500. [CrossRef]
16. Gutiérrez, J.M.; Rojas, G.; Bogarín, G.; Lomonte, B. Evaluation of the neutralizing ability of antivenoms for the treatment of snake bite envenoming in Central America. *Envenomings Treat.* **1996**, 223–231.
17. Gutiérrez, J.; Avila, C.; Rojas, E.; Cerdas, L. An alternative in vitro method for testing the potency of the polyvalent antivenom produced in Costa Rica. *Toxicon* **1988**, *26*, 411–413. [CrossRef]
18. Cardoso, J.L.; Fan, H.W.; França, F.O.; Jorge, M.T.; Leite, R.P.; Nishioka, S.A.; Avila, A.; Sano-Martins, I.S.; Tomy, S.C.; Santoro, M.L. Randomized comparative trial of three antivenoms in the treatment of envenoming by lance-headed vipers (*Bothrops jararaca*) in São Paulo, Brazil. *Q. J. Med.* **1993**, *86*, 315–325. [PubMed]
19. Lewin, M.; Samuel, S.; Merkel, J.; Bickler, P. Varespladib (LY315920) appears to be a potent, broad-spectrum, inhibitor of snake venom phospholipase A2 and a possible pre-referral treatment for envenomation. *Toxins* **2016**, *8*, 248. [CrossRef]
20. Wang, Y.; Zhang, J.; Zhang, D.; Xiao, H.; Xiong, S.; Huang, C. Exploration of the inhibitory potential of varespladib for snakebite envenomation. *Molecules* **2018**, *23*, 391. [CrossRef] [PubMed]
21. Schaloske, R.H.; Dennis, E.A. The phospholipase A2 superfamily and its group numbering system. *Biochim. Biophys. Acta-Mol. Cell Biol. Lipids* **2006**, *1761*, 1246–1259. [CrossRef] [PubMed]
22. Six, D.A.; Dennis, E.A. The expanding superfamily of phospholipase A2 enzymes: Classification and characterization. *Biochim. Biophys. Acta-Mol. Cell Biol. Lipids* **2000**, *1488*, 1–19. [CrossRef]
23. Dennis, E.A.; Cao, J.; Hsu, Y.H.; Magrioti, V.; Kokotos, G. Phospholipase A2 enzymes: Physical structure, biological function, disease implication, chemical inhibition, and therapeutic intervention. *Chem. Rev.* **2011**, *111*, 6130–6185. [CrossRef]
24. Murakami, M.; Sato, H.; Taketomi, Y. Updating Phospholipase A2 Biology. *Biomolecules* **2020**, *10*, 1457. [CrossRef] [PubMed]
25. Lee, G.-H.; Fujita, M.; Nakanishi, H.; Miyata, H.; Ikawa, M.; Maeda, Y.; Murakami, Y.; Kinoshita, T. PGAP6, a GPI-specific phospholipase A2, has narrow substrate specificity against GPI-anchored proteins. *J. Biol. Chem.* **2020**, *295*, 14501–14509. [CrossRef]
26. Aloulou, A.; Rahier, R.; Arhab, Y.; Noiriel, A.; Abousalham, A. Phospholipases: An Overview. In *Methods in Molecular Biology*; Sandoval, G., Ed.; Springer: New York, NY, USA, 2018; Volume 1835, pp. 69–105, ISBN 978-1-4939-8671-2.

27. Quach, N.D.; Arnold, R.D.; Cummings, B.S. Secretory phospholipase A2 enzymes as pharmacological targets for treatment of disease. *Biochem. Pharmacol.* **2014**, *90*, 338–348. [CrossRef]
28. Harris, J.B.; Scott-Davey, T. Secreted phospholipases A2 of snake venoms: Effects on the peripheral neuromuscular system with comments on the role of phospholipases A2 in disorders of the CNS and their uses in industry. *Toxins* **2013**, *5*, 2533–2571. [CrossRef] [PubMed]
29. Triggiani, M.; Granata, F.; Frattini, A.; Marone, G. Activation of human inflammatory cells by secreted phospholipases A2. *Biochim. Biophys. Acta Mol. Cell Biol. Lipids* **2006**, *1761*, 1289–1300. [CrossRef] [PubMed]
30. Valentin, E.; Lambeau, G. What can venom phospholipases A2 tell us about the functional diversity of mammalian secreted phospholipases A2? *Biochimie* **2000**, *82*, 815–831. [CrossRef]
31. Kudo, I.; Murakami, M. Phospholipase A2 enzymes. *Prostaglandins Other Lipid Mediat.* **2002**, *68–69*, 3–58. [CrossRef]
32. Kini, R.M. Excitement ahead: Structure, function and mechanism of snake venom phospholipase A2 enzymes. *Toxicon* **2003**, *42*, 827–840. [CrossRef] [PubMed]
33. Lambeau, G.; Barhanin, J.; Schweitz, H.; Qar, J.; Lazdunski, M. Identification and properties of very high affinity brain membrane-binding sites for a neurotoxic phospholipase from the taipan venom. *J. Biol. Chem.* **1989**, *264*, 11503–11510. [CrossRef]
34. Lambeau, G.; Schmid-Alliana, A.; Lazdunski, M.; Barhanin, J. Identification and purification of a very high affinity binding protein for toxic phospholipases A2 in skeletal muscle. *J. Biol. Chem.* **1990**, *265*, 9526–9532. [CrossRef]
35. Silliman, C.C.; Moore, E.E.; Zallen, G.; Gonzalez, R.; Johnson, J.L.; Elzi, D.J.; Meng, X.; Hanasaki, K.; Ishizaki, J.; Arita, H.; et al. Presence of the M-type sPLA 2 receptor on neutrophils and its role in elastase release and adhesion. *Am. J. Physiol. Physiol.* **2002**, *283*, C1102–C1113. [CrossRef] [PubMed]
36. Granata, F.; Petraroli, A.; Boilard, E.; Bezzine, S.; Bollinger, J.; Del Vecchio, L.; Gelb, M.H.; Lambeau, G.; Marone, G.; Triggiani, M. Activation of Cytokine Production by Secreted Phospholipase A 2 in Human Lung Macrophages Expressing the M-Type Receptor. *J. Immunol.* **2005**, *174*, 464–474. [CrossRef]
37. Hanasaki, K.; Arita, H. Phospholipase A2 receptor: A regulator of biological functions of secretory phospholipase A2. *Prostaglandins Other Lipid Mediat.* **2002**, *68–69*, 71–82. [CrossRef]
38. Ezekowitz, R.A.B.; Stahl, P.D. The structure and function of vertebrate mannose lectin-like proteins. *J. Cell Sci.* **1988**, *1988*, 121–133. [CrossRef]
39. Taylor, M.E.; Conary, J.T.; Lennartz, M.R.; Stahl, P.D.; Drickamer, K. Primary structure of the mannose receptor contains multiple motifs resembling carbohydrate-recognition domains. *J. Biol. Chem.* **1990**, *265*, 12156–12162. [CrossRef]
40. Gantzel, R.H.; Kjær, M.B.; Laursen, T.L.; Kazankov, K.; George, J.; Møller, H.J.; Grønbæk, H. Macrophage Activation Markers, Soluble CD163 and Mannose Receptor, in Liver Fibrosis. *Front. Med.* **2021**, *7*, 615599. [CrossRef]
41. Gordon, S.; Clarke, S.; Greaves, D.; Doyle, A. Molecular immunobiology of macrophages: Recent progress. *Curr. Opin. Immunol.* **1995**, *7*, 24–33. [CrossRef]
42. Stoy, N. Macrophage Biology and Pathobiology in the Evolution of Immune Responses: A Functional Analysis. *Pathobiology* **2001**, *69*, 179–211. [CrossRef]
43. Hernández, M.; Burillo, S.L.; Crespo, M.S.; Nieto, M.L. Secretory Phospholipase A2 Activates the Cascade of Mitogen-activated Protein Kinases and Cytosolic Phospholipase A2 in the Human Astrocytoma Cell Line 1321N1. *J. Biol. Chem.* **1998**, *273*, 606–612. [CrossRef] [PubMed]
44. Kinoshita, E.; Handa, N.; Hanada, K.; Kajiyama, G.; Sugiyama, M. Activation of MAP kinase cascade induced by human pancreatic phospholipase A2 in a human pancreatic cancer cell line. *FEBS Lett.* **1997**, *407*, 343–346. [CrossRef]
45. Kundu, G.C.; Mukherjee, A.B. Evidence that porcine pancreatic phospholipase A2 via its high affinity receptor stimulates extracellular matrix invasion by normal and cancer cells. *J. Biol. Chem.* **1997**, *272*, 2346–2353. [CrossRef]
46. Triggiani, M.; Calabrese, C.; Granata, F.; Gentile, M.; Marone, G. Metabolism of Lipid Mediators in Human Eosinophils. In *Human Eosinophils*; KARGER: Basel, Switzerland, 2000; pp. 77–98.
47. Hara, S.; Kudo, I.; Matsuta, K.; Miyamoto, T.; Inoue, K. Amino Acid Composition and NH2-Terminal Amino Acid Sequence of Human Phospholipase A2 Purified from Rheumatoid Synovial Fluid1. *J. Biochem.* **1988**, *104*, 326–328. [CrossRef] [PubMed]
48. Touqui, L.; Alaoui-El-Azher, M. Mammalian Secreted Phospholipases A2 and Their Pathophysiolo-gical Significance in Inflammatory Diseases. *Curr. Mol. Med.* **2001**, *1*, 739–754. [CrossRef] [PubMed]
49. Liu, N.J.; Chapman, R.; Lin, Y.; Mmesi, J.; Bentham, A.; Tyreman, M.; Abraham, S.; Stevens, M.M. Point of care testing of phospholipase A2 group IIA for serological diagnosis of rheumatoid arthritis. *Nanoscale* **2016**, *8*, 4482–4485. [CrossRef] [PubMed]
50. Feldman, M.; Ginsburg, I. A Novel Hypothetical Approach to Explain the Mechanisms of Pathogenicity of Rheumatic Arthritis. *Mediterr. J. Rheumatol.* **2021**, *32*, 112–117. [CrossRef] [PubMed]
51. Aufenanger, J.; Samman, M.; Quintel, M.; Fassbender, K.; Zimmer, W.; Bertsch, T. Pancreatic phospholipase A2 activity in acute pancreatitis: A prognostic marker for early identification of patients at risk. *Clin. Chem. Lab. Med.* **2002**, *40*, 293–297. [CrossRef] [PubMed]
52. Zhang, K.J.; Zhang, D.L.; Jiao, X.L.; Dong, C. Effect of phospholipase A2 silencing on acute experimental pancreatitis. *Eur. Rev. Med. Pharmacol. Sci.* **2013**, *17*, 3279–3284.
53. Schröder, T.; Kivilaakso, E.; Kinnunen, P.K.J.; Lempinen, M. Serum Phospholipase A 2 in Human Acute Pancreatitis. *Scand. J. Gastroenterol.* **1980**, *15*, 633–636. [CrossRef]

54. Vadas, P.; Pruzanski, W. Role of secretory phospholipases A2 in the pathobiology of disease. *Lab. Investig.* **1986**, *55*, 391–404. [PubMed]
55. Ahmad, N.S.; Tan, T.L.; Arifin, K.T.; Ngah, W.Z.W.; Yusof, Y.A.M. High sPLA2-IIA level is associated with eicosanoid metabolism in patients with bacterial sepsis syndrome. *PLoS ONE* **2020**, *15*, e0230285. [CrossRef]
56. Haapamäki, M.M.; Grönroos, J.M.; Nurmi, H.; Irjala, K.; Alanen, K.A.; Nevalainen, T.J. Phospholipase A2 in serum and colonic mucosa in ulcerative colitis. *Scand. J. Clin. Lab. Investig.* **1999**, *59*, 279–287. [CrossRef]
57. Haapamäki, M.M.; Grönroos, J.M.; Nurmi, H.; Alanen, K.; Nevalainen, T.J. Gene Expression of Group Ii Phospholipase A2 in Intestine in Crohn's Disease. *Am. J. Gastroenterol.* **1999**, *94*, 713–720. [CrossRef]
58. Woodruff, T.M.; Arumugam, T.V.; Shiels, I.A.; Newman, M.L.; Ross, P.A.; Reid, R.C.; Fairlie, D.P.; Taylor, S.M. A potent and selective inhibitor of group IIa secretory phospholipase A2 protects rats from TNBS-induced colitis. *Int. Immunopharmacol.* **2005**, *5*, 883–892. [CrossRef] [PubMed]
59. Stadel, J.M.; Hoyle, K.; Naclerio, R.M.; Roshak, A.; Chilton, F.H. Characterization of phospholipase A2 from human nasal lavage. *Am. J. Respir. Cell Mol. Biol.* **1994**, *11*, 108–113. [CrossRef]
60. Kim, D.K.; Fukuda, T.; Thompson, B.T.; Cockrill, B.; Hales, C.; Bonventre, J.V. Bronchoalveolar lavage fluid phospholipase A2 activities are increased in human adult respiratory distress syndrome. *Am. J. Physiol. Cell. Mol. Physiol.* **1995**, *269*, L109–L118. [CrossRef] [PubMed]
61. Letsiou, E.; Htwe, Y.M.; Dudek, S.M. Secretory Phospholipase A2 Enzymes in Acute Lung Injury. *Cell Biochem. Biophys.* **2021**, *79*, 609–617. [CrossRef] [PubMed]
62. Granata, F.; Staiano, R.I.; Loffredo, S.; Petraroli, A.; Genovese, A.; Marone, G.; Triggiani, M. The role of mast cell-derived secreted phospholipases A2 in respiratory allergy. *Biochimie* **2010**, *92*, 588–593. [CrossRef]
63. Sun, C.Q.; Zhong, C.Y.; Sun, W.W.; Xiao, H.; Zhu, P.; Lin, Y.Z.; Zhang, C.L.; Gao, H.; Song, Z.Y. Elevated Type II Secretory Phospholipase A2 Increases the Risk of Early Atherosclerosis in Patients with Newly Diagnosed Metabolic Syndrome. *Sci. Rep.* **2016**, *6*, 34929. [CrossRef] [PubMed]
64. Menschikowski, M.; Hagelgans, A.; Siegert, G. Secretory phospholipase A2 of group IIA: Is it an offensive or a defensive player during atherosclerosis and other inflammatory diseases? *Prostaglandins Other Lipid Mediat.* **2006**, *79*, 1–33. [CrossRef] [PubMed]
65. Cunningham, T.J.; Yao, L.; Oetinger, M.; Cort, L.; Blankenhorn, E.P.; Greenstein, J.I. Secreted phospholipase A2 activity in experimental autoimmune encephalomyelitis and multiple sclerosis. *J. Neuroinflamm.* **2006**, *3*, 26. [CrossRef] [PubMed]
66. Mirtti, T.; Laine, V.J.O.; Hiekkanen, H.; Hurme, S.; Rowe, O.; Nevalainen, T.J.; Kallajoki, M.; Alanen, K. Group IIA phospholipase A2 as a prognostic marker in prostate cancer: Relevance to clinicopathological variables and disease-specific mortality. *Apmis* **2009**, *117*, 151–161. [CrossRef] [PubMed]
67. Dong, Z.; Liu, Y.; Scott, K.F.; Levin, L.; Gaitonde, K.; Bracken, R.B.; Burke, B.; Zhai, Q.J.; Wang, J.; Oleksowicz, L.; et al. Secretory phospholipase A2-IIa is involved in prostate cancer progression and may potentially serve as a biomarker for prostate cancer. *Carcinogenesis* **2010**, *31*, 1948–1955. [CrossRef]
68. Wang, M.; Hao, F.Y.; Wang, J.G.; Xiao, W. Group IIa secretory phospholipase A2 (sPLA2IIa) and progression in patients with lung cancer. *Eur. Rev. Med. Pharmacol. Sci.* **2014**, *18*, 2648–2654. [PubMed]
69. Chen, J.; Ye, L.; Sun, Y.; Takada, Y. A Concise Update on the Relevance of Secretory Phospholipase A2 Group IIA and its Inhibitors with Cancer. *Med. Chem.* **2017**, *13*, 606–615. [CrossRef] [PubMed]
70. Schalkwijk, C.; Pfeilschifter, J.; Märki, F.; van den Bosch, H. Interleukin-1β, tumor necrosis factor and forskolin stimulate the synthesis and secretion of group II phospholipase A2 in rat mesangial cells. *Biochem. Biophys. Res. Commun.* **1991**, *174*, 268–275. [CrossRef]
71. Divchev, D.; Schieffer, B. The secretory phospholipase A2 group IIA: A missing link between inflammation, activated renin-angiotensin system, and atherogenesis? *Vasc. Health Risk Manag.* **2008**, *4*, 597–604. [CrossRef]
72. Leistad, L.; Feuerherm, A.J.; Faxvaag, A.; Johansen, B. Multiple phospholipase A2 enzymes participate in the inflammatory process in osteoarthritic cartilage. *Scand. J. Rheumatol.* **2011**, *40*, 308–316. [CrossRef] [PubMed]
73. Beck, S.; Lambeau, G.; Scholz-Pedretti, K.; Gelb, M.H.; Janssen, M.J.W.; Edwards, S.H.; Wilton, D.C.; Pfeilschifter, J.; Kaszkin, M. Potentiation of Tumor Necrosis Factor α-induced Secreted Phospholipase A2 (sPLA2)-IIA Expression in Mesangial Cells by an Autocrine Loop Involving sPLA2 and Peroxisome Proliferator-activated Receptor α Activation. *J. Biol. Chem.* **2003**, *278*, 29799–29812. [CrossRef]
74. Wu, Y.; Li, Y.; Shang, M.; Jian, Y.; Wang, C.; Bardeesi, A.S.A.; Li, Z.; Chen, T.; Zhao, L.; Zhou, L.; et al. Secreted phospholipase A2 of Clonorchis sinensis activates hepatic stellate cells through a pathway involving JNK signalling. *Parasites Vectors* **2017**, *10*, 147. [CrossRef]
75. Sarate, R.M.; Chovatiya, G.L.; Ravi, V.; Khade, B.; Gupta, S.; Waghmare, S.K. sPLA 2 -IIA Overexpression in Mice Epidermis Depletes Hair Follicle Stem Cells and Induces Differentiation Mediated Through Enhanced JNK/c-Jun Activation. *Stem Cells* **2016**, *34*, 2407–2417. [CrossRef] [PubMed]
76. Baek, S.-H.; Lim, J.-H.; Park, D.-W.; Kim, S.-Y.; Lee, Y.-H.; Kim, J.-R.; Kim, J.-H. Group IIA secretory phospholipase A2 stimulates inducible nitric oxide synthase expression via ERK and NF-κB in macrophages. *Eur. J. Immunol.* **2001**, *31*, 2709–2717. [CrossRef]
77. Beck, G.C.; Yard, B.A.; Schulte, J.; Haak, M.; van Ackern, K.; van der Woude, F.J.; Kaszkin, M. Secreted phospholipases A2 induce the expression of chemokines in microvascular endothelium. *Biochem. Biophys. Res. Commun.* **2003**, *300*, 731–737. [CrossRef]
78. Medzhitov, R. Origin and physiological roles of inflammation. *Nature* **2008**, *454*, 428–435. [CrossRef] [PubMed]

79. Barton, G.M. A calculated response: Control of inflammation by the innate immune system. *J. Clin. Investig.* **2008**, *118*, 413–420. [CrossRef]
80. Kruger, P.; Saffarzadeh, M.; Weber, A.N.R.; Rieber, N.; Radsak, M.; von Bernuth, H.; Benarafa, C.; Roos, D.; Skokowa, J.; Hartl, D. Neutrophils: Between Host Defence, Immune Modulation, and Tissue Injury. *PLOS Pathog.* **2015**, *11*, e1004651. [CrossRef] [PubMed]
81. Bonecchi, R. Chemokines and chemokine receptors: An overview. *Front. Biosci.* **2009**, *14*, 540. [CrossRef]
82. Branco, A.C.C.C.; Yoshikawa, F.S.Y.; Pietrobon, A.J.; Sato, M.N. Role of Histamine in Modulating the Immune Response and Inflammation. *Mediat. Inflamm.* **2018**, *2018*, 9524075. [CrossRef] [PubMed]
83. Bennett, M.; Gilroy, D.W. Lipid Mediators in Inflammation. *Microbiol. Spectr.* **2016**, *4*, 4–6. [CrossRef]
84. Krieglstein, C. Adhesion molecules and their role in vascular disease. *Am. J. Hypertens.* **2001**, *14*, S44–S54. [CrossRef]
85. Kameritsch, P.; Renkawitz, J. Principles of Leukocyte Migration Strategies. *Trends Cell Biol.* **2020**, *30*, 818–832. [CrossRef]
86. Rosales, C.; Uribe-Querol, E. Phagocytosis: A Fundamental Process in Immunity. *Biomed Res. Int.* **2017**, *2017*, 9042851. [CrossRef] [PubMed]
87. Gordon, S. Phagocytosis: An Immunobiologic Process. *Immunity* **2016**, *44*, 463–475. [CrossRef]
88. Collin, M.; Ehlers, M. The carbohydrate switch between pathogenic and immunosuppressive antigen-specific antibodies. *Exp. Dermatol.* **2013**, *22*, 511–514. [CrossRef]
89. Medzhitov, R. TLR-mediated innate immune recognition. *Semin. Immunol.* **2007**, *19*, 1–2. [CrossRef]
90. Aderem, A.; Ulevitch, R.J. Toll-like receptors in the induction of the innate immune response. *Nature* **2000**, *406*, 782–787. [CrossRef] [PubMed]
91. Kawai, T.; Akira, S. TLR signaling. *Semin. Immunol.* **2007**, *19*, 24–32. [CrossRef]
92. Motta, V.; Soares, F.; Sun, T.; Philpott, D.J. NOD-Like Receptors: Versatile Cytosolic Sentinels. *Physiol. Rev.* **2015**, *95*, 149–178. [CrossRef]
93. Corridoni, D.; Simmons, A. Innate immune receptors for cross-presentation: The expanding role of NLRs. *Mol. Immunol.* **2019**, *113*, 6–10. [CrossRef]
94. McGettrick, A.F.; O'Neill, L.A. Localisation and trafficking of Toll-like receptors: An important mode of regulation. *Curr. Opin. Immunol.* **2010**, *22*, 20–27. [CrossRef]
95. Kawai, T.; Akira, S. The role of pattern-recognition receptors in innate immunity: Update on toll-like receptors. *Nat. Immunol.* **2010**, *11*, 373–384. [CrossRef] [PubMed]
96. Broz, P.; Dixit, V.M. Inflammasomes: Mechanism of assembly, regulation and signalling. *Nat. Rev. Immunol.* **2016**, *16*, 407–420. [CrossRef] [PubMed]
97. Rathinam, V.A.K.; Vanaja, S.K.; Fitzgerald, K.A. Regulation of inflammasome signaling. *Nat. Immunol.* **2012**, *13*, 333–342. [CrossRef]
98. Strowig, T.; Henao-Mejia, J.; Elinav, E.; Flavell, R. Inflammasomes in health and disease. *Nature* **2012**, *481*, 278–286. [CrossRef] [PubMed]
99. Schroder, K.; Tschopp, J. The Inflammasomes. *Cell* **2010**, *140*, 821–832. [CrossRef]
100. Malik, A.; Kanneganti, T.-D. Inflammasome activation and assembly at a glance. *J. Cell Sci.* **2017**, *130*, 3955–3963. [CrossRef]
101. Lin, Y.-Z.; Yao, S.; Veach, R.A.; Torgerson, T.R.; Hawiger, J. Inhibition of Nuclear Translocation of Transcription Factor NF-κB by a Synthetic Peptide Containing a Cell Membrane-permeable Motif and Nuclear Localization Sequence. *J. Biol. Chem.* **1995**, *270*, 14255–14258. [CrossRef] [PubMed]
102. Lawrence, T. The nuclear factor NF-kappaB pathway in inflammation. *Cold Spring Harb. Perspect. Biol.* **2009**, *1*, a001651. [CrossRef]
103. Kwiatkowska, K.; Sobota, A. Signaling pathways in phagocytosis. *Bioessays* **1999**, *21*, 422–431. [CrossRef]
104. Kitaura, J.; Eto, K.; Kinoshita, T.; Kawakami, Y.; Leitges, M.; Lowell, C.A.; Kawakami, T. Regulation of Highly Cytokinergic IgE-Induced Mast Cell Adhesion by Src, Syk, Tec, and Protein Kinase C Family Kinases. *J. Immunol.* **2005**, *174*, 4495–4504. [CrossRef]
105. Yang, Q.; Langston, J.C.; Tang, Y.; Kiani, M.F.; Kilpatrick, L.E. The role of tyrosine phosphorylation of protein kinase C delta in infection and inflammation. *Int. J. Mol. Sci.* **2019**, *20*, 1498. [CrossRef]
106. Fruman, D.A.; Chiu, H.; Hopkins, B.D.; Bagrodia, S.; Cantley, L.C.; Abraham, R.T. The PI3K Pathway in Human Disease. *Cell* **2017**, *170*, 605–635. [CrossRef]
107. Stephenson, J.D.; Shepherd, V.L. Purification of the human alveolar macrophage mannose receptor. *Biochem. Biophys. Res. Commun.* **1987**, *148*, 883–889. [CrossRef]
108. Ezekowitz, R.A.; Sastry, K.; Bailly, P.; Warner, A. Molecular characterization of the human macrophage mannose receptor: Demonstration of multiple carbohydrate recognition-like domains and phagocytosis of yeasts in Cos-1 cells. *J. Exp. Med.* **1990**, *172*, 1785–1794. [CrossRef]
109. Greenberg, S. Modular components of phagocytosis. *J. Leukoc. Biol.* **1999**, *66*, 712–717. [CrossRef] [PubMed]
110. Fernández, N.; Alonso, S.; Valera, I.; Vigo, A.G.; Renedo, M.; Barbolla, L.; Crespo, M.S. Mannose-Containing Molecular Patterns Are Strong Inducers of Cyclooxygenase-2 Expression and Prostaglandin E 2 Production in Human Macrophages. *J. Immunol.* **2005**, *174*, 8154–8162. [CrossRef]
111. McNally, A.K.; DeFife, K.M.; Anderson, J.M. Interleukin-4-induced macrophage fusion is prevented by inhibitors of mannose receptor activity. *Am. J. Pathol.* **1996**, *149*, 975–985.

112. Newton, R.; Holden, N. Inhibitors of p38 Mitogen-Activated Protein Kinase. *BioDrugs* **2003**, *17*, 113–129. [CrossRef]
113. Janssen, W.J.; Henson, P.M. Cellular Regulation of the Inflammatory Response. *Toxicol. Pathol.* **2012**, *40*, 166–173. [CrossRef]
114. Buckley, C.D.; Gilroy, D.W.; Serhan, C.N. Proresolving Lipid Mediators and Mechanisms in the Resolution of Acute Inflammation. *Immunity* **2014**, *40*, 315–327. [CrossRef]
115. Serhan, C.N.; Savill, J. Resolution of inflammation: The beginning programs the end. *Nat. Immunol.* **2005**, *6*, 1191–1197. [CrossRef] [PubMed]
116. Feehan, K.T.; Gilroy, D.W. Is Resolution the End of Inflammation? *Trends Mol. Med.* **2019**, *25*, 198–214. [CrossRef]
117. Netea, M.G.; Balkwill, F.; Chonchol, M.; Cominelli, F.; Donath, M.Y.; Giamarellos-Bourboulis, E.J.; Golenbock, D.; Gresnigt, M.S.; Heneka, M.T.; Hoffman, H.M.; et al. A guiding map for inflammation. *Nat. Immunol.* **2017**, *18*, 826–831. [CrossRef] [PubMed]
118. Kini, R.M.; Chan, Y.M. Accelerated Evolution and Molecular Surface of Venom Phospholipase A2 Enzymes. *J. Mol. Evol.* **1999**, *48*, 125–132. [CrossRef] [PubMed]
119. Krizaj, I.; Bieber, A.L.; Ritonja, A.; Gubensek, F. The primary structure of ammodytin L, a myotoxic phospholipase A2 homologue from Vipera ammodytes venom. *Eur. J. Biochem.* **1991**, *202*, 1165–1168. [CrossRef] [PubMed]
120. van den Berg, B.; Tessari, M.; de Haas, G.H.; Verheij, H.M.; Boelens, R.; Kaptein, R. Solution structure of porcine pancreatic phospholipase A2. *EMBO J.* **1995**, *14*, 4123–4131. [CrossRef]
121. Díaz, C.; Gutiérrez, J.; Lomonte, B.; Gené, J. The effect of myotoxins isolated from Bothrops snake venoms on multilamellar liposomes: Relationship to phospholipase A2, anticoagulant and myotoxic activities. *Biochim. Biophys. Acta-Biomembr.* **1991**, *1070*, 455–460. [CrossRef]
122. Lomonte, B.; Gutiérrez, J.M. Phospholipases A2 from viperidae snake venoms: How do they induce skeletal muscle damage? *Acta Chim. Slov.* **2011**, *58*, 647–659.
123. Zuliani, J.P.; Fernandes, C.M.; Zamuner, S.R.; Gutiérrez, J.M.; Teixeira, C.F.P. Inflammatory events induced by Lys-49 and Asp-49 phospholipases A2 isolated from *Bothrops asper* snake venom: Role of catalytic activity. *Toxicon* **2005**, *45*, 335–346. [CrossRef]
124. Teixeira, C.F.P.; Landucci, E.C.T.; Antunes, E.; Chacur, M.; Cury, Y. Inflammatory effects of snake venom myotoxic phospholipases A2. *Toxicon* **2003**, *42*, 947–962. [CrossRef]
125. Landucci, E.C.T.; De Castro, R.C.; Toyama, M.; Giglio, J.R.; Marangoni, S.; De Nucci, G.; Antunes, E. Inflammatory oedema induced by the Lys-49 phospholipase A2 homologue piratoxin-I in the rat and rabbit. Effect of polyanions and p-bromophenacyl bromide. *Biochem. Pharmacol.* **2000**, *59*, 1289–1294. [CrossRef]
126. Chaves, F.; León, G.; Alvarado, V.H.; Gutiérrez, J.M. Pharmacological modulation of edema induced by Lys-49 and Asp-49 myotoxic phospholipases A2 isolated from the venom of the snake *Bothrops asper* (*Terciopelo*). *Toxicon* **1998**, *36*, 1861–1869. [CrossRef]
127. Daniele, J.J.; Bianco, I.D.; Fidelio, G.D. Kinetic and Pharmacological Characterization of Phospholipases A2 from Bothrops neuwiedii Venom. *Arch. Biochem. Biophys.* **1995**, *318*, 65–70. [CrossRef]
128. Landucci, E.C.; Castro, R.C.; Pereira, M.F.; Cintra, A.C.; Giglio, J.R.; Marangoni, S.; Oliveira, B.; Cirino, G.; Antunes, E.; De Nucci, G. Mast cell degranulation induced by two phospholipase A2 homologues: Dissociation between enzymatic and biological activities. *Eur. J. Pharmacol.* **1998**, *343*, 257–263. [CrossRef]
129. Ketelhut, D.F.; Homem de Mello, M.; Veronese, E.L.; Esmeraldino, L.; Murakami, M.; Arni, R.; Giglio, J.; Cintra, A.C.; Sampaio, S. Isolation, characterization and biological activity of acidic phospholipase A2 isoforms from *Bothrops jararacussu* snake venom. *Biochimie* **2003**, *85*, 983–991. [CrossRef]
130. Cogo, J.C.; Lilla, S.; Souza, G.H.M.F.; Hyslop, S.; de Nucci, G. Purification, sequencing and structural analysis of two acidic phospholipases A2 from the venom of *Bothrops insularis* (jararaca ilhoa). *Biochimie* **2006**, *88*, 1947–1959. [CrossRef]
131. de Castro, R.; Landucci, E.C.; Toyama, M.; Giglio, J.; Marangoni, S.; De Nucci, G.; Antunes, E. Leucocyte recruitment induced by type II phospholipases A2 into the rat pleural cavity. *Toxicon* **2000**, *38*, 1773–1785. [CrossRef]
132. Kanashiro, M.M.; Rita de Cássia, M.E.; Petretski, J.H.; Prates, M.V.; Alves, E.W.; Machado, O.L.; da Silva, W.D.; Kipnis, T.L. Biochemical and biological properties of phospholipases A2 from *Bothrops atrox* snake venom. *Biochem. Pharmacol.* **2002**, *64*, 1179–1186. [CrossRef]
133. Metcalfe, D.D.; Baram, D.; Mekori, Y.A. Mast cells. *Physiol. Rev.* **1997**, *77*, 1033–1079. [CrossRef] [PubMed]
134. Metz, M.; Maurer, M. Mast cells—Key effector cells in immune responses. *Trends Immunol.* **2007**, *28*, 234–241. [CrossRef]
135. Chacur, M.; Longo, I.; Picolo, G.; Gutiérrez, J.M.; Lomonte, B.; Guerra, J.L.; Teixeira, C.F.P.; Cury, Y. Hyperalgesia induced by Asp49 and Lys49 phospholipases A2 from *Bothrops asper* snake venom: Pharmacological mediation and molecular determinants. *Toxicon* **2003**, *41*, 667–678. [CrossRef]
136. Gambero, A.; Landucci, E.C.T.; Toyama, M.H.; Marangoni, S.; Giglio, J.R.; Nader, H.B.; Dietrich, C.P.; De Nucci, G.; Antunes, E. Human neutrophil migration in vitro induced by secretory phospholipases A2: A role for cell surface glycosaminoglycans. *Biochem. Pharmacol.* **2002**, *63*, 65–72. [CrossRef]
137. Gambero, A.; Thomazzi, S.M.; Cintra, A.C.O.; Landucci, E.C.T.; De Nucci, G.; Antunes, E. Signalling pathways regulating human neutrophil migration induced by secretory phospholipases A2. *Toxicon* **2004**, *44*, 473–481. [CrossRef]
138. Zuliani, J.P.; Gutiérrez, J.M.; Teixeira, C. Signaling pathways involved in zymosan phagocytosis induced by two secreted phospholipases A2 isolated from *Bothrops asper* snake venom in macrophages. *Int. J. Biol. Macromol.* **2018**, *113*, 575–582. [CrossRef]

139. de Freitas Oliveira, C.; da Silva Lopes, D.; Mendes, M.M.; Homsi-Brandeburgo, M.I.; Hamaguchi, A.; de Alcântara, T.M.; Clissa, P.B.; de Melo Rodrigues, V. Insights of local tissue damage and regeneration induced by BnSP-7, a myotoxin isolated from Bothrops (neuwiedi) pauloensis snake venom. *Toxicon* **2009**, *53*, 560–569. [CrossRef] [PubMed]
140. Menaldo, D.L.; Bernardes, C.P.; Zoccal, K.F.; Jacob-Ferreira, A.L.; Costa, T.R.; Del Lama, M.P.F.M.; Naal, R.M.Z.G.; Frantz, F.G.; Faccioli, L.H.; Sampaio, S.V. Immune cells and mediators involved in the inflammatory responses induced by a P-I metalloprotease and a phospholipase A2 from Bothrops atrox venom. *Mol. Immunol.* **2017**, *85*, 238–247. [CrossRef]
141. Cedro, R.C.A.; Menaldo, D.L.; Costa, T.R.; Zoccal, K.F.; Sartim, M.A.; Santos-Filho, N.A.; Faccioli, L.H.; Sampaio, S.V. Cytotoxic and inflammatory potential of a phospholipase A2 from *Bothrops jararaca* snake venom. *J. Venom. Anim. Toxins Incl. Trop. Dis.* **2018**, *24*, 1–14. [CrossRef] [PubMed]
142. Arend, W.P.; Gabay, C. Cytokines in the rheumatic diseases. *Rheum. Dis. Clin. N. Am.* **2004**, *30*, 41–67. [CrossRef]
143. David, B.A.; Kubes, P. Exploring the complex role of chemokines and chemoattractants in vivo on leukocyte dynamics. *Immunol. Rev.* **2019**, *289*, 9–30. [CrossRef]
144. de Oliveira, S.; Rosowski, E.E.; Huttenlocher, A. Neutrophil migration in infection and wound repair: Going forward in reverse. *Nat. Rev. Immunol.* **2016**, *16*, 378–391. [CrossRef] [PubMed]
145. Corasolla Carregari, V.; Stuani Floriano, R.; Rodrigues-Simioni, L.; Winck, F.V.; Baldasso, P.A.; Ponce-Soto, L.A.; Marangoni, S. Biochemical, Pharmacological, and Structural Characterization of New Basic Bbil-TX from *Bothriopsis bilineata* Snake Venom. *Biomed Res. Int.* **2013**, *2013*, 612649. [CrossRef]
146. Moura, A.A.D.; Kayano, A.M.; Oliveira, G.A.; Setúbal, S.S.; Ribeiro, J.G.; Barros, N.B.; Nicolete, R.; Moura, L.A.; Fuly, A.L.; Nomizo, A.; et al. Purification and Biochemical Characterization of Three Myotoxins from *Bothrops mattogrossensis* Snake Venom with Toxicity against Leishmania and Tumor Cells. *Biomed Res. Int.* **2014**, *2014*, 195356. [CrossRef]
147. Boeno, C.N.; Paloschi, M.V.; Lopes, J.A.; Pires, W.L.; Setúbal, S.D.S.; Evangelista, J.R.; Soares, A.M.; Zuliani, J.P. Inflammasome Activation Induced by a Snake Venom Lys49-Phospholipase A2 Homologue. *Toxins* **2019**, *12*, 22. [CrossRef]
148. Ranéia e Silva, P.A.; de Lima, D.S.; Mesquita Luiz, J.P.; Câmara, N.O.S.; Alves-Filho, J.C.F.; Pontillo, A.; Bortoluci, K.R.; Faquim-Mauro, E.L. Inflammatory effect of Bothropstoxin-I from Bothrops jararacussu venom mediated by NLRP3 inflammasome involves ATP and P2X7 receptor. *Clin. Sci.* **2021**, *135*, 687–701. [CrossRef] [PubMed]
149. Setúbal, S.S.; Pontes, A.S.; Furtado, J.L.; Xavier, C.V.; Silva, F.L.; Kayano, A.M.; Izidoro, L.F.M.; Soares, A.M.; Calderon, L.A.; Stábeli, R.G.; et al. Action of two phospholipases A2 purified from *Bothrops alternatus* snake venom on macrophages. *Biochemistry* **2013**, *78*, 194–203. [CrossRef]
150. Rueda, A.Q.; Rodríguez, I.G.; Arantes, E.C.; Setúbal, S.S.; Calderon, L.D.A.; Zuliani, J.P.; Stábeli, R.G.; Soares, A.M. Biochemical Characterization, Action on Macrophages, and Superoxide Anion Production of Four Basic Phospholipases A 2 from Panamanian *Bothrops asper* Snake Venom. *Biomed Res. Int.* **2013**, *2013*, 789689. [CrossRef]
151. Zuliani, J.P.; Gutiérrez, J.M.; e Silva, L.L.C.; Sampaio, S.C.; Lomonte, B.; de Fátima Pereira Teixeira, C. Activation of cellular functions in macrophages by venom secretory Asp-49 and Lys-49 phospholipases A2. *Toxicon* **2005**, *46*, 523–532. [CrossRef]
152. Furtado, J.L.; Oliveira, G.A.; Pontes, A.S.; Setúbal, S.D.S.; Xavier, C.V.; Lacouth-Silva, F.; Lima, B.F.; Zaqueo, K.D.; Kayano, A.M.; Calderon, L.A.; et al. Activation of J77A.1 macrophages by three phospholipases A2 isolated from *Bothrops atrox* snake venom. *Biomed Res. Int.* **2014**, *2014*, 683123. [CrossRef] [PubMed]
153. Brinkmann, V.; Zychlinsky, A. Beneficial suicide: Why neutrophils die to make NETs. *Nat. Rev. Microbiol.* **2007**, *5*, 577–582. [CrossRef] [PubMed]
154. Castanheira, F.V.S.; Kubes, P. Neutrophils and NETs in modulating acute and chronic inflammation. *Blood* **2019**, *133*, 2178–2185. [CrossRef]
155. Setúbal, S.D.S.; Pontes, A.S.; Nery, N.M.; Rego, C.M.A.; Santana, H.M.; de Lima, A.M.; Boeno, C.N.; Paloschi, M.V.; Soares, A.M.; Zuliani, J.P. Human neutrophils functionality under effect of an Asp49 phospholipase A2 isolated from *Bothrops atrox* venom. *Toxicon X* **2020**, *6*, 100032. [CrossRef]
156. Bon, C.; Choumet, V.; Delot, E.; Faure, G.; Robbe-Vincent, A.; Saliou, B. Different Evolution of Phospholipase A 2 Neurotoxins (Beta-Neurotoxins) from Elapidae and Viperidae Snakes. *Ann. N. Y. Acad. Sci.* **1994**, *710*, 142–148. [CrossRef] [PubMed]
157. Rossetto, O.; Morbiato, L.; Caccin, P.; Rigoni, M.; Montecucco, C. Presynaptic enzymatic neurotoxins. *J. Neurochem.* **2006**, *97*, 1534–1545. [CrossRef] [PubMed]
158. Sampaio, S.C.; Brigatte, P.; Sousa-E-Silva, M.C.C.; Dos-Santos, E.C.; Rangel-Santos, A.C.; Curi, R.; Cury, Y. Contribution of crotoxin for the inhibitory effect of Crotalus durissus terrificus snake venom on macrophage function. *Toxicon* **2003**, *41*, 899–907. [CrossRef]
159. Sampaio, S.C.; Rangel-Santos, A.C.; Peres, C.M.; Curi, R.; Cury, Y. Inhibitory effect of phospholipase A2 isolated from Crotalus durissus terrificus venom on macrophage function. *Toxicon* **2005**, *45*, 671–676. [CrossRef] [PubMed]
160. Freitas, A.P.; Favoretto, B.C.; Clissa, P.B.; Sampaio, S.C.; Faquim-Mauro, E.L. Crotoxin Isolated from Crotalus durissus terrificus Venom Modulates the Functional Activity of Dendritic Cells via Formyl Peptide Receptors. *J. Immunol. Res.* **2018**, *2018*, 7873257. [CrossRef]
161. Giannotti, K.C.; Leiguez, E.; De Carvalho, A.E.Z.; Nascimento, N.G.; Matsubara, M.H.; Fortes-Dias, C.L.; Moreira, V.; Teixeira, C. A snake venom group IIA PLA2 with immunomodulatory activity induces formation of lipid droplets containing 15-d-PGJ2 in macrophages. *Sci. Rep.* **2017**, *7*, 4098. [CrossRef] [PubMed]

162. Moreira, V.; De Castro Souto, P.C.M.; Ramirez Vinolo, M.A.; Lomonte, B.; María Gutiérrez, J.; Curi, R.; Teixeira, C. A catalytically-inactive snake venom Lys49 phospholipase A2 homolog induces expression of cyclooxygenase-2 and production of prostaglandins through selected signaling pathways in macrophages. *Eur. J. Pharmacol.* **2013**, *708*, 68–79. [CrossRef]
163. Moreira, V.; Gutiérrez, J.M.; Amaral, R.B.; Zamunér, S.R.; de Fátima Pereira Teixeira, C. Effects of *Bothrops asper* snake venom on the expression of cyclooxygenases and production of prostaglandins by peritoneal leukocytes in vivo, and by isolated neutrophils and macrophages in vitro. *Prostaglandins Leukot. Essent. Fat. Acids* **2009**, *80*, 107–114. [CrossRef]
164. Moreira, V.; Lomonte, B.; Vinolo, M.A.R.; Curi, R.; Gutiérrez, J.M.; Teixeira, C. An asp49 phospholipase A2 from snake venom induces cyclooxygenase-2 expression and prostaglandin E2 production via activation of NF- κ B, p38MAPK, and PKC in macrophages. *Mediat. Inflamm.* **2014**. [CrossRef]
165. Moreira, V.; Gutiérrez, J.M.; Amaral, R.B.; Lomonte, B.; Purgatto, E.; Teixeira, C. A phospholipase A2 from *Bothrops asper* snake venom activates neutrophils in culture: Expression of cyclooxygenase-2 and PGE2 biosynthesis. *Toxicon* **2011**, *57*, 288–296. [CrossRef]
166. Moreira, V.; Gutiérrez, J.M.; Lomonte, B.; Vinolo, M.A.R.; Curi, R.; Lambeau, G.; Teixeira, C. 12-HETE is a regulator of PGE2 production via COX-2 expression induced by a snake venom group IIA phospholipase A2 in isolated peritoneal macrophages. *Chem. Biol. Interact.* **2020**, *317*, 108903. [CrossRef] [PubMed]
167. Cristina Giannotti, K.; Leiguez, E.; Moreira, V.; Nascimento, N.G.; Lomonte, B.; Gutiérrez, J.M.; Lopes De Melo, R.; Teixeira, C. A Lys49 phospholipase A2, isolated from *Bothrops asper* snake venom, induces lipid droplet formation in macrophages which depends on distinct signaling pathways and the C-terminal region. *Biomed Res. Int.* **2013**, *2013*, 807982. [CrossRef]
168. Giannotti, K.C.; Weinert, S.; Viana, M.N.; Leiguez, E.; Araujo, T.L.S.; Laurindo, F.R.M.; Lomonte, B.; Braun-Dullaeus, R.; Teixeira, C. A secreted phospholipase A2 induces formation of smooth muscle foam cells which transdifferentiate to macrophage-like state. *Molecules* **2019**, *24*, 3244. [CrossRef]
169. Leiguez, E.; Motta, P.; Maia Marques, R.; Lomonte, B.; Sampaio, S.V.; Teixeira, C. A Representative GIIA Phospholipase A2 Activates Preadipocytes to Produce Inflammatory Mediators Implicated in Obesity Development. *Biomolecules* **2020**, *10*, 1593. [CrossRef] [PubMed]
170. Leiguez, E.; Giannotti, K.C.; Do Nascimento Viana, M.; Matsubara, M.H.; Fernandes, C.M.; Gutiérrez, J.M.; Lomonte, B.; Teixeira, C. A snake venom-secreted phospholipase A2 induces foam cell formation depending on the activation of factors involved in lipid homeostasis. *Mediat. Inflamm.* **2018**, *2018*, 2547918. [CrossRef]
171. Murakami, M.; Nakatani, Y.; Atsumi, G.I.; Inoue, K.; Kudo, I. Regulatory functions of phospholipase A2. *Crit. Rev. Immunol.* **2017**, *37*, 121–180. [CrossRef] [PubMed]
172. Yokomizo, T.; Ogawa, Y.; Uozumi, N.; Kume, K.; Izumi, T.; Shimizu, T. cDNA Cloning and Mutagenesis Study of Leukotriene B4 12-Hydroxydehydrogenase. *Adv. Exp. Med. Biol.* **1997**, 151–156. [CrossRef]
173. Sarau, H.M.; Foley, J.J.; Schmidt, D.B.; Martin, L.D.; Webb, E.F.; Tzimas, M.N.; Breton, J.J.; Chabot-Fletcher, M.; Underwood, D.C.; Hay, D.W.P.; et al. In vitro and in vivo pharmacological characterization of SB 201993, an eicosanoid-like LTB4receptor antagonist with anti-inflammatory activity. *Prostaglandins Leukot. Essent. Fat. Acids* **1999**, *61*, 55–64. [CrossRef]
174. Serhan, C.N.; Takano, T.; Maddox, J.F. Aspirin-Triggered 15-Epi-Lipoxin A4 and Stable Analogs of Lipoxin A4 are Potent Inhibitors of Acute Inflammation. *Adv. Exp. Med. Biol.* **1999**, 133–149. [CrossRef]
175. Chen, J.-K.; Wang, D.-W.; Falck, J.R.; Capdevila, J.; Harris, R.C. Transfection of an Active Cytochrome P450 Arachidonic Acid Epoxygenase Indicates That 14,15-Epoxyeicosatrienoic Acid Functions as an Intracellular Second Messenger in Response to Epidermal Growth Factor. *J. Biol. Chem.* **1999**, *274*, 4764–4769. [CrossRef]
176. Calder, P.C. Eicosanoids. *Essays Biochem.* **2020**, *64*, 423–441. [CrossRef] [PubMed]
177. Samuelsson, B.; Dahlén, S.-E.; Lindgren, J.Å.; Rouzer, C.A.; Serhan, C.N. Leukotrienes and Lipoxins: Structures, Biosynthesis, and Biological Effects. *Science* **1987**, *237*, 1171–1176. [CrossRef] [PubMed]
178. Rocca, B.; FitzGerald, G.A. Cyclooxygenases and prostaglandins: Shaping up the immune response. *Int. Immunopharmacol.* **2002**, *2*, 603–630. [CrossRef]
179. O'Neill, G.P.; Ford-Hutchinson, A.W. Expression of mRNA for cyclooxygenase-1 and cyclooxygenase-2 in human tissues. *FEBS Lett.* **1993**, *330*, 157–160. [CrossRef]
180. Merlie, J.P.; Fagan, D.; Mudd, J.; Needleman, P. Isolation and characterization of the complementary DNA for sheep seminal vesicle prostaglandin endoperoxide synthase (cyclooxygenase). *J. Biol. Chem.* **1988**, *263*, 3550–3553. [CrossRef]
181. Funk, C.D.; Funk, L.B.; Kennedy, M.E.; Pong, A.S.; Fitzgerald, G.A. Human platelet/erythroleukemia cell prostaglandin G/H synthase: cDNA cloning, expression, and gene chromosomal assignment. *FASEB J.* **1991**, *5*, 2304–2312. [CrossRef]
182. Pannunzio, A.; Coluccia, M. Cyclooxygenase-1 (COX-1) and COX-1 inhibitors in cancer: A review of oncology and medicinal chemistry literature. *Pharmaceuticals* **2018**, *11*, 101. [CrossRef]
183. Pruzanski, W.; Stefanski, E.; Vadas, P.; Kennedy, B.P.; van den Bosch, H. Regulation of the cellular expression of secretory and cytosolic phospholipases A2, and cyclooxygenase-2 by peptide growth factors. *Biochim. Biophys. Acta-Mol. Cell Res.* **1998**, *1403*, 47–56. [CrossRef]
184. Martínez-Colón, G.J.; Moore, B.B. Prostaglandin E2 as a Regulator of Immunity to Pathogens. *Pharmacol. Ther.* **2018**, *185*, 135–146. [CrossRef]
185. Minghetti, L. Cyclooxygenase-2 (COX-2) in Inflammatory and Degenerative Brain Diseases. *J. Neuropathol. Exp. Neurol.* **2004**, *63*, 901–910. [CrossRef]

186. Zidar, N.; Odar, K.; Glavac, D.; Jerse, M.; Zupanc, T.; Stajer, D. Cyclooxygenase in normal human tissues—Is COX-1 really a constitutive isoform, and COX-2 an inducible isoform? *J. Cell. Mol. Med.* **2009**, *13*, 3753–3763. [CrossRef]
187. Moreira, V.; Gutiérrez, J.M.; Soares, A.M.; Zamunér, S.R.; Purgatto, E.; de Fátima Pereira Teixeira, C. Secretory phospholipases A2 isolated from *Bothrops asper* and from Crotalus durissus terrificus snake venoms induce distinct mechanisms for biosynthesis of prostaglandins E2 and D2 and expression of cyclooxygenases. *Toxicon* **2008**, *52*, 428–439. [CrossRef]
188. Gerritsen, M.E. Physiological and pathophysiological roles of eicosanoids in the microcirculation. *Cardiovasc. Res.* **1996**, *32*, 720–732. [CrossRef]
189. Kida, T.; Sawada, K.; Kobayashi, K.; Hori, M.; Ozaki, H.; Murata, T. Diverse effects of prostaglandin e2on vascular contractility. *Heart Vessel* **2014**, *29*, 390–395. [CrossRef]
190. Gutiérrez, J.; Lomonte, B.; Chaves, F.; Moreno, E.; Cerdas, L. Pharmacological activities of a toxic phospholipase a isolated from the venom of the snake *Bothrops asper*. *Comp. Biochem. Physiol. Part C Comp. Pharmacol.* **1986**, *84*, 159–164. [CrossRef]
191. Moreira, V.; Zamuner, S.R.; Wallace, J.L.; de Fátima PereiraTeixeira, C. Bothrops jararaca and Crotalus durissus terrificus venoms elicit distinct responses regarding to production of prostaglandins E2 and D2, and expression of cyclooxygenases. *Toxicon* **2007**, *49*, 615–624. [CrossRef] [PubMed]
192. Kini, R.M.; Evans, H.J. The role of enzymatic activity in inhibition of the extrinsic tenase complex by phospholipase A2 isoenzymes from Naja nigricollis venom. *Toxicon* **1995**, *33*, 1585–1590. [CrossRef]
193. Thommesen, L.; Sjursen, W.; Gåsvik, K.; Hanssen, W.; Brekke, O.L.; Skattebøl, L.; Holmeide, A.K.; Espevik, T.; Johansen, B.; Laegreid, A. Selective inhibitors of cytosolic or secretory phospholipase A2 block TNF-induced activation of transcription factor nuclear factor-kappa B and expression of ICAM-1. *J. Immunol.* **1998**, *161*, 3421–3430. [PubMed]
194. Anthonsen, M.W.; Solhaug, A.; Johansen, B. Functional Coupling between Secretory and Cytosolic Phospholipase A2 Modulates Tumor Necrosis Factor-α- and Interleukin-1β-induced NF-κB Activation. *J. Biol. Chem.* **2001**, *276*, 30527–30536. [CrossRef]
195. Balsinde, J.; Balboa, M.A.; Dennis, E.A. Functional coupling between secretory phospholipase A2 and cyclooxygenase-2 and its regulation by cytosolic group IV phospholipase A2. *Proc. Natl. Acad. Sci. USA* **1998**, *95*, 7951–7956. [CrossRef]
196. Peng, Z.; Chang, Y.; Fan, J.; Ji, W.; Su, C. Phospholipase A2 superfamily in cancer. *Cancer Lett.* **2021**, *497*, 165–177. [CrossRef] [PubMed]
197. Naraba, H.; Murakami, M.; Matsumoto, H.; Shimbara, S.; Ueno, A.; Kudo, I.; Oh-ishi, S. Segregated coupling of phospholipases A2, cyclooxygenases, and terminal prostanoid synthases in different phases of prostanoid biosynthesis in rat peritoneal macrophages. *J. Immunol.* **1998**, *160*, 2974–2982. [PubMed]
198. Kuwata, H.; Nakatani, Y.; Murakami, M.; Kudo, I. Cytosolic Phospholipase A2 Is Required for Cytokine-induced Expression of Type IIA Secretory Phospholipase A2 That Mediates Optimal Cyclooxygenase-2-dependent Delayed Prostaglandin E2 Generation in Rat 3Y1 Fibroblasts. *J. Biol. Chem.* **1998**, *273*, 1733–1740. [CrossRef]
199. Murakami, M.; Shimbara, S.; Kambe, T.; Kuwata, H.; Winstead, M.V.; Tischfield, J.A.; Kudo, I. The Functions of Five Distinct Mammalian Phospholipase A2s in Regulating Arachidonic Acid Release. *J. Biol. Chem.* **1998**, *273*, 14411–14423. [CrossRef] [PubMed]
200. Murakami, M.; Das, S.; Kim, Y.-J.; Cho, W.; Kudo, I. Perinuclear localization of cytosolic phospholipase A 2 α is important but not obligatory for coupling with cyclooxygenases. *FEBS Lett.* **2003**, *546*, 251–256. [CrossRef]
201. Ghosh, M.; Stewart, A.; Tucker, D.E.; Bonventre, J.V.; Murphy, R.C.; Leslie, C.C. Role of Cytosolic Phospholipase A 2 in Prostaglandin E 2 Production by Lung Fibroblasts. *Am. J. Respir. Cell Mol. Biol.* **2004**, *30*, 91–100. [CrossRef]
202. Belich, M.P.; Salmerón, A.; Johnston, L.H.; Ley, S.C. TPL-2 kinase regulates the proteolysis of the NF-κB-inhibitory protein NF-κB1 p105. *Nature* **1999**, *397*, 363–368. [CrossRef]
203. Kifor, O.; MacLeod, R.J.; Diaz, R.; Bai, M.; Yamaguchi, T.; Yao, T.; Kifor, I.; Brown, E.M. Regulation of MAP kinase by calcium-sensing receptor in bovine parathyroid and CaR-transfected HEK293 cells. *Am. J. Physiol. Physiol.* **2001**, *280*, F291–F302. [CrossRef]
204. Aoki, K.; Zubkov, A.Y.; Parent, A.D.; Zhang, J.H. Mechanism of ATP-Induced [Ca^{2+}] i Mobilization in Rat Basilar Smooth Muscle Cells. *Stroke* **2000**, *31*, 1377–1385. [CrossRef]
205. Leiguez, E.; Giannotti, K.C.; Moreira, V.; Matsubara, M.H.; Gutíerrez, J.M.; Lomonte, B.; Rodríguez, J.P.; Balsinde, J.; Teixeira, C. Critical role of TLR2 and MyD88 for functional response of macrophages to a group IIA-secreted phospholipase A2from snake venom. *PLoS ONE* **2014**, *9*, e93741. [CrossRef]
206. Soto, M.E.; Guarner-Lans, V.; Herrera-Morales, K.Y.; Pérez-Torres, I. Participation of arachidonic acid metabolism in the aortic aneurysm formation in patients with Marfan syndrome. *Front. Physiol.* **2018**, *9*, 77. [CrossRef] [PubMed]
207. Rikitake, Y.; Hirata, K.; Kawashima, S.; Takeuchi, S.; Shimokawa, Y.; Kojima, Y.; Inoue, N.; Yokoyama, M. Signaling Mechanism Underlying COX-2 Induction by Lysophosphatidylcholine. *Biochem. Biophys. Res. Commun.* **2001**, *281*, 1291–1297. [CrossRef] [PubMed]
208. Hughes-Fulford, M.; Tjandrawinata, R.R.; Li, C.-F.; Sayyah, S. Arachidonic acid, an omega-6 fatty acid, induces cytoplasmic phospholipase A 2 in prostate carcinoma cells. *Carcinogenesis* **2005**, *26*, 1520–1526. [CrossRef]
209. Ruipérez, V.; Casas, J.; Balboa, M.A.; Balsinde, J. Group V Phospholipase A 2 -Derived Lysophosphatidylcholine Mediates Cyclooxygenase-2 Induction in Lipopolysaccharide-Stimulated Macrophages. *J. Immunol.* **2007**, *179*, 631–638. [CrossRef]

210. Gubern, A.; Barceló-Torns, M.; Casas, J.; Barneda, D.; Masgrau, R.; Picatoste, F.; Balsinde, J.; Balboa, M.A.; Claro, E. Lipid droplet biogenesis induced by stress involves triacylglycerol synthesis that depends on Group VIA phospholipase A2. *J. Biol. Chem.* **2009**, *284*, 5697–5708. [CrossRef] [PubMed]
211. Guijas, C.; Pérez-Chacón, G.; Astudillo, A.M.; Rubio, J.M.; Gil-de-Gómez, L.; Balboa, M.A.; Balsinde, J. Simultaneous activation of p38 and JNK by arachidonic acid stimulates the cytosolic phospholipase A2-dependent synthesis of lipid droplets in human monocytes. *J. Lipid Res.* **2012**, *53*, 2343–2354. [CrossRef] [PubMed]
212. Jarc, E.; Kump, A.; Malavašič, P.; Eichmann, T.O.; Zimmermann, R.; Petan, T. Lipid droplets induced by secreted phospholipase A2 and unsaturated fatty acids protect breast cancer cells from nutrient and lipotoxic stress. *Biochim. Biophys. Acta-Mol. Cell Biol. Lipids* **2018**, *1863*, 247–265. [CrossRef]
213. Arrese, E.L.; Saudale, F.Z.; Soulages, J.L. Lipid droplets as signaling platforms linking metabolic and cellular functions. *Lipid Insights* **2014**, *7*, 7–16. [CrossRef] [PubMed]
214. Onal, G.; Kutlu, O.; Gozuacik, D.; Dokmeci Emre, S. Lipid Droplets in Health and Disease. *Lipids Health Dis.* **2017**, *16*, 128. [CrossRef]
215. Welte, M.A. Expanding roles for lipid droplets. *Curr. Biol.* **2015**, *25*, R470–R481. [CrossRef] [PubMed]
216. Bosch, M.; Sánchez-Álvarez, M.; Fajardo, A.; Kapetanovic, R.; Steiner, B.; Dutra, F.; Moreira, L.; López, J.A.; Campo, R.; Marí, M.; et al. Mammalian lipid droplets are innate immune hubs integrating cell metabolism and host defense. *Science* **2020**, *370*, 6514. [CrossRef]
217. Karagiannis, F.; Masouleh, S.K.; Wunderling, K.; Surendar, J.; Schmitt, V.; Kazakov, A.; Michla, M.; Hölzel, M.; Thiele, C.; Wilhelm, C. Lipid-Droplet Formation Drives Pathogenic Group 2 Innate Lymphoid Cells in Airway Inflammation. *Immunity* **2020**, *52*, 620–634.e6. [CrossRef] [PubMed]
218. Marschallinger, J.; Iram, T.; Zardeneta, M.; Lee, S.E.; Lehallier, B.; Haney, M.S.; Pluvinage, J.V.; Mathur, V.; Hahn, O.; Morgens, D.W.; et al. Lipid-droplet-accumulating microglia represent a dysfunctional and proinflammatory state in the aging brain. *Nat. Neurosci.* **2020**, *23*, 194–208. [CrossRef] [PubMed]
219. Garces, F.; López, F.; Nĩo, C.; Fernandez, A.; Chacin, L.; Hurt-Camejo, E.; Camejo, G.; Apitz-Castro, R. High plasma phospholipase A 2 activity, inflammation markers, and LDL alterations in obesity with or without type 2 diabetes. *Obesity* **2010**, *18*, 2023–2029. [CrossRef]
220. Tall, A.R.; Yvan-Charvet, L. Cholesterol, inflammation and innate immunity. *Nat. Rev. Immunol.* **2015**, *15*, 104–116. [CrossRef]
221. Schaftenaar, F.; Frodermann, V.; Kuiper, J.; Lutgens, E. Atherosclerosis: The interplay between lipids and immune cells. *Curr. Opin. Lipidol.* **2016**, *27*, 209–215. [CrossRef]
222. Rodríguez, J.P.; Leiguez, E.; Guijas, C.; Lomonte, B.; Gutiérrez, J.M.; Teixeira, C.; Balboa, M.A.; Balsinde, J. A lipidomic perspective of the action of group iia secreted phospholipase a2 on human monocytes: Lipid droplet biogenesis and activation of cytosolic phospholipase a2α. *Biomolecules* **2020**, *10*, 891. [CrossRef]
223. Moujaber, O.; Stochaj, U. The Cytoskeleton as Regulator of Cell Signaling Pathways. *Trends Biochem. Sci.* **2020**, *45*, 96–107. [CrossRef]
224. Leiguez, E.; Zuliani, J.P.; Cianciarullo, A.M.; Fernandes, C.M.; Gutierrez, J.M.; Teixeira, C. A group IIA-secreted phospholipase A2 from snake venom induces lipid body formation in macrophages: The roles of intracellular phospholipases A2 and distinct signaling pathways. *J. Leukoc. Biol.* **2011**, *90*, 155–166. [CrossRef]
225. Yu, Y.H.; Liao, P.R.; Guo, C.J.; Chen, C.H.; Mochly-Rosen, D.; Chuang, L.M. PKC-ALDH2 pathway plays a novel role in adipocyte differentiation. *PLoS ONE* **2016**, *11*, e0161993. [CrossRef] [PubMed]
226. Huang, S.H.; Shen, W.J.; Yeo, H.L.; Wang, S.M. Signaling pathway of magnolol-stimulated lipolysis in sterol ester-loaded 3T3-L1 preadipocyes. *J. Cell. Biochem.* **2004**, *91*, 1021–1029. [CrossRef]
227. Le Lay, S.; Hajduch, E.; Lindsay, M.R.; Le Lièpvre, X.; Thiele, C.; Ferré, P.; Parton, R.G.; Kurzchalia, T.; Simons, K.; Dugail, I. Cholesterol-induced caveolin targeting to lipid droplets in adipocytes: A role for caveolar endocytosis. *Traffic* **2006**, *7*, 549–561. [CrossRef]
228. Than, N.G.; Sumegi, B.; Bellyei, S.; Berki, T.; Szekeres, G.; Janaky, T.; Szigeti, A.; Bohn, H.; Than, G.N. Lipid droplet and milk lipid globule membrane associated placental protein 17b (PP17b) is involved in apoptotic and differentiation processes of human epithelial cervical carcinoma cells. *Eur. J. Biochem.* **2003**, *270*, 1176–1188. [CrossRef]
229. Zhong, W.; Fan, B.; Cong, H.; Wang, T.; Gu, J. Oleic acid-induced perilipin 5 expression and lipid droplets formation are regulated by the PI3K/PPARα pathway in HepG2 cells. *Appl. Physiol. Nutr. Metab.* **2019**, *44*, 840–848. [CrossRef] [PubMed]
230. Han, X.; Wang, T.; Zhang, J.; Liu, X.; Li, Z.; Wang, G.; Song, Q.; Pang, D.; Ouyang, H.; Tang, X. Apolipoprotein CIII regulates lipoprotein-associated phospholipase A2 expression via the MAPK and NFκB pathways. *Biol. Open* **2015**, *4*, 661–665. [CrossRef] [PubMed]
231. Hu, S.-B.; Zou, Q.; Lv, X.; Zhou, R.L.; Niu, X.; Weng, C.; Chen, F.; Fan, Y.W.; Deng, Z.Y.; Li, J. 9t18:1 and 11t18:1 activate the MAPK pathway to regulate the expression of PLA2 and cause inflammation in HUVECs. *Food Funct.* **2020**, *11*, 649–661. [CrossRef] [PubMed]
232. Andersson, L.; Boström, P.; Ericson, J.; Rutberg, M.; Magnusson, B.; Marchesan, D.; Ruiz, M.; Asp, L.; Huang, P.; Frohman, M.A.; et al. PLD1 and ERK2 regulate cytosolic lipid droplet formation. *J. Cell Sci.* **2006**, *119*, 2246–2257. [CrossRef]
233. Boström, P.; Andersson, L.; Li, L.; Perkins, R.; Højlund, K.; Borén, J.; Olofsson, S.O. The assembly of lipid droplets and its relation to cellular insulin sensitivity. *Biochem. Soc. Trans.* **2009**, *37*, 981–985. [CrossRef] [PubMed]

234. Reustle, A.; Torzewski, M. Role of p38 MAPK in atherosclerosis and aortic valve sclerosis. *Int. J. Mol. Sci.* **2018**, *19*, 3761. [CrossRef] [PubMed]
235. Kavurma, M.M.; Rayner, K.J.; Karunakaran, D. The walking dead: Macrophage inflammation and death in atherosclerosis. *Curr. Opin. Lipidol.* **2017**, *28*, 91–98. [CrossRef]
236. Chakraborti, S. Phospholipase A2 isoforms: A perspective. *Cell. Signal.* **2003**, *15*, 637–665. [CrossRef]
237. Hooks, S.B.; Cummings, B.S. Role of Ca2+-independent phospholipase A2 in cell growth and signaling. *Biochem. Pharmacol.* **2008**, *76*, 1059–1067. [CrossRef]
238. Kita, Y.; Shindou, H.; Shimizu, T. Cytosolic phospholipase A 2 and lysophospholipid acyltransferases. *Biochim. Biophys. Acta-Mol. Cell Biol. Lipids* **2019**, *1864*, 838–845. [CrossRef] [PubMed]
239. Grygiel-Górniak, B. Peroxisome proliferator-activated receptors and their ligands: Nutritional and clinical implications—A review. *Nutr. J.* **2014**, *13*, 17. [CrossRef]
240. Engin, A.B.; Engin, A. *Obesity and Lipotoxicity*; Springer: Berlin/Heidelberg, Germany, 2017. [CrossRef]
241. Mei, C.-L.; He, P.; Cheng, B.; Liu, W.; Wang, Y.-F.; Wan, J.-J. Chlamydia pneumoniae induces macrophage-derived foam cell formation via PPAR α and PPAR γ-dependent pathways. *Cell Biol. Int.* **2009**, *33*, 301–308. [CrossRef] [PubMed]
242. Chistiakov, D.A.; Melnichenko, A.A.; Myasoedova, V.A.; Grechko, A.V.; Orekhov, A.N. Mechanisms of foam cell formation in atherosclerosis. *J. Mol. Med.* **2017**, *95*, 1153–1165. [CrossRef]
243. Son, S.H.; Goo, Y.H.; Chang, B.H.; Paul, A. Perilipin 2 (PLIN2)-deficiency does not increase cholesterol-induced toxicity in macrophages. *PLoS ONE* **2012**, *7*, e33063. [CrossRef]
244. Turkish, A.; Sturley, S.L. Regulation of Triglyceride Metabolism. I. Eukaryotic neutral lipid synthesis: "Many ways to skin ACAT or a DGAT". *Am. J. Physiol.-Gastrointest. Liver Physiol.* **2007**, *292*, 953–957. [CrossRef]
245. Yu, X.H.; Fu, Y.C.; Zhang, D.W.; Yin, K.; Tang, C.K. Foam cells in atherosclerosis. *Clin. Chim. Acta* **2013**, *424*, 245–252. [CrossRef] [PubMed]
246. Cruz, A.L.S.; Barreto, E.D.A.; Fazolini, N.P.B.; Viola, J.P.B.; Bozza, P.T. Lipid droplets: Platforms with multiple functions in cancer hallmarks. *Cell Death Dis.* **2020**, *11*, 105. [CrossRef] [PubMed]
247. Dias, S.S.G.; Soares, V.C.; Ferreira, A.C.; Sacramento, C.Q.; Fintelman-Rodrigues, N.; Temerozo, J.R.; Teixeira, L.; Nunes da Silva, M.A.; Barreto, E.; Mattos, M.; et al. Lipid droplets fuel SARS-CoV-2 replication and production of inflammatory mediators. *PLoS Pathog.* **2020**, *16*, e1009127. [CrossRef]
248. Jarc, E.; Petan, T. A twist of FATe: Lipid droplets and inflammatory lipid mediators. *Biochimie* **2020**, *169*, 69–87. [CrossRef]
249. De Carvalho, A.E.Z.; Giannotti, K.; Junior, E.L.; Matsubara, M.; Dos Santos, M.C.; Fortes-Dias, C.L.; Teixeira, C. Crotalus durissus ruruima Snake Venom and a Phospholipase A2 Isolated from This Venom Elicit Macrophages to Form Lipid Droplets and Synthesize Inflammatory Lipid Mediators. *J. Immunol. Res.* **2019**, *2019*, 2745286. [CrossRef]
250. Juge-Aubry, C.E.; Henrichot, E.; Meier, C.A. Adipose tissue: A regulator of inflammation. *Best Pract. Res. Clin. Endocrinol. Metab.* **2005**, *19*, 547–566. [CrossRef] [PubMed]
251. Oikonomou, E.K.; Antoniades, C. The role of adipose tissue in cardiovascular health and disease. *Nat. Rev. Cardiol.* **2019**, *16*, 83–99. [CrossRef]
252. La Cava, A.; Matarese, G. The weight of leptin in immunity. *Nat. Rev. Immunol.* **2004**, *4*, 371–379. [CrossRef]
253. Tsatsanis, C.; Zacharioudaki, V.; Androulidaki, A.; Dermitzaki, E.; Charalampopoulos, I.; Minas, V.; Gravanis, A.; Margioris, A.N. Adiponectin induces TNF-α and IL-6 in macrophages and promotes tolerance to itself and other pro-inflammatory stimuli. *Biochem. Biophys. Res. Commun.* **2005**, *335*, 1254–1263. [CrossRef] [PubMed]
254. Yamaguchi, N.; Argueta, J.G.M.; Masuhiro, Y.; Kagishita, M.; Nonaka, K.; Saito, T.; Hanazawa, S.; Yamashita, Y. Adiponectin inhibits Toll-like receptor family-induced signaling. *FEBS Lett.* **2005**, *579*, 6821–6826. [CrossRef]
255. Martínez-Sánchez, N. There and back again: Leptin actions in white adipose tissue. *Int. J. Mol. Sci.* **2020**, *21*, 6039. [CrossRef] [PubMed]

 toxins

MDPI

Review

Neglected Venomous Animals and Toxins: Underrated Biotechnological Tools in Drug Development

Guilherme Rabelo Coelho [1], Daiane Laise da Silva [1], Emidio Beraldo-Neto [1], Hugo Vigerelli [2], Laudiceia Alves de Oliveira [3], Juliana Mozer Sciani [4] and Daniel Carvalho Pimenta [1,*]

1 Laboratório de Bioquímica, Instituto Butantan, São Paulo 05503-900, Brazil; guilherme.coelho@butantan.gov.br (G.R.C.); daiane.silva@esib.butantan.gov.br (D.L.d.S.); emidio.beraldo@butantan.gov.br (E.B.-N.)
2 Laboratório de Genética, Instituto Butantan, São Paulo 05503-900, Brazil; hugo.barros@esib.butantan.gov.br
3 Laboratório de Moléstias Infecciosas—Faculdade de Medicina de Botucatu, São Paulo State University (UNESP), São Paulo 01049-010, Brazil; laudiceia.oliveira@unesp.br
4 Laboratório Multidisciplinar em Pesquisa, Universidade São Francisco, Bragança Paulista 12916-900, Brazil; juliana.sciani@usf.edu.br
* Correspondence: dcpimenta@butantan.gov.br

Abstract: Among the vast repertoire of animal toxins and venoms selected by nature and evolution, mankind opted to devote its scientific attention—during the last century—to a restricted group of animals, leaving a myriad of toxic creatures aside. There are several underlying and justifiable reasons for this, which include dealing with the public health problems caused by envenoming by such animals. However, these studies became saturated and gave rise to a whole group of animals that become neglected regarding their venoms and secretions. This repertoire of unexplored toxins and venoms bears biotechnological potential, including the development of new technologies, therapeutic agents and diagnostic tools and must, therefore, be assessed. In this review, we will approach such topics through an interconnected historical and scientific perspective that will bring up the major discoveries and innovations in toxinology, achieved by researchers from the Butantan Institute and others, and describe some of the major research outcomes from the study of these neglected animals.

Keywords: toxins; venoms; skin secretion; drug discovery

Key Contribution: This review brings up the issue of the limitations in current toxinology, that is the poor appraisal of the poisonous animals, opposed to the venomous ones. It is meant to expand the readers' perspective on venoms and toxins and the possible scientific developments associated with these thematic lines of research.

Citation: Coelho, G.R.; da Silva, D.L.; Beraldo-Neto, E.; Vigerelli, H.; de Oliveira, L.A.; Sciani, J.M.; Pimenta, D.C. Neglected Venomous Animals and Toxins: Underrated Biotechnological Tools in Drug Development. *Toxins* **2021**, *13*, 851. https://doi.org/10.3390/toxins13120851

Received: 16 October 2021
Accepted: 25 November 2021
Published: 29 November 2021

1. Introduction

"Around 1896, a modest physician that used to practice medicine in Botucatu became notorious due to his strange fascination with snakes and their venoms. It was Dr Vital Brazil that, from the tranquility of the countryside, was taking the initial steps on the brilliant research that would make him famous not only in Brazil but also all over the educated world". This free translation of the beginning of the first paragraph (Figure 1C) of the book "Memória Histórica do Instituto Butantan" (Figure 1A; Historic memory of Butantan Institute, in free translation) written by Dr Vital Brazil himself (Figure 1B) [1] refers to published news in 1914 reporting the inauguration of 'new facilities' in the Institute (Figure 1C).

Over one hundred years after the news reported above, some of the authors of this review have worked, conducted research, performed experiments, and published papers in this exact same building. Since then, a lot has changed in the Institute, including its slogan, but not the building (Figure 2). Our slogan is now "at the service of life", a humbler commitment to the institutional mission, and the research laboratories have been decommissioned from this building, which is currently undergoing restoration and will be dedicated to cultural activities only.

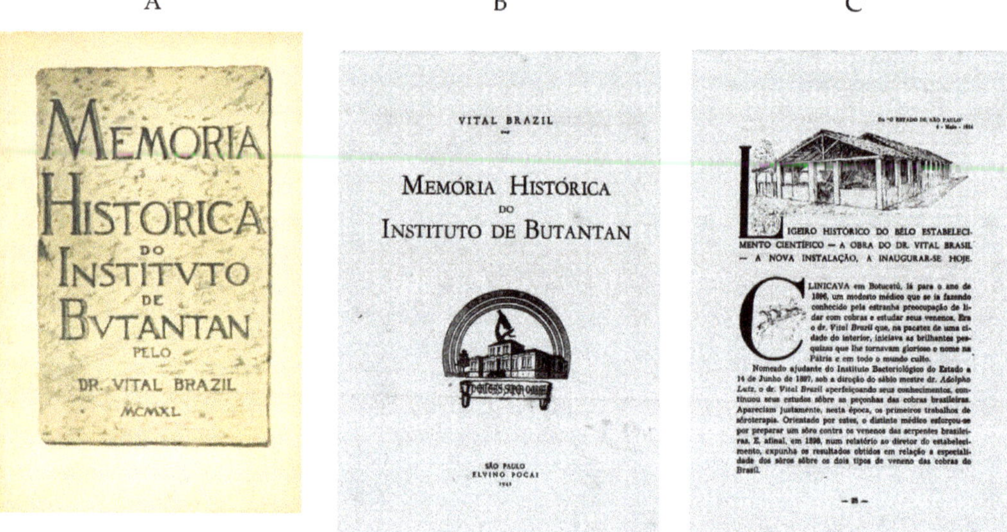

A

MEMORIA HISTORICA DO INSTITVTO DE BVTANTAN PELO DR. VITAL BRAZIL MCMXL

B

VITAL BRAZIL

MEMÓRIA HISTÓRICA DO INSTITUTO DE BUTANTAN

SÃO PAULO
ELVINO POCAI
1941

C

LIGEIRO HISTÓRICO DO BÊLO ESTABELECIMENTO CIENTÍFICO — A OBRA DO DR. VITAL BRASIL — A NOVA INSTALAÇÃO, A INAUGURAR-SE HOJE.

CLINICAVA em Botucatú, lá para o ano de 1896, um modesto médico que se ia fazendo conhecido pela estranha preocupação de lidar com cobras e estudar seus venenos. Era o dr. *Vital Brazil* que, na pacatez de uma cidade do interior, iniciava as brilhantes pesquizas que lhe tornavam glorioso o nome na Pátria e em todo o mundo culto.

Nomeado ajudante do Instituto Bacteriológico do Estado a 14 de Junho de 1897, sob a direção do sábio mestre dr. *Adolpho Lutz*, o dr. *Vital Brazil* aperfeiçoando seus conhecimentos, continuou seus estudos sôbre as peçonhas das cobras brasileiras. Apareciam justamente, nesta época, os primeiros trabalhos de sôroterapia. Orientado por estes, o distinte médico esforçou-se por preparar um sôro contra os venenos das serpentes brasileiras. E, afinal, em 1898, num relatório ao diretor do estabelecimento, expunha os resultados obtidos em relação a especialidade dos sôros sôbre os dois tipos de veneno das cobras do Brasil.

Figure 1. Selected reproductions of (**A**) the book written by Vital Brazil in 1941. This and other classic books are available at https://bibliotecadigital.butantan.gov.br/, accessed on 19 November 2021. Please note the institute logo in (**B**). It is the depiction of the main-laboratory building, underneath a microscope, bearing the motto "peritas super omnia", meaning "expert in everything" in Latin. (**C**) News advertising the inauguration of new facilities at the Institute, in 1914.

Figure 2. (**A**) A recent photograph of the main-laboratory building, in a similar framing as depicted in Figure 1. (**B**) Current Institute logotype, bearing the slogan 'A serviço da vida' (at the service of life, in free translation).

1.1. The Origins of Negligence

There are twenty (tropical) diseases that are officially classified as 'neglected tropical diseases' by the World Health Organization (WHO) [2]. Neglected tropical diseases persist under conditions of poverty and are concentrated almost exclusively in impoverished populations in the developing world. They are: Buruli ulcer, Chagas disease, Dengue and Chikungunya (only WHO, not CDC), Dracunculiasis, Echinococcosis, Yaws, Fascioliasis, African trypanosomiasis, Leishmaniasis, Leprosy, Lymphatic filariasis, Onchocerciasis, Rabies, Schistosomiasis, Soil-transmitted helminthiasis, Cysticercosis, Trachoma, Scabies and other ectoparasites, Snakebite envenoming, Mycetoma and deep mycoses. These diseases are common in 149 countries, affecting more than 1.4 billion people (including more than 500 million children) and costing developing economies billions of dollars every year.

The importance of neglected tropical diseases has been underestimated since many are asymptomatic and have long incubation periods. The connection between a death and a neglected tropical disease that has been latent for a long period of time is not often realized. Additionally, neglected tropical diseases are often associated with some kind of social stigma, making their treatment more complex.

From the toxinology perspective, one can also consider that there are 'neglected' venomous and poisonous animals by employing very similar criteria to justify such negligence: Human accidents occur with individuals who are often amongst the poorest populations, living in remote, rural areas, urban slums or conflict zones; the accident causes no rapid death of the victim and/or such animals are stigmatized (cause bad luck, carry evil spells or are cursed).

Depending on the nature and origin of the venom or toxin, one can clearly perceive that there are 'preferred' subjects and matters in the field of toxinology (Table 1). Probably due to historical and/or epidemiological factors, some animals and venoms—normally the ones that elicit acute, severe lesions due to some pronounced biological activity—were selected (or elected) as 'more relevant' to the field and have been thoroughly studied throughout the years. Endemic animals, such as spiders and scorpions that have adapted to urban environments, have also 'deserved' more attention than other species. All the consulted databases indicated that there is more literature on snakes, spiders and scorpions (the triad) than the others. Interestingly, Scopus and Web of Science present the same publications ratio for triad:neglected (7.8), whereas Google and PubMed display lower ratios (5.5 and 1.8 respectively), probably due to the differences in indexed publications queried.

Table 1. Total results retrieved according to the searched terms in different academic databases.

Term	PubMed	Scopus	Web of Knowledge	Google Scholar
Snake	29,272	56,112	44,467	771,000
Scorpion	7030	10,362	8834	91,000
Spider	15,988	42,351	39,343	1,180,000
TOTAL	52,290	108,825	92,644	2,042,000
Amphibian (skin) [1]	7714	3549	3338	134,000
Sea urchin (toxin) [2]	314	183	170	19,300
Mollusk [3]	3688	902	290	19,800
Stingray	813	1717	1817	2160
Cnidarian (toxins)	2389	913	162	17,700
Insects (toxins)	12,879	6663	6037	175,000
TOTAL	27,797	13,927	11,814	367,960
Proportion	1.8×	7.8×	7.8×	5.5×

Search performed in 11 September 2021. [1] Limited to skin, in order to exclude ecological studies; [2] Limited to toxin, in order to exclude developmental/reproductive models; [3] Limited to toxins and excluding dinoflagellates.

The aim of this review is, therefore, to shed a light upon such amazing animals and their venoms and secretions, presenting a non-anthropocentric view of their venom composition and the (few, but consistent) biomedical 'cases' derived from the study of such species, and review the literature and the biotechnological developments derived from venoms and secretions from toxic animals that have not received proper attention from the scientific community over the past years and cast a light on their unique features and interesting molecules. Afterall, just like the neglected tropical diseases, it was never about the 'importance' of these animals, only their 'relevance', i.e., their economic impact, geopolitical localization, affected population, endangerment status and profit potential, in addition to formerly listed reasons.

1.2. Biodiversity

Earth's existing biodiversity is a direct consequence of Darwin's Natural Selection, i.e., the survival of the fittest, in a constant struggle to survive [3]. With an estimated 8.7 million species inhabiting our planet, the mere 1.2 million (mostly insects) that have already been identified and described have all—or are still in the process of—adapted and evolved so that, after numerous breeding cycles, poorly suited individuals are filtered out by nature.

One particularly interesting adaptation which emerged millions of years ago was the biochemical weaponry utilized for defense and/or predation by some organisms as a means of guaranteeing survival [4]. These so called 'toxins' can be found in procaryotic species, such as *Staphylococcus aureus* and *Klebsiella pneumoniae* [5,6], plants (*Cicuta maculate* (Socrates committed suicide by drinking cicuta, circa 399 B.C.) and *Nicotiana tabacum* (homage to Jean Nicot de Villemain, who introduced snuff to the French court in 1560)) and, obviously, animals.

For animals, these toxins are believed to have originated from ancestral house-keeping genes that underwent variation and neofunctionalization [4,7], resulting in molecules displaying an 'increased' biological activity, normally targeted to major biological systems that when unbalanced may result in severe risk of death, such as the hemostatic-interfering molecules. The toxins were then specifically expressed in venom-secreting cells that eventually became specialized venom glands [8]. Such specialization became an evolutionary advantage, due to unique pharmacokinetic properties that these (typically) peptides and proteins granted to such animals [9,10].

1.3. Toxins: Snakes, Spiders and Scorpions as Classical as It Can Be

Toxinology has its origins long associated with venomous animals and not poisonous ones. There might be some controversy in this separation, but it is commonly accepted that venomous animals would possess an inoculating apparatus capable of delivering toxins into the prey/aggressor. On the other hand, poisonous animals would secrete toxins in their skin or body organs and would have to be actively eaten/beaten/attacked/poked/colonized (bacteria) in order for to the toxins exert their effect.

Nevertheless, mystical, magical, medical and/or lethal uses of some animals' venoms are well known throughout history. For example: Cleopatra may have committed suicide by letting herself be bitten by a snake (*Naja haje* probably). In the Bible there are nine verses citing scorpions (Luke 10:19 and 11:12, Kings 12:11 and 12:14, Deuteronomy 8:15, Ezekiel 2:6, Revelation 9:3, 9:5 and 9:10). Greek mythology presents us the Lernaean Hydra, a serpentine water monster with many heads (depending on the myth source) with poisonous breath and blood so virulent that even its scent was deadly, as well as the Medusa, one of the three monstrous Gorgons, generally described as winged human females with living venomous snakes in place of hair.

These venomous animals are still present in modern-day fiction, such as the famous Spiderman, whose superpowers derived from mutations resulting from the bite of a radioactive spider. Even Harry Potter was forced to deal with the Basilisk, a giant snake capable of instant kill just by gazing at the victim's eyes. There are also urban legends and local habits, such as the well-known North American arachnophobia.

On the other hand, poisonous animals share a less glamorous role in human history. They have participated, for example, in human (sacrificial) rituals and attempted pharmaceutical developments throughout history. There were Maya human bloodletting rituals that employed the sting of marine stingrays as blades, due to a 'more efficient' bleeding [11]. Hunters have long sought the Central and South American Dendrobatidae 'poison arrow frogs' (self-explanatory) to use their toxic skin secretion for hunting [12]. Traditional Chinese medicine uses the 'all healing' Chan'Su (dried *Bufo bufo* skin) for mostly any illness [13]. Amazon tribes traditionally used Kambo (or Kampum) in their purification rituals [14]. This medicine is extracted from *Phyllomedusa bicolor* skin secretion and has become known as the 'frog vaccine' in urban environments. The Bible also cites such animals in the infamous passage in Exodus 8:1–4, in which the "*great LORD says: Let my people go, so that they may worship me. If you refuse to let them go, I will plague your whole country with frogs. The Nile will teem with frogs. They will come up into your palace and your bedroom and onto your bed, into the houses of your officials and on your people, and into your ovens and kneading troughs. The frogs will go up on you and your people and all your officials".* Unfortunately, the poisonous animals are presented from a more neglected, less charming perspective, as presented above.

All this glamour associated with venomous animals has led to the establishment of what can be considered the 'greatest-hits' of (classical) toxinology: snakes, spiders, and scorpions (the triad). Undoubtedly, studying these animals' venoms has yielded a myriad of relevant scientific papers [15–19] produced by highly committed international scientific groups. The molecular dissection of the venom constituents has made it possible that effective sera could be manufactured [20–22], thus reducing mortality and morbidity associated with envenomation [23,24]. Moreover, one of the world's most administered antihypertensives (Captopril) is a direct derivative of one viper toxin [25].

Another example is a tumor-labeling molecule (tozuleristide), currently undergoing clinical phase 1 studies, that is being used in surgeries as marker and diagnostics for glioma and other tumors. This molecule is an analogue of a chlorotoxin isolated from the venom of the scorpion *Leiurus quinquestriatus* [26–28].

It is noteworthy to mention that there is young blood trying to join the party. Even though the marine mollusks of the *Conus* genus do not belong to the classic triad, they are becoming more and more famous since the discovery of ziconotide (Prialt), the strongest analgesic ever described: a calcium channel blocker, purified from the *Conus magnus* venom [29]. These animals are discussed below.

However, even for such well-studied animals there are still 'neglected' molecules present in their venoms, such as L-amino acid oxidase, crotapotin, crotamine that 'simply' for not killing or harming the animal models are put aside, turning the spotlight to the super-toxic metallopeptidases, phospholipases (A_2 and D) and ionic channel blockers.

Still, a number of other animals can (and do) cause accidents upon human encounters, displaying broad variation in terms of the clinical outcome. Marine animals are good examples: sea urchins can be solely painful [30] whereas mollusks can instantly kill [31]. Yet, for some reason, such animals have not been able to attract the attention of major research groups in toxinology, remaining in 'neglect' for the past couple of decades.

The modern reptiles are a group comprised of the Crocodila, Lepidossaura, Rhynohocephalia, Squamata, Testudines and Aves. With the exception of snakes, no other true venomous reptile (i.e., with a specialized venom inoculation apparatus) is currently known. The venomous living dinosaurs, i.e., birds pitohui, ifrita and rufous [32], and the Komodo dragon are considered to be poisonous [33,34].

However, in the end, snakes are the most classical venomous animals. Since ancient times, their behavior has been considered to be mischievous—even tempting—and their venom has been associated to magic spells and even cures. Not surprisingly, The Rod of Asclepius, i.e., the Medicine symbol (Figure 3A), is a snake serpentizing around a rod [35]. Nevertheless, the caduceus—the traditional symbol of Hermes—represented by two snakes serpentizing around a winged rod (Figure 3B) is often mistakenly used as a symbol of medicine instead of the Rod of Asclepius, especially in the United States, as a consequence of documented mistakes, misunderstandings and confusion in the late 19th and early 20th centuries. However, the two-snake caduceus design has ancient and consistent associations with trade, eloquence, negotiation, alchemy, and wisdom. Last, but not least, the current Butantan Institute logotype (created in 1983) is a clever design in which the capital 'I' and 'B' are fused and the 'B' serif becomes the snake serpentizing around the 'I', which serves as the rod (Figure 3C).

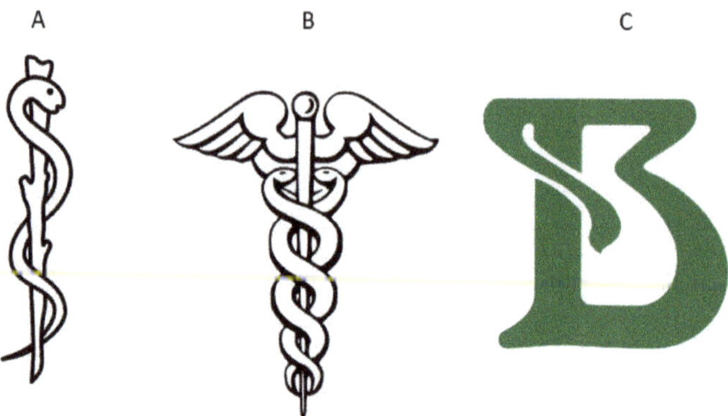

Figure 3. (**A**) Rod of Asclepius, (**B**) the caduceus and (**C**) Butantan Institute logotype.

Jumping a few centuries ahead, there is indeed current medicine based on snake venoms, such as Captopril [25,36,37], Aggrastat, Intergillin and Aggretin [30,33], proving that ancient wisdom may be old, but never outdated. Not only that, but this particular *Toxins* issue that celebrates the 120th anniversary of Butantan corroborates this. At the same time, one can easily note the iconic fascination that the snake has exerted over the local scientific community, that has—and still does—followed Vital Brazil's initial steps.

1.4. Lizards

Lizards' biting has long been discussed among the toxinology field due to the lack of an inoculating venom apparatus. *Heloderma* bites have been reported since 1882 [38,39], and the first toxic activities were described in 1900–1950. At that time, authors were aware that such lizards' toxins included neurotoxins, causing respiratory depression. Inflammation, edema and pain have also been described. However, once this animal bites 'as strong as a bulldog' according to the authors, these symptoms may not be exclusively toxin-derived [40]. Moreover, its hemolytic activity is mild, when compared to snakes, and seems to be species-specific [41].

Later, between 1950–1990, a wide range of biological activities were described, such as phospholipasic, hyaluronidasic, proteolytic [42], L-amino acid oxidase, fibrinolytic, [43] esterase, 5′-nucleotidase [44], secretagogue [45] and nerve growth factor activity [46]. Furthermore, new venom components (at the time) were isolated and identified, such as: kallikrein [47,48], Helospectins 1–2 (acting as secretagogues) [49], Gilatoxin (serine peptidase) [50], Helodermin (vasoactive peptide) [51], and Helothermine (CRISP) [52]. Hyaluronidase [53], a Phospholipase A2 [54], Helodermatin (hypotensive toxin) [55], and Exendin-3 (secretagogue) [56] were also described. Such myriad of toxins could, then, be correlated to the many established envenomation symptoms, such as hypotension and respiratory difficulties [57], smooth muscle contraction [58] and anticoagulant effect [59].

In 1992, Exendin-4 identification was a major event and *Heloderma* venom studies sky-rocketed from this year onwards [60]. Several research projects have evaluated the antidiabetic potential of this molecule, which gave rise to exenatide, a new drug for the treatment of diabetes [61]. A few years later, the inhibition of platelet aggregation by a phospholipase isolated from a Helodermatid lizard was described [62]. Even though it was already known that *Heloderma* venom presents at least five anionic phospholipases, being the most abundant similar to *Apis mellifera* phospholipase [63], it was another important event.

In the following years, Helokinestatin, a toxin that acts as an antagonist of the bradykinin B2 receptor, was described [64]. Moreover, Helofensin was identified by Fry and co-workers by genetic and functional analysis [65], and classified together with a class of lethal toxins firstly described by Komori at al. in 1988 [66]. The presence of a natriuretic peptide in *Heloderma* venom was pointed out by different authors [67,68].

A work comparing the venom proteome of *Heloderma suspectum* with the venom of *H. exasperatum* and *H. horridum* presented an interesting result. Although *H. suspectum* was evolutionarily separated from the other species 30 million before, the venom composition was basically the same for the three species, presenting the same toxins with slight differences in their relative proportions [69]. Moreover, authors could also describe two new molecules: semaphorin and a bactericidal/permeability-increasing (BPI) molecule. Another study that characterized the *H. suspectum* venom proteome relates to the presence of a neuroendocrine convertase 1 homolog, and proposes that this protein is responsible for the cleavage of the proforms of exendins. In the same study, the authors also point out the high presence of phospholipase propeptides in the venom proteome [70]. Recent works allowed access to different classes of proteins, and also new biological activities from *Heloderma* venom. The venom gland transcriptome analysis from *H. horridum horridum* revealed the presence of metalloproteases, lipases, vespryns, waspryns, lectins, cystatins and serine peptidase inhibitors, but none of these proteins were actually isolated from the venom [71]. Furthermore, *Heloderma* contains neurotoxins in its venom, and these toxins are able to bind sodium and calcium channels [72]. An important work by Fry et al. [73] evaluated phylogeny between snakes and lizards and demonstrated that the venom delivery system of these animals could have evolved from the same common ancestor. This was the first study that biochemically evaluated the venom of a lizard from the Varanidae family. The crude venom from *Varanus varius* displays a hypotensive effect and an isolated PLA2 from the venom inhibits platelet aggregation, via adrenaline pathway. The LC-MS analysis indicates the presence of natriuretic peptide, PLA2, CRISP, and Kallikrein. cDNA libraries analyses indicated the presence of AVIT, cobra venom factor, cystatin, crotamine, nerve growth factor and vespryn. Later studies demonstrated that the venom of *V. komodensis* (Komodo Dragon) also induces a hypotensive action, and that the venom is composed of toxins, such as PLA2, kallikrein, natriuretic peptide, CRISP, and AVIT [74].

A cDNA libraries analysis conducted by Fry et al. [67], comparing different lizards, was able to reveal new classes of toxins presents in the Varanidae family, such as lectin, veificolin, hyaluronidase, Cholecystotoxin (binds to CCK-A), Celestoxin (hypotensive), epididymal secretory protein and Goannatyrotoxin (hypertensive/hypotensive effect).

Then, the venoms of *Lanthanotus*, *Varanus* and *Heloderma* genus were compared through proteomic approaches and enzymatic activities profiling [69]. Interestingly, the only ubiquitous protein was Kallikrein and, different from Heloderma, which presents a conservation of venom constitution and actions in different species, the Varanus genus presents a variability in venom proteins and enzymatic activities such as serine peptidases, phospholipase activity and differential potential to cleavage alpha and beta chains from fibrinogen.

Venoms from different species of the Varanus genus were evaluated for the ability to prevent blood clotting by thromboelastography, and the venoms differ regarding the activity; the most potent effects were found in arboricole species, probably due to the selective pressure, according to the authors [75]. Similar to *Heloderma*, Varanid lizards possess neurotoxins that are able to bind sodium and calcium channels [72].

1.5. Amphibian

Although a witch's recipe benefits from venomous animals, the toe of a frog and the eye of a newt would definitely spice things up. Shakespeare's Macbeth (Act 4, Scene 1) contains a recipe for a witch's brew that goes as follows:

"Fillet of a fenny snake,
In the cauldron boil and bake;
Eye of newt and toe of frog,
Wool of bat and tongue of dog,
Adder's fork and blind-worm's sting,
Lizard's leg and owlet's wing,
For a charm of powerful trouble,
Like a hell-broth boil and bubble."

Although most of the above referred ingredients can be traced back to herbs (eye of newt = mustard seed (*Sinapis alba*); toe of frog = buttercup (*Ranunculus acris* L.); wool of bat = holly leaves (*Ilex aquifolium*); tongue of dog = gypsyflower from the genus hound's tounge (*Cynoglossum officinale* L.); adder's fork = least adder's-tongue (*Ophioglossum lusitanicum* L.); blind-worm = slowworm (*Anguis fragilis*)), a really mighty witch might as well as go on literally, seeing the true herpetological powers needed for the spell.

According to Table 1, published papers on amphibian venoms are less common than the triad. The similar figure to scorpion papers is due to two characteristics belonging to the study of the amphibian skin secretion: (i) the discovery of magainin, the first antibiotic peptide by [76], which boosted the literature by making several researchers seek other antibiotic peptides in other species, and (ii) the vast Chinese literature on Chan'Su, the all healing Chinese traditional medicine. These two events have undoubtfully contributed to these numbers. However, in general, amphibian literature on accidents is scarce in comparison to venomous animals.

The amphibian defense strategy against predators/aggressors is the "passive" defense (with the exception of *Rhaebo guttatus*, which is capable of voluntarily compressing its parotoid glands and ejecting its contents [77]), and the chemical nature of their venom is mainly protein/peptide toxins and low-molecular-mass compounds (such as alkaloids, steroids and their respective derivatives).

Some of the authors of the current review have been working with amphibian skin secretion for almost twenty years. As a consequence, they have been able to produce consistent literature on the subject that encompasses the different classes of bioactive molecules commonly found on the amphibian skin secretion. A compilation of these results will be presented below, together with the related literature.

Conceição, et al. [78] have evaluated the skin secretion of the tree frog *Phyllomedusa hypochondrialis* and described that this secretion presents proteins ranging between 68 and 14 kDa, and that proteolytic and phospholipase A2 activities could be detected in vitro. Moreover, authors also report that the injection of 0.6 ug of the venom in mice induced myotoxicity, as evaluated by the increase of creatinine-kinase activity in plasma. The same dose of the skin secretion also elicited vasoconstriction (for 20 min) and leukocyte rolling, as assayed by intravital microscopy. Edema and nociception, in a dose–response manner, could also be observed. Interestingly, the observed vascular permeability alterations displayed a different mechanism, in which the lowest tested concentration caused the most intense effect, in comparison to larger concentrations. This phenomenon is mostly likely due to the presence of different molecules, in distinct relative concentrations, acting on independent biological systems.

As a consequence of the described leukocyte rolling effect, a subsequent study was performed [79] that assessed the mechanisms involved in that effect. Experiments revealed that the toxins could lead to edema formation, within 2 h, which lasted for 24 h. Moreover, authors also described that the numbers of rolling and adherent leukocytes were augmented in post-capillary venules. Cytological analysis showed that macrophages were the main cells present 2 h after the injection, whereas neutrophils were the cells present after 24 h. The cytokine profiles indicted elevated levels of chemokines MCP-1 and KC, and also IL-6 and PGE2.

Mendes et al., 2016 [80], studied the casque-headed tree frog *Corythomantis greening*, a frog bearing a cranial bone adaptation used in phragmosis. The cutaneous secretion of this animal was able to induce inflammation (edema, for 96 h after the injection) and nociception. Moreover, relevant enzymatic activities were detected in the skin secretion, such as fibrinogenolytic, hyaluronidasic and metallopeptidasic. Enzymes presenting such activities have already been described as important toxins for snake venoms [81,82] and were also described in some amphibians from different genus, for example phospholipase in *Pithecopus azureus* [83] and serine peptidases in *Duttaphrynus melanostictus* [84]. Furthermore, Fusco et al. 2020 [85] studied the epidermal secretion of *Argenteohyla siemersi* and described both phospholipasic and hemolytic activities. They also reported that that venom is cytotoxic and capable of promoting necrosis which is independent of the proteolytic activity, a different activity pattern from *C. greeningi* (included in the same genus).

Targeting antibiotic peptides—a consequence of Zasloff's study—Conceição et al., 2006 [86], screened the skin secretion of *P. hypochondrialis* for antimicrobial peptides against Gram-positive and -negative bacteria and successfully described Phylloseptin-7 and Dermaseptin (DPh-1). These peptides were active over common pathogens, such as *Staphylococcus aureus, Escherichia coli, Pseudomonas aeruginosa* and *Micrococcus luteus*. In a complementary study, Huang and collaborators identified a new Dermaseptin from the same *P. hypochondrialis* (Dermaseptin-PH), which was active against Gram-positive/-negative bacteria and inhibited biofilm formation. This peptide was also effective against *Candida albicans*.

Other authors also reported complementary phylloseptins. For example, Wu et al., 2017 [87], isolated PNS-PC from *P. camba* the PNS-PC. This peptide displays inhibitory action against Methicillin-resistant *Staphylococcus aureus*. They also isolated PBa1–3 from *P. Burmeister*, a peptide with antibacterial and antifungal activities [88]. A recent study by Liu et al., 2020 [89], reported the antibacterial activity of PV-1, a Phylloseptin from *P. vaillantii* in vitro and in vivo. In spite of observed hemolysis (in vitro), this peptide was not toxic to hepatic and renal tissues in vivo, indicating the possible therapeutical potential of this peptide for bacterial infection.

Zhang et al., 2010 [90], isolated Phylloseptin-1 (PSN-1) from *P. sauvagei*. This peptide was active against *Staphylococcus aureus* in vitro, including bacterial biofilm formation inhibition. A few years later, Raja et al., 2013 [91], described five more Phylloseptins displaying antimicrobial activity from this species. Their work proved that the structural differences among those peptides were responsible for the different observed bactericidal potency, suggesting that the alpha-helix amphipathic conformation leads to microbial membrane disruption.

Using Zasloff's classic strategy, Conlon et al. 2007 [92] stimulated *Hylomantis lemur* skin secretion with norepinephrine and successfully purified Dermaseptin-L1 and Phylloseptin-L1, which were active against Gram-negative bacteria and *Batrachochytrium dendrobatidis*, a fungus that infect frogs.

In 2009, an unexpected antimicrobial peptide was described by Sousa et al. [93]. Leptoglycin, a peptide comprised basically by Leu and Gly (with an import Pro at the center of the sequence) was isolated from the skin secretion of *Leptodactylus pentadactylus* and was active against Gram-negative bacteria.

Bradykinin-potentiating peptides are protagonists of the most important example of drug discovery from animal venoms. Rocha e Silva's discovery of bradykinin [94] ultimately led to the discovery of the bradykinin-potentiating peptides (BBPs) from snake venoms. Such a peptide, on the other hand, led to the development of Captopril, the first drug belonging to angiotensin-converting enzyme inhibitor (ACEi) class, widely used around the world to treat arterial hypertension. In another unexpected study, Conceição et al., 2007 [78], described the first canonical BPP isolated from another source than snake venoms. Phypo-Xa, a decapeptide isolated from *P. hypochondrialis*, inhibited ACE and potentiated bradykinin both in vivo and in vitro. A few years later, those authors [95] also isolated three bradykinin-related peptides from *P. nordestina* skin secretion: two were vasodilators (Pnor3 and Pnor7) and one was a vasoconstrictor (Pnor5).

Some amphibians, particularly toads, can be considered major biological sources of low-molecular-mass compounds, such as alkaloids and steroids. Tempone et al. [96], through biomonitored assays, have isolated two bufadienolide steroids displaying antiparasitic activity from the skin secretion of *Rhinalla jimi*. Telecinobufagin and hellebrigenin were not new molecules at that time; however, the activity against *Leishmania* sp. promastigotes and amastigotes in macrophage culture (without NO production modulation) and the anti-*Trypanossoma cruzi* trypomastigotes activity were the novelties they reported in their paper. The mechanism of action of these molecules seems to be related to the disturbance of cellular membrane and mitochondrial function. Neither steroid presented hemolytic or cytotoxic activities in the tested conditions.

That same group of authors [97] later assayed the skin secretion of *P. nordestina* on antiparasitic models. They were able to demonstrate that four antimicrobial peptides (Dermaseptins 1 and 4, and Phylloseptins 7 and 8) were able to decrease the in vitro viability of *T. cruzi*, with a high theoretical therapeutic index. The proposed mechanism of action of the peptides is cell death induction, through cellular membrane permeabilization. Phylloseptin-7 was also effective against *Leishmania* sp.

Such results (selective membrane permeation) convinced Sciani et al. [98] to investigate the possible antitumor activities of the skin secretion of some Brazilian toads. MCF-7 and MDA-MB-231 lineages (breast tumor) displayed reduced proliferation and apoptosis induction when treated with eight different amphibian skin secretions. Among them, the most promising results came from *R. guttatus*, *R. margaritifera* and *P. hypochondrialis*. Moreover, *R. guttatus* and *R. marina* displayed selective antitumor activity over HL-60 (leukemia lineage), without toxicity to human leukocytes. It is believed that the observed antiproliferative effect is due to the known presence of bufadienolides in this toad secretion.

Schemda-Hirschmann in 2014 [99] related the presence of argininyl bufadienolides in *R. schneideri* dermic secretions, which were active on different tumor lineages AGS, SK-MES-1, J82 and HL-60 (gastric adenocarcinoma, lung carcinoma, bladder carcinoma and leukemia, respectively). Later, the same group showed similar activity in the Peruvian *R. marina* venom, and the mechanism of action seems related to ROS production and cell cycle arrest, for breast cancer lineages [100]. Antitumor properties were also described for the Paraguayan *Rhinella* sp. Such skin secretion is traditionally used by locals in folk medicine to treat skin lesions and tumors [101].

The crude extract of *Physalaemus nattereri* is cytotoxic for the B16F10 melanoma cell line. Carvalho et al. [102] observed that the secretion was able to induce conformational changes in cells, exposure of phosphatidylserine on cell membrane, reduction of mitochondrial membrane potential and arrest of cell cycle in S phase, indicating that apoptosis is the probable mechanism of action that explains the antitumor activity. RP-HPLC fractionated *P. nattereri* extract points out that this biological action is due to peptides

Skin venom from the Malaysian toad *B. asper* was active against HCT 116 colorectal tumor line by apoptosis induction, via caspase 3/7 activation and mitochondrial membrane potential disruption [103]. Bufadienolides also possess the ability to inhibit Na^+/K^+ ATPase and trigger caspase-induced apoptosis, being more selective to cancer cells than normal cells [104]. The venoms of two Turkish Salamandrine amphibians were tested against cancer cell lineages. The venoms, which presented proteins in their biochemical content, were active against cervix, alveolar, colon colorectal, pancreas, prostate, astrocytoma and breast carcinoma lines. However, these secretions were also toxic to human fibroblasts (HEK 293) [105].

Marinobufagin is a molecule present on *R. marina* venom displaying activity against leukemic cells without being toxic to normal blood cells. According to Machado et al. [106], this steroid induces toxicity via apoptosis, antimitotic action and cycle cell arrest at interphase in leukemia cells, without any genotoxicity.

The bufadienolides, bufotoxins, alkaloids and arginiyl derivatives from *R. jimi* cytotoxicity effects on cancer cell lineages were studied by Filho et al. [107], whereas Spinelli et al. [108] revaluated the antitumor action of 11 different Argentine amphibians: 6 *Hylidae/Microhylidae* and 5 *Leptodactylidae*. These venoms induced apoptosis and autophagy. Interestingly, *Leptodactylidae* skin secretion induced aggregation on cancer cells.

Finally, we present bufotenine: a tryptamine alkaloid found in many species and genera across nature (animals and plants), particularly in *R. crucifer*, *R. granulosa*, *R. schneideri*, *R. icteria* and *R. jimi* [109]. This molecule was selected in biomonitored assays and has the capacity to inhibit the penetration of rabies virus in mammalian cells, through an apparent competitive mechanism [110]. Complementary studies conducted by those authors [111] showed that this molecule was active in vivo, by increasing the survival rate of intracerebrally virus-infected mice from 15 to 40%. The safety of bufotenine was then evaluated [112] and no significant effects on mice could be detected at the effective antiviral dose. Interestingly, bufotenine acts synergically with ocellatin-F1—an antimicrobial peptide obtained from the frog *Leptodactylus labyrinthicus* skin secretion—in the rabies virus model [113]. Finally, recent in vitro assays showed that bufotenine has no antiviral action against canine coronavirus (CCoV), canine adenovirus type 2 (CAV-2) or herpesvirus type 1 (HSV-1), indicating some specificity against distinct types of viruses [114]. The mechanism of action of this alkaloid remains unclear (although the evaluation of its effects in the immune system is being assayed by these authors), but bufotenine is the perfect example of the potential of bioactive molecules isolated from a neglected venom, serving as biotechnological tool for a neglected disease drug development study.

1.6. Marine Animals

Oceans dominate planet Earth: approximately 70% of the Earth's surface is covered by water, and from that, 96.5% of this water is from the oceans [115]. More than 480,000 species of marine animals have been discovered and identified according to the World Register of Marine Species [116]. However, such a figure may be even larger: the Ward Appeltans of the Intergovernmental Oceanographic Commission of UNESCO (https://en.unesco.org/news/ocean-life-marine-age-discovery-0, accessed 19 November 2021) estimates that oceans may hold 700,000 species. These data represent what the ocean can become: a molecular library! Molecules that belong to an organism's physiology, act on hunting and prey digestion and/or chemical defense may ultimately lead to the discovery of new compounds with biotechnological or pharmaceutical uses.

Regarding bioprospection, marine animals have provided several molecules for a wide range of therapeutic applications. Some of them have already been approved by regulatory agencies and are being commercialized. The most known is ziconotide (Prialt), a ω-conotoxin peptide from *Conus magus*, applied by intrathecal route as analgesic for chronic and intense pain, whose mechanism of action is the selective blocking of neuronal N-type voltage-sensitive calcium channels [117,118]. Another known drug from marine animals is trabectedin (Yondelis), initially isolated from the marine ascidian *Ecteinascidia turbinata*, used to treat sarcomas and ovary cancer [119].

For cancer, other drugs have been developed, such as Ara-C (Cytarabine), a nucleoside isolated from a Caribbean sponge, *Cryptotheca crypta*. It is used for certain types of leukemia, including acute myeloid leukemia, acute lymphocytic leukemia and chronic myelogenous leukemia [120].

Brazilian sponges and cnidarians, such as *Zoanthus sociatus, Exaiptasia pallida* and *Carijoa riisei*, have yielded promising molecules active on cancer cells. Some of these authors have showed that *C. riisei* and the porifera *Tedania brasiliensis* extracts were effective in reducing the cell viability of glioblastoma, and that *C. riseii* also acts on breast and ovary cancer. Moreover, *Z. sociatus* and *E. pallida* were able to diminish leukemic cell viability [121]. Regarding the envenomation field, some of these authors have contributed for the understanding of marine animal venoms, from a biochemical and pathophysiological perspective.

1.7. Sea Urchins

Sea urchins are the most abundant animals in Brazilian shores. They are also responsible for the majority of reported marine animal accidents [122]. *Echinometra lucunter*—the rock boring urchin—can be easily found in rocky shores. Human accidents are frequent and can be associated with the animal's manipulation by bathers, or by people stepping on the animals while walking on the shore. More severe cases (in terms of the number of spines punctures) can result from people being dragged onto rocky walls by wave action. Still, the most common route that the spines penetrate the skin is through the foot or hand. This event causes local inflammatory reactions, characterized by edema, erythema and pain [123,124]. Facing this problem, the authors have wondered: is this accident solely mechanical due to the spine's penetration, or does the sea urchin have a venom that contributes to the described symptoms?

To answer that question, 'toxins' from *E. lucunter* spines were extracted, immersing the excised appendices in a physiological buffer (to avoid cell lyses by osmotic shock), followed by animal inflammation test models. Authors described that the extract induces a pro-inflammatory reaction, by increasing rolling, adhered and migrated leukocytes. Moreover, the spines extract decreased the pain threshold and induced paw edema [30]. In another study, these authors were able to isolate one molecule responsible for those effects, including its partial molecular characterization [125]. However, it was clear that there was more than one single molecule eliciting such activities; therefore, the clinical observed symptoms clearly surpass the mechanical trauma aroused by spine penetration.

This mechanism is a very successful adaptation: the venom (i.e., the 'toxins') diminishes the pain threshold—making the victim more susceptible to painful stimuli—at the same time that the spines puncture the skin. As a consequence, the mechanical accident becomes more aggressive, due to this synergism (resulting in inflammation).

E. lucunter spines do not contain typical venom glands, in the same way venomous animals do, but it is a living structure, full of granular cells, which are most likely to produce and secrete these toxins along the entire spine, particularly at in the spine tip, a region more susceptible to mechanical stress by contact (with possible predators and aggressors) [126]. Moreover, although the spine is composed mainly of calcium and/or magnesium carbonate, the myriad of cells embedded would significantly contribute to spine regeneration. It has been demonstrated that the spine secretes cathepsins B and/or X, an enzyme associated with matrix remodeling processes, contributing to the spine growth and regeneration, but also to the toxicity.

Besides spines accidents, consumption of sea urchins may elicit undesirable/toxic effects for the consumer, as they are usually eaten raw. Therefore, these authors have investigated the coelomic fluid of *E. lucunter*, searching for toxins (pro-inflammatory molecules, in particular). A bioactive peptide, termed 'echinometrin', capable of reducing rolling cells and increasing adhered and migrated ones—concomitant to edema induction—was identified. Moreover, this peptide induced mast cell degranulation, which makes us think that histamine was responsible for the observed inflammatory reaction [127]. Actually, many consumers present allergies after the consumption of raw sea urchin, and there are studies suggesting the participation of vitellogenin in such process, by increasing IgE levels [128,129]. Echinometrin is, in fact, a cryptide [130], i.e., an internal fragment of vitellogenin. Moreover, its N- and C-termini match the amino acid specificity for (the previously reported) cathepsin B/X, suggesting a local toxin generation system, in which both substrate and processing enzyme are present and ready to act.

Once the biomonitored assay reported above proved successful in the identification of one bioactive peptide, these authors decided to performed an untargeted peptidomic approach on sea urchins' peptides. The secreted peptides from *E. lucunter*, *Lytechinus variegatus* and *Arbacia lixula* were analyzed. It was possible to observe that coelomic fluids of all three species are full of peptides. On the other hand, peptides could be identified only in the spines of *L. variegatus* and *A. lixula*, whereas *E. lucunter* spines contain mainly low molecular mass compounds. Database mining suggests that some peptides may display relevant biological effects, such as antibiotic, anticancer, antiviral, phospholipase A2 inhibitor and neuroprotective properties, making sea urchin molecules a source of new therapeutic compounds [131].

1.8. Mollusks

Peptides are abundant in marine mollusks from the Gastropoda class. They are usually referred as 'conopeptides' and are responsible for prey paralysis due to their specific action on the neuromuscular ionic channels [132,133]. The genus *Conus* is a well-known source of these conopeptides. The Tox-Prot database from Uniprot/Swiss-Prot describes that 1.370 toxins are manually annotated for 117 snail species, most of them from genus *Conus* [134,135]. On the other hand, the database platform for conopeptides, ConoServer, shows that 119 *Conus* species already have at least one protein sequence/structure elucidated. Besides that, this platform shows that conopeptides can be categorized in 12 pharmacological families or in 33 cysteine frameworks. More than 2900 mature conotoxins can be found in this database [136,137].

Conus can be classified into three main groups, according to their feeding behavior: worm-hunting, molluscivorous and fish-hunting snails [138]. One of them—*C. regius*—a species that dwells the USA, Central America, and Brazil, including the Fernando de Noronha archipelago, has been studied by these authors [139]. As feeding behavior is often related with venom composition, the authors have investigated what would be the feeding habits of these animal, since they were not known at the time. They found that *C. regius* preferentially preys on fire-worms, thus being categorized as a vermivorous species. Authors have also evaluated the homogeneity of the venom and have determined that, regardless of gender, size and season of the year, there was no significant variation on venom composition (as determined by RP-HPLC peak area and similarity). Under these conditions, they have found the major peak, isolated and characterized it, which led to the identification of rg11a, a conotoxin presenting the cysteine pattern C-C-CC-CC-C-C and ~5 kDa [140]. Later, these authors also described α-RgIB: a 2.7 kDa peptide bearing the CC-C-C pattern, which is an antagonist of neuronal acetylcholine receptor and is capable of inducing hyperactivity in mice and breathing difficulties [141].

1.9. Stingrays

Stingrays accounted for 69% of aquatic animal accidents in Brazil from 2007 to 2013. Most cases (88.4%) were reported in the north region and correspond to accidents caused by freshwater stingrays [142].

In general, symptoms of freshwater stingray accidents include skin necrosis, edema, erythema and intense pain, mainly at the lower limbs, which are the most common accident site. Several studies have focused on the mechanism of action of stingray toxins. One explanation is the release of proinflammatory interleukins that lead to the inflammatory reaction and pain, besides the direct participation of mast cell degranulation and histamine release [143,144]. The presence of inflammatory cells in the necrotic tissues was reported, most lymphoid, CD3+ and CD4+ cells, as well as the presence of eosinophils [145].

Although less frequent, marine rays also cause human accidents, but few works report them. In this sense, some of these authors have studied *Hypanus americanum*'s mucus, searching for toxins [146]. It is noteworthy to mention that a marine stingray's whole body is covered by mucus produced by epithelial cells. Some animals possess a calcified spine ('sting') on their tail, which is covered by an epithelium that secretes mucus. This secretion is rich in molecules involved in the chemical defense and skin homeostasis maintenance, including establishing a barrier against microorganisms.

These authors observed that the mucus is labile, denaturating in function of the temperature and storage time after collection. Moreover, the classical scratching method for mucus collection results in the attainment of a mucus rich in cellular debris and, consequently, intracellular content that masks the 'actual' mucus. Authors were forced to develop a new method: the whole animal was submerged—for 40 s—in a tank containing only freshwater. After the animal was removed, the water was acidified (0.1% final concentration) and the solution was filtered. This large volume was directly pumped into the C18-RP-HPLC column via system pump 'A'. After total sample loading, standard chromatography was performed [146].

Nevertheless, the chemical nature of the mucus revealed itself to be more complex than initially imagined by those authors. Several proteins, peptides and low-molecular-mass compounds could be detected. The mucus elicits inflammatory reactions, such as edema and leukocyte recruitment in mice. The performed zymograms displayed proteolytic activity. Moreover, authors describe the antimicrobial effect of molecules fractionated from the mucus. The proteomic analyses revealed proteins that are involved in the immune response, and are very similar to the proteins related to the sting, and also similar to proteins described in fishes from Teleostei class, indicating that the epidermal secretions of stingrays could be more related to an innate immune system than with a venom delivery system [146]. This hypothesis was recently reinforced in a work that analyzed the genomic data of a venomous fish and associated the presence of aerolysin (considered as a toxin) with the immune system [147].

1.10. Cnidarians

The phylum Cnidaria comprises more than 10,000 species and is considered the most ancient venomous animal lineage, having emerged approximately 650 million years ago [148,149]. To the contrary of other venomous animals, cnidarians have the unique characteristic of lacking a centralized venom system [150]. Instead of a venom gland, these animals present little organelles distributed throughout their bodies, called cnidaes. Such structures are produced by the Golgi apparatus of specialized cells: the cnidoblasts [151]. It is divided into three main lineages: 1. Anthozoa, formed by Anthozoa class; 2. Medusozoa, comprised of Scyphozoa, Staurozoa, Cubozoa and Hydrozoa classes; 3. Endocnidozoa, comprising Myxozoa and Polypodiozoa classes. Cnidaria is a diverse phylum, rich in bioactive molecules, known to be used mainly for predation, defense and intraspecific competition [152].

Cnidaria early studies began in 1903 on *Anemonia sulcata* and *Actinia equina* tentacles extracts. Since then, several studies on sea anemones have been developed, leading to more than a century of research on these animals' venoms [150,153,154]. Sea anemones are exquisite sources of toxins and represent the greatest diversity in Anthozoa, having around 1200 species distributed in 46 families [150].

These cnidarians can cause envenomation through their nematocysts, specialized structures that inoculate venom. One particular case report of a human accident caused by anemones belonging to the *Stichodactyla* genus describes local skin irritations with blistering, edema and hemorrhage, mild symptoms when compared to the actual target of the toxin, prey, which is instantly killed by neuro- and cardiotoxins [155]. The anemone toxins molecular scaffolds are diverse: at least 17 different structural motifs are known [150].

The peptide neurotoxins found in sea anemones may act over different ion channels. ShK toxins, for example, bind to Kv type 1; some types of β-defensins can modify the action of Kv type 3 and Nav type 1, 2 and 4; while the inhibitor cystine-knot (ICK) can act over Kv type 5 and acid-sensing ion channels [150]. In this context, a study published in 2004, by some of these authors, investigated the differential selectivity between three sea anemones toxins against a wide range of Nav channels subtypes (Nav 1.1–1.6). The authors observed that for Nav1.3, the three toxins (ATX-II, AFT-II and Bc-III) were active only when at high concentrations. Additionally, it was observed that although ATX-II (from *A. sulcata*) and AFT-II (from *A. fuscoviridis*) exhibit similar sequences, a single amino difference was enough to alter the ion channels specificity. Lysine^{36} (ATX-II) seems to be fundamental for its action over Nav1.1 and Nav1.2 channels; meanwhile, AFT-II mainly exerts effects on Nav 1.4 and Nav 1.5. Moreover, the slight changes in amino acids between similar Nav channels can have a crucial role in toxins binding. For example, AFT-II had a more potent effect over Nav1.4 than Nav 1.5. These two channels are only marginally different and the presence of a Leucine at position 1611 in Nav1.4, instead of an Isoleucine at in Nav1.5 right after a neighboring Asparagine, may indicate the importance of these residues for the toxin binding [156].

In another evaluation of sea anemones venoms, Zaharenko et al. reported, for the first time, the proteomics analysis of the neurotoxic fraction of the sea anemone *Bunodosoma cangicum*. Authors processed by RP-HPLC such a fraction and identified at least 81 different molecules, distributed along 41 chromatographic peaks. Mass spectrometric analysis by MALDI-TOF and ESI-Q-TOF shows that that fraction is composed of low-molecular-mass (280–450 Da) as well as heavier molecules (4–5 kDa). Major fractions were purified and sequenced by Edman degradation, revealing nine novel peptides. Three peptides clearly presented the typical cysteine scaffold found in type 1 sodium channel toxins, and six of them presented new cysteine scaffolds belonging to two new classes of toxins. Additionally, when tested on extracellular crab leg nerve, the new peptides called Bcg31.16 and Bcg30.24 showed that, at very low concentrations (40–50 nM), those neurotoxins were able to diminish the amplitude of CAPs (compound action potentials) and increase its duration, showing a high potency and suggesting that these toxins target sodium channels [157].

Compared to other cnidarians, the Anthozoa (anemones included) is a well-studied group, in terms of toxins investigation. ToxProt lists 256 toxins belonging to 48 species of sea anemones (manually curated; accessed October, 2021). On the other hand, only five toxins from Cubozoa; four from Hydrozoa and one from Scyphozoa classes are deposited [134]. Of particular interest, three Cubozoa toxins (caTX-A, cqTX-A, crTX-A, cfTX-1 and cfTX-2) belong respectively to four species of box jellyfishes: *Carybdea alata*, *Chiropsoides quadrigatus*, *Carybdea rastonii*; and *Chironex fleckeri* (the Australian box jellyfish, one of the most dangerous species of cnidarians) [134,158]. Regardless of the small number of curated toxins, 327 proteins from Cubozoa—computationally analyzed and available at TrEMBL—still remain to be reviewed. The literature refers to Cubozoa toxins being enzymes (phospholipases A2, metallopeptidases and serine peptidases), CRISPs, lectins, pore-forming toxins and protease inhibitors [159]. For Hydrozoa, the four proteins manually curated and described as Hydralysin toxins belong to only two different species: *Hydra viridissima* and *H. vulgaris* [134,135].

The challenge of better knowing the toxins found in Cubozoa and Hydrozoa is not limited to the proteins and peptides; little is known about the low-molecular-mass molecules from these organisms [160]. In order to increase knowledge on the biotechnological potential of Cubozoa and Hydrozoa, two studies were recently performed. The first one, conducted by Bueno et al. [161], investigated the effects of the methanolic extracts of hydromedusa *Olindias sambaquiensis* and jellyfish *Chiropsalmus quadrumanus* over the autonomic neurotransmission. In this study, researchers employed a classical model to sympathetic co-transmission: a myographic evaluation of rat vas deferens bisected in two portions (prostatic and epididymal) for purinergic or adrenergic responses. Throughout the study, both methanolic extracts were demonstrated to be of low complexity and rich in low molecular mass molecules.

Authors report that a low concentration (0.1 µg/mL) of *C. quadrumanus* extract blocked the predominantly noradrenergic contraction of the epididymal end. On the other hand, only high concentrations (1 and 10 µg/mL) of *O. sambaquiensis* extract were capable of leading to the blockade of muscle contraction. Nevertheless, both extracts did not present significant differences concerning the phasic contractions in the prostatic portion (purinergic response), when compared to the control group. Moreover, the histological analysis showed that none of the extracts promote major tissue damage in the prostatic and epididymal vas deferens ends, showing the same unaltered morphology as the control group, which indicates their effects only on the neurotransmission, not causing toxic tissue damages [161].

Another study, published by Arruda et al. [160], focused on *C. quadrumanus* tentacles methanolic extract and its biological activity over neurite growth. In this work, the extract was tested on a human SH-SY5Y neuroblastoma cell line, a neuronal cell culture model commonly used for neurodegenerative disease investigations. Authors report alterations on neurite-related structures of neurons, without affecting cell proliferation or inducing necrosis or apoptosis [160]. The specific neurite length outgrowth observed in all cells exposed to the toxins was associated with a translin-like protein (hyccin cryptein) cryptide, as well as to small molecules acting synergically to promote the neurite/branches formation, elongation and facilitating neurotransmission. Neurite formation can happen either via microtubule and motor proteins [162] or PI4P regulation—acting on plasma membrane identity and myelin development [162]. Moreover, toxins present in the methanolic extract showed no effect on the straightness of neurite's growth or cell body area, but increased branching junctions connected to cells. More than 14 low molecular mass molecules related to neuritogenesis were found through LC-MS fingerprinting and at least 4 peptides related to neuronal function [160].

1.11. Insects

Although insects are the largest group within the arthropod phylum, and most of them are well studied due to their importance, there is always room for new research. Insects are known to create the biological foundation for most of the terrestrial ecosystems by pollinating plants, dispersing seeds, controlling populations of other organisms (including decomposing dead material to recycle nutrients) and being a major food source for other taxa. On the other hand, insects also spread diseases and can compromise a significant amount of food (grains, for example).

Among the notorious insects, there is the honeybee (*Apis mellifera*). There are reports of beekeeping as old as 10,000 years. Bee domestication started in Egypt 4500 years ago, when probably human accidents must have become more frequent, as well.

Bee stinging is mischievous: one single sting may provoke allergy and the subject may die from anaphylaxis. On the other hand, one may be stung several times and, in spite of intense pain and significant swelling, no significant harm occurs. However, when a few dozen bees sting, one may become envenomed. This event is not related to allergy and is a consequence of the bee toxins acting on the victim's body, especially in the kidneys.

The difference between poison and medicine is the dose, and apitherapy is a rather popular branch of alternative medicine which includes live bee acupuncture. Such a procedure may heal some, but is not free of risks at all! Adverse reactions to bee venom therapy are frequent. Constant exposure to the venom may lead to arthropathy, for example. In sensitized individuals, allergic reactions vary from mild, local swelling to severe systemic reactions, anaphylactic shock and even death. Yet, there are claimed cosmetic uses of the bee venom. Rumor has it that the Duchess of Cambridge has used bee venom to keep her skin looking flawless and even applied the secret ingredient to ensure a glowing complexion when she wed Prince William in 2011.

In a more practical context, a few groups have explored the possibility of developing an antiapilic serum, for treating those patients that have suffered multiple bee stings and have not suffered anaphylactic shock. Among those, authors from this group have successfully developed an efficient antiapilic serum that is currently under clinical trial (phase III). Further details can be found in the works of Ferreira Jr et al., 2010 [163], and Sciani et al., 2010 [164], who set the basis for the preclinical and clinical studies summarized by Barbosa et al., 2021 [165].

2. Conclusions

Animal venoms and toxins comprise a diverse repertoire of fascinating proteins, peptides and other bioactive molecules that have evolved through natural selection, driven by adaptive pressure and the survival of the fittest. Their biological role is—mainly—predation and defense. Mankind—and its anthropocentric perspective of nature—have always tried to develop ways to use and study these venoms and toxins as pharmacological prototypes for the research and development of novel therapeutics. Such a quest has opened new venues to the identification of an unprecedented number of new molecules and/or biological effects.

According to our view, 'classic' toxinology (as we have termed the continuous study of snakes, scorpions and spiders) will lessen in the near future and the 'new' venoms and toxins will prevail, due to subject saturation. Research of unexplored—or neglected—species of animals and their venoms and secretions should become dominant, since they contain a myriad of molecules displaying relevant biological effects on human illnesses, diseases, degenerative disorders, injuries, pain, tumors and infections (viral, bacterial and fungal), either as medicines or diagnostics tools.

Therefore, we consider that the currently reviewed literature on lizards, amphibians, and marine animals is just the beginning of a new thematic approach that we hope will become dominant in the following years. Such veiled potential currently hidden in the neglected animal venoms and toxins can set the instrumental and scientific basis for the development of new molecules with innovative potential, which could shape a "new era" in toxinology.

Author Contributions: Conceptualization, G.R.C. and E.B.-N.; writing—original draft preparation, G.R.C., D.L.d.S., E.B.-N., H.V. and L.A.d.O.; writing—review and editing, E.B.-N. and H.V.; supervision, J.M.S. and D.C.P.; funding acquisition, J.M.S. and D.C.P. All authors have read and agreed to the published version of the manuscript.

Funding: This research was funded by Conselho Nacional de Desenvolvimento Científico e Tecnológico (CNPq), grant number 301974/2019-5, Financiadora de Estudos e Projetos (FINEP), grants number 01.12.0450.00 and 01.09.0278.04, Fundação de Amparo à Pesquisa do Estado de São Paulo (FAPESP), grant number 19/19929-6. The APC was funded by Instituto Butantan.

Institutional Review Board Statement: Not applicable.

Informed Consent Statement: Not applicable.

Conflicts of Interest: The authors declare no conflict of interest.

References

1. Brazil, V. *Memória histórica do Instituto de Butantan*; Instituto Butantan: São Paulo, Brazil, 1941.
2. WHO. Investing to overcome the global impact of neglected tropical diseases. In *Third WHO Report on Neglected Tropical Diseases 2015*; World Health Organization: Geneva, Switzerland, 2015; Volume 3.
3. Darwin, C. *On the Origin of Species, 1859*; Routledge: London, UK, 2003.
4. Zhang, S.; Gao, B.; Zhu, S. Target-Driven Evolution of Scorpion Toxins. *Sci. Rep.* **2015**, *5*, 14973. [CrossRef] [PubMed]
5. Argudín, M.; Mendoza, M.C.; Rodicio, M.D.R.R. Food Poisoning and Staphylococcus aureus Enterotoxins. *Toxins* **2010**, *2*, 1751–1773. [CrossRef] [PubMed]
6. Guarino, A.; Giannella, R.; Thompson, M.R. Citrobacter freundii produces an 18-amino-acid heat-stable enterotoxin identical to the 18-amino-acid Escherichia coli heat-stable enterotoxin (ST Ia). *Infect. Immun.* **1989**, *57*, 649–652. [CrossRef]
7. Santibáñez-López, C.E.; Graham, M.R.; Sharma, P.P.; Ortiz, E.; Possani, L.D. Hadrurid Scorpion Toxins: Evolutionary Conservation and Selective Pressures. *Toxins* **2019**, *11*, 637. [CrossRef] [PubMed]
8. Hargreaves, A.D.; Swain, M.T.; Hegarty, M.J.; Logan, D.; Mulley, J.F. Restriction and Recruitment—Gene Duplication and the Origin and Evolution of Snake Venom Toxins. *Genome Biol. Evol.* **2014**, *6*, 2088–2095. [CrossRef]
9. Casewell, N.R.; Wüster, W.; Vonk, F.J.; Harrison, R.A.; Fry, B.G. Complex cocktails: The evolutionary novelty of venoms. *Trends Ecol. Evol.* **2013**, *28*, 219–229. [CrossRef]
10. Schendel, V.; Rash, L.D.; Jenner, R.A.; Undheim, E.A.B. The Diversity of Venom: The Importance of Behavior and Venom System Morphology in Understanding Its Ecology and Evolution. *Toxins* **2019**, *11*, 666. [CrossRef]
11. Haines, H.R.; Willink, P.W.; Maxwell, D. Stingray Spine Use and Maya Bloodletting Rituals: A Cautionary Tale. *Lat. Am. Antiq.* **2008**, *19*, 83–98. [CrossRef]
12. Dodd-Butera, T.; Broderick, M. Animals, Poisonous and Venomous. *Encycl. Toxicol.* **2014**, 246–251. [CrossRef]
13. Qi, J.; Tan, C.K.; Hashimi, S.M.; Zulfiker, A.H.M.; Good, D.; Wei, M.Q. Toad Glandular Secretions and Skin Extractions as Anti-Inflammatory and Anticancer Agents. *Evid. -Based Complement. Altern. Med.* **2014**, *2014*, 312684. [CrossRef]
14. Junior, V.H.; Martins, I.A. KAMBÔ: An Amazonian enigma. *J. Venom Res.* **2020**, *10*, 13. [PubMed]
15. Quintero-Hernández, V.; Jiménez-Vargas, J.; Gurrola, G.; Valdivia, H.; Possani, L. Scorpion venom components that affect ion-channels function. *Toxicon* **2013**, *76*, 328–342. [CrossRef]
16. Petricevich, V.L. Scorpion Venom and the Inflammatory Response. *Mediat. Inflamm.* **2010**, *2010*, 903295. [CrossRef]
17. Markland, F.S. Snake venoms and the hemostatic system. *Toxicon* **1998**, *36*, 1749–1800. [CrossRef]
18. Bjarnason, J.B.; Fox, J.W. Hemorrhagic metalloproteinases from snake venoms. *Pharmacol. Ther.* **1994**, *62*, 325–372. [CrossRef]
19. Gutiérrez, J.M.; Lomonte, B. Phospholipase A2 myotoxins from Bothrops snake venoms. *Toxicon* **1995**, *33*, 1405–1424. [CrossRef]
20. Kuniyoshi, A.K.; Kodama, R.T.; Moraes, L.H.F.; Duzzi, B.; Iwai, L.K.; Lima, I.F.; Carvalho, D.C.; Portaro, F.V. In vitro cleavage of bioactive peptides by peptidases from Bothrops jararaca venom and its neutralization by bothropic antivenom produced by Butantan Institute: Major contribution of serine peptidases. *Toxicon* **2017**, *137*, 114–119. [CrossRef] [PubMed]
21. Grego, K.F.; Vieira, S.E.M.; Vidueiros, J.P.; Serapicos, E.D.O.; Barbarini, C.C.; da Silveira, G.P.M.; Rodrigues, F.D.S.; Alves, L.D.C.F.; Stuginski, D.R.; Rameh-De-Albuquerque, L.C.; et al. Maintenance of venomous snakes in captivity for venom production at Butantan Institute from 1908 to the present: A scoping history. *J. Venom. Anim. Toxins Incl. Trop. Dis.* **2021**, *27*. [CrossRef] [PubMed]
22. Lucas, S.M. The history of venomous spider identification, venom extraction methods and antivenom production: A long journey at the Butantan Institute, São Paulo, Brazil. *J. Venom. Anim. Toxins Incl. Trop. Dis.* **2015**, *21*, 21. [CrossRef]
23. Heard, K.; O'Malley, G.F.; Dart, R.C. Antivenom Therapy in the Americas. *Drugs* **1999**, *58*, 5–15. [CrossRef]
24. Jorge, M.; Ribeiro, L. Dose de soro (antiveneno) no tratamento do envenenamento por serpentes peçonhentas do gênero Bothrops. *Rev. Assoc. Médica Bras.* **1997**, *43*, 74–76. [CrossRef]

25. Cushman, D.W.; Ondetti, M.A. History of the design of captopril and related inhibitors of angiotensin converting enzyme. *Hypertension* **1991**, *17*, 589–592. [CrossRef] [PubMed]
26. Patil, C.G.; Walker, D.G.; Miller, D.M.; Butte, P.; Morrison, B.; Kittle, D.S.; Hansen, S.J.; Nufer, K.L.; Byrnes-Blake, K.A.; Yamada, M.; et al. Phase 1 Safety, Pharmacokinetics, and Fluorescence Imaging Study of Tozuleristide (BLZ-100) in Adults with Newly Diagnosed or Recurrent Gliomas. *Neurosurgery* **2019**, *85*, E641–E649. [CrossRef] [PubMed]
27. Ojeda, P.G.; Wang, C.; Craik, D.J. Chlorotoxin: Structure, activity, and potential uses in cancer therapy. *Biopolymers* **2015**, *106*, 25–36. [CrossRef] [PubMed]
28. Dardevet, L.; Rani, D.; Aziz, T.A.E.; Bazin, I.; Sabatier, J.-M.; Fadl, M.; Brambilla, E.; De Waard, M. Chlorotoxin: A helpful natural scorpion peptide to diagnose glioma and fight tumor invasion. *Toxins* **2015**, *7*, 1079–1101. [CrossRef] [PubMed]
29. Wie, C.S.; Derian, A. Ziconotide. [Updated 2021 July 22]. In *StatPearls*; StatPearls Publishing: Treasure Island, FL, USA, 2021. Available online: https://www.ncbi.nlm.nih.gov/books/NBK459151/ (accessed on 19 November 2021).
30. Sciani, J.M.; Zychar, B.C.; Gonçalves, L.R.D.C.; Nogueira, T.D.O.; Giorgi, R.; Pimenta, D.C. Pro-inflammatory effects of the aqueous extract of Echinometra lucunter sea urchin spines. *Exp. Biol. Med.* **2011**, *236*, 277–280. [CrossRef]
31. Kohn, A.J. Conus envenomation of humans: In fact and fiction. *Toxins* **2019**, *11*, 10. [CrossRef]
32. Dumbacher, J.P.; Spande, T.F.; Daly, J.W. Batrachotoxin alkaloids from passerine birds: A second toxic bird genus (Ifrita kowaldi) from New Guinea. *Proc. Natl. Acad. Sci. USA* **2000**, *97*, 12970–12975. [CrossRef]
33. Diamond, J.M. Did Komodo dragons evolve to eat pygmy elephants? *Nature* **1987**, *326*, 825. [CrossRef]
34. Montgomery, J.M.; Gillespie, D.; Sastrawan, P.; Fredeking, T.; Stewart, G.L. Aerobic salivary bacteria in wild and captive komodo dragons. *J. Wildl. Dis.* **2002**, *38*, 545–551. [CrossRef] [PubMed]
35. Estevão-Costa, M.-I.; Sanz-Soler, R.; Johanningmeier, B.; Eble, J.A. Snake venom components in medicine: From the symbolic rod of Asclepius to tangible medical research and application. *Int. J. Biochem. Cell Biol.* **2018**, *104*, 94–113. [CrossRef]
36. Opie, L.H.; Kowolik, H. The discovery of captopril: From large animals to small molecules. *Cardiovasc. Res.* **1995**, *30*, 18–25. [CrossRef]
37. Ferreira, L.A.F.; Galle, A.; Raida, M.; Schrader, M.; Lebrun, I.; Habermehl, G. Isolation: Analysis and properties of three bradykinin–potentiating peptides (BPP-II, BPP-III, and BPP-V) from Bothrops neuwiedi venom. *J. Protein Chem.* **1998**, *17*, 285–289. [CrossRef]
38. Grant, M.L.; Henderson, L.J. A Case of Gila Monster Poisoning with a Summary of Some Previous Accounts. *Proc. Iowa Acad. Sci.* **1957**, *64*, 686–697.
39. Shufeldt, R.W. The bite of the Gila monster (*Heloderma suspectum*). *Am. Nat.* **1882**, *16*, 907–908.
40. Woodson, W.D. Toxicity of Heloderma venom. *Herpetologica* **1947**, *4*, 31–33.
41. Cooke, E.; Loeb, L. Hemolytic action of the venom of Heloderma suspectum. *Proc. Soc. Exp. Biol. Med.* **1908**, *5*, 104–105. [CrossRef]
42. Mebs, D. Some studies on the biochemistry of the venom gland of Heloderma horridum. *Toxicon* **1968**, *5*, 225–226. [CrossRef]
43. Stýblová, Z.; Kornalik, F. Enzymatic properties of Heloderma suspectum venom. *Toxicon Off. J. Int. Soc. Toxinolo.* **1967**, *5*, 139–140. [CrossRef]
44. Murphy, S.A.; Johnson, B.D.; Sifford, D.H. Enzymes in Heloderma horridum venom. *J. Ark. Acad. Sci.* **1976**, *30*, 61–63.
45. Raufman, J.P.; Jensen, R.T.; Sutliff, V.E.; Pisano, J.J.; Gardner, J.D. Actions of Gila monster venom on dispersed acini from guinea pig pancreas. *Am. J. Physiol. Content* **1982**, *242*, G470–G474. [CrossRef] [PubMed]
46. Banks, B.E.; Pearce, F.; Springer, C.J.; Vernon, C. On the immunology of nerve growth factor. *Neurosci. Lett.* **1985**, *61*, 127–130. [CrossRef]
47. Mebs, D. Biochemistry of Kinin-Releasing Enzymes in the Venom of the Viper Bitis Gabonica and of the Lizard Heloderma Suspectum. In *Bradykinin and Related Kinins*; Springer: Berlin/Heidelberg, Germany, 1970; pp. 107–116. [CrossRef]
48. Mebs, D. Isolierung und Eigenschaften eines Kallikreins aus dem Gift der Krustenechse *Heloderma suspectum*. *DIETRICH MEBS* **1969**, *350*, 821–826. [CrossRef]
49. Parker, D.S.; Raufman, J.P.; O'Donohue, T.L.; Bledsoe, M.; Yoshida, H.; Pisano, J.J. Amino acid sequences of helospectins, new members of the glucagon superfamily, found in Gila monster venom. *J. Biol. Chem.* **1984**, *259*, 11751–11755. [CrossRef]
50. Hendon, R.A.; Tu, A.T. Biochemical characterization of the lizard toxin gilatoxin. *Biochemistry* **1981**, *20*, 3517–3522. [CrossRef]
51. Vandermeers, A.; Vandermeers-Piret, M.-C.; Robberecht, P.; Waelbroeck, M.; Dehaye, J.-P.; Winand, J.; Christophe, J. Purification of a novel pancreatic secretory factor (PSF) and a novel peptide with VIP- and secretin-like properties (helodermin) from Gila monster venom. *FEBS Lett.* **1984**, *166*, 273–276. [CrossRef]
52. Mochca-Morales, J.; Martin, B.M.; Possani, L.D. Isolation and characterization of Helothermine, a novel toxin from Heloderma horridum horridum (Mexican beaded lizard) venom. *Toxicon* **1990**, *28*, 299–309. [CrossRef]
53. Tu, A.T.; Hendon, R.R. Characterization of lizard venom hyaluronidase and evidence for its action as a spreading factor. *Comp. Biochem. Physiol. Part B Comp. Biochem.* **1983**, *76*, 377–383. [CrossRef]
54. Sosa, B.P.; Alagon, A.C.; Martin, B.M.; Possani, L.D. Biochemical characterization of the phospholipase A2 purified from the venom of the Mexican beaded lizard (Heloderma horridum horridum Wiegmann). *Biochemistry* **1986**, *25*, 2927–2933. [CrossRef] [PubMed]
55. Alagon, A.; Possani, L.D.; Smart, J.; Schleuning, W.D. Helodermatine, a kallikrein-like, hypotensive enzyme from the venom of Heloderma horridum horridum (Mexican beaded lizard). *J. Exp. Med.* **1986**, *164*, 1835–1845. [CrossRef]

56. Eng, J.; Andrews, P.; Kleinman, W.A.; Singh, L.; Raufman, J.P. Purification and structure of exendin-3, a new pancreatic secretagogue isolated from Heloderma horridum venom. *J. Biol. Chem.* **1990**, *265*, 20259–20262. [CrossRef]
57. Patterson, R.A. Some physiological effects caused by venom from the Gila Monster, Heloderma suspectum. *Toxicon* **1967**, *5*, 5–10. [CrossRef]
58. Patterson, R.A. Smooth muscle stimulating action of venom from the Gila monster, Heloderma suspectum. *Toxicon* **1967**, *5*, 11–15. [CrossRef]
59. Patterson, R.A.; Lee, I.S. Effects of Heloderma suspectum venom on blood coagulation. *Toxicon* **1969**, *7*, 321–324. [CrossRef]
60. Eng, J.; Kleinman, W.; Singh, L.; Singh, G.; Raufman, J. Isolation and characterization of exendin-4, an exendin-3 analogue, from Heloderma suspectum venom. Further evidence for an exendin receptor on dispersed acini from guinea pig pancreas. *J. Biol. Chem.* **1992**, *267*, 7402–7405. [CrossRef]
61. Parkes, D.G.; Mace, K.F.; Trautmann, M.E. Discovery and development of exenatide: The first antidiabetic agent to leverage the multiple benefits of the incretin hormone, GLP-1. *Expert Opin. Drug Discov.* **2012**, *8*, 219–244. [CrossRef] [PubMed]
62. Huang, T.-F.; Chiang, H.-S. Effect on human platelet aggregation of phospholipase A2 purified from Heloderma horridum (beaded lizard) venom. *Biochim. Biophys. Acta (BBA) Lipids Lipid Metab.* **1994**, *1211*, 61–68. [CrossRef]
63. Vandermeers, A.; Vandermeers-Piret, M.-C.; Vigneron, L.; Rathe, J.; Stievenart, M.; Christophe, J. Differences in primary structure among five phospholipases A2 from Heloderma suspectum. *JBIC J. Biol. Inorg. Chem.* **1991**, *196*, 537–544. [CrossRef] [PubMed]
64. Kwok, H.F.; Chen, T.; O'Rourke, M.; Ivanyi, C.; Hirst, D.; Shaw, C. Helokinestatin: A new bradykinin B2 receptor antagonist decapeptide from lizard venom. *Peptides* **2008**, *29*, 65–72. [CrossRef]
65. Fry, B.G.; Winter, K.; Norman, J.A.; Roelants, K.; Nabuurs, R.J.A.; Van Osch, M.; Teeuwisse, W.M.; van der Weerd, L.; Mcnaughtan, J.E.; Kwok, H.F.; et al. Functional and Structural Diversification of the Anguimorpha Lizard Venom System. *Mol. Cell. Proteom.* **2010**, *9*, 2369–2390. [CrossRef]
66. Komori, Y.; Nikai, T.; Sugihara, H. Purification and characterization of a lethal toxin from the venom of Helodermahorridumhorridum. *Biochem. Biophys. Res. Commun.* **1988**, *154*, 613–619. [CrossRef]
67. Fry, B.G.; Roelants, K.; Winter, K.; Hodgson, W.C.; Griesman, L.; Kwok, H.F.; Scanlon, D.; Karas, J.; Shaw, C.; Wong, L.; et al. Novel Venom Proteins Produced by Differential Domain-Expression Strategies in Beaded Lizards and Gila Monsters (genus Heloderma). *Mol. Biol. Evol.* **2009**, *27*, 395–407. [CrossRef]
68. Ma, C.; Yang, M.; Zhou, M.; Wu, Y.; Wang, L.; Chen, T.; Ding, A.; Shaw, C. The natriuretic peptide/helokinestatin precursor from Mexican beaded lizard (Heloderma horridum) venom: Amino acid sequence deduced from cloned cDNA and identification of two novel encoded helokinestatins. *Peptides* **2011**, *32*, 1166–1171. [CrossRef]
69. Koludarov, I.; Jackson, T.N.W.; Sunagar, K.; Nouwens, A.; Hendrikx, I.; Fry, B.G. Fossilized Venom: The Unusually Conserved Venom Profiles of Heloderma Species (Beaded Lizards and Gila Monsters). *Toxins* **2014**, *6*, 3582–3595. [CrossRef] [PubMed]
70. Sanggaard, K.W.; Dyrlund, T.F.; Thomsen, L.R.; Nielsen, T.A.; Brøndum, L.; Wang, T.; Thøgersen, I.B.; Enghild, J.J. Characterization of the gila monster (Heloderma suspectum suspectum) venom proteome. *J. Proteom.* **2015**, *117*, 1–11. [CrossRef] [PubMed]
71. Lino-López, G.J.; Valdez-Velázquez, L.L.; Corzo, G.; Romero-Gutiérrez, M.T.; Jiménez-Vargas, J.M.; Rodríguez-Vázquez, A.; Vazquez-Vuelvas, O.F.; Gonzalez-Carrillo, G. Venom gland transcriptome from Heloderma horridum horridum by high-throughput sequencing. *Toxicon* **2020**, *180*, 62–78. [CrossRef]
72. Dobson, J.; Harris, R.; Zdenek, C.; Huynh, T.; Hodgson, W.; Bosmans, F.; Fourmy, R.; Violette, A.; Fry, B. The Dragon's Paralysing Spell: Evidence of Sodium and Calcium Ion Channel Binding Neurotoxins in Helodermatid and Varanid Lizard Venoms. *Toxins* **2021**, *13*, 549. [CrossRef] [PubMed]
73. Fry, B.G.; Vidal, N.; Norman, J.A.; Vonk, F.J.; Scheib, H.; Ramjan, S.F.R.; Kuruppu, S.; Fung, K.; Hedges, S.B.; Richardson, M.K.; et al. Early evolution of the venom system in lizards and snakes. *Nature* **2005**, *439*, 584–588. [CrossRef]
74. Fry, B.G.; Wroe, S.; Teeuwisse, W.; van Osch, M.J.P.; Moreno, K.; Ingle, J.; McHenry, C.; Ferrara, T.; Clausen, P.; Scheib, H.; et al. A central role for venom in predation by Varanus komodoensis (Komodo Dragon) and the extinct giant Varanus (Megalania) priscus. *Proc. Natl. Acad. Sci. USA* **2009**, *106*, 8969–8974. [CrossRef]
75. Dobson, J.S.; Zdenek, C.N.; Hay, C.; Violette, A.; Fourmy, R.; Cochran, C.; Fry, B.G. Varanid Lizard Venoms Disrupt the Clotting Ability of Human Fibrinogen through Destructive Cleavage. *Toxins* **2019**, *11*, 255. [CrossRef] [PubMed]
76. Zasloff, M. Magainins, a class of antimicrobial peptides from Xenopus skin: Isolation, characterization of two active forms, and partial cDNA sequence of a precursor. *Proc. Natl. Acad. Sci. USA* **1987**, *84*, 5449–5453. [CrossRef]
77. Mailho-Fontana, P.L.; Antoniazzi, M.M.; Toledo, L.F.; Verdade, V.K.; Sciani, J.M.; Barbaro, K.C.; Pimenta, D.C.; Rodrigues, M.T.; Jared, C. Passive and active defense in toads: The parotoid macroglands in Rhinella marina and Rhaebo guttatus. *J. Exp. Zool. Part A Ecol. Genet. Physiol.* **2014**, *321*, 65–77. [CrossRef] [PubMed]
78. Conceição, K.; Konno, K.; de Melo, R.L.; Antoniazzi, M.M.; Jared, C.; Sciani, J.M.; Conceição, I.M.; Prezoto, B.C.; de Camargo, A.C.M.; Pimenta, D.C. Isolation and characterization of a novel bradykinin potentiating peptide (BPP) from the skin secretion of Phyllomedusa hypochondrialis. *Peptides* **2007**, *28*, 515–523. [CrossRef] [PubMed]
79. Conceição, K.; Bruni, F.M.; Pareja-Santos, A.; Antoniazzi, M.M.; Jared, C.; Lopes-Ferreira, M.; Lima, C.; Pimenta, D.C. Unusual profile of leukocyte recruitment in mice induced by a skin secretion of the tree frog Phyllomedusa hypochondrialis. *Toxicon* **2007**, *49*, 625–633. [CrossRef] [PubMed]

80. Mendes, V.A.; Barbaro, K.C.; Sciani, J.M.; Vassão, R.C.; Pimenta, D.C.; Jared, C.; Antoniazzi, M.M. The cutaneous secretion of the casque-headed tree frog Corythomantis greeningi: Biochemical characterization and some biological effects. *Toxicon* **2016**, *122*, 133–141. [CrossRef] [PubMed]
81. Assakura, M.T.; Silva, C.A.; Mentele, R.; Camargo, A.C.; Serrano, S.M. Molecular cloning and expression of structural domains of bothropasin, a P-III metalloproteinase from the venom of Bothrops jararaca. *Toxicon* **2002**, *41*, 217–227. [CrossRef]
82. Leme, A.F.P.; Prezoto, B.C.; Yamashiro, E.T.; Bertholim, L.; Tashima, A.K.; Klitzke, C.F.; Camargo, A.C.M.; Serrano, S.M.T. BothropsproteaseA, a unique highly glycosylated serine proteinase, is a potent, specific fibrinogenolytic agent. *J. Thromb. Haemost.* **2008**, *6*, 1363–1372. [CrossRef]
83. Souza, B.B.P.; Fh, J.L.C.; Murad, A.M.; Prates, M.V.; Coura, M.M.A.; Brand, G.D.; Barbosa, E.A.; Bloch, C., Jr. Identification and characterization of phospholipases A2 from the skin secretion of Pithecopus azureus anuran. *Toxicon* **2019**, *167*, 10–19. [CrossRef]
84. Mariano, D.; Messias, M.D.G.; Neto, J.P.P.; Spencer, P.; Pimenta, D.C. Biochemical Analyses of Proteins from Duttaphrynus melanostictus (Bufo melanostictus) Skin Secretion: Soluble Protein Retrieval from a Viscous Matrix by Ion-Exchange Batch Sample Preparation. *Protein J.* **2018**, *37*, 380–389. [CrossRef]
85. Fusco, L.S.; Cajade, R.; Piñeiro, J.M.; Torres, A.M.; Da Silva, I.R.F.; Hyslop, S.; Leiva, L.C.; Pimenta, D.C.; Bustillo, S. Biochemical characterization and cytotoxic effect of the skin secretion from the red-spotted Argentina frog Argenteohyla siemersi (Anura: Hylidae). *J. Venom. Anim. Toxins Incl. Trop. Dis.* **2020**, *26*, e20190078. [CrossRef]
86. Conceição, K.; Konno, K.; Richardson, M.; Antoniazzi, M.M.; Jared, C.; Daffre, S.; Camargo, A.C.M.; Pimenta, D.C. Isolation and biochemical characterization of peptides presenting antimicrobial activity from the skin of Phyllomedusa hypochondrialis. *Peptides* **2006**, *27*, 3092–3099. [CrossRef]
87. Wu, X.; Pan, J.; Wu, Y.; Xi, X.; Ma, C.; Wang, L.; Zhou, M.; Chen, T. PSN-PC: A Novel Antimicrobial and Anti-Biofilm Peptide from the Skin Secretion of Phyllomedusa-camba with Cytotoxicity on Human Lung Cancer Cell. *Molecules* **2017**, *22*, 1896. [CrossRef] [PubMed]
88. Wu, Y.; Wang, L.; Zhou, M.; Chen, T.; Shaw, C. Phylloseptin-PBa1,-PBa2,-PBa3: Three novel antimicrobial peptides from the skin secretion of Burmeister's leaf frog (Phyllomedusa burmeisteri). *Biochem. Biophys. Res. Commun.* **2019**, *509*, 664–673. [CrossRef] [PubMed]
89. Liu, Y.; Shi, D.; Wang, J.; Chen, X.; Zhou, M.; Xi, X.; Cheng, J.; Ma, C.; Chen, T.; Shaw, C.; et al. A Novel Amphibian Antimicrobial Peptide, Phylloseptin-PV1, Exhibits Effective Anti-staphylococcal Activity without Inducing Either Hepatic or Renal Toxicity in Mice. *Front. Microbiol.* **2020**, *11*, 565158. [CrossRef]
90. Zhang, R.; Zhou, M.; Wang, L.; McGrath, S.; Chen, T.; Chen, X.; Shaw, C. Phylloseptin-1 (PSN-1) from Phyllomedusa sauvagei skin secretion: A novel broad-spectrum antimicrobial peptide with antibiofilm activity. *Mol. Immunol.* **2010**, *47*, 2030–2037. [CrossRef] [PubMed]
91. Raja, Z.; Andre, S.; Piesse, C.; Sereno, D.; Nicolas, P.; Foulon, T.; Oury, B.; Ladram, A. Structure, antimicrobial activities and mode of interaction with membranes of bovel phylloseptins from the painted-belly leaf frog, Phyllomedusa sauvagii. *PLoS ONE* **2013**, *8*, e70782. [CrossRef]
92. Conlon, J.M.; Woodhams, D.C.; Raza, H.; Coquet, L.; Leprince, J.; Jouenne, T.; Vaudry, H.; Rollins-Smith, L.A. Peptides with differential cytolytic activity from skin secretions of the lemur leaf frog Hylomantis lemur (Hylidae: Phyllomedusinae). *Toxicon* **2007**, *50*, 498–506. [CrossRef] [PubMed]
93. Sousa, J.C.; Berto, R.F.; Gois, E.A.; Fontenele-Cardi, N.C.; Honório-Júnior, J.E.; Konno, K.; Richardson, M.; Rocha, M.F.; Camargo, A.A.; Pimenta, D.C.; et al. Leptoglycin: A new Glycine/Leucine-rich antimicrobial peptide isolated from the skin secretion of the South American frog Leptodactylus pentadactylus (Leptodactylidae). *Toxicon* **2009**, *54*, 23–32. [CrossRef]
94. Silva, M.R.E.; Beraldo, W.T.; Rosenfeld, G. Bradykinin, a hypotensive and smooth muscle stimulating factor released from plasma globulin by snake venoms and by trypsin. *Am. J. Physiol. Content* **1949**, *156*, 261–273. [CrossRef] [PubMed]
95. Conceição, K.; Bruni, F.; Sciani, J.; Konno, K.; Melo, R.; Antoniazzi, M.; Jared, C.; Lopes-Ferreira, M.; Pimenta, D. Identification of bradykinin: Related peptides from Phyllomedusa nordestina skin secretion using electrospray ionization tandem mass spectrometry after a single-step liquid chromatography. *J. Venom. Anim. Toxins Incl. Trop. Dis.* **2009**, *15*, 633–652. [CrossRef]
96. Tempone, A.G.; Pimenta, D.C.; Lebrun, I.; Sartorelli, P.; Taniwaki, N.N.; de Andrade, H.F., Jr.; Antoniazzi, M.M.; Jared, C. Antileishmanial and antitrypanosomal activity of bufadienolides isolated from the toad Rhinella jimi parotoid macrogland secretion. *Toxicon* **2008**, *52*, 13–21. [CrossRef] [PubMed]
97. Pinto, E.G.; Pimenta, D.; Antoniazzi, M.M.; Jared, C.; Tempone, A. Antimicrobial peptides isolated from Phyllomedusa nordestina (Amphibia) alter the permeability of plasma membrane of Leishmania and Trypanosoma cruzi. *Exp. Parasitol.* **2013**, *135*, 655–660. [CrossRef] [PubMed]
98. Sciani, J.M.; De-Sá-Júnior, P.L.; Ferreira, A.K.; Pereira, A.; Antoniazzi, M.M.; Jared, C.; Pimenta, D.C. Cytotoxic and antiproliferative effects of crude amphibian skin secretions on breast tumor cells. *Biomed. Prev. Nutr.* **2013**, *3*, 10–18. [CrossRef]
99. Schmeda-Hirschmann, G.; Quispe, C.; Theoduloz, C.; de Sousa, P.T., Jr.; Parizotto, C. Antiproliferative activity and new argininyl bufadienolide esters from the "cururú" toad Rhinella (Bufo) schneideri. *J. Ethnopharmacol.* **2014**, *155*, 1076–1085. [CrossRef] [PubMed]
100. Schmeda-Hirschmann, G.; Quispe, C.; Arana, G.V.; Theoduloz, C.; Urra, F.A.; Cárdenas, C. Antiproliferative activity and chemical composition of the venom from the Amazonian toad Rhinella marina (Anura: Bufonidae). *Toxicon* **2016**, *121*, 119–129. [CrossRef]

101. Schmeda-Hirschmann, G.; Gomez, C.V.; de Arias, A.R.; Burgos-Edwards, A.; Alfonso, J.; Rolon, M.; Brusquetti, F.; Netto, F.; Urra, F.A.; Cárdenas, C. The Paraguayan Rhinella toad venom: Implications in the traditional medicine and proliferation of breast cancer cells. *J. Ethnopharmacol.* **2017**, *199*, 106–118. [CrossRef]
102. Carvalho, A.C. Avaliação dos Efeitos Citotóxicos e Antiproliferativos da Secreção Cutânea e de Peptídeos do Anuro Physalaemus Nattereri (Steindachner, 1863). 2015. Available online: https://repositorio.unb.br/handle/10482/18589 (accessed on 19 November 2021).
103. Dahham, S.; Asif, M.; Tabana, Y.; Sandai, D.; Majid, A.; Harn, G. Antiproliferative and apoptotic activity of Crude Skin Secretion from Malaysian Toad (Bufo asper) on in vitro colorectal cancer cells. *J. Appl. Pharm. Sci.* **2017**, *7*, 1–6. [CrossRef]
104. de Sousa, L.Q.; Machado, K.; Oliveira, S.F.D.C.; Araújo, L.D.S.; Monção-Filho, E.D.S.; Cavalcante, A.; Junior, G.M.V.; Ferreira, P.M.P. Bufadienolides from amphibians: A promising source of anticancer prototypes for radical innovation, apoptosis triggering and Na+/K+-ATPase inhibition. *Toxicon* **2017**, *127*, 63–76. [CrossRef]
105. Karış, M.; Şener, D.; Yalçın, H.T.; Nalbantsoy, A.; Göçmen, B. Major biological activities and protein profiles of skin secretions of Lissotriton vulgaris and Triturus ivanbureschi. *Turk. J. Biochem.* **2018**, *43*, 605–612. [CrossRef]
106. Machado, K.; de Sousa, L.Q.; Lima, D.J.B.; Soares, B.M.; Cavalcanti, B.C.; Maranhão, S.S.A.; de Noronha, J.D.C.; de Jesus Rodrigues, D.; Militão, G.C.G.; Chaves, M.H.; et al. Marinobufagin, a molecule from poisonous frogs, causes biochemical, morphological and cell cycle changes in human neoplasms and vegetal cells. *Toxicol. Lett.* **2018**, *285*, 121–131. [CrossRef]
107. Filho, E.S.M.; Pio, Y.P.F.; Chaves, M.H.; Ferreira, P.M.P.; Fonseca, M.G.; Pessoa, C.; Lima, D.J.B.; Araújo, B.Q.; Vieira, G.M., Jr. Chemical Constituents and Cytotoxic Activity of Rhinella jimi (Anura: Bufonidae). *J. Braz. Chem. Soc.* **2021**. [CrossRef]
108. Spinelli, R.; Guevara, L.A.B.; López, J.A.; Camargo, C.M.; De Restrepo, H.G.; Siano, A.S. Cytotoxic and antiproliferative activities of amphibian (anuran) skin extracts on human acute monocytic leukemia cells. *Toxicon* **2020**, *177*, 25–34. [CrossRef]
109. Sciani, J.M.; Angeli, C.B.; Antoniazzi, M.M.; Jared, C.; Pimenta, D.C. Differences and Similarities among Parotoid Macrogland Secretions in South American Toads: A Preliminary Biochemical Delineation. *Sci. World J.* **2013**, *2013*, 937407. [CrossRef] [PubMed]
110. Vigerelli, H.; Sciani, J.M.; Jared, C.; Antoniazzi, M.M.; Caporale, G.M.M.; Silva, A.D.C.R.D.; Pimenta, D.C. Bufotenine is able to block rabies virus infection in BHK-21 cells. *J. Venom. Anim. Toxins Incl. Trop. Dis.* **2014**, *20*, 45. [CrossRef]
111. Vigerelli, H.; Sciani, J.; Pereira, P.M.C.; Lavezo, A.A.; Silva, A.C.R.; Collaço, R.; Rocha, T.; Bueno, T.C.; Pimenta, D.C. Bufotenine, a tryptophan-derived alkaloid, suppresses the symptoms and increases the survival rate of rabies-infected mice: The development of a pharmacological approach for rabies treatment. *J. Venom. Anim. Toxins Incl. Trop. Dis.* **2020**, *26*, e20190050. [CrossRef]
112. Vigerelli, H.; Sciani, J.M.; Eula, M.A.C.; Sato, L.A.; Antoniazzi, M.M.; Jared, C.; Pimenta, D.C. Biological Effects and Biodistribution of Bufotenine on Mice. *BioMed Res. Int.* **2018**, *2018*, 1032638. [CrossRef]
113. Neto, R.D.S.C.; Vigerelli, H.; Jared, C.; Antoniazzi, M.M.; Chaves, L.B.; Silva, A.D.C.R.D.; De Melo, R.L.; Sciani, J.M.; Pimenta, D.C. Synergic effects between ocellatin-F1 and bufotenine on the inhibition of BHK-21 cellular infection by the rabies virus. *J. Venom. Anim. Toxins Incl. Trop. Dis.* **2015**, *21*, 50. [CrossRef] [PubMed]
114. Barboza, C.M.; Pimenta, D.C.; Vigerelli, H.; Silva, A.D.C.R.D.; Garcia, J.G.; Zamudio, R.M.; Castilho, J.G.; Montanha, J.A.; Roehe, P.M.; Batista, H.B.D.C.R. In vitro effects of bufotenine against RNA and DNA viruses. *Braz. J. Microbiol.* **2021**, *52*, 2475–2482. [CrossRef] [PubMed]
115. Santoro, F.; Selvaggia, S.; Scowcroft, G.; Fauville, G.; Tuddenham, P. *Ocean Literacy for All: A Toolkit*; UNESCO Publishing: Paris, France, 2017; Volume 80.
116. WoRMS Editorial Board. World Register of Marine Species. 2021. Available online: https://www.marinespecies.org (accessed on 19 November 2021).
117. McIntosh, M.; Cruz, L.; Hunkapiller, M.; Gray, W.; Olivera, B. Isolation and structure of a peptide toxin from the marine snail Conus magus. *Arch. Biochem. Biophys.* **1982**, *218*, 329–334. [CrossRef]
118. Miljanich, G.P. Ziconotide: Neuronal Calcium Channel Blocker for Treating Severe Chronic Pain. *Curr. Med. Chem.* **2004**, *11*, 3029–3040. [CrossRef] [PubMed]
119. Larsen, A.K.; Galmarini, C.M.; D'Incalci, M. Unique features of trabectedin mechanism of action. *Cancer Chemother. Pharmacol.* **2015**, *77*, 663–671. [CrossRef] [PubMed]
120. Schwartsmann, G.; da Rocha, A.B.; Berlinck, R.G.; Jimeno, J. Marine organisms as a source of new anticancer agents. *Lancet Oncol.* **2001**, *2*, 221–225. [CrossRef]
121. Longato, G.B.; De Barros, H.V.; de Lima, C.; Picolo, G.; Zambelli, V.O.; Morandini, A.; Marques, A.C.; Sciani, J.M. Screening of Brazilian marine animals extracts on tumor cell line panel. *Toxicon* **2019**, *168*, S31. [CrossRef]
122. Haddad, V., Jr.; Lupi, O.; Lonza, J.P.; Tyring, S.K. Tropical dermatology: Marine and aquatic dermatology. *J. Am. Acad. Dermatol.* **2009**, *61*, 733–750. [CrossRef]
123. Cracchiolo, A.; Goldberg, L. Local and Systemic Reactions to Puncture Injuries by the Sea Urchin Spine and the Date Palm Thorn. *Arthritis Rheum.* **1977**, *20*, 1206–1212. [CrossRef] [PubMed]
124. Rossetto, A.L.; Mora, J.D.; Haddad Junior, V. Sea urchin granuloma. *Rev. Do Inst. Med. Trop. São Paulo* **2006**, *48*, 303–306. [CrossRef]
125. Sciani, J.M.; Zychar, B.; Gonçalves, L.R.; Giorgi, R.; Nogueira, T.; Pimenta, D.C. Preliminary molecular characterization of a proinflammatory and nociceptive molecule from the Echinometra lucunter spines extracts. *J. Venom. Anim. Toxins Incl. Trop. Dis.* **2017**, *23*, 43. [CrossRef]

126. Sciani, J.M.; Antoniazzi, M.M.; Neves, A.D.; Pimenta, D.C. Cathepsin B/X is secreted by Echinometra lucunter sea urchin spines, a structure rich in granular cells and toxins. *J. Venom. Anim. Toxins Incl. Trop. Dis.* **2013**, *19*, 1–8. [CrossRef]
127. Sciani, J.M.; Sampaio, M.C.; Zychar, B.C.; Gonçalves, L.R.D.C.; Giorgi, R.; Nogueira, T.D.O.; De Melo, R.L.; Teixeira, C.D.F.P.; Pimenta, D.C. Echinometrin: A novel mast cell degranulating peptide from the coelomic liquid of Echinometra lucunter sea urchin. *Peptides* **2014**, *53*, 13–21. [CrossRef]
128. Yamasaki, A.; Higaki, H.; Nakashima, K.; Yamamoto, O.; Hein, K.; Takahashi, H.; Chinuki, Y.; Morita, E. Identification of a Major Yolk Protein as an Allergen in Sea Urchin Roe. *Acta Derm. Venereol.* **2010**, *90*, 235–238. [CrossRef]
129. Rodriguez, V.; Bartolomé, B.; Armisen, M.; Vidal, C. Food allergy to Paracentrotus lividus (sea urchin roe). *Ann. Allergy Asthma Immunol.* **2007**, *98*, 393–396. [CrossRef]
130. Pimenta, D.C.; Lebrun, I. Cryptides: Buried secrets in proteins. *Peptides* **2007**, *28*, 2403–2410. [CrossRef]
131. Sciani, J.M.; Emerenciano, A.K.; Silva, J.R.; Pimenta, D.C. Initial peptidomic profiling of Brazilian sea urchins: Arbacia lixula, Lytechinus variegatus and Echinometra lucunter. *J. Venom. Anim. Toxins Incl. Trop. Dis.* **2016**, *22*. [CrossRef] [PubMed]
132. Olivera, B.M.; Cruz, L.J. Conotoxins, in retrospect. *Toxicon* **2001**, *39*, 7–14. [CrossRef]
133. Kaas, Q.; Westermann, J.-C.; Craik, D.J. Conopeptide characterization and classifications: An analysis using ConoServer. *Toxicon* **2010**, *55*, 1491–1509. [CrossRef]
134. Jungo, F.; Bairoch, A. Tox-Prot, the toxin protein annotation program of the Swiss-Prot protein knowledgebase. *Toxicon* **2005**, *45*, 293–301. [CrossRef]
135. The UniProt Consortium, UniProt: The universal protein knowledgebase in 2021. *Nucleic Acids Res.* **2021**, *49*, D480–D489. [CrossRef] [PubMed]
136. Kaas, Q.; Westermann, J.-C.; Halai, R.; Wang, C.; Craik, D. ConoServer, a database for conopeptide sequences and structures. *Bioinformatics* **2008**, *24*, 445–446. [CrossRef] [PubMed]
137. Kaas, Q.; Yu, R.; Jin, A.-H.; Dutertre, S.; Craik, D.J. ConoServer: Updated content, knowledge, and discovery tools in the conopeptide database. *Nucleic Acids Res.* **2011**, *40*, D325–D330. [CrossRef]
138. Terlau, H.; Olivera, B.M. ConusVenoms: A Rich Source of Novel Ion Channel-Targeted Peptides. *Physiol. Rev.* **2004**, *84*, 41–68. [CrossRef]
139. Eston, V.R.; Migotto, A.E.; Oliveira Filho, E.C.; Rodrigues, S.D.; Freitas, J.C. Vertical distribution of benthic marine organisms on rocky coasts of the Fernando de Noronha Archipelago (Brazil). *Bol. Do Inst. Oceanográfico* **1986**, *34*, 37–53. [CrossRef]
140. Braga, M.C.V.; Konno, K.; Portaro, F.C.; de Freitas, J.C.; Yamane, T.; Olivera, B.M.; Pimenta, D.C. Mass spectrometric and high performance liquid chromatography profiling of the venom of the Brazilian vermivorous mollusk Conus regius: Feeding behavior and identification of one novel conotoxin. *Toxicon* **2005**, *45*, 113–122. [CrossRef]
141. Braga, M.C.V.; Nery, A.A.; Ulrich, H.; Konno, K.; Sciani, J.M.; Pimenta, D.C. α-RgIB: A Novel Antagonist Peptide of Neuronal Acetylcholine Receptor Isolated from Conus regius. *Venom. Int. J. Pept.* **2013**, *2013*, 543028. [CrossRef] [PubMed]
142. Reckziegel, G.C.; Dourado, F.S.; Garrone, D.; Haddad, V. Injuries caused by aquatic animals in Brazil: An analysis of the data present in the information system for notifiable diseases. *Rev. Soc. Bras. Med. Trop.* **2015**, *48*, 460–467. [CrossRef] [PubMed]
143. Magalhães, M.R.; da Silva, N.J., Jr.; Ulhoa, C.J. A hyaluronidase from Potamotrygon motoro (freshwater stingrays) venom: Isolation and characterization. *Toxicon* **2008**, *51*, 1060–1067. [CrossRef]
144. Kimura, L.; Neto, J.P.P.; Távora, B.C.; Faquim-Mauro, E.; Pereira, N.A.; Antoniazzi, M.M.; Jared, S.G.; Teixeira, C.F.; Santoro, M.L.; Barbaro, K.C. Mast cells and histamine play an important role in edema and leukocyte recruitment induced by Potamotrygon motoro stingray venom in mice. *Toxicon* **2015**, *103*, 65–73. [CrossRef]
145. Germain, M.; Smith, K.; Skelton, H. The cutaneous cellular infiltrate to stingray envenomization contains increased TIA+ cells. *Br. J. Dermatol.* **2000**, *143*, 1074–1077. [CrossRef]
146. Coelho, G.R.; Neto, P.P.; Barbosa, F.C.; Dos Santos, R.S.; Brigatte, P.; Spencer, P.J.; Sampaio, S.C.; D'Amélio, F.; Pimenta, D.C.; Sciani, J.M. Biochemical and biological characterization of the Hypanus americanus mucus: A perspective on stingray immunity and toxins. *Fish Shellfish. Immunol.* **2019**, *93*, 832–840. [CrossRef]
147. Lima, C.; Disner, G.R.; Falcão, M.A.P.; Seni-Silva, A.C.; Maleski, A.L.A.; Souza, M.M.; Tonello, M.C.R.; Lopes-Ferreira, M. The Natterin Proteins Diversity: A Review on Phylogeny, Structure, and Immune Function. *Toxins* **2021**, *13*, 538. [CrossRef]
148. Jouiaei, M.; Yanagihara, A.A.; Madio, B.; Nevalainen, T.J.; Alewood, P.F.; Fry, B.G. Ancient Venom Systems: A Review on Cnidaria Toxins. *Toxins* **2015**, *7*, 2251–2271. [CrossRef]
149. Van Iten, H.; Leme, J.M.; Pacheco, M.L.A.F.; Simões, M.G.; Fairchild, T.R.; Rodrigues, F.; Galante, D.; Boggiani, P.C.; Marques, A.C. Origin and Early Diversification of Phylum Cnidaria: Key Macrofossils from the Ediacaran System of North and South America. In *The Cnidaria, Past, Present and Future*; Springer: Berlin/Heidelberg, Germany, 2016; pp. 31–40. [CrossRef]
150. Madio, B.; King, G.F.; Undheim, E.A.B. Sea Anemone Toxins: A Structural Overview. *Mar. Drugs* **2019**, *17*, 325. [CrossRef]
151. Beckmann, A.; Özbek, S. The Nematocyst: A molecular map of the Cnidarian stinging organelle. *Int. J. Dev. Biol.* **2012**, *56*, 577–582. [CrossRef]
152. Ashwood, L.M.; Norton, R.S.; Undheim, E.A.B.; Hurwood, D.A.; Prentis, P.J. Characterising Functional Venom Profiles of Anthozoans and Medusozoans within Their Ecological Context. *Mar. Drugs* **2020**, *18*, 202. [CrossRef]
153. Richet, C. De l'action de la congestine (virus des Actinies) sur les lapins et de ses effects anaphylactiques. *CR Soc. Biol. Paris* **1905**, *58*, 109–112.
154. Richet, C. De la thalassine toxine cristalliseé et pruritogène. *CR Soc. Biol. Paris* **1903**, *55*, 707–710.

155. Tsai, H.-S.; Niu, K.-Y. Acute Skin Manifestation of Sea Anemone Envenomation. *J. Emerg. Med.* **2021**, *60*, 536–537. [CrossRef]
156. Oliveira, J.S.; Redaelli, E.; Zaharenko, A.J.; Cassulini, R.R.; Konno, K.; Pimenta, D.C.; Freitas, J.C.; Clare, J.J.; Wanke, E. Binding specificity of sea anemone toxins to Nav 1.1-1.6 sodium channels: Unexpected contributions from differences in the IV/S3-S4 outer loop. *J. Biol. Chem.* **2004**, *279*, 33323–33335. [CrossRef]
157. Zaharenko, A.J.; Ferreira, W.A., Jr.; Oliveira, J.S.; Richardson, M.; Pimenta, D.C.; Konno, K.; Portaro, F.C.V.; de Freitas, J.C. Proteomics of the neurotoxic fraction from the sea anemone Bunodosoma cangicum venom: Novel peptides belonging to new classes of toxins. *Comp. Biochem. Physiol. Part D Genom. Proteom.* **2008**, *3*, 219–225. [CrossRef]
158. Yanagihara, A.A.; Shohet, R.V. Cubozoan Venom-Induced Cardiovascular Collapse Is Caused by Hyperkalemia and Prevented by Zinc Gluconate in Mice. *PLoS ONE* **2012**, *7*, e51368. [CrossRef]
159. Jaimes-Becerra, A.; Chung, R.; Morandini, A.C.; Weston, A.J.; Padilla, G.; Gacesa, R.; Ward, M.; Long, P.; Marques, A.C. Comparative proteomics reveals recruitment patterns of some protein families in the venoms of Cnidaria. *Toxicon* **2017**, *137*, 19–26. [CrossRef]
160. Arruda, G.L.M.; Vigerelli, H.; Bufalo, M.C.; Longato, G.B.; Veloso, R.V.; Zambelli, V.O.; Picolo, G.; Cury, Y.; Morandini, A.C.; Marques, A.C. Box jellyfish (Cnidaria, Cubozoa) extract increases neuron's connection: A possible neuroprotector effect. *BioMed Res. Int.* **2021**, *2021*, 8855248. [CrossRef] [PubMed]
161. Bueno, T.C.; Collaço, R.D.C.; Cardoso, B.A.; Bredariol, R.F.; Escobar, M.L.; Cajado, I.B.; Gracia, M.; Antunes, E.; Zambelli, V.O.; Picolo, G.; et al. Neurotoxicity of Olindias sambaquiensis and Chiropsalmus quadrumanus extracts in sympathetic nervous system. *Toxicon* **2021**, *199*, 127–138. [CrossRef]
162. Stein, J.M.; Bergman, W.; Fang, Y.; Davison, L.; Brensinger, C.; Robinson, M.B.; Hecht, N.B.; Abel, T. Behavioral and Neurochemical Alterations in Mice Lacking the RNA-Binding Protein Translin. *J. Neurosci.* **2006**, *26*, 2184–2196. [CrossRef]
163. Junior, R.S.F.; Sciani, J.M.; Marques-Porto, R.; Junior, A.L.; Orsi, R.D.O.; Barraviera, B.; Pimenta, D.C. Africanized honey bee (Apis mellifera) venom profiling: Seasonal variation of melittin and phospholipase A2 levels. *Toxicon* **2010**, *56*, 355–362. [CrossRef] [PubMed]
164. Sciani, J.M.; Marques-Porto, R.; Lourenço, A.; Orsi, R.D.O.; Junior, R.S.F.; Barraviera, B.; Pimenta, D.C. Identification of a novel melittin isoform from Africanized Apis mellifera venom. *Peptides* **2010**, *31*, 1473–1479. [CrossRef] [PubMed]
165. Barbosa, A.N.; Ferreira, R.S., Jr.; de Carvalho, F.C.T.; Schuelter-Trevisol, F.; Mendes, M.B.; Mendonça, B.C.; Batista, J.N.; Trevisol, D.J.; Boyer, L.; Chippaux, J.-P. Single-arm, multicenter phase I/II clinical trial for the treatment of envenomings by massive africanized honey bee stings using the unique apilic antivenom. *Front. Immunol.* **2021**, *12*, 860. [CrossRef]

 toxins

Review

Lonomia obliqua Envenoming and Innovative Research

Miryam Paola Alvarez-Flores [1,2], Renata Nascimento Gomes [1,2], Dilza Trevisan-Silva [1,2], Douglas Souza Oliveira [1,2], Isabel de Fátima Correia Batista [1,2], Marcus Vinicius Buri [1,2], Angela Maria Alvarez [1,2], Carlos DeOcesano-Pereira [1,2], Marcelo Medina de Souza [1,2] and Ana Marisa Chudzinski-Tavassi [1,2,*]

1 Centre of Excellence in New Target Discovery (CENTD), Butantan Institute, Butantã 05503-900, SP, Brazil; miryam.flores@butantan.gov.br (M.P.A.-F.); renata.gomes@esib.butantan.gov.br (R.N.G.); dilza.silva@butantan.gov.br (D.T.-S.); douglas.oliveira@butantan.gov.br (D.S.O.); isabel.batista@butantan.gov.br (I.d.F.C.B.); marcus.buri@butantan.gov.br (M.V.B.); angela.alvarez@esib.butantan.gov.br (A.M.A.); carlos.ocesano@butantan.gov.br (C.D.-P.); marcelo.souza@butantan.gov.br (M.M.d.S.)

2 Development and Innovation Centre, Butantan Institute, Butantã 05503-900, SP, Brazil

* Correspondence: ana.chudzinski@butantan.gov.br; Tel.: +55-11-2627-9738

Abstract: As a tribute to Butantan Institute in its 120th anniversary, this review describes some of the scientific research efforts carried out in the study of *Lonomia* envenoming in Brazil, a country where accidents with caterpillars reach over 42,000 individuals per year (especially in South and Southeast Brazil). Thus, the promising data regarding the studies with *Lonomia*'s toxins contributed to the creation of new research centers specialized in toxinology based at Butantan Institute, as well as to the production of the antilonomic serum (ALS), actions which are in line with the Butantan Institute mission "to research, develop, manufacture, and provide products and services for the health of the population". In addition, the study of the components of the *Lonomia obliqua* bristle extract led to the discovery of new molecules with peculiar properties, opening a field of knowledge that could lead to the development and innovation of new drugs aimed at cell regeneration and inflammatory diseases.

Keywords: *Lonomia*; envenoming; innovation

Key Contribution: This review highlights the current knowledge of *Lonomia* envenoming and treatment, in addition to already identified bioactive molecules and future perspectives on innovative research with new derived molecules.

Citation: Alvarez-Flores, M.P.; Gomes, R.N.; Trevisan-Silva, D.; Oliveira, D.S.; Batista, I.d.F.C.; Buri, M.V.; Alvarez, A.M.; DeOcesano-Pereira, C.; de Souza, M.M.; Chudzinski-Tavassi, A.M. *Lonomia obliqua* Envenoming and Innovative Research. *Toxins* **2021**, *13*, 832. https://doi.org/10.3390/toxins13120832

Received: 16 October 2021
Accepted: 9 November 2021
Published: 23 November 2021

1. Introduction

Although lepidopteran species are widely distributed around the world, only a few of them cause severe damage to humans or animals that have had contact with adult animal hairs (lepidopterism) or with the bristles of caterpillars (erucism) [1]. Locally, accidental contact with hair or bristles leads to a skin reaction, and systemic symptoms can be treated using oral antipruritic and antihistamines [2]. However, some caterpillar species of the *Lonomia* genus cause serious injuries, which are sometimes irreversible, leading to death. Patients that develop clinical manifestations of disseminated intravascular coagulation (DIC) and consumptive coagulopathy can progress to hemorrhagic syndrome with serious consequences if the antilonomic serum (ALS) produced by the Butantan Institute (SP/Brazil) is not administered in due time [3–8]. Although treatment with ALS is effective for *Lonomia*'s envenoming, deaths resulting from contact with caterpillars are still a public health problem in Brazil [9,10]. The literature reports that, between 2007 and 2017 a total of 42,264 accidents were caused by caterpillars in Brazil, among them 248 were severe cases and five evolved to deaths. Most accidents occurred in the states of south and southern Brazil between December and April, a period corresponding to an increase in temperature and rainfall [10].

Over the years, Brazil has gained significant knowledge in the field of toxinology that benefits the politics of public health. One example is the existence of Toxicological Information and Assistance Centers (CIATox), created for the Brazilian Unified Health System (SUS) to provide specific information on poisoning and treatment to health professionals and to the community. Furthermore, the creation of special programs and centers for research in the study of animal toxins contributed to the innovation in the development of new molecules derived from animal toxins or secretions, accelerating the interaction between science and industry. Therefore, this review highlights the current knowledge about *Lonomia* envenoming, as well as its treatment and already identified bioactive molecules, approaching the future perspectives on innovative research with new derived compounds as potential drugs for the treatment of inflammatory diseases (Figure 1).

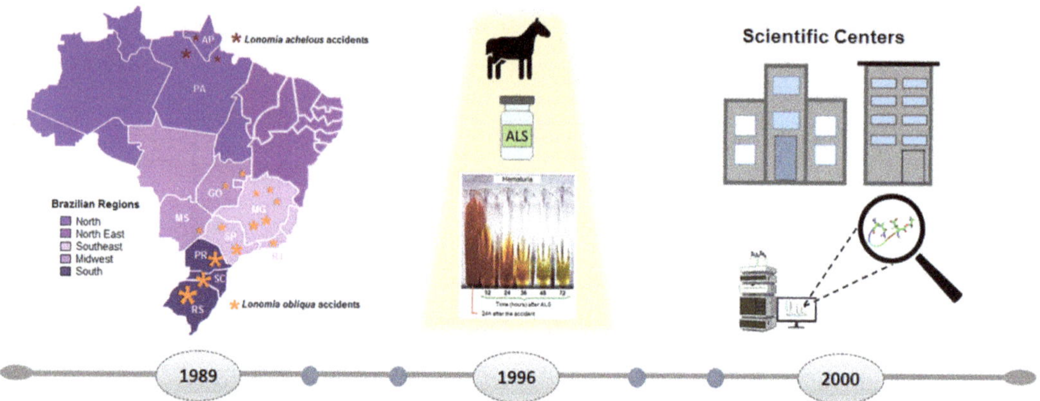

Figure 1. Overview of *Lonomia obliqua* epidemiology, treatment, and research over the years. Since 1989, a burst of accidents with hemorrhagic manifestations were reported in Brazil to be caused by *L. obliqua* (Walker, 1855) (orange asterisks), mainly in Santa Catarina (SC), Rio Grande do Sul (RS), and Paraná (PR). In Venezuela and northern Brazil [Amapá (AP) and Pará (PA)], caterpillars were identified as *L. achelous* (Cramer) (red asterisks). Other cases were registered in the states of Goiás (GO), Minas Gerais (MG), São Paulo (SP), Mato Grosso do Sul (MS), and Rio de Janeiro (RJ) (orange asterisks). In 1996, an antivenom against *L. obliqua* toxins was developed [3]. Today, the treatment of patients is based on the administration of ALS, produced at the Butantan Institute, which has been shown to be effective in reversing hemostatic and hemorrhagic disorders. Photograph (Dr. Marlene Zannin) showing the reduction in hematuria of urine samples of a patient with treatment started with ALS after 24 h of having an accident with *L. obliqua* caterpillars. In 2000, the State of São Paulo Research Foundation (FAPESP) started a program to create Research, Innovation, and Dissemination Centers (RIDC) leading to the creation of the "Center for Applied Toxinology (CAT)", the "Center of Toxins, Immune Response, and Cell Signaling (CETICS)", and the "Center of Excellence in New Target Discovery (CENTD)", with the latter aiming at not only the study of toxins from poisons and animal secretions, but also the development of new molecules based on toxins and, in public–private partnerships, their use as tools for studying molecular targets for several diseases.

2. *Lonomia* spp. Epidemiology and Impact on Public Health

The occurrence of hemorrhage after contact with South American caterpillars was first reported by Alvarenga and collaborators [11]. Although 26 species of the genus *Lonomia* (Saturniidae family) are distributed in the American continent, the most studied species are *L. obliqua* (Figure 2a) and *Lonomia achelous* caterpillars; both are capable of inducing hemorrhagic effects in humans after contact with their broken spines [8,12–20].

Accidents involving *L. achelous* caterpillars have been reported in Venezuela since 1967 [21] and in Brazil since 1982, in the state of Amapá and Ilha de Marajó (state of Pará) [15,22]. Accidents related to *L. achelous* have been characterized by a hemorrhagic syndrome attributed to the fibrinolytic activity of the venom [23]. On the other hand, accidents involving *L. obliqua* specimens have been reported since 1989 in southern Brazil in Santa Catarina, Rio Grande do Sul, and Paraná, also affecting states in the southeast region [10,15,16,24–26].

In Brazil, accident notifications are registered and informed by the Brazilian Ministry of Health through the Information System for Notification of Diseases (SINAN). Data from SINAN between 2007 and 2017 [10] indicate that most cases of accidents involving caterpillars occurred in southern and southeast states during the warm and rainy season [9,10], representing ideal conditions for the hatching of eggs and subsequent development of larvae. A study performed with 105 patients in the State of Santa Catarina (between December 1998 and June 2000), showed that most accidents occur in rural areas (85%), during work activities (55%) [8]. Envenoming by *L. obliqua* caterpillars is considered a public health problem in southern Brazil [10,27,28]. The relevance of the accident is due not only to the increase in the number of accidents, but also to the expansion of the caterpillar population to other areas of the country and to the hemorrhagic syndrome that affects the victims.

3. Clinical Manifestations and Complications

Immediately after contact with the caterpillar bristles (Figure 2b), an urticating dermatitis occurs, accompanied by pain and swelling. Some general and nonspecific manifestations may appear later, such as holocranial headache, general malaise, nausea and vomiting, anxiety, myalgia, and, less frequently, abdominal pain, hypothermia, and hypotension. After a period that can vary from 1 to 48 h, blood dyscrasia appears, accompanied or not by hemorrhagic manifestations that usually appear 8 to 72 h after contact [7,8,27–30]. Ecchymoses can be found, which can reach extensive hemorrhagic dysfunctions, hematomas caused by trauma or healed lesions, hemorrhages from mucosal cavities (gingivorrhagia, epistaxis, hematemesis, and enterorrhagia), macroscopic hematuria, bleeding from recent wounds, intraarticular, abdominal (intra- and extraperitoneal), pulmonary, and glandular (thyroid, salivary glands) hemorrhages, and intraparenchymal cerebral hemorrhage [15–19,27,28] (Figure 2b). Lonomism is the term used to designate the severe hemorrhagic disease related to *Lonomia* accidents [31].

The main complication of *L. obliqua* envenomation is acute renal failure, which can occur in up to 12% of the cases, being frequent in patients over 45 years old and in those with heavy bleeding [24,28–30,32]. Moreover, some deaths related to hemorrhage and renal failure have been reported [9,10,15,16,29]. However, the early diagnosis and proper treatment with ALS within 12 h of contact can prevent severe coagulopathy and hemorrhage events [8,15,19,33–35].

Considering that caterpillars have gregarious habits, the severity of symptoms may be influenced by the number of caterpillars crushed on contact, the extent of the exposed body area, the depth of wound, and the amount of venom inoculated [31].

According to the intensity of the hemostatic disturbances [8,27,28], accidents can be classified as follows:

(a) Mild: patient with local envenomation and without coagulation changes or bleeding within 12 h after the accident, confirmed with the identification of the agent.

(b) Moderate: patient with local manifestations, alterations in global coagulation tests, or hemorrhagic manifestations in the skin and/or mucous membranes (gingivorrhagia, ecchymosis, hematoma), and hematuria, without hemodynamic alterations (hypotension, tachycardia, or shock) (see Figure 2b).

(c) Severe: patient with impaired coagulation, hemorrhagic manifestations in the viscera (hematemesis, hypermenorrhagia, pulmonary bleeding, and intracranial hemorrhage), hemodynamic changes, and/or failure of multiple organs or systems.

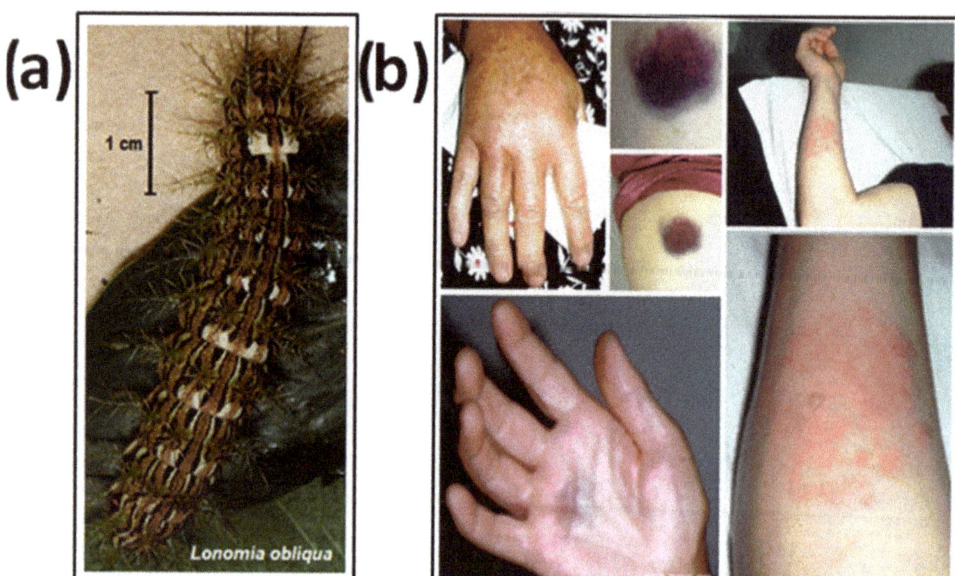

Figure 2. *L. obliqua* and clinical manifestations. (**a**) *L. obliqua* caterpillar. This photograph shows a caterpillar at the sixth stage or instar; and (**b**) Initial symptoms. This photograph shows some clinical manifestation that begin 12 to 24 h after the accident involving contact with broken bristles. Edema (hands), erythema, heat, and blisters (arm), in addition to systemic symptoms, have been reported. Ecchymosis, after 3 days of contact, of variable intensity and hematuria (abdominal bruises, after 24 h), may occur (Photographs: (**a**) Dr. Miryam P. Alvarez-Flores; (**b**) Dr. Marlene Zannin).

4. The Importance of Butantan Institute in Antilonomic Serum Treatment

The hemostatic disturbances observed in the envenoming by *L. obliqua* caterpillars result in a consumption coagulopathy (resembling a DIC) and secondary fibrinolysis, which can lead to the hemorrhagic syndrome [8,19,28,36]. Treatments with antifibrinolytic drugs such as aprotinin and ε-aminocaproic acid associated with whole blood, fresh-frozen plasma, or cryoprecipitates were initially used to treat patients; however, rather than reverting symptoms, the treatment exacerbated them [3,6,7,14,16,28]. In 1996, Da Silva and collaborators [3] developed the ALS from horses immunized with four doses of *Lonomia obliqua* bristle extract (LOCBE), producing antibodies capable of neutralizing the components responsible for typical noncoagulated blood induced by contact with *Lonomia* caterpillars in rats. The developed antivenom is composed of specific immunoglobulin F(ab′)2 fragments purified from horse plasma [4].

Clinical studies showed that hemostasis alteration in this kind of accident could be severe within the first 6 h, with intense fibrinogen reduction [8,19]. The patients with the occurrence of abnormal coagulation in screening tests, such as thrombin time (TT), fibrinogen (Fg) concentration, and whole-blood clotting time (WBCT) or hemorrhage manifestations must be hospitalized to receive treatment with ALS according to the Guidelines of the Ministry of Health of Brazil. After the introduction of ALS therapy, the number of deaths resulting from lonomism was reduced. Currently, ALS is the only specific therapy used to treat victims of *L. obliqua* envenomation and is widely distributed by the Ministry of Health in Brazil. Fibrinolytic agents are not recommended for treatment as the venom mainly contains procoagulant agents [7,8,27,28,36]. The correction of anemia due to blood loss should be instituted through the administration of packed red blood cells [27]. Whole blood or fresh plasma are contraindicated, since intravascular coagulation can be accentuated. The recommended doses of ALS in the treatment, according to severity, are shown in Table 1.

Table 1. Classification of the severity of accidents by *L. obliqua* and therapeutic guidance according to the Brazilian Ministry of Health [27,28].

Manifestations of Severity	Clinical Local Picture	Coagulation Time	Bleeding	Treatment
Mild	Present	Normal	Absent	No ALS required
Moderate	Present or absent	Altered	Absent or present in skin/mucous membranes	Serotherapy: 5 ampoules of ALS
Severe	Present or absent	Altered	Present in viscera. Life-threatening	Serotherapy: 10 ampoules of ALS

5. Blood Coagulation Alterations

One of the main clinical manifestations of *Lonomia* envenoming is the consumption of coagulopathy due to depletion of coagulation factors, as well as a secondary activation of fibrinolysis accompanied by bleeding into the skin or mucosa, as described by Zannin and collaborators [8]. Hemostatic changes were observed in patients who have had contact with larvae of 5 cm or more corresponding to the last larval instars shortly before entering the pupal stage [36].

The hemorrhagic syndrome developed by patients is a consequence of a type of DIC [8,28]. DIC is defined as a pathological syndrome that results in thrombin formation, activation and consumption of some coagulation factors, and fibrin clot formation [37].

According to Zannin and collaborators [8], the global clotting times (thrombin time—TT, prothrombin time—PT, and activated partial thromboplastin time—aPTT) were prolonged in most cases and were related to an intense reduction in plasma fibrinogen. The levels of von Willebrand factor (vWF), Protein S, tissue plasminogen activator (tPA), and urokinase were not altered, while the levels of factors V and VIII and prekallikrein (PK) were reduced, and this reduction can be attributed to consumption suffered in the activation of coagulation. Factors XII, II, and X levels were unchanged. These results indicate that the consumption coagulopathy developed in this envenomation is different from that observed in DIC associated with other clinical conditions, in which these factors are usually reduced [37–39]. On the other hand, activation of the contact phase of coagulation is unlikely since factor XII levels were normal, although PK levels were shown to be reduced. Zannin and collaborators [8] suggested that PK could be activated by some component of the venom. A subtle reduction in factor XIII was observed in patients envenomed by *L. obliqua*, different from *L. achelous* envenoming where a drastic reduction in this factor is observed, attributed to a factor that degrades FXIII, present in the hemolymph of *L. achelous*, which was called "Lonomin V" [40]. The generation of fragments 1 and 2 of prothrombin (F1+2) and the thrombin/antithrombin complex (TAT) was also observed, like DIC. Although the generation of F1+2 and TAT confirms the formation of thrombin, the number of platelets was not altered in the blood of the patients. Regarding coagulation inhibitors, there was a marked reduction in the levels of Protein C, and there was no significant consumption of antithrombin (AT), as observed in other cases of DIC. Moreover, the formation of thrombin and the TAT complex was observed mainly in patients with severe coagulopathy. These results suggest that AT activity does not depend on coagulation activation. Thrombocytopenia in patients is rare, and its absence can be explained by the high generation of fibrin degradation products (PDFn) suggested by the extremely high levels of D-dimers (DD) [8,19,41,42]. In patients with high levels of DD, a reduction in proteins involved in the fibrinolytic system such as plasminogen, plasminogen activator inhibitor (PAI), and α2-antiplasmin (α2-AP) was observed. In blood coagulation, DD is generated by the action of plasmin on crosslinked fibrin, while the action of plasmin on fibrinogen is observed by the generation of another class of degradation products (PDFg) [43]. Thus, the high increase in DD in patients envenomed by *L. obliqua* suggests that the observed fibrinolysis is an event secondary to the formation of intravascular fibrin [8,19].

In summary, LOCBE induces consumption coagulopathy, depletion of some coagulation factors and inhibitors, and secondary fibrinolysis. It induces a special form of DIC, different from that observed in other clinical situations such as trauma, neoplasia, and sepsis [8,19,28,37,39].

6. Toxins Identified in LOCBE

In vitro studies with LOCBE revealed that it has mainly procoagulant activity [43–48]. However, several studies have identified in bristles or hemolymph many biological activities that could be associated with the effects observed in patients such as inflammation, leukocyte migration, degradation of extracellular matrix, or even pain. Table 2 lists the main biological activities, toxins, or transcripts identified in *L. obliqua* tissues and secretions.

Table 2. Complexity of *L. obliqua* caterpillar venom and secretions.

Toxin/Biological Activity	Source	Method of Detection/Characteristic	References
Lipocalin	LOCBE Hemolymph	Edman sequencing cDNA library More abundant	[47,49]
	LOCBE	Proteome	[50]
	Tegument Cryosecretion	Edman sequencing	[49]
Hemolin	LOCBE	Edman sequencing cDNA library	[47,49]
Serpin	LOCBE	Edman sequencing cDNA library	[49,51]
		Proteome	[50]
	Tegument	Edman sequencing cDNA library	[49]
	Hemolymph Cryosecretion	Edman sequencing	[49]
Kininogen	LOCBE	cDNA library	[49,51]
Trypsin	LOCBE	Edman sequencing	[49,51]
Lectin	LOCBE	Edman sequencing cDNA library	[49,51]
	Tegument	cDNA library	[49]
Transferrin	LOCBE Tegument Hemolymph Cryosecretion	Edman sequencing	[49]
Laminin	LOCBE	Edman sequencing	[49]
Protease inhibitor	Hemolymph	Edman sequencing	[49]
	Tegument	cDNA Library	[49]
Serine proteases	LOCBE	cDNA Library	[49]
Phospholipase A2 (PLA-2)/hemolytic activity	LOCBE	cDNA Library	[49,51]
		Purified protein with Indirect hemolytic activity	[52–55]
	Tegument	cDNA Library	[49]

Table 2. *Cont.*

Toxin/Biological Activity	Source	Method of Detection/Characteristic	References
Lopap	LOCBE	Proteoma cDNA library Native and recombinant protein Lipocalin-like Prothrombin activator Cytoprotector	[45–47,56,57]
Bilin-binding proteins (BBP)	LOCBE	Recombinant Lipocalin Similar to Lopap No prothrombinase activity	[58]
Losac	LOCBE	Native and recombinant protein Hemolin-like Factor X activator Neuroprotection Antiapoptotic	[59–63]
Factor Xa-like	LOCBE	Sequence similar to Lopap Enzymatic activity on S-2222 chromogenic substrate	[64]
Lonofibrase	Hemolymph	Fibrinogenolytic activity	[48,65]
Lonoglyases	LOCBE	Hyaluronidase Degradation of ECM	[66]
Antiapoptotic/proliferative	LOCBE	Activity on several cell cultures	[56,60,62,63,67]
	Hemolymph	Activity on *Spodoptera frugiperda* (Sf-9)	[68]
Antiviral	Hemolymph	Recombinant protein Effect of several virus	[69,70]
Nociceptive and edematogenic	LOCBE	Prostaglandins favors nociception Kallikrein inhibitor or bradykinin B2 receptor antagonists reduces the edema and hypotension	[35,71]
Kallikrein–kinin system activation	LOCBE	In vitro and in vivo studies showed activation of kinin system	[32,35,71]
Proinflammatory response	LOCBE	Proinflammatory phenotype in macrophages and endothelial cells NF-κB pathway activation	[72–74]
Modulation of cell adhesion/cytoskeleton dynamics	LOCBE	Changes in cell–ECM interaction	[74,75]
Platelet adhesion and aggregation	LOCBE	Platelet aggregation in vitro inhibited by PLA-2 inhibitors	[76–78]

Transcriptome and proteome approaches have shown the complexity of LOCBE at the molecular level. However, these approaches did not provide a clear identification of procoagulant molecules or toxins equivalent to well-known coagulation activators [31,48]. The most studied toxins isolated from LOCBE are two procoagulant proteins: the *Lonomia obliqua* Stuart factor activator (Losac) and the *Lonomia obliqua* prothrombin activator protease (Lopap). Both proteins, far from being similar to known clotting factors, were identified as belonging to the hemolin and lipocalin families, respectively [43,47].

6.1. Expressed Genes in LOCBE

Strategies based on expressed sequence tags (EST) were commonly used for identifying a large number of genes in species of interest until the development of high-throughput methods [79]. This approach was used to identify and characterize the major transcripts present in LOCBE [51,53]. Reis and collaborators [47] identified and submitted to GenBank in 2004 sequences from 1270 independent clones assembled into 702 clusters of distinct genes and corresponding proteins such as lipocalins, hemolins, serpins, and other proteins (Table 2). The Lopap whole sequence is identified in GenBank by access number AY908986 [47,51].

Aiming to maximize the identification of putative toxins in *L. obliqua* tissues, Veiga and collaborators constructed separate cDNA libraries from both bristle and tegument mRNAs [49]. These libraries correspond to mRNA isolated from *L. obliqua* bristles and from tegument. A catalog for the transcripts from *L. obliqua* structures showed that lipocalin is the most abundant transcript in this genome. Both cDNA libraries of *L. obliqua* contain sequences with homology to lipocalins. A total of 1152 independent clones from the tegument library and 960 from the bristle library were identified as expressed, yielding 938 and 730 sequences, respectively [49].

In the *L. obliqua* bristles library, over 50% of cDNAs code for a lipocalin, followed by kininogen (16.5%), serine proteases (14.7%). and lectin (5.5%) [49]. Concerning the tegument library, the number of clusters found for serpins is the most abundant (25.8%), followed by serine proteases (16.1%), lipocalin (16%). and lectin (12.9%). These gene sequences from both cDNA libraries were applied independently in GenBank. and they are complementary [47,49,51]. Sequence analysis also showed that Lopap is a member of the lipocalin family of proteins, since it presents an identity of 20% to 59% with other lipocalins [47,51]. Lopap has a serine protease-like activity and acts on prothrombin, such as FXa, in the absence of prothrombinase components [47]. Characterization of the transcripts present in LOCBE showed several kinds of components that distinctly take part in the envenoming. However, the exact toxins involved in the envenomation are not entirely clear so far.

6.2. Proteomic Analysis of LOCBE

Proteomic analyses of LOCBE are scarce, and studies are focused on specific bioactive molecules related to the hematological disturbances observed during envenomation rather than an in-depth description of the bristles extract protein content [50]. A protein profile of LOCBE analyzed through 2D electrophoresis revealed the presence of 159 to 129 spots under nonreducing or reducing conditions, respectively. Most of the spots were detected at acidic to neutral isoelectric point values ($4 < pI > 7$) distributed in a wide molecular mass range (<10 to 105 kDa). This complexity was predominantly diminished at low molecular mass range under reducing conditions, suggesting the presence of dimers or oligomers, and its monomers were not retained on the acrylamide gel mesh used. According to Coomassie blue staining and immunogenic potential, 25 spots were submitted to mass spectrometry analysis, and three protein categories were identified: lipocalins (eight spots), cuticle proteins (five spots), and serpin (one spot). Twelve spots were described as unknown proteins; some of them were immunodetected using ALS or anti-Lopap rabbit serum, suggesting the presence of interesting immunogenic molecules to be further investigated [50].

6.3. Procoagulant Toxins from LOCBE

Losac is the first factor X activator purified as a monomer of 45 kDa from LOCBE [43,63]. The cloning, heterologous expression, and characterization of recombinant Losac (rLosac) was described by Alvarez-Flores and collaborators [43,48]. rLosac specifically activates factor X in the absence of calcium and phospholipids, although the presence of these cofactors accelerates its activity. Its enzymatic characterization was performed, revealing that this protein has no homology to known procoagulant proteases. Instead, Losac belongs to the hemolin family of proteins, a group of multifunctional proteins exclusively expressed by Lepidoptera order insects involved in several cell interactions, but mainly in immunity [43,80]. The tertiary structure model of rLosac was built through homology modeling using the crystal structure of *H. crecopia* hemolin as a template, and it shares the multidomain structure D1–D4 and its conserved motifs. In addition, the multiple amino-acid sequence alignment of rLosac showed up to 76% identity with other hemolin protein families [43].

Studies were conducted to evaluate its effects on endothelium. rLosac inhibited apoptosis in serum-deprived human umbilical vein endothelial cells (HUVECs) and induced cell proliferation [63]. This reduction in cell death under nutrient deprivation conditions was also observed in mouse cortical neurons and human dermal fibroblasts, showing an effective prevention of reactive oxygen species generation and loss of mitochondrial membrane potential, suggesting an antioxidant activity [60–62]. An in vivo experimental model in rats demonstrated that rLosac improved wound healing by increasing the epidermal proliferation, as well as by preserving the extracellular matrix organization through collagen type I, fibronectin, and laminin expression. Thus, rLosac was indicated as a very promising molecule, potentially useful as a bioactive agent to develop new formulations for wound healing [59].

Lopap is a 69 kDa prothrombin activator that shares with LOCBE its role in inflammatory processes and belongs to the lipocalin protein family, being the most abundantly studied isolated toxin from the LOCBE [47,48,51,57]. The recombinant form of Lopap (rLopap) recognizes and hydrolyzes prothrombin, which in turn leads to an active thrombin generation, showing a proteolytic activity similar to native Lopap [47,57]. Lopap displayed a Ca^{2+}-activating serine protease activity that was included into the group I of prothrombin activators [36,45,47]; the infusion of native Lopap produced intravascular coagulation and thrombosis in the post capillary vessels of mice [46].

Properties of rLopap were evaluated in an in vivo model of leukocyte–endothelial cell interaction, revealing that rLopap as the native Lopap induced NO production and ICAM-1 expression in both neutrophils and endothelial cells. In addition, it induced antiapoptotic effects mediated by NO production [81]. The study of its effects in human platelets showed that there is not a direct effect on platelet function, since Lopap showed no effect on platelet aggregation induced by collagen ADP or thrombin. On the other hand, Lopap induced the expression of adhesion molecules ICAM-1 and E-selectin of human endothelial cells [56]. In those cells, Lopap promotes survival mechanisms since it induces the release of nitric oxide and prostaglandin I2, along with the release of inflammatory cytokine IL-8 and t-PA. Moreover, synthetic peptides based on lipocalin motif 2, which is found in a primary sequence of Lopap, showed cytoprotective and antiapoptotic activity in vitro and in vivo approaches, suggesting the involvement of that lipocalin motif in cell protection [82–85]. This knowledge brought new perspectives on the use of these synthetic molecules since they do not exhibit hemostatic functions.

7. Effect of LOCBE and Toxins in the Inflammatory Response

Dermatitis and skin reactions such as urticaria are well-known signs after accidental contact with the spines and bristles of venomous lepidopteran caterpillars [1]. Generally, the consequences of these reactions are limited to local inflammation, with no systemic or tissue damage. In the case of the envenomation caused by *L. obliqua*, the process is characterized by triggering an intense inflammatory response in victims followed by coagulation, complement, and kallikrein–kinin systems [36,48,51]. In recent years, several studies have been carried out aiming to clarify and describe the role of the venom-induced inflammatory response in the clinical symptoms characteristic of lonomism.

L. obliqua proinflammatory effects are first manifested by pain, burning sensation, edema, and erythema formation [8,19,28]. The first pharmacological studies showed that venom-induced nociception in animal models is largely facilitated by the production of prostaglandins, and later edematogenic symptoms are induced by prostanoids and histamines [48,71]. The kallikrein–kinin system is also involved in the edematogenic and hypotensive responses triggered by the venom. Bohrer and collaborators demonstrated that administration, prior to treatment with LOCBE, of plasma kallikrein inhibitor reduces the volume of venom-induced edema in a mouse paw model [35].

Envenomed patients presented low levels of plasma prekallikrein [8,19], indicating that kallikrein was activated and released into the blood circulation. The kallikrein–kinin system is composed of proteolytic enzymes and their substrates, being able to generate potent vasoactive and proinflammatory molecules that are involved in the control of blood pressure, vascular permeability, vascular smooth muscle cell contraction or relaxation, and pain [86]. One of the common consequences of lonomism is the sudden loss of basic renal functions. Kidneys and urine from envenomed animals were enriched with proteins related to inflammatory stress, tissue damage, oxidative stress, coagulation, complement system activation, and kinin system [87,88]. When simultaneously treated with kallikrein inhibitors and antilonomic serum, envenomed rats showed improvements in renal and vascular function, reducing tubular necrosis and renal inflammation [32]. The mechanism underlying these effects was associated with lowering renal inflammation, with a decrease in proinflammatory cytokines and matrix metalloproteinase expression, reduced tubular degeneration, and protection against oxidative stress.

An increase in the permeability of the endothelium allows greater infiltration of cells of the immune system, effectors, and regulators of acute inflammation into tissues [89]. Increased vascular tissue permeability is a characteristic event of the inflammatory response and can be induced by several proinflammatory and vasoactive substances such as bradykinin, histamine, thrombin, cytokines, prostaglandins, and free radicals [74,90]. Activation of the vascular tissue was observed after a single subcutaneous injection of LOCBE in rats [91]. Envenomed animals demonstrated neutrophilic leukocytosis in several tissues, where their histological sections provided evidence of inflammatory cell infiltrates in the heart, lungs, and kidneys, characterizing a systemic acute inflammatory response induced by the venom [92]. Furthermore, an increase was observed in leukocyte rolling and adhesion of these circulating blood cells to the endothelium of hamster cheek pouch tissue that was previously incubated with low doses of LOCBE [72]. The ability of LOCBE to induce an increase in the permeability of the vasculature and immune cell infiltration may provide a favorable environment for hemorrhages, especially in microvessels in the brain.

Due to the important relationship between inflammation and vasculature, studies were carried out seeking to elucidate the direct effect of LOCBE on vascular tissue. In vitro studies showed that non-hemorrhagic concentrations of LOCBE modify the cytoskeleton dynamics and increase focal adhesion in endothelial cells [72]. Furthermore, low doses of the LOCBE can induce activation of the nuclear transcription factor κB (NF-κB) pathway in these cells [73]. The NF-κB pathway is a critical signaling in several events associated with triggering acute inflammation and immune system cell recruitment [93]. Consequently, LOCBE also induced significant increases in the expression of COX-2, NOS-2, HO-1, MMP-2, and MMP-9, enzymes related to prostaglandin production, oxidative stress, and extracellular matrix degradation [72,74].

Additionally, the LOCBE was also shown to be a potent activator of vascular smooth muscle cells, being able to induce cell chemotaxis, exacerbated proliferation, and production of reactive oxygen species. Smooth muscle cell dysfunction is characterized by increased cell migration and proliferation, events that are amplified by the release of inflammatory mediators [91]. Furthermore, researchers also carried out a broad analysis of the gene expression profile of fibroblasts treated with LOCBE. The results show an upregulation of several proinflammatory mediator genes, such as IL-8, IL-6, and CCL2, as well as the adhesion molecule ICAM-3 and COX-2 [74]. Recently, our group showed a direct effect of LOCBE upon macrophage activation. The LOCBE directly induces THP-1 macrophages to a proinflammatory phenotype by activating NF-κB pathway, leading the cells to release proinflammatory cytokines and chemokines such TNFα, IL-1β, IL-6, IL-8, and CXCL10 [73].

The role of isolated toxins present in the venom triggering inflammatory responses is still to be investigated. In addition to its procoagulant activity, in vivo studies show that injection of a high concentration of recombinant *L. obliqua* prothrombin activator protease (rLopap) in rats promotes neutrophil and monocyte infiltration in pulmonary microcirculation vessels. In HUVECs, rLopap stimulates the increase of IL-8, ICAM-1, and E-selectin, proteins involved in the recruitment of immune cells to the tissue [56,75]. In contrast, there is no evidence that Losac presents proinflammatory activity beyond its cytoprotective and proliferative effects.

Taken together, the evidence indicates that LOCBE can induce a local acute inflammatory response that can evolve into a systemic response. Many studies have characterized the role of the kinin–kallikrein system and the liberation of other proinflammatory mediators by the affected tissues, related to several clinical stomps. The isolated effect of the toxins present in the LOCBE, such as Lopap, and their roles in the activation of prekallikrein, the ability to directly induce cell responses, and the molecular mechanism underlying these effects still need to be clarified.

8. Innovation in Toxinology: Research Centers and New Molecule Development

8.1. Research Centers Specialized in Toxinology

Created in 1962, the State of São Paulo Research Foundation (FAPESP) has robust programs of fellowships and research aiming to promote scientific research in the state of São Paulo. Focusing on the production of multidisciplinary scientific knowledge with high-impact science, in 2000, FAPESP started a program to create Research, Innovation, and Dissemination Centers (RIDC). The "Center for Applied Toxinology (CAT)", dedicated to studying animal and microbial toxins, based at Butantan Institute in São Paulo, Brazil was one of the 17 RIDC selected projects (https://fapesp.br/cepid/pasta_cepid.pdf?t=1; (accessed on 31 August 2001). The CAT's scientific findings enrolled in toxin-based drugs affecting the blood clotting, cardiovascular system, pain perception, antiproliferative compounds, and immune suppression resulted in the filing of eight patents, and the technology was transferred to three Brazilian pharmaceutical companies [94].

Motivated by the knowledge and technological innovation resulted from CAT experience along its 10 years, the assembled researchers started another ambitious project and created the "Center of Toxins, Immune Response, and Cell Signaling (CETICS)", a center of excellence in toxins, immune response, and cell signaling, also based at Butantan Institute and supported by FAPESP (https://bv.fapesp.br/pt/auxilios/58571/cetics-centro-de-toxinas-imuno-resposta-e-sinalizacao-celular/; (accessed on 31 August 2001).

Again, according to the rules of an FAPESP program, Butantan Institute initiated a long-term research program enrolling Research Institutes and companies. Created in the public–private partnership (PPP) model, the "Center of Excellence in New Target Discovery (CENTD)" resulted from a partnership among the Butantan Institute, FAPESP, and GlaxoSmithKline Pharmaceutical (GSK). As a multidisciplinary center inaugurated in 2017, the CENTD focus is to use venoms and secretions to validate new therapeutic targets for inflammatory diseases, such as rheumatoid arthritis, metabolic syndrome, and neurodegenerative diseases. Recently renewed, CENTD has so far achieved eight patents and many scientific publications.

8.2. New Molecular Entities for the Study of Anti-Inflammatory Diseases

Lopap is the most studied lipocalin isolated from *L. obliqua*, and the activity of its derived peptides has been the aim of several studies.

Lipocalins are known by their role as protective factors with participation in development, regeneration, and tissue repair [95–97]. Moreover, several lipocalins have been involved in the mediation of cell regulation, such as cell growth and differentiation [98,99]. Lipocalins share three main structures: structurally conserved region 1 (SCR1) is the highest conserved region with the 3–10 helix and adjacent strand leading into a loop. SCR2 is formed by two strands, and the loop that connects them varies across proteins; however, the essential features of the loop are retained, despite the unfavorable main chain. Lastly, SCR3 contains a strand with arginine as the last residue and part of the loop linking it to the C-terminal α-helix [100].

Synthetic peptides based on these three conserved motifs found in Lopap—pM1, pM2a/pM2b, and pM3 (the numbers are related to the motif)—were studied to determine their capability to promote cell survival. Motif 2-related peptides showed a cytoprotective activity, in which pM2b was found to be the shortest peptide with similar activity to Lopap, despite not containing the residues supposedly involved in the catalytic activity of Lopap. This peptide triggers cell survival of neutrophils and endothelial cells via nitric oxide [82]. The effect of peptide pM2b was also evaluated in the modulation of extracellular matrix (ECM) proteins using both in vitro and in vivo approaches and showed an increased production of ECM, as well as a modulation of mediators involved in apoptosis, anti-apoptosis, and proliferation in human fibroblasts. Additionally, an increased production of collagen was observed in vivo [83]. Peptide pM2b also displayed modulating effects in an in vivo model of wound healing. Furthermore, the increased production of collagen, glycosaminoglycans, and metalloproteinases along with improvement of wound closure was observed [84]. Those studies suggest an involvement of motif 2 in cell protection. Moreover, using a peptide mapping approach and tertiary structure comparison, a lipocalin sequence signature able to modulate cell survival was identified [82]. A computational analysis of the peptide sequence signature YAIGYSC, later called pM2c, revealed great similarity with antiapoptotic lipocalins [85]. Alignment of the Lopap sequence with other lipocalins showed that this peptide sequence was highly conserved, and yet few different patterns could be observed. Ten aligned peptides, along with pM2c, were modeled in 3D structures, followed by the analysis of molecular descriptors. This study showed that, even with the amino-acid modifications, the calculated molecular properties were generally maintained, especially the molecular shape and electronic density distribution, highlighting the importance of these two properties for molecular recognition process, while the lipophilicity was more related to the pharmacokinetic profile, validating the lipocalin sequence signature previously reported [85].

Other peptides have been isolated from Lopap (Table 3). The peptide CNF011.05D, along with three peptides comprising an amino-acid sequence of at least 70% identity, was described and patented for being capable of stimulating the production of ECM proteins in human fibroblast cells (Patent No. P4US88374OB2 in the USA and No. EP2245149B1 in Europe). The ECM proteins evaluated were fibronectin, tenascin, procollagen, and collagen. In addition, it was demonstrated that this peptide induced the production of nitric oxide, and a wound healing model in rats and pigs showed a faster rate of reduced wound size.

Regarding Losac, the structural analysis revealed that it shares the conserved motifs and immunoglobulin-like domains scattered throughout their four domains (D1 to D4) with hemolins from different Lepidoptera species [43]: *N*-glycosylation and protein kinase C in D3 and cAMP/cGMP in D2; furthermore, the conserved Lys–Gly–Asp (KDG) motif is found in domains D1 and D3 [101]. The KDG motif is related to cell adhesion, and previous studies in hemocytes revealed that hemolin could be involved in cell differentiation or regeneration of wounded tissue, acting as a cell attachment component, suggesting its participation in cellular immune response and morphogenesis [102,103]. This also resembles the function of cell adhesion molecules such as selectins and cadherins; they are important in lymphocyte homing during the inflammatory response and morphogenesis, as well as in the development of neural systems [103–105].

Table 3. Patents applied and granted regarding *Lonomia* toxins and derived peptides.

Molecule	Patent Number	Claim
Lopap	INPI (PI0200269-8, Brazil, 26 July 2016). WO03/070746 (WIPO) AU2003208190 (Australia) CAN2,471,410 (Canada) EP1482969 (German) JP2003-569653 (Japan) MX04007344 (Mexico) US10/501,238 (USA)	Lopap obtention process
P4	INPI (PCT/IB2009/05023, Brazil, 21 July 2010) 101965398B, 2015 (China) JP5777886B2, 2015 (Japan) EP2245149B1, 2015 (German) US8,883,740 B2, 2014 (USA)	Tissue remodeling and tissue repair
Losac-derived peptides	INPI (BR10201901866, Brazil, 9 September 2019)	Anti-inflammatory agents

Despite advances involving the pathogenesis of inflammatory diseases and new therapeutic strategies, the progression of inflammatory diseases has become a public health problem in recent years, mainly due to new viruses and resistant bacteria [106], along with inflammatory diseases related to aging [107]. Therefore, studies on new molecular targets are needed for development of more effective drugs and therapeutic approaches that can benefit the health and quality of life of patients. Accordingly, allied with the promising results from *Lonomia* peptides, several other peptides were engineered and assayed in several in vitro models of inflammatory diseases, such as arthritis. Some of these peptides had anti-inflammatory effects on several cellular models developed at CENTD (Table 3), reversing the expression of molecules associated with the inflammatory mechanism of cells, in addition to decreasing pain markers in the neuron model and inducing regenerative proteins in different cell models. Those peptides were chosen as a powerful tool for the discovery of new inflammatory targets against arthritis, associating omics approaches with the developed cell models and LOCBE.

9. Concluding Remarks

The hemorrhagic syndrome is one of the most serious complications in patients who have contact with the *L. obliqua* caterpillar bristles. After the burst of accidents with hemorrhagic manifestations in 1989 in the states of south and southeast Brazil, many efforts were put into finding a solution to the public health problem that reached alarming proportions. Fortunately, the treatment with ALS produced at the Butantan Institute drastically reduced the consequences of the envenoming. All these efforts have resulted in substantial knowledge about the pathophysiology of the envenoming and the toxins contained in the venom.

Over the years, research has diversified, addressing not only the hematological alterations of the venom but also the role of toxins in each observed manifestation (Table 2). New molecules identified demonstrate peculiarities, such as hemolin and lipocalin structures with enzymatic or cellular activities that have never previously been reported for these types of proteins. This opened new perspectives for the use of those molecules in several applications, as diagnostic agents, for example, to detect dysprothrombinemias, as in the case of Lopap, which resulted in several patents granted, or as therapeutic agents for use in defibrinogenating and antithrombotic therapy (Lopap) or for promoting wound healing (Lopap- and Losac-derived peptides) (Table 3).

The creation of multidisciplinary research centers specialized in the study of animal envenomation and animal toxins promoted the interaction of partnerships with the pharmaceutical industry, e.g., CENTD, taking the study of molecules derived from venoms and toxins to another level, favoring the development and innovation of new molecules. High-throughput screening technologies for new drugs or target discovery are extensively used at CENTD using toxin-derived peptides as tools for identifying new targets for inflammatory diseases.

Therefore, it is clear that the efforts of distinct groups at Butantan Institute have not only contributed to accumulated knowledge about the toxinology of *Lonomia*, but also created the basis for the development of ALS, essential in reducing the deaths due to lononism, and prompted the institute to develop innovative initiatives with toxin-derived peptides.

Author Contributions: Conceptualization, M.P.A.-F., D.T.-S. and A.M.C.-T.; writing—original draft preparation, M.P.A.-F., R.N.G., D.T.-S., D.S.O., I.d.F.C.B., M.V.B., A.M.A., C.D.-P., M.M.d.S. and A.M.C.-T.; writing—review and editing, M.P.A.-F., R.N.G., D.T.-S., D.S.O., I.d.F.C.B., M.V.B., A.M.A. and A.M.C.-T.; funding acquisition, A.M.C.-T. All authors have read and agreed to the published version of the manuscript.

Funding: This research was funded by "Grant 2015/50040-4, São Paulo Research Foundation and GlaxoSmithKline", "Grant 2020/13139-0, São Paulo Research Foundation and GlaxoSmithKline", and Grant number 2013/07467-1 (FAPESP-CETICs). A.M.C.-T. is a recipient of CNPq-PQ grant number 303197/2017-0 and Fundação Butantan-PQ grants. R.N.G. is a recipient of FAPESP grant number 2018/20469-7. A.M.A is a recipient of FAPESP grant number 2018/10937-3.

Institutional Review Board Statement: Not applicable.

Informed Consent Statement: Not applicable.

Data Availability Statement: The data presented in this study are available on request from the corresponding author.

Acknowledgments: The authors thank Rafaella Lafraia for research on SINAN.

Conflicts of Interest: The authors declare no conflict of interest.

References

1. Villas-Boas, I.M.; Alvarez-Flores, M.P.; Chudzinski-Tavassi, A.M.; Tambourgi, D.V. Envenomation by caterpillars. In *Clinical Toxinology*; Gopalakrishnakone, P., Faiz, S., Gnanathasan, C., Habib, A., Fernando, R., Yang, C.C., Eds.; Springer: Dordrecht, The Netherlands, 2016. [CrossRef]
2. Seldeslachts, A.; Peigneur, S.; Tytgat, J. Caterpillar Venom: A Health Hazard of the 21st Century. *Biomedicines* **2020**, *8*, 143. [CrossRef] [PubMed]
3. Da Silva, W.D.; Campos, C.M.; Gonçalves, L.R.; Sousa-E-Silva, M.C.; Higashi, H.G.; Yamagushi, I.K.; Kelen, E.M. Development of an antivenom against toxins of Lonomia obliqua caterpillars. *Toxicon* **1996**, *34*, 1045–1049. [CrossRef]
4. Rocha-Campos, A.C.; Gonçalves, L.R.; Higashi, H.G.; Yamagushi, I.K.; Fernandes, I.; Oliveira, J.E.; Ribela, M.T.; Sousa-E-Silva, M.C.; da Silva, W.D. Specific heterologous F(ab')2 antibodies revert blood incoagulability resulting from envenoming by Lonomia obliqua caterpillars. *Am. J. Trop. Med. Hyg.* **2001**, *64*, 283–289. [CrossRef]
5. Caovilla, J.J.; José Guardão Barros, E. Efficacy of two different doses of antilonomic serum in the resolution of hemorrhagic syndrome resulting from envenoming by Lonomia obliqua caterpillars: A randomized controlled trial. *Toxicon* **2004**, *43*, 811–818. [CrossRef]
6. Gonçalves, L.R.; Souza-E-Silva, M.C.; Tomy, S.C.; Sano-Martins, I.S. Efficacy of serum therapy on the treatment of rats experimentally envenomed by bristle extract of the caterpillar Lonomia obliqua: Comparison with epsilon-aminocaproic acid therapy. *Toxicon* **2007**, *50*, 349–356. [CrossRef]
7. Sano-Martins, I.S.; González, C.; Anjos, I.V.; Díaz, J.; Gonçalves, L.R.C. Effectiveness of Lonomia antivenom in recovery from the coagulopathy induced by Lonomia orientoandensis and Lonomia casanarensis caterpillars in rats. *PLoS Negl. Trop. Dis.* **2018**, *12*, e0006721. [CrossRef]
8. Zannin, M.; Lourenço, D.M.; Motta, G.; Dalla Costa, L.R.; Grando, M.; Gamborgi, G.P.; Noguti, M.A.; Chudzinski-Tavassi, A.M. Blood coagulation and fibrinolytic factors in 105 patients with hemorrhagic syndrome caused by accidental contact with Lonomia obliqua caterpillar in Santa Catarina, southern Brazil. *Thromb. Haemost.* **2003**, *89*, 355–364. [CrossRef]
9. Favalesso, M.M.; Cuervo, P.F.; Casafús, M.G.; Guimarães, A.T.B.; Peichoto, M.E. Lonomia envenomation in Brazil: An epidemiological overview for the period 2007–2018. *Trans. R. Soc. Trop. Med. Hyg.* **2021**, *115*, 9–19. [CrossRef]
10. Brazilian Ministry of Health. SINAN, Sistema de Informação de Agravos de Notificação (Information System for Notification of Diseases). Available online: http://tabnet.datasus.gov.br/cgi/menu_tabnet_php.htm (accessed on 31 August 2021).

11. Alvarenga, Z. A taturana. In *VII Congresso Brasileiro de Medicina Cirúrgica;* Anais da Sociedade de Medicina e Cirurgia: Belo Horizonte, Brazil, 1912; pp. 132–135.
12. Fraiha-Neto, H.; Ballarini, A.J.; Leão, R.N.Q.; Costa, J.R.D.; Dias, L.B. Síndrome hemorrágica por contato com larvas de mariposa (Lepidoptera, Saturniidae). In *Instituto Evandro Chagas: 50 Anos de Contribuição às Ciências Biológicas e à Medicina Tropical;* Fundação Serviços de Saúde Pública: Belém, Brazil, 1986; Volume 2, pp. 811–820.
13. Fan, H.W.; Cardoso, J.L.C.; Olmos, R.D.; Almeida, F.J.; Viana, R.P.; Martinez, A.P.P. Hemorrhagic syndrome and Acute renal failure in a pregnant woman after contact with Lonomia caterpillars: A case report. *Rev. Inst. Med. Trop. São Paulo* **1998**, *40*, 119–120. [CrossRef]
14. Arocha-Piñango, C.L.; de Bosch, N.B.; Torres, A.; Goldstein, C.; Nouel, A.; Argüello, A.; Carvajal, Z.; Guerrero, B.; Ojeda, A.; Rodriguez, A. Six New Cases of a Caterpillar-Induced Bleeding Syndrome. *Thromb. Haemost.* **1992**, *67*, 402–407. [CrossRef] [PubMed]
15. Kelen, E.M.A.; Picarelli, Z.P.; Duarte, A.C. Hemorrhagic syndrome induced by contact with caterpillars of the genus Lonomia (saturniidae, hemileucinae). *J. Toxicol.-Toxin Rev.* **1995**, *14*, 283–308. [CrossRef]
16. Duarte, A.C.; Crusius, P.S.; Pires, C.A.; Schilling, M.A.; Fan, H.W. Intracerebral haemorrhage after contact with Lonomia caterpillars. *Lancet* **1996**, *348*, 9033. [CrossRef]
17. Arocha-Piñango, C.L.; Guerrero, B. Lonomia obliqua and haemorrhagic syndrome. *Lancet* **1999**, *354*, 9186. [CrossRef]
18. Pineda, D.; Amarillo, A.; Becerra, J.; Montenegro, G. Síndrome hemorrágico por contacto con orugas del género Lonomia (Saturniidae) en Casanare, Colombia: Informe de dos casos. *Biomédica* **2001**, *21*, 328. [CrossRef]
19. Sano-Martins, I.S.; Duarte, A.C.; Guerrero, B.; Moraes, R.H.P.; Barros, E.J.G.; Arocha-Piñango, C.L. Hemostatic disorders induced by skin contact with Lonomia obliqua (Lepidoptera, Saturniidae) caterpillars. *Rev. Inst. Med. Trop. Sao Paulo* **2017**, *59*. [CrossRef] [PubMed]
20. Spadacci-Morena, D.D.; Soares, M.A.; Moraes, R.H.; Sano-Martins, I.S.; Sciani, J.M. The urticating apparatus in the caterpillar of Lonomia obliqua (Lepidoptera: Saturniidae). *Toxicon* **2016**, *119*, 218–224. [CrossRef]
21. Arocha-Piñango, C.L. Fibrinólisis producida por contacto con orugas. *Acta Cient. Venez.* **1967**, *18*, 136–139.
22. Cardoso, A.E.C.; Junior, V.H. Accidents caused by lepidopterans (moth larvae and adult): Study on the epidemiological, clinical and therapeutic aspects. *An. Bras. Dermatol.* **2005**, *80*, 571–578. [CrossRef]
23. Arocha-Piñango, C.L.; Layrisse, M. Fibrinolysis Produced by contact with a caterpillar. *Lancet* **1969**, *293*, 810–812. [CrossRef]
24. Duarte, A.C.; Caovilla, J.; Lorini, I.; Lorini, D.; Mantovani, G.; Sumida, J.; Manfre, P.C.; Silveira, R.C.; Moura, S.P. Insuficiência renal aguda por acidentes com lagartas. *J. Bras. Nefrol.* **1990**, *12*, 1033–1187.
25. Secretária da Saúde do Paraná. SESA/PR. Available online: https://www.saude.pr.gov.br/ (accessed on 31 August 2001).
26. CIT/RS. Centro de Informação Toxológica do Rio Grande do Sul. 2005. Available online: http://www.cit.rs.gov.br/ (accessed on 31 August 2001).
27. Ministério da Saúde. Acidentes por lepidópteros. In *Manual de Diagnóstico e Tratamento de Acidentes por Animais peçoNhentos*, 2nd ed.; Assessoria de Comunicação e Educação em Saúde/Ascom/Pre/FUNASA, Ed.; Fundação Nacional de Saúde (FUNASA): Brasilia, Brazil, 2001; pp. 67–76.
28. Chudzinski-Tavassi, A.M.; Zannin, M. Aspectos clínicos y epidemiológicos del envenenamiento por Lonomia obliqua. In *Emergencias por Animales Ponzoñosos en las Américas;* D'Suze, G., Corzo Burguete, G.A., Paniagua Solis, J.F., Eds.; Instituto Bioclon: Mexico City, Mexico, 2011; pp. 303–321.
29. Burdmann, E.A.; Antunes, I.; Saldanha, L.B.; Abdulkader, R.C. Severe acute renal failure induced by the venom of Lonomia caterpillars. *Clin. Nephrol.* **1996**, *46*, 337–339. [PubMed]
30. Schmitberger, P.A.; Fernandes, T.C.; Santos, R.C.; de Assis, R.C.; Gomes, A.P.; Siqueira, P.K.; Vitorino, R.R.; de Mendonça, E.G.; de Almeida Oliveira, M.G.; Siqueira-Batista, R. Probable chronic renal failure caused by Lonomia caterpillar envenomation. *J. Venom. Anim. Toxins Incl. Trop. Dis.* **2013**, *19*, 14. [CrossRef] [PubMed]
31. Chudzinski-Tavassi, A.M.; Alvarez Flores, M.P. South American Lonomia Obliqua Caterpillars: Morphologic Aspects and Venom Biochemistry. In *Lepidoptera;* Guerritore, E., DeSare, J., Eds.; Nova Science Publishers: New York, NY, USA, 2013; pp. 169–185.
32. Berger, M.; de Moraes, J.A.; Beys-da-Silva, W.O.; Santi, L.; Terraciano, P.B.; Driemeier, D.; Cirne-Lima, E.O.; Passos, E.P.; Vieira, M.; Barja-Fidalgo, T.C.; et al. Renal and vascular effects of kallikrein inhibition in a model of Lonomia obliqua venom-induced acute kidney injury. *PLoS Negl. Trop. Dis.* **2019**, *13*, e0007197. [CrossRef] [PubMed]
33. Kowacs, P.A.; Cardoso, J.; Entres, M.; Novak, E.M.; Werneck, L.C. Fatal intracerebral hemorrhage secondary to Lonomia obliqua caterpillar envenoming: Case report. *Arq. Neuro-Psiquiatr.* **2006**, *64*, 1030–1032. [CrossRef] [PubMed]
34. Santos, J.H.A.; Oliveira, S.S.; Alves, E.C.; Mendonça-da-Silva, I.; Sachett, J.A.G.; Tavares, A.; Ferreira, L.C.; Fan, H.W.; Lacerda, M.; Monteiro, W.M. Severe Hemorrhagic Syndrome After Lonomia Caterpillar Envenomation in the Western Brazilian Amazon: How Many More Cases Are There? *Wilderness Environ. Med.* **2017**, *28*, 46–50. [CrossRef] [PubMed]
35. Bohrer, C.B.; Reck Junior, J.; Fernandes, D.; Sordi, R.; Guimarães, J.A.; Assreuy, J.; Termignoni, C. Kallikrein–kinin system activation by Lonomia obliqua caterpillar bristles: Involvement in edema and hypotension responses to envenomation. *Toxicon* **2007**, *49*, 663–669. [CrossRef]
36. Alvarez Flores, M.P.; Zannin, M.; Chudzinski-Tavassi, A.M. New Insight into the Mechanism of Lonomia obliqua envenoming: Toxin involvement and molecular approach. *Pathophysiol. Haemost. Thromb.* **2010**, *37*, 1–16. [CrossRef]
37. Gando, S.; Levi, M.; Toh, C.-H. Disseminated intravascular coagulation. *Nat. Rev. Dis. Primers* **2016**, *2*, 16037. [CrossRef]

38. Wada, H.; Sakuragawa, N.; Mori, Y.; Takagi, M.; Nakasaki, T.; Shimura, M.; Hiyoyama, K.; Nisikawa, M.; Gabazza, E.C.; Deguchi, K.; et al. Hemostatic molecular markers before the onset of disseminated intravascular coagulation. *Am. J. Hematol.* **1999**, *60*, 273–278. [CrossRef]
39. Spero, J.A.; Lewis, J.H.; Hasiba, U. Disseminated intravascular coagulation. Findings in 346 patients. *Thromb. Haemost.* **1980**, *43*, 28–33.
40. Guerrero, B.A.; Arocha-Piñango, C.L.; San Juan, A.G. Lonomia achelous caterpillar venom (LACV) selectively inactivates blood clotting factor XIII. *Thromb. Res.* **1997**, *87*, 83–93. [CrossRef]
41. Bauer, K.A. Activation markers of coagulation. *Best Pract. Res. Clin. Haematol.* **1999**, *12*, 387–406. [CrossRef]
42. Delabranche, X.; Helms, J.; Meziani, F. Immunohaemostasis: A new view on haemostasis during sepsis. *Ann. Intensive Care* **2017**, *7*, 117. [CrossRef]
43. Alvarez-Flores, M.P.; Furlin, D.; Ramos, O.H.P.; Balan, A.; Konno, K.; Chudzinski-Tavassi, A.M. Losac, the First Hemolin that Exhibits Procogulant Activity through Selective Factor X Proteolytic Activation. *J. Biol. Chem.* **2011**, *286*, 6918–6928. [CrossRef] [PubMed]
44. Donato, J.L.; Moreno, R.A.; Hyslop, S.; Duarte, A.; Antunes, E.; Le Bonniec, B.F.; Rendu, F.; de Nucci, G. Lonomia obliqua caterpillar spicules trigger human blood coagulation via activation of factor X and prothrombin. *Thromb. Haemost.* **1998**, *79*, 539–542. [CrossRef] [PubMed]
45. Reis, C.V.; Kelen, E.; Farsky, S.; Portaro, F.; Sampaio, C.; Fernandes, B.; Camargo, A.C.; Chudzinski-Tavassi, A.M. A Ca++ activated serine protease (LOPAP) could be responsible for the haemorrhagic syndrome caused by the caterpillar Lonomia obliqua. *Lancet* **1999**, *353*, 1942. [CrossRef]
46. Reis, C.V.; Farsky, S.H.P.; Fernandes, B.L.; Santoro, M.L.; Oliva, M.L.V.; Mariano, M.; Chudzinski-Tavassi, A.M. In Vivo Characterization of Lopap, a Prothrombin Activator Serine Protease from the Lonomia obliqua Caterpillar Venom. *Thromb. Res.* **2001**, *102*, 437–443. [CrossRef]
47. Reis, C.V.; Andrade, S.A.; Ramos, O.H.P.; Ramos, C.R.R.; Ho, P.L.; de Batista, I.F.C.; Chudzinski-Tavassi, A.M. Lopap, a prothrombin activator from Lonomia obliqua belonging to the lipocalin family: Recombinant production, biochemical characterization and structure–function insights. *Biochem. J.* **2006**, *398*, 295–302. [CrossRef]
48. Chudzinski-Tavassi, A.M.; Alvarez-Flores, M.P.; Carrijo-Carvalho, L.C.; Ricci-Silva, M.E. Toxins from Lonomia obliqua: Recombinant production and molecular approach. In *An Integrated View of the Molecular Recognition and Toxinology—From Analytical Procedures to Biomedical Applications*; Baptista, G.R., Ed.; InTech: Rijeka, Croatia, 2013; pp. 1–32.
49. Veiga, A.B.G.; Ribeiro, J.M.C.; Guimarães, J.A.; Francischetti, I.M.B. A catalog for the transcripts from the venomous structures of the caterpillar Lonomia obliqua: Identification of the proteins potentially involved in the coagulation disorder and hemorrhagic syndrome. *Gene* **2005**, *355*, 11–27. [CrossRef] [PubMed]
50. Ricci-Silva, M.E.; Valente, R.H.; León, I.R.; Tambourgi, D.V.; Ramos, O.H.P.; Perales, J.; Chudzinski-Tavassi, A.M. Immunochemical and proteomic technologies as tools for unravelling toxins involved in envenoming by accidental contact with Lonomia obliqua caterpillars. *Toxicon* **2008**, *51*, 1017–1028. [CrossRef]
51. Chudzinski-Tavassi, A.M.; Alvarez Flores, M.P. Exploring new molecules and activities from Lonomia obliqua caterpillars. *Pathophysiol. Haemost. Thromb.* **2005**, *34*, 228–233. [CrossRef]
52. Seibert, C.S.; Shinohara, E.M.G.; Sano-Martins, I.S. In vitro hemolytic activity of Lonomia obliqua caterpillar bristle extract on human and Wistar rat erythrocytes. *Toxicon* **2003**, *41*, 831–839. [CrossRef]
53. Seibert, C.S.; Oliveira, M.R.L.; Gonçalves, L.R.C.; Santoro, M.L.; Sano-Martins, I.S. Intravascular hemolysis induced by Lonomia obliqua caterpillar bristle extract: An experimental model of envenomation in rats. *Toxicon* **2004**, *44*, 793–799. [CrossRef]
54. Seibert, C.S.; Santoro, M.L.; Tambourgi, D.V.; Sampaio, S.C.; Takahashi, H.K.; Peres, C.M.; Curi, R.; Sano-Martins, I.S. Lonomia obliqua (Lepidoptera, Saturniidae) caterpillar bristle extract induces direct lysis by cleaving erythrocyte membrane glycoproteins. *Toxicon* **2010**, *55*, 1323–1330. [CrossRef]
55. Seibert, C.S.; Tanaka-Azevedo, A.M.; Santoro, M.L.; Mackessy, S.P.; Soares Torquato, R.J.; Lebrun, I.; Tanaka, A.S.; Sano-Martins, I.S. Purification of a phospholipase A2 from Lonomia obliqua caterpillar bristle extract. *Biochem. Biophys. Res. Commun.* **2006**, *342*, 1027–1033. [CrossRef]
56. Fritzen, M.; Flores, M.P.A.; Reis, C.V.; Chudzinski-Tavassi, A.M. A prothrombin activator (Lopap) modulating inflammation, coagulation and cell survival mechanisms. *Biochem. Biophys. Res. Commun.* **2005**, *333*, 517–523. [CrossRef] [PubMed]
57. Reis, C.V.; Portaro, F.C.V.; Andrade, S.A.; Fritzen, M.; Fernandes, B.L.; Sampaio, C.A.; Camargo, A.C.; Chudzinski-Tavassi, A.M. A Prothrombin Activator Serine Protease from the Lonomia obliqua Caterpillar Venom (Lopap). *Thromb. Res.* **2001**, *102*, 427–436. [CrossRef]
58. Veiga, A.B.G.; Ribeiro, J.M.C.; Francischetti, I.M.B.; Xu, X.; Guimarães, J.A.; Andersen, J.F. Examination of the Ligand-Binding and Enzymatic Properties of a Bilin-Binding Protein from the Poisonous Caterpillar Lonomia obliqua. *PLoS ONE* **2014**, *9*, e95424. [CrossRef]
59. Sato, A.C.; Bosch, R.V.; Will, S.E.; Alvarez-Flores, M.P.; Goldfeder, M.B.; Pasqualoto, K.F.; da Silva, B.A.; de Andrade, S.A.; Chudzinski-Tavassi, A.M. Exploring the in vivo wound healing effects of a recombinant hemolin from the caterpillar Lonomia obliqua. *J. Venom. Anim. Toxins Incl. Trop. Dis.* **2016**, *22*, 36. [CrossRef] [PubMed]
60. Bosch, R.V.; Alvarez-Flores, M.P.; Maria, D.A.; Chudzinski-Tavassi, A.M. Hemolin triggers cell survival on fibroblasts in response to serum deprivation by inhibition of apoptosis. *Biomed. Pharmacother.* **2016**, *82*, 537–546. [CrossRef]

61. Chudzinski Tavassi, A.M.; Alvarez-Flores, M.P. Lonomia obliqua brasileña: Bioquímica y fisiopatologia del veneno. In *Emergencias por Animales Ponzoñosos en las Américas*; D'Suze, G., Corzo Burguete, G.A., Paniagua Solis, J.F., Eds.; Instituto Bioclon: Mexico City, Mexico, 2011; pp. 323–369.
62. Alvarez-Flores, M.P.; Hébert, A.; Gouelle, C.; Geller, S.; Chudzinski-Tavassi, A.M.; Pellerin, L. Neuroprotective effect of rLosac on supplement-deprived mouse cultured cortical neurons involves maintenance of monocarboxylate transporter MCT2 protein levels. *J. Neurochem.* **2019**, *148*, 80–96. [CrossRef] [PubMed]
63. Alvarez Flores, M.P.; Fritzen, M.; Reis, C.V.; Chudzinski-Tavassi, A.M. Losac, a factor X activator from Lonomia obliqua bristle extract: Its role in the pathophysiological mechanisms and cell survival. *Biochem. Biophys. Res. Commun.* **2006**, *343*, 1216–1223. [CrossRef]
64. Lilla, S.; Pereira, R.; Hyslop, S.; Donato, J.L.; Le Bonniec, B.F.; de Nucci, G. Purification and initial characterization of a novel protein with factor Xa activity from Lonomia obliqua caterpillar spicules. *J. Mass Spectrom.* **2005**, *40*, 405–412. [CrossRef] [PubMed]
65. Pinto, A.F.; Dobrovolski, R.; Veiga, A.B.; Guimarães, J.A. Lonofibrase, a novel alpha-fibrinogenase from Lonomia obliqua caterpillars. *Thromb. Res.* **2004**, *113*, 147–154. [CrossRef]
66. da CB Gouveia, A.I.; da Silveira, R.B.; Nader, H.B.; Dietrich, C.P.; Gremski, W.; Veiga, S.S. Identification and partial characterisation of hyaluronidases in Lonomia obliqua venom. *Toxicon* **2005**, *45*, 403–410. [CrossRef] [PubMed]
67. Heinen, T.E.; de Farias, C.B.; Abujamra, A.L.; Mendonça, R.Z.; Roesler, R.; da Veiga, A.B.G. Effects of Lonomia obliqua caterpillar venom upon the proliferation and viability of cell lines. *Cytotechnology* **2014**, *66*, 63–74. [CrossRef]
68. Souza, A.P.; Peixoto, C.C.; Maranga, L.; Carvalhal, A.V.; Moraes, R.H.; Mendonça, R.M.; Pereira, C.A.; Carrondo, M.J.; Mendonça, R.Z. Purification and characterization of an anti-apoptotic protein isolated from Lonomia obliqua hemolymph. *Biotechnol. Prog.* **2005**, *21*, 99–105. [CrossRef]
69. Carmo, A.C.V.; Yamasaki, L.H.T.; Figueiredo, C.A.; da Silva Giovanni, D.N.; de Oliveira, M.I.; Dos Santos, F.C.P.; Curti, S.P.; Rahal, P.; Mendonça, R.Z. Discovery of a new antiviral protein isolated Lonomia obliqua analysed by bioinformatics and real-time approaches. *Cytotechnology* **2015**, *67*, 1011–1022. [CrossRef]
70. Sousa, Á.P.B.; Moraes, R.H.P.; Mendonça, R.Z. Improved replication of the baculovirus Anticarsia gemmatalis nucleopolyhedrovirus (AgMNPV) in vitro using proteins from Lonomia obliqua hemolymph. *Cytotechnology* **2015**, *67*, 331–342. [CrossRef]
71. de Castro Bastos, L.; Veiga, A.B.; Guimarães, J.A.; Tonussi, C.R. Nociceptive and edematogenic responses elicited by a crude bristle extract of Lonomia obliqua caterpillars. *Toxicon* **2004**, *43*, 273–278. [CrossRef]
72. Nascimento-Silva, V.; Rodrigues da Silva, G.; Moraes, J.A.; Cyrino, F.Z.; Seabra, S.H.; Bouskela, E.; Almeida Guimarães, J.; Barja-Fidalgo, C. A pro-inflammatory profile of endothelial cell in Lonomia obliqua envenomation. *Toxicon* **2012**, *60*, 50–60. [CrossRef]
73. Oliveira, D.S.; de Souza, J.G.; Alvarez-Flores, M.P.; Cunegundes, P.S.; DeOcesano-Pereira, C.; Lobba, A.M.; Gomes, R.N.; Chudzinski-Tavassi, A.M. Lonomia obliqua Venom Induces NF-κB Activation and a Pro-Inflammatory Profile in THP-1-Derived Macrophage. *Toxins* **2021**, *13*, 462. [CrossRef]
74. Pinto, A.F.M.; Dragulev, B.; Guimarães, J.A.; Fox, J.W. Novel perspectives on the pathogenesis of Lonomia obliqua caterpillar envenomation based on assessment of host response by gene expression analysis. *Toxicon* **2008**, *51*, 1119–1128. [CrossRef]
75. Bernardi, L.; Pinto, A.F.M.; Mendes, E.; Yates, J.R.; Lamers, M.L. Lonomia obliqua bristle extract modulates Rac1 activation, membrane dynamics and cell adhesion properties. *Toxicon* **2019**, *162*, 32–39. [CrossRef]
76. Chudzinski-Tavassi, A.M.; Schattner, M.; Fritzen, M.; Pozner, R.G.; Reis, C.V.; Lourenço, D.; Lazzari, M.A. Effects of Lopap on Human Endothelial Cells and Platelets. *Pathophysiol. Haemost. Thromb.* **2001**, *31*, 257–265. [CrossRef]
77. Berger, M.; Reck, J.; Terra, R.M.S.; Pinto, A.F.M.; Termignoni, C.; Guimarães, J.A. Lonomia obliqua caterpillar envenomation causes platelet hypoaggregation and blood incoagulability in rats. *Toxicon* **2010**, *55*, 33–44. [CrossRef] [PubMed]
78. Berger, M.; Reck, J.; Terra, R.M.S.; Beys da Silva, W.O.; Santi, L.; Pinto, A.F.; Vainstein, M.H.; Termignoni, C.; Guimarães, J.A. Lonomia obliqua venomous secretion induces human platelet adhesion and aggregation. *J. Thromb. Thrombolysis* **2010**, *30*, 300–310. [CrossRef]
79. Lowe, R.; Shirley, N.; Bleackley, M.; Dolan, S.; Shafee, T. Transcriptomics technologies. *PLoS Comput. Biol.* **2017**, *13*, e1005457. [CrossRef]
80. Schmidt, O.; Söderhäll, K.; Theopold, U.; Faye, I. Role of adhesion in arthropod immune recognition. *Ann. Rev. Entomol.* **2010**, *55*, 485–504. [CrossRef] [PubMed]
81. Waismam, K.; Chudzinski-Tavassi, A.M.; Carrijo-Carvalho, L.C.; Fernandes Pacheco, M.T.; Farsky, S.H.P. Lopap: A non-inflammatory and cytoprotective molecule in neutrophils and endothelial cells. *Toxicon* **2009**, *53*, 652–659. [CrossRef] [PubMed]
82. Chudzinski-Tavassi, A.M.; Carrijo-Carvalho, L.C.; Waismam, K.; Farsky, S.H.P.; Ramos, O.H.P.; Reis, C.V. A lipocalin sequence signature modulates cell survival. *FEBS Lett.* **2010**, *584*, 2896–2900. [CrossRef] [PubMed]
83. Carrijo-Carvalho, L.C.; Maria, D.A.; Ventura, J.S.; Morais, K.L.P.; Melo, R.L.; Rodrigues, C.J.; Chudzinski-Tavassi, A.M. A Lipocalin-Derived Peptide Modulating Fibroblasts and Extracellular Matrix Proteins. *J. Toxicol.* **2012**, *2012*, 325250. [CrossRef]
84. Wlian, L.; Carrijo-Carvalho, L.C.; Andrade, S.A.; Lourenço, S.V.; Rodrigues, C.J.; Maria, D.A.; Chudzinski-Tavassi, A.M. Wound Healing Effects of a Lipocalin-Derived Peptide. *J. Clin. Toxicon* **2014**, *4*, 2–6.
85. Mesquita Pasqualoto, K.F.; Carrijo-Carvalho, L.C.; Chudzinski-Tavassi, A.M. Rational development of novel leads from animal secretion based on coagulation and cell targets: 1. In silico analysis to explore a peptide derivative as lipocalins' signature. *Toxicon* **2013**, *69*, 200–210. [CrossRef] [PubMed]

86. Björkqvist, J.; Jämsä, A.; Renné, T. Plasma kallikrein: The bradykinin-producing enzyme. *Thromb. Haemost.* **2013**, *110*, 399–407. [CrossRef]
87. Zanon, P.; Pizzato, S.B.; da Rosa, R.L.; Terraciano, P.B.; Moraes, J.A.; Beys-da-Silva, W.O.; Santi, L.; Yates, J.R., 3rd; Passos, E.P.; Barja-Fidalgo, C.; et al. Urine proteomic analysis reveals alterations in heme/hemoglobin and aminopeptidase metabolism during Lonomia obliqua venom-induced acute kidney injury. *Toxicol. Lett.* **2021**, *341*, 11–22. [CrossRef] [PubMed]
88. Berger, M.; Santi, L.; Beys-da-Silva, W.O.; Oliveira, F.M.S.; Caliari, M.V.; Yates, J.R., 3rd; Vieira, M.A.; Guimarães, J.A. Mechanisms of acute kidney injury induced by experimental Lonomia obliqua envenomation. *Arch. Toxicol.* **2015**, *89*, 459–483. [CrossRef]
89. Nourshargh, S.; Alon, R. Leukocyte Migration into Inflamed Tissues. *Immunity* **2014**, *41*, 694–707. [CrossRef]
90. Claesson-Welsh, L. Vascular permeability—The essentials. *Upsala J. Med. Sci.* **2015**, *120*, 135–143. [CrossRef]
91. Moraes, J.; Rodrigues, G.; Nascimento-Silva, V.; Renovato-Martins, M.; Berger, M.; Guimarães, J.; Barja-Fidalgo, C. Effects of Lonomia obliqua Venom on Vascular Smooth Muscle Cells: Contribution of NADPH Oxidase-Derived Reactive Oxygen Species. *Toxins* **2017**, *9*, 360. [CrossRef]
92. Berger, M.; Beys-da-Silva, W.O.; Santi, L.; de Oliveira, I.M.; Jorge, P.M.; Henriques, J.A.; Driemeier, D.; Vieira, M.A.; Guimarães, J.A. Acute Lonomia obliqua caterpillar envenomation-induced physiopathological alterations in rats: Evidence of new toxic venom activities and the efficacy of serum therapy to counteract systemic tissue damage. *Toxicon* **2013**, *74*, 179–192. [CrossRef]
93. Chen, L.; Deng, H.; Cui, H.; Fang, J.; Zuo, Z.; Deng, J.; Li, Y.; Wang, X.; Zhao, L. Inflammatory responses and inflammation-associated diseases in organs. *Oncotarget* **2018**, *9*, 7204–7218. [CrossRef] [PubMed]
94. De Camargo, A.C.M. Perspective for pharmaceutical innovation in Brazil—center for applied toxinology (CEPID- center for research, innovation and dissemination—FAPESP). *J. Venom. Anim. Toxins Incl. Trop. Dis.* **2005**, *11*, 384–390. [CrossRef]
95. Kim, H.J.; Je, H.J.; Cheon, H.M.; Kong, S.Y.; Han, J.; Yun, C.Y.; Han, Y.S.; Lee, I.H.; Kang, Y.J.; Seo, S.J. Accumulation of 23kDa lipocalin during brain development and injury in Hyphantria cunea. *Insect Biochem. Mol. Biol.* **2005**, *35*, 1133–1141. [CrossRef]
96. Spreyer, P.; Schaal, H.; Kuhn, G.; Rothe, T.; Unterbeck, A.; Olek, K.; Müller, H.W. Regeneration-associated high level expression of apolipoprotein D mRNA in endoneurial fibroblasts of peripheral nerve. *EMBO J.* **1990**, *9*, 2479–2484. [CrossRef]
97. Playford, R.J.; Belo, A.; Poulsom, R.; Fitzgerald, A.J.; Harris, K.; Pawluczyk, I.; Ryon, J.; Darby, T.; Nilsen-Hamilton, M.; Ghosh, S.; et al. Effects of Mouse and Human Lipocalin Homologues 24p3/lcn2 and Neutrophil Gelatinase–Associated Lipocalin on Gastrointestinal Mucosal Integrity and Repair. *Gastroenterology* **2006**, *131*, 809–817. [CrossRef]
98. Flower, D.R. The lipocalin protein family: Structure and function. *Biochem. J.* **1996**, *318*, 1–14. [CrossRef]
99. Rassart, E.; Desmarais, F.; Najyb, O.; Bergeron, K.-F.; Mounier, C. Apolipoprotein D. *Gene* **2020**, *756*, 144874. [CrossRef] [PubMed]
100. Flower, D.R.; North, A.C.T.; Attwood, T.K. Structure and sequence relationships in the lipocalins and related proteins. *Protein Sci.* **1993**, *2*, 753–761. [CrossRef]
101. Li, W.; Terenius, O.; Hirai, M.; Nilsson, A.S.; Faye, I. Cloning, expression and phylogenetic analysis of Hemolin, from the Chinese oak silkmoth, Antheraea pernyi. *Dev. Comp. Immunol.* **2005**, *29*, 853–864. [CrossRef] [PubMed]
102. Bettencourt, R.; Lanz-Mendoza, H.; Lindquist, K.R.; Faye, I. Cell Adhesion Properties of Hemolin, an Insect Immune Protein in the Ig Superfamily. *Eur. J. Biochem.* **1997**, *250*, 630–637. [CrossRef]
103. Bettencourt, R.; Gunne, H.; Gastinel, L.; Steiner, H.; Faye, I. Implications of hemolin glycosylation and Ca^{2+}-binding on homophilic and cellular interactions. *Eur. J. Biochem.* **1999**, *266*, 964–976. [CrossRef] [PubMed]
104. Maurer, P.; Hohenester, E.; Engel, J. Extracellular calcium-binding proteins. *Curr. Opin. Cell Biol.* **1996**, *8*, 609–617. [CrossRef]
105. Tedder, T.F.; Steeber, D.A.; Chen, A.; Engel, P. The selectins: Vascular adhesion molecules. *FASEB J. Off. Publ. Fed. Am. Soc. Exp. Biol.* **1995**, *9*, 866–873.
106. WHO. Ten Threats to Global Health in 2019. 2019. Available online: https://www.who.int/news-room/spotlight/ten-threats-to-global-health-in-2019 (accessed on 31 August 2021).
107. Greene, M.A.; Loeser, R.F. Aging-related inflammation in osteoarthritis. *Osteoarthr. Cartil.* **2015**, *23*, 1966–1971. [CrossRef]

MDPI
St. Alban-Anlage 66
4052 Basel
Switzerland
Tel. +41 61 683 77 34
Fax +41 61 302 89 18
www.mdpi.com

Toxins Editorial Office
E-mail: toxins@mdpi.com
www.mdpi.com/journal/toxins

www.ingramcontent.com/pod-product-compliance
Lightning Source LLC
LaVergne TN
LVHW071612170726
843515LV00009B/2374